Methods in
Neurobiology
VOLUME 1

Methods in Neurobiology

VOLUME 1

Edited by
Robert Lahue
University of Waterloo
Waterloo, Ontario, Canada

PLENUM PRESS · NEW YORK AND LONDON

Library of Congress Cataloging in Publication Data

Main entry under title:

Methods in neurobiology.

Includes index.
1. Neurophysiology—Technique. 2. Neurobiology—Technique. I. Lahue, Robert.
QP357.M47   591.1'88'028   80-15623
ISBN 0-306-40517-2 (v. 1)

© 1981 Plenum Press, New York
A Division of Plenum Publishing Corporation
233 Spring Street, New York, N. Y. 10013

Printed in the United States of America

Preface

Rapid advances in knowledge have led to an increasing interest in neurobiology over the last several years. These advances have been made possible, at least in part, by the use of increasingly sophisticated methodology. Furthermore, research in the most rapidly advancing areas is essentially multidisciplinary and is characterized by contributions from many investigators employing a variety of techniques. While a grasp of fundamental neurobiological concepts is an obvious prerequisite for those who wish to follow or participate in this field, critical awareness and evaluation of neurobiological research also requires an understanding of sophisticated methodologies.

The objective of *Methods in Neurobiology* is the development of such critical abilities. The reader is exposed to the basic concepts, principles, and instrumentation of key methodologies, and the application of each methodology is placed in the special context of neurobiological research. The reader will gain familiarity with the terminology and procedures of each method and the ability to evaluate results in light of the particular features of neurobiological preparations and applications.

Robert Lahue

Waterloo

Contributors

P. Kontro, Department of Biomedical Sciences, University of Tampere, Tampere, Finland

Robert Lahue, Department of Psychology, Renison College, University of Waterloo, Waterloo, Ontario, Canada

E. Marani, Laboratory of Anatomy and Embryology, University of Leiden, Leiden, The Netherlands

S. S. Oja, Department of Biomedical Sciences, University of Tampere, Tampere, Finland

Neville N. Osborne, Nuffield Laboratory of Ophthalmology, University of Oxford, Oxford OX2 6AW, England

Karl H. Pfenninger, Department of Anatomy, Columbia University, New York, New York 10032

Kedar N. Prasad, Department of Radiology, University of Colorado Medical Center, Denver, Colorado 80220

Philip Rosenberg, Section of Pharmacology and Toxicology, University of Connecticut School of Pharmacy, Storrs, Connecticut 06268

Werner T. Schlapfer, Western Research and Development Office, Veterans Administration Medical Center, Livermore, California 94550

R. J. Thompson, Department of Clinical Biochemistry, University of Cambridge, Cambridge, England

J. S. Woodhead, Department of Medical Biochemistry, Welsh National School of Medicine, Cardiff CF4 4XN, Wales, and Department of Clinical

Contents

Chapter 3
Tissue and Organ Culture
Werner T. Schlapfer

Chapter 4
Cell Culture
Kedar N. Prasad

Chapter 5
Enzyme Kinetics
P. Kontro and S. S. Oja

Chapter 6
Spectrophotometry and Fluorometry
Robert Lahue

Chapter 7
Immunological Techniques in Biochemical Investigation
J. S. Woodhead and R. J. Thompson

Chapter 8
Freeze–Fracturing
Karl H. Pfenninger

Chapter 9
Enzyme Histochemistry
E. Marani

The Squid Giant Axon: Methods and Applications

Philip Rosenberg

1. Introduction

The unique properties of the squid giant axon cannot be ignored by any neurobiologist with a serious interest in understanding axonal function. The neurochemists and neuropharmacologists are now following the lead of the electrophysiologists in applying their ingenuity to exploiting the natural advantages to be gained by using the largest single nerve fiber that nature has created.

I had not expected that it would be necessary to re-review the squid giant axon so soon after my earlier effort (Rosenberg, 1973). I have been convinced otherwise, however, by the great increase in the number of articles dealing with the squid axon and by the nature of these articles, several of which have opened up entirely new fields of study, such as, for example, the "optical spike." In addition, certain older areas of research, such as the electrophysiological basis of drug action and the use of chemicals as tools in understanding the excitable membrane, are now being intensively exploited using very sophisticated techniques. During the intervening years, an excellent guide to the laboratory use of the squid, *Loligo pealei*, has appeared (Arnold *et al.*, 1974), although the emphasis in that brief text is on the natural history, population dynamics, reproductive behavior, and embryonic development of the squid. A brief description of the dissection and biochemistry of the squid axon is, however, also provided.

In this chapter, I shall attempt to provide detailed methods for dissecting the squid giant axon, which until recently has been an art passed from one

Philip Rosenberg • Section of Pharmacology and Toxicology, University of Connecticut School of Pharmacy, Storrs, Connecticut 06268.

investigator to another with techniques rarely being recorded. While no substitute for practice and consultation, this section should decrease the time required to dissect functioning intact axons. The fear of learning a new technique, especially a dissection that is not very difficult, should not stand in the way of using this useful preparation. It is essential in most studies, even of a nonelectrophysiological nature, that the investigator ascertain that he is working with "living" axons. A simple method for recording action potentials with extracellular electrodes is therefore presented. This technique will show whether the axon can conduct an action potential, and should be an adequate correlate for many biochemical studies. Where a more detailed comparison between electrical and biochemical properties of the axon is desired, or where the primary interest is in its electrophysiological properties, it is necessary to use intracellular and voltage-clamping techniques, the details of which are beyond the scope of this chapter. I do, however, provide an introduction to the methods and usefulness of these techniques along with references for further study. Methods for extrusion of axoplasm and perfusion of the axon are considered in some detail, since the ability to expose the "naked" inner aspect of the axonal membrane (axolemma) to any desired solution is unique to the squid axon and this technique has not been adequately exploited in studies of a nonelectrophysiological nature. As a result of comments received on my earlier review (Rosenberg, 1973), I have greatly expanded the section on the applications of and results obtained using the techniques described in this chapter. It is hoped that this will overcome yet another hurdle in the way of investigators who have until now been hesitant to use the squid axon. This section should allow them to rapidly catch up on "squid axonology," no matter what their particular area of interest. Established investigators may also find some important "missed" references. Results obtained since 1968 are considered in special detail, although the major earlier references are also included. A summary table detailing the effects of chemical agents on the electrical properties of the squid giant axon is also provided in this section. Finally, a word of caution has been added in reference to the applicability and evaluation of data obtained using this preparation. This review was prepared in 1977 and includes literature up to 1976.

1.1. Uniqueness of the Squid Giant Axon

There are very few invertebrate preparations used in biomedical research. Among the molluscs in addition to the squid, the sea hare (Aplysia) is also used in neurological research. Arthropods extensively studied would include the horseshoe crab (Limulus), lobsters, and the brine shrimp (Artemia). An echinoderm widely used in embryological research is the sea urchin. What unique properties does the squid giant axon have that place it in such exclusive company? Why should those of us who are interested in the general, non-species-specific, properties of axons or in the human nervous system in particular use a cephalopod axon in preference to an axon from a mouse or monkey?

Table 1. Size of Squid Giant Axons

Species	Diameter (μm)	Ref. No.[a]
Architeuthis dux	137–210	1
Doryteuthis bleekeri	400–900	2
Doryteuthis plei	200–500	3
Dosidicus gigas	600–1500	4
Ilex illecebrosus	200	5
Loligo edulis	400–800	6
Loligo forbesi	500–1040	7
Loligo opalescens	250–350	8
Loligo pealei[b]	270–800	9
	(mostly 350–500)	
	1080 (May) to 320 (July)	10
Loligo vulgaris	530–975	11
Lolliguncula brevis	75–150	12
Sepioteuthis sepioidea	200–500	13
Todarodes pacificus	140–260	14
Todarodes sagittatus	400	15

[a] References: (1) Steele and Aldrich, reported in Arnold *et al.* (1974); (2) Tasaki *et al.* (1962, 1968a); Matsumoto *et al.* (1970; (3) R. Villegas and G.M. Villegas (1960), G.M. Villegas and R. Villegas (1968), R. Villegas *et al.* (1971), DiPolo, reported in Arnold *et al.* (1974); (4) G.M. Villegas and R. Villegas (1968), G.M. Villegas (1969), Tasaki and Luxoro (1964), Tasaki *et al.* (1965a), Huneeus-Cox and Fernandez (1967), Huneeus-Cox *et al.* (1966); (5) Stallworthy and Fensom (1966); (6) Tasaki *et al.* (1968a); (7) Baker *et al.* (1962a), L.B. Cohen *et al.* (1970, 1972), Kerkut (1967), Meves (1966b, 1974), Meves and Vogel (1973), Levinson and Meves (1975), Chandler and Hodgkin (1965), Caldwell *et al.* (1964); (8) Llinas, reported in Arnold *et al.* (1974), Hagiwara *et al.* (1972); (9) Kerkut (1967), Gilbert (1971), Wang *et al.* (1972), Tasaki *et al.* (1961, 1965b, 1966b, 1967a, 1968a, 1971), Tasaki (1963), Tasaki and Spyropoulos (1961), Takashima *et al.* (1975), Seyama and Narahashi (1973), Mauro *et al.* (1972), Fishman (1970, 1975a), Freeman (1971), Narahashi *et al.* (1967a, 1969b), Narahashi and Anderson (1967), Begenisich and Lynch (1974), Binstock and Lecar (1969), Blaustein and Russell (1975); (10) Fishman *et al.* (1975a); (11) Spyropoulos (1965); (12) Joiner and DeGroob, reported in Arnold *et al.* (1974); (13) J. Villegas (1974, 1975), G.M. Villegas and R. Villegas (1968), R. Villegas *et al.* (1962, 1963), J. Villegas *et al.* (1976); (14) Chailaklyou (1961); (15) Mauro *et al.* (1970).
[b] Less preferably spelled *Loligo pealii* (Arnold *et al.*, 1974).

The ability to use certain specialized techniques on the squid axon and on almost no other preparation is a result of the large size of the axon. As shown in Table 1, squid giant axons range in size from 100 to over 1000 μm in diameter. In comparison, most other giant axons are smaller; for example, those of *Lumbricus* (earthworm), *Myxicola* (marine polychaete), *Canbarus* (crayfish), and *Periplaneta* (cockroach) are approximately 100, 100, 250, and 40 μm, respectively, in diameter (Kerkut, 1967). Young (1952) has contrasted the size of a squid giant axon with that of the multifibered rabbit sciatic nerve (Fig. 1). The significance of the giant axons was not appreciated until their rediscovery by Young (1936a, b, 1939), even though Williams (1909) reported on their presence.

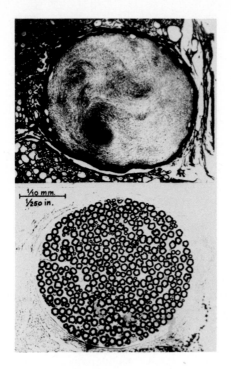

Fig. 1. Cross section of the giant axon of squid (*top*) compared with that of rabbit sciatic nerve (*bottom*) at the same magnification. Reprinted from Young (1952) by courtesy of Oxford University Press.

The large size of the giant axon allows us to perform the following procedures:

1. Insert microelectrodes for stimulating and recording of electrical activity, passing current, voltage-clamping axon, measuring pH, and other electrophysiological measures.

2. Obtain samples of pure axoplasm for measurements of its composition, enzymatic activity, penetration (influx) of radioactive materials, metabolism, and other biological properties.

3. Inject drugs, ions, fluorescent dyes, and other substances into axoplasm and then measure their effects on electrical activity, radioactive efflux, active transport, metabolism, and other functions. By comparing and contrasting effects of agents applied externally to the squid axon with results obtained by internal application, it is sometimes possible to check for asymmetries in the properties of the internal and external faces of the axonal membrane. It has, for example, been suggested that the "gate" for the sodium channel is on the external aspect of the axolemma, whereas the "gate" for the potassium channel is on its inner aspect, because tetrodotoxin (Nakamura *et al.*, 1965; Narahashi *et al.*, 1966, 1967*a, b*) blocks only conduction when externally applied, whereas the reverse is true for tetraethylammonium (Armstrong, 1966; Armstrong and Binstock, 1965; Tasaki and Hagiwara, 1957).

4. Perfuse the interior of the squid giant axon, following removal of axoplasm, for long periods of time with any solution we desire while

monitoring electrical activity, active transport, metabolism, penetration, and other activities.

5. Obtain samples of membranal sheath, essentially free of axoplasm. The sheath will consist of axolemma, Schwann cell, and connective tissue.

6. Correlate biochemical and physiological effects. We need not be concerned that our results merely represent average values, as is the case when multifibered preparations are used. Most mammalian axons consist of various-sized sensory, motor, and autonomic fibers, each with different properties. By correlating the chemical properties of the squid axon with its well-known electrophysiological properties, we may finally achieve understanding of how axons function at the molecular level.

1.2. A Brief History of Squid and Squid Users

Squid belong to the suborder Decapoda of the order Dibranchiata of the class Cephalopoda of the phylum Mollusca. They are characterized by a reduced shell and ten arms. The other suborder of Dibranchiata is Octopoda (ex. octopus), which have no shell and only eight arms. Decapods can be subdivided into the true squids (tribes Myopsida and Oegopsida) and the cuttlefish (tribe Sepioidea). Most genera of squid used in biological research (Table 1) belong to the Myopsida tribe, which inhabit coastal areas, whereas the large oceanic squid of the Oegopsida tribe are less well known (Borradaile *et al.*, 1967; Arnold *et al.*, 1974).

Information concerning the natural history, population dynamics, reproductive behavior, and embryonic development of *L. pealei* has been well presented in several articles and reviewed and summarized in a laboratory text (Arnold, 1965; McMahon and Summers, 1971; Arnold *et al.*, 1974). *Loligo pealei* is the commonest squid in the east coast of North America, found primarily between Cape Cod and Cape Hatteras, but also as far north as the Bay of Fundy and as far south as Colombia (Mercer, 1970; Summers, 1969; Arnold *et al.*, 1974). Although mainly available only during the summer months at the Marine Biological Laboratory on Cape Cod, the winter population of *Loligo* in the mid-Atlantic region has been described (Summers, 1969). Tropical species of squid in the Caribbean region include *D. plei* and *S. sepioidea*, while *D. gigas* is found in the Pacific Ocean, *D. bleekeri* off the coast of Japan, *L. vulgaris* in the Mediterranean, and *L. forbesi* near the English coast. *Loligo pealei* has been reported to grow at a rate of approximately 1.8 cm dorsal mantle length per month (Summers, 1968). Mean sizes (mantle length) of 1- and 2-year-old squid are reported as about 16 and 27 cm, respectively, for females and 18 and 32 cm for males. Maximum longevity was also estimated as 36 months for males and 19 for females (Summers, 1971). The mantle length for the large squid *D. gigas* has been reported to be about 100 cm (Deffner and Hafter, 1959a). The growth and distribution of the squids *D. bleekeri* (Araya and Ishii, 1974) and *Bathothauna lyromma* (Alfred, 1974) have also been described (for compilation of reported growth rates, see also Summers, 1968). *Loligo pealei* avoids water temperatures below

8°C, which explains its presence off the coast of Cape Cod only between May and November (Summers, 1969).

Because of the difficulty, cost, and effort involved in collecting and maintaining adult squid (Section 2.1), there have been attempts at laboratory rearing of *L. pealei* that have been unsuccessful (Arnold *et al.*, 1974), although the embryos are readily maintained for a period of time in flowing seawater, with the rate of development being related to temperature (12–23°C). Arnold *et al.* (1972) were able to rear a Hawaiian sepiolid squid *Euprymna scolopes*. A technique for rearing *S. sepioidea* and *D. plei* has been reported (LaRoe, 1973). The former were reared from eggs to mature adults in 146 days.

The ability of squid to survive in their natural environment is dependent to a great extent on their ability to escape from predators, which explains the development of giant axons in this species. The giant axons innervate the muscle of the body wall, that is, the mantle. A concomitant of the large size of the giant axon is a high speed of conduction (25 m/sec), which is as great as that in myelinated fibers (for references, see Hodgkin, 1965). It is most fortunate for biologists that the problem of developing high conduction speed in the squid was solved by the evolving of giant axons rather than myelination as in mammalian nerves. The high speed of conduction allows a rapid contraction of the mantle muscle, which serves as a compression chamber, thereby ejecting water through the narrow funnel (siphon) of the squid, providing the energy for the rapid jetlike movements that are characteristic for cephalopods (Johnson *et al.*, 1972). The squid can repeat this process by expanding the mantle, thereby drawing water into the cavity through a wide inlet, the mantle aperture, and then producing a number of jet pulses in rapid succession by squeezing the water out through a narrow funnel. The giant fiber system ensures that the mantle cavity is emptied of all water it contains in a simultaneous and maximal contraction of all muscle elements. The volume of seawater expelled is about 200 ml in *L. vulgaris*, with a theoretical pressure of 300 g/cm^3 and a pulse duration of about 0.2 sec (Johnson *et al.*, 1972). The theoretical values these authors calculated for maximum squid velocity are in good agreement with those of Packard (1969) measured in an adult squid during a backward escape reaction (208 cm/sec and an acceleration of 3.3 g). With *L. pealei*, speeds of about 200 cm/sec were also estimated, while with *D. gigas*, a top estimated speed of 700 cm/sec was calculated (Cole and Gilbert, 1970). The feeding and locomotion of the squid and functioning of the mantle have also been described in the short-finned squid *Ilex illecerebrosus illecerebrosus* (Bradbury and Aldrich, 1969a, b).

Much faster moving than the squid are the squid users, which are found, for example, every summer at the Marine Biological Laboratory in Woods Hole, Massachusetts, where all work must be done during the few months when squid are available. Many outstanding scientists were quick to recognize the usefulness of the squid giant axon, and I would be remiss not to mention their contributions. Our sophisticated understanding of the electrophysiological properties of the squid axon would not have been possible without the pioneering studies of Cole on the development and applications in the squid of the voltage-clamp techniques (Cole, 1968). Utilizing this technique, Hodgkin and Huxley mathematically described and demonstrated in squid the ionic basis for the resting and action potentials (for references, see Hodgkin, 1965), work for which they received the Nobel Prize in Physiology and Medicine in 1963. Hodgkin's and Tasaki's groups were the first to demonstrate the methodology and applications of the perfused giant axon (Baker *et al.*, 1961, 1962*a,b*; Oikawa *et al.*, 1961). This technique allowed Tasaki to continue his own studies on the ion-exchange properties of the nerve membrane leading to his two-stable-state theory (Tasaki, 1968, 1975), which in some aspects extends and in others contradicts the ion-pore theory. Cohen, Keynes, and Tasaki have led the way in studying the optical properties of the squid axon, that is, its light-scattering, birefringent, and fluorescent properties (for references, see Von Muralt, 1975). Major credit for our understanding of the electrophysical action of pharmacological agents is due to an extensive series of studies on the squid giant axon by Narahashi and co-workers, who have also shown the usefulness of chemicals, toxins, and enzymes as tools in the study of excitable membranes (Narahashi, 1974), a field of study in which I have also been particularly interested (Rosenberg, 1971, 1976). The Villegases and their group have studied in detail the ultrastructure of the squid giant axon, and have demonstrated an operative cholinergic system in the Schwann cell surrounding the axon (G.M. Villegas and R. Villegas, 1968; G. Villegas, 1975; J. Villegas, 1975). The extensive and continuing contributions of the aforenamed authors are indicated in the reference list of this chapter, in which their contributions are documented in almost 200 of a total of approximately 500 references. While I will not at this time mention the excellent and extensive contributions of a much larger group of squid workers, their contributions will be referred to throughout the remainder of this chapter. The usefulness of the squid giant axon is limited only by the ingenuity of the investigator, and my goal in writing this chapter will have been fulfilled if a future revision of this work finds many new names in the reference list.

2. Methods

2.1. Obtaining and Maintaining Squid

Great efforts have been made to find optimal conditions for transport and maintenance of squid. Nevertheless, even under ideal conditions, most squid do not survive in the laboratory for prolonged periods of time (Arnold *et al.*, 1974). *Sepioteuthis sepioidea* was reared from eggs to adults in the laboratory, while adult *D. plei* were maintained for 38 days. In both cases, the choice and quantity of food were most critical (LaRoe, 1973). Opaque tanks with a seminatural bottom substrate and UV illumination were found to increase survival. Young Pacific squid (*Todarodes pacificus*) lived up to 35 days and fed well with regular changes of water in 5 × 5 × 1.6 m basins (Mikulich and Kozak, 1971). Matsumoto (1976) noted that *D. bleekeri*, a squid widely used in Japan, can be transported for 3–5 hr and maintained in a closed-system aquarium for up to 3 weeks, with the survival for the first week being 90%. Filtration of the seawater was crucial for survival. Neill (1971) obtained an average survival rate of 2 weeks for captive Mediterranean squid (*L. vulgaris*). During 12 summers of research at the Marine Biological Laboratory (MBL) and the direct observation of over 1000 *L. pealei* squid, I observed survival over 4 days only rarely, even though each laboratory has large tanks (1.20 × 0.6 m) that are provided with a drain and a source of running seawater. We usually inserted syringe needles into the rubber tubing leading from the seawater spigots to the holding tanks to increase aeration of the seawater; however, this did not appear to increase survival time significantly. Summers and McMahon (1970) reported that *L. pealei* kept in a 600-liter aquarium with running seawater and constant illumination showed a survival rate of 71% per day over a 1-week period. Crowding and confinement appeared to increase mortality; however, the trauma of capture and starvation of the squid did not appear to obviously increase mortality. Improvements in the methods for maintaining *L. pealei* have since been described (Summers and McMahon, 1974; Summers *et al.*, 1974). Under the best conditions, a mean survival time of 248 hr was noted, with a maximum survival of 1400 hr. Survival was better in rectangular than in square containers, and younger squid did not survive as well in the presence of older squid. Within natural limits, the water temperature, size of container, sex, level of crowding, and conditions of feeding did not seem to alter survival significantly. In contrast, ample water exchange, the use of opaque tanks with lids, and illumination were important for survival. Collision with aquarium walls seemed to be a likely cause of mortality; therefore, a special bumper guard system was devised (Fig. 2) that increased survival.

Because of the limited duration of survival, it is still best to carry out squid studies at marine stations that can provide a regular supply. Some stations with available squid are: MBL, Woods Hole, Massachusetts (*L. pealei*); Friday Harbor Laboratories, Washington State (*L. forbesi*); Laboratory of the Marine Biological Association, Plymouth, England (*L. forbesi*); Marine Station of Tokyo University, Misaki, Japan (*D. bleekeri*); Marine Biological Station of the University of Chile at Vina del'Mar and the Laboratorio di Fisiologia Celular, Montemar, Chile (*D. gigas*); and the Zoological Stations in Naples and in Calmogli, Italy (*L. vulgaris*).

By writing to the MBL, one can obtain their annual announcement booklets, which describe the operation of the laboratory, fee schedules, courses available for students, and other particulars. One should also request application blanks for research space, library desk, or housing by late fall of the year prior to your intended residence. Woods Hole is one of the world's leading centers for biological and oceanographic research, having, in addition to the MBL, the Woods Hole Oceanographic Institute and the National Marine Fisheries Service. Live squid are collected by groundfish otter trawls from a chartered fishing vessel, and large numbers of squid (over 1000 per week) are usually available from late May until early September, although

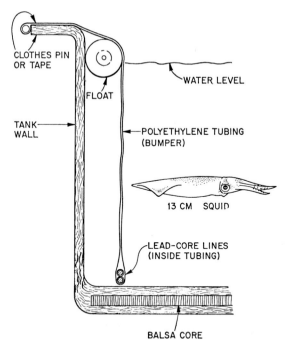

Fig. 2. Typical cross section of the bumper system at a tank wall. Squid colliding with the bumper must express seawater from the space between the polyethylene tubing and the tank wall before striking a solid obstacle; this provides viscous damping of their kinetic energy. The bumper returns to its original position after a collision. Reprinted from Summers *et al.* (1974) by courtesy of the Marine Biological Laboratory, Woods Hole, Massachusetts.

erratic "dry" periods do occur. It has been my experience, and also noted by Arnold *et al.* (1974), that the largest squid and largest-diameter axons are available early in the season. The annual report of the laboratory appears in the August issue of *Biological Bulletin*, a journal published by the MBL (the 81st report, for the year 1978, appeared in *Biol. Bull.* **157**:1, 1978).

2.2. Dissection of the Giant Axon

Where large amounts of axon are required, as for certain biochemical analyses or for internal-perfusion experiments, one should try to obtain large squid, which in the case of *L. pealei* would mean mantle lengths of at least 20 cm. The mantle is about two thirds of the total length of the squid (Fig. 3). "Fat" squid should also be used if available, since it was my impression that axonal diameter correlated better with mantle circumference at the olfactory crest than with mantle length, although with the longer squid one can obtain up to an 8-cm length of giant axon.

In addition to my earlier description of the dissection of the giant axon (Rosenberg, 1973), there have been two other recent descriptions (Arnold *et al.*, 1974; Tasaki and Sisco, 1975). Squid should be handled as little as

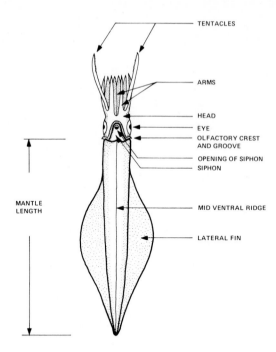

Fig. 3. Ventral external view of *Loligo*. Reprinted (with modifications) from Bullough (1950, p. 380) by courtesy of St. Martin's Press, Inc., Macmillan and Co., Ltd.

possible, since they become disturbed when trapped and can readily injure themselves. With the aid of a net (at least 25 cm in diameter), a squid should be carefully removed from the holding tank. This should be done rapidly without exciting the other squid in the tank, so that they will not injure themselves or discharge a cloud of black ink from their ink sacs. Hold the squid with the head end pointed away from you to avoid being squirted and about midway down the body, never at the head end, since the sharp beak can inflict a painful pinch. Some of the external anatomy of the squid (*Loligo*) is shown in Fig. 3. The skin, especially dorsally, contains many chromatophores that can expand and contract, changing the animal from almost white to deep brownish-red.

The squid should be rapidly decapitated with a large scissors in the region of the olfactory crest (Fig. 3), and the time recorded, if it is desired to keep records on dissection time, time until beginning of experiment, or total survival time of the axon. The body of the squid should then be placed on a dissection table that is near the holding tank and incorporates a drain and raised edge so that seawater can be continuously circulated during dissection. The center of the table should have a window (approximately 10 × 15 cm) into which a glass plate has been cemented. Directly underneath the glass should be a bright light that will allow dissection to be carried out because of the translucent nature of the mantle musculature. While a 100-W lamp plugged into the electrical circuit can be used, there is a hazard near seawater so that a battery-powered light is safer. The mantle cavity is then opened up by cutting through the mantle muscle along the full length of the midventral ridge, beginning at the olfactory crest. The mid-ventral ridge (Fig. 3) is readily visible in the intact squid. At this time, the mantle length and circumference of the mantle in the region of the olfactory crest can be recorded, as well as the sex of the squid. The presence of the ovary, large nidamental glands, and, often, bright red accessory nidamental glands in the female and spermatophores in the male makes identification relatively easy (Fig. 4). For convenience, a centimeter scale can be marked directly on the dissecting table.

With forceps or fingers, the two layers of skin can now be peeled off the surface of the mantle. Although the skin need be removed only over the area of the giant axon, in practice the skin over the entire mantle will usually peel off readily. The subsequent observation of the giant axon by transmitted light is greatly aided by removal of the skin. At all stages of dissection, great care should be taken not to unnecessarily cut or injure the mantle tissue, since wherever the mantle is injured it changes from translucent to opaque. If the opacities are in the region of the axon, its subsequent dissection will be much more difficult.

The mantle should now be pinned dorsal side down on the wooden dissecting tray with the region containing the giant axon over the glass. The silvery ink sac is readily visible and can be removed at this time to avoid the

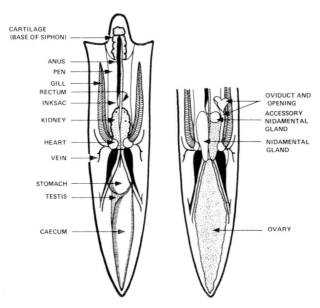

Fig. 4. Ventral view of internal organs of a male (*left*) and a female (*right*) from which the left nidamental gland has been removed to expose the left accessory nidamental gland. Reprinted (with modifications) from Bullough (1950, p. 382) by courtesy of St. Martin's Press, Inc., Macmillan and Co., Ltd.

possibility of leakage of ink or accidentally cutting into it. The gills can easily
be cut away from their underlying connections to the mantle musculature
and removed. All the remaining visceral organs (Fig. 4) lie over the pen (a
transparent hard chitinous structure extending the entire length of the
mantle in the middorsal region) and may be cut away from their underlying
connections and removed.

Parts of the siphon and cartilaginous material adhering to the siphon
may still be attached at the anterior end of the mantle and should be
removed. Lift up this tissue and carefully observe the position of the two
stellate ganglia (Fig. 5). These are on the inner surface of the mantle
musculature, slightly to the left and right of the pen. To remove this tissue,
it will be necessary to cut the mantle nerve, which innervates the stellate
ganglia. It should be cut as far as possible from the ganglia, at which time a

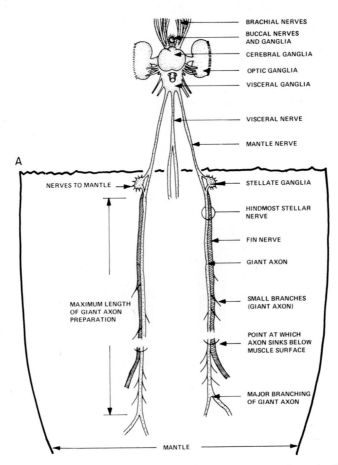

Fig. 5. Nervous system of the squid. That part of the nervous system above A will be removed
when the squid is decapitated. Reprinted (with modifications) from Bullough (1950, p. 388) by
courtesy of St. Martin's Press, Inc., Macmillan and Co., Ltd.

twitch of the mantle may be noted. After the removal of this tissue, the pen can be removed by carefully cutting the connections between the pen and muscle. While it is faster to simply grasp the pen near the posterior portion of the mantle and pull it out, this has on occasion led to the rupture and loss of the giant axon, which at the anterior portion of the mantle is close to the left and right margins of the pen.

For the further steps in dissection, it is convenient to use a good dissecting stereomicroscope with a magnification of 5–10×. You can now remove any strands of tissue that are close to the giant axon, usually slightly in toward the midline, tissue that had been underneath or slightly to the side of the pen. The giant axons as they appear at this stage of dissection are shown in Fig. 5. There are many giant axons that radiate out from the stellate ganglion. I shall describe the dissection of the giant axon contained in the hindmost stellar nerve, which is the largest, longest, and easiest to dissect, and the one usually referred to in all studies utilizing giant axons. One therefore obtains two giant axons per squid (one per ganglion). For further stages of dissection, it is convenient to use Du Mont No. 5 forceps and very fine spring-action iris scissors (available from Roboz Surgical Instrument Co., Washington, D.C., or from Clay Adams Co.). At this time, one can remove the fin nerve and most of the small nerve fibers of the stellar nerve that surround the giant axon. The giant axon is clearer in appearance than the fin nerve and usually appears closer to the midline. Carefully cut into the fin nerve slightly below the ganglion and pull up carefully on the cut end of the fin nerve with the tweezers. With the scissors, you can now free the giant axon of most but not all of the small nerve fibers surrounding it. At a certain distance posterior to the ganglion, the giant axon is no longer at the surface of the muscle, but dips below the muscle. For a maximum length of giant axon, the dissection should be continued until the major bifurcation of the axon (Fig. 5). The giant axon can be observed below the surface of the mantle, and I have found further dissection most easily accomplished by making longitudinal cuts in the mantle musculature slightly to the left and right of the axon. The muscle tissue that overlies the giant axon can then be peeled back and cut away. In this region, the fin nerve no longer adheres closely to the giant axon. For some studies, you may wish to save the fin nerve and compare results with those obtained on the giant axon. To avoid leakage from the axoplasm, and to simplify removal from the squid, you can now, with the aid of needle and thread, tie knots around the axon near its most distal point and also just below the stellate ganglion, in this way obtaining a maximum length of axon. You may wish to use different-color threads to differentiate the wider end of the axon near the ganglion from the narrower peripheral portion. This is important for certain procedures, as, for example, extrusion of axoplasm (Section 2.4). After making the knots, it is convenient to make small loops with the thread at both ends of the giant axon to mount it for further dissection or for recording of electrical activity. By cutting distal to the knots, you can now pull up slightly on the loops with tweezers and free the length of the giant axon from any connections still adhering to the mantle. Always hold on to at least one of the loops, since I have observed all the work of dissection literally going down the drain, especially if the flow of seawater is rapid. If the axon is not to be used immediately, you can grasp the loops and transfer the axon to a 9-cm or larger Petri dish containing filtered seawater, and place it in the refrigerator until use. The axon should not be allowed to curl or kink, nor should it be stretched or pinched. The loops of thread may be placed over syringe needles that are firmly mounted either in the Petri dish or on a glass slide placed in the Petri dish. The axon should be kept at about its resting length or slightly less and submerged below the surface of the seawater. The preparation at this stage free of the fin nerve but still containing some adhering small nerve fibers may be referred to as a "crudely dissected" giant axon, the relative crudeness depending on how many of the small nerves were removed during the course of dissection. The axons should be identified by date, squid number, and whether they came from the left or the right side of the squid.

To obtain a "finely dissected" axon free of all small nerve fibers, further dissection is performed with the axon firmly mounted in the Petri dish and submerged slightly below the surface of the water, although running seawater is not needed. The Petri dish can be surrounded with crushed ice to keep the nerve cool during the further dissection. Once again, illumination can be from below with the light off-center, making the axon more visible, although dark-field illumination is often of great aid in visualizing the very fine nerve fibers that encircle the giant axon (Fig. 6). Higher magnification (40–100×) than previously will be required to visualize all the small nerve fibers. These fibers are carefully peeled away and pulled off with fine tweezers and dissecting needles. Scissors should never be used very close to the giant axon, since they are likely to cause injury. There are several very small branches of the giant axon, especially in the region of the axon that had been below the mantle. Be careful not to mistake these for adhering small nerve fibers. They should be cut no closer than 0.5 mm away from the surface of the giant axon, or injury to the axon may result. These branches are more readily visible at the sides of the giant axon as compared to its upper or lower surface. Any injury to the giant axon is readily detected within half an hour by the appearance of opaque or constricted regions in the axon. The opaque regions are probably produced by the entry of $CaCl_2$ into the axoplasm.

After practice, two crudely dissected axons may be obtained in 20–30 min, and finely dissected axons in 40–60 min. A crudely dissected axon usually weighs between 10 and 40 mg, depending on the length and the number of small nerve fibers adhering to the giant axon. A finely dissected preparation usually weighs between 3 and 9 mg, of which 80–90% is axoplasm, with the remainder being envelope (sheath, i.e., connective tissue, basement membrane, Schwann cell, axolemma).

Axons (*L. pealei*) carefully dissected in the manner described above will routinely maintain full-sized action potentials in an air-conditioned laboratory (18–20°C) for at least 4 hr. At 5–10°C, it has been reported that axons maintain full-sized action potentials for over 10 hr (Tasaki and Sisco, 1975).

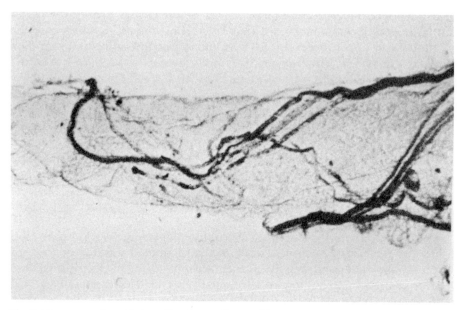

Fig. 6. Giant axon of squid plus adhering small nerve fibers stained for cholinesterase. The giant axon membrane stains only lightly, whereas the small nerve fibers show a dense deposit of precipitate. Reprinted from Brzin *et al.* (1965*a*) by courtesy of Rockefeller University Press, New York.

Even longer survival times were reported by Segal (1968*a*, *b*) for the Italian squid (Table 2).

The ionic composition of the seawater as it comes from the tap at the MBL in Woods Hole, Massachusetts, has been reported to be equivalent to the following (g/liter): NaCl, 24.72; KCl, 0.62; CaCl$_2$, 1.36; MgCl$_2$·6H$_2$O, 4.66; MgSO$_4$·7H$_2$O, 6.29; NaHCO$_3$, 0.18; KBr, 0.089; NaF, 0.003; SrCl$_2$·6H$_2$O,

Table 2. Survival Time of Squid Giant Axons[a]

Immersion fluid	Survival time (hr) mean ± S.E.	Number of experiments
ASW	13.3 ± 0.4	22
ASW, 9°C	39.0 ± 2.4	3
ASW + 10^{-3} M CN$^-$	6.5 ± 0.6	6
ASW + 5 × 10^{-4} M DNP	6.5 ± 0.3	7
ASW + 10^{-4} M ouabain	11.8 ± 0.3	4
Choline SW	6.0 ± 0.1	8
Lithium SW	17.3 ± 1.2	4

[a] Modified from Segal (1968*a*). The excitability of Italian squids (not specified, probably *L. vulgaris*) was measured with extracellular recordings. Artificial seawater (ASW) contained (mM): NaCl, 460; MgCl$_2$, 55; CaCl$_2$, 11; KCl, 10; KHCO$_3$, 0.6. Choline and lithium seawater had NaCl replaced by 460 mM choline chloride or lithium chloride, respectively. The pH of all solutions was 7.3. The temperature was 22.0 ± 0.5°C except where otherwise noted.

0.037; H_3BO_3, 0.024 (Cavanaugh, 1975). A satisfactory artificial seawater can be prepared without the last four trace constituents, in which case the amount of KCl can be increased to 0.67. One may also use a final concentration of about 2 mM Tris [(hydroxymethyl)amino methane] as a buffer. The pH of the seawater solutions is usually kept between 7.5 and 8.0, although I have found that electrical functioning of the squid axon can be maintained for considerable periods of time over a wide range of external pH from approximately 5.5 to 9.5. "Instant Ocean," a mixture of seawater salts (8 lb/ 25 gal water) containing trace elements would also be suitable for use, if natural seawater is not available. It may be obtained from Aquarium Systems Inc., 33208 Lakeland Blvd., East Lake, Ohio 44094.

2.3. Recording of Electrical Activity

While a detailed discussion of electrophysiology and electrical recording techniques is clearly beyond the scope of this chapter, I will discuss some of the most important points relative to accurate monitoring of electrical activity. Prior to actually performing experiments, texts on electrophysiology and neurophysiology should be consulted (e.g., Bureš *et al.*, 1967; Hodgkin, 1965; Katz, 1966; Nastuk, 1964; Tasaki, 1968; Woodbury, 1965). In addition, the manufacturers' instruction manuals for the various pieces of equipment should be carefully studied. If you find it necessary to use intracellular or voltage-clamping techniques, it would be worthwhile to consult with an electrophysiologist prior to undertaking your experiments.

A suitable artificial seawater for external bathing of the squid giant axon was noted in Section 2.2. This solution and three additional solutions that have been used are shown in Table 3. You will note that either a mixture of calcium chloride plus magnesium salts can be used or the magnesium salts can be left out as long as the calcium concentration is increased. Bicarbonate, Tris, or other agents that have effective buffering capacity in the pH 7–8 range can be used. I have found that 1–3 mM Tris is usually adequate for routine work in which no strong acids or bases are used.

Since axonal preparations survive much better at low temperatures (see Table 2), the nerve chambers containing the axon can be surrounded with

Table 3. Solutions for External Bathing of the Squid Giant Axon

NaCl (mM)	KCl (mM)	CaCl$_2$ (mM)	MgCl$_2$ (mM)	Other (mM)	pH	Reference
430	9.2	9.5	23.4 (+26.4 MgSO$_4$)	2.19 NaHCO$_3$	7.9	Adelman and Moore (1961)
449	10	50	0	30 Tris	8.0	Frazier *et al.* (1975)
460	10	11	55	0.6 KHCO$_3$		Pepe *et al.* (1975)
423	9	9.3	22.9 (+25.5 MgSO$_4$)	2.15 NAHCO$_3$		Cavanaugh (1975)

a crushed ice water solution to bring the temperature down to about 4–10°C. It has been reported with *L. forbesi* that the resting potential is independent of temperature between 3 and 20°C (Hodgkin and Katz, 1949*a*), although in *D. gigas* the resting potential decreases linearly as the temperature is raised from 3 to 20°C (Latorre and Hidalgo, 1969). The velocity of impulse propagation increases as the temperature rises, the Q_{10} being 1.7–2.21 (Chapman, 1967; Easton and Swenberg, 1975). Block of conduction occurs below −3.4°C in *L. pealei* and at 0°C in *L. vulgaris* (Easton and Swenberg, 1975).

It has been my experience that the squid will conduct about 400,000 impulses when stimulated at a rate of 100/sec, and after a brief respite can conduct more impulses. The axon must of course be kept moist during this extended period of stimulation. This ability to respond to high rates of stimulation for prolonged periods of time is very useful if you wish to monitor biochemical changes associated with stimulation. In addition, squid giant axons exposed to seawater containing 25% of the usual concentration of Ca^{2+} and Mg^{2+} will show spontaneous or repetitive electrical activity at rates up to 300/sec (Rosenberg and Bartels, 1967). This activity may last 40–80 min, after which the evoked action potential may be decreased about 50%, although this decrease is partially reversible in normal seawater.

2.3.1. Extracellular Recordings

In certain biochemical studies, the only electrical recording you may feel to be necessary is that which will confirm that you are starting out with functionally active axons. In other experiments, you may wish to know only whether a drug blocks electrical activity, without concern as to its electrophysiological mechanism. In these cases, the recording of the action potential with external electrodes may be sufficient. This is the simplest method of recording; however, it does not provide information about the resting potential (difference in steady electrical potential between the inside and outside of the membrane), nor does it reveal the absolute height of the action potential (wave of electronegativity passing along the axon, which consists of a transient reversal in membrane potential). The action potential height you record with external electrodes will be influenced by many factors including the number of small fibers surrounding the giant axon. Because there is no absolute "correct" height for the externally recorded action potential, it is essential that several measurements of the size of the action potential and voltage required for stimulation be made over a period of at least one half hour prior to beginning any experiment to make sure that the preparation is not spontaneously deteriorating. With external recordings, decreases of 5% or less in action potential size in any 30-min period are probably not significant and may represent a slow spontaneous deterioration of the preparation.

I shall describe the equipment that I have used, although obviously other manufacturers' equipment may be substituted. A suitable nerve chamber can

Fig. 7. Bionix nerve chamber for extracellular recording of nerve action potential. (S) Stimulating electrodes; (G) ground electrode; (R) recording electrodes. The axon lies over these electrodes, and the loops of thread, which tie the ends of the axon, are placed over the bent ends of the needles (N) and hold the axon in place. The level of the solution is lowered, for recording of action potential, by pulling on the plunger (P). Reprinted from Rosenberg (1973) by courtesy of Marcel Dekker, Inc.

be purchased from Harvard Apparatus Co. (150 Dover Road, Millis, Massachusetts 02054) or Bionix Industries (10601 San Pablo Ave., El Cerrito, California 94530). The Bionix chamber (Fig. 7) was modified by inserting stainless steel syringe needles with bent tips through each end of the chamber. The loops of threads can be placed over these hooks and the needles moved to adjust for differing lengths of axons. The preparation can be oxygenated by a syringe needle dipping into the bathing solution; however, in experiments lasting up to 3 hr, I never found any difference in survival with or without oxygen. The axon successively passes over two stimulating, one ground, and two recording electrodes, all of Ag–AgCl (for method and importance of coating silver electrodes, see Bureš *et al.*, 1967). Other types of nerve chambers that are suitable for extracellular recording are described by Bureš *et al.* (1967). The axon is submerged during incubation in seawater, or a solution of drugs or radioactive compounds dissolved in seawater. For recording the action potential, the seawater must be removed; otherwise, extensive short-circuiting of the preparation will not allow recording of the electrical activity. For stimulating the axon, a square-wave stimulator such as the model S4 manufactured by Grass Instrument Co. (Quincy, Massachusetts) is connected to a Grass model S105 stimulus isolation unit, which is connected to the two stimulating electrodes in the chamber. Since the impulse originates at the cathode, and to avoid depression at the anode, the cathode should be closest to the recording electrodes. The recording and ground electrodes are connected to a Grass model P 15 AC preamplifier that is in turn connected to a Tektronix 502A dual-beam oscilloscope or 564 storage oscilloscope (Tektronix Inc., Portland, Oregon). Plug-in units suitable for the 564 oscilloscope are a 2B67 time base and a 3A3 dual-trace differential amplifier. A connection is made between the external output of the stimulator and the trigger input of the oscilloscope to synchronize the sweep of the oscilloscope trace to the time of stimulation. If necessary, the action potentials may be photographed with a Tektronix Model C-12 camera. A block diagram of the apparatus used for extracellular monitoring of the action potential is shown in Fig. 8.

Appropriate parameters for stimulation are a frequency of once per second with a pulse duration of 0.1 msec. Under these conditions, it usually requires between 0.1 and 0.5 V to stimulate the giant axon. The small nerve fibers of a "crudely dissected" preparation will not respond to the minimal voltage required for stimulation of the giant axon; however, they will be stimulated by higher voltages. The action potential of the small nerve fibers appears to increase with increasing stimulating voltage, because of the large numbers of fibers present that do not all have the same threshhold of stimulation. In contrast, the response of the giant axon is all or none; that is, as the stimulating voltage reaches threshhold, a maximal-sized action potential is observed. The effects of externally applied chemicals on the action potential may be checked every 5 or 10 min for as long as desired

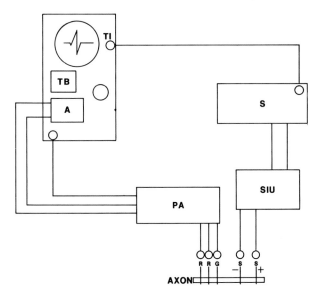

Fig. 8. Simplified block diagram of the electrophysiological apparatus used for external recordings of action potential. (A) Oscilloscope amplifier (Tektronix 3A3); (G) ground; (O) oscilloscope; (PA) preamplifier (Grass P 15 AC); (R) recording electrodes; (S) stimulating electrodes; (S) stimulator (Grass S4); (SIU) stimulus isolation unit (Grass SIU5); (TB) time base (Tektronix 2B67); (TI) trigger input. See the text for further details.

(usually 30 or 60 min). Reversibility of any observed decrease in the action potential is checked in control seawater during a subsequent 30 or 60 min. An example of the action potentials recorded with extracellular electrodes is shown in Fig. 9. The heights of the recorded action potentials in millivolts may then be plotted against time as shown in Fig. 10. In crudely dissected axons, the spike height is usually between 5 and 15 mV, whereas in finely dissected axons, the spike height recorded with extracellular electrodes varies between 25 and 45 mV (Rosenberg and Podleski, 1962).

The "sucrose-gap" technique is an extracellular recording method that allows the observation of nearly full-sized resting and action potentials in axons too small in diameter to allow the use of intracellular microelectrodes. This is achieved by perfusing a short segment of the axon with a high-resistance isotonic sucrose solution that penetrates throughout the extracellular fluid of the preparation. This minimizes the electrical shunt, which otherwise would make the observed potentials much less than their true values (Fig. 11). This technique has been used with the squid giant axon

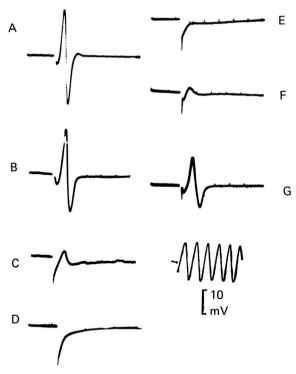

Fig. 9. Restoration by pyridine-2-aldoxime methiodide (PAM) of the extracellularly recorded action potential blocked by Paraoxon in venom-treated squid giant axon. (A) Control; (B) after exposure to 25 μg/ml cottonmouth moccasin venom for 30 min; (C, D) after exposure to 0.01 M Paraoxon for 2 and 5 min; (E) 30 min after return to seawater; (F, G) after exposure to 0.05 M PAM for 5 and 20 min. Time signal is 750 cycles/sec. Reprinted from Rosenberg and Dettbarn (1967) by courtesy of Pergamon Press, Inc.

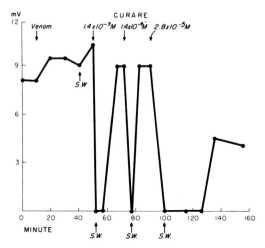

Fig. 10. Extracellular recordings of the effects of curare on the giant axon of squid following pretreatment with 15 µg/ml cottonmouth moccasin venom. (S.W) Indicates return to seawater. Reprinted from Rosenberg and Podleski (1962) by courtesy of Williams and Wilkins Co., Baltimore.

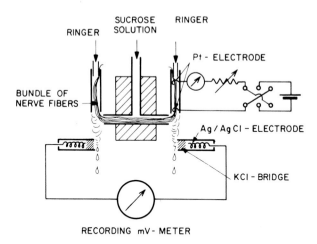

Fig. 11. "Sucrose-gap" apparatus. The nerve is threaded through three polyethylene tubes that are arranged to form a vertical U-shaped system. The horizontal part of the U is filled from its middle with isotonic sucrose. The Ringer solution flowing down in the vertical tubes at a rapid and constant rate sweeps away the dripping sucrose solution. There is thus a sharp drop of longitudinal resistance of the external medium at both ends of the horizontal tube, thereby preventing the shunting of current. At the right-angled junctions formed by the horizontal and vertical tubes, a cotton wick carries away the constantly flowing solutions. These wicks connect to Ag–AgCl electrodes by 3 M KCl bridges. A Pt electrode is used for stimulating the preparation. Drugs may be added to the flowing Ringer solution in one of the vertical tubes. Reprinted from Rosenberg (1973) by courtesy of Marcel Dekker, Inc.

(Dettbarn and Davis, 1962; J.W. Moore *et al.*, 1964*a*, *b*). Intracellular recording techniques, however, are preferred in a large axon such as the squid giant axon.

Fishman has described an elegant technique for electrically isolating a small area or patch (10^{-4} to 10^{-5} cm^2) of the external surface of the squid axon (Fishman, 1975*a*; Fishman *et al.*, 1975*a*). This is achieved by using concentric glass pipettes; the inner one, which is filled with seawater, makes contact with the axon, while the outer pipette has flowing sucrose to the area of membrane surrounding the patch. The patch remains in good condition for about 30 min, and resistance and capacitance measurements indicate that good electrical control and response times are achieved. This method may be especially useful for and should be tested on other preparations where limited areas of membrane are available, provided that the preparation is not damaged by the flowing sucrose.

2.3.2. Intracellular Recordings

The best method for obtaining the absolute values of the resting and action potentials is to record the millivolt potential difference between an electrode inside the axon (intracellular electrode) and another in the external solution bathing the squid axon. The first intracellular electrodes were introduced along the longitudinal axis of the squid giant axon (Curtis and Cole, 1940, 1942; Hodgkin and Huxley, 1939, 1945). The use of these types of electrodes is limited to very large axons such as the squid giant axon, whereas transverse capillary microelectrodes can be used with the squid giant axon or with smaller axons, soma, or muscle cells down to about 20 μm. A comparison of the two types of microelectrodes is shown in Fig. 12. Either the preparation can be stimulated with external electrodes, as previously described, or one can insert a fine metal wire electrode (insulated except at its tip) directly into the axon while the other stimulating electrode remains on the outside of the preparation, thereby stimulating across the membrane.

The axonal preparation should be held rigidly in place to allow easier penetration of the microelectrode. We have routinely used in our studies a lucite chamber with a narrow well (5-ml volume) containing a center rod around which the axon is supported (Fig. 13). The entire preparation is then mounted on a stage that allows transillumination and microscopic observation. Either seawater can be kept flowing continuously and slowly through the chamber at a rate that does not disturb the axon or the flow can be interrupted as desired. Adhering small nerve fibers should be carefully removed from the area, about 2 cm, over which the recording microelectrode is to be inserted. In our studies, we have used the transverse glass capillary microelectrodes of the Ling–Gerard type. These glass microelectrodes can be drawn with any electromechanical puller, such as supplied by Industrial Science Associates, Inc. (63-15 Forest Avenue, Ridgewood, New York 11227). These electrodes can be pulled from carefully cleaned glass capillary tubing of 0.7- to 1.0-mm outside diameter, and are best filled with 3 M KCl. We

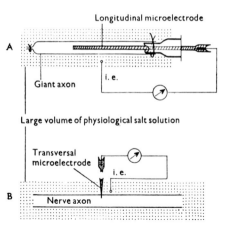

Fig. 12. Two types of capillary microelectrodes. (A) Hodgkin–Huxley type introduced longitudinally into the axon for approximately 30 mm (capillary diameter about 50 μm); (B) Ling–Gerard type introduced transversely into the axon (tip diameter less than 0.5 μm, i.e., indifferent electrode). Reprinted from Bureš *et al.* (1967) by courtesy of Academic Press, Inc.

have found that the simplest method of filling is to allow the microelectrodes to stay overnight with their tips dipping into a filtered solution of 3 M KCl. A convenient holder can be purchased from W-P Instruments Inc. (Hamden, Connecticut). By the next day, the fine tips should be filled, and it is then necessary to fill the shanks using a microhypodermic needle. Any trapped air bubbles can be removed with fine wisps of glass drawn with the

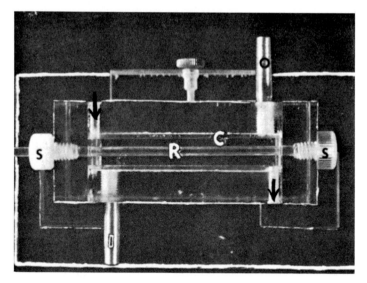

Fig. 13. Chamber suitable for intracellular recording of electrical activity. (C) Center well (5 ml) containing seawater that bathes the axon; (R) rod that supports the axon, allowing insertion of microelectrode; (S) screw that tightens up on rod and on thread used for tying off ends of axon; (I, O) inlets and outlets for controlled-temperature solutions that flow in jacketed area surrounding center well and allow maintenance of desired temperature; (↓) inlets and outlets for seawater bathing axon in center well. Reprinted (with modifications) from Rosenberg (1973) by courtesy of Marcel Dekker, Inc.

electromechanical puller from capillary tubing of about 0.4 mm outside diameter. The completed microelectrode should have a resistance between 7 and 12 MΩ and a tip diameter of about 0.5 μm. Further details concerning preparation, filling, and electrical measurements with microelectrodes can be found in many electrophysiology texts (e.g., Bureš *et al.*, 1967).

The microelectrode is mounted so as to allow electrical contact with an amplifier and rigid attachment to a micromanipulator. Bioelectric Instruments (155 Marine St., Farmingdale, New York 11735) supplies a convenient probe assembly that allows contact between the 3 M KCl within the microelectrode and an Ag–AgCl wire, which is in turn connected to a high-impedance amplifier such as the NFI or P series made by Bioelectric Instruments. The probe assembly is held firmly in place on a micromanipulator such as provided by Brinkman Instruments (Cantiague Road, Westbury, New York 11590). The other recording electrode is extracellular and can conveniently be a simple Ag–AgCl wire dipping into the seawater bathing solution of the chamber. This electrode is also connected to the high-impedance amplifier. The amplifier is connected to one channel of a dual-beam oscilloscope. The two beams of the oscilloscope are initially set at

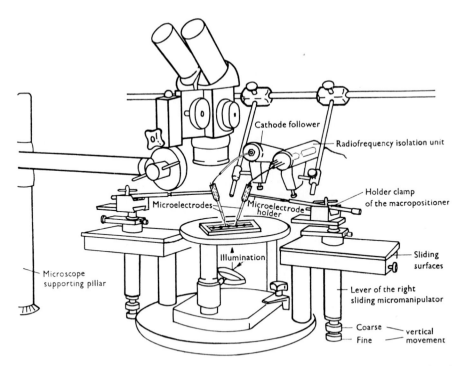

Fig. 14. Experimental setup using intracellular capillary microelectrodes. The preparation is illuminated from below using a mirror. The preparation is observed through the microscope from above. The recording electrode is manipulated with the left, the stimulating electrode with the right, sliding micromanipulator. The recording chamber is on the object plate of the microscope. Reprinted from Bureš *et al.* (1967) by courtesy of Academic Press, Inc.

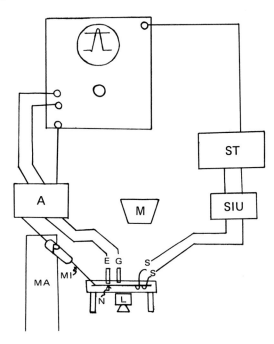

Fig. 15. Block diagram of units needed for intracellular recording with transverse microelectrodes. (A) Amplifier; (E) extracellular recording electrode; (G) ground electrode; (L) lamp; (M) microscope; (MA) micromanipulator; (MI) intracellular recording microelectrode; (N) nerve; (O) oscilloscope; (P) probe assembly; (S) stimulating electrodes; (SIU) stimulus isolation unit; (ST) stimulator. See the text for further details.

identical positions. The reference beam from the inactive channel will remain at this position, whereas penetration of the axon with the aid of the fine-movement knob of the micromanipulator will cause the other beam trace to move rapidly to a new position. After the oscilloscope is calibrated using the built-in calibration control, the distance in millivolts between the initial position (indicated by the inactive beam) and the final position of the beam represents the resting potential of the axon. It is also possible to lead the electrical signal from the amplifier into both the oscilloscope and an ink-writing recorder, thereby allowing action potentials to be recorded on an oscilloscope and continuous records of the resting potential to be obtained on a chart-recorder. Schematics of experimental setups for intracellular recording are shown in Figs. 14 and 15.

The squid giant axon should have control resting potentials of 55–65 mV and action potentials of 90–110 mV. An example of reversible block of the intracellularly recorded action potential by diphenylhydantoin is shown in Fig. 16. The effect of a component of snake venoms, cardiotoxin, is shown in Fig. 17. The action potentials were recorded on an oscilloscope and resting potentials on an ink-writing recorder.

Resting and action potentials of the squid giant axon have also been measured *in situ* in the animal, where blood oxygenation and circulation were maintained, and compared to those obtained in axons removed from the animal (J.W. Moore and Cole, 1960). After an hour, the parameters in the *in situ* axon were the same as those in the excised axon; however, at shorter intervals, the resting potentials were larger (up to 73 mV) in the *in situ*

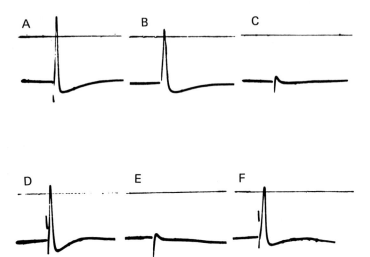

Fig. 16. Effect of diphenylhydantoin on the action potential. (A) Control in seawater; (B) 6 and (C) 14 min after application of 1×10^{-3} M diphenylhydantoin; (D) 7 min after return to seawater; (E) 22 min after addition of 1×10^{-3} M diphenylhydantoin; (F) 55 min after return to seawater. Note that the resting potential remains unchanged throughout the experiment. Calibration: vertical bar, 50 mV; horizontal bar, 2 msec. Reprinted from Rosenberg and Bartels (1967) by courtesy of Williams and Wilkins Co., Baltimore.

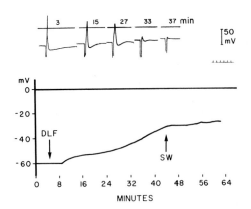

Fig. 17. Effects of direct lytic factor [(DLF) cardiotoxin] on the resting and action potentials of the squid giant axon. Intracellularly recorded action potentials are shown at the top of the figure as photographed from the oscilloscope, and the time course change of the resting potential is shown below as recorded on a Varian ink-writer. Time-signal calibration (in msec) is shown beneath the calibration for mV. DLF (1000 µg/ml) applied at 5 min induced a slow depolarization of the membrane leading to block of conduction at 33 min. Reprinted from Condrea and Rosenberg (1968) by courtesy of Elsevier Publishing Co., New York.

preparation. Changes in the excised preparation are probably due to ionic leakages.

2.3.3. Voltage-Clamping

For a detailed analysis of the ionic currents flowing across the membrane during an action potential and to analyze the effects of drugs on these currents, it is necessary to use the voltage-clamp technique. For a description of the development of this technique and the early results obtained with it, the following references may be consulted: Cole (1949, 1968), Cole and Moore (1960), Hodgkin and Huxley (1952, 1952a–d).

The problem that electrophysiologists faced in attempting to measure the individual ionic currents associated with an action potential was that these currents could not be sorted out from the total current. In addition, current flow changes the membrane potential, which changes the conductance. It was necessary, therefore, to hold the system steady and thus be able to systematically analyze the manner in which conductance changes at different membrane potentials. The voltage clamp holds the membrane potential at any desired value so that the current flow for that particular potential can be monitored. Briefly, the method is as follows: A silver wire for measuring membrane potentials is inserted longitudinally into the axon. The signal is led to an amplifier that can deliver current to the inside of the axon through another longitudinal electrode. Whenever the voltage (potential) set by the experimenter in an external circuit is not the same as that of the membrane potential as determined by the internal electrode, current will flow. The current flow is such as to drive the membrane potential to the value set by the investigator. The experimenter can thus send in a command voltage changing the membrane potential, for example, from its resting level (perhaps -60 mV) to, let us say, -15 mV. The voltage clamp will then measure the current needed to keep the membrane potential constant. The current will be exactly equal (opposite in direction) to the current carried by ions going across the membrane as a result of this potential change. In this manner, the ionic currents that flow during every stage of the action potential can be analyzed. The circuit diagram for a voltage-clamped squid axon is shown in Fig. 18. The following description of this figure is taken from Wu and Narahashi (1973): A glass capillary of 75-µm diameter filled with 0.6 M KCl solution is the internal potential electrode. A bare platinum wire of 25 µm diameter is inserted in the capillary to reduce high-frequency impedance. A 0.6 M KCl–agar and an Ag–AgCl wire connect the capillary electrode to the input of a high-input-impedance preamplifier. For the internal delivery of current, a 75-µm-diameter platinum wire with a 12-mm exposed tip is used. The external reference electrode is a glass capillary (100-µm diameter) filled with seawater. This reference electrode is connected to another high-input-impedance preamplifier by a 0.6 M KCl–agar and an Ag–AgCl wire. The membrane potential could then be held, for example, at -70 mV and step depolarizing or hyperpolarizing pulses applied. The membrane currents

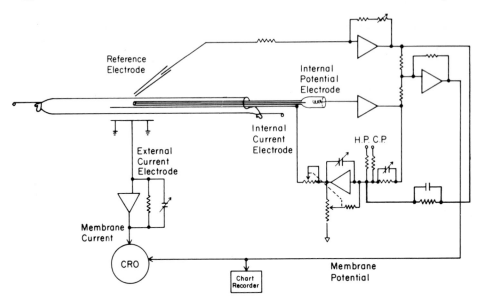

Fig. 18. Circuit diagram for voltage-clamping of squid axons. (H.P.) Holding potential; (C.P.) command pulse. See the text for explanation. Reprinted from Wu and Narahashi (1973) by courtesy of Williams and Wilkins Co., Baltimore.

associated with these pulses could then be recorded. Results obtained with the voltage-clamped squid axon are shown in Figs. 19 and 20. A detailed numerical simulation and evaluation of the voltage-clamp technique in the squid axon has recently appeared (J.W. Moore *et al.*, 1975*a, b*). By voltage-clamp analysis and other electrophysiological experiments, it was shown that the resting potential is primarily a potassium concentration potential, with the resting membrane being more permeable to potassium than to sodium. Associated with an action potential, there is an initial transient inward current carried by sodium ions, the membrane becoming briefly preferentially permeable to sodium ions. A sodium-inactivation process (even with a maintained depolarization) turns off this current flow. The increased sodium conductance is followed by a delayed steady-state increase in potassium conductance, which is maintained as long as the membrane is depolarized. Since sodium ions are normally more concentrated in the external seawater than within the axoplasm, while the reverse is true for potassium, these ions flow down their concentration gradients during an action potential. An active sodium–potassium coupled transport serves to restore the selective ionic concentration gradient. It has been suggested that there are sodium pores or channels on the outside of the squid axolemma, whereas there are potassium channels on the inside. These selectively change their configurations, allowing ionic flow. The idea of spatially separate ionic channels is based partially on the fact that tetrodotoxin selectively blocks the peak transient sodium current associated with an action potential when applied from the outside, but not when injected into the squid giant axon. In contrast,

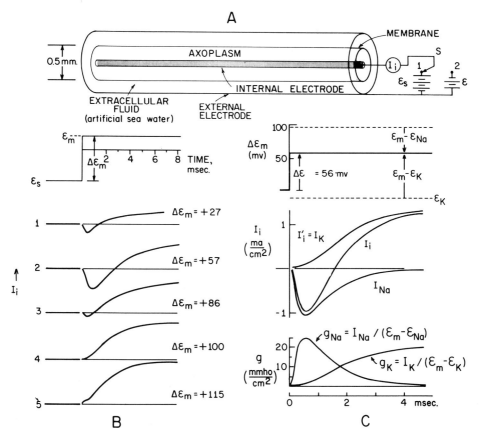

Fig. 19. Voltage-clamping in squid giant axon. (A) Transmembrane voltage (ϵ_m) is held constant over a considerable length of membrane by connecting the internal and external media to a battery through long electrodes. The ϵ_m can be changed suddenly from the resting membrane potential ($I_i = 0$) to any other value by flipping switch (S) to position 2. Total current (I_i) through the membrane is measured as a function of time by an ammeter (cathode-ray oscilloscope). (B) Transmembrane current flow as a function of time after a sudden change in ϵ_m. The topmost curve is ϵ_m as a function of time. Curves 1–5 show the membrane current that flows after the membrane is depolarized, increasing amounts (in mV) shown at the right. In curve 4, depolarization was near the sodium equilibrium potential (ϵ_{Na}), and in curve 5, ϵ_m was greater than ϵ_{Na}. Thus, for all but the largest depolarizations, the early component of current flows in a direction opposite to that expected from change in ϵ_m and the late current flows in the same direction. The time scale at the top applies to all records in (B). (C) Components of total membrane current and conductance. *Top curve:* ϵ_m as a function of time; (———) ϵ_{Na} and potassium equilibrium potentials (ϵ_K). *Middle curve:* Total membrane ionic current (I_i) broken up into its two components, I_{Na} and I_K (I_{Cl} is constant, small, and neglected here). Separation was made by reducing the external concentration of sodium to a value at which a depolarization of 56 mV equaled ϵ_{Na}. Since $I_{Na} = 0$ under these conditions, the total ion current (I_i) is equal to I_K as labeled. *Bottom curve:* Conductance of sodium (g_{Na}) and g_K as functions of time for the step change in ϵ_m shown in the top curve. Conductances are the same shape as the current curves because they are calculated, as shown, by dividing ionic current by effective voltage driving ion (indicated in top curves). The time scale at the bottom applies to all records in (C). Reprinted from Ruch and Patton (1966) by courtesy of W.B. Saunders Co.

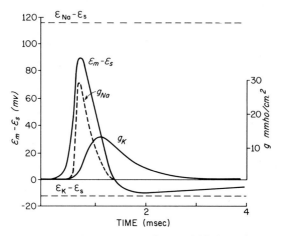

Fig. 20. Calculated time courses of membrane voltage, which in order to make all voltages positive is expressed as $\epsilon_m - \epsilon_s$, that is, membrane voltage (potential) minus resting voltage. Sodium conductance (g_{Na}) and potassium conductance (g_K) in squid giant axon are also shown. Note the time relationships between the upstroke of the action potential and g_{Na} and between g_K and the downstroke and after hyperpolarization. (ϵ_{Na}, ϵ_K) Sodium and potassium equilibrium potentials; (ϵ_s) resting membrane potential (voltage). Reprinted from Ruch and Patton (1966) by courtesy of W.B. Saunders Co.

tetraethylammonium specifically blocks the delayed potassium steady-state current when injected into the squid giant axon, but has no effect when externally applied. A voltage-clamp analysis of drug action will allow the experimenter to state whether the drug affects the magnitude or time duration of the transient peak sodium current or the steady-state potassium equilibrium current, whether the drug alters the rate of sodium activation (increase in conductance) and inactivation (turnoff of increased conduction) or potassium activation, or whether the drug induces a potassium inactivation. While currents are measured, results of voltage-clamp analyses are often reported as conductance changes. The conductances for sodium or potassium (g_{Na} and g_K) equal their corresponding currents (I_{Na} and I_K) divided by the sum of the membrane potential (ϵ_m) minus the reversal or equilibrium potential for sodium or potassium (ϵ_{Na} and ϵ_K). The equilibrium potentials for sodium and potassium ions in the squid axon are approximately $+55$ mV and -70 mV, respectively. An example of a drug effect (propranolol) on the current–voltage curve obtained from a voltage-clamp analysis of the squid giant axon is shown in Fig. 21. It was possible to conclude from this figure that propranolol suppresses the peak transient sodium conductance with a dissociation constant of 2.07×10^{-4} M. The drug shifts the curve relating the peak conductance to the membrane potential in the direction of depolarization, and the time to peak current is shortened. The steady-state potassium conductance is increased by low concentrations of propranolol, whereas it is suppressed by higher concentrations. Voltage-clamp analysis is thus a powerful tool for understanding at an electrophysiological level how drugs act. It of course cannot directly indicate how drugs act at the molecular or biochemical level.

It is hoped that this brief review of the voltage-clamp technique will allow the reader to appreciate the significance of the electrophysiological data and drug effects reported in Section 3. While electrophysiologists have satisfactorily explained the ionic currents and movements associated with an action potential, the underlying molecular and biochemical basis of electrical excitability is not agreed on even though theories have been proposed (Nachmansohn, 1971; Nachmansohn and Neumann, 1975; Tasaki, 1968).

2.4. Extrusion of Axoplasm and Measurement of Penetration

It has been known for a long time that axoplasm can be extruded from the cut end of a squid giant axon (Bear *et al.*, 1937*a,b*). I shall describe a simple and reliable procedure that we have used for obtaining samples of axoplasm after exposure of the giant axon of *L. pealei* to radioactive drug or enzyme containing external incubation media (Hoskin and Rosenberg, 1964, 1965; Rosenberg, 1975; Rosenberg and Hoskin, 1963, 1965; Rosenberg and Khairallah, 1974).

The giant axon, tied off at both ends, is placed into the incubation solution, which would be in a nerve chamber if it is desired to monitor electrical activity at the same time that penetration is being checked. The end

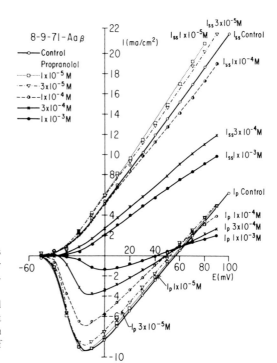

Fig. 21. Current–voltage relationships for peak transient current (I_p) and for steady-state current (I_{ss}) before and during application of 1×10^{-5} M, 3×10^{-5} M, 1×10^{-4} M, 3×10^{-4} M, and 1×10^{-3} M propranolol. See the text for a description. Reprinted from Wu and Narahashi (1973) by courtesy of Williams and Wilkins Co., Baltimore.

of the axon (about 10 mm) that was closest to the stellate ganglion is kept suspended out of the incubation solution to decrease the possibility of contamination during the subsequent cutting of the axon. After the desired incubation time, the axon is removed and passed rapidly (total time 1 min) through three washings using large volumes of seawater. The axon is then blotted on filter paper and the larger end of the axon, the end that had been suspended out of the incubation solution (the end closest to the ganglion), is cut open. The axon is placed on a microscope slide and held in a vertical position with about 5 mm of the cut end of the axon hanging free below the slide (Fig. 22). A roller is then pressed with enough force to cause axoplasm to extrude from the end of the axon. The roller can be a simple piece of polyethylene tubing or can be prepared by forcing tygon tubing over a 2-cm-long piece of thick-walled glass tubing. A piece of heavy wire can be inserted through the bore of the glass tubing and fashioned into a convenient handle. As the droplets of axoplasm appear at the cut end, they are rapidly sucked up into a capillary (preweighed if the weight of the axoplasm is desired). The tips of disposable Pasteur pipettes are convenient to use and can be attached to rubber tubing to better control the amount of suction exerted (Fig. 22). The capillary tubing is then reweighed and the axoplasm blown out into any desired container, care being taken to rinse the pipette with distilled water, seawater, or other desired solution if quantitative transferral of axoplasm is required. Using this procedure, we have usually obtained 2–8 mg axoplasm per axon. *Dosidicus gigas* is reported to provide up to 30 times as much axoplasm (Deffner and Hafter, 1959*a*). The axoplasm has been estimated to be about 94% of the weight of a finely dissected *L. pealei* giant axon (Hoskin, 1976). If necessary, axoplasm from several axons can be collected into one preweighed capillary. The droplets of axoplasm can also be collected onto a weighed microscope cover glass; however, evaporation is

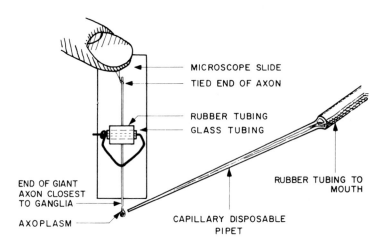

Fig. 22. Method for extrusion and collection of axoplasm. See the text for details. Reprinted from Rosenberg (1973) by courtesy of Marcel Dekker, Inc.

more rapid and quantitative transfers less convenient. This method is, however, convenient if you wish to collect axoplasm without measuring penetration or radioactivity, since the axon can then be dipped for several seconds into distilled water prior to cutting the end open. The axoplasm will then be more gelatinous, which simplifies collection on the cover slide. This method cannot be used if penetration is to be measured, since the structure of the tissue will be affected by the distilled water, and there may be movements of material between axoplasm and sheath of the giant axon. After extrusion of axoplasm, the remaining sheath (envelope), consisting of axolemma, Schwann cell, and connective tissue, may also be studied. After extrusion of a finely dissected axon, only the dry weight of the envelope can be determined, since it immediately dries, appearing as a fine "hair" (0.5–1.5 mg), and must be handled carefully. A procedure similar to that described above has been used for penetration and permeability experiments with *D. plei* (R. Villegas *et al.*, 1966). Of course, more sophisticated techniques of microinjection and perfusion of the interior of the axon can also be used for the measurements of permeability and ionic fluxes (Baker *et al.*, 1961; Oikawa *et al.*, 1961; Tasaki and Spyropoulos, 1961; Tasaki *et al.*, 1961). The perfusion technique is discussed in the next section.

If radioactive experiments are to be performed, it is important that the final mixture of radioactive material in seawater contain at least 5×10^5 cpm/ml, so that even a few tenths of one percent penetration can be detected. The axoplasm can be conveniently blown out from the capillary tubing into 0.1 ml distilled water in a scintillation vial. The capillary pipette should then be rinsed with at least two additional 0.1-ml samples of distilled water, which are added to the same scintillation vial. Bray's solution (Bray, 1960) or other suitable scintillator fluid is added and the samples shaken and counted. Using this technique for measuring the penetration of radioactive materials known to be impermeable, such as sucrose and acetylcholine, we have found less than 1% apparent penetration (Hoskin and Rosenberg, 1964, 1965; Rosenberg and Hoskin, 1963, 1965). Whether this low level of apparent penetration indicates real penetration, contamination of the radioactive material, or contamination during the process of extrusion is uncertain. In any case, these experiments show that contamination during the extrusion described above, if it occurs at all, is very slight.

2.5. Perfusion of the Axon

Possibly the earliest experiments on altering the internal composition of the squid axon were those carried out by Grundfest *et al.* (1954). They determined the effects of ions microinjected into the axon. Although the microinjection technique has been found adequate for some studies, its usefulness is limited by the presence of axoplasm remaining inside the axon, the uncertainty as to the chemical concentrations internally, and the inability to make rapid and repeated changes in the internal medium. In 1961, an exciting new technique was described that allowed the internal aspect of the

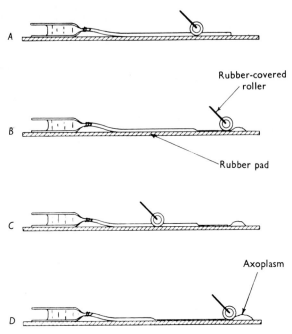

Fig. 23. Extrusion of axoplasm. Reprinted from Baker *et al.* (1962*a*) by courtesy of Cambridge University Press, New York.

axolemma to be perfused with any desired solution (Baker *et al.*, 1961, 1962*a*; Oikawa *et al.*, 1961). By 1965, an entire conference was devoted to the new observations made using perfused squid axons (Mullins, 1965). These methods have been successfully and extensively applied only to squid giant axons. Studies with perfused axons clearly show that the majority of the axoplasm is not needed for conduction, and that the process of excitation is localized in the axolemmal membrane.

There are two major different methods of internal perfusion, which I shall describe, one developed in England and the other in the United States. The English method (Baker *et al.*, 1961, 1962*a*) was reviewed by Meves

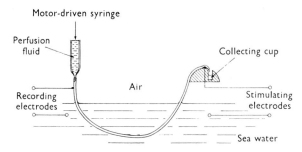

Fig. 24. Standard arrangement for recording external action potential from perfused axons. Reprinted from Baker *et al.* (1962*a*) by courtesy of Cambridge University Press, New York.

(1966a), and I shall summarize the technique. A giant axon (*L. forbesi*), 6–8 cm long, was crudely dissected, and a small glass cannula tied into its distal end. Most of the axoplasm was squeezed out by passing a rubber-covered roller about four times over the axon (Fig. 23). The axon was then suspended vertically in a beaker of seawater, and perfusion fluid (K_2SO_4) was forced into the axon at a rate of 6 μl/min with a mechanically driven syringe. Of slightly over 200 axons, about 75% were excitable and about 45% continued to give action potentials for 1–5 hr. Later, it was reported (Baker *et al.*, 1964) that 90% of the extruded fibers are excitable. An arrangement used for externally recording action potentials in the perfused axon is shown in Fig. 24. For intracellular recording of resting and action potential, a 100-μm microelectrode was inserted into the cannula and moved about 20 mm into the axon. The fiber could be stimulated either externally or internally through the cannula. There have also been many studies showing that the

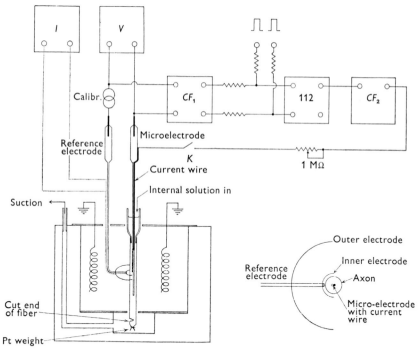

Fig. 25. Diagram of apparatus for voltage-clamp experiments on the perfused axon. The main parts of the apparatus are: glass cannula with the perfused axon hanging vertically; Perspex cell filled with artificial seawater and covered with lids; microelectrode for measuring the internal potential with platinum wire for sending current attached; external reference electrode with two C-shaped Ag–AgCl wires; feedback amplifier [Tektronix d.c. amplifier (112) with cathode followers (CF$_1$, CF$_2$)]; 1 MΩ protective resistance; switch (K) to connect amplifier output to the current wire; pulse generators (⎍); dual-beam oscilloscope for recording membrane current (I) and membrane potential (V). Inset: Horizontal section through the axon and electrodes at the level of the tip of the internal capillary. Reprinted from Chandler and Meves (1965) by courtesy of Cambridge University Press, New York.

perfused axon can be voltage-clamped (Adelman and Gilbert, 1964; Chandler and Meves, 1965; J.W. Moore *et al.*, 1964*b*). A diagram of one of these procedures is shown in Fig. 25. The microscopic appearance of the perfused axon is shown in Fig. 26. It was estimated that 95% of the axoplasm had been removed by extrusion and perfusion.

Tasaki and co-workers described a different procedure in which the giant axon is perfused internally with two glass cannulas (Oikawa *et al.*, 1961; Tasaki and Sisco, 1975; Tasaki *et al.*, 1962). The axon is mounted horizontally on a lucite chamber, care being taken to make sure that the incision site is not immersed in the external medium, since the seawater would diffuse into and destroy the axon. The experimental setup is shown in Fig. 27. There are electrodes in the chamber for extracellular stimulation and recording. The inlet and outlet cannulas, which have outside diameters of about 90 and 300 μm, are both connected to micromanipulators, and must be perfectly aligned. The outlet cannula is inserted in the membrane, and as it is advanced, the axoplasm is continuously aspirated by mouth so that the diameter of the axon is not altered and the axon damaged. The inlet cannula connected to the reservoir of perfusing fluid is introduced at the opposite end. The smaller pipette is pushed into the lumen of the larger pipette and perfusion fluid forced from the smaller into the larger. The two pipettes are then separated by 7–10 mm. With a flow rate of 10–25 μl/min, propagation could be maintained for a prolonged period of time using a satisfactory perfusing medium. A slight modification of this technique is to use only one cannula instead of two (Adelman and Gilbert, 1964; Fishman, 1970). This perfusion technique is shown in Fig. 28. Adelman and Gilbert state that this method of perfusion removes about 75% of the axoplasm.

Narahashi (1963) first showed that the perfusion previously carried out only on large axons (700–900 μm) could be extended to *L. pealei* (300–500 μm), and that the squeezing method is successful with smaller axons. In *D. gigas*, however, it was noted that the compression and roller technique of Baker *et al.* (1961) led to irreversible blocking of electrical activity, so that the method of Tasaki *et al.* (1962) had to be used (Huneeus-Cox *et al.*, 1966). Since this method did not remove all the axoplasm, these workers studied means of removing the axoplasm. They found that the protein gel in the axoplasm could be dissociated by 3–10 min of perfusion with reducing agents such as a high concentration of cysteine. The liquefaction of the axoplasm is apparently due to cleavage of the disulfide bridges stabilizing the gel structure of the axoplasm. This method was adopted to remove the axon of the bulk of its gel-like layer of axoplasm. The flow rate was greatly enhanced, although some axoplasm may have still been present. With internal perfusion, action potentials were recorded for 11 hr in *D. gigas* and for over 6 hr in *Loligo*. The resting potential was −50 to −60 mV for *Loligo* and −40 to −50 mV for *D. gigas*, while corresponding values for action potentials were 105–120 and 90–130 mV. Pronase (0.05–0.1 mg/ml) perfusion has also been used as a method to liquefy and remove remaining axoplasm prior to perfusion studies (Tasaki *et al.*, 1965*b*, 1966*b*). Squid axons perfused with

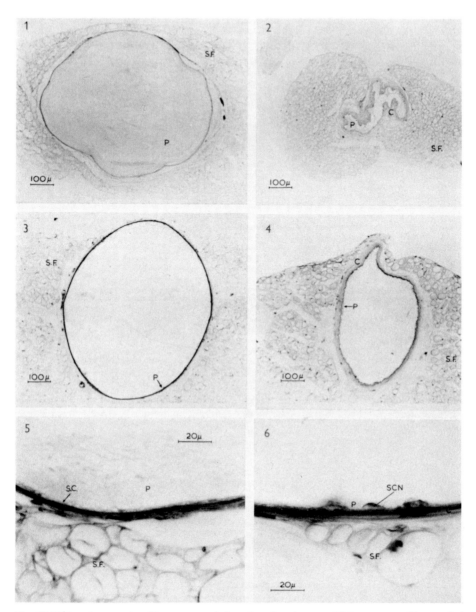

Fig. 26. Light micrographs of intact, extruded, and perfused axons. (1) Intact axon; (2) extruded axon; (3) axon perfused and fully inflated with isotonic KCl; (4) similar to (3) but not fully inflated; (5) surface of intact axon at higher magnification; (6) surface of perfused axon at higher magnification. Residual axoplasm reaches maximum thickness of 10 μm near a Schwann cell nucleus (SCN). (C) Connective tissue sheath; (P) protoplasm (axoplasm) of giant axon; (S.C.) Schwann cell; (S.F.) small nerve fiber. Reprinted from Baker *et al.* (1962a) by courtesy of Cambridge University Press, New York.

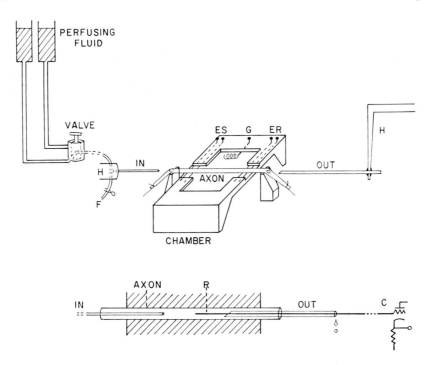

Fig. 27. Schematic diagram illustrating the experimental setup used for intracellular perfusion of squid giant axon. The inlet cannula (IN) is connected to two reservoirs of the perfusion fluid through a valve. The outlet cannula (OUT), the internal recording electrode (R), and the IN are held with separate micromanipulators. (ES) A pair of stimulating electrodes; (ER) external recording electrodes; (F) outflow; (G) ground electrode; (H) holders held by micromanipulators; (C) cathode-follower. Reprinted from Tasaki (1968) by courtesy of Charles C. Thomas.

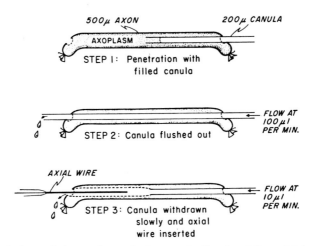

Fig. 28. A perfusion technique using a single cannula. Reprinted from Adelman and Gilbert (1964) by courtesy of Wistar Institute Press, Philadelphia.

Bacillus protease, strain N', retained excitability for several hours, while electron microscopy studies showed that after 8–30 min, all the axoplasm was removed from inside the plasma membrane. The internal structure of the Schwann cell was affected, whereas plasma membrane was not affected (Takenaka *et al.*, 1968; Takenaka and Yamagishi, 1969). Prozyme and bromelain also removed axoplasm without blocking conduction. Trypsin, in contrast, rapidly blocked the action potential.

Brinley and Mullins (1967, 1968; Mullins, 1968) were concerned that the ionic permeabilities and other properties of the perfused axon might be different from those of the intact axon, perhaps due to the loss of protein. They therefore devised a method to dialyze a length of axoplasm internally against a flowing fluid of known composition. They thereby controlled the internal concentration of ions while at the same time retaining the proteins of the axoplasm.

Survival of the perfused axon depends on the choice of perfusing fluid. Baker *et al.* (1961) used isotonic potassium sulfate, methylsulfate, chloride, or isethionate plus 30 mM phosphate buffer, pH 7. Replacing potassium sulfate with potassium chloride decreased the resting potential by 5 mV; replacing with isotonic sodium chloride decreased it to near zero. With isotonic potassium chloride outside and isotonic sodium chloride inside, the resting potential was reversed in sign, that is, $+40$ to 60 mV. The effects of internal pH, Ca^{2+}, and Mg^{2+} on survival are shown in Fig. 29. It is obvious that a low internal divalent cation concentration and a pH between about 7 and 8 are required for prolonged survival. Tasaki and co-workers (Tasaki *et al.*, 1962) first used 250 mM potassium sulfate plus 500 mM sucrose and adjusted the pH to 7.3 with dibasic potassium phosphate. *Dosidicus gigas* axons maintained relatively large action potentials even when perfused with a 350 mM sodium-rich solution (Tasaki and Luxoro, 1964; Tasaki *et al.*, 1965*a*).

The effects of the internal perfusion of many cations and anions on the survival of the perfused axon were then studied (Tasaki and Luxoro, 1964; Tasaki and Takenaka, 1964; Tasaki *et al.*, 1965*b*). Their orders of effectiveness in prolonging survival agreed with their lyotropic sequence:

Anions: fluoride > phosphate > glutamate = aspartate > citrate > tartrate > propionate = butyrate > sulfate > chloride > nitrate > bromide > iodide > thiocyanate

Cations: Cesium > rubidium > potassium > sodium > lithium ≫ barium, strontium, magnesium, calcium

In addition, it was noted that dilution of the internal perfusion fluid with nonelectrolytes such as 1.1 M sucrose or 12% glycerol always had a favorable effect on excitability. The more impermeable ions appear more effective in maintaining a high resting potential and excitability. Axons perfused with rubidium or cesium fluoride maintained excitability in a sodium-free external medium containing suitable polyatomic univalent cations. The sodium-substituting ability decreases with the number and length of hydrocarbon chains

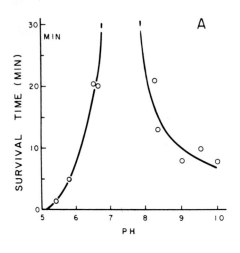

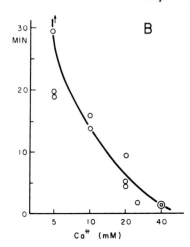

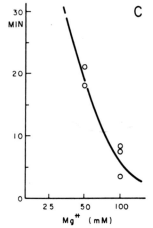

Fig. 29. Dependence of survival time of perfused axons on pH (A), Ca^{2+} concentration (B), and Mg^{2+} concentration (C) of perfusing fluid. Data from 11 axons were incorporated in (A), from 12 axons in (B), and from 9 axons in (C). Room temperature was approximately 15°C. Reprinted from Tasaki *et al.* (1962) by courtesy of the National Academy of Sciences U.S.A.

attached to the nitrogen atom [examples: (1) $NH_4 > NH_3(CH_3) > NH_2(CH_3)_2 > NH(CH_3)_3 > (CH_3)_4N$; (2) hydroxyl > amino > hydrogen > alkyl > phenyl] (Tasaki *et al.*, 1965c). Regardless of the perfusing medium used, excitability cannot be maintained in the absence of external divalent cations (Tasaki *et al.*, 1967a). The following are some examples of satisfactory internal perfusing solutions (concentrations in mM): (1) KF, 400; K_2HPO_4, 26.6; KH_2PO_4 3.4 (Adelman *et al.*, 1965b); (2) K glutamate, 370; sucrose, 333; KH_2PO_4, 15; pH 7.3 (Frazier *et al.*, 1975); (3) K glutamate, 320; NaF, 50; KH_2Po_4, 15; sucrose 333; pH 7.3 (Narahashi and Anderson, 1967); (4) K glutamate, 370; sucrose, 333; Tris, 5; pH 7.3 (Narahashi *et al.*, 1970); (5) KF, 440; glycerol, 2.4%; K phosphate buffer (small amount); pH 7.2–7.4 (Tasaki and Sisco, 1975).

Using perfused axons, it is possible in many cases to critically determine whether a compound acts from the inside or outside of the axolemma, by selective application of a drug and determination of the time for its effects.

Tetrodotoxin, for example, affects only conduction when applied externally, but has no effect from inside the nerve membrane (Narahashi *et al.*, 1966, 1967*a*; Nakamura *et al.*, 1965). Since tetrodotoxin cannot readily penetrate biological membranes, this would indicate that the receptor for tetrodotoxin is on the external surface of the axolemma and that the Schwann cell and connective tissue are not major permeability barriers. By controlling the pH of the external and internal incubation solution, it may also be possible to determine whether a compound acts in its charged or neutral form. For example, Narahashi *et al.* (1970, 1972*b*) determined that tertiary amine local anesthetics related to lidocaine penetrate in their uncharged form and block the action potential from inside the membrane in the charged form.

The effects of other ions and drugs used in the perfused-axon preparation are discussed in the next section.

3. Applications

I have summarized in this section the many studies that have been carried out in the squid giant axon, including those of a structural, chemical, electrophysiological, optical, biochemical, and pharmacological nature. It is not possible to discuss in detail the significance of these studies; however, extensive references are provided for the interested reader.

3.1. The Giant Axon: A Model for Nonmyelinated Axons

The structure, composition, and permeability properties of the squid giant axon have been extensively studied. These results are of general interest insofar as they are representative of properties of nonmyelinated axons. Except for the fact that the giant axons are much larger, their structure and composition are similar to those of other nonmyelinated axons. For those planning to work with the squid giant axon, it is essential to understand those aspects of structure, composition, and permeability that may influence the results that are obtained and their interpretation.

3.1.1. Structure

Many giant axons arise from cell bodies in the stellate ganglia and innervate the mantle musculature. I described in Section 2.2 the dissection of the largest of the giant axons that run parallel to the pen. The axon arises by a fusion of the processes of several hundred small nerve cell bodies (Young, 1936*a,b*) and is one of the few examples of neurons that are a syncytium. The squid stellar nerve (*L. pealei*) consists of hundreds of small nerve fibers (50 μm or less in diameter), connective tissue (endoneurial layer consisting of fibroblasts and collagen), basement membrane, Schwann cell, and the giant axon. In *D. plei*, there are large endoneurium cells (2 × 10 μm) in which are embedded bundles of connective tissue fibrils. The

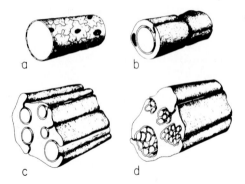

Fig. 30. Scheme representing the four different types of nerve fibers found in the stellar nerve of the squid. (a) Giant fiber with the axon surrounded by several Schwann cells; (b) medium-sized fiber with the axon, 1.5–10 μm in diameter, ensheathed by one Schwann cell; (c) small fibers with several axons, 0.5–1 μm in diameter, surrounded by one Schwann cell; (d) minute fibers with bundles of axons, less than 0.5 μm in diameter, ensheathed by one Schwann cell. Reprinted from G.M. Villegas and R. Villegas (1968) by courtesy of Rockefeller University Press, New York.

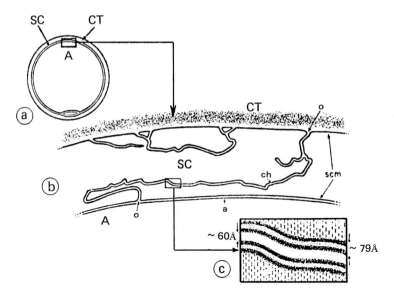

Fig. 31. Three-dimensional diagram of the giant nerve fiber of the squid. (a) A segment of the nerve fiber showing the axon (A) covered by the Schwann cell (SC); the latter is covered by the connective tissue (CT). (b) Enlarged portion of the fiber in which channels (ch) are shown as slits crossing the Schwann cell from outer surface to axonal surface. Special attention is called to the openings (o) of the channels (ch). Some of the channels that in ultrathin sections appear to end in a blind alley are found to be continuous at different levels, as has been observed in serial electron micrographs. (c) Highly enlarged view of one of the channel openings (o) in which the continuity of the channel walls with the Schwann cell membrane (scm) is demonstrated. The fine structure of the axolemma (a) is shown. No difference can be appreciated between this structure (a), the Schwann cell membrane (scm), and the channel wall structure (ch). Reprinted from R. Villegas and G.M. Villegas (1960) by courtesy of Rockefeller University Press, New York.

relationship of the Schwann cells to different-sized axons (including the giant axon) is shown in Fig. 30. A diagram of the giant axon is shown in Fig. 31, and electron micrographs in Figs. 32 and 33. External to the axoplasm is the axolemma (80–120 Å), which is the site of origin of axonal bioelectricity, a gap of 50–150 Å, and the Schwann sheath (0.2–5 μm), which is a single cellular layer composed of a mosaic of several Schwann cells and their interdigitated processes (Geren and Schmitt, 1954; Martin and Rosenberg, 1968; R. Villegas and G.M. Villegas, 1960; G.M. Villegas and R. Villegas, 1960, 1963; R. Villegas *et al.*, 1963). Several Schwann cell nuclei may be found in a transverse section of the axon. External to the Schwann cell, there is a prominent basement membrane (0.1–0.3 μm thick), and more externally there is the connective tissue endoneurium. Three simultaneous efflux processes of [^{14}C]glycerol with rate constants of 21, 4, and 0.33×10^{-3}/sec were related to tissue compartments in the squid axon. The slowest is the efflux from the giant axon, and the fastest is efflux from the extracellular space. The middle component of efflux, inhibited by copper, is from the periaxonal (external to the giant axon) cellular space (4–16 μm wide). The general tissue arrangements are as described above in all squid giant axons, although the sizes of the different structural components may vary in different squid [for example, see Geren and Schmitt (1954) (*L. pealei*), G.M. Villegas and R. Villegas (1960) (*D. plei*), R. Villegas *et al.* (1963) (*S. sepioidea*), Baker *et al.* (1962a,b) (*L. forbesi*)]. The Schwann sheath is between 0.2 and 0.8 μm thick in *D. plei*, 1–3 μm in *L. pealei*, and 1.5–6 μm in *D. gigas* and *S. sepioidea*. On an average, it was reported that a 500-μm-diameter *L. pealei* axon had a Schwann cell layer 1.2 μm thick, a basement membrane of 0.3 μm, and a layer of connective tissue of about 5.5 μm (Hoskin, 1976).

The axolemma is probably the only barrier, between the axoplasm and the outside of the axon, that is capable of maintaining the ionic concentration differences necessary for normal axonal functioning (R. Villegas and Barnola, 1961; R. Villegas and G.M. Villegas, 1960; R. Villegas *et al.*, 1963). The axolemma in *D. gigas*, *D. plei*, and *S. sepioidea* had a laminated pattern with repeated globular units and local thickenings (Fig. 33) (G.M. Villegas, 1969). There are long tortuous channels about 60 Å wide and 50,000 Å long that go through the Schwann cell (Figs. 31–33). The interior of the channels is extracellular in nature, and large enough to allow water and ions to reach the axolemmal surface (R. Villegas and G.M. Villegas, 1960; G.M. Villegas and R. Villegas, 1960, 1963, 1968). Thorium particles penetrated through the endoneurial space, through the channels of the Schwann cell, and into the axolemma–Schwann cell space (G.M. Villegas and R. Villegas, 1968). The Schwann cell in *S. sepioidea* is so large and so arranged that it is possible to insert a microelectrode and measure resting potentials in the Schwann cell as well as in the axon and in the endoneurial cells that are external to the Schwann cell (R. Villegas *et al.*, 1963). Action potentials, however, were found only in the axon, thereby conclusively demonstrating that the axolemma is the electrically excitable membrane. On the basis of studies on the water permeability of the axons and of changes in volume to small nonelectrolyte

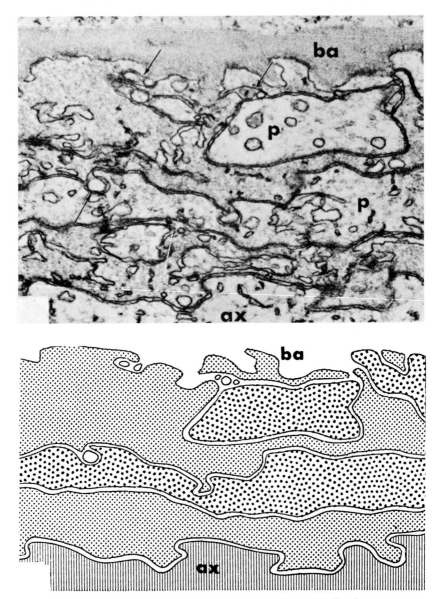

Fig. 32. *Top:* Schwann sheath of a control giant axon (*L. pealei*) fixed in permanganate immediately after dissection. (ax) Axoplasm. The sheath is formed by several processes (p) and surrounded by a prominent basement membrane (ba). At the arrows, there are either transversely cut finger-like extensions or globular postmortem artifacts (×50,000). *Bottom:* Diagram of the top micrograph, which separates the different layers of the sheath by different patterns. It is not intended to indicate that the elements marked by one pattern originate from the same cell. Reprinted from Martin and Rosenberg (1968) by courtesy of Rockefeller University Press, New York.

molecules such as methanol, ethanol, urea, and glycerol, it was proposed that the axolemma has equivalent pores of 4.25 ± 0.25 Å in diameter. Other studies gave a range of 4- to 5-Å pores spaced 1000 Å apart. The reflection coefficients were calculated for ten penetrating nonelectrolytes in the resting and stimulated (100/sec) axon, giving estimates of 4.7 and 6.2 Å, respectively (R. Villegas and Barnola, 1960, 1961; R. Villegas and G.M. Villegas, 1960; R. Villegas *et al.*, 1966, 1968). The presence of water-filled pores in the membrane is thought to be primarily responsible for the electroosmosis that was observed when a current was passed lengthwise through *L. forbesi* and *I. illecebrosus* giant axons. The direction of water flow was always toward the negative terminal, with a magnitude of 16–28 molecules of water per positive charge (Stallworthy, 1970; Stallworthy and Fenson, 1966). These estimates of pore size are compatible with the known restricted diffusion of sodium observed in the resting squid membrane. Similar pores have been proposed on the basis of ionic-permeability studies (Mullins, 1956, 1960). The axolemma is about 50 times as permeable to potassium as to sodium and chloride (Baker *et al.*, 1964; Freeman *et al.*, 1966). Functionally, the axolemma is asymmetric, indicating that there are differences in the structure or composition, or both, of the inside and outside of the axolemma. As noted before, tetrodotoxin will block conduction when applied to the outside of the squid giant axon, whereas it is inert when microinjected or perfused into the axon interior (Nakamura *et al.*, 1965; Narahashi *et al.*, 1966, 1967a). In contrast, tetraethylammonium, cesium ion, and trypsin are active only when applied to the interior of the giant axon (Adelman and Senft, 1966; Armstrong, 1966, 1969; Armstrong and Binstock, 1965; Chandler and Meves, 1965; Narahashi and Tobias, 1964; Pickard *et al.*, 1964; Rojas and Luxoro, 1963; Rosenberg and Ehrenpreis, 1961; Tasaki and Hagiwara, 1957; Tasaki and Takenaka, 1964).

It has been suggested by Tasaki and co-workers (for references, see Tasaki and Sisco, 1975) that the outer surface of the axolemma has a fixed negative charge that can bind calcium ions. Gilbert and Ehrenstein (1969) measured the influence of divalent cations on potassium conductance, finding that a decrease in divalent cations shifts the conductance–voltage curve in the hyperpolarized direction. These results were also explained by assuming that the outer surface of the membrane has a fixed negative charge of one electronic charge per 120 Å^2. Segal (1968b) also concluded that the surface charge is negative (-1.9×10^{-8} coulombs/cm^2) on the basis of electrophoretic-mobility measurements in single isolated axons. Rojas and Atwater (1968) have described methods for determining membrane charges in the squid giant axon, and they estimated the negative charge density as one negative charge per 1600 Å^2. It is well known that squid axons are more permeable to cations then to anions, which is further evidence of fixed negative charges.

Electron-dense structures have been found along the axonal plasma membrane, in the mitochondria, and along the basal plasma membrane of the Schwann cell, but not in the axoplasm of the giant axon. These densities contain high concentrations of calcium and phosphorus, and may represent

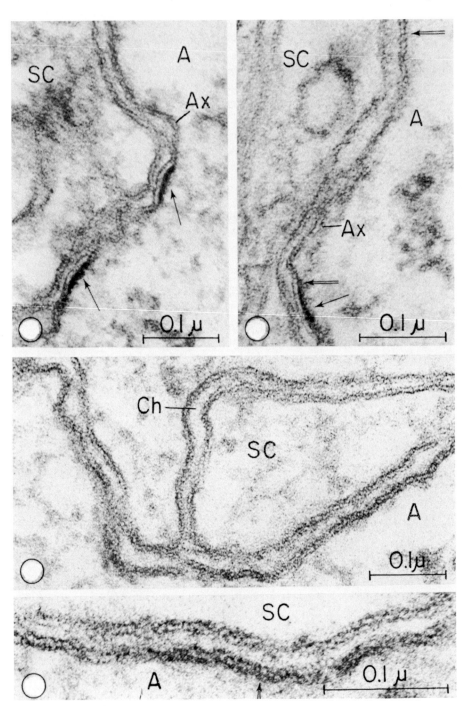

Fig. 33. High-magnification electron micrographs of the boundary between giant axon (A) and Schwann cell (SC). Schwann cell plasma membrane, channel wall (Ch), and axolemma (Ax) show three-layered continuous pattern and areas with septa placed among external and internal dense

sites where the calcium-binding protein is saturated with calcium. The electron-dense structure may also localize sites of ATPase activity.

I would expect in the future that greater use will be made of free radical probes such as spin labels to obtain information on the structure of the squid axon membrane. Yeh *et al.* (1975) have used a spin-labeled local anesthetic, and compared its effects with those of its non-spin-labeled analogue. The spin-labeled analogue was more potent and had some effects that were different from those of its analogue.

The structural properties of the axoplasm have been extensively studied both *in situ* and after extrusion. The axoplasm is liquefied by divalent cations, whereas monovalent cations tend to preserve the jellied nature of the axoplasm (Chambers and Kao, 1952). The axoplasm itself is not uniform, but contains large numbers of protein fibers over 2000 Å in length and 50–80 Å in diameter (Davison and Taylor, 1960). It is now known (Davison and Huneeus, 1970; Huneeus and Davison, 1970) that there are both neurofilaments and microtubules in the axoplasm (Fig. 34). The protein subunits of the neurofilaments in the axoplasm of *D. gigas* were isolated by selective extraction. The neurofilaments were dissociated by dilute urea and guanidine hydrochloride. The protein subunit was an acidic protein of molecular weight 80,000 and is different from the microtubule protein in size, immunologically, in chemical characteristics, in amino acids, and in other properties (Davison and Huneeus, 1970; Huneeus and Davison, 1970). Microtubules were identified in negatively stained freshly extruded axoplasm. Sodium citrate or EDTA (0.01 M in D_2O) at pH 6 were best for preservation. Mercaptoethanol increased the instability of the microtubules and stabilized the neurofilaments. $MgCl_2$ dissociated the tubule structure but gelled the neurofilaments (Davison and Huneeus, 1970). Axoplasm from *L. pealei* was studied by differential interference microscopy and electron microscopy (Metuzals, 1969; Metuzals and Izzard, 1969). A continuous three-dimensional network of threadlike elements was in axoplasm, being more parallel and densely packed in the peripheral as compared to the central axoplasmic region. Filaments from 30 to 250 Å in width to threads of 1–3 μm were found. The larger microtubules were surrounded by filaments. Calcium ions caused a loosening and disintegration of the network of filaments that explains the ability of calcium to liquefy the axoplasm. It was suggested by the aforenamed workers that folding and unfolding of the filaments may cause changes in the physicochemical properties of the axoplasm. Isolated microtubules were reassembled in isolated axoplasm obtained from *L. pealei* (Witman and Rosenbaum, 1973). The neurofilaments, negatively stained, appeared as long straight tubes 60–80 Å in diameter having a lumen of

---

zones of membrane ($\Leftarrow$). In some regions, favorable orientation allows easy identification of globular units in septated regions. Cytoplasmic densities ($\leftarrow$) closely apposed to internal dense zone of axolemma are observed spaced along the membrane. OsO_4-fixed, Epon-embedded material. Reprinted from G.M. Villegas and R. Villegas (1968) by courtesy of Rockefeller University Press, New York.

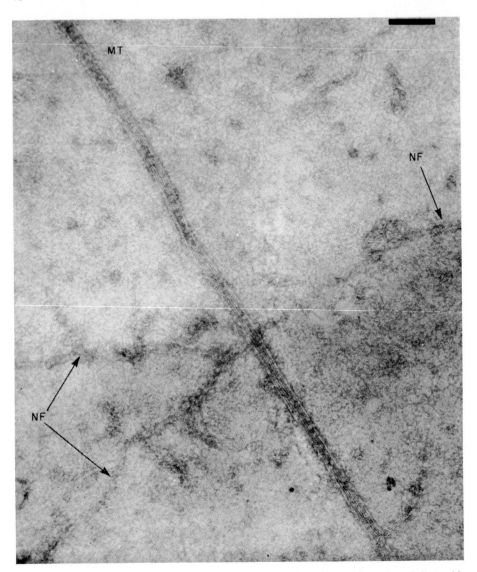

Fig. 34. Microtubules (MT) and neurofilaments (NF) in axoplasm from a giant axon of the squid *L. pealii*; dispersed in 0.01 M EDTA, pH 6.3, in D_2O and negatively stained by uranyl acetate. the bar represents 100 nm. In regions where the tubule did not appear swollen, the diameter was 26–29 nm. Neurofilaments are also detectable, but they are largely obscured by adsorbed cellular material. Reprinted from Davison and Huneeus (1970) by courtesy of Academic Press, Inc., New York.

15–20 Å diameter and walls 25–30 Å thick. They appeared to be composed of four strands, and purified preparations were enriched in two proteins of molecular weight 78,000 and 170,000 (Witman and Rosenbaum, 1973). It would be of interest to determine whether these fibrillar proteins are involved in the transport processes of the axon. It was reported, however, that Drs.

Twomey and Lasek were not able to find suitable conditions for studying axonal transport from the stellate ganglia (Arnold *et al.*, 1974). Scattered among the protein fibrils in the axoplasm are peculiar clusters of vesicles and large cisterns of endoplasmic reticulum (G.M. Villegas, 1969). Mitochondria in the axoplasm of *D. gigas* are usually elongated bodies, 0.2–0.4 μm thick and 3 μm long, of which there are about 40 per 100 μm^2 in the peripheral axoplasm, near the axolemma, and about 30 per 100 μm^2 in the deeper regions of the axoplasm. Near the axolemma, rounded (> 0.6-μm diameter) mitochondria were also found, while the elongated type were found throughout the axoplasm (G.M. Villegas, 1969; G.M. Villegas and R. Villegas, 1968).

Carpenter *et al.* (1971) used metal microelectrodes to measure the intracellular conductance of squid axoplasm. They reported that its conductance is equal to that of seawater, suggesting that there is no extensive binding of water or small ions. Cole (1975), however, notes that the resistivity (reciprocal of conductance) of squid axoplasm has been reported to be anywhere from 1.0 to 6.9 times that of seawater. His own measurements on extruded axoplasm from *L. pealei* ranged from 1.2 to 1.6 times that of seawater. In other studies, Carpenter *et al.* (1975) also found values higher than 1.0; using metal microelectrodes and *in situ* measurements, they reported resistivity values of 1.55 times (*L. pealei*) and 1.30 times (*L. opalescens*) that of seawater.

3.1.2. Composition

Most studies on composition of the squid giant axon have emphasized the axoplasm, since pure samples can easily be obtained. In contrast, after one extrudes the axoplasm from the giant axon, the remaining "sheath" or "envelope" is still a heterogeneous complex containing axolemma, Schwann cell, and connective tissue as previously discussed (Section 3.1.1). There is thus a great need for procedures that will separate the structural components of the envelope after axoplasmic extrusion. Two membranal fractions have been isolated from the first stellar nerve of *D. gigas* by a series of differential discontinuous and gradient centrifugations (Camejo *et al.*, 1969). On the basis of various criteria including morphological appearance, yield, and distribution of Na$^+$-K$^+$-ATPase, one fraction was thought to derive primarily from the axolemma membrane, and one from the Schwann cell plasma membrane. For absolute identification, however, more detailed studies will be required. Similar methods were used by Marcus *et al.* (1972) for isolating membrane fractions from the squid giant axon (*D. gigas*), which has a low axolemma/ Schwann cell ratio ($\approx$ 1:5), and from the squid retinal nerve, which has a high ratio ($\approx$ 5:1). A light membrane fraction, equilibrating at 12–25% sucrose, was ascribed to the axolemma, and a heavy membrane fraction, equilibrating at 35–50% sucrose, to the Schwannlemma and basement membrane. The giant axons were richer in the heavier membranes. Protein, Na$^+$-K$^+$-ATPase and NADH-ferricyanide oxido-reductase were measured along a linear sucrose gradient. Between 50 and 100 finely dissected axons were used for each batch of membranes. The original paper should be consulted for details of the isolation procedure. A high priority should be given to these types of studies so that the membranes of the giant axon can be characterized in detail as great as has been accomplished for brain tissue. It will be necessary that micromethods of separation and analysis be utilized since, for example, even in the most sensitive method employed (Marcus *et al.*, 1972), it required about 100 of the large *D. gigas* axons. It would require many more *L. pealei* axons to obtain an equivalent weight of tissue. A brief handbook of micromethods in the biological sciences has recently been published (Keleti and Lederer, 1974).

The axoplasm of the finely dissected squid giant axon has been reported to represent between 73 and 88% of the total weight of the fibers (Bear and Schmitt, 1939), with the water content of the axoplasm being 77.3 ± 1.1% (R. Villegas and G.M. Villegas, 1960) or 86–89% (see Hodgkin, 1965; Koechlin, 1955) of the total axoplasmic weight. Using intracellular pH microelectrodes, the pH of the axoplasm (*L. pealei*) was estimated to be 7.0 ± 0.2, and not to be influenced by changes in the resting or action potential (Bicher and Okhi, 1972). Other estimates were 7.28 ± 0.02, with external exposure to 5% CO_2, sodium cyanide, azide, or dinitrophenol decreasing the pH and NH_4Cl increasing the pH (Boron and DeWeer, 1976); 7.35 (Spyropoulos, 1960); and about 7.0 in *L. forbesi* (Caldwell, 1958). The nitrogen content of the axoplasm is reported to be 6.5–7.8 mg nitrogen/g wet weight.

The ionic constituents of the axoplasm and sheath have been studied, and some representative values are shown in Table 4. For comparison, values in seawater and squid blood are also shown. The high concentration reported for sodium in the Schwann cell is unexpected, and may be related to the maintenance of the sodium concentration in the axolemma–Schwann cell space. The Schwann cell values are calculated values indicating that the total cellular volume of the nerve fiber sheaths in *S. sepioidea* is about 87% Schwann cell (G. Villegas *et al.*, 1965). Electron-microscopic studies also indicate that the Schwann cell is rich in sodium (J. Villegas, 1968). The Schwann cell was found to have the ability to concentrate potassium, extrude sodium, and still maintain a high internal sodium concentration. The values in Table 4 represent total concentrations; however, it should be realized that a fraction of these electrolytes may be bound to phosphate esters, nucleic acids, and proteins. Using cation-selective glass microelectrodes, it was found that if one assumes potassium to be 100% free, then sodium is only 76% free (Hinke, 1961), while chloride is bound hardly at all (Keynes, 1963). The free ionized divalent cation concentration within the axoplasm must be low, since divalent cations such as Mg^{2+} and especially Ca^{2+} are quite toxic when perfused in the giant axon (Section 2.5). The ionized calcium level in *L. pealei* was measured by recording light produced by injected aequorin and gave a value of 0.02 μM. With the use of an arsenazo dye, the value was 0.05 μM (Dipolo *et al.*, 1976). Baker *et al.* (1971) measured the level of ionized calcium in axoplasm of *L. forbesi*, also using aequorin. The internal concen-

Table 4. Concentration of Ions in Axoplasm, Sheath, and Schwann Cells of the Squid Giant Axon, and in Squid Blood and Seawater[a]

Tissue	Na^+	K^+	Ca^{2+}	Mg^{2+}	Cl^-	Ref. No.[b] and (squid)
			(mmol/liter)			
Seawater	445	10	—	—	580	1
	460	10	10	53	540	2
	423	8.3	9.3	48	506	3
Squid blood	440	20	10	54	560	2
Axoplasm	52 ± 10	335 ± 25	7 ± 2	20 ± 6	135 ± 14	1 (*S. sepioides*)
	65	344	3.5	10	151	4 (*L. pealei*)
	81	409	—	—	150	4 (*D. gigas*)
				6.4 ± 0.8		5 (*L. forbesi*)
	65 ± 10	344 ± 20	7 ± 5	20 ± 10	140 ± 20	6[c] (*L. pealei*)
					108 ± 2	7[d] (*L. forbesi*)
	50	400	—	10	(40–150)	2
	44	360	—	—	36–75	8 (*L. pealei*)
	46	323	—	—	—	9 (*L. forbesi*)
Schwann cell	312	220	—	—	167	1 (*S. sepioidea*)
	(241–404)	(157–308)			(138–208)	
Sheath		353	—	—	—	10 (*L. pealei*)

[a] Concentrations in parentheses are ranges. Some values are presented as means ± S.E.
[b] References: (1) G. Villegas *et al.* (1965); (2) for references from which these summary data were obtained, see Hodgkin, (1965); (3) Cavanaugh (1975); (4) Deffner (1961b); (5) Baker and Crawford (1972); (6) Koechlin (1955); (7) Keynes (1963); (8) Steinbach and Spiegelman (1943), Steinbach (1941); (9) Keynes and Lewis (1951); (10) Coelho *et al.* (1960).
[c,d] Results are shown as: [c] μeq/g axoplasm; [d] mmol/kg axoplasm.

tration of ionized calcium was about 0.3 μM. Cyanide released internal calcium and ATP antagonized the action of cyanide. Stimulation increased the level of ionized calcium in proportion to the frequency of stimulation. The ionized Mg^{2+} level has been reported to be much higher, 2.0–3.5 mM. (Baker and Crawford, 1972; Brinley and Scarpa, 1975). This would agree with the 20-fold greater diffusion constant of injected radioactive Mg^{2+} (2×10^{-6} cm²/sec) than of Ca^{2+} (Baker and Crawford, 1972) and the finding that the axoplasm is about 9 times richer than the optic lobes of the squid in a calcium-binding "brain-specific" acidic protein that is highly concentrated in the nervous system of cephalopods (Alema *et al.*, 1973).

Deffner and Hafter (Deffner and Hafter, 1959a,b,1960a,b; Deffner, 1961a,b) have studied the axoplasmic constituents in *L. pealei* and *D. gigas*. Some earlier studies are reviewed by Schmitt and Geschwind (1957). Using paper electrophoresis, Deffner and Hafter (1959a) separated the dialyzable portion of the axoplasm (about 70% of the total axoplasm) into 11 fractions. Sixteen free amino acids were detected in amounts ranging from 0.02 to 0.06% for methionine to about 10% for aspartic acid and taurine (percentage of dry weight of dialyzable material from axoplasm). The concentrations of taurine and neutral amino acids were lower in *D. gigas* than in *L. pealei* axoplasm. A sulfonic acid derivative of taurine, isethionic acid ($HOCH_2CH_2SO_3$), had been found to be the main organic anion present in squid axoplasm (Koechlin, 1954, 1955), that is, about 20% of the dry weight of dialyzed axoplasm (Deffner and Hafter, 1960*b*), equivalent to a concentration of 165 mM (Deffner, 1961*a*), which is in fair agreement with the value of 220 ± 20 found by Koechlin (1955) and the value of 143 ± 22 found by Hoskin and Brande (1973). The metabolic pathway for isethionate formation is not known. A small amount of the transported cysteine is metabolized via a "taurine pathway" to hypotaurine but not to isethionate (Hoskin *et al.*, 1975; Hoskin and Kordik, 1975). Some other values for axoplasmic constituents were reported by Deffner (1961*a*), Lewis (1952), and Koechlin (1955) for aspartate, alanine, glutamate, taurine, glycine, homarine, and betaine. A large fraction of the anions of squid axoplasm are thus organic, with isethionic acid having a key role in the maintenance of acid–base balance. Cysteine and cystine are absent in the free state, while ornithine is present instead of histidine. Glycocoll-betaine, glycerol, myoinositol, and homarine (*N*-methyl picolinic acid) are found in the free state in the axoplasm and in total account for 23% of the dry weight of dialyzed axoplasm (Deffner and Hafter, 1960*a*). Small amounts of glucose, sucrose, fructose, and hypoxanthine were also present along with traces of quaternary ammonium bases (Deffner, 1961*a*; Deffner and Hafter, 1960*a*). The concentrations of amino acids and other constituents in the blood of *L. pealei* have been reported (Deffner, 1961*a*). The concentrations of isethionic acid, taurine, betaine, and homarine are, respectively, 104, 23, 17, and 6 times as great in axoplasm as in blood. The axoplasm of *D. gigas* was studied by disk electrophoresis; 14 protein bands were resolved and antibodies were produced in rabbits (Huneeus-Cox, 1964; Huneeus-Cox and Fernandez, 1967). Internal, but not external, perfusion with the antiaxoplasmic antibody preparation blocked conduction in 3 hr with no marked change in the resting potential. This may be due to a structural change in the macromolecules lining the inner wall of the axolemma, due to the binding of the antibodies. Antineurofilament and nonspecific antibodies had no effect.

Complete analyses of phospholipids and free amino acids in the squid

Table 5. Phospholipids in Squid Axons (L. pealei)[a]

Phospholipid	GA +	GA	ENV	AX
TLP (μg/mg wet wt)	0.38 ± 0.02	0.23 ± 0.04	0.5–2.0	0.09 ± 0.01
PE	27 ± 1	31 ± 1	28 ± 1	31 ± 2
PS	7 ± 0.3	8 ± 0.3	11 ± 1	7 ± 1
PI	2 ± 0.3	5 ± 0.2	5 ± 0.4	7 ± 1
PC	52 ± 1	45 ± 1	40 ± 2	56 ± 1
SM	12 ± 1	11 ± 0.2	16 ± 1	0 ± 0

[a] Modified from Condrea and Rosenberg (1968). Individual phospholipid values are expressed as percentages of the total lipid phosphorus (TLP). Results are recorded as means ± S.E. Abbreviations: (GA +) giant axon with adhering small nerve fibers (crudely dissected giant axon); (GA) finely dissected giant axon (no adhering small nerve fibers; (ENV) envelope (sheath) of finely dissected giant axon after extrusion of axoplasm; (AX) axoplasm; (PE) phosphatidylethanolamine; (PS) phosphatidylserine; (PI) phosphatidylinositol; (PC) phosphatidylcholine (lecithin); (SM) sphingomyelin.

Table 6. Free Amino Acids in Axoplasm and Envelope of Squid
Giant Axon (L. pealei)[a]

Amino acids	Incubation solutions (nmol/mg axon exposed to sol'n for 60 min)	Envelope (nmol/mg)	Axoplasm (nmol/mg)
Tau	9.5–21.1	37.6–47.9	81.3–137.7
Asp	3.4–7.5	16.6–28.2	33.1–39.1
Glu	1.0–2.4	4.2–8.0	6.2–8.4
Ala	1.3–2.0	1.5–2.1	1.7–2.0
Gly	0.56–1.09	1.1–2.1	4.5–5.4
Arg	0.61–0.74	0.40–1.7	0.36–2.2
Gln	0.58–0.72	0.47–0.61	0
Ser	0.18–0.48	0.30–0.44	0.34–0.35
Leu	0.21–0.22	0.18–0.18	+ to 0.23
Tyr	+	0.07–0.10	0 to +
Ile	0.08–0.09	0.05–0.10	0 to +
Orn	0.18–0.21	+ to 0.06	+ to 0.15
Lys	0.08–0.14	0–0.04	0–0
Thr	0.03–0.05	0 to +	0–0
Phe	+ to 0.09	+ to +	0–0
Val	0–0.11	0–0	0–0
Met	+ to +	0–0	0 to +
NH$_3$	3.2–3.7	1.4–3.6	8.6–9.5
TOTAL:	21.3–40.1	64.3–94.8	137.4–203.7

[a] Modified from Rosenberg and Khairallah (1974). Crudely dissected squid axons were incubated in seawater for 60 min, after which free amino acids were determined in extruded axoplasm, envelope, and incubation solutions. (+) Detected, but below level of quantitation. No histidine, proline, or half cystines were detected. The duplicate determinations for each sample are shown as ranges. The crude axons as dissected consisted of about 20% axoplasm and 80% envelope.

giant axon as determined by thin-layer chromatographic and amino acid analyzer techniques, respectively, are shown in Tables 5 and 6 (Condrea and Rosenberg, 1968; Rosenberg and Khairallah, 1974). No doubt these constituents are partly responsible for the fixed negative charges present on the axolemmal membrane (Section 3.1.1). The axon is rich in charged phospholipids (e.g., PE, PS) and in polar, free amino acids (e.g., taurine, aspartic acid, glutamic acid, arginine). The results in Table 6 are similar but not identical to what Deffner and Hafter (1959a) found; however, their methods were less sensitive and they had to pool many more axons together for a single determination. It has been suggested that a protein-bound reactive sulfhydryl group in the axolemma may be involved in the action potential mechanism, since oxidizing agents and *N*-ethylmaleimide irreversibly blocked conduction (Huneeus-Cox *et al.*, 1966). Disulfide groups were not thought to have a significant role in overall membrane structure. It should be noted, however, that stimulation of the giant axon in the presence of thiol reagents (e.g., mercuric chloride, *p*-chloromercuribenzoate) markedly potentiated their ability to block conduction (Marquis and Mautner, 1974). It was suggested that the results may be due to increased permeability to the thiols, unmasking of buried sulfhydryl groups, or reduction of disulfides to thiols. Only molecules capable of reacting with thiol groups produced effects that were potentiated by stimulation.

The fibrous proteins in the axoplasm of squid have also been studied (see Section 3.1.1) (Bear *et al.*, 1937*a,b*; Davison and Taylor, 1960; Maxfield, 1953; Maxfield and Hartley, 1957). The intact fibers in the axoplasm are thought to consist of one or several continuous protein backbones, probably as chains of globular units. These axoplasmic fibrils comprise about 70% of the axoplasmic protein. Conditions for the preservation of axoplasmic microtubules and for the isolation from *D. gigas* of the acidic protein tubulin (molecular weight $\approx$ 60,000), which has the property of binding colchicine, have been described (Davison and Huneeus, 1970). The binding of colchicine has been taken as a useful marker for the presence of microtubules. The binding of colchicine to squid axoplasm was the highest of 20 tissues tested (Borisy and Taylor, 1967). The neurofilaments did not bind heavy mero-myosin; however, the axoplasm does contain filaments that react with heavy meromyosin to form typical arrowhead complexes. A protein that comigrated with rabbit actin was also observed in sodium dodecyl sulfate (SDS)–poly-acrylamide gels of isolated axoplasm, suggesting that an actin-like protein is present in squid axoplasm (Witman and Rosenbaum, 1973).

Other information concerning lipids, proteins, and enzymes present in the squid giant axon will be found in Section 3.4.

3.1.3. Membranal Permeability

The permeability properties of the squid giant axon membrane seem similar to those of other biological membranes, and may be summarized as follows: The ability of a compound to penetrate from the external medium into the axoplasm of the squid giant axon depends on the extent of the nonpolar lipophilic character of the compound (Hoskin and Rosenberg, 1965). Water-soluble ionized compounds do not penetrate at all and un-ionized water-insoluble compounds penetrate readily, while partially ionized compounds show intermediate behavior. Certain compounds are exceptions to this general rule, for example, glucose, which, although lipid-insoluble, penetrates readily, no doubt because of an active transport process (Hoskin and Rosenberg, 1965). Extensive studies by Narahashi and his group (see Sections 2.5 and 3.5) clearly show that the free base (uncharged) form of tertiary amines, such as local anesthetics, is the form that penetrates across the squid membrane. Dettbarn *et al.* (1972) also found that the free base form of procaine and atropine penetrates the squid axon membrane, whereas the cationic form cannot penetrate. Levorphanol penetrates into the axoplasm of the squid giant axon much better at pH 8, at which there is a higher concentration of uncharged free base molecules, than at pH 6 (Simon and Rosenberg, 1970). Cations, such as sodium, potassium, and calcium, are more penetrant than are anions. The results obtained with ionic fluxes suggest that there are pores in the membrane that specifically allow the aforementioned impenetrant ions to penetrate both at rest and during electrical activity, although as discussed elsewhere the relative permeabilities and the equivalent radii of the pores will differ greatly under different conditions (see Section 3.1.1).

Extensive studies have shown that phospholipase A_2 is the component present in certain snake venoms, such as cottonmouth moccasin venom, that increases the permeability of the squid giant axon to any poorly penetrating compound (Brzin *et al.*, 1965*b*; Condrea and Rosenberg, 1968; Hoskin and Rosenberg, 1964, 1965; Rosenberg, 1970, 1976; Rosenberg and Condrea, 1968; Rosenberg and Hoskin, 1963, 1965). The effects are, however, not directly due to disruption of membranal phospholipids but are rather due to the detergent action of lysophosphatides liberated as a result of phospholipase action. A summary of some of our results on the penetration of various compounds in the absence and in the presence of cottonmouth moccasin venom is shown in Table 7. While all these experiments were carried out on crudely dissected giant axons, we found that the penetration of acetylcholine (4.5×10^{-3} M) is identical in the finely dissected squid giant axon (1.1–1.6%) (Rosenberg and Hoskin, 1965). This would indicate that the connective tissue and small nerve fibers do not constitute the major permeability barrier to the penetration of externally applied compounds. The low levels of apparent penetration ($<2\%$) of lipid-insoluble compounds (in the absence of venom pretreatment) may represent impurities in the radioactive material, contamination during extrusion of the axoplasm, or actual penetration. While measuring cholinesterase activity of the squid giant axon, it was observed that even very small pieces of axon have permeability barriers to the penetration of the substrate (acetylcholine) (Brzin *et al.*, 1965*a*), and that cottonmouth moccasin venom or phospholipase A_2 will increase the apparent cholinesterase activity of the intact squid giant axon by allowing greater access of the substrate (acetylcholine) to the enzyme of the axon (Rosenberg and Dettbarn, 1964). Choline, neostigmine, and physostigmine decrease the penetration of acetylcholine and the effects of acetylcholine on the action potential of axons pretreated with cottonmouth moccasin venom (Hoskin and Rosenberg, 1964). These results indicate that there may be a competition for pathways of penetration.

The permeabilities of urea, thiourea, ethylene glycol, urethane, and toluene were measured in perfused axons, and permeability constants were found that ranged from 0.8×10^{-6} cm/sec for urea to 0.8×10^{-4} cm/sec for toluene and tritiated water. Decreasing the temperature from 18° to 5°C gave a 12–50% decrease in permeability of the aforenamed agents. It was concluded that the axonal membrane has a nonhomogeneous composition (Hidalgo and Latorre, 1970*a*). Penetration of sugars into the axoplasm was studied by Krolenko and Nikol'skii (1967). Insulin and raffinose could not penetrate, while the permeabilities of arabinose, fructose, and sucrose were about 2×10^{-7}, 2×10^{-7}, and 0.35×10^{-7} cm/sec, respectively. The Q_{10}

Table 7. Penetration of Compounds into the Axoplasm of the Squid Giant Axon (L. pealei) Following External Application[a]

Compound	Concentration (M)	Control	Penetration (%)[b] following pretreatment with cottonmouth moccasin venom	
			15–25 µg/ml	50–100 µg/ml
Acetylcholine	9×10^{-2}–9×10^{-5}	0.5–2.1	5.5–13	15–34
Acetylcholine +	4.4×10^{-2}	0.3–0.8	1.1–3.4	12–60
Physostigmine	2.4×10^{-4}	—	—	—
Acetylsalicylic acid	4.5×10^{-3}	26–47	—	—
γ-Aminobutyric acid	4.5×10^{-3}	2.8–3.9	5.2–16.5	22–24
Aspartate	4.5×10^{-3}	1.8–2.6	—	—
Choline	5.4×10^{-2}	0.2–0.5	3.8–14.2	32–48
Cortisol	2.5×10^{-4}	93–127	—	—
Dehydroepiandrosterone	2.5×10^{-4}	3.7–4.3	—	—
Dieldrin	1.9×10^{-6}	43–131	—	—
Diphenylhydantoin	2.5×10^{-4}	56–81	—	—
DOPA	4.5×10^{-3}	0.4–2.4	—	8.8–67
Dopamine	4.5×10^{-3}	2.5–3.9	—	62–78
Glucose	4.5×10^{-3}	19–23	—	—
Glutamate	4.5×10^{-3}	1.7	7.1	11–45
Glutamine	4.5×10^{-3}	0.8–3.6	—	18–72
Indoleacetic acid	1×10^{-3}	11–30	—	—
Mannitol	4.5×10^{-3}	0.6–1.1	—	—
Neostigmine	5×10^{-2}	0.02–0.08	0.02–0.10	0.5–1.0
Paraoxon	1×10^{-2}	5–6	3–4	—
Physostigmine	1×10^{-5}	30–90	30–90	—
Serotonin	4.5×10^{-3}	5.0–8.0	—	50–58
Sucrose	4.5×10^{-3}	0.3–2.7	—	36–41
Trimethylamine	1.1×10^{-2}	113–117	—	95–115
D-Tubocurarine (dimethyl)	1.1×10^{-3}	0.2–0.6	3.1–4.7	3–57

[a] Based on studies of Rosenberg and Hoskin (1963, 1965) and Hoskin and Rosenberg (1965).
[b] The percentage penetration is the percentage of the external concentration prevailing within the axoplasm. For purposes of calculation, it is assumed that 1 mg axoplasm contains 1 µl solution into which the drug can diffuse. Crudely dissected axons were exposed to the externally applied radioactive compound for 1 hr, and the axoplasm was then extruded and radioactivity counted.

for arabinose penetration is about 1.9. Mullins (1966) studied the efflux of sucrose after injection into the axoplasm, observing a permeability of about 1 Å/second that was uninfluenced by stimulation or temperature change. It was concluded that the membrane is impermeable to sucrose and that its efflux is through leakage pathways, which have previously been described electrophysiologically. Tasaki and Spyropoulos (1961) determined the following "time constants" for the efflux of radioactive compounds injected into the axoplasm (hr): sucrose, 50–150; starch, 150–200; cesium, 5–15; thiourea, 0.5–1.5; urea, 0.5–1.5; guanidine, 4–15; choline, 10–25. Guanidine, cesium, and choline efflux were increased by stimulation. R. Villegas *et al.* (1965*b*, 1966) studied the penetration of erythritol, mannitol, and sucrose at rest and during stimulation (in parentheses), finding values of 2.9 (5.2), 2.3

(4.0), and 0.9 (1.8) $\times$ 10^{-7} cm/sec, respectively. Increase in nonelectrolyte penetration caused by stimulation might be due to an increased area for diffusion in the axolemma or a drug effect of sodium ions entering the axon, since sodium ions and nonelectrolytes may share a common pathway across the axolemma. The magnitude of the increase due to stimulation decreased as the axon diameter increased, which correlated with an increased thickness of the axolemma (R. Villegas *et al.*, 1971). Sodium influx and electrical properties were independent of axon size, suggesting that there is a specific reduction in the size of pathways used by nonelectrolytes. This explains why Mullins (1966) and Hidalgo and Latorre (1970*a,b*) found no increase in nonelectrolyte permeability of stimulated axons, since they used large-diameter axons.

Rojas (1965) suggested that protein on the inside of the axolemmal membrane is essential for the maintenance of cell-membrane permeability and properties. This was based on the well-known findings that trypsin and other proteases have no effect when externally applied to the squid giant axon, but block conduction, decrease membrane resistance, and increase sodium and potassium effluxes when applied internally. I should caution, however, that these results may merely mean that externally applied trypsin and other proteases cannot reach the axolemmal membrane and may therefore not signify anything as regards the relative importance of the inner or outer aspects of the axolemmal membrane.

The great majority of phospholipids in the envelope of the finely or crudely dissected squid giant axon are not essential for the permeability or conduction properties of the squid giant axon. For example, phospholipases A_2 and C, when externally applied, penetrate into the axoplasm (Rosenberg, 1975), hydrolyze the great majority of phospholipids both in the axoplasm and in the axon sheath (Condrea and Rosenberg, 1968; Rosenberg, 1970; Rosenberg and Khairallah, 1974), and yet have no effect on conduction or permeability. Apparent effects of phospholipase A_2 were actually due to detergent properties of liberated lysophosphatides. After phospholipase C pretreatment, the great majority of the polar head groups of the envelope leave the axonal sheath, and yet its permeability properties are not altered. While the axolemmal phospholipids make only a small contribution to the total phospholipids measured in the axonal sheath, the fact that axoplasmic phospholipids are hydrolyzed and that these enzymes penetrate into the axoplasm would suggest that axolemmal phospholipids were also affected. The ability of proteins such as phospholipase A_2 and C to penetrate into the axoplasm is surprising, although their hydrolytic activity may aid such penetration (Rosenberg, 1975). It has also been reported, however, that ^{125}I-labeled human serum albumin can penetrate into the axoplasm. Electrophoretic behavior indicated that the intact molecule had penetrated. Electrical stimulation increased uptake about 2- to 3-fold (Giuditta *et al.*, 1971). The penetration of these high-molecular-weight proteins may proceed through pinocytotic and exocytotic mechanisms from Schwann cell to the giant axon.

In testing the effects of drugs on the squid giant axon, it is important to realize that a lack of effect may not be due simply to lack of an adequate "receptor," but may also be due to permeability barriers that prevent access of the compound to the appropriate site of action. It is therefore not surprising that lipid-insoluble compounds such as acetylcholine and tubo-curarine (curare) have no effect on conduction in the squid giant axon, and this lack of effect cannot be used as evidence against the proposal that acetylcholine is essential for axonal conduction (Nachmansohn and Neumann, 1975). In fact, reversible block of conduction by acetylcholine and curare is observed (see Fig. 10) after the permeability of the preparation is increased. See Section 3.4.4 for further details concerning the possible nature of a cholinergic system in the squid giant axon.

3.2. The Electrical Spike

The property of the squid giant axon, and indeed of other axons, that has been most extensively studied is its electrical spike. This is not surprising, considering that electrical impulses are the most appropriate stimuli for axons and the major function of the axon is to transmit excitation (an electrical stimulus) from one point to another. For reasons that have already been discussed, studies on the squid giant axon have contributed more to our knowledge of the electrical properties of axons than studies of any other preparation. These studies have been noted in many texts and reviews, including those by Cole (1968), Hodgkin (1965), and Tasaki (1968, 1975), and some of the most important references, especially those by Hodgkin and Huxley on the ionic basis of the action potential, are included in *Cellular Neurophysiology: A Source Book* (Cooke and Lipkin, 1972). Some additional references that may be consulted for background information on the electrophysiology of axons (although not always specifically the squid giant axon) are those by Nastuk (1964) on electrophysiological methods of stimulation and recording; by Camougis and Takman (1971) on the use of isolated nerve preparations; by Katz (1966) on the initiation and transmission of signals within nerve, muscle, and synapse; by Aidley (1971) on the physiology of excitable cells, including methods of recording and theories of nerve conduction; by Bureš *et al.* (1967) on the electrophysiological methods in biological research; and by Kuffler and Nicholls (1976), who have recently co-authored an excellent new text on conduction and transmission in axon and synapse, in the periphery, and in the central nervous system.

The squid giant axon is unique not only for its large size, but also because in certain axons such as those of *S. sepioidea*, the Schwann cell (up to 5 μm thick) and endoneurial cell layer (up to 20 μm thick) are large enough to allow the simultaneous measurement of three different and independent potentials: (1) intracellular potential of endoneurium cells (-10 to -26 mV); (2) Schwann cell potential (-33 to -46 mV); (3) axon resting potential (-50 to -65 mV). When the nerve was stimulated, action potentials were registered only from the axon, clearly showing that the axolemma is the site of bioelectricity (R. Villegas *et al.*, 1962, 1963). The Schwann cell is, however, hyperpolarized following a train of impulses in the axon or following a series of depolarizing (but not hyperpolarizing) pulses. The hyperpolarizing response is blocked by external trypsin (J. Villegas, 1972).

An exciting recent development in the electrophysiology of the squid axon is the observation of the gating current, which may be related to the Hodgkin–Huxley "m" system. The sodium and potassium permeabilities in the axonal membrane increase greatly when the membrane is depolarized. This permeability increase must involve the movement of charged structures within the membrane that change position or conformation. These membrane molecules may therefore serve as "gates" for the hydrophilic ionic pores. Armstrong and Bezanilla (1973) first detected small gating currents using internal-perfusion, voltage-clamp, and signal-averaging techniques. This work was then confirmed in several later studies. The gating currents are best observed when the sodium and potassium currents are blocked [e.g., tetrodotoxin (TTX) externally and Cs internally] and the axon is subjected to exactly equal positive and negative voltage-clamp pulses. The gating currents are seen as an asymmetric transient outward current at the beginning of the pulse and a transient inward current at the end of the pulse. The gating currents observed seem at present to be associated with the sodium channel. This exciting development may lead to a better understanding of the membranal events associated with control of ionic conductance.

Some of the major studies, primarily the more recent ones, dealing with the electrical properties of the squid giant axon are listed below chronologically by topic. Although space does not allow a critical review of these findings, the reader is referred to the original references for additional information. Some major studies, such as the papers by Hodgkin and Huxley (1952*a–d*), that are mentioned elsewhere in this chapter are not included in this section. The references are separated into those concerned primarily (but not necessarily exclusively) with: (1) calcium and magnesium; (2) cesium and rubidium; (3) gating currents; (4) noise measurements; (5) passive properties and leakage current; (6) potassium; (7) sodium; (8) temperature effects; and (9) miscellaneous studies.

3.2.1. Calcium and Magnesium

Hodgkin and Keynes (1957): No change in calcium efflux during stimulation, indicating that calcium movement is only inward. Calcium influx is 0.08 $pmol/cm^2$ per impulse.

Tasaki and Shimamura (1962): Membrane potential depends on external divalent cations when interior is perfused with sodium chloride, sodium sulfate, choline, and other compounds. Outward current replaces divalent with univalent ions bound to membrane.

Tasaki *et al.* (1967*b*): Increased calcium efflux occurs during excitation of perfused axon. Internal calcium decreased action potentials and membrane resistance.

Luxoro and Yañez (1968): Calcium efflux is 0.072–0.44 $pmol/cm^2$ per sec.

Rojas and Hidalgo (1968): Calcium efflux was decreased by decreased temperature and uncouplers of oxidative phosphorylation, but not by ouabain.

Baker *et al.* (1969*a*): There is coupling between inward calcium movement and outward sodium movement.

Blaustein and Hodgkin (1969): Calcium efflux (0.2 pmol/cm^2 per sec) decreased if external calcium was replaced with magnesium, or if sodium was replaced with lithium, choline, or dextrose. Calcium efflux is increased by cyanide, not affected by ouabain, and decreased by EGTA. Part of the calcium efflux is coupled with sodium entry.

Tasaki *et al.* (1969*a*): A minimal condition for excitability in perfused axon is a divalent cation externally and a univalent cation internally. Many univalent cations are good. Addition of univalent cation to external calcium chloride gives increased height of the action potentials and increased membrane conductance during excitation.

Van Breemen and DeWeer (1970): In the presence of 10 mM external calcium, 50% of the calcium efflux is blocked by 0.23 mM lanthanum, while removal of external calcium reduced the value to 0.10 mM. Lanthanum has a high attraction for negative calcium-binding sites.

Baker *et al.* (1971): The early component of calcium entry with depolarizing pulse is blocked by tetrodotoxin, while the late component is not affected. Early is due to leak of calcium through sodium channels (1% of sodium permeability).

Baker and Crawford (1972): Influx and efflux of magnesium are about 1 pmol/cm^2 per sec. Efflux is decreased by cyanide, dinitrophenol, and choline seawater. Removal of external potassium or calcium had little effect, and removal of external magnesium increased efflux. Magnesium influx is unaffected by cyanide. Extra entry of magnesium during action potentials same as extra calcium entry.

Hallett and Carbone (1972): Calcium influx studied with injected aequorin. Inward current gave increased influx (luminescence). Time course of luminescence is discussed.

Baker *et al.* (1973*a*): Calcium influx was monitored with intracellular aequorin. Late calcium and potassium channels are distinct.

Baker *et al.* (1973*b*): Tetraethylammonium ion has no effect on tetrodotoxin-insensitive outward calcium current even though it blocks outward potassium current. Calcium does not pass through potassium permeability channels.

Dipolo (1973, 1974): Calcium efflux from axons dialyzed with 0.3 μM calcium and 5 mM ATP was 0.26 pmol/cm^2 per sec. Replacement of external calcium and sodium by Tris and magnesium decreased calcium efflux by 80%. Sodium–calcium exchange component of calcium efflux is not dependent on ATP in axoplasm. In the presence of ATP, cyanide increases calcium efflux. With ATP at less than 5 μM, the efflux of calcium was 0.11 ± 0.01. Magnitude of the ATP-dependent calcium efflux varies with external sodium calcium, and magnesium concentrations.

Inoue *et al.* (1973, 1974): Sodium fluoride or cesium fluoride was present internally. The addition of potassium chloride to 100 mM calcium chloride externally gave a change from the resting to the depolarized state. Cooling to 5°C gave a 70-mV depolarization. Addition of sodium chloride to the external calcium chloride gave increased membrane conductance and rate of propagation. There is a nonuniformity of the axon membrane.

Meves and Vogel (1973): Calcium inward currents in voltage-clamped perfused axons can produce action potentials. The preparation was perfused with cesium fluoride and sucrose and placed in 100 mM calcium chloride plus sucrose. Tetrodotoxin blocked inward and part of outward current. Slow inactivation of calcium inward current gives long-lasting action potentials.

Baker and Glitsch (1975): Review of voltage-dependent changes in permeability to calcium and other divalent cations. Depolarization increases calcium influx by two routes, one of which is blocked by tetrodotoxin. Sensitive route is sodium channel, but insensitive route is neither sodium nor potassium channel. Tetrodotoxin-insensitive route is blocked by magnesium, manganese, cobalt, and the calcium antagonist D-600.

Blaustein and Russell (1975): Fluxes of calcium and sodium in internally dialyzed axon. Calcium efflux is dependent on external sodium and calcium. There is a mobile carrier that can exchange one calcium from axoplasm with either three sodiums or one calcium (plus alkali metal ion) from external media.

Brinley *et al.* (1975): Calcium efflux is linear with internal calcium (5–100 mM). Calcium pump favors calcium over magnesium by 10^6-fold. Calcium efflux decreased by increase in internal sodium.

Brown *et al.* (1975*a,b*): The study used arsenazo III as an indicator of ionized intracellular calcium. Confirmed results with aequorin of an early calcium entry associated with early inward sodium current. Tetrodotoxin decreases calcium entry by 80%.

Meves (1975*a*): The calcium current was measured in voltage-clamped and perfused axons. External medium contained 100 mM calcium chloride and 0 mM sodium, while the interior had cesium or rubidium fluoride plus sucrose. With depolarizing steps, inward current is carried by calcium ions going through sodium channel (blocked by tetrodotoxin). The outward current at large depolarizations was partly blocked by tetrodotoxin. Voltage-clamp experiments are compared with direct measurements of calcium entry.

Mullins and Brinley (1975): The preparation was internally dialyzed with calcium. At concentrations greater than 1.0 μM, efflux was 1–3 pmol/ cm^2 per sec. There is a coupled sodium–calcium efflux. At concentrations less than 1.0 μM, the sensitivity of efflux to membrane potential is greater. Hyperpolarization increased and depolarization decreased calcium efflux.

Rojas and Taylor (1975): Resting calcium influx (10 mM external calcium chloride) of 0.016 pmol/cm^2 per sec is a linear function of external concentration. Tetrodotoxin has no effect. Extra calcium influx during activity is in two phases; the early phase is blocked by tetrodotoxin.

Rojas and Taylor (1975): Resting magnesium influx (55 mM magnesium chloride external) is 0.105 pmol/cm^2 per sec. Tetrodotoxin had no effect on the influx. Extra magnesium influx during activity was not affected by external tetrodotoxin or internal tetraethylammonium ion.

Dipolo *et al.* (1976): Stimulation increased only ionized calcium (aequorin glow) to an extent of about 0.1% of that expected by known calcium influx, suggesting calcium buffering.

3.2.2. Cesium and Rubidium

Baker *et al.* (1962*a*): Membrane discriminates between potassium and cesium during an action potential. Rubidium perfusion gave long-duration action potentials and excitability was maintained.

Pickard *et al.* (1964): Cesium applied externally has little effect; does not carry membrane currents.

Chandler and Meves (1965): Cesium was perfused in voltage-clamped axon. The delayed steady-state current decreased to 2–5% of control.

Adelman and Senft (1966): Internal cesium delays sodium conductance turnoff.

Sjodin (1966): Internally applied cesium sulfate gave long-duration action potentials. Extra potassium efflux during action potentials reduced to 7–22% of normal.

Tasaki *et al.* (1966*a,b*, 1968*a*): Internally, cesium fluoride or cesium phosphate was present, and the external medium had divalent cations. The action potentials had a long duration. Cesium displaces calcium in the membrane. Addition of external univalent cations displaces divalent ions and gives depolarization and decreased membrane resistance.

3.2.3. Gating Currents

Armstrong and Bezanilla (1974): The gating current is outward during opening of sodium pores and inward when pores close. Both are unaffected by tetrodotoxin. Gating current is due to a reorientation of charged or dipole molecules. Procedures that block sodium current also block gating current.

Keynes and Rojas (1974): Movement of mobile charges in membrane identified with gating particles responsible for controlling sodium conductance. Procaine reduced the total charge transfer and halved the time constant. Sodium current was blocked by tetrodotoxin and removal of sodium. Potassium current was blocked by perfusion with cesium fluoride.

Meves (1974, 1975*b*): The asymmetry current was measured in the perfused axon. It apears to be related to the opening and closing of sodium gates. Currents are blocked by internal glutaraldehyde, which also blocks sodium current. These currents may not simply reflect "m" system.

Ulbricht (1974): Separate structures for gating and ion discrimination. Veratridine and pronase change gating without influencing selectivity.

Armstrong and Bezanilla (1975): Methods for measuring gating current are described. Evidence that gating current is associated with the sodium channel is reviewed.

Bezanilla and Armstrong (1975): Kinetic properties and inactivation of the gating currents are discussed. The capacitative current is 50 times smaller than the sodium current.

Henderson and Gilbert (1975): Helium pressure (204 atm) prolongs early and late current. It affects kinetics of channel opening and closing.

Rojas and Keynes (1975): The relationship between displacement gating current and activation of sodium conductance is discussed. Gating current involves displacement of 3 charged particles from a blocking to an open position in the sodium channel.

Keynes and Rojas (1976): Each sodium channel behaved as though it incorporated 3 gating particles. Gating particles are Hodgkin–Huxley's hypothetical "m" particles.

3.2.4. "Noise" Measurements

Fishman (1973): Electrical noise ascribed to potassium-ion passage. Two-state open–closed conductance model not correct.

Conti *et al.* (1975): Space and voltage-clamp conditions used. Calculated that there are 330 sodium channels/μm^2, each with conductance of $4 \times 10^{-12}\ \Omega^{-1}$ and 40–70 potassium channels each with a conductance of $12 \times 10^{-12}\ \Omega^{-1}$.

DeFelice *et al.* (1975): There are three components of current noise (1/f, potassium, sodium). The potassium system noise is larger in magnitude than that of the sodium system.

Fishman *et al.* (1975a): Noise measurement of fluctuations in membrane potantial and current using "patch" voltage clamp.

Fishman (1975a), Fishman *et al.* (1975b): Conductance channels produce noise characteristic of a relaxation process. Sodium-conductance noise lower than potassium noise.

3.2.5. Passive Properties and Leakage Current

Cole and Curtis (1939), Cole (1941, 1968), Cole and Baker (1941), Cole and Marmont (1942), Takashima and Schwann (1974): Studies on the passive electrical and linear properties of the squid axon, including conductance and capacitance.

Adelman and Taylor (1961): Leakage current cannot be represented as a constant conductance. Time constant of rectification is less than 100 μsec. Lowering external divalent ions decreases leak. Leakage current is caused by the outward movement of an internal ion. Inward movement of potassium, chloride, or sodium could account for only a small fraction of the leakage current.

Rojas *et al.* (1969): Leakage current depends on external calcium and magnesium concentrations.

Takashima *et al.* (1975): Passive properties of perfused axon are similar to those of intact axon. Phospholipase and pronase increase capacitance. Presence of a frequency-dependent capacitance indicates membrane is in liquid crystalline state.

3.2.6. Potassium

Narahashi (1963): The resting potential and action potential were 16–33 and 100 mV, respectively, at 6 mM internal potassium concentration and 60 and 100 mV at 538 or 1016 mM internal potassium. With 6 mM potassium inside and 0 mM outside, low resting potential and long-duration action potentials are seen. With potassium-free media on both sides, resting potential was low or reversed, but action potential was still produced. With low internal potassium, conductance and potential relations shift along the potential axis.

Brinley and Mullins (1965): Axons were injected with radioactive potassium and isethionate. Potassium carried the depolarizing current, and sodium and potassium fluxes were measured. The resting membrane has high selectivity for potassium as compared to sodium and chloride.

Sjodin and Mullins (1967): Efflux data with potassium indicate two compartments; one is a rapidly equilibrating superficial compartment.

Adelman and Senft (1968): Resting potential changes with metal cations in the order potassium > rubidium > cesium > sodium > lithium. Site controlling late conductance is internal, in contrast to early-conductance site, which is external. There is a dynamic asymmetry of the membrane.

Mullins and Brinley (1969): With no sodium inside and with 1 μM or 4 mM ATP inside, the potassium influx was 6 and 8 pmol/cm^2 per sec, respectively. The influx increased from 8 to 19 when 80 mM sodium was added to the inside. Other phosphorylated compounds were not effective energy sources.

Armstrong (1971): Derivatives of tetraethylammonium ion injected into the axon could enter only potassium channels that had open gates, which they would then occlude. Hyperpolarization or increased external potassium helps remove them from channel. Wide inner mouth of potassium pore can accept hydrated potassium ion or the tetraethylammonium ion derivatives, while narrower inner portion can accept only potassium, not the derivatives.

Bezanilla and Armstrong (1972): With internal cesium, external sodium and lithium interfere with outward potassium current through potassium pores. Potassium pores have a wide (8 Å) and nonselective inner mouth. Remainder of pore is 2.6–3.0 Å.

Hagiwara *et al.* (1972): Cation permeability in resting state: thallium, 1.8; potassium, 1.0; rubidium, 0.72; cesium, 0.16; sodium, <0.08; lithium, <0.08.

Adam (1973): Potassium equilibrium potential may be changed during potassium current flow by accumulation of potassium in the intermembranous spaces of the Schwann cell. There is an increased magnitude and slower time rise of potassium conductance.

Ulbricht (1974): Selectivity of potassium channel: thallium > potassium > rubidium > ammonium ≫ sodium > cesium. Dissociation constant for tetraethylammonium ion and potassium channel is 4×10^{-4} M.

Landowne (1975b): Potassium efflux and membrane potential studied. Efflux rises to 8 times normal and potential decreases from 55 to 32 mV when external divalent cations are below one tenth of normal. Fluxes of thallium and potassium are compared. Different mechanisms underlie ion permeability at rest and during activity. Thallium is a useful probe for studying interaction of potassium with membrane.

3.2.7. Sodium

Hodgkin and Katz (1949b): Action potential reversibly blocked by sodium-free solutions. Action-potential height increased by hypertonic sodium, but not by sucrose. Sodium hypothesis of action potential proposed.

Keynes (1951), Keynes and Lewis (1951): Extra sodium influx of 3.5–3.7 pmol/cm^2 per impulse and resting efflux of 31 pmol/cm^2 per sec.

Hodgkin and Keynes (1955): Resting sodium influx is 32 pmol/cm^2 per sec (*L. forbesi*).

Adelman and Moore (1961): A decrease in the external calcium and magnesium increased sodium influx from 97 to 186 pmol/cm^2 per sec.

Hinke (1961): Sodium gain is 3.8 ± 0.1 pmol/cm^2 per impulse. Potassium loss is 5.6 ± 0.8 pmol/cm^2 per impulse.

J. W. Moore and Adelman (1961): Voltage-clamp used to measure intracellular sodium concentration, which increased in 1 hr from 38 to 50 mM, giving a net inward flux of 40 pmol/cm^2 per sec. Stimulation increased sodium accumulation. Additional flux associated with excitation is 1.5 pmol/cm^2 per impulse. Others found values of 3 or 4.

Tasaki (1963), Shaw (1966): With chloride or sulfate as internal anion, sodium permeability increases with influx values of 131 ± 25 (*L. pealei*) and 300 pmol/cm^2 per sec (*L. forbesi*).

Tasaki and Takenaka (1963), Tasaki and Luxoro (1964): With the same sodium concentrations inside and outside the axon, all-or-none action potentials are produced. The resting potential was not changed by replacing internal sodium with potassium. With a high internal sodium concentration, differences in anions changed electrophysiological behavior of the axon.

Adelman and Taylor (1964): A decreased level of sodium chloride (50, 25, and 10% of normal) decreased transient peak inward current and delayed steady-state current following a decrease in resting potential. Potassium current is independent of the sodium current.

Baker *et al.* (1964): Perfusion was maintained with isotonic sucrose plus 6 mM sodium chloride in an external medium containing potassium-free artificial seawater. Resting potential was 0, but full-sized long-duration action potentials (2 sec) were observed. Sodium conductance and inactivation curve shifted along voltage axis in positive direction. There is a decrease in the delayed steady-state current, and inactivation of the sodium-carrying system is retarded.

Frumento and Mullins (1964): When internal sodium concentration is one-half normal, efflux is insensitive to the external potassium concentration.

Tasaki and Luxoro (1964), Tasaki *et al.* (1965*a*): Amplitude of action potential overshoot does not agree with Nernst equation applied to sodium. Data do not agree with those of Hodgkin and Chandler (1965).

Chandler and Hodgkin (1965): Overshoot in perfused axon never exceeds sodium equilibrium potential provided low-impedance electrodes are used.

Chandler *et al.* (1965): Kinetics of sodium inactivation are the same in perfused and intact axons.

Chandler and Meves (1965): Outward sodium current decreased when internal potassium was replaced with rubidium, cesium, or sucrose. Selectivity of the sodium channel to cations is: lithium (1.1) > sodium (1.0) > potassium (1/12) > rubidium (1/40) > cesium (1/61).

J. W. Moore *et al.* (1966): Sodium and lithium traverse early transient pathway equally well, but potassium and rubidium do so only sparingly. Rubidium and potassium pass steady-state pathway, but sodium and lithium are excluded. Cesium cannot enter either pathway.

Tasaki *et al.* (1966*b*): In sodium-free external medium, excitability is maintained with organic and inorganic sodium substitutes. Tetrodotoxin depressed excitability. Sodium is not essential for excitation.

Sjodin and Beauge (1967): Radioactive sodium efflux is least in the absence of external potassium, rubidium, and cesium. Effectiveness in increasing sodium efflux was potassium > rubidium > cesium (1:0.84:0.22). Two sodium ions are actively extruded for each potassium ion taken in.

Taskaki *et al.* (1967*b*): Anesthetics acted the same in sodium-free and sodium-containing external media.

Tasaki *et al.* (1967*a*): Current, conductance, and impedance in sodium-free and sodium-containing media were measured. Product of membrane resistance and flux is a constant.

Watanabe *et al.* (1967*b*): Sodium chloride (5–50 mM) was present internally and calcium chloride (30–100 mM) externally. Gradient for sodium was reversed, but the action potential was present even if no sodium difference was present across membrane. The two-stable-state theory of excitation is discussed.

Watanabe *et al.* (1967*a*): Action potential is blocked by tetrodotoxin with or without sodium in external medium.

Canessa *et al.* (1968): Resting sodium influx is 52 pmol/cm^2 per sec (*D. gigas*) and efflux is 67.

Rojas and Canessa-Fischer (1968): As sodium fluoride inside is increased from 2 to 200 mM, the sodium efflux increases from 0.09 to 34 and influx from 42.9 to 64.5 pmol/cm^2 per sec. Net extra sodium influx at any sodium concentration is 5 pmol/cm^2 per impulse.

Sjodin and Beauge (1969): The sensitivity of sodium efflux to removal of external potassium or sodium is discussed. Magnitude of sodium efflux varies in potassium-free seawater. Efflux-stimulation sites are 98% selective to potassium ion and 2% selective to sodium or lithium.

Atwater *et al.* (1970): Time course of ionic fluxes during nonpropagated action potential agree with the Hodgkin–Huxley equation. Sodium influx is 7.13 pmol/cm^2 per impulse (12°C).

Chandler and Meves (1970*a–d*): Fibers perfused with 300 mM sodium fluoride show cardiac-like action potential and two types of sodium conductance. One is maintained with depolarization giving steady level of sodium conductance. Sodium current, potassium current, rate constants, and permeabilities were studied.

Hidalgo and Latoree (1970*b*): Stimulation of perfused axon increased sodium influx but not influx of urethane, water, ethylene glycol, or urea. Sodium and nonelectrolyte fluxes are not coupled.

Meves and Vogel (1973): Permeabilities of sodium channel to ions: sodium (10) > calcium (1) > cesium (0.6).

L.B. Cohen and Landowne (1974): Temperature dependence of sodium fluxes during action potential is lower (Q_{10} of 1.2–1.6) than predicted by Hodgkin–Huxley analysis.

Ulbricht (1974): Selectivity of the sodium channel is: sodium $\approx$ lithium > thallium > ammonium > potassium > rubidium > cesium. The dissociation constant for tetrodotoxin and the sodium channel is 3×10^{-9} M. There are estimated to be 100 channels/μm^2.

M. Cohen *et al.* (1975): Sodium-current inward density in voltage-clamped axons fits a positive cooperative homotropic reaction with at least two allosteric sites. External potassium inhibited sodium current.

Goldman (1975): Pharmacological treatments cannot distinguish between coupled and independent models for sodium conductance.

Hille (1975): An essential negatively charged acidic group is within the sodium channel (pK_a 5–6). It is part of the tetrodotoxin receptor and responsible for block of sodium current by metal and organic cations.

Hodgkin (1975): Sodium channels vary from 2.5/μm^2 in garfish olfactory nerve to 500/μm^2 in squid. Maximum conduction velocity should be observed at 500–1000 sites/μm^2.

Keynes *et al.* (1975): Number of tetrodotoxin-binding sites based on experimental and computer models is 300–600/μm^2.

Landowne (1975*a*): Radioactive sodium injected into voltage-clamped axon. In the presence of ouabain, 0.1% of sodium leaves axon per minute. Increased efflux with repetitive depolarization is blocked by tetrodotoxin. Inward sodium current becomes outward and extra efflux decreases when sodium is removed from the external medium. Extra efflux exhibits sodium–sodium exchange properties.

Levinson and Meves (1975): Number of [^{3}H]tetrodotoxin-binding sites is 553 $\pm$ 119/μm^2. Larger than in most other nerves.

Rojas and Taylor (1975): Resting sodium influx with 430 mM sodium chloride in the external solution is 27.7 $\pm$ 4.5 pmol/cm^2 per sec and is decreased by tetrodotoxin to 25.1 $\pm$ 6.2 in *D. gigas* and decreased from 50.5 to 20 in *Loligo*.

J.W. Moore and Cox (1976): A kinetic model for the sodium-conductance system more nearly matches experimental results than does the Hodgkin–Huxley model.

3.2.8. Temperature Effects

Guttman and Barnhill (1968*a*): The effects of temperature variation, tetrodotoxin, and low sodium concentration on rheobase threshold were measured.

Guttman and Barnhill (1968*b*): The Q_{10} for accommodation is 44% higher than for excitation in space-clamped axons.

Guttman (1969): Temperature dependence and frequency of oscillation near threshold were studied.

Guttman and Barnhill (1970): Repetitive firing by low calcium strongly dependent on temperature (Q_{10} of 2.7).

Guttman and Hachmeister (1971): The effects of temperature, calcium, and currents on the excitation threshold of space-clamped axons were studied. The threshold is minimal at 120 Hz. Low calcium increases optimal frequency and decreases current threshold. Increased temperature increases optimum frequency (Q_{10} of 1.8) and increases threshold current.

Wang *et al.* (1972): Allethrin has a negative temperature coefficient, blocking the action potential, and decreasing the potassium steady-state conductance and the sodium conductance, more effectively at 8° than at 23°C.

3.2.9. Miscellaneous Studies

Mullins (1959): Pore model of conductance changes analyzed to provide a mechanism for changes in ionic permeability.

Tasaki *et al.* (1961): Radioactive tracers injected into the axoplasm. The time constant for loss at rest is (hr): potassium, 8; sodium, 3; water, 0.02; chloride, 25–55; sulfate, 50–85; phosphate, 50; calcium, 0.33–0.5. Stimulation increased efflux of potassium ($5–10\times$) and sodium ($2–7\times$), but had no effect on water, chloride, sulfate, phosphate, or calcium efflux.

Tasaki (1963): Effluxes of sodium, potassium, calcium, cesium, and bromine were measured after injection and influxes measured during perfusion with potassium sulfate. Permeability to anions is smaller than to sodium and potassium.

Hirsch (1965): Repetitive response to alternating current is observed at a critical membrane potential.

Tasaki and Singer (1966): Two-stable-state theory discussed. Ion-exchange process and negative fixed charges are basis for action potential.

Agin and Schauf (1968): Mathematical description of negative conductance.

Vargas (1968): Water flux, electrokinetic phenomena, streaming potentials, and filtration coefficients were studied.

Binstock and Lecar (1969): Voltage-clamp was used in intact and perfused axons. Ammonium ion carries transient peak inward current with 0.3 times sodium permeability and delayed steady-state current with 0.3 times potassium permeability. Tetrodotoxin externally and tetraethylammonium ion internally block early and late current, respectively, even when both are carried by the same ion (ammonium ion).

Mauro *et al.* (1970, 1972): Subthreshold responses studied.

Kootsey (1975): Voltage-clamp simulation and evaluation of quality.

Adelman and Fitzhugh (1975): Hodgkin–Huxley equations modified to include properties of the external diffusion barrier.

3.3. The Optical Spike

I am following the terminology of Von Muralt (1975) in applying the term "optical spike" to include changes in light-scattering, birefringence, and induced fluorescence observed during the passage of an action potential. In the past few years, optical spikes have been studied intensively with the hope that an analysis of their properties would aid our understanding of the physicochemical processes and macromolecular changes in the membrane that must underlie bioelectricity. While studies of the electrical spike have greatly increased our understanding of the electrical manifestations and ionic changes associated with bioelectricity, they have not elucidated the molecular basis of bioelectricity. It was not until 1968 that optical spikes associated with rapidly reversible changes in membrane macromolecules were reported (L.B. Cohen *et al.*, 1968; Tasaki *et al.*, 1968*b*). there are many problems and technical difficulties associated with making these measurements because usually small changes must be sorted out from a strong background light ($1:10^4$ or $1:10^5$). A high time resolution is also required, since measurements must be made in less than 10 msec. There have been two very good recent reviews on the properties of the optical spike and the problems associated with making these optical measurements (Von Muralt, 1975; Tasaki and Sisco, 1975). Tasaki and Sisco (1975) discuss various physical problems associated with making optical measurements, including random noise, fluctuations in source-light intensity, suppression of 60 cps disturbances in averaging computer recordings, and mechanical disturbances.

It has been noted (Von Muralt, 1975) that the squid giant axon is best suited for studying the optical spike, since intracellular electrodes can be inserted, microinjections made, and perfusion of the axon interior performed.

The true value of the studies of the optical spike and whether they will enable us to pinpoint those macromolecular changes in the membrane responsible for the initiation of the action potential is yet to be demonstrated. It is, however, disappointing that voltage-clamp studies of the optical spikes show a dependence on and a correlation with the potential of the membrane, rather than with conductance. Even in those few cases in which a current dependence has been shown, there is no conductance dependence. These studies are described in the sections that follow. If the optical spikes are merely manifestations of or responses to the potential changes that occur in the membrane, they will not be useful tools for probing the membranal changes responsible for the potential changes.

3.3.1. Light-Scattering

Under light-scattering, I would include those processes that disperse light from the transmitted beam so that it can be detected by photodetector placed outside the path of the transmitted beam, and might also include reflected and refracted light. Light-scattering during the passage of an action potential may either increase or decrease, depending on the angle of observation. The squid scatters more light in the forward direction than at right angles. These turbidity changes during nerve excitation are recorded as $\Delta I/I$, where ΔI is the size of optical spike and I is the resting light-scattering. Because of the small changes, it is usually necessary to average many sweeps. Signals are usually observed with monochromatic light at all wavelengths between about 365 and 650 nm. The experimental apparatus used and an example of the results obtained are shown in Figs. 35 and 36.

L.B. Cohen *et al.* (1972*a,b*) measured light-scattering changes during the

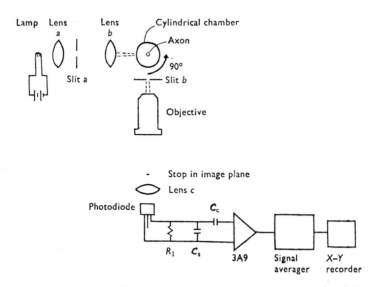

Fig. 35. A schematic diagram of the experimental apparatus for measuring light-scattering, shown in the position for measuring 90° scattering. Two Balzers B_1/K_1 heat filters were placed between lens *a* and slit *a*. The cylindrical chamber was equipped with a positionable gate about 20 mm below the illuminated portion of the axon. For action-potential experiments, the stimulus currents were passed between electrodes on either side of this gate. Reprinted from L.B. Cohen *et al.* (1972*a*) by courtesy of Cambridge University Press, New York.

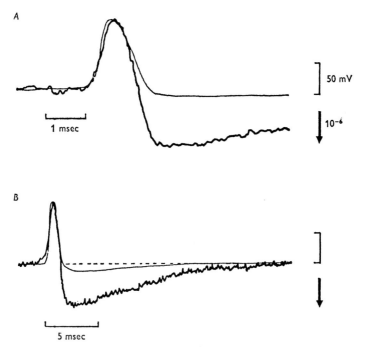

Fig. 36. Forward-angle scattering changes (heavy lines) compared with action potentials (thin lines). The changes had time courses similar to those of the action potentials at the faster (A) and slower (B) sweeps. The scattering traces were inverted to facilitate comparison with the action potential. Potential-measuring electrode. *L. forbesi:* time constant, 50 μsec; 23,000 sweeps averaged. Reprinted from L.B. Cohen *et al.* (1972*a*) by courtesy of Cambridge University Press, New York.

squid axon action potential using voltage-clamp techniques, and they observed both a light-scattering signal that varied with the potential and one that varied with the current. At forward angles, there was a decrease in the light-scattering associated with the action potential, which was shown to be potential-dependent and have a time course similar to that of the action potential (Fig. 36). At right angles, there was one scattering increase during the action potential, and a second longer-lasting increase. These were partially current-dependent. Since we are interested primarily in changes in the conformation of membrane molecules responsible for opening and closing "pores," we would be most interested in a change that could be correlated with the change in ionic conductances associated with the action potential. Unfortunately, the current-related response is not conductance-related. For example, if Na in the seawater is replaced with choline or conduction is blocked by tetraethylammonium, a depolarizing current will still give a large increase in conductance; however, no axonal current or optical spike will be produced. It must be concluded, therefore, that until we understand the molecular basis of the processes that give rise to the light-scattering spike, we cannot evaluate whether this method is of much usefulness.

Recently, the technique of light-scattering spectroscopy has been applied to the squid axons (J.W. Moore *et al.*, 1975c). In this technique, using monochromatic polarized and coherent light, the frequency distribution of the scattered light as well as its intensity can be measured. The amplitude of the spectra increased with depolarization and decreased with hyperpolarization. The authors suggest that the optical signals originate in the axon membrane, and that they reflect fluctuations in the local electrical field. Further studies will be required to evaluate this technique.

3.3.2. Birefringence

Any tissue that has oriented structural elements (e.g., nerve, hair, bone) will show birefringence. An analyzer or compensator is set in the crossed position so that a polarized light beam is completely blocked. If a birefringent material is placed between the polarizer and analyzer, some of the light will reach the detector. This can be expressed in terms of retardation of the light with the nerve introducing a varying retardation that can be offset by the compensator. Maximal light intensity is obtained when the long axis of the nerve is at about 45° to the polarizer and analyzer. An action potential induces a decrease in birefringence in the squid axon. This birefringence

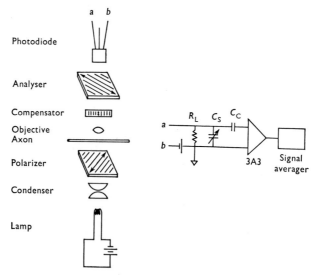

Fig. 37. Schematic diagram of the experimental arrangement for measuring retardation changes in nerve. Parallel light from a tungsten-halogen bulb was passed through a Glan-Thompson prism polarizer and focused by a long-working-distance condenser onto the nerve. An image of the nerve was formed by a 10× strain-free objective, above which was a slot for introduction of a Brace-Kohler compensator or a quarter-wave plate, and then a Polaroid analyzer. The output from the photodetector was measured across the load resistor (R_L). The d.c. and low-frequency components of the signal were eliminated by the coupling capacitor (C_c). The high-frequency response was regulated by the smoothing capacitor (C_s). Reprinted from L.B. Cohen *et al.* (1970) by courtesy of Cambridge University Press, New York.

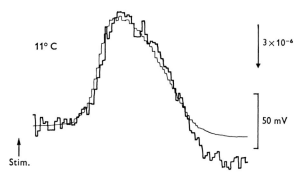

11° C

3 × 10⁻⁶

50 mV

Stim.

Fig. 38. Comparison of the intensity changes (thick lines) and the action potentials (thin lines) recorded at the same point in a squid giant axon at two temperatures, and scaled to coincide at the peak of the spike. The intensity traces have been inverted, so that an increase is downward. Light-response time constant, 16 μsec; number of sweeps averaged was 10,000 at 11°C, 11,000 at 25°C. Reprinted from L.B. Cohen *et al.* (1970) by courtesy of Cambridge University Press, New York.

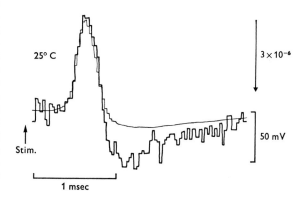

25° C

3 × 10⁻⁶

50 mV

Stim.

1 msec

signal during excitation is associated with a decrease in the difference between the refractive index of the nerve in the longitudinal and in the transverse directions. A schematic of the method used and an example of the results obtained are shown in Figs. 37 and 38. Improvements in method should now allow the recording of the optical birefringent spike in a single sweep without averaging.

Studies on birefringence were initiated by L.B. Cohen *et al.* (1968) and Tasaki *et al.* (1968a). The origin of the birefringent change was suggested as being in a thin cylinder surrounding the axon, probably the excitable membrane, and possibly due to changes in membrane thickness under the influence of an electrical field (L.B. Cohen *et al.*, 1970, 1971). The optical spike followed the time course of the action potential in fresh axons, in perfused axons, and at different temperatures (Fig. 38). There appear to be at least three potential dependent components in the birefringent response (L.B. Cohen *et al.*, 1971). Tetrodotoxin and high calcium slow the fast phase of the birefringent response, increase its size, and produce a new slow component. Tasaki and his group (Sato *et al.*, 1973; Tasaki and Sisco, 1975) also suggest that these signals originate in longitudinally oriented fibrous material near the axonal membrane, material that was shown to be digested with pronase.

As pointed out by Von Muralt (1975), membranes consist of very few sodium-channel sites but have many oriented phospholipid and other molecules that may contribute to the birefringent response that is observed during an action potential. It would seem, therefore, that there is little hope of obtaining direct information on what is happening during activation and inactivation of channels, unless the conductance change affects a large area of the membrane.

3.3.3. Fluorescence

Flourescent probes may be used to "label" squid axons, and can be applied either externally or internally. When an axon is exposed to any of a wide variety of these probe molecules and then electrically stimulated, there is a brief change in the intensity of the fluorescent light. A schematic diagram of the apparatus and an example of a result obtained with one of the most intensely fluorescent dyes are shown in Figs. 39 and 40. The optical setup must allow an adequate excitation of probe molecules at an appropriate wavelength, and an arrangement to collect the emitted fluorescent light following stimulation.

Fluorescent probes were first tried successfully by Tasaki *et al.* (1968*b*). A great number of fluorescent probes have been tested and their appropri-

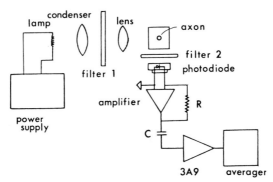

Fig. 39. Schematic drawing of apparatus used to measure fluorescence changes in axons. The light source was a 100-W quartz-halogen tungsten-filament lamp in a Schoeffel LH 150 Lamp housing. The lamp was powered by a Kepco JQE 75-15 (M) HS power supply (KEPCO, Inc., Flushing, New York) operated in the constant-voltage mode. The output of the lamp was focused onto a 5-mm length of axon by the condenser and a second quartz lens. The incident radiant flux density (intensity) at the axon was 0.1 W/cm^2 with filter 1 of 570 ± 15 nm. Between the two lenses were two heat filters (G-776-7100, Oriel Optics Corp., Stamford, Connecticut) and an interference filter, filter 1 (Oriel Optics Corp.). The barrier filters (filter 2) were made by Schott Optical Glass, Inc. (Duryea, Pennsylvania). The photodiode was an SGD 444 (E.G. & G., Inc., Salem, Massachusetts), and the amplifier used to convert current to voltage was a battery-powered Q25AH operational amplifier (Philbrick Researches, Inc., Dedham, Massachusetts). Additional amplification and high-frequency filtration were provided by a Tektronix 3A9. The output of the 3A9 amplifier was fed into one of two signal averagers, a Biomac 1000 (Data Laboratories, Ltd., Mitchum, Surrey, England) or a TDH-9 waveform eductor (Princeton Applied Research Corp., Princeton, New Jersey). Reprinted from L.B. Cohen *et al.* (1974) by courtesy of Springer-Verlag, New York.

Dye I

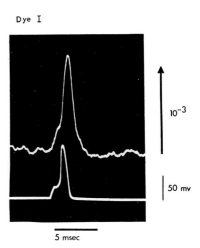

Fig. 40. Fluorescence change (*top trace*) observed with a merocyanine dye during an action potential (*bottom trace*). A single oscilloscope sweep was photographed; no signal-averaging was required. The rising phase of the electrically recorded action potential was retouched. The excitation wavelength was 570 ± 15 nm, and filter 2 was a Schott RG 610. The time constant of the light-measuring system was 600 μsec. Reprinted from L.B. Cohen *et al.* (1974) by courtesy of Springer-Verlag, New York.

ateness for these types or studies evaluated (Conti and Tasaki, 1970; Tasaki *et al.*, 1971; Tasaki *et al.*, 1969*c*; L.B. Cohen *et al.*, 1974). In these studies of Tasaki and his group, it was found that the internal application of acridinic orange and 8-anilinonaphthalene 1-sulfonate (ANS) gave, respectively, the largest increase and decrease in fluorescence during an action potential. When applied internally, a volume of about 1 mm^3 (0.1–1.0 mg/ml) is injected. The optimal concentration has to be determined, since the optical spike may decrease if the concentration is either too high or too low. It is also necessary to ascertain that the dye has no deleterious effect on the axon, and that the exciting UV light does not damage the axon. Later studies have shown that certain merocyanine dyes have the most intense fluorescence known, and the optical spike as a result of a single action potential can be recorded (Davila *et al.*, 1973). In these studies, the axon was illuminated with wavelengths of 540 ± 25 nm, and the emitted wavelengths greater than 590 nm were measured with photodiodes at right angle to the incident beam. Tasaki has studied and extensively used aminonaphthalene derivatives, such as 2-*p*-toluidinylnaphthalene 6-sulfonate (2,6-TNS) (Tasaki *et al.*, 1971, 1972, 1973*b*). The dissociation constant of 2,6-TNS binding is 0.22 mM. Of 5 × 10^{14} molecules bound/cm^2 to sites in or near the axonal membrane, it is estimated that 2 × 10^{10} molecules/cm^2 contribute to the transient decrease in fluorescence during nerve excitation (Tasaki *et al.*, 1973*a*). Another interesting approach they employed was to make a spectral analysis of the emitted fluorescent light. During the optical spike, there is a shift toward shorter wavelengths and a sharpening of the spectrum. The emission spectra of 2-*p*-chloro-ANS and other probes was studied and found to be sensitive to changes in solvent polarity. The polarity of the binding of 2-*p*-chloro-ANS was high, whereas that of a merocyanine dye was low.

The fluorescence response is entirely dependent on potential for all the dyes studied (L.B. Cohen *et al.*, 1970, 1971; Davila *et al.*, 1972, 1973, 1974).

Voltage-clamp experiments showed that the fluorescence change does not result from ionic currents or conductance increases during the action potential. These observations give rise to the suggestion that the dyes may be used for visualizing action potentials, since changes of fluorescence occur with the same time course as the action potential (Davila *et al.*, 1973). This would be a nondestructive method of monitoring membrane potentials while an experiment is being performed.

It is of course essential to know where these probes bind, and how their fluorescence would be affected by different binding sites. It would be expected that the intensity of fluorescence and its spectral distribution would be sensitive to viscosity, pH, polarity, and other physiochemical factors. Conformational changes within the membrane might alter one or more of these properties. It was suggested that the binding site of 2,6-TNS is partially exposed to surrounding water, and that relaxation of polar groups in a microenvironment of high viscosity is responsible for fluorescence emission (Carbone *et al.*, 1974). In some experiments, the exciting UV light was polarized, and the degree of polarization of the fluorescent light contributing to decrease in intensity was higher for 2,6-ANS and 2,6-TNS than for 1,8-ANS (Tasaki *et al.*, 1974). Results indicate that there is, near the membrane, a longitudinally oriented molecular structure that causes a high degree of alignment with 2,6-ANS and 2,6-TNS. Since TNS, like ANS, is a hydrophobic probe, it has been suggested that decrease in fluorescence associated with an action potential is due to a membranal transition from a hydrophobic to a more hydrophilic state (Tasaki *et al.*, 1971). These types of conclusions are suggested on the basis of the physicochemical properties of 2,6-TNS, which have been studied in various solvents and bound to various macromolecules (Tasaki *et al.*, 1973*b*). The peak wavelength, lifetime, and band width were reported for a number of mixtures. The Schwann cell and connective tissue are not thought to be involved in the fluorescent response when the probes are injected, since the membrane is impermeable to probe molecules. It is still difficult at this stage, however, to predict whether the fluorescence changes are related to conformational changes in the membrane and what will ultimately be learned about axonal structure from fluorescence changes. Von Muralt (1975) points out four trivial possibilities to explain the fluorescent spike, including a direct effect of membrane potential on fluorescence, an alignment of dye dipoles in a changing potential gradient, a concentration change of the dye during excitation, and electrophoresis of the dye into another environment.

If fluorescent dyes were available with specific ionic sensitivities, it might be possible to monitor ionic changes. In this connection, Hallett *et al.* (1972) have used tetracycline dyes (e.g., chlortetracycline) as probes of membrane-associated calcium. When stained internally, the axons showed a small increase in fluorescence during the action potential, which the authors suggest may be monitoring calcium associated with the inner surface of the nerve membrane.

3.4. Biochemical Studies

Biochemists have come to realize the advantages to be gained by studies on the squid giant axon, especially as a source of cellular cytoplasm (axoplasm). In summarizing those biochemical studies that have been applied to the squid giant axon, I shall not describe the techniques employed, since they are usually standard procedures that have also been used with other biological preparations. Because of the limited amounts of axoplasm and sheath tissue that may be available for any particular study, it is wise to select sensitive biochemical assay procedures. A reasonable estimate of the amount of tissue (wet weight) you might expect to obtain from one day's dissection by an experienced dissector would be 80 mg of axoplasm and 20 mg of finely dissected sheath or 140 mg of finely dissected whole axon, or greater than 400 mg of crudely dissected giant axon (depending on how "crude"). Relatively large amounts of fin nerve and small nerve fibers surrounding the giant axon could also readily be obtained for comparative studies. The foregoing estimates are based on the dissection of 20 axons (10 squid) of the size of those obtained from *L. pealei*, that is, axons of 400- to 500-μm diameter.

3.4.1. Energy Metabolism and Active Transport

Both cyanide and dinitrophenol decrease the survival time of the squid giant axon (Segal, 1968*a*) (see Table 2). This was observed both in regular seawater and in artificial solutions free of sodium, suggesting that the slow decrease in excitability produced by metabolic poisons is not due solely to the block of active transport. Various metabolic processes are no doubt essential for the long- and medium-term survival of the squid giant axon. While the creation and maintenance of the electrochemical gradient necessary for excitation may be one of the essential contributions of metabolism to nerve function, it is not the only function that axonal metabolism may subserve, and thence arises the necessity for studying all aspects of metabolism. It has been reported that the sodium pump in the squid giant axon does not contribute to its membrane potential, in contrast to the squid nerve cell bodies, in which such a contribution was found. This was related to the much lower membrane resistance in the axon (Carpenter, 1973). In contrast to these results, DeWeer and Geduldig (1973) found a hyperpolarizing electrogenic sodium pump that is stimulated by internal sodium and external potassium. Strophanthidin, which has effects similar to those of ouabain, depolarized the membrane (1.4–4.7 mV) and enhanced the depolarizing effect of external potassium. These effects of strophanthidin were abolished by cyanide.

ATP provides the squid nerve with the energy required for the active transport (efflux) of sodium, which is coupled to the influx of potassium (Baker and Shaw, 1965; Caldwell and Keynes, 1957; Caldwell *et al.*, 1960;

Hodgkin and Keynes, 1955). Information on ionic fluxes in the resting and stimulated state can be found in Section 3.2, although in many cases these fluxes are passive rather than involving active transport. Just as in other tissues, the cardiac glycoside ouabain, an inhibitor of Na-K-activated ATPase, inhibits active transport in the squid giant axon (Caldwell and Keynes, 1959; J. Villegas *et al.*, 1968), and is a useful tool for differentiating between passive and active fluxes. Baker *et al.* (1969*b*) observed that 50–90% of sodium efflux from *L. forbesi* axons is blocked by ouabain. The properties of the ouabain-sensitive component were different from those of the insensitive component. While TTX decreased sodium flux, it had no effect on the ouabain-sensitive flux. The inhibition of the sodium pump by ouabain was examined in greater detail by Baker and Willis (1972). The onset of inhibition was concentration-dependent and irreversible, and had a roughly exponential time course. Replacement of external sodium by choline, dextrose, or potassium slowed the rate of inhibition. Measurements of [^{3}H]ouabain-binding indicate that there are between 10^3 and 10^4 sodium-pumping sites/μm^2. Baker and Manii (1968) observed 50% inhibition of sodium efflux by 10^{-5} ouabain in about 37 sec. Baker (1968) has reviewed the similarities and differences between the ouabain-sensitive and-insensitive components of sodium flux in the squid axon. Canessa-Fischer *et al.* (1968) found that sodium efflux was inhibited 82, 57, and 64% by 10^{-5} M ouabain, 6 mM amytal, and 3 mM cyanide. Sodium efflux in natural seawater was 46 ± 10 pmol/cm^2 per sec with an influx of 52 ± 8. A complex sodium transport was demonstrated. Potassium fluoride perfused internally inhibited 92% of the sodium efflux and inhibited sodium-potassium ATPase. When glutamate and aspartate were perfused, the survival of the axons was good, but sodium efflux was only 20% of normal and not dependent on internal sodium or potassium.

Sabatini *et al.* (1968) have attempted to locate the sites of ATPase activity histochemically by reacting lead with the inorganic phosphate liberated from the splitting of ATP by the enzyme ATPase. Na-K-activated ATPase is the enzyme known to be essential for active transport processes. ATPase activity, as judged by lead deposits, was found in the axoplasmic side of the axolemma, in the Schwann cell channels, and in mitochondria (Fig. 41). In contrast to the deposits found when ATP was used as substrate, there were no deposits when AMP, ADP, glycerophosphate, or GTP was used as substrate, indicating the specificity of ATP. The authors note, however, that the histochemical procedure is essentially qualitative and that some of the ATPase activity may be lost during fixation or inhibited by the lead salts.

ATP, ADP, GTP, arginine-phosphate, and phosphoenolpyruvate all increased sodium efflux from giant axons previously poisoned with cyanide (Caldwell *et al.*, 1960). The phosphorus compounds in the axoplasm and their turnover were also studied by these authors (Caldwell *et al.*, 1964). In cyanide-poisoned axons, the injection of arginine-phosphate restored the ATP levels, and the injection of phosphoenolpyruvate restored the levels of both ATP and arginine-phosphate.

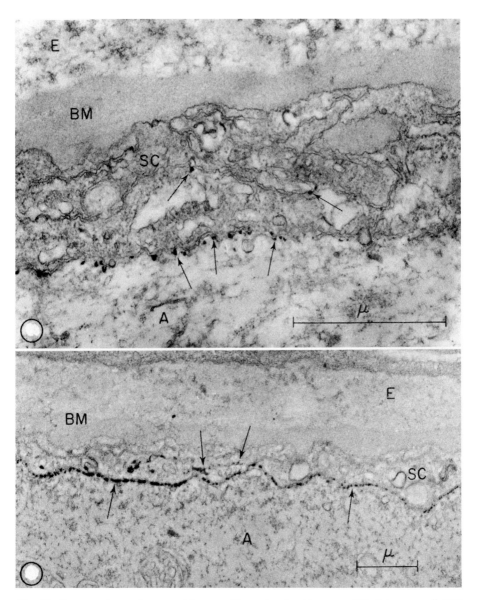

Fig. 41. Cross sections of two squid nerve fibers showing the axon (A), Schwann cell (SC), basement membrane (BM), and endoneurium (E). The arrows indicate some of the sites of ATPase activity (deposits of lead phosphate) at the axolemma and Schwann cell channels. These nerve fibers were fixed in glutaraldehyde. Frozen slices were incubated in Wachstein–Meisel medium containing ATP and $Pb(NO_3)_2$, postfixed in O_sO_4, and embedded in Epon. The sections were stained with uranyl acetate. Reprinted from Sabatini *et al.* (1968) by courtesy of Rockefeller University Press, New York.

DeWeer (1968, 1970) has shown that the dependence of sodium efflux on external potassium requires low ADP levels (not high phosphoarginine) inside the axon. Procedures that raised the ADP level, such as application of cyanide, ADP, AMP, arginine, or creatine plus creatine-kinase, all rendered sodium efflux less dependent on external potassium, even in the cases in which the axons were depleted of arginine and phsophoarginine. The sodium pump can switch from a sodium–potassium exchanging mode of operation to a sodium–sodium exchanging one, and the pump's behavior is governed by the ATP/ADP ratio. The sodium–sodium exchange is therefore seen only in the partially poisoned squid axon (Caldwell *et al.*, 1960; DeWeer, 1968, 1970). Ouabain (10^{-5} M) also rapidly inhibits the sodium–sodium exchange induced by dinitrophenol (Baker and Manii, 1968). The activation of the sodium pump by external potassium is inhibited by external sodium or lithium. Ouabain probably combines with a group that normally has an affinity for both sodium and lithium and that is involved in both the sodium–potassium and the sodium–sodium exchange. After microinjection of ^{24}Na, it was found that two to three sodium ions are transported outward for every potassium ion transported inward. Under conditions of low sodium, the coupling ratio is 1:1. Sodium efflux is approximately proportional to the sodium concentration up to concentrations of 220 mM. Total potassium influx is 20 pmol/cm^2 per sec when internal sodium is 22 mM and 36 pmol when internal sodium is 122 mM, of which about two thirds is coupled to sodium efflux, and inhibited by cyanide (Sjodin and Beauge, 1968). The relative selectivity of external ions for stimulation of sodium extrusion is K = Rb = Cs (1:0.84:0.22) (Sjodin and Beauge, 1967).

Only passive sodium movements are seen in axons exposed to seawater containing cyanide and internally dialyzed with solutions free of ATP (Mullins and Brinley, 1967). The normal ATP content in axons was estimated to be 4.4 mM and its consumption to be 43 pmol/cm^2 per sec, and ATP was shown to be responsible for more than 96% of the sodium efflux. Also measured in the axoplasm were ATPase, adenylate kinase, and arginine kinase. In contrast to the results reported with ATP, nine other naturally occurring high-energy compounds including ADP, AMP, GTP, CTP, and UTP could not support sodium extrusion (Brinley and Mullins, 1968). The relationship between internal ATP concentration and sodium efflux was nonlinear, rising most steeply in the range between 1 and 10 µM ATP, although saturation was not observed even at 10,000 µM. The relationship between sodium efflux and internal sodium concentration was linear in the range of 2–240 mM sodium. Sodium influx was reduced 70% by removal of ATP and sodium from inside the axon. A calcium-activated ATPase has been reported in the sheath of the squid giant axon (Bonting and Caravaggio, 1962; Canessa, 1965). The ATPase system in the sheath of the giant axon from *D. gigas* has been analyzed (Canessa-Fischer *et al.*, 1967). From the 100,000*g* pellets obtained after ultracentrifugation of homogenates of the giant axon sheath, a sodium-potassium-activated ATPase of high specific activity was obtained.

The effects of cations, anions, and ouabain on the activity of this preparation were also determined.

If radioactive phosphorus is to be used in studies on the metabolism of phosphorus compounds, it should be realized that the membrane may be a barrier to the penetration of aqueous soluble phosphorus or phosphorus compounds. For example, a very slow rate of ^{32}P efflux is observed after the injection of orthophosphate into the axoplasm (Caldwell *et al.*, 1964; Tasaki *et al.*, 1961). The influx of ^{32}P was 20.9 fmol/cm^2 per sec at an external orthophosphate concentration of 0.02–0.5 mM. Influx is reduced by cyanide, dinitrophenol, and ouabain, and by the absence of external potassium. Influx, although low, appears to be mediated by an active transport process that may have some connection with active transport of sodium and potassium (Caldwell and Lowe, 1970).

A microfluorometric procedure was used by Doane (1967) to follow changes in the oxidation–reduction state of the giant axon from *L. pealei* that are caused by anoxia, cyanide, azide, and other factors. The method used by the authors to measure the fluorescence emission intensity of NADH in the isolated axon is described in detail. Oxygen depletion reduced 90% of the NAD within 1–2 min. Cyanide and amobarbital also reduced NAD, while ouabain and strophanthidin had no effect. The NAD–NADH system seems to be largely confined to the mitochondria. Doane also assayed NAD, NADH, NADP, and NADPH in the axoplasm with the enzymatic cycling technique (Lowry *et al.*, 1961).

The greater production of $^{14}CO_2$ from glucose-1 than from glucose-6-[^{14}C] by whole axon and axonal sheath indicates some participation of a pentose phosphate pathway in the membrane. No activity in the axoplasm could be detected, indicating little or no activity of the citric acid cycle in the axoplasm (Hoskin and Rosenberg, 1965). In contrast, anaerobic glycolysis is much greater in the axoplasm than in the envelope, although under aerobic conditions the glycolysis rate in the axoplasm would be expected to be much lower (Hoskin, 1966).

3.4.2. Lipids and Their Metabolism

Lipids in the central ganglia and axoplasm of the giant axon were first analyzed by McColl and Rossiter (1951). They concluded that the distribution of lecithin, cholesterol, cephalins, and sphingomyelin is similar to that in human nonmyelinated axons. It was necessary, however, to reevaluate the lipid distribution using more specific methods of analysis. R. Villegas and Camejo (1968) determined the lipid composition of a fraction rich in membranes that they isolated from giant axons of *D. gigas*. The fraction consisted of 33% protein and 67% lipid. The lipid distribution (percentage of total lipids) was as follows: cholesterol, 35; fatty acids, 6; hydrocarbons, 9; sphingomyelin, 6; choline phospholipids, 18; other polar lipids, 3. Larrabee and Brinley (1968) determined the incorporation of inorganic phosphate

(^{32}P) into the axoplasm and sheath of the giant axons when ^{32}P was applied to the external bathing solution, injected into the axon, or added to separated axoplasm. The following phospholipids were labeled in both sheath and axoplasm: phosphatidylinositol, phosphatidylethanolamine, phosphatidic acid, and lysophosphatidylethanolamine, whereas, surprisingly, phosphatidylcholine (lecithin) was not labeled. Lipid metabolism was observed even in extruded axoplasm. Levorphanol increases the overall incorporation of ^{32}P into phospholipids (Dole and Simon, 1974). However, the significance of this observation is questionable, since, as is discussed in Section 4, the squid axon is not a good model for studying the analgesic properties of morphine-like compounds (Frazier *et al.*, 1973*b*; Simon and Rosenberg, 1970).

The individual phospholipids of the squid giant axon have been measured in crudely and finely dissected giant axons and in separated axoplasm and envelope. The data are shown in Table 5 (Condrea and Rosenberg, 1968). Phospholipase A$_2$ was found to be the component of snake venoms responsible for their ability to increase permeability, cause vesiculation of the Schwann cell, render the squid axon sensitive to the blocking action of cholinergic compounds (see Fig. 10), and, in higher concentrations directly and irreversibly block conduction (Condrea and Rosenberg, 1968; Hoskin and Rosenberg, 1964; Martin and Rosenberg, 1968; Rosenberg and Condrea, 1968; Rosenberg and Ehrenpreis, 1961; Rosenberg and Hoskin, 1963; Rosenberg and Podleski, 1962; 1963). Extensive splitting (25–50%) of the major phospholipids, by the phospholipase A$_2$ enzyme was found not only in the sheath but also in the axoplasm (Rosenberg and Condrea, 1968; Condrea and Rosenberg, 1968). This agrees with the finding that both phospholipase A$_2$ and phospholipase C readily penetrate in their active form from the external bathing solution into the axoplasm (Rosenberg, 1975). While these results might suggest that phospholipids are essential for axonal functioning, this may not be the case, since all the effects of phospholipase A$_2$ are not directly due to phospholipid splitting, but are secondary effects due to the detergent action of lysophosphatides liberated as a result of phospholipase A$_2$ action. Lysolecithin and a mixture of lysophospholipids mimicked all the actions of crude venom or purified phospholipase A$_2$ prepared from the venom (Rosenberg and Condrea, 1968). Further support for this conclusion was obtained from a comparison of phospholipase action on crudely and finely dissected axons. In crudely dissected axons, phospholipase A$_2$ and snake venoms had all the activities enumerated above, whereas on finely dissected giant axons they were completely inert, even though the percentage of phospholipid splitting was indentical in the two preparations (Condrea and Rosenberg, 1968; Rosenberg and Condrea, 1968). However, the amount of substrate and therefore the amount of lysophosphatides produced in the finely dissected preparation was estimated to be much lower than in the crudely dissected axon. Lysophosphatides acted equally well on crudely or finely dissected giant axons, as did a lysophosphatide mixture prepared by incubating phospholipase A$_2$ with crudely dissected axons. Further support for the suggestion that extensive splitting of phospholipids is compatible with axonal functioning, unless the hydrolytic products are themselves toxic, was obtained using phospholipase C (Rosenberg, 1970; Rosenberg and Condrea, 1968). Phospholipase C, the action of which does not give rise to toxic hydrolytic products, had no effect on conduction or permeability of the squid axon even though it split up to 100% of sphingo-myelin, 84% of lecithin, and 50% of phosphatidylethanolamine, which are the three major phospholipids in the squid axon (Table 5). Indeed, 84–100% of the phosphorylated bases liberated as a result of phospholipase C action leave the membrane, appearing in the external bathing solution (the digly-ceride remained attached to the membrane). These results would suggest that the polar head groups of the phospholipids are at the surface of the membrane rather than in its interior. This would agree with the newer theories of membrane structure (Singer and Nicolson, 1972; Vanderkooi and Green, 1970), which suggest that the polar head groups of the phospholipids are at the surface of the membrane and bound to protein primarily by hydrophobic interactions. These theories are in contrast to the unit membrane hypothesis, which suggests that phospholipids are in the membrane interior and covered by protein to which they bind primarily by electrostatic inter-actions (Robertson, 1960). We also found that disruption of hydrophobic bonding by phospholipase A_2 has a greater effect on the release of free amino acids from the squid giant axon than does disruption of hydrophilic interactions with phospholipase C (Rosenberg and Khairallah, 1974). These results also support the theory that binding between membranal proteins and phospholipids is primarily hydrophobic in nature. Our results would also appear to cast some doubts on theories that attribute vital functions to axonal phospholipids (Goldman, 1964; Blaustein and Goldman, 1966; Haw-thorne and Kai, 1970). The effects of phospholipases on the squid giant axon have been reviewed recently (Rosenberg, 1976), and the pharmacology of snake venom phospholipase A_2 has been thoroughly analyzed in a recent chapter (Rosenberg, 1977).

Abbott *et al.* (1972) reported that an enzyme preparation having phos-pholipase A and phospholipase B activity had no effect on external appli-cation, but blocked conduction and had other electrophysiological effects following internal perfusion. They concluded, therefore, that the phospho-lipids in the squid axonal membrane may have a decisive role in excitability, in contradistinction to our conclusions. They also criticize our results on the basis that while we showed splitting of "envelope" phospholipids, we could not measure axolemmal phospholipids, which would be only a very small percentage of the total phospholipids we measure in the envelope. While we did point out this criticism in our manuscripts, we also noted that phospho-lipids are split by phospholipase A_2 not only in the sheath but also in the axoplasm, and that phospholipase A_2 and C readily penetrate into the axoplasm. To therefore suppose that axolemmal phospholipids were not hydrolyzed, it would be necessary that the axolemmal phospholipids are preferentially protected against or resistant to the action of phospholipase A_2, a supposition for which there is no evidence. It is unfortunate that in the studies of Abbott *et al.* (1972), only one concentration of their phospholipase preparation was tested externally, which was found to be inactive. That concentration was equivalent to the minimal concentration that had an effect internally, so that what they interpreted as a qualitative difference between the effects of external and internal application of phospholipase may in fact have been only a small quantitative difference. They did not check the effects of lysophosphatides produced as a result of the internal application of phospholipase; therefore, their results cannot be used to support any theory as to the essentiality of phospholipids. The suggestion (Narahashi, 1974) that rapid internal perfusion would wash away lysophosphatides so they would not affect the axolemma in unlikely. The lysophosphatides would be produced right in the axolemma and would be ideally situated for exerting their detergent action. It would have been much simpler to directly test the effects of added lysophosphatides. All their results may have been due to the detergent action of evolved lysophosphatides, as we found in our studies that were described previously in this section.

It has recently been reported that the squid giant axon (*L. opalescens*) has endogenous phospholipase activity (Eyrich *et al.*, 1976). Activity was detected in centrifuged homogenate supernatants, and in extracts, although the type of phospholipase (A, A_2, B) was not definitely established.

Tetrodotoxin was shown to expand monolayers prepared from a non-polar lipid extract of squid axons, whereas the polar extract (phospholipids) did not interact with tetrodotoxin (R. Villegas and Camejo, 1968). Later studies showed that both tetrodotoxin and saxitoxin would expand cholesterol monolayers from about 39 $Å^2$ per molecule to over 41 $Å^2$ per molecule (R. Villegas and Barnola, 1972; R. Villegas *et al.*, 1970). It was suggested that cholesterol may be part of the sodium channels in the axon or of the toxin receptors. If this were so, however, then these "active" cholesterol spots should be in some ways different from the bulk of cholesterol that is a ubiquitous constituent of all membranes. The interaction of [^{14}C]-DDT with isolated membranes of the squid axon and their lipid and protein components was studied (Barnola *et al.*, 1971). DDT accumulated in the plasma membrane and reduced the fluorescence observed when protein is bound to lipid.

3.4.3. Proteins and Their Metabolism

The free amino acid composition of the squid giant axon is shown in Table 6 (Rosenberg and Khairallah, 1974). Other details of amino acids in the squid giant axon can be found in Section 3.1.2. Korey (1950) reported that amino acids such as glycine, alanine, and aspartic acid can penetrate the axonal sheath of the squid giant axon. In contrast, we found that glutamate, aspartate, and glutamine show only a low level of penetration (see Table 7) (Hoskin and Rosenberg, 1965). A sodium-dependent component of glutamate influx was found, which was sensitive to changes in the metabolic state of the cell (Baker and Potashner, 1973).

Various studies have shown that amino acids are incorporated into proteins of the squid giant axon. For example, radioactive serine, valine, and leucine were injected into the axoplasm, and radioactivity was then measured in the trichloroacetic acid (TCA)-insoluble fractions of the axoplasm and sheath (Fischer and Litvak, 1967; Fischer *et al.*, 1968). Most of the incorporation was in the sheath, the rate of incorporation was higher in stimulated than in nonstimulated axons, and the incorporation was inhibited by chloramphenicol and actinomycin D. When a mixture of radioactive amino acids was incubated with giant axons, most of the radioactive proteins were found in the particulate fraction of the envelope and in the soluble fraction of the axoplasm (Giuditta *et al.*, 1968). Puromycin, cycloheximide, and chloramphenicol inhibited the incorporation. After microinjection, tritiated uridine and ^{14}C hydrolyzed protein are incorporated into RNA (White-Oritz, 1967). However, there is no increase in uridine incorporation in stimulated axons unless stimulation precedes the injection of the uridine (Fischer *et al.*, 1969). About 60% of the incorporation was in the sheath and the remainder in the axoplasm. Incorporation at low temperatures was greatly decreased. Nevertheless, Fischer *et al.* (1969) suggest that the observed apparent incorporation may be due only to physical absorption of the uridine or to terminal addition of the tracer to RNA molecules. Pepe *et al.* (1975) checked the effect of 100 mM potassium chloride on the synthesis of protein following incubation with labeled amino acids. The salt induced about a 17% inhibition in the axon sheath and a 77% inhibition in the axoplasm. No difference was seen in the TCA-soluble radioactivity. Ouabain did not alter the potassium effect, indicating that it does not act through activation of an ionic pump. The migration patterns in gel electrophoresis of newly synthesized soluble proteins from the axoplasm and sheath of the giant axon were examined (Alema and Giuditta, 1976).

The RNA content of the axoplasm and sheath was reported as 0.42 and 2.90 mg/100 mg protein, while corresponding values for DNA were 0 and 0.64 (Fischer *et al.*, 1969). Base composition and density-gradient centrifugations of the RNA were also performed. For each experiment on RNA content, it was necessary to pool approximately 100 finely dissected axons. Similar results were found by Lasek (1970); however, he expressed the results in several different ways. If the amount of RNA was related to a 5-mm length of axon, then the amount in axoplasm (0.805 μg) was the same as that in the sheath (0.828 μg). Expressed per dry weight or per milligram protein, however, the amount in the axoplasm (0.18% and 7.39 μg) is only about one fourth that in the sheath (0.76% and 29.3 μg). No DNA was detected in the axoplasm, while the sheath had an activity of 26 μg/mg protein. While it is thought that axoplasmic proteins are primarily synthesized in the soma and moved into the axon by axoplasmic transport, the possibility was also noted that direct axoplasmic synthesis or transfer from the Schwann cell could be the source of the axoplasmic protein. Lasek noted that he and S. Tuomey have attempted to study axonal transport in the giant axon after injection of [^{3}H]leucine into the stellate ganglia (Arnold *et al.*, 1974). The sheath of the axon, however, was more heavily labeled than the axoplasm, which made interpretation of the data difficult and raised the possibility of artifactual diffusion of the label. The RNA in the axoplasm is of a low molecular weight, and this was suggested by Lasek *et al.* (1973) as being the reason for the inability of Fischer *et al.* (1968) to find RNA in the axoplasm, since their method would probably not have detected this RNA. Since the RNA in the axoplasm is more than 95% 4 S (probably transfer) RNA with little or no 28 + 18 S (ribosomal) RNA or ribosomes being found, it is doubtful that any protein synthesis occurs in the axoplasm (Lasek *et al.*, 1973). About 85% of the squid axoplasmic RNA was associated with the 27,000*g* supernatant, and only about 15% in the mitochondria-containing pellet. Using polyacrylamide gel electrophoresis, these authors found that the squid ganglia and the sheath of the giant axon had 28 and 18 S RNA in addition to smaller amounts of 4 S RNA. The function of the 4 S RNA in the axoplasm is unknown, although suggestions have been made (Arnold *et al.*, 1974). Lasek *et al.* (1974, 1975) showed that some axoplasmic proteins are synthesized in the sheath (Schwann cell) and transported into the axoplasm. Perfusion, extrusion, and radioautography techniques were used to analyze the protein synthesis that occurred subsequent to incubation of giant axons in [³H]leucine. Labeled protein was found in the Schwann cell 15–30 min prior to its appearance in the axoplasm. The synthesis was inhibited by puromycin and cycloheximide, but not by chloramphenicol, suggesting that neither bacteria nor mitochondria are the source of the synthesis. The transferred proeins, which represent about 50% of the newly synthesized Schwann cell proteins, are soluble and of 12,000–200,000 daltons. They may have some sort of regulatory role in neuronal functioning. It was suggested that there is exocytosis from the Schwann cell and pinocytosis into the giant axon. The presence of vesicles in both tissues and the decrease in appearance of axoplasmic protein under conditions of zero calcium were taken as evidence in favor of this suggestion. The observation that large molecules such as albumin (Giuditta *et al.*, 1971) and phospholipases (Rosenberg, 1975) appear in the axoplasm after external application supports the suggestion of a pinocytotic mechanism. As pointed out in the text by Arnold *et al.* (1974), the squid giant axon may be a very useful preparation for studying cell-to-cell transfer because the axon is large and incorporates amino acids into proteins, while in the axoplasm little synthesis occurs and perfusion can be used to collect labeled proteins transferred from the Schwann cell to the axoplasm.

The lactoperoxidase technique of iodinating accessible tyrosine and histidine residues was used as a probe to search for proteins in the outer and inner membranes of the squid giant axon (Gainer *et al.*, 1974). After incubation in the lactoperoxidase and Na¹²⁵I, two 13-mm perfused segments of axons were pooled, homogenized, reduced with β-mercaptoethanol, and analyzed by SDS disk electophoresis. The iodinated proteins from the external suface were probably primarily from the Schwann cell and showed a wide range of molecular weights, mostly greater than 200,000. Iodination of the internal membrane showed two major peaks of activity with molecular weights of 12,000 and 68,000. These peaks are not derived from the axoplasm, since axoplasmic proteins showed a different profile with 16 major bands. Potassium-induced depolarization decreased the iodination of the 12,000-molecular-weight peak, which may indicate a conformational change. These promising studies should be continued.

The axoplasm from *D. gigas* axons had a low level of neutral protease-like activity equivalent to about 4×10^{-6} mg chymotrypsin/mg axoplasmic protein; the activity was increased by calcium (Orrego, 1971). An acid proteinase activity (pH 4.8) was also present. These activities were also in the axonal sheath. The protease may be involved in the degradation of proteins that came down from the cell body or proteins transported from the Schwann cell.

3.4.4. Cholinergic System

The squid giant axon has been widely used for studies on the components of the cholinergic system, including acetylcholine (ACh), cholinesterase (ChE), choline acetyltransferase (ChA), and the ACh receptor. These studies are of direct relevance to the theory proposed by Nachmansohn (Nachmansohn and Neumann, 1975), that ACh and the other components of the cholinergic system are essential for all bioelectrical activity including that of the giant axon. There is also the most interesting recent observation that there may be an operative cholinergic system in the Schwann cell surrounding the giant axon (G. Villegas, 1975).

The content in the squid giant axon of ACh (μmol/g), ChA (μmol ACh formed/g per hr), and ChE (μmol Ach split/g per hr) were reported to be about 0.02, 1.8, and 2.1, respectively (Rosenberg et al., 1966a; Webb *et al.*, 1966). These values are lower than those for the squid ganglia. Histochemical staining of the ChE in the small nerve fibers surrounding the giant axon is shown in Fig. 6. The ChE activity in the axoplasm is low (0.14–0.56 μmol ACh hydrolyzed/g per hr) compared to that in the envelope (2.1–2.5 μmol/g per hr), while the reverse is true for ChA (0.055 μmol ACh formed/g per hr in envelope and 0.75 in axoplasm) (Boell and Nachmansohn, 1940; Brzin *et al.*, 1965a; Rosenberg *et al.*, 1972; Kremzner and Rosenberg, 1971). Treatment of the intact crudely dissected giant axon with snake venoms or phospholipase A_2 prior to determination of ChE activity increases the apparent activity (Rosenberg and Dettbarn, 1964). The treatment, by disrupting permeability barriers, allows higher concentrations of the substrate (ACh) to reach the enzyme (ChE). Using a microgasometric technique with a magnetic diver, the ChE activity per square millimeter of surface of the finely dissected extruded giant axon has been estimated to be 9.5×10^{-5} μmol/hr and 2.1 μmol/g per hr (Brzin *et al.*, 1965b). There were permeability barriers even in small fragments of the cell wall, as evidenced by the significantly lower values (1.05 μmol/g per hr) that were obtained if the bits of tissue were not exposed to sonic disintegration.

In attempting to determine the essentiality of ChE for conduction as proposed in the theory of Nachmansohn (Nachmansohn and Neumann,

1975), various ChE inhibitors were applied to the squid giant axon, and effects on electrical and ChE activity were measured. Diisopropylfluorophosphate (DFP), a potent inhibitor of ChE, blocks conduction irreversibly only if applied in high concentrations (10^{-2} M) for long periods of time (Bullock *et al.*, 1946; Hoskin *et al.*, 1966). High amounts of a DFPase, and enzyme that hydrolyzes DFP, was found in the axon, which could explain the high concentrations required for block of electrical activity (Hoskin *et al.*, 1966). This so-called DFPase enzyme was purified from the squid 1300-fold and is markedly different from mammalian and bacterial DFPases (Hoskin and Long, 1972). Virtually all the DFPase was recently reported to be located in the axoplasm, not in the envelope of the giant axon (Hoskin, 1976). This would suggest that DFPase is not a barrier to the penetration of intact DFP into the conducting membrane of nerves. Other ChE inhibitors such as Sarin, Tabun, and Soman are also hydrolyzed, although at a lower rate than DFP, by the squid giant axon (Hoskin, 1971). High concentrations of Tabun had little effect on the action potential, even though it readily penetrates in its active inhibitory form into the axoplasm, is a more potent inhibitor of ChE than DFP, and is detoxified at only one tenth the rate of DFP. Other potent inhibitors of ChE such as the tertiary anaologue of Phospholine, Selenophos, and Tetriso are also very weak in their effects on conduction, even though they are not hydrolyzed by the axon and readily penetrate into the axoplasm in their active inhibitory form when externally applied (Hoskin and Rosenberg, 1967; Hoskin *et al.*, 1969).

To evaluate whether ChE is essential for axonal conduction, it is necessary to accurately measure the ChE activity following exposure to the aforementioned inhibitors, since it is possible that they penetrated without inhibiting the enzyme. It is difficult to accurately measure enzymatic activity such as ChE activity following exposure to an inhibitor. If the tissue is homogenized, excess inhibition may occur *in vitro*, while if intact tissue is used, a substrate must be selected for the *in vitro* enzymatic assay that will penetrate to all the enzyme (Dettbarn and Rosenberg, 1962; Hoskin *et al.*, 1969). In addition, a single-fiber preparation such as the squid giant axon should be used, since otherwise only average values are obtained and it becomes difficult to accurately relate electrical and ChE activity. Since reversible inhibition may also be reversed during the enzyme assay, on the addition of substrate, it is preferable to use irreversible ChE inhibitors. Recently, we found that acetyl-β-methylcholine, an ester that is hydrolyzed by acetyl-ChE (the specific enzyme as contrasted to pseudo-ChE), can penetrate and apparently reach all the ChE in the intact giant axon (Kremzner and Rosenberg, 1971). Using this substrate, it was found that reversible conduction block after exposure to 10^{-2} M DFP was associated with 70% ChE inhibition, whereas a long exposure and irreversible conduction block were associated with 99% inhibition of the enzyme. When ChE activity was measured, however, after a brief exposure to 5×10^{-2} M DFP, which also caused irreversible conduction block, there was only about 70% inhibition. The tertiary analogue of Phospholine had no effect on conduction even though we found 98–99% inhibition of ChE. From these results, it was not possible to demonstrate any direct relationship between electrical and ChE activity.

In about half the experiments, pyridine aldoxime methiodide (PAM), a specific reactivator of organophosphate-inhibited ChE, restored the irreversible conduction block produced by DFP and Paraoxon (see Fig. 9) (Rosenberg and Dettbarn, 1967; Kremzner and Rosenberg, 1971). These experiments would seem to indicate an essential role for ChE in conduction; however, we could find no significant difference in ChE activity between those experiments in which PAM restored activity and those in which it had no effect (Kremzner and Rosenberg, 1971).

Certain styrylpyridine analogues are potent inhibitors of ChA, and it was hoped that they could be used to evaluate the essentiality of this enzyme for axonal functioning. The inhibitors did block conduction at a concentration of about 5×10^{-3} M; however, it was shown that this was a direct action on the nerves not related to their inhibition of ChA (Rosenberg *et al.*, 1972). Hemicholinium is a potent inhibitor of the synthesis of ACh that blocks conduction when internally perfused in the giant axon (Frazier *et al.*, 1969, 1970). This action has, however, been ascribed to a direct noncholinergic mechanism, since neither choline nor acetylcholine is able to block the action of hemicholinium.

As noted elsewhere (Sections 3.4.2 and 3.5.4 and Fig. 10), cholinergic compounds such as ACh and curare block conduction after snake venom or phospholipase A_2 treatment, whereas they are inactive without treatment due to their inability to penetrate. We attempted, therefore, to determine whether the "receptor" for these compounds in the axon has properties similar to that in the junction of the isolated single electroplax, a typical cholinergic preparation. A large series of selenium isologues of benzoylcholine and its tertiary analogue were compared for their ability to block conduction in control and venom-treated giant axons (Rosenberg *et al.*, 1966*b*; Rosenberg and Mautner, 1967). Their relative potencies on the axon were quite similar to those on the electroplax junction, suggesting that the "receptors" in the junction and axon might be similar. In contrast to this conclusion are the experiments of Yeh and Narahashi (1974*b*), who perfused cholinergic compounds such as ACh, curare, decamethonium, and hexamethonium into the voltage-clamped axon. In most cases, only weak effects were seen (Section 3.5.4), and the effects that were seen were different from those on the postsynaptic membrane. It was concluded, therefore, that their action is noncholinergic in nature, and that typical ACh synaptic-type receptors are not present in the axon. The results seem fairly conclusive, although it could be argued that the "real" cholinergic receptor of the axon is imbedded in the membrane, and accessible from neither the outside nor the inside to exogenously applied cholinergic compounds, whereas the receptor would be accessible to physiologically released ACh, assuming that this occurs.

Recently, evidence has been accumulated by Villegas that an operative cholinergic system is present in the Schwann cell of the giant axon (G.

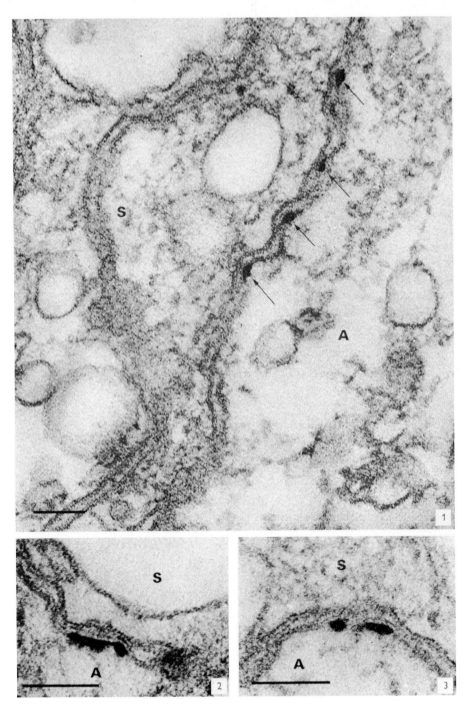

Fig. 42. (1) Electron micrograph of squid giant nerve fiber incubated in medium lacking inhibitors. The thiocholine end product appears as dense deposits focally distributed along the axolemma and attached to the inner aspect of this membrane (←). No precipitate is present in

Villegas, 1975; J. Villegas, 1975). Following conduction of a train of impulses in the giant axon (*S. sepioidea*), there is a long-lasting hyperpolarization of the Schwann cell, which is measured by inserting microelectrodes into the cell. Physostigmine at a concentration of 10^{-9} M prolongs while 10^{-7} and 10^{-4} M decreases and blocks this hyperpolarization (J. Villegas, 1973). d-Tubocurarine at 10^{-9} M blocked this hyperpolarizing response of the Schwann cell. These compounds had no direct effect on the action potential of the giant axon. ACh (10^{-7} M) and carbamylcholine (10^{-6} M) directly hyperpolarized the Schwann cell, and the ACh effect was prolonged by 10^{-9} M physostigmine (J. Villegas, 1974). The carbamylcholine effect was not blocked by a 100-fold decrease in the external sodium or by 5×10^{-8} M tetrodotoxin. Carbamylcholine did, however, increase the permeability of the Schwann cell to potassium. α-Bungarotoxin (10^{-6} to 10^{-9} M), which binds to the ACh receptor, blocked the hyperpolarization induced by carbamylcholine or a train of impulses in the giant axon. d-Tubocurarine (10^{-5} M) protected against the irreversible α-bungarotoxin action. Nicotine (10^{-6} M) induced a direct hyperpolarization of the resting fiber, whereas muscarine had no effect. None of the drugs had any effect on the resting or action potentials of the giant axon. On the basis of these results, it was concluded that there is a nicotinic-type ACh receptor at the junction between the axon and the Schwann cell. The mechanism of hyperpolarization may involve an increased permeability to potassium; sodium ions do not appear to contribute to the hyperpolarization. In contrast, sodium pathways did seem to be involved in a depolarization of the Schwann cell induced by grayanotoxin and veratrine (J. Villegas *et al.*, 1976). Tetrodotoxin or removal of external sodium reversed the depolarization. It was concluded that there are non-voltage-dependent sodium pathways in the plasma membrane of the Schwann cell that are blocked by tetrodotoxin. These pathways are different from those in the axolemma, which show a voltage-dependent conductance.

Using the electron microscope, ChE was histochemically localized using acetylthiocholine as substrate (G.M. Villegas and J. Villegas, 1974). Deposits were found focally distributed along the axoplasmic side of the axolemma. No definite deposits were seen in the axoplasm or Schwann cell. By use of selective inhibitors, the enzyme in the axolemma was shown to be true acetylcholinesterase. Most interestingly, there were at the points of cholinesterase localization trilaminar substructures of the axolemma, an undercoating of dense material, and a narrowing of the Schwann cell space (Fig. 42). It was suggested that these may be the sites of the intercellular junctions between axon and Schwann cell. These structural complexes (≈ 0.1 μm long

---

the axoplasm (A) or in the Schwann cell (S). The solid line indicates 0.1 μm. (2, 3) High-magnification electron micrographs showing the exact localization of the thiocholine end product at the level of the inner leaflet of the axolemma. The trilaminar substructure of this membrane and the narrowing of the axon (A)–Schwann cell (S) interspace at the level of the positive reaction are observed in both micrographs. Tissue was incubated without inhibitors. Reprinted from G.M. Villegas and J. Villegas (1974) by courtesy of Academic Press, Inc., New York.

and 70–170 Å thick) show a frequency of about 137 per 1000 μm of axon perimeter (G.M. Villegas and J. Villegas, 1976). The number of complexes decreased when the external sodium, potassium, or magnesium was decreased or when calcium was increased in the presence of magnesium. The complexes may be involved in active ion transport.

3.5. Pharmacological Studies

Many different toxins, venoms, enzymes, and drugs have been applied to the squid giant axon. In some cases, the primary interest was in analyzing the electrophysiological mechanism of action of these materials, which would include many of the studies using local anesthetics, barbiturates, and other drugs. In other studies, the primary interest was in using the agents as tools for obtaining a better understanding of how electrically excitable membranes work, which would include many of the studies with toxins and enzymes. Venoms and enzymes have also been used for quite specific purposes such as to remove axoplasm or to increase permeability. Regardless of the intended use, I have grouped together many of the electrophysiological effects produced by toxins, venoms, enzymes, drugs, and poisons in Table 8. The original references should be consulted for further details concerning the electrophysiological effects and for information on the duration of application, mechanism of action, and other properties.

I would like to summarize briefly the usefulness of these pharmacological studies (Table 8 should be consulted for the appropriate references). An excellent review on chemicals as tools in the study of excitable membranes, including but not limited to the squid giant axon, has appeared (Narahashi, 1974).

3.5.1. Enzymes

"Explaining" the action potential in terms of ionic conductance changes does not help our understanding of the membranal proteins or lipids that undergo specific conformational changes to account for these conductance changes. Various enzymes have therefore been studied to see the effect of specific modifications of membrane components, for example, proteases and phospholipases to hydrolyze proteins and phospholipids, respectively. In other experiments, enzymes were used to help remove axoplasm in the perfused axon or to increase permeability of the axonal preparation.

Unfortunately, some of the data in the literature are difficult to interpret because of the following sources of error.

1. *Impure enzyme preparations.* In many experiments, impure or poorly characterized enzyme preparations were used. It then becomes difficult to be certain that the effects are not due to a contaminant. When impure preparations are used, the reported concentrations (mg/ml) have little meaning, and different studies cannot be compared. In many of the studies using enzymes, the authors have not clearly recorded the purity of their

Table 8. Effects of Enzymes, Toxins, Venoms, Poisons, and Drugs on the Squid Giant Axon[a]

Enzymes (EC No.)[b]	Effects[c]	References
Alkaline phosphatase (3.1.3.1)	– AP	Rosenberg and Ehrenpreis (1961)
L-Amino acid oxidase (1.4.3.2)	– AP	Rosenberg and Ng (1963)
Arginase (3.5.3.1)	*↓ AP; ↓ RP; *– AP	Tasaki and Takenaka (1964); DeWeer (1968)
Bromelain (3.4.24.)	*– AP	Takenaka and Yamagishi (1969)
Carboxypeptidase A (3.4.2.1)	*– g_P; ↓ g_{ss}; *– Na_{in}; ↑ RP; ↕ g_l; *↓ AP; ↓ RP	Sevcik and Narahashi (1975); Tasaki and Takenaka (1964)
Carboxypeptidase B (3.4.2.2)	*– g_P; ↓ g_{ss}; *– Na_{in}; ↑ RP; ↕ g_l; *↓ AP; ↓ RP; – AP; – RP	Sevcik and Narahashi (1975); Tasaki and Takenaka (1964)
Chymotrypsin A (3.4.4.5)	*↑ g_l; ↓ Na_{in}; *↑ I_P; ↓ I_{ss}; *↓ AP; ↓ RP; SF	Sevcik and Narahashi (1975); Tasaki and Takenaka (1964); Tasaki and Takenaka (1964)
Clostridiopeptidase A (collagenase) (3.4.4.19)	– AP; – RP	DeWeer (1968)
Creatine kinase (2.7.3.2)	*– AP	Tasaki and Takenaka (1964)
Deoxyribonuclease (DNAase) (3.1.4.5)	*– AP; – RP	Tasaki and Takenaka (1964)
Ficin (3.4.12)	– AP; – RP; *↓ AP; ↓ RP; SF; ↕ AP	Tasaki and Takenaka (1964); Rosenberg and Ehrenpreis (1961)
Hyaluronidase (3.2.1.35)		
Leucine aminopeptidase (3.4.1.1.)	*↓ AP; ↓ RP	Tasaki and Takenaka (1964)

(Continued)

Table 8. (Continued)

Enzymes (EC No.)[b]	Effects[c]	References
Lipase (3.1.1.3)	– AP; – RP	Rosenberg and Ehrenpreis (1961)
Lysozyme (3.2.1.17)	*↓ AP; ↓ RP	Tasaki and Takenaka (1964)
	– AP	Rosenberg and Ehrenpreis (1961)
NAD nucleosidase (DPNase) (3.2.2.5)	*– AP; – RP	Tasaki and Takenaka (1964)
Neuraminidase (3.2.1.18)	– AP	Rosenberg (1965)
	*– AP; – RP	Tasaki and Takenaka (1964)
Papain (3.4.4.10)	– AP	Rosenberg and Ehrenpreis (1961)
	*↓ AP; ↓ RP; SF	Tasaki and Takenaka (1964)
Phospholipase A (3.1.1.4)	↓ AP; ↓ RP[d]	Condrea and Rosenberg (1968), Rosenberg and Condrea (1968)
Phospholipase A + B	– AP; – RP	Abbott et al., (1972)
Phospholipase B (3.1.1.5)	*↓ AP; ↓ RP; ↓ I_P; ↕ I_{ss}; ↑ I_i; – gNa_{in}	Rosenberg and Condrea (1968)
	– AP	Rosenberg and Condrea (1968)
Phospholipase C (3.1.4.3)	– AP; – RP	Rosenberg (1970), Rosenberg and Ng (1963), Tasaki and Takenaka (1964)
	*↓ AP; ↓ RP	Rosenberg and Condrea (1968)
Phospholipase D (3.1.4.4)	– AP	Rosenberg and Ng (1963), Tasaki and Takenaka (1964)
	*↓ AP; ↓ RP	
Pronase	*↑ AP; – RP; ↓ I_{ss}	Rojas and Atwater (1967)
	*↑ gK; ↓ gNa_{in}	Armstrong et al. (1973)
	*– gNa_{act}; ↓ K_E	Takashima et al. (1975)
	– gNa_{in}	Takenaka et al. (1968)
Protease (BPN')	*– AP; – RP	Takenaka and Yamagishi (1969)
Prozyme	*– AP	Takenaka and Yamagishi (1969)
Ribonuclease (RNase)	*– AP; – RP	Tasaki and Takenaka (1964)
	– AP; – RP	Rosenberg and Ehrenpreis (1961)
Trypsin (3.4.4.4)	*↓ AP; ↓ RP; ↓ R_m	Rojas and Luxoro (1963), Rojas (1965), Tasaki and Takenaka (1964)

Toxins and venoms	Conc. [M] (mg/ml)	Effects[c]	References
Acanthophis Antarcticus venom (death adder)	(0.1)	$\downarrow$AP	Rosenberg (1965)
Agkistrodon contortrix mokeson venom (copperhead moccasin)	(1)	$\downarrow$AP	Rosenberg and Podleski (1963)
Agkistrodon piscivorus venom (cottonmouth moccasin)	(0.05)	$\downarrow$AP; $\downarrow$RP	Rosenberg and Podleski (1962, 1963)
Apis mellifera venom (bee)	(0.01)	$\downarrow$AP	Rosenberg and Podleski (1963)
Batrachotoxin	10^{-8}	$\downarrow$RP	Albuquerque *et al.* (1973)
	10^{-6}	$\downarrow$AP; $\downarrow$RP; $\uparrow$PNa$_r$	Narahashi *et al.* (1971b)
	10^{-6}	*$\downarrow$AP; $\downarrow$RP; $\uparrow$PNa$_r$	
Bitis arietans venom (puff adder)	(2)	$\downarrow$AP	Rosenberg (1965)
Bothrops atrox venom (fer de lance)	(1)	$\downarrow$AP	Rosenberg and Podleski (1963)
Bufotenin	5×10^{-4}	$-$AP	Rosenberg (1965)
α-Bungarotoxin	10^{-6}	$-$AP; $-$RP	J. Villegas (1975)
Bungarus coeruleus venom (krait)	(0.5)	$\downarrow$AP	Rosenberg (1965)
Buthus tamulus venom (scorpion)	(0.1)	$\downarrow$RP; $\downarrow$AP; SF; $\downarrow$I$_s$; $-$I$_{Na}$	Narahashi *et al.* (1972a)
	(0.1)	*$-$RP; $\uparrow$AP	
Cardiotoxin (direct lytic factor)	(1.4)	$\downarrow$AP; $\downarrow$RP	Condrea and Rosenberg (1968)
Centruroides sculpturatus venom (scorpion)	(0.1)	$-$AP	Rosenberg (1965)
Cobrotoxin	(0.25)	$-$AP	Rosenberg (1965)
Condylactis toxin		$-$g$_p$; $-$g$_{ss}$	Narahashi *et al.* (1969a)
Crotalus adamanteus venom (Eastern diamondback rattlesnake)	(2)	$-$AP	Rosenberg and Podleski (1962)
Crotalus atrox venom (Western diamondback rattlesnake)	(0.2)	$\downarrow$AP	Rosenberg and Podleski (1963)
Crotalus h. horridus venom (timber rattlesnake)	(0.5)	$-$AP	Rosenberg (1965)

(Continued)

Table 8. (Continued)

Toxins and venoms	Conc. [M] (mg/ml)	Effects[c]	References
Dendroaspis polylepis venom (black mamba)	(0.1)	$\downarrow$AP	Rosenberg (1965)
Enhydrina schistosa venom (sea snake)	(0.15)	$\downarrow$AP	Rosenberg (1965)
Grayanotoxin	10^{-5}	$\downarrow$AP; $\downarrow$RP	J. Villegas *et al.* (1976)
Grayanotoxin II (dihydro α)	10^{-6}	$\downarrow$RP; $\uparrow$PNa$_r$	Seyama and Narahashi (1973)
		g_{ss}; $-g_p$	
	10^{-6}	*$\downarrow$RP; $\uparrow$PNa$_r$	
Grayanotoxin II (dihydro β)	10^{-4}	$-$RP; $-I_p$; $-I_{ss}$	Seyama and Narahashi (1973)
	10^{-5}	*$-$RP	
Hemachatus haemachatus venom (ringhals)	(0.05)	$\downarrow$AP	Condrea and Rosenberg (1968)
Holothurin A	10^{-4}	$\downarrow$AP; $\downarrow$RP; $\uparrow g_l$	de Groof and Narahashi (1974)
		*$\downarrow$AP; $\downarrow$RP; $\uparrow g_l$	
Latrodectus geometricus venom (spider)	(0.1)	$-$AP	Rosenberg (1966)
Latrodectus mactans venom (spider)	(0.001)	$\downarrow$AP; $\updownarrow$RP	Gruener (1973)
		$\downarrow^I$Na$_d$; $\uparrow^I$K$_o$	
	(0.001)	*$\downarrow$AP; $-$RP	
		*$\downarrow^I$Na$_d$; $\uparrow^I$K$_o$	
Latrodectus varidus venom (spider)	(0.1)	$-$AP	Rosenberg (1966)
Naja naja venom (Indian cobra)	(0.05)	$\downarrow$AP; $\downarrow$RP	Rosenberg and Ehrenpreis (1961), Rosenberg and Podleski (1962)
Notechis scutatus venom (tiger snake)	(0.1)	$\downarrow$AP	Rosenberg (1965)
Ophiophagus hannah venom (king cobra)	(0.4 mg/ml)	$\downarrow$AP	Rosenberg and Podleski (1962)
Tetrodotoxin	3×10^{-7}	$\uparrow$RP	Narahashi *et al.* (1966, 1967a,b)
	10^{-7}	$\downarrow$AP; $-$RP; $\downarrow I_P$; $-I_G$	Nakamura *et al.* (1965)
	10^{-9}	$\downarrow I_P$	
	10^{-6}	*$-$AP; $-$RP; $-I_p$	Rosenberg and Ehrenpreis (1961), J. W. Moore *et al.* (1967), Freeman (1971), Armstrong and Bezanilla (1974), Cuervo and Adelman (1970)
Vespula arenaria venom (yellow hornet)	(0.1)	$-$AP	Rosenberg (1965)
	(1 droplet/ml)	$\downarrow^g$Na; $\downarrow^g$K	Parmentier and Narahashi (1975)
Vipera palestinae venom (Palestinian viper)	(0.4)	$\downarrow$AP	Condrea and Rosenberg (1968)
Vipera russellii venom (Russells viper)	(1)	$-$AP	Rosenberg and Podleski (1962)

Drugs and poisons	Conc. [M]	Effects[c]	References
Acetylcholine	10^{-1}	−AP; −RP	Rosenberg and Podleski (1962, 1963)
	10^{-3}	+AP; ↓RP; CMV	Rosenberg and Hoskin (1965)
Acetylsalicylic acid (aspirin)	10^{-2}	*−RP; ↓AP; ↓g_p; ↓g_{ss}	Frazier et al. (1969), Yeh and Narahashi (1974a)
	10^{-3}	*−RP; −AP	
Aconitine	5×10^{-3}	−AP	Hoskin and Rosenberg (1965)
	10^{-6}	−AP; −RP; SF	Herzog et al. (1964)
Alcohols C_1–C_5	10^{-1}–5×10^{-1}	↓AP; ↓RP; ↓g_m; ↓g_p	Armstrong and Binstock (1964)
C_8	10^{-3}	↓AP; ↓RP; ↓g_m; ↓g_p −C_m; −Na_{in}	
Aldrin (transdiol)	5×10^{-5}	↓AP; →RP	Van Den Bercken and Narahashi (1974)
Allethrin	10^{-4}	↓I_p; ↑I_{ss}; ↑I_l ↑Na_{in}; ↓g_p; ↓g_{ss} ↓AP; ↑RP *AP; ↓RP	Narahashi and Anderson (1967) Wang et al. (1972) Yeh et al. (1976)
2-Aminopyridine 3-Aminopyridine 4-Aminopyridine	10^{-3} 10^{-3} $> 10^{-3}$	*I_{ss}; −I_p; SF	
Arsenite	2×10^{-2}	−AP	Rosenberg (1965)
Aspartate	5×10^{-3}	−AP	Hoskin and Rosenberg (1965)
ATP	2×10^{-3}	−AP	Rosenberg (1965)
Atropine	5×10^{-3}	↓AP; ↓RP	Rosenberg and Ehrenpreis (1961)
	2×10^{-4}	↓AP; CMV	Rosenberg and Bartels (1967) Rosenberg and Podleski (1963)
Barbital	2×10^{-1}	+AP	Rosenberg and Bartels (1967)
	4×10^{-2}	↓SF	
Benzilic acid	1×10^{-2}	−AP	Rosenberg and Bartels (1967)
Benzoylcholine	1×10^{-2}	−AP; CMV	
	1×10^{-1}	−AP	
	2×10^{-2}	↓AP; CMV	Rosenberg and Podleski (1963)

(Continued)

Table 8. (Continued)

Drugs and poisons	Conc. [M]	Effects[c]	References
Benztropine	2 × 10⁻⁴	$-g_P$; $-g_{ss}$ ↓AP; $-$RP	Rosenberg and Bartels (1967)
	10⁻⁵	→SF	Rosenberg (1965)
Bretylium tosylate	5 × 10⁻³	$-$AP	Rosenberg and Bartels (1967)
Caramiphen	2 × 10⁻⁴	+AP	
	10⁻⁵	→SF	
Carbamylcholine	3 × 10⁻²	$-$AP	Rosenberg and Podleski (1963)
	10⁻²	*↑AP; $-$RP	Yeh and Narahashi (1974a), J. Villegas (1974)
Cevadine	10⁻⁴	↑g_P; ↓g_{ss}	Ohta et al. (1973)
Cetyltrimethyl-ammonium bromide	6 × 10⁻⁵	↓AP; ↓RP ↓AP; ↓RP	Rosenberg and Ehrenpreis (1961), Kishimoto and Adelman (1964)
Chloranil	10⁻³	*↓AP; $-$RP ↑AP_d	Huneeus-Cox and Smith (1965) Huneeus-Cox et al. (1966)
Chloroform	1 × 10⁻²	↓AP	Rosenberg and Bartels (1967)
Chlorpheniramine	5 × 10⁻⁴	↓AP	Rosenberg and Bartels (1967)
	10⁻⁴	→SF	
Chlorpromazine	10⁻⁴	↓AP	Rosenberg and Bartels (1967)
	10⁻⁵	$-$RP; ↓SF; ↓I_P	Rosenberg and Ehrenpreis (1961)
	10⁻⁵	*↓AP; $-$RP	Gruener and Narahashi (1972)
		↓I_P; ↓I_{ss}	Lakshminarayanaiah and Bianchi (1975)
Choline	10⁻¹	$-$AP	Rosenberg and Podleski (1962)
	10⁻¹	$-$AP; CMV	
	10⁻³	*$-$AP; $-$RP	
Cortisol	3 × 10⁻⁴	$-$AP	Frazier et al. (1969)
Cysteine	3 × 10⁻¹	*↓AP	Hoskin and Rosenberg (1965)
	10⁻¹	*$-$AP; $-$RP	Huneeus-Cox and Smith (1965)
	10⁻²	*$-$AP; ↑RP	Huneeus-Cox et al. (1966)
Decamethonium	10⁻¹	$-$AP	Rosenberg and Podleski (1963)
	2 × 10⁻³	↓AP; CMV	
	10⁻²	*↓AP; $-$RP →g_P; ↓g_{ss}	Yeh and Narahashi (1974a)

Compound	Concentration	Effect	Reference
Deoxycholate	1.3×10^{-5}	↑ AP	Rosenberg and Ehrenpreis (1961)
Deuterium oxide		↑ AP; − RP; − I_{ss}; ↓ I_p	Stillman et al. (1968)
Dextrorphan	10^{-3}	↓ AP	Simon and Rosenberg (1970)
Dibucaine	3×10^{-5}	↓ AP; ↓ g_p; ↓ g_{ss}	Rosenberg and Ehrenpreis (1961)
	5×10^{-4}	*↓ AP; ↑ RP	Narahashi et al. (1969a), Lakshminarayanaiah and Bianchi (1975)
Dieldrin	10^{-4}	− AP; − RP	Van Den Bercken and Narahashi (1974)
Digitonin	2×10^{-5}	↑ AP	Rosenberg and Ehrenpreis (1961)
Diisopropylfluorophosphate	10^{-2}	↓ AP	Rosenberg and Ehrenpreis (1961)
	10^{-3}	↓ AP; CMV	Hoskin et al. (1966), Kremzner and Rosenberg (1971)
Dimethylammonium chloride (N,N-bis-phenylcarbamoylmethyl)	10^{-3}	→ AP; − RP	Frazier et al. (1970)
N,N-Dimethylformamide	10^{-4}	*↓ AP; − RP	Rosenberg (1965)
Dimethylsulfoxide	1.0	− AP	Rosenberg (1965)
2,4-Dinitrobenzene (1,5-difluoro)	10^{-1}	− AP	Cooke et al. (1968)
	2×10^{-3}	↓ I_p; ↓ I_{ss}; → RP; ↓ I_p	Cooke et al. (1968)
2,4-Dinitrobenzene (1-fluoro)	2×10^{-3}	↓ AP; ↓ RP	
	2×10^{-3}	*↓ I_{ss}; ↑ I_p; ↓ RP	
Dinitrophenol	2×10^{-3}	− AP	Simon and Rosenberg (1970)
Diphenhydramine	4×10^{-4}	↓ AP	Rosenberg and Bartels (1967)
Diphenylhydantoin	10^{-3}	− AP; − RP	Rosenberg and Bartels (1967)
	5×10^{-4}	↓ AP; CMV	Korey (1951)
	5×10^{-5}	− RP; ↓ g_p; ↑ g_{ss}	Lipicky et al. (1972)
DOPA	5×10^{-3}	→ SF	Hoskin and Rosenberg (1965)
DOPAmine	5×10^{-3}	− AP	Hoskin and Rosenberg (1965)
5,5'-Dithiobis-2-nitrobenzoate (DTNB)	10^{-4}	− AP	Marquis and Mautner (1974)
Dithiothreitol (DTT)	2×10^{-2}	− AP	Marquis and Mautner (1974)
Epinephrine	10^{-2}	− AP	Rosenberg and Bartels (1967), Rosenberg (1965)
Ethanol	1.5	↓ AP; ↓ RP	Rosenberg and Bartels (1967)
		↓ g_p; g_{ss}	J. W. Moore et al. (1964b)
Ether (diethyl)	3×10^{-1}	↓ AP	Rosenberg and Bartels (1967)
	10^{-1}	↓ AP; CMV; ↓ Na_E	Schwartz (1968)

(Continued)

Table 8. (Continued)

Drugs and poisons	Conc. [M]	Effects[c]	References
Ethopropazine	10^{-4}	$\downarrow$ AP	Rosenberg and Bartels (1967)
	10^{-5}	$\downarrow$ SF	
N-Ethylmaleimide (NEM)	10^{-3}	$\uparrow$ AP; $-$ RP	Marquis and Mautner (1974)
	10^{-3}	* $\downarrow$ AP	Huneeus-Cox et al. (1966)
Etorphine	10^{-3}	* $\downarrow$ I$_p$; $\rightarrow$ I$_{ss}$	Frazier et al. (1973b)
Ferricyanide	10^{-3}	* $\downarrow$ AP; $\rightarrow$ AP$_d$	Huneeus-Cox et al. (1966)
Fluorescein mercuric acetate	10^{-4}	* $\downarrow$ AP; $-$ RP	Huneeus-Cox and Smith (1965)
	10^{-4}	$\downarrow$ AP	Huneeus-Cox et al. (1966)
γ-Aminobutyric acid (GABA)	5×10^{-3}	$-$ AP	Hoskin and Rosenberg (1965)
Glutamine	5×10^{-3}	$-$ AP	Hoskin and Rosenberg (1965)
Hemicholinium-3	10^{-2}	$-$ AP; $-$ RP	Frazier et al. (1969, 1970)
	5×10^{-3}	* $\downarrow$ AP; $-$ RP	
		* $\downarrow$ I$_p$; $\downarrow$ I$_{ss}$	
Hexamethonium	10^{-2}	* $\uparrow$ AP; $-$ RP	Yeh and Narahashi (1974a)
		$\uparrow$ g$_p$; $\uparrow$ g$_{ss}$	
Hexobarbital	10^{-2}	$-$ RP; $\uparrow$ AP	Frazier et al. (1975)
	7×10^{-3}	* $-$ RP; $\uparrow$ AP	
Histamine	10^{-1}	$-$ AP; CMV	Rosenberg and Bartels (1967)
	10^{-1}	$-$ AP; CMV	Rosenberg (1965)
	10^{-3}	* $\uparrow$ AP; $\downarrow$ RP; SF	Scuka (1971)
Homocysteic acid	5×10^{-3}	$-$ AP	Rosenberg (1965)
Hydrazine	10^{-1}	* $\downarrow$ AP	Huneeus-Cox and Smith (1965)
	10^{-1}	$\downarrow$ AP	Huneeus-Cox et al. (1966)
Hydrogen peroxide	10^{-3}	* $\downarrow$ AP; $-$ RP	Huneeus-Cox and Smith (1965)
		$\uparrow$ AP$_d$	Huneeus-Cox et al. (1966)
Hydroquinone	10^{-3}	* $\downarrow$ AP; $\uparrow$ AP$_d$	Huneeus-Cox et al. (1966)
p-Hydroxymercuribenzoate	10^{-3}	* $\downarrow$ AP; $\downarrow$ RP	Huneeus-Cox et al. (1966)
	10^{-4}	$\downarrow$ AP; $\downarrow$ RP	
Indoleacetic acid	10^{-3}	$-$ AP	Hoskin and Rosenberg (1965)
Iodoacetamide	10^{-2}	$\downarrow$ AP	Marquis and Mautner (1974)
	10^{-2}	* $-$ AP	Huneeus-Cox et al. (1966)

Compound	Concentration	Effect	Reference
Iodoacetate	10^{-2}	$* - AP; - RP$	Huneeus-Cox and Smith (1965), Huneeus-Cox et al. (1966)
	10^{-2}	$\downarrow AP$	Marquis and Mautner (1974), Huneeus-Cox et al. (1966)
O-Iodosobenzoate	10^{-3}	$* \downarrow AP; \uparrow AP_d$	Huneeus-Cox et al. (1966)
Isosystox	5×10^{-3}	$\downarrow AP$	Hoskin et al. (1969)
Levallorphan	1×10^{-3}	$\downarrow AP$	Simon and Rosenberg (1970)
Levorphanol	10^{-3}	$\downarrow AP; - RP$	Simon and Rosenberg (1970)
Lobeline	5×10^{-5}	$\downarrow SF$	Dole and Simon (1974)
	2×10^{-5}	$\downarrow AP; \downarrow g_P$ $\uparrow K_{in}$	Yeh and Narahashi (1974a)
Lysergic acid diethylamide (LSD)	1×10^{-4}	$* \downarrow AP; \downarrow g_P; g_{ss}$ $\uparrow K_{in}$	Henkin et al. (1974)
Lysolecithin	3×10^{-6}	$- I_P; - I_{ss}$	Rosenberg and Condrea (1968)
	5×10^{-3}	$\downarrow AP$	
Mephenytoin	5×10^{-3}	$- AP$	Rosenberg and Bartels (1967)
	2×10^{-3}	$\downarrow SF$	
2-Mercaptoethanol	10^{-1}	$* - AP; - RP$	Huneeus-Cox and Smith (1965)
	2×10^{-1}	$* \downarrow AP$	
Mercuric chloride	2×10^{-1}	$\downarrow AP; \uparrow RP$	Huneeus-Cox et al. (1966)
	10^{-4}	$* \downarrow AP; \downarrow RP$	Huneeus-Cox and Smith (1965)
Mersalyl (sodium)	10^{-4}	$\downarrow AP; \downarrow RP$	Huneeus-Cox et al. (1966)
Methantheline	2×10^{-3}	$* \downarrow AP; \downarrow RP$ $\downarrow AP$	Rosenberg and Ehrenpreis (1961)
Monactin	10^{-6}	$- I_{ss}$	Stillman et al. (1970)
Morphine	10^{-3}	$\uparrow AP$	Simon and Rosenberg (1970)
	10^{-3}	$* \updownarrow AP; - RP$	
	10^{-2}	$* \downarrow AP; - RP$	
Muldamine	10^{-4}	$\downarrow g_P; \downarrow g_{ss}$	Frazier et al. (1972)
Muscarine	10^{-6}	$\downarrow AP; - RP$	Ohta et al. (1973)
Naloxone	10^{-3}	$- AP; - RP$	G. Villegas (1975)
Neostigmine	5×10^{-2}	$* \downarrow I_P; \downarrow I_{ss}$ $- AP$	Frazier et al. (1973b)
			Rosenberg and Ehrenpreis (1961), Rosenberg and Podleski (1962), Brzin et al. (1965b)

(Continued)

Table 8. (Continued)

Drugs and poisons	Conc. [M]	Effects[c]	References
Nicotine	10^{-2}	$-AP$; $\downarrow g_{ss}$; $-g_P$	Rosenberg and Podleski (1963)
	10^{-3}	$-g_{ss}$	
	10^{-2}	$*\downarrow g_{ss}$; $\downarrow g_P$	Frazier et al. (1973a)
	10^{-3}	$*\downarrow g_{ss}$; $-g_P$	G. Villegas (1975)
Nor-acetylcholine	5×10^{-4}	$\rightarrow AP$	Rosenberg and Ehrenpreis (1961)
Oxotremorine	10^{-2}	$-AP$	Rosenberg and Bartels (1967)
p-Chloromercuribenzoate (PCMB)	10^{-4}	$*\downarrow AP$; $\downarrow RP$	Huneeus-Cox and Smith (1965)
	5×10^{-6}	$\updownarrow AP$; $-AP$	Marquis and Mautner (1974)
Paraoxon	10^{-2}	$\rightarrow AP$	Brzin et al. (1965b)
Pentobarbital	3×10^{-3}	$\rightarrow AP$; $-RP$	Narahashi et al. (1969a)
	3×10^{-4}	$*\downarrow AP$; $-RP$	
		$\downarrow g_P$; $\downarrow g_{ss}$	Narahashi et al. (1971a)
Pentylenetetrazol (Metrazol)	3×10^{-3}	$*\downarrow AP$; $-RP$	Frazier et al. (1975)
Phenobarbital	10^{-1}	$\rightarrow AP$	Rosenberg and Bartels (1967)
	10^{-1}	$\rightarrow AP$	
	10^{-2}	$\updownarrow AP$; $-RP$	Rosenberg and Bartels (1967)
	3×10^{-2}	$\rightarrow AP$; CMV	Frazier et al. (1975)
	3×10^{-3}	SF	
	3×10^{-3}	$*\downarrow AP$; $-RP$	
Phenylmercuric acetate	10^{-4}	$*\downarrow AP$	Huneeus-Cox and Smith (1965), Huneeus-Cox et al. (1966), Hoskin and Rosenberg (1967)
Phospholine (217AO)	10^{-2}	$-AP$	Kremzner and Rosenberg (1971)
	10^{-2}	$\rightarrow AP$; CMV	Rosenberg and Ehrenpreis (1961)
Physostigmine	5×10^{-3}	$\rightarrow AP$	Rosenberg and Bartels (1967)
	5×10^{-4}	$\rightarrow AP$; CMV	Rosenberg and Podleski (1962), Bullock et al. (1946). J. Villegas (1973)
	10^{-4}	$-AP$; $-RP$	
Picrotoxin	10^{-3}	$-AP$	Rosenberg and Bartels (1967)
	6×10^{-2}	$*\downarrow AP$	Huneeus-Cox and Smith (1965)
Potassium ferricyanide	3×10^{-3}	$\downarrow AP$; $-RP$	Rosenberg and Ehrenpreis (1961)
	5×10^{-4}	$\downarrow I_P$; $\downarrow I_{ss}$; $\downarrow SF$	Narahashi et al. (1967a)
Procaine	3×10^{-3}	$*\downarrow AP$; $\uparrow RP$	Rosenberg and Bartels (1967)
		$\downarrow I_P$; $\downarrow I_{ss}$	Taylor (1959)

Compound	Concentration	Effect	Reference
Propranolol	10^{-5}	↑g_{ss}	Wu and Narahashi (1973)
	10^{-3}	↓AP; −RP; ↓I_p; ↓I_{ss}	
2-Pyridine aldoxime methiodide (2-PAM)	10^{-1}	−AP	Rosenberg and Podleski (1963)
	10^{-2}	↓AP; CMV	
Quinidine HCl	2×10^{-4}	*↓AP; −RP; ↓g_p; ↓g_{ss}; $^g K_{in}$	Yeh and Narahashi (1976)
	2×10^{-4}	↓AP; −RP; ↓g_p; ↓g_{ss}; $^g K_{in}$	
Quinone	10^{-3}	*↓AP; −RP; ↑AP_d	Huneeus-Cox and Smith (1965)
		→AP	Huneeus-Cox et al. (1966)
		−AP	Rosenberg (1965)
			Hoskin et al. (1969)
Saponin	(3 mg/ml)	+AP; CMV	Hoskin and Rosenberg (1965)
Selenophos	5×10^{-3}	−AP	Kishimoto and Adelman (1964)
Serotonin	5×10^{-3}	−AP	Rosenberg and Ehrenpreis (1961)
Sodium lauryl sulfate	5×10^{-3}	↑AP; ↕RP	Ohta and Narahashi (1973)
	10^{-4}	−AP	
Span 2°	(5 mg/ml)	↕AP_d; ↓g_{ss}; ↕g_p	
Spartein	10^{-3}	−AP	
Spermidine	2×10^{-3}	−AP	Simon and Rosenberg (1970)
Spermine	5×10^{-3}	*↓AP	Simon and Rosenberg (1970)
Strychnine	10^{-3}	↓AP; −RP; ↓I_p; ↓I_{ss}; ↑K_{in}	Rosenberg and Ehrenpreis (1961), Rosenberg and Bartels (1967)
			Tasaki et al. (1965b)
			Shapiro et al. (1974)
Tetracaine	2×10^{-4}	*↓AP; −RP	Rosenberg and Bartels (1967)
	5×10^{-3}	↓AP; −RP	Tasaki et al. (1965b)
	10^{-4}	SF	
Tetrathionate (sodium)	3×10^{-5}	*↓AP; ↑AP_d	Huneeus-Cox et al. (1966)
Tetraethylammonium (TEA)	10^{-3}	−I_p; −I_{ss}	Oikawa (1962)
	10^{-4}	* ↓g_{ss}	Tasaki and Hagiwara (1957)
	10^{-2}	↓I_{ss}; −I_p; SF; ↑AP_d	Armstrong and Binstock (1965)
			Armstrong (1966, 1969, 1971)

(Continued)

Table 8. (Continued)

Drugs and poisons	Conc. [M]	Effects[c]	References
Triethylammonium − butyl − methyl − pentyl − propyl	10^{-4}–3×10^{-3}	*↓g_{ss}; AR	Armstrong (1969, 1971)
Triethylammonium (N-2,6-dimethylphenyl carbamoylmethyl)	10^{-2} 10^{-3}	−AP; −RP *↓AP; −RP	Frazier et al. (1970)
Tetrahydrocannabinol (delta 9)	3×10^{-3} 3×10^{-3}	−AP *−AP	Brady and Carbone (1973)
Tetrahydrocannabinol (11-hydroxy delta 9)	10^{-8}	↓AP	Brady and Carbone (1973)
Tetriso	5×10^{-3}	−AP	Hoskin et al. (1969)
Tricaine	5×10^{-3} 3×10^{-3}	↓AP; CMV *↓I_p; ↕I_{ss}	Frazier and Narahashi (1975)
Trihexyphenidyl	3×10^{-4} 3×10^{-5}	↓AP; −RP; −g_p; −g_{ss} ↓AP; CMV	Rosenberg and Bartels (1967)
Trimethadione	3×10^{-6} 10^{-1}	↓SF ↓AP	Wu and Narahashi (1976) Rosenberg and Bartels (1967)
2,4,6-Trinitrophenol	5×10^{-3} 2×10^{-3}	↓SF ↓AP; ↕RP ↓I_p; −I_{ss}	Cooke et al. (1968)
Tropine-p-tolyl acetate	10^{-3} 10^{-3}	↓AP *↓AP; ↓g_p; ↓g_{ss}	Narahashi et al. (1969a)

Compound	Concentration	Effect	Reference
d-Tubocurarine	10^{-2}	$-$ AP; $-$ RP	Tasaki et al. (1965b)
	3×10^{-5}	$\downarrow$ AP; $-$ RP; CMV	Rosenberg and Podleski (1962, 1963)
	5×10^{-3}	$*\downarrow$ AP; $\uparrow$ RP	Yeh and Narahashi (1974a), J. Villegas (1973)
Tween 80	10^{-1}	$\downarrow g_p$; $\downarrow g_{ss}$; $\downarrow Na_{in}$	Kishimoto and Adelman (1964)
Veratramine	10^{-4}	$\updownarrow$ AP; $-$ RP	Ohta et al. (1973)
5-Veratranine (3β 11α diol)	10^{-4}	$\downarrow$ AP; $-$ RP	Ohta et al. (1973)
		$\downarrow$ AP; $\updownarrow$ RP	
Veratridine	2×10^{-5}	$\downarrow g_p$; $\downarrow g_{ss}$	Rosenberg (1965)
	2×10^{-5}	$-$ AP; $-$ RP	Meves (1966b)
	10^{-4}	$*-$AP; $\updownarrow$ RP	Shanes et al. (1953), Shanes (1952), Ohta et al. (1973)
		$\downarrow$ AP; SF; $\downarrow$ RP	
		$\uparrow PNa_r$	
Veratrine-proto	5×10^{-5}	$\updownarrow$ AP; $\downarrow$ RP	Ohta et al. (1973)
Veratrosine	10^{-4}	$-$ AP; $-$ RP	Ohta et al. (1973)

[a] Unless otherwise stated, all compounds were applied externally to squid (mostly *L. pealei*) giant axons. The concentrations listed are the lowest that caused reported effect or the highest tested that had no effect.

[b] No concentrations for the enzymes are shown because of uncertainties as to the purity of the preparations that were used. To avoid ambiguity concerning which enzyme is being referred to, their enzyme commission (EC) numbers are given in parentheses.

[c] Symbols: (*) Compound was perfused or microinjected into the axoplasm; ($\downarrow$) marked decrease or block; ($\downdownarrows$) slight or less marked decrease; ($\uparrow$) marked increase; ($\upuparrows$) less marked increase; ($-$) no effect; (act) activation; (AP) action potential; (AR) anomalous (ingoing) rectification; (C) capacitance; (CMV) after pretreatment of axon with 15–35 μg cottonmouth moccasin (*Agkistrodon piscivorus*) venom/ml; (d) duration; (E) efflux; (g) conductance; (G) gating; (in) inactivation; (K) potassium; (l) leakage; (m) membrane; (Na) sodium; (o) onset; (p) peak transient (sodium); (r) resting; (R) resistance; (RP) resting potential; (SF) induces spontaneous or repetitive firings; ($\downarrow$ SF) blocks spontaneous or repetitive firings of axon induced by decreased divalent cations in seawater; (ss) steady state (potassium).

[d] These effects were on crudely dissected giant axons. On finely dissected axons, phospholipase A_2 and the snake venoms had no effect.

preparation or its absolute activity in International or other recognized units. I have therefore not recorded the mg/ml concentrations of enzymes used, since they have little meaning.

2. *Neglecting effects of hydrolytic products.* It is necessary to differentiate between the direct effects of the enzymes due to modification of the membranal constituent (substrate) and effects due to the products of the reaction between enzyme and substrate. For example, lysophosphatides have detergent actions and are liberated as a result of the reaction of phospholipase A_2 with phospholipids. Lysophosphatides are responsible for the effects of phospholipase A_2 on the squid axon. Incorrect conclusions may therefore be reached if reaction products are not tested.

3. *Failure to quantitate extent of enzymatic reaction.* If you want to critically relate an observed effect on the squid axon to an enzyme's action, it is important to quantitate the actual extent of substrate hydrolysis in the axoplasm and in the envelope. No *in vitro* assay can completely suffice, nor can the extent of activity on the squid axon action be predicted. The effects of pH, ionic conditions, permeability barriers, and other factors may all modify an enzyme's action in the tissue. Even after making these measurements, caution is necessary for their interpretation, since, for example, the envelope contains mostly Schwann cell and connective tissue, and results cannot be directly assumed to represent the results on the axolemma. In the case of phospholipase A_2, however, it was suggested that the splitting in the envelope did indicate hydrolysis of axolemmal phospholipids, since equal splitting was also observed in the axoplasm following the external application of the phospholipase. It was also unequivocally demonstrated that phospholipases penetrate into the axoplasm following external application. Nevertheless, it would ideally be best to prepare a fraction enriched in axolemmal membranes and then measure the extent of hydrolysis on this isolated fraction.

4. *Permeability barriers to the penetration of enzymes.* If substrate hydrolysis is not measured, then the lack of effect of an enzyme on the action potential, permeability, and other properties cannot be related to a nonessentiality of the membranal constituents on which the enzyme acts. It may be that the enzyme simply cannot penetrate to the appropriate site of action. The ability to perfuse the squid axon provides the opportunity of avoiding external permeability barriers such as Schwann cell and connective tissue. There may, however, be internal barriers if, for example, the reactive substrate site is oriented on the external aspect of the axolemma or is buried within the axolemma.

A variety of enzymes have been used on the squid giant axon, although almost all the studies suffer from one or more of the sources of error described above. Proteases have no effect on the action potential when externally applied. However, they have dramatic effects when internally perfused. It is not known whether the externally applied proteases are unable to reach the axolemma because of the Schwann cell and connective tissue or whether the external aspect of the axolemma has chemical bonds not susceptible to the proteases, in contrast to the inner aspect of the axolemma. Phospholipases have varied effects on the squid axon, including increasing permeability, rendering the axon sensitive to the action of cholinergic compounds, disrupting the Schwann cell, depolarizing, and blocking conduction. As noted above (also see Section 3.4.2), these effects appear to be due to the production of lysophosphatides. Additional studies on the squid giant axon using enzymes should be most rewarding, especially if it is possible to measure hydrolysis in subfractions of the axon after external and internal exposure to enzymes.

To avoid any ambiguity as to the enzyme being referred to in Table 8, I have listed their enzyme commission (EC) numbers. These numbers specifically identify the enzymes and are listed in *Enzyme Nomenclature* (International Union of Biochemistry, 1973) along with the enzymes' systematic names, recommended trivial names, and other names (not recommended), reactions in which the enzymes participate, and other notes.

3.5.2. Toxins and Venoms

Toxins and venoms have been used mainly as tools in analyzing bioelectricity because many toxins have very specific and potent effects, while venoms are storehouses of many enzymes and toxins. Tetrodotoxin, which is very widely used and obtained from pufferfish and certain newts, blocks the transient channel through which sodium normally flows in response to a depolarization. Since the toxin blocks fluxes through the transient channel regardless of the ion involved, it has been used to determine whether any particular ionic flux is through the channel normally used by sodium. It has been very useful for voltage-clamp studies in which it is desired to eliminate the contribution of the transient current. Since the toxin is inactive on internal perfusion, it has been suggested that the receptor for tetrodotoxin and perhaps the opening for the sodium channel are on the external face of the axolemma. The fact that tetrodotoxin does not affect the gating current would suggest that the "gate" and the "channel" have important molecular differences and are separate entities. Tetrodotoxin also decreases resting membrane sodium conductance and causes a slight hyperpolarization of the resting potential. Radioactive-tetrodotoxin-binding has been used to estimate the density of sodium channel sites, and the toxin may be useful for attempts to extract toxin-binding material, that is, fractions enriched in sodium channels. Saxitoxin has actions similar to those of tetrodotoxin. Batrachotoxin, contained in the skin secretion of certain frogs, is quite interesting because it specifically increases the resting sodium permeability without affecting the voltage-dependent transient sodium increase associated with the initiation of the action potential. This suggests perhaps a biochemical or structural difference in the resting and in the activated sodium channel. Grayanotoxins, obtained from the leaves of various plants, have effects similar to those of batrachotoxin except that they are less potent and much more reversible. Condylactis toxin obtained from an anemone is most interesting because it has no effect on the squid axon, while it markedly prolongs the falling phase of the transient current in lobster and crayfish axons. It may eventually be useful as a tool to discover in squid and other axons membrane ultrastructural differences that could explain the marked differences in sensitivity to this toxin.

Snake venoms are rich sources of many enzymes including phospholipase A_2, proteases, hyaluronidase, l-amino acid oxidase, phosphodiesterase, and others. In addition, these venoms have potent toxins such as α- and β-bungarotoxin, cobrotoxin, and others that have specific effects either pre- or postsynaptically. These synaptic toxins do not, however, affect the squid giant axon. Crude venoms (e.g., snake, spider, bee) may be used on axons to relate effects on conduction to those experienced clinically, or in preliminary experiments to see whether this "pharmaceutical factory" has any actions on a particular function of the axon. The potencies of different crude venoms on the squid axon vary greatly, which may be due to differences in their phospholipase A_2 activities or to differences in the physicochemical properties of the enzyme (e.g., pH optima, ionic requirements). Phospholipase and cardiotoxin (direct lytic factor) are the only known snake venom components to have effects on axons (Rosenberg, 1977). For most studies, however, it would be necessary to isolate the components of interest and test them in their pure state.

3.5.3. Drugs and Poisons

Interest in the action of drugs on the squid giant axon is not so directly related to their use as tools to accomplish specific purposes or to understand biological processes as is the case with enzymes and toxins. Some drugs, such as tetraethylammonium, are, however, extremely useful tools. Studies with most drugs, however (including, for example, local anesthetics, barbiturates, antiarrhythmic agents, and cholinergic agents), have evaluated whether the effects on the electrophysiological properties of the squid axon could explain or at least aid in understanding the drugs' mechanisms of action. It would also be hoped that eventually the squid axon might have predictive value as a biological screen for various types of pharmacological activity. As discussed elsewhere (Section 4.2), one must be very cautious in extending results obtained on the squid giant axon to other nonaxonal systems. There is evidence that the effects of narcotics, barbiturates, and perhaps cholinergic compounds are quite different on the squid axon as compared to the synapse or central nervous system.

Tetraethylammonium (TEA) is quite effective in suppressing steady-state potassium conductance, without affecting transient sodium conductance, thereby causing the development of prolonged cardiac-type action potentials. Since TEA acts only from the inside of the squid axon, it has been suggested that the "gate" for the potassium channel is on the inner aspect of the axolemma. It is probable, however, that this charged lipid-insoluble compound cannot even reach the axolemma when externally applied. The usefulness of TEA in blocking the potassium channel is analogous to the usefulness of tetrodotoxin in blocking the sodium channel.

Drugs have been applied both externally and internally to the squid giant axon while at the same time the pH of the external and internal solutions was varied. This protocol allowed the active form of the drug (charged or uncharged) and the site of drug action (external or internal) to be estimated. These studies, mainly by Narahashi and his group, have indicated that nicotine, morphine, barbiturates, and local anesthetics all act at the internal aspect of the axolemmal surface. It was suggested that in addition to tetrodotoxin, scorpion venom acted on the external axolemmal surface. The charged cationic form of local anesthetics and chlorpromazine was shown to be the active form of these drugs.

Sulfhydryl reagents such as dithiothreitol (DTT), dithiobisnitrobenzoate (DTNB), N-ethylmaleimide (NEM), p-chloromercuribenzoate (PCMB), and others offer promise as tools to modify proteins of the axonal membrane. While they have been extensively used on the isolated eel electroplax (Narahashi, 1974), there have been only a few papers concerned with their effects on the squid giant axon, and their effects have not been thoroughly evaluated. Dinitrophenol derivatives that form stable bonds with proteins have been found to irreversibly suppress steady-state current and the action potential.

Metabolic inhibitors such as cyanide and inhibitors of active transport such as ouabain have been extensively used to determine which functions or properties of the squid axon require metabolic energy or specifically which ionic fluxes involve active transport. They have also been used to evaluate whether a portion of the resting membrane potential may be due to an active-transport electrogenic pump.

The use of aequorin, a photoprotein isolated from a jellyfish, in the determination of ionized calcium levels and the use of fluorescence probes in studying membranal architecture have been mentioned in other sections of this chapter, wherein appropriate references may be found. The usefulness of these compounds depends on the fact that they do not have direct actions of their own on the squid axon, and they are therefore not listed in Table 8.

4. Concluding Comments

To avoid the appearance of recommending the squid axon as a salve for all neurobiological ills, and of being overly sanguine about its usefulness, a few brief critical commentaries would now appear appropriate.

4.1. Evaluation of Data

During the course of your "life with the squid," you will have to critically analyze and interpret published literature data. You will not be able to do this if you are overly specialized. For example, if you are a biochemist using the squid giant axon in biochemical experiments, how can you be certain of the reliability of your data if you never tested the electrical activity of the axons you have used? You would not think of missing a biochemical control; do not miss this essential electrical control. Make sure that you are starting your experiments with a functioning axon. I hope that it is clear from the discussion of methods (Section 2) that the extracellular recording of electrical activity is both simple and adequate for showing that you have a conducting axon. If you are to appreciate the relevance of your studies, and how they relate to the main function of the squid axon, which is to generate and conduct an electrical signal, then you must learn the significance of electro-physiological studies, and what types of information they provide. Even if you never do a voltage-clamp experiment, you should be able to understand a current–voltage diagram. Electrophysiological data can sometimes provide clues as to what types of macromolecules or what types of conformational changes are necessary to account for the electrophysiological data. For example, knowing that tetrodotoxin specifically blocks the early transient sodium current may tell you something about the properties that the membrane constituents in the region of the sodium "pore" must have to interact with the chemical groupings of tetrodotoxin.

Likewise, we should look at the reverse of what we have discussed above. If you are an electrophysiologist, devote special attention to understanding the significance of the biochemical studies being carried out on the squid giant axon. You are not working with an electrical wire, and your work will achieve added interest for you and significance for others if you can relate your electrical findings to the real axon of proteins, lipids, enzymes, and other substances. Unfortunately, because of different backgrounds and jargon, the biochemist and the electrophysiologist often seem to be separated by a broad gap. I hope that this chapter helps provide a bridge and encourage communication. In my own studies on the squid giant axon, I have found it necessary to alternate between electrical measurements and phospholipid measurements, between intracellular microelectrodes and electrophoretic electrodes, between cutting axons and cutting thin-layer plates.

The point that I am trying to make is that if you are to be a serious squid user, you must be able to critically evaluate, interpret, and integrate data from more than your single field of specialization.

4.2. Generalization of Results

There is always the tendency to generalize results, to hope that our findings are of broad interest, to speak of the unity of nature. In other words, we would not like to think that our results are unique to this "abnormally" large axon. The electrophysiological data clearly indicate that the electrical properties of the giant axon are not unique but in fact can be generalized to all nonmyelinated axons, and its more fundamental properties to all axons. For example, the ionic basis of the action potential seems to be similar for all axons. Other electrical properties are greatly influenced by anatomical considerations; for example, the high speed of conduction of the giant axon is clearly related to its large size and cannot be generalized to other, smaller nonmyelinated axons. The presence or absence of myelin will also modify some parameters such as the speed and mechanism of electrical conduction. I feel confident, however, that the great majority of biochemical and electrical properties we deduce using the giant axon will be generally shown to be true for all axons.

Some caution will be necessary, however, in extending data obtained with pharmacological agents. For example, TEA ion acts only from the inside of the squid giant axon, whereas it appears to act from both the inside and the outside of the Ranvier node in myelinated preparations (Hille, 1967; Koppenhofer, 1967; Armstrong, 1969; Armstrong and Hille, 1972). The block of activity by external application to the node is, however, quite different from (non-time- and -voltage-dependent) and the block by internal application quite similar to that produced in the squid axon. Typical local anesthetics such as procaine seem to act on the squid giant axon in manner similar to that in which they affect mammalian axons. The greatest caution must be exercised in attempting to generalize data obtained on the squid axon to what might be expected in a junctional or central nervous system preparation. There are, for example, many drugs well characterized in their central and synaptic actions that do not show comparable actions on the squid axon. This, of course, merely means that the most sensitive site of action of these drugs is elsewhere than on the axon. The effects on the squid axon may still be typical if what axonal effects would be observed *in vivo* on mammalian preparations if adequate concentrations were reached. Nevertheless, the squid axon has little predictive value as to what results might be expected at synapses and on the central nervous system. For example, narcotics and narcotic antagonists such as morphine, levallorphan, and naloxone act quite differently on the squid than on the central mammalian narcotic receptor (Simon and Rosenberg, 1970; Frazier *et al.*, 1972, 1973*b*). Both narcotics and narcotic antagonists have similar effects on the squid axon, and extremely potent narcotic analgetics do not show any great degree of potency on the squid axon. The duration of action of barbiturates on the squid axon also has no bearing on their duration of action in man (Frazier *et al.*, 1975). This, of course, is not surprising, since the metabolism by the liver and excretion by the kidney are the main factors responsible for determining their duration of action. The squid results are generalizable to the extent that they show that the division of short- intermediate- and long-acting barbiturates is not based on differences in their fundamental duration of action at the receptor level. There are many other examples of drugs having effects on the squid axon that are not typical of those observed on nonaxonal bioelectrically excitable tissue such as brain and synapse, for example, tetrahydrocannabinol derivatives (Brady and Carbone, 1973) and cholinergic drugs (Yeh and Narahashi, 1974*a*).

In contrast, it has been suggested that some of the effects of antiarrhythmic agents on the squid (e.g., increased steady-state potassium conductance) can explain their antiarrhythmic activity (Narahashi and Wang, 1973; Wu and Narahashi, 1973; Yeh and Narahashi, 1976). It has thus been suggested on the basis of effects observed on the squid axon that trihexyphenidyl, an antiparkinsonian agent, should be tried as an antiarrhythmic agent (Wu and Narahashi, 1976). Various treatments, such as the injection of TEA (Tasaki and Hagiwara, 1957) or the manipulation of internal and external ionic concentrations (Adelman, 1965; Adelman *et al.*, 1965*a*), can produce long-duration, cardiac-like action potentials from the squid giant axon (Fig. 43). It would be of great interest to know whether cardioactive drugs modify these long-duration responses from the squid in a manner similar to that observed on cardiac action potentials. If this were the case, the squid axon might have advantages over heart muscle for biological screening and studies on mechanism of action of cardiac drugs. The repetitively firing squid giant axon shows some promise also as a test object for detecting antiepileptic and antiparkinsonian activity. Well-known drugs having these two types of activities were shown to block spontaneous or repetitive activity (induced by low external divalent cations) of the squid giant axons in much lower concentrations than required to block the evoked action potential (Rosenberg and Bartels, 1967). In contrast, local anesthetics and other compounds showed little specificity in their relative activities against repetitive and evoked activity.

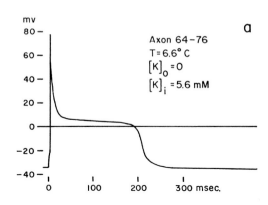

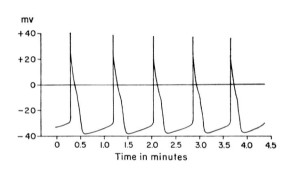

Fig. 43. (a) Membrane action potentials recorded from an axon internally perfused with 5.6 mM potassium acid phosphates in isosmotic dextrose. Response was obtained after 16 min in K-free artificial seawater. (b) Spontaneous discharge recorded from an axon internally perfused with isosmotic dextrose having a Tris-Cl concentration of 5 mM. The external solution was K-free artificial seawater. Reprinted from Adelman *et al.* (1965*a*) by courtesy of Rockefeller University Press, New York.

These few examples clearly indicate that there are numerous areas for further exploration before we can reach decisions on the usefulness of the squid axon as a biological screen or for understanding the actions of drugs having their major actions at sites other than the axon. It is obvious, however, that caution must be exercised and each case carefully investigated before results obtained with the squid axon can be generalized.

4.3. Some Final Thoughts

If you have gained an insight into how the squid giant axon can be of assistance in your research studies, then I have succeeded in the goals that prompted me to write this chapter. In emphasizing both "methods" and "applications," I have attempted to overcome the two greatest barriers we all experience when moving into new areas of research or when we are considering learning new techniques or using a new biological preparation. One barrier we find difficult to overcome is that of summoning the energy necessary to attack the veritable mountain of literature in the new field, literature that may seem unorganized and unitelligible except to the few devotees of the subject. I have therefore attempted to organize the literature, showing the many areas of studies for which the squid axon has been found useful, and giving an indication of the results obtained and their significance so that you will be able to narrow down the number of references that have to be consulted for detailed study. I have included the great majority of all studies on the squid giant axon since 1968, as well as many of the important references before that date. I hope that the chapter is therefore also useful to the experienced investigator who may find some "missing" references. The other barrier, which tends to keep us doing the same thing perhaps for longer than we should, is the hesitancy to learn a new technique that at least at first glance seems difficult and strange. I have therefore described the methods in detail, even though I realize that no amount of description can substitute for going into the laboratory and practicing. I hope, however, that my description of "methods" has given you useful hints that will smooth the path of your apprenticeship.

The giant axon of the squid should be useful to all neurobiologists interested in the electrophysiological and biochemical properties of axon. In addition, the opportunity to carry out structure–activity relationship studies with drugs on this axonal preparation have not yet been fully exploited. It is a unique preparation for any studies requiring pure axoplasm or the selective modification of the internal environment of the axolemmal membrane. For some biochemical studies, however, a limitation may be the amount of material that can be obtained, since for most studies with the axonal envelope, the finely dissected axon should be employed and techniques of membranal separation used for obtaining fractions enriched in various components such as axolemma.

The studies on the optical spike show some promise, although I would expect that in the future, advancements in the techniques of nuclear magnetic resonance, electron spin resonance, and other physical techniques will allow their application to the squid axon. We will then perhaps begin to really understand the conformational changes that occur in the axonal membrane and what "pores" and "gates" are at the molecular and biochemical level.

Our understanding of axonal functioning at the electrophysiological level was greatly aided by making use of the unique properties of the squid giant axon. It is hoped that other neurobiologists will be no less ingenious in using this preparation.

ACKNOWLEDGMENTS. I am grateful to the many collaborators whose work I have had the pleasure to quote. Grateful thanks are also given to the publishers for allowing me to reprint copyrighted material. I am pleased to acknowledge the National Institutes of Health, the National Science Foundation, the National Multiple Sclerosis Society, and the University of Connecticut Research Foundation, which have supported some of my personal research. A personal thank you is extended to my wife Sybil, son Stuart, and daughters Gail and Rachelle for assistance in various phases of preparation of materials for this chapter. The excellent secretarial assistance of Linda Kelsey was most helpful.

References

Abbott, N.J., Deguchi, T., Frazier, D.T., Murayama, K., Narahashi, T., Ottolenghi, A., and Wang, C.M., 1972, The action of phospholipases on the inner and outer surface of the squid giant axon membrane, *J. Physiol.* **220**:73.

Adam, G., 1973, The effect of potassium diffusion through the Schwann cell layer on potassium conductance of the squid axon, *J. Membrane Biol.* **13**:353.

Adelman, W.J., 1965, Cardiac like responses from internally perfused squid axons, *Excerpta Medica International Congr. Ser.*, No. 87, Proceedings of the XXIIIrd International Congress of Physiological Sciences (Tokyo, Sept. 1965), p. 542.

Adelman, W.J., and Fitzhugh, R., 1975, Solutions of the Hodgkin–Huxley equations modified for potassium accumulation in a periaxonal space, *Fed. Proc. Fed. Am. Soc. Exp. Biol.* **34**:1322.

Adelman, W.J., and Fok, Y.B., 1964, Internally perfused squid axons studied under voltage clamp conditions. II. Results: The effects of internal potassium and sodium on membrane electrical characteristics, *J. Cell. Comp. Physiol.* **64**:429.

Adelman, W.J., and Gilbert, D.L., 1964, Internally perfused squid axons studied under voltage clamp conditions. I. Method, *J. Cell. Comp. Physiol.* **64**:423.

Adelman, W.J., and Moore, J.W., 1961, Action of external divalent ion reduction on sodium movement in the squid giant axon, *J. Gen. Physiol.* **45**:93.

Adelman, W.J., and Senft, J.P., 1966, Voltage clamp studies on the effect of internal cesium ion on sodium and potassium currents in the squid giant axon, *J. Gen. Physiol* **50**:279.

Adelman, W.J., and Senft, J.P., 1968, Dynamic assymetries in the squid axon membrane, *J. Gen. Physiol.* **51**:102S.

Adelman, W.J., and Taylor, R.E., 1961, Leakage current rectification in the squid giant axon, *Nature (London)* **190**:883.

Adelman, W.J., and Taylor, R.E., 1964, Effects of replacement of external sodium chloride with sucrose on membrane currents of the squid giant axon, *Biophys. J.* **4**:451.

Adelman, W.J., Dyro, F.M., and Senft, J., 1965a, Long duration responses obtained from internally perfused axons, *J. Gen. Physiol.* **48**:1.

Adelman, W.J., Dyro, F.M., and Senft, J.P., 1965*b*, Internally perfused axons: Effects of two different anions on ionic conductance, *Science* **151**:1392.

Agin, D., and Schauf, C., 1968, Concerning negative conductance in the squid axon, *Proc. Natl. Acad. Sci. U.S.A.* **59**:1201.

Aidley, D.J., 1971, *The Physiology of Excitable Cells*, Cambridge University Press, London.

Albuquerque, E.X., Seyama, I., and Narahashi, T., 1973, Characterization of batrachotoxin-induced depolarizations of the squid giant axons, *J. Pharmacol. Exp. Ther.* **184**:308.

Alema, S., and Giuditta, A., 1976, Site of biosynthesis of brain-specific proteins in the giant fibre system of the squid, *J. Neurochem.* **26**:995.

Alema, S., Calissano, P., Rusca, G., and Giuditta, A., 1973, Identification of a calcium-binding brain specific protein in the axoplasm of squid giant axons, *J. Neurochem.* **20**:681.

Alfred, R.G., 1974, Structure, growth and distribution of the squid *Bathothauma lyromma*, Chun., *J. Mar. Biol. Assoc. U.K.* **51**:995.

Araya, H., and Ishii, M., 1974, Information on the fishery and the ecology of the squid, *Doryteuthis bleekeri* Keferstein, in the waters of Hokkaido, *Bull. Hokkaido Reg. Fish Res. Lab.* **40**:1.

Armstrong, C.M., 1966, Time course of TEA$^+$-induced anomalous rectification in squid giant axons, *J. Gen. Physiol.* **50**:491.

Armstrong, C.M., 1969, Inactivation of the potassium conductance and related phenomena caused by quaternary ammonium ion injection in squid axons, *J. Gen. Physiol.* **54**:553.

Armstrong, C.M., 1971, Interaction of tetraethylammonium ion derivatives with the potassium channels of giant axons, *J. Gen. Physiol.* **58**:413.

Armstrong, C.M., and Bezanilla, F., 1973, Currents related to the movement of the gating particles of the sodium channels, *Nature (London)* **242**:459.

Armstrong, C.M., and Bezanilla, F., 1974, Change movement associated with the opening and closing of the activation gates of the Na channels, *J. Gen. Physiol.* **63**:533.

Armstrong, C.M., and Bezanilla, F., 1975, Currents associated with the ionic gating structures in nerve membrane, in: Carriers and channels in biological systems (A.E. Shamoo, ed.), *Ann. N.Y. Acad. Sci.* **264**:265.

Armstrong, C.M., and Binstock, L., 1964, The effects of several alcohols on the properties of the squid giant axon, *J. Gen. Physiol.* **48**:265.

Armstrong, C.M., and Binstock, L., 1965, Anomalous rectification in the squid axon injected with tetraethylammonium chloride, *J. Gen. Physiol.* **48**:859.

Armstrong, C.M., and Hille, B., 1972, The inner quaternary ammonium ion receptor in potassium channels of the node of Ranvier, *J. Gen. Physiol.* **59**:388.

Armstrong, C.M., Bezanilla, F., and Rojas, E., 1973, Destruction of sodium conductance inactivation in squid axons perfused with pronase, *J. Gen. Physiol.* **62**:375.

Arnold, J.M., 1965, Normal embryonic stages of the squid, *Loligo pealii* (Lesuer), *Biol. Bull.* **128**:24.

Arnold, J.M., Singley, C.T., and Williams-Arnold, L.D., 1972, Embryonic development and post-hatching survival of the sepiolid squid *Euprymma scolopes* under laboratory conditions, *Veliger* **14**:361.

Arnold, J.M., Summers, W.C., Gilbert, D.L., Manalis, R.S., Daw, N.W., and Lasek, R.J., 1974, A guide to the laboratory use of the squid *Loligo pealei*, Marine Biological Laboratory, Woods Hole, Massachusetts.

Atwater, I., Bezanilla, F., and Rojas, E., 1970, Time course of the sodium permeability change during a single membrane action potential, *J. Physiol.* **211**:753.

Baker, P.F., 1968, Recent experiments on the properties of the Na efflux from squid axons, *J. Gen. Physiol.* **51**:172S.

Baker, P.F., and Crawford, A.C., 1972, Mobility and transport of magnesium in squid giant axons, *J. Physiol.* **227**:855.

Baker, P.F., and Glitsch, H.G., 1975, Voltage-dependent changes in the permeability of nerve membranes to calcium and other divalent cations, *Philos. Trans. R. Soc. London B Ser.* **270**:389.

Baker, P.F., and Manil, J., 1968, The rates of action of K$^+$ and ouabain on the sodium pump in squid axons, *Biochim. Biophys. Acta* **150**:328.

Baker, P.F., and Potashner, S.J., 1973, The role of metabolic energy in the transport of glutamate by invertebrate nerve, *Biochim. Biophys. Acta* **318:**123.

Baker, P.F., and Shaw, T.I., 1965, A comparison of the phosphorus metabolism of intact squid nerve with that of isolated axoplasm and sheath, *J. Physiol.* **180:**424.

Baker, P.F., and Willis, J.S., 1972, Inhibition of the sodium pump in squid giant axons by cardiac glycosides: Dependence of extracellular ions and metabolism, *J. Physiol.* **224:**463.

Baker, P.F., Hodgkin, A.L., and Shaw, T.I., 1961, Replacement of the protoplasm of a giant nerve fibre with artificial solutions, *Nature (London)* **190:**885.

Baker, P.F., Hodgkin, A.L., and Shaw, T.I., 1962a, Replacement of the axoplasm of giant nerve fibres with artificial solutions, *J. Physiol.* **164:**330.

Baker, P.F., Hodgkin, A.L., and Shaw, T.I., 1962b, The effects of changes in internal ionic concentrations on the electrical properties of perfused giant axons, *J. Physiol.* **164:**355.

Baker, P.F., Hodgkin, A.L., and Meves, H., 1964, The effect of diluting the internal solution on the electrical properties of a perfused giant axon, *J. Physiol.* **170:**541.

Baker, P.F., Blaustein, M.P., Hodgkin, A.L., and Steinhardt, R.A., 1969a, The influence of calcium on sodium efflux in squid axons, *J. Physiol.* **200:**431.

Baker, P.F., Blaustein, M.P., Keynes, R.D., Manil, J., Shaw, T.I., and Steinhardt, R.A., 1969b, The ouabain sensitive fluxes of sodium and potassium in squid giant axons, *J. Physiol.* **200:**459.

Baker, P.F., Hodgkin, A.L., and Ridgeway, E.B., 1971, Depolarization and calcium entry in squid giant axons, *J. Physiol.* **218:**709.

Baker, P.F., Meves, H., and Ridgeway, E.B., 1973a, Calcium entry in response to maintained depolarization of squid axons, *J. Physiol.* **231:**527.

Baker, P.F., Meves, H., and Ridgeway, E.B., 1973b, Effects of manganese and other agents on the calcium uptake that follows depolarization of squid axons, *J. Physiol.* **231:**511.

Barnola, F.V., Camejo, G., and Villegas, R., 1971, Ionic channels and nerve membrane lipoproteins: DDT–nerve membrane interaction, *Int. J. Neurosci.* **1:**309.

Bear, R.S., and Schmitt, F.O., 1939, Electrolytes in the axoplasm of the great nerve fibers of the squid, *J. Cell. Comp. Physiol.* **14:**205.

Bear, R.S., Schmitt, F.O., and Young, J.Z., 1937a, The ultrastructure of nerve axoplasm, *Proc. R. Soc. London Ser. B* **123:**505.

Bear, R.S., Schmitt, F.O., and Young, J.Z., 1937b, Investigations on the protein constituents of nerve axoplasm, *Proc. R. Soc. London Ser. B* **123:**520.

Begenisich, T., and Lynch, C., 1974, Effects of internal divalent cations on voltage-clamped squid axons, *J. Gen. Physiol.* **63:**675.

Bezanilla, F., and Armstrong, C.M., 1972, Negative conductance caused by entry of sodium and cesium ions into the potassium channels of the squid axons, *J. Gen. Physiol.* **60:**588.

Bezanilla, F., and Armstrong, C.M., 1975, Kinetic properties and inactivation of the gating currents of sodium channels in squid axon, *Philos. Trans. R. Soc. London Ser. B* **270:**449.

Bicher, H.I., and Ohki, S., 1972, Intracellular pH electrode: Experiments on the giant squid axon, *Biochim. Biophys. Acta* **255:**900.

Binstock, L., and Lecar, H., 1969, Ammonium ion currents in the squid giant axon, *J. Gen. Physiol.* **53:**342.

Blaustein, M.P., and Goldman, D.E., 1966, Action of anionic and cationic nerve blocking agents: Experiment and interpretation, *Science* **153:**429.

Blaustein, M.P., and Hodgkin, A.L., 1969, The effect of cyanide on the efflux of calcium from squid axons, *J. Physiol.* **200:**497.

Blaustein, M.P., and Russell, J.M., 1975, Sodium–calcium exchange and calcium–calcium exchange in internally dialyzed squid giant axons, *J. Membrane Biol.* **22:**285.

Boell, E.J., and Nachmansohn, D., 1940, Localization of choline esterase in nerve fibers, *Science* **92:**513.

Bonting, S.L., and Caravaggio, L.L., 1962, Sodium–potassium-activated adenosine triphosphatase in the squid giant axon, *Nature (London)* **194:**1180.

Borisy, G.G., and Taylor, E.W., 1967, The mechanism of action of colchicine: Binding of colchicine-^{3}H to cellular protein, *J. Cell Biol.* **34:**525.

Boron, W.F., and DeWeer, P., 1976, Intracellular pH transients in squid giant axons caused by CO_2, NH_3 and metabolic inhibitors, *J. Gen. Physiol.* **67**:91.

Borradaile, L.A., Potts, F.A., Eastham, L.E.S., and Saunders, J.T., 1967, *The Invertebrata*, 4th ed., Revised by G.A. Kerkut, pp. 636–647, Cambridge University Press, London.

Bradbury, H.E., and Aldrich, F.A., 1969a, Observations on locomotion of the short-finned squid, *Illex illecerebrosus illecerebrosus* Lesueur, 1821), in captivity, *Can. J. Zool.* **47**:741.

Bradbury, H.E., and Aldrich, F.A., 1969b, Observations on feeding of the squid *Illex illecerebrosus illecerebrosus* (Lesueur, 1821) in captivity, *Can. J. Zool.* **47**:913.

Brady, R.O., and Carbone, E., 1973, Comparison of the effects of delta[9]-tetrahydrocannabinol, 11-hydroxy-delta[9]-tetrahydrocannabinol and ethanol on the electrophysiological activity of the giant axon of the squid, *Neuropharmacology* **12**:601.

Bray, G.A., 1960, A simple efficient liquid scintillator for counting aqueous solutions in liquid scintillation counter, *Anal. Biochem.* **1**:279.

Brinley, F.J., Jr., and Mullins, L.J., 1965, Ion fluxes and transference number in squid axons, *J. Neurophysiol.* **28**:526.

Brinley, F.J., Jr., and Mullins, L.J., 1967, Sodium extrusion by internally dialyzed squid axons, *J. Gen. Physiol.* **50**:2303.

Brinley, F.J., Jr., and Mullins, L.J., 1968, Sodium fluxes in internally dialyzed squid axons, *J. Gen. Physiol.* **52**:181.

Brinley, F.J., Jr., and Scarpa, A., 1975, Ionized magnesium concentration in axoplasm of dialyzed squid axons, *FEBS Lett.* **50**:82.

Brinley, F.J., Jr., Spangler, S.G., and Mullins, L.J., 1975, Calcium and EDTA fluxes in dialyzed squid axons, *J. Gen. Physiol.* **66**:223.

Brown, J.E., Cohen, L.B., DeWeer, P., Pinto, L.H., Ross, W.N., and Salzberg, B.M., 1975a, Arsenazo III, an indicator of rapid changes of intracellular ionized calcium in squid giant axons, *Biol. Bull.* **149**:421.

Brown, J.E., Cohen, L.B., DeWeer, P., Pinto, L.H., Ross, W.N., and Salzberg, B.M., 1975b, Rapid changes of intracellular free calcium concentration, *Biophys. J.* **15**:1155.

Brzin, M., Dettbarn, W.-D., Rosenberg, P., and Nachmansohn, D., 1965a, Cholinesterase activity per unit surface area of conducting membranes, *J. Cell Biol.* **26**:353.

Brzin, M., Dettbarn, W.-D., and Rosenberg, P., 1965b, Penetration of neostigmine, physostigmine and paraoxon into the squid giant axon, *Biochem. Pharmacol.* **14**:919.

Bullock, T.H., Nachmansohn, D., and Rothenberg, M.A., 1946, Effects of inhibitors of choline esterase on the nerve action potential, *J. Neurophysiol.* **9**:9.

Bullough, W.S., 1950, *Practical Invertebrate Anatomy*, Macmillan, London.

Bureš, J., Petráň, M., and Zachar, J., 1967, *Electrophysiological Methods in Biological Research*, 3rd ed., Academic Press, New York.

Caldwell, P.C., 1958, Studies on the internal pH of large muscle and nerve fibres, *J. Physiol.* **142**:22.

Caldwell, P.C., and Keynes, R.D., 1957, The utilization of phosphate bond energy for sodium extrusion from giant axons, *J. Physiol.* **137**:12P.

Caldwell, P.C., and Keynes, R.D., 1959, The effect of ouabain on the efflux of sodium from squid giant axon, *J. Physiol.* **148**:8P.

Caldwell, P.C., and Lowe, A.G., 1970, The influx of orthophosphate into squid giant axons, *J. Physiol.* **207**:271.

Caldwell, P.C., Hodgkin, A.L., Keynes, R.D., and Shaw, T.I., 1960, The effects of injecting "energy rich" phosphate compounds on the active transport of ions in the giant axons of *Loligo, J. Physiol.* **152**:561.

Caldwell, P.C., Hodgkin, A.L., Keynes, R.D., and Shaw, T.I., 1964, The rate of formation and turnover of phosphorus compounds in squid giant axons, *J. Physiol.* **171**:119.

Camejo, G., Villegas, G.M., Barnola, F.V., and Villegas, R., 1969, Characterization of two different membrane fractions isolated from the first stellar nerves of the squid *Dosidicus gigas, Biochim. Biophys. Acta* **193**:247.

Camougis, G., and Takman, B.H., 1971, Nerve and nerve muscle preparations, in: *Methods in Pharmacology*, Vol. 1 (A. Schwartz, ed.) Chapt. 1, pp. 1–40, Appleton-Century-Crofts, New York.

Canessa, M., 1965, Properties of ATPase activities of membrane fractions from the sheath of squid giant axons, *J. Cell. Comp. Physiol.* **66:**165.

Canessa-Fischer, M., Zambrano, F., and Riveros-Moreno, V., 1967, Properties of the ATPase system from the sheath of squid giant axons, *Arch. Biochem. Biophys.* **122:**658.

Canessa-Fischer, M., Zambrano, F., and Rojas, E., 1968, The loss and recovery of the sodium pump in perfused giant axons, *J. Gen. Physiol.* **51:**162S.

Carbone, E., Sisco, K., and Warashima, A., 1974, Physicochemical properties of 2,6 TNS binding sites in squid giant axons: Involvement of water molecules in the excitable process, *J. Membrane Biol.* **18:**263.

Carpenter, D.O., 1973, Electrogenic sodium pump and high specific resistance in nerve cell bodies of the squid, *Science* **179:**1336.

Carpenter, D.O., Hovey, M.M., and Bak, A.F., 1971, Intracellular conductance of Aplysia neurons and squid axon as determined by a new technique, *Int. J. Neurosci.* **2:**35.

Carpenter, D.O., Hovey, M.M., and Bak, A.F., 1975, Resistivity of axoplasm. II. Internal resistivity of giant axons of squid and Myxicola, *J. Gen. Physiol.* **66:**139.

Cavanaugh, G.M. (ed.), 1975, *Formulae and Methods VI of the Marine Biological Laboratory Chemical Room,* pp. 67–68, Marine Biological Laboratory, Woods Hole, Massachusetts.

Chailaklyou, L.M., 1961, Measurement of the resting and action potentials of the giant fibre of the squid in various conditions of recording, *Biophysics* **6:**344.

Chambers, R., and Kao, C.-Y., 1952, The effect of electrolytes on the physical state of the nerve axon of the squid and of Stentor, a protozoon, *Exp. Cell Res.* **3:**564.

Chandler, W.K., and Hodgkin, A.L., 1965, The effect of internal sodium on the action potential in the presence of different internal anions, *J. Physiol.* **181:**594.

Chandler, W.K., and Meves, H., 1965, Voltage clamp experiments on internally perfused giant axons, *J. Physiol.* **180:**788.

Chandler, W.K., and Meves, H., 1970a, Sodium and potassium currents in squid axons perfused with fluoride solutions, *J. Physiol.* **211:**623.

Chandler, W.K., and Meves, H., 1970b, Evidence for two types of sodium conductance in axons perfused with sodium fluoride solution, *J. Physiol.* **211:**653.

Chandler, W.K., and Meves, H., 1970c, Rate constants associated with changes in sodium conductance in axons perfused with sodium fluoride, *J. Physiol.* **211:**679.

Chandler, W.K., and Meves, H., 1970d, Slow changes in membrane permeability and long-lasting action potentials in axons perfused with fluoride solutions, *J. Physiol.* **211:**707.

Chandler, W.K., Hodgkin, A.L., and Meves, H., 1965, The effect of changing the internal solution on sodium inactivation and related phenomena in giant axons, *J. Physiol.* **180:**821.

Chapman, R.A., 1967, Dependence on temperature of the conduction velocity of the action potential of the squid giant axon, *Nature (London)* **213:**1143.

Coelho, R.R., Goodman, J.W., and Bowers, M.B., 1960, Chemical studies of the satellite cells of the squid giant nerve fibre, *Exp. Cell Res.* **20:**1.

Cohen, L.B., and Landowne, D., 1974, The temperature dependence of the movement of sodium ions associated with nerve impulses, *J. Physiol.* **236:**95.

Cohen, L.B., Keynes, R.D., and Hilla, B., 1968, Light scattering and birefringence changes during nerve activity. *Nature (London)* **218:**438.

Cohen, L.B., Hille, B., and Keynes, R.D., 1970, Changes in axon birefringence during the action potential, *J. Physiol.* **211:**495.

Cohen, L.B., Hille, B., Keynes, R.D., Landowne, D., and Rojas, E., 1971, Analysis of the potential-dependent changes in optical retardation in the squid giant axon, *J. Physiol.* **218:**205.

Cohen, L.B., Keynes, R.D., and Landowne, D., 1972a, Changes in light scattering that accompany the action potential in squid giant axons: Potential dependent components, *J. Physiol.* **224:**701.

Cohen, L.B., Keynes, R.D., and Landowne, D., 1972b, Changes in axon light scattering that accompany the action potential: Current dependent components, *J. Physiol.* **224:**727.

Cohen, L.B., Salzberg, B.M., Davita, H.V., Ross, W.N., Landowne, D., Waggoner, A.S., and Wang, C.H., 1974, Changes in axon fluorescence during activity: Molecular probes of membrane potential, *J. Membrane Biol.* **19:**1.

Cohen, M., Palti, Y., and Adelman, W.J., 1975, Ionic dependence of sodium currents in squid axons analyzed in terms of specific ion "channel" interactions, *J. Membrane Biol.* **24:**201.

Cole, K.S., 1941, Rectification and inductance in the squid axon membrane, *J. Gen. Physiol.* **25:**29.

Cole, K.S., 1949, Dynamic electrical characteristics of the squid axon membrane, *Arch. Sci. Physiol.* **3:**253.

Cole, K.S., 1968, *Membrane, Ions and Impulses*, University of California Press, Berkeley and Los Angeles.

Cole, K.S., 1975, Resistivity of axoplasm. 1. Resistivity of extruded squid axoplasm, *J. Gen. Physiol.* **66:**133.

Cole, K.S., and Baker, R.F., 1941, Longitudinal impedance of the squid giant axon, *J. Gen. Physiol.* **24:**771.

Cole, K.S., and Curtis, H.J., 1939, Electrical impedance of the squid giant axon during activity, *J. Gen. Physiol.* **22:**649.

Cole, K.S., and Gilbert, D.L., 1970, Jet propulsion of squid, *Biol. Bull.* **138:**245.

Cole, K.S., and Marmont, G., 1942, The effect of ionic environment upon the longitudinal impedance of the squid giant axon, *Fed. Proc. Fed. Am. Soc. Exp. Biol.* **1:**15.

Cole, K.S., and Moore, J.W., 1960, Ionic current measurements in the squid giant axon membrane, *J. Gen. Physiol.* **44:**123.

Condrea, E., and Rosenberg, P., 1968, Demonstration of phospholipid splitting as the factor responsible for increased permeability and block of axonal conduction induced by snake venom. II. Study on squid axons, *Biochim. Biophys. Acta* **150:**271.

Conti, F., and Tasaki, I., 1970, Changes in extrinsic fluorescence in squid axons during voltage-clamp, *Science* **169:**1322.

Conti, F., DeFelice, L.J., and Wanke, E., 1975, Potassium and sodium ion current noise in the membrane of the squid giant axon, *J. Physiol.* **248:**45.

Cooke, I., and Lipkin, M., Jr., 1972, *Cellular Neurophysiology: A Source Book,* Holt, Rinehart and Winston, New York.

Cooke, I.M., Diamond, J.M., Grinnell, A.D., Hagiwara, S., and Sakata, H., 1968, Suppression of the action potential in nerve by nitrobenzene derivatives, *Proc. Natl. Acad. Sci. U.S.A.* **60:**470.

Cuervo, L.A., and Adelman, W.J., 1970, Equilibrium and kinetic properties of the interaction between tetrodotoxin and the excitable membrane of the squid giant axon, *J. Gen. Physiol.* **55:**309.

Curtis, H.J., and Cole, K.S., 1940, Membrane action potentials from the squid giant axon, *J. Cell. Comp. Physiol.* **15:**147.

Curtis, H.J., and Cole, K.S., 1942, Membrane resting and action potentials from the squid giant axon, *J. Cell. Comp. Physiol.* **19:**135.

Davila, H.V., Salzberg, B.M., Cohen, L.B., and Waggoner, A.S., 1972, Changes in fluorescence of squid axons during activity, *Biol. Bull.* **134:**457.

Davila, H.V., Salzberg, B.M., Cohen, L.B., and Waggoner, A.S., 1973, A large change in axon fluorescence that provides a promising method for measuring membrane potential, *Nature (London) New Biol.* **241:**159.

Davila, H., Cohen, L.B., Salzberg, B.M., and Shrivastav, B.B., 1974, Changes in ANS and TNS fluorescence in giant axons from *Loligo, J. Membrane Biol.* **15:**29.

Davison, P.F., and Huneeus, F.C., 1970, Fibrillar proteins from squid axons. II. Microtubule protein, *J. Mol. Biol.* **52:**429.

Davison, P.F., and Taylor, E.W., 1960, Physical–chemical studies of proteins of squid nerve axoplasm with special reference to the axon fibrous protein, *J. Gen. Physiol.* **43:**801.

DeFelice, L.J., Wanke, E., and Conti, F., 1975, Potassium and sodium current noise from squid axon membranes, *Fed. Proc. Fed. Am. Soc. Exp. Biol.* **34:**1338.

Deffner, G.G.J., 1961a, Chemical investigations of the giant nerve fibers of the squid. V. Quaternary ammonium ions in axoplasm, *Biochim. Biophys. Acta* **50:**555.

Deffner, G.G.J., 1961b, The dialyzable free organic constituents of squid blood: A comparison with nerve axoplasm, *Biochim. Biophys. Acta* **47:**378.

Deffner, G.G.J., and Hafter, R.E., 1959a, Chemical investigations of the giant nerve fibers of the squid. I. Fractionation of dialyzable constituents of axoplasms and quantitative determination of the free amino acids, *Biochim. Biophys. Acta* **32:**362.

Deffner, G.G.J., and Hafter, R.E., 1959*b*, Chemical investigations of the giant nerve fibers of the squid. II. Detection and identification of cysteic acid amide in squid nerve axoplasm, *Biochim. Biophys. Acta* **35**:334.

Deffner, G.G.J., and Hafter, R.E., 1960*a*, Chemical investigations of the giant nerve fibers of the squid. III. Identification and quantitative estimation of free organic ninhydrin-negative constituents, *Biochim. Biophys. Acta* **42**:189.

Deffner, G.G.J., and Hafter R.E., 1960*b*, Chemical investigations of the giant nerve fibers of the squid. IV. Acid–base balance in axoplasm, *Biochim. Biophys. Acta* **42**:200.

de Groof, R.C., and Narahashi, T., 1974, Effects of holothurin A on squid axon membranes, *Fed. Proc. Fed. Am. Soc. Exp. Biol.* **33**:319.

Dettbarn, W.-D., and Davis, F.A., 1962, "Sucrose gap" technique applied to single-nerve-fiber preparation, *Biochim. Biophys. Acta* **60**:648.

Dettbarn, W.-D., and Rosenberg, P., 1962, Sources of error in relating electrical and acetylcholinesterase activity, *Biochem. Pharmacol.* **11**:1025.

Dettbarn, W.-D., Heilbronn, E., Hoskin, F.C.G., and Katz, R., 1972, The effects of pH on penetration and action of procaine ^{14}C, atropine ^{3}H, *n* butanol ^{14}C and halothane ^{14}C in single giant axons of the squid, *Neuropharmacology* **11**:727.

DeWeer, P., 1968, Restoration of a potassium requiring sodium pump in squid giant axons poisoned with CN and depleted of arginine, *Nature (London)* **219**:730.

DeWeer, P., 1970, Effects of intracellular adenosine-5'-diphosphate and orthophosphate on the sensitivity of sodium efflux from squid axon to external sodium and potassium, *J. Gen. Physiol.* **56**:583.

DeWeer, P., and Geduldig, D., 1973, Electrogenic sodium pump in squid giant axon, *Science* **179**:1326.

Dipolo, R., 1973, Calcium efflux from internally dialyzed squid giant axons, *J. Gen. Physiol.* **62**:575.

Dipolo, R., 1974, Effect of ATP on the calcium efflux in dialyzed squid giant axons, *J. Gen. Physiol.* **64**:503.

Dipolo, R., Requena, J., Brinley, F.J., Jr., Mullins, L.J., Scarpa, A., and Tiffert, T., 1976, Ionized calcium concentration in squid axons, *J. Gen. Physiol.* **67**:433.

Doane, M.G., 1967, Fluorometric measurement of pyridine nucleotide reduction in the giant axon of the squid, *J. Gen. Physiol.* **50**:2603.

Dole, W.P., and Simon, E.J., 1974, Effects of levorphanol on phospholipid metabolism in the giant axon of the squid, *J. Neurochem.* **22**:183.

Easton, D.M., and Swenberg, C.E., 1975, Temperature and impulse velocity in giant axon of squid, *Loligo pealei, Am. J. Physiol.* **229**:1249.

Eyrich, T.L., Barrett, D., and Rock, P.A., 1976, Phospholipase activity in squid and frog axons, *J. Neurochem.* **26**:1079.

Fischer, S., and Litvak, S., 1967, The incorporation of microinjected ^{14}C-amino acids into TCA insoluble fractions of the giant axon of the squid, *J. Cell. Physiol.* **70**:69.

Fischer, S., Cellino, M., Gariglio, P., and Tellez-Nagel, M.I., 1968, Protein and RNA metabolism of squid axons (*Dosidicus gigas*), *J. Gen. Physiol.* **51**:72S.

Fischer, S., Gariglio, P., and Tarifeno, E., 1969, Incorporation of H$_3$-uridine and the isolation and characterization of RNA from squid axon, *J. Cell. Physiol.* **74**:155.

Fishman, H.M., 1970, Direct and rapid description of the individual ionic currents of squid axon membrane by ramp potential control, *Biophys. J.* **10**:799.

Fishman, H.M., 1973, Relaxation spectra of potassium channel noise from squid axon membranes, *Proc. Natl. Acad. Sci. U.S.A.* **70**:876.

Fishman, H.M., 1975*a*, Patch voltage clamp of squid axon membrane, *J. Membrane Biol.* **24**:26S.

Fishman, H.M., Poussart, D.J.M., and Moore, L.E., 1975*a*, Noise measurements in squid axon membrane, *J. Membrane Biol.* **24**:281.

Fishman, H.M., Moore, L.E., and Poussart, D.J.M., 1975*b*, Potassium-ion conduction noise in squid axon membrane, *J. Membrane Biol.* **24**:305.

Frazier, D.T., and Narahashi, T., 1975, Tricaine (MS-222): Effects on ionic conductances of squid axon membranes, *Eur. J. Pharmacol.* **33**:313.

Frazier, D.T., Narahashi, T., and Moore, J. W., 1969, Hemicholinium 3: Noncholinergic effects on squid axons, *Science* **163**:820.

Frazier, D.T., Narahashi, T., and Yamada, M., 1970, The site of action and active form of local anesthetics. II. Experiments with quaternary compounds, *J. Pharmacol. Exp. Ther.* **171**:45.

Frazier, D.T., Murayama, K., Abbott, N.J., and Narahashi, T., 1972, Effects of morphine on internally perfused squid giant axons, *Proc. Soc. Exp. Biol. Med.* **139**:434.

Frazier, D.T., Sevcik, C., and Narahashi, T., 1973a, Nicotine: Effect on nerve membrane conductances, *Eur. J. Pharmacol.* **22**:217.

Frazier, D.T., Ohta, M., and Narahashi, T., 1973b, Nature of the morphine receptor present in the squid axon, *Proc. Soc. Exp. Biol. Med.* **142**:1209.

Frazier, D.T., Murayama, K., Abbott, N.J., and Narahashi, T., 1975, Comparison of the action of different barbiturates on squid axon, *Eur. J. Pharmacol.* **32**:102.

Freeman, A.R., 1971, Electrophysiological activity of tetrodotoxin on the resting membrane of the squid giant axon, *Comp. Biochem. Physiol.* **40A**:71.

Freeman, A.R., Reuben, J.P., Brandt, P.W., and Grundfest, H., 1966, Osmometrically determined characteristics of the cell membrane of squid and lobster giant axons, *J. Gen. Physiol.* **50**:423.

Frumento, A.S., and Mullins, L.J., 1964, Potassium free effect in squid axons, *Nature (London)* **204**:1312.

Gainer, H., Carbone, E., Singer, I., Sisco, K., and Tasaki, I., 1974, Depolarization-induced changes in the enzymatic radio-iodination of a protein on the internal surface of the squid giant axon membrane, *Comp. Biochem. Physiol.* **47A**:477.

Geren, B.B., and Schmitt, F.O., 1954, The structure of the Schwann cell, and its relation to the axon in certain invertebrate nerve fibers, *Proc. Natl. Acad. Sci. U.S.A.* **40**:863.

Gilbert, D.L., 1971, Internal perfusion of squid giant axon, in: *Biophysics and Physiology of Excitable Membranes* (W.J. Adelman, Jr., ed.), pp. 264–273, Van Nostrand Reinhold, New York.

Gilbert, D.L., and Ehrenstein, G. 1969, Effect of divalent cations on potassium conductance of squid axons: Determination of surface charge, *Biophys. J.* **9**:447.

Giuditta, A., Dettbarn, W.-D., and Brzin, M., 1968, Protein synthesis in the isolated giant axon of the squid, *Proc. Natl. Acad. Sci. U.S.A.* **59**:1284.

Giuditta, A., D'Udine, B., and Pepe, M., 1971, Uptake of protein by the giant axon of the squid, *Nature (London) New Biol.* **229**:29.

Goldman, D.E., 1964, A molecular structural basis for the excitation properties of axons, *Biophys. J.* **4**:167.

Goldman, L., 1975, Pronase and models for the sodium conductance, *J. Gen. Physiol.* **65**:551.

Gruener, R., 1973, Excitability blockade of the squid giant axon by the venom of *Latrodectus mactans* (black widow spider), *Toxicon* **11**:155.

Gruener, R., and Narahashi, T., 1972, The mechanism of excitability blockade by chlorpromazine, *J. Pharmacol. Exp. Ther.* **181**:161.

Grundfest, H., Kao, C.Y., and Altamirano, M., 1954, Bioelectric effects of ions microinjected into the giant axon of *Loligo*, *J. Gen. Physiol.* **38**:245.

Guttman, R., 1969, Temperature dependence of oscillation in squid axons: Comparison of experiments with computations, *Biophys. J.* **9**:269.

Guttman, R., and Barnhill, R., 1968a, Effect of low sodium, tetrodotoxin and temperature variation upon excitation, *J. Gen. Physiol.* **51**:621.

Guttman, R., and Barnhill, R., 1968b, Temperature dependence of accommodation and excitation in space-clamped axons, *J. Gen. Physiol.* **51**:759.

Guttman, R., and Barnhill, R., 1970, Oscillation and repetitive firing in squid axons: Comparison of experiments with computations, *J. Gen. Physiol.* **55**:104.

Guttman, R., and Hachmeister, L., 1971, Effect of calcium, temperature and polarizing currents upon alternating current excitation of space-clamped squid axons, *J. Gen. Physiol.* **58**:304.

Hagiwara, S., Eaton, D.C., Stuart, A.E., and Rosenthal, N.P., 1972, Cation selectivity of the resting membrane of squid axon, *J. Membrane Biol.* **9**:373.

Hallett, M., and Carbone, E., 1972, Studies of calcium influx into squid giant axons with aequorin, *J. Cell. Physiol.* **80**:219.

Hallett, M., Schneider, A.S., and Carbone, E., 1972, Tetracycline fluorescence as probe for nerve membrane with some model studies using erythrocyte ghosts, *J. Membrane Biol.* **10**:31.

Hawthorne, J.N., and Kai, M., 1970, Metabolism of phosphoinositides, in: *Handbook of Neuro-chemistry*, Vol. 3 (A. Lajtha, ed.), pp. 491–508, Plenum Press, New York.

Henderson, J.V., Jr., and Gilbert, D.L., 1975, Slowing of ionic currents in the voltage-clamped squid axon by helium pressure, *Nature (London)* **258**:351.

Henkin, R.I., Stillman, I.S., Gilbert, D.L., and Lipicky, R.J., 1974, Ineffectiveness of lysergic acid diethyl amide -25 (LSD) on altering Na–K currents in squid giant axon, *Experientia* **30**:916.

Herzog, W.H., Feibel, R.M., and Bryant, S.H., 1964, The effect of aconitine on the giant axon of the squid, *J. Gen. Physiol.* **47**:719.

Hidalgo, C., and Latorree, R., 1970a, Temperature dependence of non-electrolyte and sodium permeability in giant axon of squid, *J. Physiol.* **211**:173.

Hidalgo, C., and Latorree, R., 1970b, Effect of stimulation and hyperpolarization on non-electrolyte and sodium permeability in perfused axons of squid, *J. Physiol.* **211**:193.

Hille, B., 1967, The selective inhibition of delayed potassium currents in nerve by tetraethylam-monium ion, *J. Gen. Physiol.* **50**:1287.

Hille, B., 1975, An essential ionized acid group in sodium channels, *Fed. Proc. Fed. Am. Soc. Exp. Biol.* **34**:1318.

Hinke, J.A.M., 1961, The measurement of sodium and potassium activities in the squid axon by means of cation-selective glass microelectrodes, *J. Physiol.* **156**:314.

Hirsch, H.R., 1965, Squid giant axon: Repetitive responses to alternating current stimulation, *Nature (London)* **208**:1218.

Hodgkin, A.L., 1965, *The Conduction of the Nervous Impulse*, Liverpool University Press, England.

Hodgkin, A., 1975, The optimum density of sodium channels in an unmyelinated nerve, *Philos. Trans. R. Soc. London Ser. B* **270**:297.

Hodgkin, A.L., and Chandler, W.K., 1965, Effects of changes in ionic strength on inactivation and threshold in perfused nerve fibers of *Loligo, J. Gen. Physiol.* **48**:27.

Hodgkin, A.L., and Huxley, A.F., 1939, Action potentials recorded from inside a nerve fibre, *Nature (London)* **144**:710.

Hodgkin, A.L., and Huxley, A.F., 1945, Resting and action potentials in single nerve fibres, *J. Physiol.* **104**:176.

Hodgkin, A.L., and Huxley, A.F., 1952a, Currents carried by sodium and potassium ions through the membrane of the giant axon of *Loligo, J. Physiol.* **116**:449.

Hodgkin, A.L., and Huxley, A.F., 1952b, The components of membrane conductance in the giant axons of *Loligo, J. Physiol.* **116**:473.

Hodgkin, A.L., and Huxley, A.F., 1952c, The dual effect of membrane potential on sodium conductance in the giant axon of *Loligo, J. Physiol.* **116**:497.

Hodgkin, A.L., and Huxley, A.F., 1952d, A quantitative description of membrane current and its application to conduction and excitation in nerve, *J. Physiol.* **117**:500.

Hodgkin, A.L., and Katz, B., 1949a, The effect of temperature on the electrical activity of the giant axon of squid, *J. Physiol.* **109**:240.

Hodgkin, A.L., and Katz, B., 1949b, The effect of sodium ions on the electrical activity of the giant axon of the squid, *J. Physiol.* **108**:37.

Hodgkin, A.L., and Keynes, R.D., 1955, Active transport of cations in giant axons from *Sepia* and *Loligo, J. Physiol.* **128**:28.

Hodgkin, A.L., and Keynes, R.D., 1957, Movements of labelled calcium in squid giant axons, *J. Physiol.* **138**:253.

Hodgkin, A.L., Huxley, A.F., and Katz, B., 1952, Measurement of current–voltage relations in the membrane of the giant axon of *Loligo, J. Physiol.* **116**:424.

Hoskin, F.C.G., 1966, Anaerobic glycolysis in parts of the giant axon of squid, *Nature (London)* **210**:856.

Hoskin, F.C.G., 1971, Diisopropylphosphofluoridate and tabun: Enzymatic hydrolysis and nerve function, *Science* **172**:1243.

Hoskin, F.C.G., 1976, Distribution of diisopropylphosphofluoridate-hydrolyzing enzyme between sheath and axoplasm of squid giant axon, *J. Neurochem.* **26**:1043.

Hoskin, F.C.G., and Brande, M., 1973, An improved sulphur assay applied to a problem of isethionate metabolism in squid axon and other nerves, *J. Neurochem.* **20:**1317.

Hoskin, F.C.G., and Kordik, E.R., 1975, Rhodanese and DFPase in relation to isethionate in squid nerve, *Biol. Bull.* **149:**429.

Hoskin, F.C.G., and Long, R.J., 1972, Purification of a DFP-hydrolyzing enzyme from squid head ganglion, *Arch. Biochem. Biophys.* **150:**548.

Hoskin, F.C.G., and Rosenberg, P., 1964, Alteration of acetylcholine penetration into, and effects on, venom-treated squid axons by physostigmine and related compounds, *J. Gen. Physiol.* **47:**1117.

Hoskin, F.C.G., and Rosenberg, P., 1965, Penetration of sugars, steroids, amino acids and other organic compounds into the interior of the squid giant axon, *J. Gen. Physiol.* **49:**47.

Hoskin, F.C.G., and Rosenberg, P., 1967, Penetration of an organophosphorus compound into squid axon and its effects on metabolism and function, *Science* **156:**966.

Hoskin, F.C.G., Rosenberg, P., and Brzin, M., 1966, Re-examination of the effect of DFP on electrical and cholinesterase activity of squid giant axon, *Proc. Natl. Acad. Sci. U.S.A.* **55:**1231.

Hoskin, F.C.G., Kremzner, L.T., and Rosenberg, P., 1969, Effects of some cholinesterase inhibitors on the squid giant axon, *Biochem. Pharmacol.* **18:**1727.

Hoskin, F.C.G., Pollock, M.L., and Prusch, R.D., 1975, An improved method for the measurement of $^{14}CO_2$ applied to a problem of cysteine metabolism in squid nerve, *J. Neurochem.* **25:**445.

Huneeus, F.C., and Davison, P.F., 1970, Fibrillar proteins from squid axons. I. Neurofilament protein, *J. Mol. Biol.* **52:**415.

Huneeus-Cox, F., 1964, Electrophoretic and immunological studies of squid axoplasm proteins, *Science* **143:**1036.

Huneeus-Cox, F., and Fernandez, H.L., 1967, Effect of specific antibodies on the excitability of internally perfused squid axons, *J. Gen. Physiol.* **50:**2407.

Huneeus-Cox, F., and Smith, B.H., 1965, The effects of oxidizing, reducing and sulfhydryl reagents on the resting and action potentials of the internally perfused axon of *Loligo pealeii*, *Biol. Bull.* **129:**408.

Huneeus-Cox, F., Fernandez, H.L., and Smith, B.H., 1966, Effects of redox and sulfhydryl reagents on the bioelectric properties of the giant axon of the squid, *Biophys. J.* **6:**675.

Inoue, I., Kobatake, Y., and Tasaki, I., 1973, Excitability, instability and phase transitions in squid axon membrane under internal perfusion with dilute salt solutions, *Biochim. Biophys. Acta* **307:**471.

Inoue, I., Tasaki, I., and Kobatake, Y., 1974, A study of the effects of externally applied sodium-ions and detection of spatial non-uniformity of the squid axon membrane under internal perfusion, *Biophys. Chem.* **2:**116.

International Union of Biochemistry, 1973, *Enzyme Nomenclature*, 3rd ed., Elsevier, New York.

Johnson, W., Soden, P.D., and Trueman, E.R., 1972, A study in jet propulsion: An analysis of the motion of the squid, *Loligo vulgaris*, *J. Exp. Biol.* **56:**155.

Katz, B., 1966, *Nerve, Muscle and Synapse*, McGraw-Hill, New York.

Keleti, G., and Lederer, W.H., 1974, *Handbook of Micromethods for the Biological Sciences*, Van Nostrand Reinhold, New York.

Kerkut, G.A., 1967, Biochemical aspects of invertebrate nerve cells, in: *Invertebrate Nervous System* (C.A.G. Wiersma, ed.), pp. 5–37, The University of Chicago Press.

Keynes, R.D., 1951, The ionic movements during nervous activity, *J. Physiol.* **114:**119.

Keynes, R.D., 1963, Chloride in the squid giant axon, *J. Physiol.* **169:**690.

Keynes, R.D., and Lewis, P.R., 1951, The sodium and potassium content of cephalopod nerve fibres, *J. Physiol.* **114:**151.

Keynes, R.D., and Rojas, E., 1974, Kinetics and steady-state properties of the charged system controlling sodium conductance in the squid giant axon, *J. Physiol.* **239:**393.

Keynes, R.D., and Rojas, E., 1976, The temporal and steady-state relationship between activation of the sodium conductance and movement of the gating particles in the squid giant axon, *J. Physiol.* **255:**157.

Keynes, R.D., Bezanilla, F., Rojas, E., and Taylor, R.E., 1975, The rate of action of tetradotoxin on sodium conductance in the squid giant axon, *Philos. Trans. R. Soc. London Ser. B* **270**:365.

Kishimoto, U., and Adelman, W.J., Jr., 1964, Effect of detergent on electrical properties of squid axon membrane, *J. Gen. Physiol.* **47**:975.

Koechlin, B.A., 1954, The isolation and identification of the major anion fraction of the axoplasm of squid giant nerve fiber, *Proc. Natl. Acad. Sci. U.S.A.* **40**:60.

Koechlin, B.A., 1955, On the chemical composition of the axoplasm of squid giant nerve fibers with particular reference to its ion pattern, *Biophys. Biochem. Cytol.* **1**:511.

Kootsey, J.M., 1975, Voltage clamp simulation, *Fed. Proc. Fed. Am. Soc. Exp. Biol.* **34**:1343.

Koppenhofer, E., 1967, Die Wirkung von Tetraathylammoniumchlorid auf die Membranstrome Ranvierscher Schnurringe von *Xenopus laevis, Pfluegers Arch. Gesamte Physiol. Menschen Tiere* **293**:34.

Korey, S.R., 1950, Permeability of axonal surface membranes to amino acids, *Fed. Proc. Fed. Am. Soc. Exp. Biol.* **9**:191.

Korey, S.R., 1951, Effect of Dilantin and Mesantoin on the giant axon of the squid, *Proc. Soc. Exp. Biol. Med.* **76**:297.

Kremzner, L.T., and Rosenberg, P., 1971, Relationship of acetylcholinesterase activity to axonal conduction, *Biochem. Pharmacol.* **20**:2953.

Krolenko, S.A., and Nikol'skii, N.N., 1967, Penetration of sugars into the squid giant axon (*Loligo-vulgaris*), *Tsitologiya* **9**:273.

Kuffler, S.W., and Nicholls, J.G., 1976, *From Neuron to Brain*, Sinauer, Sunderland, Massachusetts.

Lakshminarayanaiah, N., and Bianchi, C.P., 1975, Ca^{2+} concentration and interaction of long-lasting local anesthetics with the squid axon membrane, *J. Pharm. Pharmacol.* **27**:787.

Landowne, D., 1975a, Sodium efflux from voltage clamped squid axons, *Biol. Bull.* **149**:434.

Landowne, D., 1975b, A comparison of radioactive thallium and potassium fluxes in the giant axon of the squid, *J. Physiol.* **252**:79.

LaRoe, E.T., 1973, Laboratory culture of squid, *Fed. Proc. Fed. Am. Soc. Exp. Biol.* **32**:2212.

Larrabee, M.G., and Brinley, F.J., Jr., 1968, Incorporation of labelled phosphate into phospholipids in squid giant axons, *J. Neurochem.* **15**:533.

Lasek, R.J., 1970, The distribution of nucleic acids in the giant axon of the squid (*Loligo pealii*), *J. Neurochem.* **17**:103.

Lasek, R.J., Dabrowski, C., and Nordlander, R., 1973, Analysis of axoplasmic RNA from invertebrate giant axons, *Nature (London) New Biol.* **244**:162.

Lasek, R.J., Gainer, H., and Przybylski, R.J., 1974, Transfer of newly synthesized proteins from Schwann cells to the squid giant axon, *Proc. Natl. Acad. Sci. U.S.A.* **71**:1188.

Lasek, R.J., Gainer, H., and Barker, J., 1975, Transfer of newly synthesized proteins between glia and neurons: The squid giant axon as a model, *Trans. Am. Soc. Neurochem.* **6**:74.

Latorre, R., and Hidalgo, M.C., 1969, Effect of temperature on resting potential in giant axons of squid, *Nature (London)* **221**:962.

Levinson, S.R., and Meves, H., 1975, The binding of tritiated tetrodotoxin to squid giant axons, *Philos. Trans. R. Soc. London Ser. B* **270**:349.

Lewis, P.R., 1952, The free amino acids of invertebrate nerve, *Biochem J.* **52**:330.

Lipicky, R.J., Gilbert, D.L., and Stillman, I.M., 1972, Diphenylhydantoin inhibition of sodium conductance in squid giant axon, *Proc. Natl. Acad. Sci. U.S.A.* **69**:1758.

Lowry, O.H., Passonneau, J.V., Schulz, D.W., and Rock, M.K., 1961, The measurement of pyridine nucleotides by enzymatic cycling, *J. Biol. Chem.* **236**:2746.

Luxoro, M., and Yañez, E., 1968, Permeability of the giant axon of *Dosidicus gigas* to calcium ions, *J. Gen. Physiol.* **51**:115S.

Marcus, D., Canessa-Fischer, M., Zampighi, G., and Fischer, S., 1972, The molecular organization of nerve membranes. VI. The separation of axolemma from Schwann cell membranes of giant and retinal squid axons by density gradient centrifugation, *J. Membrane Biol.* **9**:209.

Marquis, J.K., and Mautner, H.G., 1974, The effect of electrical stimulation on the action of sulfhydryl reagents in the giant axon of squid: Suggested mechanisms for the role of thiol and disulfide groups in electrically-induced conformational changes, *J. Membrane Biol.* **15**:249.

Martin, R., and Rosenberg, P., 1968, Fine structural alterations associated with venom action on squid giant nerve fibers, *J. Cell Biol.* **36**:341.

Matsumoto, G., 1976, Transportation and maintenance of adult squid (*Doryteuthis bleekeri*) for physiological studies, *Biol. Bull.* **150**:279.

Matsumoto, N., Inoue, I. and Kishimoto, V., 1970, The electrical impedance of the squid axon membrane measured between internal and external electrodes, *Jpn. J. Physiol.* **20**:516.

Mauro, A., Conti, F., Dodge, F., and Schor, R., 1970, Subthreshold behavior and phenomenological impedance of the squid giant axon, *J. Gen. Physiol.* **55**:497.

Mauro, A.R., Freeman, A.R., Cooley, J.W., and Ross, A., 1972, Propagated subthreshold oscillatory response and classical electrotonic response of squid giant axon, *Biophysik* **8**:118.

Maxfield, M., 1953, Axoplasmic proteins of the squid giant nerve fiber with particular reference to the fibrous proteins, *J. Gen. Physiol.* **37**:201.

Maxfield, M., and Hartley, R.W., 1957, Dissociation of the fibrous protein of nerve, *Biochim. Biophys. Acta* **24**:83.

McColl, J.D., and Rossiter, R.J., 1951, Lipids of the nervous system of the squid *Loligo pealii*, *J. Exp. Biol.* **28**:116.

McMahon, J.J., and Summers, W.C., 1971, Temperature effects on the developmental rate of squid (*Loligo pealei*) embryos, *Biol. Bull.* **141**:561.

Mercer, M.C., 1970, Sur la limite septentrionale du calmar *Loligo pealei* Lesueur, *Nat. Can. (Ottawa)* **97**:823.

Metuzals, J., 1969, Configuration of a filamentous network in the axoplasm of the squid (*Loligo pealii L.*) giant nerve fiber, *J. Cell Biol.* **43**:480.

Metuzals, J., and Izzard, C.S., 1969, Spatial patterns of threadlike elements in the axoplasm of the giant nerve fiber of the squid (*Loligo pealii L.*) as disclosed by differential interference microscopy and by electron microscopy, *J. Cell Biol.* **43**:456.

Meves, H., 1966*a*, Experiments on internally perfused squid giant axons, *Ann. N. Y. Acad. Sci.* **137**:807.

Meves, H., 1966*b*, The effects of veratridine on internally perfused giant axons, *Arch. Ges. Physiol.* **290**:211.

Meves, H., 1974, The effect of holding potentials on the assymetry currents in squid giant axons, *J. Physiol.* **243**:847.

Meves, H., 1975*a*, Calcium currents in squid giant axon, *Philos. Trans. R. Soc. London Ser. B* **270**:377.

Meves, H., 1975*b*, Assymetry currents in intracellularly perfused squid giant axons, *Philos. Trans. R. Soc. London Ser. B* **270**:493.

Meves, H., and Vogel, W., 1973, Calcium inward currents in internally perfused giant axons, *J. Physiol.* **235**:225.

Mikulich, L.V., and Kozak, L.P., 1971, Experiment in the keeping of the Pacific squid under artificial conditions, *Ekologiya* **2**:94.

Moore, J.W., and Adelman, W.J., Jr., 1961, Electronic measurement of the intracellular concentration and net flux of sodium in the squid axon, *J. Gen. Physiol.* **45**:77.

Moore, J.W., and Cole, K.S., 1960, Resting and action potentials of the squid giant axon *in vivo*, *J. Gen. Physiol.* **43**:961.

Moore, J.W., and Cox, E.B., 1976, A kinetic model for the sodium conductance system in squid axon, *Biophys. J.* **16**:171.

Moore, J.W., Narahashi, T., and Ulbricht, W., 1964*a*, Sodium conductance shift in an axon internally perfused with a sucrose and low-potassium solution, *J. Physiol.* **172**:163.

Moore, J.W., Ulbricht, W., and Takata, M., 1964*b*, Effect on ethanol on the sodium and potassium conductances of the squid axon membrane, *J. Gen. Physiol.* **48**:279.

Moore, J.W., Anderson, N., Blaustein, M., Takata, M., Lettvin, J.Y., Pickard, W.F., Bernstein, T., and Pooler, J., 1966, Alkali cation selectivity of squid axon membrane, *Ann. N. Y. Acad. Sci.* **137**:818.

Moore, J.W., Blaustein, M.P., Anderson, N.C., and Narahashi, T., 1967, Basis of tetrodotoxin's selectivity in blockage of squid axons, *J. Gen. Physiol.* **50**:1401.

Moore, J.W., Ramon, F., and Joyner, R.W., 1975a, Axon voltage-clamp simulations. II. Double sucrose-gap method, *Biophys. J.* **15**:25.

Moore, J.W., Ramon, F., and Joyner, R.W., 1975b, Axon voltage-clamp simulations. I. Methods and tests, *Biophys. J.* **15**:11.

Moore, L.E., Tufts, M., and Soroka, M., 1975c, Light scattering spectroscopy of the squid axon membrane, *Biochim. Biophys. Acta* **382**:286.

Mullins, L.J., 1956, The structure of nerve cell membranes, in: *Molecular Structure and Functional Activity of Nerve Cells* (R.G. Grenell and L.J. Mullins, eds.), American Institute of Biological Science, Washington, D.C.

Mullins, L.J., 1959, An analysis of conductance changes in squid axon, *J. Gen. Physiol.* **42**:1013.

Mullins, L.J., 1960, An analysis of pore size in excitable membranes, *J. Gen. Physiol.* **43**:105.

Mullins, L.J. (ed.), 1965, A conference on newer properties of perfused squid axons, *J. Gen. Physiol.* **48**:1.

Mullins, L.J., 1966, Ion and molecular fluxes in squid axons, *Ann. N. Y. Acad. Sci.* **137**:830.

Mullins, L.J., 1968, Ion fluxes in dialyzed squid axons, *J. Gen. Physiol.* **51**:146S.

Mullins, L.J., and Brinley, F.J., Jr., 1967, Some factors influencing sodium extrusion by internally dialyzed squid axons, *J. Gen. Physiol.* **50**:2333.

Mullins, L.J., and Brinley, F.J., Jr., 1969, Potassium fluxes in dialyzed squid axons, *J. Gen. Physiol.* **53**:704.

Mullins, L.J., and Brinley, F.J., Jr., 1975, Sensitivity of calcium efflux from squid axons to changes in membrane potential, *J. Gen. Physiol.* **65**:135.

Nachmansohn, D., 1971, Proteins in bioelectricity: Acetylcholine esterase and receptor, in: *Handbook of Sensory Physiology*, Vol. 1 (W.R. Loewenstein, ed.), Springer-Verlag, Berlin.

Nachmansohn, D., and Neumann, E., 1975, *Chemical and Molecular Basis of Nerve Activity* (revised), Academic Press, New York.

Nakamura, Y., Nakajima, S., and Grundfest, H., 1965, The action of tetrodotoxin on electrogenic components of squid giant axons, *J. Gen. Physiol.* **48**:985.

Narahashi, T., 1963, Dependence of resting and action potentials on internal potassium in perfused squid giant axons, *J. Physiol.* **169**:91.

Narahashi, T., 1974, Chemicals as tools in the study of excitable membranes, *Physiol. Rev.* **54**:813.

Narahashi, T., and Anderson, N.C., 1967, Mechanism of excitation block by the insecticide allethrin applied externally and internally to squid giant axons, *Toxicol. Appl. Pharmacol.* **10**:529.

Narahashi, T., and Tobias, J.M., 1964, Properties of axon membranes as affected by cobra venom, digitonin and proteases, *Am. J. Physiol.* **207**:1441.

Narahashi, T., and Wang, C.M., 1973, Effects of antiarrhythmic drugs on ionic conductances of squid axon membranes, *Pharmacologist* **15**:178.

Narahashi, T., Anderson, N.C., and Moore, J.W., 1966, Tetrodotoxin does not block excitation from inside the nerve membrane, *Science* **153**:765.

Narahashi, T., Anderson, N.C., and Moore, J.W., 1967a, Comparison of tetrodotoxin and procaine in internally perfused squid giant axon, *J. Gen. Physiol.* **50**:1413.

Narahashi, T., Moore, J.W., and Poston, R.N., 1967b, Tetrodotoxin derivatives: Chemical structure and blockage of nerve membrane conductance, *Science* **156**:976.

Narahashi, T., Moore, J.W., and Poston, R.N., 1969a, Anesthetic blocking of nerve membrane conductances by internal and external applications, *J. Neurobiol.* **1**:3.

Narahashi, T., Moore, J.W., and Shapiro, B.I., 1969b, Condylactis toxin: Interaction with nerve membrane ionic conductances, *Science* **163**:680.

Narahashi, T., Frazier, D.T., and Yamada, M., 1970, The side of action and active form of local anesthetics. I. Theory and pH experiments with tertiary compounds, *J. Pharmacol. Exp. Ther.* **171**:32.

Narahashi, T., Frazier, D., Deguchi, T., Cleaves, C.A., and Ernau, M.C., 1971a, The active form of pentobartital in squid giant axons, *J. Pharmacol. Exp. Ther.* **177**:25.

Narahashi, T., Albuquerque, E.X., and Deguchi, T., 1971b, Effects of batrachotoxin on membrane potential and conductance of squid giant axons, *J. Gen. Physiol.* **58**:54.

Narahashi, T., Shapiro, B.I., Deguchi, T., Scuka, M., and Wang, C.M., 1972*a*, Effects of scorpion venom on squid axon membranes, *Am. J. Physiol.* **222**:850.

Narahashi, T., Frazier, D.T., and Moore, J.W., 1972*b*, Comparison of tertiary and quaternary amine local anesthetics in their ability to depress membrane ionic conductances, *J. Neurobiol.* **3**:267.

Nastuk, W.L. (ed.), 1964, *Physical Techniques in Biological Research,* Vol. 5, Academic Press, New York.

Neill, S.S.J., 1971, Notes on squid and cuttlefish: Keeping, handling and colour patterns, *Pubbl. St. Zool. Napoli* **39**:64.

Ohta, M., and Narahashi, T., 1973, Sparteine interaction with nerve membrane potassium conductance, *J. Pharmacbl. Exp. Ther.* **187**:47.

Ohta, M., Narahashi, T., and Keeler, R.F., 1973, Effects of veratrum alkaloids on membrane potential and conductance of squid and crayfish giant axons, *J. Pharmacol. Exp. Ther.* **184**:143.

Oikawa, T., 1962, Electrical interactions between normal and TEA-treated zones of squid axon, *Am. J. Physiol.* **202**:865.

Oikawa, T., Spyropoulos, C.S., Tasaki, I., and Teorell, T., 1961, Methods for perfusing the giant axon of *Loligo pealii, Acta Physiol. Scand.* **52**:195.

Orrego, F., 1971, Protein degradation in squid giant axons, *J. Neurochem.* **18**:2249.

Oschman, J.L., Hall, T.A., Peters, P.D., and Wall, B.J., 1974, Association of calcium with membranes of squid giant axon, *J. Cell Biol.* **61**:156.

Packard, A., 1969, Jet propulsion and the giant fibre response of *Loligo, Nature (London)* **221**:875.

Parmentier, J.L., and Narahashi, T., 1975, Effects of hornet venom on squid axon membranes, *Biol. Bull.* **149**:440.

Pepe, I.M., Giuditta, A., and Cimarra, P., 1975, Inhibition of neuronal protein synthesis in the giant fibre system of the squid by a high potassium concentration, *J. Neurochem.* **24**:1271.

Pickard, W.F., Lettvin, J.Y., Moore, J.W., Takata, M., Pooler, J., and Bernstein, T., 1964, Caesium ions do not pass the membrane of the giant axon, *Proc. Natl. Acad. Sci. U.S.A.* **52**:1177.

Robertson, J.D., 1960, The molecular structure and contact relationships of cell membranes, *Prog. Biophys. Biophys. Chem.* **10**:343.

Rojas, E., 1965, Membrane potentials, resistance and ion permeability in squid giant axons injected or perfused with proteases, *Proc. Natl. Acad. Sci. U.S.A.* **53**:306.

Rojas, E., and Atwater, I., 1967, Blocking of potassium currents by pronase in perfused giant axons, *Nature (London)* **215**:850.

Rojas, E., and Atwater, I., 1968, An experimental approach to determine membrane charges in squid giant axons, *J. Gen. Physiol.* **51**:131S.

Rojas, E., and Canessa-Fischer, M., 1968, Sodium movements in perfused squid giant axons, *J. Gen. Physiol.* **52**:240.

Rojas, E., and Hidalgo, C., 1968, Effect of temperature and metabolic inhibitors on Ca^{45} outflow from squid giant axons, *Biochim. Biophys. Acta* **163**:550.

Rojas, E., and Keynes, R.D., 1975, On the relation between displacement currents and activation of the sodium conductance in the squid giant axon, *Philos. Trans. R. Soc. London Ser. B* **270**:459.

Rojas, E., and Luxoro, M., 1963, Micro-injection of trypsin into axons of squid, *Nature (London)* **199**:78.

Rojas, E., and Taylor, R.E., 1975, Simultaneous measurements of magnesium, calcium and sodium influxes in perfused giant axons under membrane potential control, *J. Physiol.* **252**:1.

Rojas, E., Taylor, R.E., Atwater, I., and Bezanilla, F., 1969, Analysis of the effects of calcium or magnesium on voltage clamp currents in perfused squid axons bathed in solutions of high potassium, *J. Gen. Physiol.* **54**:532.

Rosenberg, P., 1965, Effects of venoms on the squid giant axon, *Toxicon* **3**:125.

Rosenberg, P., 1966, Use of venoms in studies on nerve excitation, *Mem. Inst. Butantan, Sao Paulo* **33**(2):477.

Rosenberg, P., 1970, Function of phospholipids in axons: Depletion of membrane phosphorus by treatment with phospholipase C, *Toxicon* **8**:235.

Rosenberg, P., 1971, The use of snake venoms as pharmacological tools in studying nerve activity, in: *Neuropoisons: Their Pathophysiological Actions*, Vol. 1 (L.L. Simpson, ed.), pp. 111–137, Plenum Press, New York.

Rosenberg, P., 1973, The giant axon of the squid: A useful preparation for neurochemical and pharmacological studies, in: *Methods of Neurochemistry*, Vol. 4 (R. Fried, ed.), pp. 97–160, Marcel Dekker, New York.

Rosenberg, P., 1975, Penetration of phospholipase A_2 and C into the squid (*Loligo pealii*) giant axon, *Experientia* **31**:1401.

Rosenberg, P., 1976, Bacterial and snake venom phospholipases: Enzymatic probes in the study of structure and function in bioelectrically excitable tissues, in: *Animal, Plant and Microbial Toxins*, Vol. 2 (A. Ohsaka, K. Hayashi, and Y. Sawai, eds.), pp. 229–261, Plenum Press, New York.

Rosenberg, P., 1977, Pharmacology of phospholipase A_2 from snake venoms, in: *Handbook of Experimental Pharmacology*, Vol. 52, *Snake Venoms* (C.Y. Lee, ed.), pp. 403–447, Springer-Verlag, New York.

Rosenberg, P., and Bartels, E., 1967, Drug effects on the spontaneous electrical activity of the squid giant axon, *J. Pharmacol. Exp. Ther.* **155**:532.

Rosenberg, P., and Condrea, E., 1968, Maintenance of axonal conduction and membrane permeability in presence of extensive phospholipid splitting, *Biochem. Pharmacol.* **17**:2033.

Rosenberg, P., and Dettbarn, W.-D., 1964, Increased cholinesterase activity of intact cells caused by snake venoms, *Biochem. Pharmacol.* **13**:1157.

Rosenberg, P., and Dettbarn, W.-D., 1967, Use of venoms in testing for essentiality of cholinesterase in conduction, in: *Animal Toxins* (Russell and Saunders, eds.), p. 379, Pergamon, New York.

Rosenberg, P., and Ehrenpreis, S., 1961, Reversible block of axonal conduction by curare after treatment with cobra venom, *Biochem. Pharmacol.* **8**:192.

Rosenberg, P., and Hoskin, F.C.G., 1963, Demonstration of increased permeability as a factor in the effect of acetylcholine on the electrical activity of venom-treated axons, *J. Gen. Physiol.* **46**:1065.

Rosenberg, P., and Hoskin, F.C.G., 1965, Penetration of acetylcholine into squid giant axons, *Biochem. Pharmacol.* **14**:1765.

Rosenberg, P., and Khairallah, E., 1974, Effect of phospholipases A and C on free amino acid content of the squid axon, *J. Neurochem.* **23**:55.

Rosenberg, P., and Mautner, H.G., 1967, Acetylcholine receptor: Similarity in axons and junctions, *Science* **155**:1569.

Rosenberg, P., and Ng, K.Y., 1963, Factors in venom leading to block of axonal conduction by curare, *Biochim. Biophys. Acta* **75**:116.

Rosenberg, P., and Podleski, T.R., 1962, Block of conduction by acetylcholine and D-tubocurarine after treatment of squid axon with cottonmouth mocassin venom, *J. Pharmacol. Exp. Ther.* **137**:249.

Rosenberg, P., and Podleski, T.R., 1963, Ability of venoms to render squid axons sensitive to curare and acetylcholine, *Biochim. Biophys. Acta* **75**:104.

Rosenberg, P., Dettbarn, W.-D., and Brzin, M., 1966a, Acetylcholine and choline acetylase in squid giant axon, ganglia and retina, *Nature (London)* **210**:858.

Rosenberg, P., Mautner, H.G., and Nachmansohn, D., 1966b, Similarity in effects of oxygen, sulfur and selenium isologs on the acetylcholine receptor in excitable membranes of junctions and axons, *Proc. Natl. Acad. Sci. U.S.A.* **55**:835.

Rosenberg, P., Kremzner, L.T., McCreery, D., and Willette, R.E., 1972, Inhibition of choline acetyltransferase activity in squid giant axon, *Biochim. Biophys. Acta* **268**:49.

Ruch, T.C., and Patton, H.D. (eds.), 1966, *Physiology and Biophysics*, W.B. Saunders, Philadelphia.

Sabatini, M.T., Dipolo, R., and Villegas, R., 1968, Adenosine triphosphatase activity in the membranes of the squid nerve fibre, *J. Cell Biol.* **38**:176.

Sato, H., Tasaki, I., Carbone, E., and Hallett, M., 1973. Changes in axon birefringence associated with excitation: Implications for the structure of the axon membrane, *J. Mechanochem. Cell Motil.* **2**:209.

Schmitt, F.O., and Geschwind, N., 1957, The axon surface, *Prog. Biophys. Biophys. Chem.* **8**:165.

Schwartz, E.A., 1968, Effect of diethyl ether on sodium efflux from squid axons, *Curr. Mod. Biol.* **2**:1.

Scuka, M., 1971, Effects of histamine on resting and action potentials of squid giant axons, *Life Sci.* **10**:355.

Segal, J.R., 1968*a*, Effect of metabolism on the excitability of the squid giant axon, *Am. J. Physiol.* **215**:467.

Segal, J.R., 1968*b*, Surface charge of giant axons of squid and lobster, *Biophys. J.* **8**:470.

Sevcik, C., and Narahashi, T., 1975, Effects of proteolytic enzymes on ionic conductances of squid axon membranes, *J. Membrane Biol.* **24**:329.

Seyama, I., and Narahashi, T., 1973, Increase in sodium permeability of squid axon membranes by α-dihydro-grayanotoxin II, *J. Pharmacol. Exp. Ther.* **184**:299.

Shanes, A.M., 1952, The ultraviolet spectra and neurophysiological effects of "veratrine" alkaloids, *J. Pharmacol. Exp. Ther.* **105**:216.

Shanes, A.M., Grundfest, H., and Freygang, W., 1953, Low level impedance changes following the spike in the giant axon before and after treatment with "veratrine" alkaloids, *J. Gen. Physiol.* **37**:39.

Shapiro, B.I., Wang, C.M., and Narahashi, T., 1974, Effects of strychnine on ionic conductances of squid axon membrane, *J. Pharmacol. Exp. Ther.* **188**:66.

Shaw, T.I., 1966, Cation movements in perfused giant axons, *J. Physiol.* **182**:209.

Simon, E.J., and Rosenberg, P., 1970, Effects of narcotics on the giant axon of the squid, *J. Neurochem.* **17**:881.

Singer, S.J., and Nicolson, G.L., 1972, The fluid mosaic model of the structure of cell membranes, *Science* **175**:720.

Sjodin, R.A., 1966, Long duration responses in squid giant axons injected with cesium sulfate solution, *J. Gen. Physiol.* **50**:269.

Sjodin, R.A., and Beauge, L.A., 1967, The ion selectivity and concentration dependence of cation coupled active sodium transport in squid giant axons, *Curr. Mod. Biol.* **1**:105.

Sjodin, R.A., and Beauge, L.A., 1968, Coupling and selectivity of sodium and potassium transport in squid giant axons, *J. Gen. Physiol.* **51**:152S.

Sjodin, R.A., and Beauge, L.A., 1969, The influence of potassium- and sodium-free solutions on sodium efflux from squid giant axons, *J. Gen. Physiol.* **54**:664.

Sjodin, R.A., and Mullins, L.J., 1967, Tracer and nontracer potassium fluxes in squid giant axons and the effects of changes in external potassium concentration and membrane potential, *J. Gen. Physiol.* **50**:533.

Spyropoulos, C.S., 1960, Cytoplasmic pH of nerve fibers, *J. Neurochem.* **5**:185.

Spyropoulos, C.S., 1965, The role of temperature, potassium and divalent ions in the current–voltage characteristics of nerve membranes, *J. Gen. Physiol.* **48**:49.

Stallworthy, W.B., 1970, Electro-osmosis in squid axons, *J. Mar. Biol. Assoc. U.K.* **50**:349.

Stallworthy, W.B., and Fensom, D.S., 1966, Electroosmosis in axons of freshly killed squid, *Can. J. Physiol. Pharmacol.* **44**:866.

Steinbach, H.B., 1941, Chloride in the giant axons of the squid, *J. Cell. Comp. Physiol.* **17**:57.

Steinbach, H.B., and Spiegelman, S., 1943, The sodium and potassium balance in squid nerve axoplasm, *J. Cell. Comp. Physiol.* **22**:187.

Stillman, I.M., Binstock, L., and Taylor, R.E., 1968, Effect of D_2O upon the neural activity of the squid giant axon, *Fed. Proc. Fed. Am. Soc. Exp. Biol.* **27**:703.

Stillman, I.M., Gilbert, D.L., and Robbins, M., 1970, Monactin does not influence potassium permeability in the squid axonal membrane, *Biochim. Biophys. Acta* **203**:338.

Summers, W.C., 1968, The growth and size distribution of current year class *Loligo pealei*, *Biol. Bull.* **135**:366.

Summers, W.C., 1969, Winter population of *Loligo pealei* in the mid-Atlantic bight, *Biol. Bull.* **137**:202.

Summers, W.C., 1971, Age and growth of *Loligo pealii*, a population study of the common Atlantic coast squid, *Biol. Bull.* **141**:189.

Summers, W.C., and McMahon, J.J., 1970, Survival of unfed squid, *Loligo pealei*, in an aquarium, *Biol. Bull.* **138**:389.

Summers, W.C., and McMahon, J.J., 1974, Studies on the maintenance of adult squid (*Loligo pealei*). I. Factorial survey, *Biol. Bull.* **146:**279.

Summers, W.C., McMahon, J.J., and Ruppert, G.N.P.A., 1974, Studies on the maintenance of adult squid (*Loligo pealei*). II. Empirical extensions, *Biol. Bull.* **146:**291.

Takashima, S., and Schwan, H.P., 1974, Passive electrical properties of squid axon membrane, *J. Membrane Biol.* **17:**51.

Takashima, S., Yantorno, R., and Pal, N.C., 1975, Electrical properties of squid axon membrane. II. Effect of partial degradation by phospholipase A and pronase on electrical characteristics, *Biochim. Biophys. Acta* **401:**15.

Takenaka, T., and Yamagishi, 1969, Morphology and electrophysiological properties of squid giant axons perfused intracellularly with protease solution, *J. Gen. Physiol.* **53:**81.

Takenaka, T., Hirakow, R., and Yamagishi, S., 1968, Ultrastructural examination of the squid giant axons perfused intracellularly with protease, *J. Ultrastruct. Res.* **25:**408.

Tasaki, I., 1963, Permeability of squid axon membrane to various ions, *J. Gen. Physiol.* **46:**755.

Tasaki, I., 1968, *Nerve Excitation*, Charles C. Thomas, Springfield, Illinois.

Tasaki, I., 1970, Effects of ultraviolet and visible light on nerve fibers and changes in optical properties during nervous activity, *Adv. Biol. Med. Phys.* **13:**307.

Tasaki, I., 1975, Evolution of theories of nerve excitation, in: *The Nervous System*, Vol. 1, *The Basic Neurosciences* (D.B. Tower, ed.), pp. 177–195, Raven Press, New York.

Tasaki, I., and Hagiwara, S., 1957, Demonstration of two stable potential states in the squid giant axon under tetraethylammonium chloride, *J. Gen. Physiol.* **40:**859.

Tasaki, I., and Luxoro, M., 1964, Intracellular perfusion of Chilean giant squid axons, *Science* **145:**1313.

Tasaki, I., and Shimamura, M., 1962, Further observations on resting and action potential of intracellularly perfused squid axon, *Proc. Natl. Acad. Sci. U.S.A.* **48:**1571.

Tasaki, I., and Singer, I., 1966, Membrane macromolecules and nerve excitability: A physicochemical interpretation of excitation in squid giant axons, *Ann. N. Y. Acad. Sci.* **137:**792.

Tasaki, I., and Sisco, K., 1975, Electrophysiological and optical methods for studying the excitability of the nerve membrane, in: *Methods in Membrane Biology*, Vol. 5 (E.D. Korn, ed.) Chapt. 4, pp. 163–194, Plenum Press, New York.

Tasaki, I., and Spyropoulos, C.S., 1961, Permeability of the squid axon membrane to several organic molecules, *Am. J. Physiol.* **201:**413.

Tasaki, I., and Takenaka, T., 1963, Resting and action potential of squid giant axons intracellularly perfused with sodium-rich solutions, *Proc. Natl. Acad. Sci. U.S.A.* **50:**619.

Tasaki, I., and Takenaka, T., 1964, Effects of various potassium salts and proteases upon excitability of intracellularly perfused squid giant axons, *Proc. Natl. Acad. Sci. U.S.A.* **52:**804.

Tasaki, I., Teorell, T., and Spyropoulos, C.S., 1961, Movement of radioactive tracers across squid axon membrane, *Am. J. Physiol.* **200:**11.

Tasaki, I., Watanabe, A., and Takenaka, T., 1962, Resting and action potentials of intracellularly perfused squid giant axon, *Proc. Natl. Acad. Sci. U.S.A.* **48:**1177.

Tasaki, I., Luxoro, M., and Ruarte, A., 1965a, Electrophysiological studies of Chilean squid axons under internal perfusion with sodium-rich media, *Science* **150:**899.

Tasaki, I., Singer, I., and Takenaka, T., 1965b, Effects of internal and external ionic environment on excitability of squid giant axon—A macromolecular approach, *J. Gen. Physiol.* **48:**1095.

Tasaki, I., Singer, I., and Watanabe, A., 1965c, Excitation of internally perfused squid giant axons in sodium-free media, *Proc. Natl. Acad. Sci. U.S.A.* **54:**763.

Tasaki, I., Watanabe, A., and Singer, I., 1966a, Excitability of squid giant axons in the absence of univalent cations in the external medium, *Proc. Natl. Acad. Sci. U.S.A.* **56:**1116.

Tasaki, I., Singer, I., and Watanabe, A., 1966b, Excitation of squid giant axons in sodium-free external media, *Am. J. Physiol.* **211:**746.

Tasaki, I., Singer, I., and Watanabe, A., 1967a, Cation interdiffusion in squid giant axons, *J. Gen. Physiol.* **50:**989.

Tasaki, I., Watanabe, A., and Lerman, R., 1967b, Role of divalent cations in excitation of squid giant axons, *Am. J. Physiol.* **213:**1465.

Tasaki, I., Takenaka, T., and Yamagishi, S., 1968a, Abrupt depolarization and bi-ionic action potentials in internally perfused squid giant axons, *Am. J. Physiol.* **215:**152.

Tasaki, I., Watanabe, A., Sandlin, R., and Carnay, L., 1968*b*, Changes in fluorescence, turbidity, and birefringence associated with nerve excitation, *Proc. Natl. Acad. Sci. U.S.A.* **61**:883.

Tasaki, I., Lerman, L., and Watanabe, A., 1969*a*, Analysis of excitation process in squid giant axons under bi-ionic conditions, *Am. J. Physiol.* **216**:130.

Tasaki, I., Carnay, L., Sandlin, R., and Watanabe, A., 1969*b*, Fluorescence changes during conduction in nerves stained with acridine orange, *Science* **163**:683.

Tasaki, I., Carnay, L., and Watanabe, A., 1969*c*, Transient changes in extrinsic fluorescence of nerve produced by electrical stimulation, *Proc. Natl. Acad. Sci. U.S.A.* **64**:1362.

Tasaki, I., Watanabe, A., and Hallett, M., 1971, Properties of squid axon membrane as revealed by a hydrophobic probe, 2-*p*-toluidinylnaphthalene-6-sulfonate, *Proc. Natl. Acad. Sci. U.S.A.* **68**:938.

Tasaki, I., Watanabe, A., and Hallett, M., 1972, Fluorescence of squid axon membrane labeled with hydrophobic probes, *J. Membrane Biol.* **8**:109.

Tasaki, I., Hallett, M., and Carbone, E., 1973*a*, Further studies of nerve membranes labeled with fluorescent probes, *J. Membrane Biol.* **11**:353.

Tasaki, I., Carbone, E., Sisco, K., and Singer, I., 1973*b*, Spectral analysis of extrinsic fluorescence of the nerve membrane labeled with aminonaphthalene derivatives, *Biochim. Biophys. Acta* **323**:220.

Tasaki, I., Sisco, K., and Warashima, A., 1974, Alignment of anilinonaphthalene-sulfonate and related fluorescent probe molecules in squid axon membrane and in synthetic polymers, *Biophys. Chem.* **2**:316.

Tasaki, I., Warashima, A., and Pant, H., 1976, Studies of light emission, absorption and energy transfer in nerve membranes labeled with fluorescent probes, *Biophys. Chem.* **4**:1.

Taylor, R.E., 1959, Effect of procaine on electrical properties of squid axon membranes, *Am. J. Physiol.* **196**:1071.

Ulbricht, W., 1974, Ionic channels through the axon membrane (a review), *Biophys. Struct. Mech.* **1**:1.

Van Breemen, C., and DeWeer, P., 1970, Lanthanum inhibition of ^{45}Ca efflux from the squid giant axon, *Nature (London)* **226**:760.

Van Den Bercken, J., and Narahashi, T., 1974, Effects of aldrin-transdiol, a metabolite of the insecticide dieldrin, on nerve membrane, *Eur. J. Pharmacol.* **27**:255.

Vanderkooi, G., and Green, D.E., 1970, Biological membrane structure. 1. The protein crystal model for membranes, *Proc. Natl. Acad. Sci. U.S.A.* **66**:615.

Vargas, F.F., 1968, Water flux and electrokinetic phenomena in the squid axon, *J. Gen. Physiol.* **51**:123S.

Villegas, G.M., 1969, Electron microscopy of the giant nerve fiber of the giant squid *Dosidicus gigas*, *J. Ultrastruct. Res.* **26**:501.

Villegas, G., 1975, Effects of cholinergic compounds on the axon–Schwann cell relationship in the squid nerve fiber, *Fed. Proc. Fed. Am. Soc. Exp. Biol.* **34**:1370.

Villegas, G.M., and Villegas, R., 1960, The ultrastructure of the giant nerve fibre of the squid axon–Schwann cell relationship, *J. Ultrastruct. Res.* **3**:362.

Villegas, G.M., and Villegas, R., 1963, Morphogenesis of the Schwann channels in the squid nerve, *J. Ultrastruct. Res.* **8**:197.

Villegas, G.M., and Villegas, R., 1968, Ultrastructural studies on the squid nerve fibers, *J. Gen. Physiol.* **51**:44S.

Villegas, G.M., and Villegas, J., 1974, Acetylcholinesterase localization in the giant nerve fiber of the squid, *J. Ultrastruct. Res.* **46**:149.

Villegas, G.M., and Villegas, J., 1976, Structural complexes in the squid giant axon membrane sensitive to ionic concentrations and cardiac glycosides, *J. Cell. Biol.* **69**:19.

Villegas, G., Villegas, L., and Villegas, R., 1965, Sodium, potassium and chloride concentrations in the Schwann cell and axon of the squid nerve fiber, *J. Gen. Physiol.* **49**:1.

Villegas, J., 1968, Transport of electrolytes in the Schwann cell and location of sodium by electron microscopy, *J. Gen. Physiol.* **51**:61S.

Villegas, J., 1972, Axon–Schwann cell interactions in the squid nerve fibre, *J. Physiol.* **225**:275.

Villegas, J., 1973, Effects of tubocurarine and eserine on the axon–Schwann cell relationship in the squid nerve fibre, *J. Physiol.* **232**:193.

Villegas, J., 1974, Effects of acetylcholine and carbamylcholine on the axon and Schwann cell electrical potentials in the squid nerve fibre, *J. Physiol.* **242**:647.

Villegas, J., 1975, Characterization of acetylcholine receptors in the Schwann cell membrane of the squid nerve fibre, *J. Physiol.* **249**:679.

Villegas, J., Villegas, R., and Gimenez, M., 1968, Nature of the Schwann cell electrical potential: Effect of ions and a cardiac glycoside, *J. Gen. Physiol.* **51**:47.

Villegas, J., Sevcik, C., Barnola, F.V., and Villegas, R., 1976, Grayanotoxin, veratrine, and tetrodotoxin-sensitive sodium pathways in the Schwann cell membrane of squid nerve fibers, *J. Gen. Physiol.* **67**:369.

Villegas, R., and Barnola, F.V., 1960, Equivalent pore radius in the axolemma of the giant axon of the squid, *Nature (London)* **188**:762.

Villegas, R., and Barnola, F.V., 1961, Characterization of the resting axolemma in the giant axon of the squid, *J. Gen. Physiol.* **44**:963.

Villegas, R., and Barnola, F.V., 1972, Ionic channels and nerve membrane constituents: Tetrodotoxin-like interaction of saxitoxin with cholesterol monolayers, *J. Gen. Physiol.* **59**:33.

Villegas, R., and Camejo, G., 1968, Tetrodotoxin interaction with squid nerve membrane lipids, *Biochim. Biophys. Acta* **163**:421.

Villegas, R., and Villegas, G.M., 1960, Characterization of membranes in the giant nerve fiber of the squid, *J. Gen. Physiol.* **43**:73.

Villegas, R., Gimenez, M., and Villegas, L., 1962, The Schwann-cell electrical potential in the squid nerve, *Biochim. Biophys. Acta* **62**:610.

Villegas, R., Villegas, L., Gimenez, M., and Villegas, G.M., 1963, Schwann cell and axon electrical potential difference: Squid nerve structure and excitable membrane location, *J. Gen. Physiol.* **46**:1047.

Villegas, R., Blei, M., and Villegas, G.M., 1965a, Penetration of non-electrolyte molecules in resting and stimulated squid nerve fibers, *J. Gen. Physiol.* **48**:35.

Villegas, R., Herrera, F.C., Villegas, G.M., and Blei, M., 1965b, Sodium influx and non-electrolyte penetration in stimulated squid axons, *J. Cell. Comp. Physiol.* **66**:155.

Villegas, R., Villegas, G.M., Blei, M., Herrera, F.C., and Villegas, J., 1966, Nonelectrolyte penetration and sodium fluxes through the axolemma of resting and stimulated medium sized axons of the squid *Doryteuthis plei*, *J. Gen. Physiol.* **50**:43.

Villegas, R., Bruzual, I.B., and Villegas, G.M., 1968, Equivalent pore radius of the axolemma of resting and stimulated squid axons, *J. Gen. Physiol.* **51**:81S.

Villegas, R.F., Barnola, V., and Camejo, G., 1970, Ionic channels and nerve membrane lipids: Cholesterol–tetrodotoxin interaction, *J. Gen. Physiol.* **55**:548.

Villegas, R., Villegas, G.M., DiPolo, R., and Villegas, J., 1971, Nonelectrolyte permeability, sodium influx, electrical potentials, and axolemma ultrastructure in squid axons of various diameters, *J. Gen. Physiol.* **57**:623.

Von Muralt, A., 1975, The optical spike, *Philos. Trans. R. Soc. London Ser. B* **270**:411.

Wang, C.M., Narahashi, T., and Scuka, M., 1972, Mechanism of negative temperature coefficient of nerve blocking action of allethrin, *J. Pharmacol. Exp. Ther.* **182**:442.

Watanabe, A., Tasaki, I., Singer, I., and Lerman, L., 1967a, Effect of tetrodotoxin on excitability of squid giant axons in sodium-free media, *Science* **155**:95.

Watanabe, A., Tasaki, I., and Lerman, L., 1967b, Bi-ionic action potentials in squid giant axons internally perfused with sodium salts, *Proc. Natl. Acad. Sci. U.S.A.* **58**:2246.

Webb, G.D., Dettbarn, W.-D., and Brzin, M., 1966, Biochemical and pharmacological aspects of the synapses of the squid stellate ganglion, *Biochem. Pharmacol.* **15**:1813.

White-Ortiz, A., 1967, Metabolismo de RNA en azones de *Dosidicus gigas*, *An. Fac. Quim. Farm. Univ. Chile* **19**:138.

Williams, L.W., 1909, *The Anatomy of the Common Squid Loligo pealii, Lesueur*, E.J. Brill, Leiden.

Witman, G., and Rosenbaum, J., 1973, Filamentous components of isolated squid axoplasm, *Biol. Bull.* **145**:460.

Woodbury, J.W., 1965, Biophysics of the cell membrane (Chapt. 1) and Nerve and muscle (Chapt. 2), in: *Physiology and Biophysics* (T.C. Ruch and H.D. Patton, eds.), pp. 1–72, W.B. Saunders, Philadelphia.

Wu, C.H., and Narahashi, T., 1973, Mechanism of action of propranolol on squid axon membranes, *J. Pharmacol. Exp. Ther.* **184:**155.

Wu, C.H., and Narahashi, T., 1976, Actions of trihexyphenidyl and benztropine on squid axon membranes, *J. Pharmacol. Exp. Ther.* **197:**135.

Yeh, J.Z., and Narahashi, T., 1974*a*, Effects of lobeline on ionic conductances of squid axon membranes, *Fed. Proc. Fed. Am. Soc. Exp. Biol.* **33:**272.

Yeh, J.Z., and Narahashi, T., 1974*b*, Noncholinergic mechanism of action of cholinergic drugs on squid axon membranes, *J. Pharmacol. Exp. Ther.* **189:**697.

Yeh, J.Z., and Narahashi, T., 1976, Mechanism of action of quinidine on squid axon membranes, *J. Pharmacol. Exp. Ther.* **196:**62.

Yeh, J.Z., Takeno, K., Rosen, G.M., and Narahashi, T., 1975, Ionic mechanism of action of a spin-labeled local anesthetic on squid axon membranes, *J. Membrane Biol.* **25:**237.

Yeh, J.Z., Oxford, G.S., Wu, C.H., and Narahashi, T., 1976, Interactions of aminopyridines with potassium channels of squid axon membranes, *Biophys. J.* **16:**77.

Young, J.Z., 1936*a*, The structure of nerve fibers in cephalopods and crustacea, *Proc. R. Soc. London Ser. B* **121:**319.

Young, J.Z., 1936*b*, The giant nerve fibers and epistellar body of cephalopods, *Q. J. Microsc. Sci.* **78:**367.

Young, J.Z., 1939, Fused neurons and synaptic contacts in the giant nerve fibers of cephalopods, *Philos. Trans. R. Soc. London Ser. B* **229:**465.

Young, J.Z., 1952, *Doubt and Certainty in Science*, Oxford University Press, New York.

2

Single-Cell Isolation and Analysis

Neville N. Osborne

1. Introduction

The human brain has approximately 10^{10} nerve cells, and this vast population of neurons presents a formidable challenge to the biologist trying to understand how the nervous system works. The great structural complexity of the nervous system and the consequent difficulty in interpreting gross observations were enough to stimulate numerous early attempts to study isolated individual units. In fact, Deiters (1865), more than 100 years ago, published excellent drawings of neurons he dissected from the anterior horn of the spinal cord. It is now clear, from the mass of electrophysiological and electron-microscopic data that has accumulated, that nerve cells are independent units that are interrelated in complex ways (see, for example, Bullock, 1967; Bullock and Horridge, 1965; Eccles, 1964; Segundo, 1970; Horridge, 1968). Thus, one classic approach by the biochemist trying to elucidate the complex structure of the brain is to separate the component parts (e.g., neurons, glia, myelin, nuclei, synaptosomes, synaptic vesicles) and study them in isolation (see, for example, Rose, 1967; Whittaker, 1968, 1973; Poduslo and Norton, 1972). Studies of this kind by the biochemist have many advantages, but they can suffer from certain drawbacks such as the possibility that changes in the constituents may be caused by the elaborate separation or fractionation procedures employed. Moreover, any differences there may be in the properties of similar structures obtained from the brain cannot be observed. Another approach is to analyze small defined areas of the nervous system or individual hand-dissected neurons. In recent years, many studies have been carried out using this approach (see Osborne, 1974), the most

Neville N. Osborne • Nuffield Laboratory of Ophthalmology, University of Oxford, Oxford OX2 6AW, England.

prominent developers being Lowry (1952, 1953, 1963), Giacobini (1956, 1959, 1964), and Hydèn (1959, 1960, 1964). The purpose of this chapter is to highlight various procedures used to isolate, by dissection, defined components of the nervous system and to analyze them with suitable microbiochemical methods. More detailed information on microbiochemical procedures and their application in the analyses of biological materials is presented elsewhere (Giacobini, 1968, 1969, 1970, 1975; Hydèn, 1960, 1972; Lowry, 1953, 1963; Lowry and Passoneau, 1972; Neuhoff, 1973; Osborne, 1974; Osborne and Neuhoff, 1973; Johnston and Roots, 1972; Peterson, 1972).

2. Single-Cell Isolation and Removal of Discrete Tissue Regions

2.1. Choice and Validity of Isolated Neurons

Certain invertebrate nervous systems (Table 1) offer a number of advantages over vertebrate nervous systems for the analysis of isolated neurons in that the neurons are organized in an orderly manner, fewer nerve cells are present, and specialized giant neurons exist that can be individually characterized from one animal to another (see Figs. 1 and 2). There are certain vertebrate preparations that do contain populations of giant neurons (see Table 1), though they are almost impossible to characterize individually and thus identify in different individuals of the same species. Another important advantage of the invertebrate neurons is that they can retain their fundamental activity after dissection and survive for several hours or even days (Strummwasser, 1967). This, therefore, makes it possible to perform *in vitro* experiments on invertebrate nervous systems, monitoring the activity of individual neurons by means of intra- or extracellular recording, while the environment of the cell can be controlled or changed by adding or substituting ions, inhibitors, drugs, or other substances. These and other advantages in using invertebrate neurons in the analysis of individual cells are summarized elsewhere (Osborne, 1974). Two questions have to be considered in the

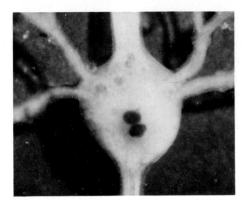

Fig. 1. Photograph showing a single ganglion from the leech, *Hirudo medicinalis*. The connective tissue was removed and the ganglion was then slightly, and rapidly, stained with methylene blue. As can be seen, the two Retzius cells (each cell has a diameter of about 80 μm) quickly accumulate the dye, which allows them to be easily identified.

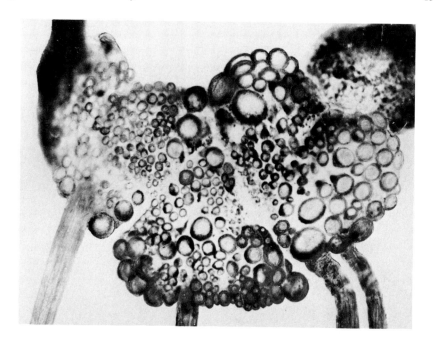

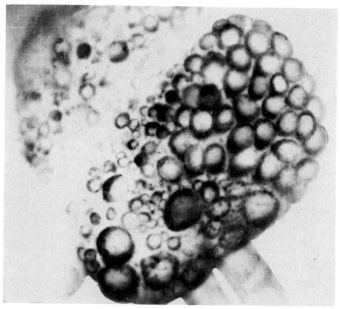

Fig. 2. Photographs of the subesophageal ganglia (top) and the right parietal ganglion (bottom) of *Helix aspersa* that demonstrate the localization and distribution of the different-sized neurons. The large or giant neurons (measuring between 90 and 150 μm) can be repeatedly identified in different preparations. The preparations were slightly stained with methylene blue to assist in the identification process. Photographs by courtesy of G.A. Kerkut (taken from Kerkut *et al.*, 1975).

Table 1. Invertebrate and Vertebrate Cell Bodies Suitable for Microanalysis

Species	Preparation	Cell-body diameter (μm)
Crustacea (invertebrate)		
Astacus fluviatilis (crayfish)	Stretch receptor (slow-adapting)	50–80
Astacus astacus (crayfish)	Stretch receptor (slow-adapting)	50–85
Homarus americanus (lobster)	Stretch receptor (slow-adapting)	75–120
Insecta (invertebrate)		
Periplaneta americana (cockroach)	Thoracic ganglion	50–120
Mollusca (invertebrate)		
Aplysia californica (sea slug)	Visceral ganglion	400–800
Helix aspersa (snail)	Visceral ganglion	40–320
Helix pomatia (snail)	Visceral ganglion	60–360
	Cerebral ganglion	40–180
	Parietal ganglion	260–400
Tritonia gilberti (sea slug)	Pleural ganglia	500–800
	Parietal ganglia	500–1000
Loligo (squid)	Pedal ganglion	150–750
Octopus (octopus)	Cerebral ganglion	20–80
Annelida (invertebrate)		
Lumbricus (earthworm)	Ventral nerve cord	30–60
Hiruda medicinalis (leech)	Ventral nerve cord	30–65
Fish (vertebrates)		
Goldfish	Mauthner's cells	30–40
Puffer fish	Supramedullary cells	200–400
Amphibians (vertebrate)		
Frog	Spinal ganglion	15–20
	Sympathetic ganglion	10–35
Mammals (vertebrate)		
Rabbit	Deiters's cells	50–100
	Cortical cells	20–40
	Cells of nucleus supraopticus	30–45
	Spinal ganglion cells	60–150
	Anterior horn cells	20–50
	Granular cells of cerebellum	10–20
	Hippocampus cells	10–30
Cat	α-Motoneurons of lumbar spinal cord region	40–80

analysis of hand-dissected isolated neurons from fresh tissue, thus placing such studies in a better perspective. First, how much contamination of glia, blood vessels, or other tissue is associated with the neurons and, second, can the isolated neurons be considered as "normal neurons" insofar as they are free of significant damage and alteration to their characteristics? The answer to both questions depends largely on the experimenter; the more experienced should avoid damage to the cell and at the same time dissect the neuron free from most sources of contamination. In reality, it is impossible to dissect a neuron totally free of glia without appreciable damage to the neurons because of the close interrelationship displayed between glia and neurons (Johnston and Roots, 1972). It is now well established that both vertebrate and invertebrate neurons are penetrated by glia processes, so that a percentage of glia will always be associated with the isolated neurons. Whether an isolated neuron can be considered a "normal neuron" is, of course, a problem. Hillman and Hydèn (1965) clearly demonstrated that hand-dissected vertebrate neurons maintained a membrane potential, though it was later shown that the neurons also displayed some ultrastructural damage (Bondareff and Hydèn, 1969). Nevertheless, the neurons are still viable, judged by the uptake of dyes such as lissamine green, trypan blue, and fluorescein diacetate (Hydèn and Rönnbäch, 1975). In reality, the problem is not whether the isolated neuron is "still normal," but whether the parameter to be analyzed has changed in some way during the isolation process. From this point of view, it is obvious that invertebrate neurons have the advantage over vertebrate neurons in that they can overcome slight environmental changes (e.g., pH, temperature) that often prove disastrous for vertebrate cells. When working with vertebrate material, the most hazardous period can frequently be the moment the blood supply is cut off and the neuron dissected. The question of isolated neurons being "abnormal" should thus not be over-stressed, but rather viewed with caution, depending on the parameter studied. In practice, clearly, everything should be done to maintain the functional integrity of the nerve cell after dissection, or the metabolism of the neuron should be stopped abruptly by means of rapid freezing.

2.2. Isolation of Neurons

Some basic dissection instruments are usually necessary for the dissection of neurons (Neuhoff, 1973; Osborne, 1974). These include microscalpels, which can easily be made by crushing halved razor blades between the teeth of a pair of pliers and mounting them on suitable handles with cement. A variety of glass needles can also be prepared, and these are best made from Pyrex glass rods. Another requisite is an assortment of nylon loops and hair points (Fig. 3), which can easily be manufactured from soft glass tubing by heating the tubing, pulling it out to form a Pasteur-like pipette, and then inserting into it the nylon (9–15 denier) loop or hair point (nostril hair is especially good) and sealing the loop or point into place with a drop of cement (e.g., Plexiglas or post-office sealing wax).

2.2.1. Isolation of Characterized Invertebrate Neurons

The location of the characterized neurons will depend on the familiarity of the experimenter with the ganglia, though a number of published maps of the ganglia are available, e.g., *Aplysia californica* (Coggeshall *et al.*, 1966; Frazier *et al.*, 1957; Kandel *et al.*, 1957), *Helix pomatia* (Kerkut, 1969; Osborne, 1973), *Helix aspersa* (Kerkut *et al.*, 1975), the cockroach (Cohen and Jacklet, 1965), *Hirudo medicinalis* (Nicholls and Baylor, 1969), and *Tritonia* (Willows,

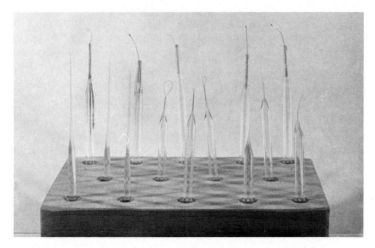

Fig. 3. Examples of various hair points, loops, glass needles, and microknives necessary for the dissection of individual neurons.

1967, 1968). Generally, the individual cells can be identified by morphology alone (see Figs. 1 and 2), and it is usually preferable to characterize the neurons before cutting through the final connective tissue sheath. It is, however, often necessary to use neurophysiological or neuropharmacological methods, or both, to confirm the identification of the neurons, since variations in the morphology of a neuron frequently occur with age.

Optimum illumination is a prerequisite for the identification and removal of a neuron, intact, from a ganglion. It is essential to pin the ganglion down initially (through the connective tissues) in a relatively stretched-out position. In the instance of the snail (*Helix*) circumesophageal ganglia, a small bath (vol., 0.7 ml) containing a nylon sheet at the bottom and filled with snail saline is most suitable. Either transillumination, reflected light, or dark-field illumination (which is often the most useful) can be used for the identification and dissection of individual neurons, which is carried out under microscopic vision (Fig. 2). One simple but very good approach is to focus a pencil of light on the cell concerned. This can be done by attaching a tapered glass rod to a microscope lamp and bringing the glass tip close to the preparation, thus illuminating the cell alone.

A couple of pairs of fine forceps, together with a microscalpel, are all that is required to free a characterized cell from the surrounding nervous tissue. A constant check should be made that the tips of the forceps fit exactly and are sharply pointed, this being best achieved by filing them with emery stone. With a pair of forceps in each hand, one can gently tease away the connective tissue surrounding the cell until the cell 'pops out.' The axon can then be cut and freed entirely with a microscalpel and a single pair of forceps. The cell can now be gently transferred either with a scalpel tip (not a method advocated for the beginner) or by means of a constricted pipette attached to the mouth by rubber tubing and thus allowing complete control of the transfer process. Another procedure is not to cut the axon of the cell, but to drop a loop of fine wire over the cell and tie it tightly around the axon at the cell base; the axon is then severed and the cell and wire transferred together to a microtube (McCaman and Dewhurst, 1970).

2.2.2. Isolation of Neurons from Fresh Tissue

In this method, pieces of tissue approximately 1 mm square and containing the cells to be dissected are rapidly removed after animals have been killed, usually by bleeding, and are placed in a drop of suitable physiological solution (Ringer's, 0.32 M sucrose, or buffered HEPES-salt solution, pH 7.2) on the top of a paraffin-wax-covered slide. The uppermost surface of the piece of tissue should be observed under a stereomicroscope and differences in the color and texture noted so that it is possible to identify the cell bodies. For even clearer identification of the cells, a drop of solution of methylene blue in isotonic Ringer's solution (1 part in 10,000) can be applied to the surface. As soon as the dye is taken up by the neurons (Fig. 4), the methylene blue solution is washed away and substituted by Ringer's alone. Single cells can then be carefully dissected with various microtools (e.g., steel needles, hair points, microknives), which should not come into contact with the cell. The nerve cells that float in the drop of Ringer's are clearly visible under the stereomicroscope and can be transferred by use of hair loops. Once the technique has been mastered, nerve cells with many

Fig. 4. Thick section of the hippocampus of the rabbit. The section, while being kept cool, was stained for 3 min with 1% methylene blue solution to observe the neurons. The dark regions show areas of nerve perikarya.

dendrites can be isolated individually, without the use of methylene blue, in less than a minute. The method described above is similar to the original techniques developed by Hydèn (1959) and Giacobini (1956). Some neurons isolated by Hydèn and his group from the rat are shown in Fig. 5, from which it can be seen that these authors have mastered the technique to such an extent that it is even possible for them to dissect the isolated neuron further and expose the inner and outer surfaces of the membrane.

2.2.3. Isolation of Neurons from "Fixed" Tissue

This method was originally developed by Edström (1964). Cells can be isolated either from paraffin-embedded or from deparaffinized sections (Fig. 6). In this procedure (see, for example, Edström and Neuhoff, 1973), a cover slip with a section of material on it is inverted and placed on a thick glass slide that has a wide groove running across it. The groove forms the floor and ends of the chamber while the cover slip with the section on its underside forms the roof. The sides are left open. The chamber is then filled with paraffin oil, which is retained in the open-sided chamber by surface tension.

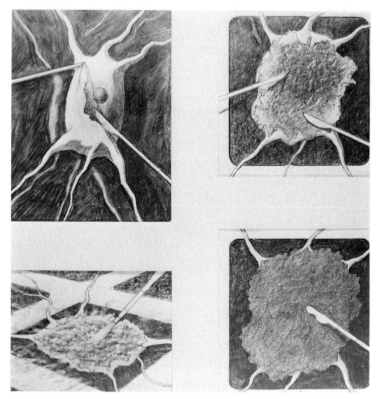

Fig. 5A. Diagrammatic representation of the procedure used for the preparation of nerve-cell membranes from single cells using the technique of freehand microsurgery.

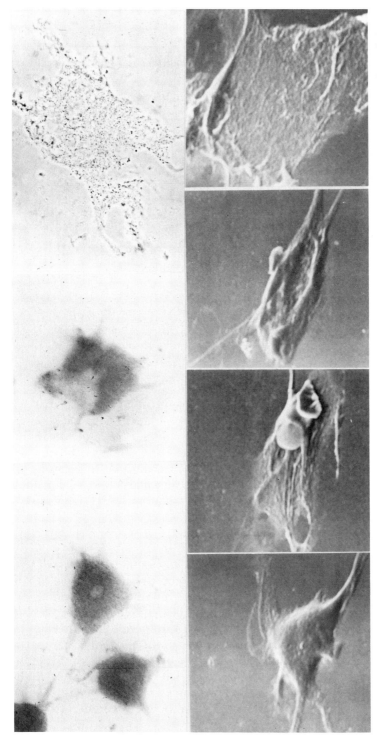

Fig. 5B. *Top row:* (Left and center) Isolated Deiters's neurons. (Right) Membrane preparation of a cell slightly stained with methylene blue. All three photographs are stereomicroscopic views using incident light. *Bottom row:* Scanning electron micrographs. The first picture is of an isolated Deiters's neuron, the second and third of partially dissected Deiters's neurons, and the fourth of a dissected membrane preparation of a Deiters's neuron. Magnification: ×500. Photographs by courtesy of H. Hyden.

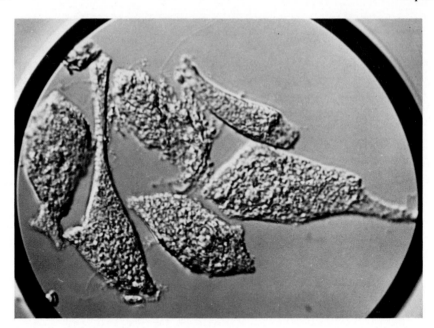

Fig. 6. Isolated nerve cells from a section of nervous tissue (human medulla oblongata) embedded in paraffin and dissected as described in the text. Magnification: × 1500. Photograph by courtesy of V. Neuhoff.

This is placed on the stage of a phase-contrast microscope that has its focusing device associated with the tube of the microscope rather than the stage. Fine glass needles, bent upward, are mounted on a micromanipulator and inserted into the chamber through the open sides. Small neurons and cell parts can then be isolated and pushed away while still sticking to the cover slip (Fig. 7), and a micropipette, also attached to a micromanipulator, can be used to suck up the cell and transfer it.

2.2.4. Isolation of Neurons from Freeze–Dried Tissue

Small pieces of nervous tissue (or whole "brain" in the case of invertebrates) are placed on a small metal microtome chuck and frozen in a liquid gas. Sections (20–150 μm) of the tissue are then prepared at about $-15°C$ and freeze–dried. Individual cell bodies may be dissected from the freeze–dried sections by teasing out freehand with the use of fine glass needles under microscopic vision. The cells can then be transferred with hair points or nylon loops or both. It is advisable to carry out the dissection procedure in a room free of drafts because the freeze–dried sections can easily blow away. This procedure was developed by Lowry (see Lowry and Pasonneau, 1972) and has even been used to dissect different, defined cell parts from a single neuron.

2.2.5. Isolation of Neurons from Frozen Impregnated Tissue

This method (Giller and Schwartz, 1971) has the virtue of stabilizing the nervous tissue, thus facilitating the dissection while allowing the enzyme activities to be retained. The nervous tissue is covered with 70% ethylene glycol and then frozen with powdered solid carbon dioxide. The ethylene glycol solution does not solidify under these conditions, although the nervous tissue is hardened to the consistency of solid cheese. The cells can easily be seen with dark-field light or transillumination under optical magnification, and can then be dissected as already described. Though this method has been used to isolate cells only from invertebrate ganglia (Giller and Schwartz, 1971; Osborne *et al.*, 1975), it can also be applied to pieces of vertebrate nerve.

2.2.6. Isolation of Invertebrate Neurons by Digestion and Mechanical Treatment

Chen *et al.* (1971, 1973) isolated neurons from *Aplysia* by classic trypsinization and showed them to be capable of maintaining a spontaneous, normal electrical activity for at least 4 hr. A similar method has also been used by Kostenko (Kostenko, 1972; Kostenko *et al.*, 1974) and Bocharova *et al.* (1975) to isolate neurons from *Helix* and *Lymnea*. Their method is basically as follows: Either pronase, trypsin, or hyaluronidase in physiological solution (0.25–0.5%) is incubated with the ganglia for a defined time, depending on the texture of the connective tissue. The ganglia are then washed rapidly in pure physiological solution, the connective tissue sheath removed with needles, and the cells extracted with special micropipettes. A typical cell

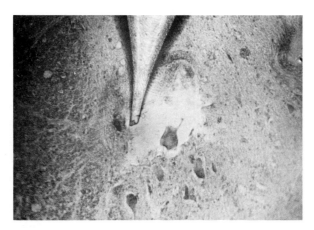

Fig. 7. A 20-μm-thick histological section of nervous tissue from the medulla of the rat that has been stained with methylene blue. The microscope slide containing the section is placed atop the paraffin oil. Cells from the section are then dissected free from the surrounding glia with a dissecting glass needle that is manipulated by a De Fonbrune attachment. This figure clearly illustrates the dissecting needle used to free the neurons, which are darkly stained. Magnification: ×200.

produced by this procedure is shown in Fig. 8, in which it is also demonstrated that the fine structure of the neuron is not damaged. Viable isolated neurons appear easily obtainable by this method, but whether the method allows them to be individually and reliably characterized is not certain. Moreover, the precise effect of the enzymes used on the biochemical constituents of the neurons is not known.

2.2.7. Removal of Discrete Nervous-Tissue Regions

In this procedure, tissue containing the area to be investigated is first frozen with solid carbon dioxide and placed in a refrigerator so that the temperature of the tissue slice reaches $-20°C$. A slice (about 1 mm thick) is then placed on a cool microscope slide and the precise area punched out. The punched-out frozen tissue sample ($<$ 1 mg tissue) is then pushed out rapidly with a metal plunger (for further details, see Neuhoff, 1973; Osborne, 1974). A micropunch system based on the method described has also been developed by Palkovits (1973).

Recently, Jacobowitz (1974) has described a modification of the method described above that has an advantage in that it enables the removal of small regions from fresh unfrozen rat brains.

3. Microbiochemical Procedures

3.1. Microanalysis of Transmitter Substances

For the biochemical analysis of isolated neurons (and also of definite discrete brain areas, biopsy material, and in other circumstances in which the

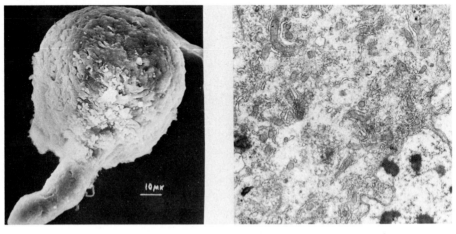

Fig. 8. (Left) Scanning electron micrograph of an isolated molluscan neuron. The neuron was isolated by fermentative digestion followed by mechanical treatment. (Right) Electron micrograph of an isolated neuron (as shown on left) demonstrating that the cytoplasm is not damaged through the isolation procedure. From Bocharova *et al.* (1975) by courtesy of the authors.

amount of tissue available is limited), it is necessary to use suitable micro-procedures. The term "microprocedure" implies a method with sufficient inherent sensitivity to measure minute quantities of a component. At this point, perhaps, it is advisable to draw a distinction between a microprocedure and normal (macro-) and ultraprocedures. In theory, they are a scale continuum, though in practice the change from the macroscale (i.e., brain homogenates) to the microscale (i.e., one very large neuron or microquantities of nervous tissue) is often simply a matter of scaling down and a slight modification of the macromethods. However, the next step, the ultramicro-procedure (parts of a single minute nerve cell), is a quantum jump and often requires elaborate apparatus and new approaches. The following account will summarize some recently developed microprocedures with special reference to the analysis of transmitter molecules. In practice, micromethods are no more difficult to carry out than normal methods and have the added advantage of being less time-consuming and less expensive with respect to the cost of materials. However, a simple scaling-down of the normal method is often not enough, and extensive adaptations are necessary to produce microprocedures. Although micromethods are theoretically no more inac-curate than normal methods, the sources of error caused by contamination, dirty glassware, efficiency of procedure, efficiency of experimenter, and other factors can be more significant—a "microerror" in the "microscale" is in effect similar to a "macroerror" in the "normal scale." Finally, the point should be made that the following basic techniques are used in the chemical analysis of tissues independent of the scale: electrophoresis, colorimetry, chromatography, fluorometry, radiometry, manometry (Cartesian diver), magnetometry (magnetic diver), fluorometry-cycling, and mass spectometry, allowing between 10^{-8} and 10^{-16} mol of substance to be measured. These methods, used alone or in combinations, most often do not provide the requirements for unambiguous identification of a substance, however, and it is fair to say that only a combination of chromatography (gas or thin-layer) and electrophoresis with mass spectrometry can achieve this. The technique of mass spectrometry (not discussed in this chapter) is a most powerful procedure, allowing, for example, femtomole levels of transmitter substances to be unambiguously identified (Abramson *et al.*, 1974).

3.1.1. Screening of Transmitters

A rapid screening method for identifying possible transmitter substances at the cellular level has been developed by Hildebrand *et al.*, (1971), though the procedure does not lend itself to the detection of new transmitter molecules. The method is based on a selective synthesis and localization of transmitters in those cells that are presumed to release them. The procedural steps involve (1) incubation of nervous tissue in a physiological medium with one or more radioactive precursors or related transmitters (choline for acetylcholine, glutamate for GABA, tyrosine for catecholamines, tyramine for octopamine, tryptophan for serotonin); (2) isolation of nervous-tissue components (e.g., cell bodies, parts of cells) and the chemical extraction of the transmitter substances under conditions that preserve them; (3) high-voltage paper electrophoresis of the extract under conditions that separate most of the compounds of interest (Fig. 9); (4) quantification of the isotope in each transmitter candidate. This whole procedure is very sensitive, allowing 10^{-12} mol of material to be detected. Experiments with rat superior cervical ganglia have shown them to produce noradrenaline and acetylcholine from their precursors; both substances are known to function as transmitter substances in the sympathetic nervous system. Hildebrand *et al.* (1971) also showed identified excitatory lobster neurons to take up glutamate and convert it to GABA, while the inhibitory neurons lacked this capacity. Moreover, only certain sensory lobster neurons could form acetylcholine from its precursors (Barker *et al.*, 1972). The usefulness of this procedure has also been demonstrated by Lam *et al.* (1974), who showed the likely transmitter substances in the cephalopod retina to be acetylcholine and dopamine.

Recently, a suitable chromatographic procedure to separate a number of precursors and transmitter molecules on 5×5 cm silica gel plates (Fig. 10) has been developed (Osborne, unpublished data), and this has been used successfully to screen a number of transmitters. In the screening of transmitters, the procedural steps involved are the same as those used by Hildebrand *et al.* (1971), except that two-dimensional chromatography instead of high-voltage paper electrophoresis is used. This method has been used to

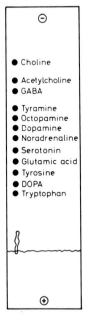

Fig. 9. Diagram showing the relative positions of some possible transmitter substances, precursors, and neutral metabolites after high-voltage electrophoresis. After Hildebrand *et al.* (1971).

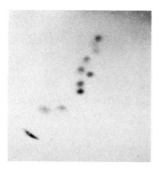 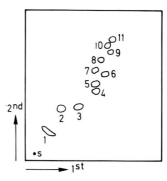

Fig. 10. Photograph of a chromatogram (silica gel, 5 × 5 cm) and a corresponding map to show the relative positions of various transmitter substances and precursors separated in two solvent systems. The directions of chromatography are indicated by the arrows, and the solvent systems used were butanol–acetic acid–water (15 : 3 : 5 by vol.) in the 1st direction and butanol–pyridine–acetic acid–water (15 : 2 : 3 : 5 by vol.) in the 2nd direction. The substances were visualized on the chromatogram by initial exposure to paraformaldehyde vapor for 30 min at 80° C followed by spraying with 1% potassium ferricyanide in ammonium hydroxide solution. After drying, the chromatogram was sprayed with ninhydrin solution. (S) Starting point; (1) glutamic acid; (2) GABA; (3) DOPA; (4) 5-hydroxytryptophan; (5) tyrosine; (6) tryptophan; (7) noradrenaline; (8) dopamine; (9) 5-hydroxytryptamine (serotonin); (10) octopamine; (11) tyramine.

show that a cell in the water snail (*Planorbis*) brain that is known to contain serotonin can produce [^{14}C]serotonin from [^{14}C]tryptophan, while a dopamine-containing cell and a cell that contains neither dopamine nor serotonin (Fig. 11) lack this capacity. The same method has demonstrated that the Retzius cells of the leech, known to contain serotonin, can produce

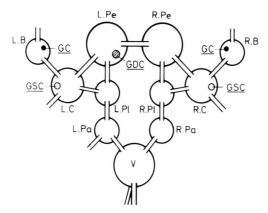

Fig. 11. Diagram of the dorsal surface of the subesophageal ganglionic mass together with the cerebral and buccal ganglia of *Planorbis*. The intercerebral and interbuccal connections have been cut to show the positions of the "giant serotonin cells" (GSCs) in each cerebral ganglion, the "giant cells" (GCs) in each buccal ganglion, and the single "giant dopamine cell" (GDC) in the left pedal ganglia. (L.B, R.B) Left and right buccal ganglia; (L.pe, R.Pe) left and right pedal ganglia; (L.C, R.C) left and right cerebral ganglia; (L.Pl, R.Pl) left and right pleural ganglia; (L.Pa, R.Pa) left and right parietal ganglia; (V) visceral ganglia. From Osborne *et al.* (1975).

[^{14}C]serotonin from [^{14}C]tryptophan, but could not metabolize [^{14}C]choline or [^{14}C]tyrosine to form either [^{14}C]acetylcholine or [^{14}C]dopamine.

3.1.2. Microchromatography of Amines

A microchromatographic procedure, based on the method of Bell and Somerville (1966), has been described by Osborne (1971) for the detection of monoamines in the nanogram range. Standard amounts of amines (1–1000 ng), dissolved in 50% acetone in 0.01 N HCl and applied to the bottom edge of a 3 × 3 cm sheet of polyamide with an ultrathin capillary, can be fractionated by ascending chromatography in a 50-ml beaker that is covered to prevent evaporation of the solvent and contains just enough of the solvent to cover the base. As single separation of the various amines can be effected by one of two solvent systems, viz., methyl acetate–isopropanol–ammonia 25% (9 : 7 : 5 by vol.) or butanol–chloroform–acetic acid (4 : 1 : 1 by vol.). When the solvent reaches the upper edge of the chromatogram, the plate is dried and placed in a sealed jar (500 ml) with about 3 g paraformaldehyde. The paraformaldehyde must have been stored at a relative humidity of 60% for at least 5 days to produce optimum fluorescence. After the jar has been heated in an oven for 3 hr at 80°C, the amines are located as fluorescent spots by viewing the chromatogram under UV light. From Table 2, which records the minimum amounts of biogenic amines detectable by this technique, it can be seen that this procedure is particularly sensitive for the detection of the transmitter substances dopamine, noradrenaline, and serotonin.

This very simple procedure for the analysis of amines can be applied to a number of neurochemical problems. The applicability of the method has been demonstrated by analyzing the serotonin content in snail neurons known to contain the amine (see Osborne, 1973). Analysis of chromatograms from extracts of as few as six neurons revealed a fluorescent spot to correspond in both color and position to pure serotonin. Moreover, examination of the excitation and emission spectra demonstrated that the substance was serotonin. A semiquantitative estimation of the serotonin content in the neurons was made by simply comparing the intensity of fluorescence in the neuronal extract with that in standard amounts of amine. From 12 different experiments, it was estimated that a "giant serotonin cell" (GSC) contains 0.9 ng serotonin. This figure is slightly higher than that obtained by other procedures (see Cottrell and Osborne, 1970) in which the serotonin content was calculated to be 0.7 ng per cell.

Microchromatographic procedures [on 5 × 5 cm silica precoated plates (Merck 60)] have also been developed to fractionate serotonin and its metabolites (Fig. 12) and dopamine and its metabolites (Fig. 13). Both procedures are, however, comparatively insensitive for the detection of endogenous levels of substances in isolated neurons, though they have proved most useful for metabolic studies. Paraformaldehyde vapor alone was not sufficient to detect the various metabolites (compare the relative insensitivity

Table 2. *Minimum Amounts of Biogenic Amines and
Other Compounds Detectable after Formaldehyde
Treatment*[a]

Compound	Detectable amount (ng)
Adrenaline	100
Bufotenine	80
n,n-Dimethyltryptamine	1000
3,4-Dimethoxyphenethylamine	15
3,4-Dihydroxyphenylalanine	50
Dopamine	6
5-Hydroxytryptophan	50
5-Hydroxytryptamine	5
3-Hydroxykynurenine	80
3-Hydroxy-4-methoxyphenylethylamine	20
Kynurenine	100
Kynurenic acid	100
Kynuramine	1000
Melatonin	90
Metadrenaline	1000
Mescaline	90
α-Methyl-*m*-tyrosine	1000
3-Methoxy-4-hydroxyphenylethylamine	50
Noradrenaline	7
Normetadrenaline	100
Octopamine	1000

[a] From Osborne (1971).

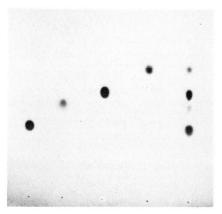 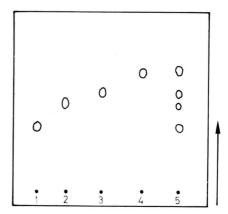

Fig. 12. Chromatogram (silica gel, 5 × 5 cm) and map to show the separation of 5-hydroxytryptamine and metabolites following one-dimensional chromatography. The direction of chromatography is indicated by the arrow, and the solution used was butanol–pyridine–acetic acid–water (15:2:3:5 by vol.) (1) 5-Hydroxytryptophan; (2) tryptophan; (3) 5-hydroxytryptamine; (4) 5-hydroxyindoleacetic acid; (5) mixture. Substances were visualized as described for Fig. 10. From Osborne and Pentreath (1976).

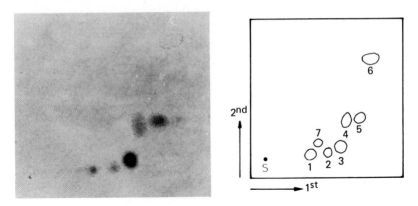

Fig. 13. Chromatogram (silica gel, 5 × 5 cm) and map to show the separation of dopamine and metabolites following two-dimensional chromatography. The solvent in the 1st dimension was *n*-butanol–pyridine–acetic acid–water (15:2:3:5 by vol.). In the 2nd dimension, the solvent was the organic phase of chloroform–acetic acid–water (2:2:1 by vol.). The substances were visualized as described for Fig. 10. (S) Starting point; (1) DOPA; (2) noradrenaline; (3) dopamine; (4) methoxytyramine; (5) dihydroxyphenylacetic acid; (6) homovanillic acid; (7) tyrosine. From Osborne *et al.* (1975).

of certain metabolites to form fluorescent products as shown in Table 3) and other detecting agents had to be used in addition. In one study, the effect of 5,7-dihydroxytryptamine on the incorporation of radioactivity from [³H]tryptophan into 5-hydroxytryptophan and serotonin in a snail giant neuron was successfully investigated (Table 4), it being shown that 5,7-dihydroxytryptamine not only slightly influenced the accumulation of [³H]tryptophan by the neuron but also inhibited the metabolism of the

Table 3. Minimum Amounts of Various Substances Detectable on Silica Gel Thin-Layer Plates Using Paraformaldehyde Vapor

Compound	Color of fluorescence	Amount detectable as judged by eye (μg)
Methoxytryptamine	Yellow	0.005
Serotonin	Yellow	0.003
5-Hydroxytryptophan	Yellow	0.003
Tryptophan	Blue-green	0.1
5-Hydroxyindolacetic acid	Yellow	4.0
Melatonin	Yellow	5.0
Tryptamine	Green-yellow	0.05
Homovanillic acid	Blue-yellow	10.00
Dopacetic acid	Yellow-orange	5.00
3-Methoxytyramine	Blue-green	0.05
Adrenaline	Green-yellow	0.5
Noradrenaline	Green-yellow	0.01
Dopamine	Green-yellow	0.03
Dihydroxypenylalanine	Green-yellow	0.5

Table 4. *Effect of 5,7-Dihydroxytryptamine on the Incorporation of Radioactivity from [³H]Tryptophan into 5-Hydroxytryptophan and 5-Hydroxytryptamine in "Giant Serotonin Cells"*[a]

Time (min)	Condition	Tryptophan	5-HTP	5-HT
30	Saline alone	14,000 ± 200	641 ± 34	582 ± 34
	Saline + 5,7-DHT	13,942 ± 240	632 ± 38	641 ± 32
45	Saline alone	26,000 ± 240	689 ± 38	640 ± 40
	Saline + 5,7-DHT	25,210 ± 230	700 ± 37	630 ± 33
60	Saline alone	38,120 ± 439	754 ± 39	820 ± 48
	Saline + 5,7-DHT	32,000 ± 384	714 ± 36	742 ± 68
90	Saline alone	39,800 ± 880	810 ± 40	1000 ± 68
	Saline + 5,7-DHT	34,840 ± 600	765 ± 38	800 ± 44[b]
120	Saline alone	47,700 ± 600	820 ± 45	1840 ± 54
	Saline + 5,7-DHT	39,120 ± 594[b]	700 ± 50	1000 ± 50[b]
150	Saline alone	50,200 ± 398	829 ± 39	1840 ± 60
	Saline + 5,7-DHT	42,400 ± 520[b]	681 ± 48	900 ± 39[b]

[a] From Osborne and Pentreath (1976). Incorporation of radioactivity in dpm/mg protein was measured after incubation of tissue in [³H]tryptophan for varying periods of time. Results are mean values of 4 experiments ± S.E. (5,7-DHT) 5,7-Dihydroxytryptamine; (GSC) "giant serotonin cell"; (5-HT) 5-hydroxytryptamine; (5-HTP) 5-hydroxytryptophan.
[b] $P < 0.05$ and therefore significantly different by Student's t test.

amino acid to form serotonin (Osborne and Pentreath, 1976). In another study (Osborne *et al.*, 1975) on defined dopamine-, serotonin-, and another non-amine-containing neuron (see Fig. 11), it was demonstrated, among other things, that only the dopamine cell contained the enzyme tyrosine-hydroxylase and that this cell type metabolized some dopamine to form methoxytyramine, homovanillic acid, and dihydroxyphenylacetic acid (see Table 5).

3.1.3. *Enzymatic–Isotopic Micromethods for Amines*

A number of very sensitive enzymatic–isotopic methods have recently been developed (Saavedra and Axelrod, 1972, 1973; Coyle and Henry, 1973; Saavedra *et al.*, 1973; Snyder *et al.*, 1966; Brownstein *et al.*, 1973; Saavedra, 1974; Taylor and Snyder, 1972). These methods are based on the incubation of amines with methyltransferase enzymes together with the donor of radioactive methyl groups, *S*-adenosylmethionine. The radioactive methyl derivatives formed are then separated by means of solvent extraction and the radioactivity counted. Table 6 shows the sensitivity obtained for the various amines. Specificity of the assays is achieved by (1) use of methylating enzymes with marked specificity for a given amine or group of amines; (2) use of solvents of different degrees of polarity to separate radioactive products formed in the reaction; and (3) use of selective evaporating techniques to eliminate radioactive but volatile contaminating substances. The procedural steps used for the enzymatic–isotopic assays are as follows: (1) extraction of amine from tissue by acid or buffer; (2) incubation with

Table 5. *Percentage Distribution of Radioactivity Associated with Catecholamine Metabolites in the "Giant Dopamine Cells," "Giant Serotonin Cells," and "Giant Cells" following Incubation in [³H]Tyrosine, [¹⁴C]-DOPA, and [¹⁴C]Dopamine*[a]

Incubation substance	[³H]Tyrosine		[¹⁴C]-DOPA		[¹⁴C]Dopamine	
GDC	[³H]Tyrosine	100%	[¹⁴C]-DOPA	100%	[¹⁴C]-DA	100%
	[³H]-DOPA	25 ± 3%	[¹⁴C]-DA	33 ± 5%	[¹⁴C]-DOPAC	25 ± 4%
	[³H]-DA	23 ± 6%	[¹⁴C]-DOPAC	14 ± 6%	[¹⁴C]-HVA	23 ± 7%
	[³H]-NA	6 ± 3%[b]	[¹⁴C]-HVA	16 ± 4%	[¹⁴C]-MTA	11 ± 2%
			[¹⁴C]-MTA	8 ± 3%[b]	[¹⁴C]-NA	6 ± 3%[b]
			[¹⁴C]-NA	6 ± 3%[b]		
GSC	[³H]Tyrosine	100%	[¹⁴C]-DOPA	100%	[¹⁴C]-DA	100%
	[³H]-DOPA	8 ± 4%[b]	[¹⁴C]-DA	27 ± 9%	[¹⁴C]-DOPAC	6 ± 3%
	[³H]-DA	7 ± 3%[b]	[¹⁴C]-DOPAC	7 ± 4%[b]	[¹⁴C]-HVA	5 ± 3%
	[³H]-NA	6 ± 3%[b]	[¹⁴C]-HVA	6 ± 4%[b]	[¹⁴C]-MTA	6 ± 2%[b]
			[¹⁴C]-MTA	6 ± 3%[b]	[¹⁴C]-NA	6 ± 2%[b]
			[¹⁴C]-NA	5 ± 2%[b]		
GC	[³H]Tyrosine	100%	[¹⁴C]-DOPA	100%	[¹⁴C]-DA	100%
	[³H]-DOPA	8 ± 3%[b]	[¹⁴C]-DA	8 ± 4%[b]	[¹⁴C]-DOPAC	5 ± 3%[b]
	[³H]-DA	7 ± 4%[b]	[¹⁴C]-DOPAC	7 ± 3%[b]	[¹⁴C]-HVA	6 ± 2%[b]
	[³H]-NA	6 ± 3%[b]	[¹⁴C]-HVA	6 ± 2%[b]	[¹⁴C]-MTA	5 ± 2%[b]
			[¹⁴C]-MTA	6 ± 3%[b]	[¹⁴C]-NA	6 ± 3%[b]
			[¹⁴C]-NA	6 ± 3%[b]		

[a] The substances identified are as in Fig. 11. Each value is the mean of 4 experiments ± S.E.M. (DA) Dopamine; (DOPAC) dihydroxyphenylacetic acid; (GSC) "giant cell"; (GDC) "giant dopamine cell"; (GSC) "giant serotonin cell"; (HVA) homovanillic acid; (MTA) methoxytyramine; (NA) noradrenaline.
[b] Not significant by Student's *t* test.

corresponding *N*-methyltransferase and radioactive methyl donor; (3) extraction of radioactive *N*- or *O*-methylated product found in the reaction with organic solvent; (4) elimination of radioactive contaminants by selective procedures; and (5) counting the radioactivity.

The enzymatic–isotopic methods have several advantages in that they are simple to carry out, relatively inexpensive, and very sensitive, and 50–100 assays can be undertaken simultaneously.

Table 6. *Sensitivity of Various Enzymatic–Isotopic Methods for the Determination of Biogenic Monoamines*

Amine	Enzyme required	Sensitivity (pg)
Noradrenaline	Catechol-*O*-methyltransferase	10
Histamine	Histamine-*N*-methyltransferase	25
Octopamine	Phenylethanolamine-*N*-methyltransferase	25
Phenylethanolamine	Phenylethanolamine-*N*-ethyltransferase	25
N-Acetylserotonin	Hydroxyindole-*O*-methyltransferase	50
Serotonin	*N*-acetyltransferase and hydroxyindole-*O*-methyltransferase	50
Dopamine	Catechol-*O*-methyltransferase	100
β-Phenylethylamine	Dopamine-β-hydroxylase and phenylethanolamine-*N*-methyltransferase	200
Tryptamine	Nonspecific-*N*-methyltransferase	1000

These assays have already been used successfully for the analysis of a number of brain tissues, e.g., individual nuclei in the hypothalamus and limbic systems (Brownstein *et al.*, 1973; Palkovits *et al.*, 1974; Saavedra *et al.*, 1974). Isolated neurons of *Aplysia* have also been analyzed with these techniques (Saavedra *et al.*, 1974; Brownstein *et al.*, 1974; Weinreich *et al.*, 1975; Milinoff *et al.*, 1969), and it has been demonstrated that (1) octopamine exists only in certain neuron types, (2) more than one transmitter type amine often occurs in the same neuron, and (3) a cell-specific distribution of histamine exists, providing indirect support for histamine's being a transmitter in *Aplysia*.

3.1.4. Determination of Amines and Amino Acids as Dansyl Derivatives

5-Dimethylaminonaphthalene-1-sulfonylchloride (dansyl chloride) is increasingly used as a reagent in different fields of biological research. The application of dansyl chloride in biochemical analysis and certain aspects of this topic, e.g., end-group determinations of peptides, amino acid separations, estimation and identification of amines, and others, have been reviewed by several authors (Seiler, 1970; Seiler and Wiechmann, 1970; Gray, 1972; Niederwieser, 1972; Rosmus and Deyl, 1971; Neuhoff, 1973; Osborne, 1974). It is the sulfonyl group of the dansyl chloride molecule that can react with $-NH_2$ or $-OH$ groups, and it is the conjugated ring system with the addition of the 1-dimethylamine group that makes the reagent highly fluorescent and therefore highly suitable for detecting minute amounts of peptides, imidazoles, phenolic compounds, sulfhydryl groups, amines and amino acids, and cyclic AMP (Osborne and Neuhoff, 1976, unpublished data). A number of derivatives of dansyl chloride, e.g., bansyl chloride (Seiler and Knödgen, 1974) or mansyl chloride (Osborne *et al.*, 1976), can also be used to analyze substances in the same way as dansyl chloride, since preliminary experiments have shown them to have the same fundamental characteristics. However, the chromatographic mobilities of the various mansyl amino acids are different from those of the dansyl derivatives, so that different chromatographic systems have to be used (see Fig. 14). In this respect, the mansyl reagent can be of particular value in that it allows one to verify, chromatographically data that have been obtained through the use of dansyl chloride. It should, however, be pointed out here that mansyl derivatives are all blue, and this, in conjunction with the lack of data concerning chromatographic mobilities of various derivatives, identification, and other properties, makes the reagent not an alternative to dansyl chloride (the derivatives of which vary among yellow, green, orange, and blue, and on which vast amounts of information are available concerning identification, chromatographic mobilities, and other properties), but rather a useful addition.

Though dansyl chloride has found an application in biochemistry to compare with Fischer's naphthalene sulfonyl chloride or Sanger's 2,4-dinitrofluorobenzene, it nevertheless has drawbacks. A major drawback lies essentially in the problem of quantification and in particular renders the quantitative microanalysis of dansyl derivatives difficult. It is now clear

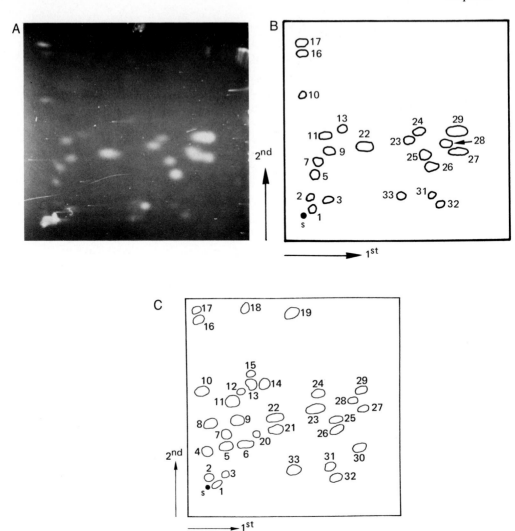

Fig. 14. Photographs of a 3 × 3 cm polyamide microchromatogram (A) and a corresponding map (B) to show the separation of various amino acids that were reacted with mansyl chloride and fractionated in two solvent systems. The chromatographic positions of some other mansyl derivatives in relation to those shown on the microchromatogram (A) are diagrammatically represented in (C). The directions of the chromatographic development using benzene–acetic acid (1st dimension) and ethanol–water–acetic acid (2nd dimension) are indicated by the arrows. (S) Starting point; (1) mansyl-OH; (2) mansyl-taurine; (3) mansyl-histidine; (4) mansyl-5-hydroxytryptophan; (5) mansyl-serotonin; (6) mansyl-tryptophan; (7) mansyl-aspartic-acid; (8) mansyl-tyrosine; (9) mansyl-glutamic acid; (10) mansyl-cystine; (11) mansylserine; (12) mansyl-threonine; (13) mansyl-glutamine; (14) mansyl-cystine; (15) mansyl-GABA; (16) mansyl-ornithine; (17) mansyl-arginine; (18) mansyl-lysine; (19) mansyl-histidine; (20) mansyl-ornithine; (21) mansyl-5-hydroxyindole acetic acid; (22) mansyl-glycine; (23) mansyl-alanine; (24) mansyl-NH$_2$; (25) mansyl-methionine; (26) mansyl-phenylalanine; (27) mansyl-leucine and mansyl-isoleucine; (28) mansyl-valine; (29) mansyl-proline; (30) mansyl-putrescine; (31) mansyl-histidine; (32) mansyl-tyrosine; (33) mansyl-lysine. Note only single individual mansyl derivatives of serotonin, histidine, tyrosine, and ornithine were characterized.

(Seiler, 1970; Neadle and Pollitt, 1965) that in the reaction of substances, dansyl chloride reacts easily with water to form a blue compound. Furthermore, the carboxyl groups of already-formed dansyl amino acids can react with dansyl chloride to give dansyl amines and, consequently, the degradation of the dansyl amino acids (see Fig. 15). Excess of dansyl chloride is, however, utterly essential in the reaction mixture, so that the recovery of the products in reality depends on the concentrations of dansyl chloride and the derivative to be dansylated, the nature of the reactive groups, pH, time of reaction, and also temperature. These conditions are difficult to standardize, particularly on the microscale. If consistent results are to be obtained for any given tissue to be analyzed, it is necessary that optimal reaction conditions be ascertained by initial trial-and-error experiments and then maintained for the series of experiments.

3.1.4a. General Procedures. A summary of the method used (after Briel *et al.*, 1972) to obtain microchromatograms for the analysis of nerve tissue is shown below:

1 mg tissue, 20 μl 0.05 M NaHCO$_3$ (pH 10)
Homogenized thoroughly
$\downarrow$

Centrifuged for 15 min at 20,000 g
$\downarrow$

Supernatant transferred to a clean tube
An equal volume of acetone added
$\downarrow$

Dried *in vacuo*
$\downarrow$

Redissolved in 4 μl acetone-acetic acid (3 : 2 vol./vol.)
$\downarrow$

0.4 μl applied to a single 3 × 3 cm microchromatogram
    The microchromatogram is then developed in an ascending way in two or three solvent systems and viewed under UV light. Since each dansylated substance is radioactive and fluoresces, the content can be quantified or autoradiograms prepared.

The principle for analyzing isolated cells is similar to that used for the analysis of small amounts of nervous tissue. However, instead of using small reaction tubes for homogenization, centrifugation, and other steps, the entire procedure is carried out in 10-μl Drummond capillaries with an internal diameter of 560 μm (see Fig. 16). This complicates the analysis and necessitates more careful handling in all the steps involved.

The problem in analyzing isolated neurons lies essentially in keeping the volume of solution as small as is practically possible, since the whole of the contents will finally be applied to a single microchromatogram. The amount of 0.05 M NaHCO$_3$ used is of particular importance when one bears in mind that dansyl chloride reacts with water and also that the salt content has to be kept low to achieve a good separation of substances by chromatography.

The procedure employed is as follows: A 10-μl Drummond capillary (Microcap) is carefully heat-sealed at one end and all but 0.05 cm of the tubing removed. The minute tube, with a volume of about 1 μl and measuring

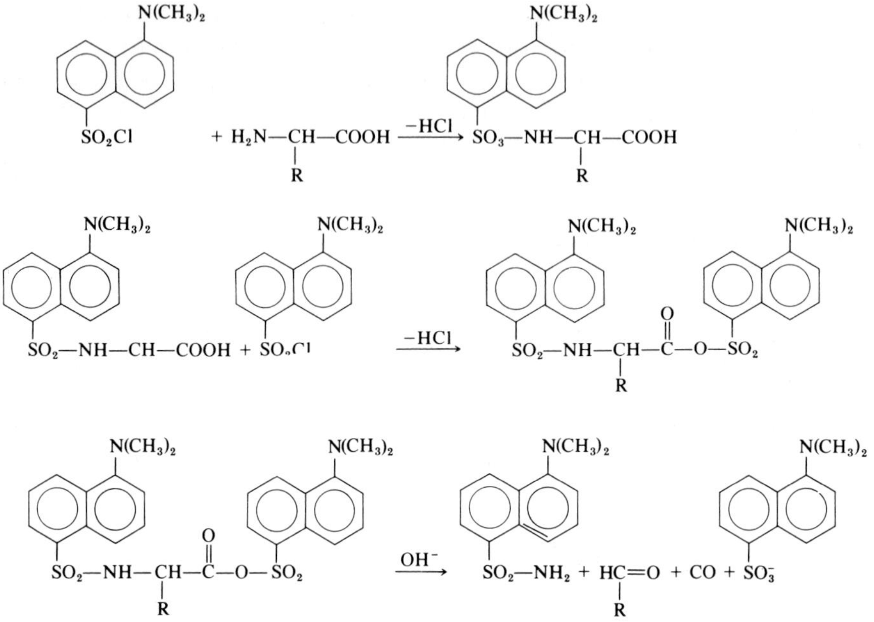

$$N(CH_3)_2 \quad SO_2Cl \quad + \quad H_2N-CH-COOH \quad \xrightarrow{-HCl} \quad N(CH_3)_2 \quad SO_3-NH-CH-COOH$$
$$\text{R} \qquad\qquad\qquad\qquad \text{R}$$

$$N(CH_3)_2 \quad N(CH_3)_2 \quad SO_2-NH-CH-COOH + SO_2Cl \quad \xrightarrow{-HCl} \quad N(CH_3)_2 \quad N(CH_3)_2$$

$$SO_2-NH-CH-\overset{O}{\overset{\|}{C}}-O-SO_2$$
$$\text{R}$$

$$N(CH_3)_2 \quad N(CH_3)_2 \quad SO_2-NH-CH-\overset{O}{\overset{\|}{C}}-O-SO_2 \quad \xrightarrow{OH^-} \quad N(CH_3)_2 \quad N(CH_3)_2 \quad SO_2-NH_2 + HC=O + CO + SO_3^-$$
$$\text{R} \qquad\qquad\qquad\qquad\qquad\qquad \text{R}$$

Fig. 15. The dansyl reaction according to Seiler (1970).

0.5 cm in length with a diameter of 560 μm, is then filled with 0.05 M NaHCO₃ (pH 10) using a pipette capillary and kept upright in a piece of plasticine. The cells to be analyzed are placed on the meniscus of the "buffer" with a pipette capillary, and a piece of 10-μl Drummond glass tubing is attached to the open end with polythene tubing. Any attempt to immerse the cells in the buffer must be avoided, since this often results in the loss of either cells or buffer and so produces inaccuracies. The capillary is centri-

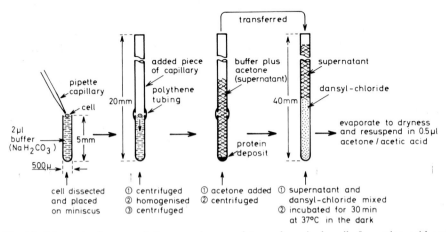

Fig. 16. Schematic diagram of the procedure used to analyze single cells for amino acid and amine content using the microdansyl procedure.

fuged, and 2 μl acetone is added to precipitate the proteins. To increase the precipitating effect, the capillary and contents are placed in a freezer ($-20°C$) for 30 min. The capillary is centrifuged again, and the supernatant is carefully transferred to a 10-μl Microcap, heat-sealed at one end. This transference is critical and depends solely on the kind of pipette capillary used and the way in which it is manipulated (see Osborne, 1974). A specific amount of dansylchloride (determined by trial experiments) is added to the 10-μl Microcap, and the contents are incubated in the dark at 37°C for 30 min. After the incubation period, the Microcap contents are vacuum-dried, and 0.5 μl acetone–acetic acid (3 : 2, vol./vol.) is used not only to wash down the walls of the Microcap, but also to dissolve the dansylated contents. The whole of the solution (0.5 μl) is then applied to the corner of a single 3 × 3 cm polyamide layer.

The reasons for using $NaHCO_3$ to extract amines and amino acids in nervous tissues and the procedure for dansylating are discussed elsewhere (Osborne, 1974; Neuhoff, 1973). It should, however, be pointed out here that one must always check the concentration of radioactive dansyl chloride carefully and analyze the purity of the particular batch. Following dansylation, the dansylated products are quantitatively applied to the corner of a 3 × 3 cm polyamide layer (Schleicher and Schüll F1700, Mikropolyamid), taking great care not to damage the polyamide sheets, and then chromatographed using different solvent systems (for details, see Osborne, 1974; Neuhoff, 1973). To separate the important dansyl amino acids on polyamide layers, water–formic acid is usually used for the first dimension (100 : 3, vol./vol.) and benzene–acetic acid (9 : 1, vol./vol.) in the second dimension (see Fig. 17). A third chromatography in the second dimension (ethyl acetate–methanol–acetic acid, 20 : 1 : 1 by vol.) is, however, necessary to separate certain groups of substances further (see Fig. 17). The chromatographic positions of amino acids and of some other important amines and precursors using the first two chromatographic systems are shown in Fig. 18.

3.1.4b. Quantification of Dansyl Products. As mentioned earlier, the complexity of the reactions between dansyl chloride and other substances in tissue extracts makes it very difficult to determine the absolute content of individual substances in an extract. It is, however, possible if the nature of all substances that react with dansyl chloride is known. This can be achieved when analyzing the tissue extracts simply by selectively extracting certain substances before dansylating. However, this greatly decreases the sensitivity of the method, and such a decrease is fundamentally opposed to the concept of microprocedure. From experience, it has been found useful to express the quantity of each substance as the percentage of the total dansylated substance removed from a chromatogram. This is particularly expedient when employing [^{14}C]dansyl chloride (see, for example, Neuhoff, 1973; Osborne, 1973). Nevertheless, this way of expressing results is unsatisfactory for detecting changes in amino acids or amines, since a change in any substance of a mixture will obviously affect the apparent concentration of the other substances, and of course it gives no information on the content.

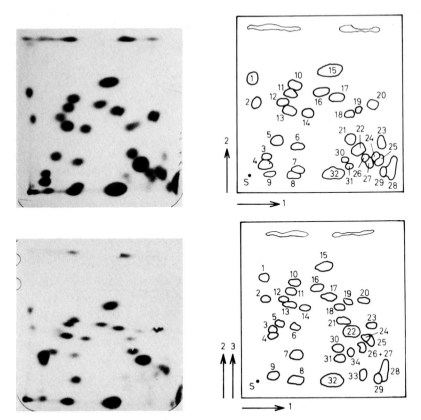

Fig. 17. Autoradiograms of 3 × 3 cm microchromatograms and corresponding maps to show
the positions of various amines and amino acids that were reacted with [^{14}C]dansyl/chloride and
separated in two (A) and three (B) solvent systems. The directions of chromatography are
indicated by the arrows, and the solvent systems used were water–formic acid (10:3 by vol.) in
the 1st direction, benzene–acetic acid (9:1 by vol.) in the 2nd direction, and ethyl
acetate–methanol–acetic acid (20:1:1 by vol.) in the 3rd direction. (S) Starting point; (1) dansyl-
bis-serotonin; (2) dansyl-*bis*-tyrosine; (3) dansyl-tryptophan; (4) dansyl-*N*-serotonin; (5) dansyl-
bis-lysine; (6) dansyl-*bis*-ornithine; (7) dansyl-*N*-tyrosine; (8) dansyl-taurine; (9) dansyl-*bis*-cystine;
(10) dansyl-isoleucine; (11) dansyl-leucine; (12) dansyl-*bis*-histidine; (13) dansyl-phenylalanine;
(14) dansyl-methionine; (15) dansyl-proline; (16) dansyl-valine; (17) dansyl-GABA; (18) dansyl-
alanine; (19) dansyl-NH$_2$; (20) dansyl-ethanolamine; (21) dansyl-glycine; (22) dansyl-hydroxy-
proline; (23) not identified; (24) dansyl-glutamine; (25) dansyl-asparagine; (26) dansyl-threonine;
(27) dansyl-serine; (28) dansyl-arginine, dansyl-α-amino histidine, and dansyl-α-lysine; (29)
dansyl-cystine; (30) dansyl-glutamic acid; (31) dansyl-aspartic acid; (32) dansyl-OH; (33, 34) not
identified.

It is generally inadvisable to use internal standards because it is difficult to
include minute but exact amounts of exogenous substances with small
amounts of material to be dansylated. Also, the inclusion of exogenous
substances alters the optimal dansylation of all substances dansylated. However, where the content of substances in a series of tissue extracts is not
significantly different, then internal standards (e.g., norvaline or sarcosine)

can be used successfully. In certain circumstances, where one substance in an extract may be considered altered, use of external standards is practicable. In this case, different amounts of standard amino acid mixtures (0.01–1.0 M) are dansylated and chromatographically separated. Standard curves can then be made of the relationship between the concentration of the substance and the radioactivity, and these can be referred to, to determine the concentration of substances in the biological extract.

Another method for quantifying dansyl substances is by double analysis. In this method, a known amount of radioactive amino acids is added to the sample before dansylating with unlabeled dansyl chloride. The fluorescence of the individual substances can then be measured either by direct scanning of the chromatogram or by eluting the spots and then measuring by fluorometry. The concentrations of the individual substances can then be calculated by taking into account the fluorescence and the amount of radioactivity associated with the standards. When radioactivity is first incorporated into tissue amines and amino acids (e.g., by incubating brain slices with [^{14}C]glucose), quantification can also be obtained by double labeling. Here, the tissue extracts can be reacted with [^{3}H]dansyl chloride and the ^{3}H/^{14}C ratio for the individual substance analyzed and corrected to produce a reliable method for quantification. Alternatively, it is possible to add known amounts of ^{14}C-labeled amino acids to a mixture, dansylate with [^{3}H]dansyl chloride, and quantitatively analyze the ^{3}H/^{14}C ratio of the individual substances (Casola and di Matteo, 1972). Such double-isotope-derivative

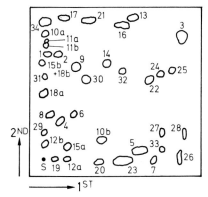

Fig. 18. Positions of some dansylated substances after chromatography on 3 × 3 cm polyamide layers using two solvent systems as described for Fig. 17. (S) starting point; (1, 2) normetanephrine; (3) adenosine; (4) 4-hydroxy-3-methoxymandelic acid; (5) 3',5'-cAMP; (6) *p*-chlorophenylalanine; (7) 5-AMP; (8) 3,4-dihydroxymandelic acid; (9) putrescine; (10a,b) octopamine; (11a,b) spermidine; (12a,b) 5-hydroxytryptophan; (13) acetylhistamine; (14) norvaline; (15a,b) 5-hydroxytryptamine; (16) histidinol; (17) histamine; (18a,b) hydroxymandelic acid; (19) ATP; (20) ADP; (21) adenine; (22) dansyl-NH$_2$; (23) dansyl-OH; (24) ethanolamine; (25) choline; (26) galactosamine; (27) α-amino-β-hydroxybutyric acid; (28) α-methyl histidine; (29) homocarnosine; (30) bufotenin; (31) *N*-acetyl-serotonin; (32) cytosine; (33) homoserine; (34) dihydroxyphenylacetic acid, tyramine, dopamine, adrenaline, metanephrine, 3,4-dihydroxymandelic acid, 6-hydroxydopamine, *p*-hydroxyphenylacetic acid, methoxytyramine.

procedures have recently been described by Snodgrass and Iversen (1973), Brown and Perman (1973), and Joseph and Halliday (1975). Theoretically, it is also possible to use a triple-labeling procedure with dansyl chloride by labeling the dansyl chloride molecule with ^{35}S. This would allow the quantitative determination of the turnover (^{3}H) and endogenous content (^{14}C) of a substance in microquantities of tissue.

Two further general techniques can be employed for the quantitative analysis of dansyl compounds. The first is mass spectrometry, which has been found excellent and enables one to analyze or confirm the identity of the dansyl derivative (see Boulton and Majer, 1972; Seiler and Knödgen, 1973; Leonard and Osborne, 1974). The other general technique is fluorometry, in which the chromatogram is scanned directly or the individual spots are eluted and evaluated by fluorometry (see Boulton, 1968; Seiler and Wiechmann, 1970; Pataki and Wang, 1968). It is worth noting that recently an automatic fluorescence scanning procedure has been developed in this laboratory that allows all the dansyl substances on a microchromatogram to be directly and quantitatively analyzed (Kronberg *et al.*, 1978).

3.1.4c. Application of the Microdansyl Procedure to Analyze Isolated Neurons. The procedure described has been used successfully to obtain good chromatograms from single *Aplysia*, *Planorbis*, and *Helix* neurons. Figure 19 shows two neuron types that were analyzed and demonstrates that one neuron type contains serotonin and the other does not. In a study concerned with analyses of six different giant neurons in the subesophageal ganglionic mass of *Helix pomatia* (see Fig. 20), Osborne *et al.* (1973) showed that the amino acids in each of the neuron types were similar. The most predominant amino acids were glutamic acid, glycine, alanine, and aspartic acid (see Table 7). Most interesting was the fact that two of the cell types analyzed (A and B) contained serotonin and less phenylalanine but more tryptophan than the other neurons. A number of other isolated neurons and defined nervous tissues have been analyzed by means of the microdansyl procedure, not only for amino acid and amine content but also for their metabolism, and this has been reviewed elsewhere (Osborne, 1973, 1974).

3.1.5. Fluorometry of Amines

Quantitative fluorometric procedures for the analysis of serotonin and dopamine at the cellular level have been developed by McCaman *et al.* (1973). The principle of the procedures involves extraction of the amines from the tissues with a special liquid cation-exchange system so that the amines free of their precursors are recovered and measured by a micromodification of the methods of Maickel and Miller (1966) for serotonin and of Shellenberger and Gordon (1971) for dopamine. The sensitivity limits of the methods are 2 pmol for serotonin and 4 pmol for dopamine.

The applicability of the methods described above has been demonstrated by analysis of the serotonin content in individual Retzius cells of the leech *Hirudo medicinalis*, in which the content was shown to be 2.5 pmol per cell.

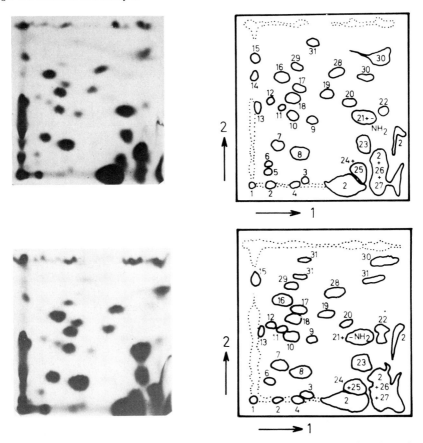

Fig. 19. Autoradiograms of dansylated substances in cell A and cell B present in the subesophageal ganglionic mass of *Helix pomatia*. The position of each neuron is indicated in Fig. 20. The content of individual amines and amino acids can be seen by comparing the numbers on the corresponding map with those in Table 7. Dansyl peptides are indicated by the dotted outlines on the maps. The direction of the chromatography is shown by the arrows and is as described for Fig. 17. From Osborne *et al.* (1973).

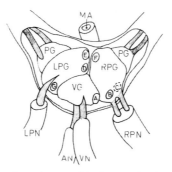

Fig. 20. Diagram of the subesophageal mass of *Helix pomatia* (dorsal view) to show the position of six identifiable giant cells analyzed for their amine and amino acid contents (see Table 7 for the actual analysis).

Table 7. Percentage Composition of Dansylated Substances in Six Identifiable Giant Neurons in the Subesophageal Ganglionic Mass of Helix pomatia[a]

Spot No.	Substance	Cell					
		A	B	C	D	E	F
1	Starting point	—	—	—	—	—	—
2	Dansyl-OH	—	—	—	—	—	—
3	Dansyl-N-tyrosine	0.60 ± 0.06	0.57 ± 0.04	0.65 ± 0.03	0.79 ± 0.06	0.89 ± 0.06	0.58 ± 0.04
4	Dansyl-taurine	0.24 ± 0.03	0.34 ± 0.06	0.26 ± 0.02	0.30 ± 0.03	0.38 ± 0.06	0.38 ± 0.03
5	Dansyl-N-serotonin	0.46 ± 0.04	0.48 ± 0.07	Abs.	Abs.	Abs.	Abs.
6	Dansyl-tryptophan	1.29 ± 0.19	0.98 ± 0.17	0.86 ± 0.14	0.88 ± 0.15	0.88 ± 0.09	0.81 ± 0.11
7	Dansyl-bis-lysine	2.65 ± 0.32	3.02 ± 0.42	2.50 ± 0.45	2.61 ± 0.49	2.98 ± 0.55	2.51 ± 0.61
8	Dansyl-bis-ornithine	7.20 ± 0.66	10.58 ± 0.71	6.47 ± 0.73	6.57 ± 0.59	8.46 ± 0.61	10.12 ± 0.63
9	Dansyl-methionine	0.50 ± 0.04	1.06 ± 0.09	0.76 ± 0.05	0.81 ± 0.06	0.70 ± 0.06	1.44 ± 0.10
10	Dansyl-phenylalanine	3.10 ± 0.36	3.88 ± 0.46	5.59 ± 0.39	5.03 ± 0.41	5.82 ± 0.31	4.98 ± 0.32
11	Dansyl-bis-histidine	0.37 ± 0.04	0.41 ± 0.06	0.33 ± 0.07	0.36 ± 0.09	0.43 ± 0.05	0.33 ± 0.05
12	Dansyl-unknown (U₁)	0.69 ± 0.05	0.60 ± 0.04	0.76 ± 0.09	0.64 ± 0.11	0.85 ± 0.12	0.55 ± 0.05
13	Dansyl-bis-tyrosine	1.93 ± 0.11	2.40 ± 0.21	2.40 ± 0.17	2.27 ± 0.21	1.79 ± 0.09	2.13 ± 0.17
14	Dansyl-bis-serotonin	0.42 ± 0.05	0.47 ± 0.05	Abs.	Abs.	Abs.	Abs.

15	Dansyl-5-hydroxyindole	1.13 ± 0.09	1.83 ± 0.11	1.59 ± 0.14	1.90 ± 0.13	1.66 ± 0.17	2.27 ± 0.11
16	Dansyl-unknown (U_2)	8.90 ± 1.11	8.47 ± 1.39	7.12 ± 0.91	7.50 ± 1.12	8.87 ± 1.23	8.10 ± 0.99
17	Dansyl-isoleucine	1.75 ± 0.13	2.19 ± 0.12	2.74 ± 0.09	2.39 ± 0.09	2.20 ± 0.07	1.66 ± 0.09
18	Dansyl-leucine	3.91 ± 0.31	6.13 ± 0.29	4.98 ± 0.31	4.22 ± 0.33	3.61 ± 0.36	6.59 ± 0.32
19	Dansyl-valine	2.81 ± 0.28	2.23 ± 0.22	2.16 ± 0.36	3.01 ± 0.28	2.85 ± 0.22	2.38 ± 0.27
20	Dansyl-GABA	0.31 ± 0.05	0.29 ± 0.06	0.37 ± 0.05	0.30 ± 0.04	0.34 ± 0.07	0.27 ± 0.03
21	Dansyl-alanine	19.57 ± 1.43	18.66 ± 1.72	17.74 ± 1.33	16.88 ± 1.27	17.44 ± 1.81	16.32 ± 1.39
22	Dansyl-ethanolamine	2.47 ± 0.24	1.11 ± 0.27	1.83 ± 0.28	1.91 ± 0.31	1.62 ± 0.22	1.41 ± 0.21
23	Dansyl-glycine	10.49 ± 1.08	9.10 ± 1.12	10.95 ± 1.31	10.19 ± 0.98	10.34 ± 0.97	8.83 ± 1.21
24	Dansyl-aspartic acid	2.85 ± 0.41	2.33 ± 0.46	2.46 ± 0.51	2.66 ± 0.38	2.90 ± 0.39	2.10 ± 0.41
25	Dansyl-glutamic acid	7.90 ± 1.01	6.43 ± 1.06	9.13 ± 0.07	9.79 ± 1.1	9.20 ± 0.91	7.21 ± 0.81
26	Dansyl-glutamine, dansyl-serine, dansyl-asparagine, and dansyl-threonine	5.41 ± 0.92	5.25 ± 1.10	5.95 ± 0.98	6.60 ± 0.99	5.42 ± 1.02	6.80 ± 1.11
27	Dansyl-arginine, dansyl-ε-lysine, dansyl-α-aminohistidine, and dansyl-cystine	9.12 ± 1.01	7.58 ± 1.12	7.93 ± 1.21	7.80 ± 0.91	6.17 ± 1.01	6.80 ± 1.04
28	Dansyl-proline	3.73 ± 0.34	3.99 ± 0.42	4.57 ± 0.61	4.61 ± 0.59	4.14 ± 0.72	3.65 ± 0.61
29	Dansyl-unknown (U_3)	0.61 ± 0.05	0.32 ± 0.05	0.65 ± 0.07	0.79 ± 0.04	0.89 ± 0.06	0.58 ± 0.05
30	Excess [14C]dansyl chloride	—	—	—	—	—	—
31	Impurities of [14C]dansyl chloride	—	—	—	—	—	—

[a] From Osborne *et al.* (1973). The position of each neuron is shown in Fig. 20. Data are means for 6 experiments ± S.E.M.

This value agrees with that found by Rude *et al.* (1969), who analyzed groups of Retzius cells. An analysis of various nervous components from the molluscs *Aplysia* and *Tritonia* was also made, though none of the cells analyzed had measurable levels of dopamine (see McCaman *et al.*, 1973). In a recent study, Osborne and Pentreath (1976) used the method described by McCaman *et al.* (1973) to analyze the serotonin content in the GSC of *Helix* and showed the cell to contain 700 ± 50 pg serotonin. A 50% drop in the serotonin content of the GSCs was recorded on treating the animals with 5,7-dihydroxytryptamine.

Many other fluorometric procedures exist that are sensitive enough to analyze transmitter molecules at the cellular level (see, for example, Schlumpf *et al.*, 1974). Perhaps mention should be made here of the use of fluorescamine, which is known to form fluorescent products with amino acids, amines, and peptides, allowing their fluorometric detection in the picomole range (Udenfriend *et al.*, 1972; Stein *et al.*, 1973).

The enzymatic recycling technique developed by Lowry (see Lowry and Passoneau, 1972) in combination with fluorometry is another very elegant procedure for analyzing transmitter molecules in single neurons. Otsuka *et al.* (1971) were able to measure as little as 10^{-14} M GABA, thus making possible the isolation of single nerve cells from the cat CNS and their analysis. These workers showed that GABA is concentrated within single axon terminals, probably belonging to Purkinje neurons, synapsing with dorsal Deiters's cells. The results of this investigation are consistent with the idea that GABA is specifically concentrated in inhibitory neurons of the mammalian CNS. The same technique has also been used by Otsuka *et al.* (1973) to measure adrenaline, noradrenaline, dopamine, and DOPA in discrete minute parts of the brain.

3.1.6. Transmitter Enzymes and Enzymes in General

When measuring enzyme activities, the handling of the nervous-tissue component prior to the analysis is a very critical stage in the procedure, since the accuracy of the results obtained depends on the localization of the enzyme in the cell. The nature of the enzyme and the binding of the enzyme molecule to specific cellular units are of prime importance. The studies of Giacobini (1959, 1969) on acetylcholinesterase (AChE), cholineacetylase (ChAc), and monoamine oxidase (MAO) have demonstrated the necessity for careful attention to these points.

A number of methods exist for analyzing enzyme activities at the cellular level. The elegant technique of the Cartesian diver respirometer of Linderström-Lang (1937) and Holter (1961) has been brought to the single-cell level by the ampulla diver of Zeuthen (1961), resulting in a number of studies at the cellular level (e.g., Giacobini, 1968; Hydèn and Pigon, 1960). The procedure does have disadvantages. It is rather an involved and time-consuming technique requiring repeated manual adjustments by the experimeter. Furthermore, only a limited number of samples can be examined in one single experiment, and the more sensitive type of instrument, which may have a self-recording magnetic diver to automatically register both temperature and volume variations, can be expensive.

The two methods now particularly used for enzyme analysis are fluorometry, in which the technique of cycling and recycling increases the sensitivity (for details, see Lowry and Passoneau, 1972), and isotopic procedures. The latter method has certain unique advantages in quantitative microstudies: (1) radioactive methods are specific, and their high sensitivity can be increased by using substrates of higher specific activities; (2) any drugs or other substances that may be present in the incubation medium often have no effect on the assay; (3) rapid and simultaneous analysis of a relatively large number of samples (50–200) can be made; (4) special instruments are mostly not required, except for the conventional scintillation counter; and (5) isotopically labeled substrates may be used to develop new and specific methods for analyzing almost all enzyme activities. Microprocedures in which isotopes are used usually amount to measurements of the labeled product, labeled gaseous release, or labeled products of a secondary chemical reaction. For each of these procedures, special substrates have to be employed. The practicality of these isotope methods depends generally on their sensitivity and efficiency. Both these criteria are related to adequate separation techniques of the labeled reaction product from the labeled substrate. Furthermore, the procedure involved should allow quantitative recovery of the labeled product.

A number of authors have described in detail the procedural steps involved when isotopes are used to analyze transmitter enzymes at the cellular level (see, for example, McCaman, 1968, 1971; Nagatsu, 1973; Coyle, 1975; Snyder and Taylor, 1972). The levels of some transmitter enzyme activities in single molluscan neurons are shown in Table 8.

3.1.7. Acetylation of Transmitters

A very good method for analyzing small amounts of amines and metabolites is by acetylation. Laverty and Sharman (1965) originally described a method using acetic anhydride as an acetylating agent for the analysis of dopamine in nervous tissues. In the original method, the acetylated catecholamines were extracted into dichloromethane and then separated chromatographically to detect as little as 10 ng dopamine. With different solvent systems, not only dopamine, noradrenaline, and adrenaline, but also a number of metabolites such as normetanephrine, metanephrine, and methoxytyramine, can be separated from one another (Sharman, 1971).

The acetylation technique has since been developed on the microscale (Rentzhog, 1970). This method is very elegant in that an internal standard of [^{14}C]amine is added to the extracted amines from a tissue sample and the complete contents acetylated with acetic-[^{3}H]anhydride. The labeled acetylated products are then chromatographically separated and the amount of catecholamines in the tissue sample calculated from the ^{3}H/^{14}C ratios

Table 8. Enzyme Activity in Single Molluscan Neurons

Aplysia cell	Enzyme	Activity *in vitro* (pmol/cell per hr)
McCaman and Dewhurst (1970)		
R_1	Choline acetyltransferase	0
L_1	Choline acetyltransferase	0
R_2	Choline acetyltransferase	588 ± 44
LPGC	Choline acetyltransferase	507 ± 57
R_{14}	Choline acetyltransferase	0
R_{15}	Choline acetyltransferase	0
L_{10}	Choline acetyltransferase	322
L_{11}	Choline acetyltransferase	199
McCaman and Dewhurst (1971)		
R_1	Acetylcholinesterase	2.6
L_1	Acetylcholinesterase	3.7
L_7	Acetylcholinesterase	6.5
L_{11}	Acetylcholinesterase	6.5
R_{14}	Acetylcholinesterase	7.2
R_{15}	Acetylcholinesterase	7.7 ± 1.4
R_2	Acetylcholinesterase	12.8 ± 1.2
LPGC	Acetylcholinesterase	14.1
Dewhurst (1972)		
L_7	Catechol-*O*-methyltransferase	6.62
L_{10}	Catechol-*O*-methyltransferase	2.66
L_{11}	Catechol-*O*-methyltransferase	5.7 ± 1.1
R_{14}	Catechol-*O*-methyltransferase	4.01
R_{15}	Catechol-*O*-methyltransferase	4.90
R_2	Catechol-*O*-methyltransferase	34 ± 4.1
LPGC	Catechol-*O*-methyltransferase	34.0 ± 3.3
L_1	Choline phosphokinase	120
L_2	Choline phosphokinase	439 ± 188
L_3	Choline phosphokinase	1123
L_4	Choline phosphokinase	670
L_6	Choline phosphokinase	468
L_7	Choline phosphokinase	583 ± 127
LPGC	Choline phosphokinase	1863
R_1	Choline phosphokinase	238 ± 80
R_2	Choline phosphokinase	1344 ± 200
R_{14}	Choline phosphokinase	568
R_{15}	Choline phosphokinase	491

obtained. This double-isotope-dilution-derivative technique allows about 1 ng catecholamine to be detected, the limits of sensitivity being determined by the specific activity of the radioactive chemicals.

3.2. Microanalysis of Other Substances

A number of other sensitive procedures, in addition to those already described, are available for the study of various biochemical parameters at the cellular level. Some of these methods and detailed information on them and other microprocedures can be found in the articles by Hydèn (1960, 1972), Giacobini (1968, 1969), Lowry (1953, 1963), Lowry and Passoneau (1972), Zeuthen (1961), Edström (1964), Neuhoff (1973), and Osborne (1974). The work by Keleti and Lederer (1974) describes the methodology involved in a number of chemical microanalytical procedures (e.g., protein estimation), though the sensitivity of most of the procedures described is limited. A brief description of some of the biochemical microprocedures will now be presented to demonstrate their applicability.

3.2.1. Nucleic Acids: Content and Base Analysis

Though a number of microprocedures can be scaled down to allow measurement of nucleic acid content in the various components of the nervous system, the methods of Edström and his collaborators (Edström, 1956; Pigon and Edström, 1959; Edström and Kawiak, 1961) are best known. Essentially, the procedure is as follows: For the analysis of RNA content, ribonuclease is used to extract the nucleic acids. The extract is then collected on a specially prepared quartz glass, placed in an "oil chamber," treated with glycerol-containing buffer, and eventually photographed in UV light at 257 μm with a reference system of the different optical densities. The photographs, magnified if necessary, are then investigated by photometry, and the RNA content is ultimately determined by means of a device for integration of the absorption. This technique also works for DNA extracted with deoxyribonuclease (Edström and Kawiak, 1961), though the extraction is slightly different from that used for RNA determination. Edström (1956, 1964) described a procedure for the base analysis of RNA in individual neurons and axons. The original method utilizes an alkaline-treated fiber (made from cellulose, for example), impregnated with a strongly acidic, viscous buffer to which hydrolysates containing 500–1000 pg RNA are applied. After high-voltage electrophoresis, the absorbing zones on the fiber are photographed in a UV microscope at 257 μm, and the photographic plate is scanned in a double-beam automatic recording densitometer. The molar base proportions are determined by measuring each area under the density curve and are expressed as percentages of the sum of the base areas. This technique, with a simple modification (see Edström, 1964), also allows DNA base content and composition to be determined. The fibers normally employed for electrophoresis or microphoresis, as termed by Edström (1956), are made from cellulose or rayon silk, or, when radioactivity is used, from cellophane (Koenig and Brattgård, 1963).

3.2.2. Sodium, Potassium, Calcium, and Magnesium

Several different techniques are available for the determination of ions at the cellular level. Techniques such as activation analysis, X-ray fluorescence, or microprobe analysis necessitate the use of fairly expensive equipment and the time-consuming preparation of the samples. Ordinary emission flame photometry or atomic absorption spectrophotometry does not yet have a sensitivity sufficient for work with single cells. Cation-sensitive glass electrodes have been used to measure certain ions, though it is questionable whether their sensitivity is specific for ions alone. Probably the best approach for analyzing ions at the cellular level is through the use of integrating micro-flame photometers. For a discussion of the problems peculiar to micro- and ultramicro-flame photometric instruments, and for the application of this technique to the analysis of ions at the cellular level, the reader is referred to the article by Haljamäe and Waldman (1972). Since the first integrating ultramicro-flame photometer designed by Ramsay *et al.* (1951), several improved instruments have been described, e.g., by Müller (1958), Malnic *et al.* (1964), Carlsson *et al.* (1967), Haljamäe and Larsson (1968), and Katz (1968). All these instruments are used mainly for Na^+ and K^+ analysis. Because of the relatively low flame temperature, the emission of Na^+ and K^+, both of which have low excitation potentials, is sufficient for analytical work, allowing as little as 10^{-14} mol of ion to be measured.

The analysis of other cations of interest in biological systems, such as Ca^{2+} or Mg^{2+}, is not possible with the aforementioned ultramicro-flame photometric techniques due to the much higher excitation potentials of these bivalent cations. Since much higher flame temperatures are needed, the introduction of the sample onto the tip of a wire is not possible. The wire would melt. Two different approaches to the analysis of Ca^{2+} and Mg^{2+} have been outlined: an emission microphotometric instrument developed by Vurek and Bowman (1965) and then further improved by Vurek (1967) and an ultramicro-flame spectroscopic technique (Haljamäe and Waldman, 1972) in which a Beckmann DN spectroscope was adapted. This latter procedure enabled Haljamäe and Waldman (1972) to detect as little as 10^{-13} mol of chloride ions.

3.2.3. Phospholipids and Lipids

The normal procedure for analyzing phospholipids in nervous tissue can be divided into the following methodological steps: extraction, separation of lipids from nonlipid contaminants, chromatographic separation, identification, and quantification of each phospholipid. For the extraction procedure, chloroform–methanol (2 : 1, vol./vol.) is generally used (Folch *et al.*, 1957). For the microdetermination of phospholipids, the procedural steps are, in principle, the same. Kleinig and Lambert (1970) and Svetashev and Vaskovsky (1972) have described methods in which the thin-layer plates (silica gel) are simply reduced to 6 × 6 cm so that nonogram quantities of the individual phospholipids can be detected. Recently, even smaller plates (cover glasses 24 × 48 × 32 mm) covered with thin layers of silica gel either alone, containing some plaster of Paris, or impregnated with water glass have been used for the chromatographic separation of lipids and phospholipids, a technique that reveals minute amounts of the individual substances (Schiefer and Neuhoff, 1971; Althaus and Neuhoff, 1973; Hubmann, 1973). Normally, the individual lipids on microchromatograms can be detected with iodine vapor, rhodamine spray, phosphate sprays, or Dragendorf reagent (see Osborne, 1974). However, problems arise in the quantification of the small amounts of substances. One method used is that described by Schiefer and Neuhoff (1971), in which the individual phospholipids are scraped from the microchromatograms, eluted from the silica particles, and analyzed by fluorometry after reaction with definite amounts of cyclohexane, rhodamine 6G0 in chloroform, and acetic acid. This method has been used by Althaus *et al.* (1973) to determine the phospholipid content in isolated giant serotonin neurons from *Helix* after separating the phospholipids by one-dimensional chromatography (see Fig. 21). An alternative to elution of substances from microchromatograms and their fluorometric analysis is direct fluorometric determination on the chromatograms. Heyneman *et al.* (1972) used such a method to identify the phospholipids with the dye 1-aniline-8-naphthalene sulfonate and then scan them directly. This efficient and simple technique, named "*in situ* measurement," in which the substances can be identified on the microchromatograms by absorption photometry in remission allows nanogram levels of a substance to be analyzed (see also Hezel, 1973).

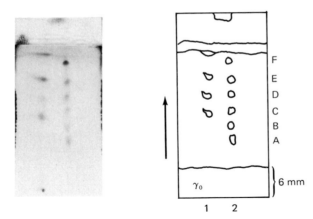

Fig. 21. Microchromatogram (A) and a map (B) to show the chromatographic extraction and fractionation of lipids from three nerve cell bodies (GSCs) of the snail *Helix pomatia* (1) and a fractionation of a control mixture of phospholipids (2). The thin-layer chromatogram measured 48 × 24 mm, and the cells were placed directly on the microchromatogram in position Y. Thereafter, the lipids were extracted from the cells by ascending chromatography in chloroform–methanol (2:1 by vol.). The chromatography terminated when the front had migrated a distance of 6 mm as indicated by the line across the map. After drying, the microplate was developed and the individual phospholipids were visualized with either iodine vapor or Hane's reagent. It is clear that the GSCs contain phosphatidylcholine (B), phosphoinositides (C), phosphatidylserine (D), and phosphatidylethanolamine (E). Spots A and F in the standard solutions belong to sphingomyelin and cardiolipin, respectively. From two different experiments, in which the individual phospholipids were removed from the microchromatograms and analyzed by spectrophotometry, it was deduced that a single GSC contains 0.68 μg phosphatidylcholine, 0.57 μg phosphatidylethanolamine, 0.07 μg phosphoinositides, and 0.12 μg phosphatidylserine. After Althaus *et al.* (1973).

3.2.4. Microelectrophoresis on Polyacrylamide: Proteins and RNA

Numerous papers have been written on polyacrylamide gel electrophoresis, and it is not my aim here even to attempt to quote the best examples.
However, these techniques are reviewed in Volume 2, Chapter 4. The theory
of polyacrylamide gel electrophoresis, whether on the normal or on the
microscale, is a separation of particles based on molecular size and charge
and is a simple procedure. A gel is formed by polymerization of acrylamide
in a tube or slab mold, and the gel tube or slab mold is then positioned
between two buffer chambers. The sample is placed on the gel and an
electrical field applied across them. Most of the components migrate as
individual bands, each with a characteristic electrophoretic migration rate
depending on its size and net charge. Fixation and staining or slicing and
assay of the gel subsequent to electrophoresis reveal characteristic band
positions. An original limitation of disk electrophoresis was the limited
number of systems available due to the fact that the mobilities of relatively
few ions were known. The limitation has been somewhat eased since Jovin
et al. (1970) compiled the characteristics of a larger number of disk systems.
The first adaptation of disk electrophoresis to the microscale was in 1964
when Pun and Lombrozo (1964) analyzed brain proteins. Since then, a
number of microelectrophoretic systems have been described to fractionate
minute amounts of proteins on cylindrical gel, Grossbach, 1965; Hydèn *et
al.*, 1966; Neuhoff, 1968; Rüchel *et al.*, 1973, 1974) or flat gels (Maurer and
Dati, 1972; Been and Rasch, 1972; Hazama and Uchimura, 1972). The
recent adaptation of microelectrophoresis in cylindrical gradient gels (Rüchel
et al., 1973, 1974) has proved of great value. The procedure is easy to carry
out, the separation of the different proteins is sharply defined, and the
molecular weights of the proteins can be determined. For this reason, the
method has been adapted for the analysis of proteins from single neurons
isolated from the central nervous system of the water snail *Planorbis corneus*
(Osborne and Rüchel, 1975). A scheme for the "single-step procedure" used
is shown in Fig. 22. A single giant neuron was placed directly on the meniscus
of the "sample solution" above the gradient gel. Once the neuron had
penetrated the "sample solution" due to gravity, the capillary, standing in a
small amount of buffer, was covered with a beaker to create a moist
atmosphere and put in an oven for 2 hr at 55°C in the presence of nitrogen.
Thereafter, any space at the top of the capillary due to a slight evaporation
of the sample solution was filled with a small amount of concentrated sucrose
solution containing a little bromophenol red as a marker dye, and then
subjected to electrophoresis as shown in Fig. 22. Electrophoresis was terminated after 30–45 min when the bromophenol red reached the lower end of
the gel. The gel was then removed from the capillary with a thin wire, stained
in Coomassie blue, and examined after destaining by densitometry. Figure
23 shows two microgels and electrophoretic patterns of the protein distribution in two types of giant neurons in *Planorbis*; it can be seen that (1)
reproducible patterns are produced by the procedure and (2) a variation

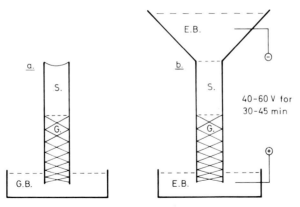

Fig. 22. Scheme of the separation procedure discussed in the text. (a) A polymerized gradient gel (G.) overlaid with the "sample solution" (S.) in a small container holding gel buffer (G.B.). The dissected cell is placed directly on the meniscus of the "sample solution" and, following incubation (2 hr at 55°C in the presence of nitrogen) and the addition of a small amount of sucrose solution containing bromophenol red, is subjected to electrophoresis as shown in (b.). Compositions: "sample solution" (S.), 20 mM Tris-sulfate buffer (pH 8.4) + 1% mercaptoethanol; electrophoresis buffer (E.B.), 50 mM Tris-glycine buffer (pH 8.4) + 0.1% SDS; gel buffer (G.B.), 350 mM Tris-sulfate buffer (pH 8.4). The procedure for casting the microgradient polyacrylamide gels containing Tris-sulfate buffer is described by Rüchel *et al.* (1973). From Osborne and Rüchel (1975).

occurs in the protein pattern between two different cell types. This method was also adapted to analyze only the water-soluble proteins of a cell by eliminating the use of sodium dodecyl sulfate (SDS) (see Osborne and Rüchel, 1975). Microdisk electrophoretic techniques—SDS–polyacrylamide gels—have also been used by a number of other groups (in particular, Wilson, 1971; Gainer, 1971, 1972, 1973) for the analysis of proteins in isolated molluscan neurons, though in these instances either pooled neurons were used or radioactivity was initially incorporated into the proteins.

Microelectrophoresis on gradient gels has many advantages, as discussed elsewhere (Neuhoff, 1973; Osborne, 1974; Rüchel *et al.*, 1973, 1974; Dames and Maurer, 1974), and has been modified for the fractionation of RNA species (Wolfrum *et al.*, 1974). A number of electrophoretic buffer systems were tested to obtain a good separation of the different RNA species in different tissue types (see Wolfrum *et al.*, 1974), and Fig. 24 demonstrates the RNA separation from a nervous-tissue sample using a discontinuous buffer system.

Several authors have also developed electrophoretic procedures for micro-isoelectric focusing (Dale and Latner, 1968; Wrigley, 1968; Riley and Coleman, 1968; Catsimpoolas, 1968; Grossbach, 1971; Quentin and Neuhoff, 1972; Gainer, 1973). Isoelectric focusing is basically a method whereby proteins are made to migrate in a pH gradient until they reach a pH at which their net charge is zero. In theory, this method is the best potential criterion for purity, since the pH range in the gels can be narrowed to increase the resolving power.

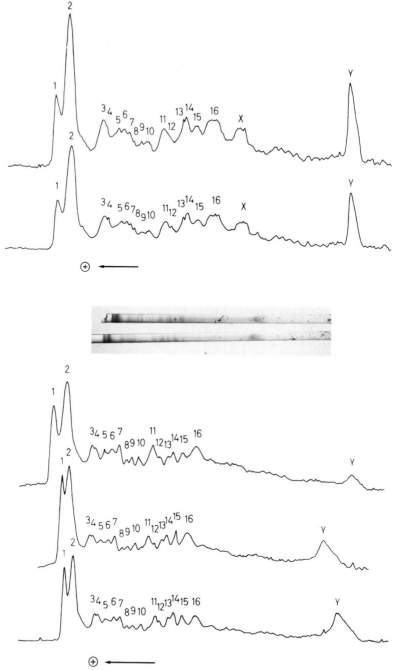

Fig. 23. Densitometric traces of micropolyacrylamide gradient gels showing the separation of total proteins from single neurons of *Planorbis corneus* using the conditions of electrophoresis as shown in Fig. 22 and staining with Coomassie blue. (*Top*) Two electropherograms of different gels from the same cell type. *Inset:* Photograph of two stained microgels that were cast in 2-μl capillaries. (*Bottom*) Three electropherograms of different gels from the same cell type (although a different cell type than that analyzed in A). A constant fractionation of protein patterns from a single cell type is achieved by the method used. Variations in protein compositions occur between the different cell types (e.g., protein band "X" occurs predominantly in the cell type shown in A).

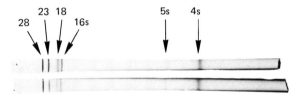

Fig. 24. Fractionation of RNA (a mixture extracted from *E. coli* and from rat brain) in gradient gels cast in 5-µl capillaries using a discontinuous buffer system. A number of RNA species are clearly fractionated by this procedure. These species are observed after staining with toluidine blue. Photograph by courtesy of D. Wolfrum.

The procedure of microelectrophoresis or microelectrofocusing, in which capillaries are used to cast the gels, can also be advantageously used for the analysis of enzyme activities (see, for example, Cremer *et al.*, 1972). Since the diameter of the microgels is so small, substrates diffuse into the gels almost instantaneously when the latter are placed in an incubation mixture. It is therefore possible to consider the microgel as having a "microcuvette" property and containing the enzyme in a volume of 0.003–0.015 µl, since a single protein band in a 5-µl gel measures between 20 and 100 µm. Because of this "microcuvette" property, it is possible to record the binding of a dye (e.g., tetrazolium mixture for analyzing glucose-6-phosphate-dehydrogenase)

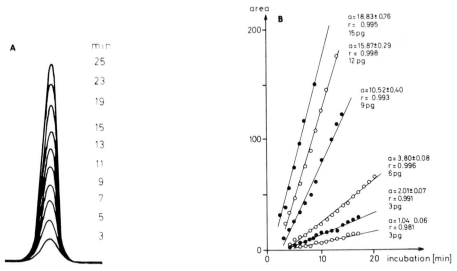

Fig. 25. (A) Microdensitometric tracings of glucose-6-phosphate dehydrogenase activity at different time intervals in the same microgel. After a liver extract was fractionated in a 5-µl capillary containing a 20% polyacrylamide gel, the gel was placed in a special chamber filled with a tetrazonium mixture and at intervals of 2 min scanned with a microdensitometer in the darkroom. The figure shows that the formazan peak increases linearly with the time of incubation. (B) Enzyme kinetics of glucose-6-phosphate dehydrogenase from pure yeast. Extracts of yeast corresponding to between 3 and 15 pg enzyme were fractionated on 20% polyacrylamide microgels and then incubated with a tetrazonium mixture as in (A). When the peak areas are plotted against the incubation time at intervals of 1 min, linear plots are achieved as seen in (B).

with the enzymatically active site in the gel at different time intervals, thus allowing the kinetics of the enzyme to be studied (see, for example, Fig. 25).

ACKNOWLEDGMENTS. The author wishes to acknowledge Prof. Dr. V. Neuhoff, in whose laboratory the author has worked for a number of years, Miss E. Priggemeier for her technical assistance, Mr. H. Ropte for producing the photographs and diagrams, and Mrs. J.S. Osborne for proofreading the chapter. Thanks are also due to Mrs. E. Zilken for secretarial help.

References

Abramson, R.P., McCaman, M.W., and MaCaman, R.E., 1974, Fentamole level analysis of biogenic amines and amino acids using functional group mass spectrometry, *Anal. Biochem.* **51**:482–499.

Althaus, H.H., and Neuhoff, V., 1973, One-dimensional microchromatography of phospholipids and neutral lipids on sodium silicate-impregnated silica gel layers, *Hoppe-Seyler's Z. Physiol. Chem.* **354**:1073–1076.

Althaus, H.H., Osborne, N.N., and Neuhoff, V., 1973, Mikrochromatographische Extraktion und Fraktionierung von Lipiden einzelner Nervenzellen von *Helix pomatia*, *Naturwissenschaften* **60**:553–554.

Barker, D.L., Herbert, E., Hildebrand, J.G., and Kravitz, E.A., 1972, Acetylcholine and lobster sensory neurones, *J. Physiol.* **226**:205–209.

Been, A.C., and Rasch, E.M., 1972, A vertical microsystem for discontinuous electrophoresis of insect tissue proteins using thin sheets of polyacrylamide gel, *J. Histochem. Cytochem.* **20**:368–384.

Bell, C.E., and Sommerville, A.R., 1966, A new fluorescence method for detection and possible quantitative assay for some catecholamine and tryptamine derivatives on paper, *Biochem. J.* **98**:1c–3c.

Bocharova, L.S., Kostenko, M.A., Veprintov, B.N., and Allachverov, B.L., 1975, Completely isolated molluscan neurons: An ultrastructural study, *Brain Res.* **101**:185–198.

Bondareff, W., and Hydén, H., 1969, Submicroscopic structure of single neurons isolated from rabbit lateral vestibular nucleus, *J. Ultrastruct. Res.* **26**:399–411.

Boulton, A.A., 1968, The automated analysis of absorbent and fluorescent substances separated on paper strips, in: *Methods in Biochemical Analysis*, Vol. 16 (D. Glick, ed.), pp. 327–363, Interscience, New York.

Boulton, A.A., and Majer, J.R., 1972, Detection and quantitative analysis of some noncatecholic primary aromatic amines, in: *Research Methods in Neurochemistry*, Vol. 1 (N. Marks and R. Rodnight, eds.), pp. 341–356, Plenum Press, New York.

Briel, G., Neuhoff, V., and Meier, M., 1972, Microanalysis of amino acids and their determination in biogenic material using dansyl chloride, *Hoppe-Seyler's Z. Physiol. Chem.* **253**:540–553.

Brown, J.P., and Perman, R.N., 1973, A highly sensitive method for amino acid analysis by a double isotope labelling technique using dansyl chloride, *Eur. J. Biochem.* **39**:69–75.

Brownstein, M.J., Saavedra, J.M., and Axelrod, J., 1973, Control of N-acetylserotonin by a β-adrenergic receptor, *Mol. Pharmacol.* **9**:605–611.

Brownstein, M.J., Saavedra, J.M., Axelrod, J., Zeman, G.H., and Carpenter, D.O., 1974, Coexistence of several putative neurotransmitters in single identified neurons of *Aplysia*, *Proc. Natl. Acad. Sci. U.S.A.* **71**:4662–4685.

Bullock, T.H., 1967, Signals and neuronal coding, in: *The Neurosciences: A Study Program* (G.C. Quarton, T. Melnechuk, and F.O. Schmitt, eds.), pp. 347–452, Rockefeller University Press, New York.

Bullock, T.H., and Horridge, G.A., 1965, *Structure and Function in the Nervous Systems of Invertebrates*, W.H. Freeman, San Francisco.

Carlsson, B., Giacobini, E., and Hovmark, S., 1967, An instrument for simultaneous determination of sodium and potassium in microsamples of biological material, *Acta Physiol. Scand.* **71:**379–390.

Casola, L., and di Matteo, G., 1972, Studies on the dansylation reaction by use of ^{14}C-dansyl chloride application to the analysis of free amino acids in rat optic nerve, *Anal. Biochem.* **38:**316–321.

Catsimpoolas, N., 1968, Micro-isolectric focusing in polyacrylamide gel columns, *Anal. Biochem.* **26:**480–482.

Chen, C.F., von, Baumgarten, R., and Tandeda, K., 1971, Pacemaker properties of completely isolated neurons in *Aplysia californica, Nature (London)* **233:**27–29.

Chen, C.F., von Baumgarten, R., and Harth, O., 1973, Metabolic aspects of the rythmogenisis in *Aplysia* pacemaker neurons, *Pfluegers Arch. Gesamte Physiol.* **345:**179–193.

Coggeshall, R.E., Kandel, E.R., Kupferman, I., and Waziri, R., 1966, A morphological and functional study on a cluster of identifiable neurosecretory cells in the abdominal ganglia of *Aplysia californica, J. Cell Biol.* **31:**363–368.

Cohen, M.J., and Jacklet, J.W., 1965, Neurons of insects: RNA changes during injury regeneration. *Science* **148:**1237–1239.

Cottrell, G.A., and Osborne, N.N., 1970, Subcelluar localisation of serotonin in an indentified serotonin-containing neuron, *Nature (London)* **225:**470–472.

Coyle, J.T., 1975, A practical introduction to radiometric enzymatic assays in psychopharmacology, in: *Handbook of Psychopharmacology*, Vol. 1 (L.L. Iversen, S.D. Iversen, and S.H. Snyder, eds.), pp. 71–100, Plenum Press, New York.

Coyle, J.T., and Henry, D., 1973, Catecholamines in the fetal and newborn rat brain, *J. Neurochem.* **21:**61–68.

Cremer, T., Dames, W., and Neuhoff, V., 1972, Microdisc electrophoresis and quantitative assay of glucose-6-phosphate dehydrogenase at the cellular level, *Hoppe-Seyler's Z. Physiol. Chem.* **353:**1317–1329.

Dale, G., and Latner, A., 1968, Isoelectric focusing in polyacrylamide gels, *Lancent* **1:**847–848.

Dames, W., and Maurer, H.R., 1974, Simultaneous preparation for electrophoresis of a large number of micropolyacrylamide gels with continuous concentration gradients, in: *Electrophoresis and Isoelectric Focusing in Polyamide Gel* (R.C. Allen and H.R. Maurer, eds.), pp. 221–231, de Gryter, Berlin and New York.

Deiters, O., 1865, *Untersuchungen über Gehirn und Rückenmark des Menschen und der Säugetiere*, von M. Schulze, Braunschweig (Brunswick), Germany.

Dewhurst, S.A., 1972, Choline phospokinase activities in ganglia and neurons of *Aplysia, J. Neurochem.* **19:**2217–2219.

Eccles, J.C., 1964, *The Physiology of Synapses*, Academic Press, New York.

Edström, J.E., 1956, Separation and determination of purines and pyridine nucleotides in picogram amounts, *Biochim. Biophys. Acta* **22:**378–388.

Edström, J.E., 1964, Microextraction and microelectrophoresis for determination and analysis of nucleic acids in isolated cellular units, in: *Methods in Cell Physiology*, Vol. I (D.M. Prescott, ed.), pp. 417–447, Academic Press, New York.

Edström, J.E., and Kawiak, J., 1961, Microchemical deoxyribonucleic acid determination in individual cells, *J. Biophys. Biochem. Cytol.* **9:**619–616.

Edström, J.E., and Neuhoff, V., 1973, Micro-electrophoresis for RNA and DNA base analysis, in: *Micromethods in Molecular Biology* (V. Neuhoff, ed.), pp. 215–256. Springer-Verlag, Berlin.

Folch, J., Lees, M., and Sloane-Stanley, G.H., 1957, A simple method for the isolation and purification of total lipids from animal tissues, *J. Biol. Chem.* **226:**497–509.

Frazier, W.T., Kandel, E.R., Kupferman, I., Waziri, R., and Coggeshall, R.E., 1957, Morphological and functional properties of identified neurons in the abdominal ganglion of *Aplysia californica, J. Neurophysiol.* **30:**1287–1351.

Gainer, H., 1971, Microdisc electrophoresis in sodium dodecyl sulphate: An application to the study of protein synthesis in individual, identified neurons, *Anal. Biochem.* **44:**589–605.

Gainer, H., 1972, Patterns of protein synthesis in individual, identified molluscan neurons, *Brain Res.* **39:**369–385.

Gainer, H., 1973, Isoelectric focusing of proteins at the 10^{-10} and 10^{-9}g level, *Anal. Biochem.* **51:**646–650.

Giacobini, E., 1956, Histochemical demonstration of AChE activity in isolated nerve cells, *Acta Physiol. Scand.* **36:**276–290.

Giacobini, E., 1959, The distribution and localisation of cholinesterase in nerve cells, *Acta Physiol. Scand. Suppl.* **156:**1–54.

Giacobini, E., 1964, in: *Morphological and Biochemical Correlates of Neural Activity* (M.M. Cohen and R.S. Snider, eds.), pp. 15–31, Harper and Row, New York.

Giacobini, E., 1968, Chemical studies on Individual Neurons: Part I, in: *Neurosciences Research*, Vol. 1 (S. Ehrenpreis and O.C. Solnitzky, eds.), pp. 1–66, Academic Press, New York.

Giacobini, E., 1969, Chemical studies on individual neurons: Part II, in: *Neurosciences Research*, Vol. 2 (S. Ehrenpreis and O.C. Solnitzky, eds.), pp. 112–198, Academic Press, New York.

Giacobini, E., 1970, Biochemistry of single neuronal models, in: *Biochemical Psychopharmacology*, Vol. 2 (E. Costa and E. Giacobini, eds.), pp. 9–64. Raven Press, New York.

Giacobini, E., 1975, The use of microchemical techniques for the identification of new transmitter molecules in neurons, *J. Neurosci. Res.* **1:**1–18.

Giller, E., and Schwartz, J.H., 1971, Choline acetyltransferase in identified neurons of the abdominal ganglion of *Aplysia californica, J. Neurophysiol.* **34:**108–115.

Gray, W.R., 1972, End-group analysis using dansyl chloride, *Methods Enzymol.* **25:**121–138.

Grossbach, U., 1965, Acrylamide gel electrophoresis in capillary columns, *Biochim. Biophys. Acta* **107:**180–182.

Grossbach, U., 1971, Chromosomen-Struktur und Zell-Funktion, *Mitt. Max-Planck-Ges.* **2:**93–108.

Haljamäe, H., and Larsson, S., 1968, An ultramicroflame photometer for K and Na analysis of single cells and nanoliter quantities of biological fluids, *Chem. Instrum.* **1:**131–144.

Haljamäe, H., and Waldman, A.A., 1972, Flame photometry at the cell level, in: *Techniques of Biochemical and Biophysical Morphology*, Vol. 1 (D. Glick and R.M. Rosenbaum, eds.), pp. 233–268. John Wiley, New York.

Hazama, H., and Uchimura, H., 1972, Separation of lactate dehydrogenase isoenzymes of nerve cells in the central nervous system by micro-disc electrophoresis on polyacrylamide gels, *Biochim. Biophys. Acta* **200:**414–417.

Heyneman, R.A., Bernard, D.M., and Vercauteren, R.E., 1972, Direct fluorometric microdetermination of phospholipids on thin-layer chromatograms, *J. Chromatogr.* **68:**285–288.

Hezel, U., 1973, Direkte quantitätive Photometrie und Dünnschicht-Chromatogrammen, *Angew. Chem.* **85:**334–342.

Hildebrand, J.G., Barker, D.L., Herbert, E., and Kravitz, E.A., 1971, Screening for neutrotransmitters: A rapid radiochemical procedure, *J. Neurobiol.* **2:**231–246.

Hillman, H., and Hydèn, H., 1965, Membrane potentials in isolated neurones *in vitro* from Deiters' nucleus of rabbit, *J. Physiol.* **177:**398–410.

Holter, H., 1961, The Cartesian diver, in: *General Cytochemical Methods*, Vol. 2 (J. Danieli ed.), pp. 93–128. Academic Press, New York.

Horridge, G.A., 1968, *Interneurons*, W.H. Freeman, San Francisco.

Hubmann, F.-H., 1973, Two-step, two-dimensional development thin-layer chromatography of lipids on a microscale, *J. Chromatogr.* **86:**197–199.

Hydèn, H., 1959, Quantitative assay of compounds in isolated, fresh nerve cells and glial cells from control and stimulated animals, *Nature (London)* **184:**433–435.

Hydèn, H., 1960, The neuron, in: *The Cell*, Vol. IX (J. Brachet and A. Mirsky, eds.), pp. 215–323. Academic Press, New York.

Hydèn, H., 1964, Biochemical and functional interplay between neuron and glia, in: *Recent Advances in Biological Psychiatry*, Vol. VI (J. Wortis, ed.), pp. 31–52, Plenum Press, New York.

Hydèn, H., 1972, Macromolecules and behavior, in: *Arthur Thomson Lectures* (G.B. Ansell and P.B. Bradley eds.), pp. 3–75, Macmillan, London.

Hydèn, H., and Pigon, A., 1960, A cytophysiological study of the functional relationship between oligodendroglial cells and nerve cells of Deiters' nucleus, *J. Neurochem.* **6:**57–72.

Hydèn, H., and Rönnbäch, L., 1975, Membrane-bound S-100 protein on nerve cells and its distribution, *Brain Res.* **100:**615–628.

Hydèn, H., Bjurstam, K., and McEwen, B., 1966, Protein at the cellular level by microdisc electrophoresis, *Anal. Biochem.* **17**:1–15.

Jacobowitz, D.M., 1974, Removal of discrete fresh regions of the rat brain, *Brain Res.* **80**:111–115.

Johnston, P.V., and Roots, B.I., 1972, *Nerve Membranes*, Vol. 36, Pergamon Press, Oxford and New York.

Joseph, M.H., and Halliday, J., 1975, A dansylation microassay for some amino acids in brain, *Anal. Biochem.* **64**:389–402.

Jovin, T.M., Dante, L.M., and Chrambach, A., 1970, Multiphorese buffer systems output, Publ. Nos. 196085–196091 and 203016, National Information Service, Springfield, Virginia.

Kandel, E.R., Frazier, W.T., Waziri, R., and Coggeshall, R.E., 1957, Direct and common connections among identified neurons in *Aplysia*, *J. Neurophysiol.* **30**:1352–1376.

Katz, G.M., 1968, Another look at ultramicro integrative flame photometry, *Anal. Biochem.* **26**:381–397.

Kerkut, G.A., 1969, Neurochemistry of invertebrates, in: *Handbook of Neurochemistry*, Vol. II (A. Lajthe, ed.), pp. 539–562, Plenum Press, New York.

Kerkut, G.A., Lambert, J.D.C., Gayton, R.J., Loker, J.E., and Walker, R.J., 1975, Mapping of nerve cells in the suboesophageal ganglia of *Helix espera*, *Comp. Biochem. Physiol.* **50A**:1–25.

Keleti, G., and Lederer, W.H., 1974, *Micromethods for the Biological Sciences*, Van Nostrand Reinhold, New York.

Kleinig, H., and Lempert, U., 1970, Phospholipid analysis on a micro scale, *J. Chromatogr.* **53**:595–597.

Koenig, E., and Brattgård, S.O., 1963, A quantitative micromethod for determination of specific radioactivity of ^{3}H-purines and ^{3}H-pyrimidines, *Anal. Biochem.* **6**:424–434.

Kostenko, M.A., 1972, The isolation of single nerve cells of the brain of the mollusc *Lymnaea stagnalis* for their further cultivation *in vitro*, *Tsitologia* **14**:1274–1279 (in Russian).

Kostenko, M.A., Geletyuk, V.I., and Veprintsev, B.N., 1974, Completely isolated neurons in the mollusc *Lymnaea stagnalis*: A new objective for nerve cell biology investigation, *Comp. Biochem. Physiol.* **49A**:89–100.

Kronberg H., Zimmer H.-G., and Neuhoff, V., 1978, Automatische Fluorimetrik von Mikro-Dünnschicht-Chromatogrammen, *Z. Anal. Chem.* **290**:2145–2150.

Lam, D.M.K., Wiesel, T.N., and Kaneko, A., 1974, Neurotransmitter synthesis in cephalopod retina, *Brain Res.* **82**:365–368.

Laverty, R., and Sharman, D.F., 1965, The estimation of small quantities of 3,4-dihydroxyphenylethylamine in tissues, *Br. J. Pharmacol.* **24**:538–548.

Leonard, B.E., and Osborne, N.N., 1974, The use of dansyl-chloride for the detection of amino acids and serotonin in nervous tissue, in: *Research Methods in Neurochemistry*, Vol. 3 (N. Marks and R. Rodnight, eds.), pp. 443–462, Plenum Press, New York.

Linderström-Lang, K., 1973, Principle of Cartesian diver applied to gasometric technique, *Nature (London)* **140**:108.

Lowry, O.H., 1952, The quantitative histochemistry of the brain, *Science* **116**:526.

Lowry, O.H., 1953, The quantitative histochemistry of the brain, *J. Histochem. Cytochem.* **1**:420–428.

Lowry, O.H., 1963, The chemical study of single neurons, *Harvey Lect.* **58**:1–19.

Lowry, O.H., and Passonneau, J.V., 1972, *A Flexible System of Enzymatic Analysis*, Academic Press, New York.

Maickel, R.P., and Miller, F.P., 1966, Fluorescent products formed by reaction of indole derivatives with *o*-phthaldehyde, *Anal. Chem.* **38**:1937–1938.

Malnic, G., Klose, R.M., and Giebisch, G., 1964, Micropuncture study of renal potassium excretion in the rat, *Am. J. Physiol.* **206**:674–686.

Maurer, H.R., and Dati, F.A., 1972, Polyacrylamide gel electrophoresis on microslabs, *Anal. Biochem.* **46**:19–32.

McCaman, R.E., 1968, Application of tracers to quantitative biochemical and cytochemical studies, in: *Advances in Tracer Methodology*, Vol. 4 (S. Rothchild, ed.), pp. 137–202, Plenum Press, New York.

McCaman, R.E., 1971, Quantitative isotopic methods for measuring enzyme activities and endogenous substrate levels, in: *International Encyclopedia of Pharmacology and Therapeutics,* Sect. 78, pp. 275–314, Pergamon Press, New York.

McCaman, R.E., and Dewhurst, S.A., 1970, Choline acetyltransferase in individual neurons of *Aplysia californica, J. Neurochem.* **17:**1421–1426.

MacCaman, R.E., and Dewhurst, S.A., 1971, Metabolism of putative transmitters in individual neurons of *Aplysia california, J. Neurochem.* **18:**1329–1335.

McCaman, M.W., Weinreich, D., and McCaman, R.E., 1973, The determination of picomole levels of 5-hydroxytryptamine and dopamine in *Aplysia, Tritonia* and leech nervous tissues, *Brain Res.* **53:**129–137.

Milinoff, P.C., Landsberg, L., and Axelrod, J., 1969, An enzymatic assay for octopamine and other β-hydroxylated phenylethylamines, *J. Pharmacol. Exp. Ther.* **170:**253–261.

Müller, P., 1958, Experiments on current flow and ionic movements in single myelinated nerve fibres, *Exp. Cell Res. Suppl.* **5:**118–152.

Nagatsu, T., 1973, *Biochemistry of Catecholamines,* University Park Press, Baltimore, London, and Tokyo.

Needle, D.J., and Pollitt, R.J., 1965, The formation of 1-dimethylaminonaphthalene-5-sulphon-amide during the preparation of 1-dimethylaminonaphthalene-5-sulphonylamino acids, *Biochem. J.* **97:**607–608.

Neuhoff, V., 1968, Micro-disc-electrophorese von Hirnproteinen, *Arzneim.-Forsch.* **18:**35–38.

Neuhoff, V (ed.), 1973, *Micromethods in Molecular Biology,* Springer-Verlag, Berlin.

Niederwieser, A., 1972, Thin layer chromatography of amino acids and derivatives, *Methods Enzymol.* **25:**60–99.

Nicholls, J.G., and Baylor, D.A., 1969, The specificity and functional role of individual cells in a simple central nervous system, *Endeavor* **29:**3–7.

Osborne, N.N., 1971, A micro-chromatographic method for the detection of biologically active monoamines from isolated neurons, *Experientia* **25:**1502–1513.

Osborne, N.N., 1973, The analysis of amines and amino acids in microquantities of tissue, in: *Progress in Neurobiology,* Vol. 1, Part 4 (G.A. Kerkut and J.W. Phillis, eds.), pp. 299–329, Pergamon Press, Oxford.

Osborne, N.N., 1974, *Microchemical Analysis of Nervous Tissue,* Pergamon Press, Oxford and New York.

Osborne, N.N., and Neuhoff, V., 1973, Neurochemical studies on characterised neurons, *Naturwissenschaften* **60:**78–87.

Osborne, N.N., Szczepaniak, A.C., and Nenhoff, V., Amines and amino acids in identified neurons of *Helixpomatia, Int. J. Neurosci.* **5:**125–131.

Osborne, N.N., and Pentreath, V.W., 1976, Effects of 5,7-dihydroxytryptamine on an identified 5-hydroxytryptamine-containing neurone in the central nervous system of the snail *Helix pomatia, Br. J. Pharmacol.* **56:**29–38.

Osborne, N.N., and Rüchel, R., 1975, Fractionation of proteins from single neurons of *Planorbis corneus* by microelectrophoresis on SDS-gradient polyacrylamide gels, *J. Chromatogr.* **105:**197–200.

Osborne, N.N., Priggemeier, E., and Neuhoff, V., 1975, Dopamine metabolism in characterised neurons of *Planorbis corneus, Brain Res.* **90:**261–271.

Osborne, N.N., Stahl, W.L., and Neuhoff, V., 1976, Separation of amino acids as mansyl derivatives on polyamide layers *J. Chromatogr.* **123:**212–215.

Otsuka, M., Obata, K., Migata, Y., and Tanaka, T., 1971, Measurement of γ-aminobutyric acid in isolated nerve cells of cat central neurons, *J. Neurochem.* **18:**287–295.

Otsuka, M., Migara, T., Konishi, S., and Takahashi, T., 1973, A study of neurotransmitters in the spinal cord, Proceedings of International Society of Neurochemistry Meeting 52-2 (Tokyo), p. 23.

Palkovits, M., 1973, Isolated removal of hypothalamic or other brain nuclei of the rat, *Brain Res.* **59:**449–450.

Palkovits, M., Brownstein, M., Saavedra, J.M., and Axelrod, J., 1974, Norepinephrine and dopamine content of hypothalamic nuclei of the rat, *Brain Res.* **77:**137–149.

Pataki, G., and Wang, K.-T., 1968, Quantitative thin-layer chromatography. VII. Further investigations of direct fluorometric scanning of amino acid derivatives, *J. Chromatogr.* **37**:499–507.

Peterson, R.P., 1972, Biochemical methods used to study single neurons of *Aplysia californica* (see hare), in: *Methods of Neurochemistry*, Vol. 2 (R. Fried, ed.), pp. 73–99, Marcel Dekker, New York.

Pigon, A., and Edström, J.E., 1959, Nucleic changes during starvation and encystment in a ciliate (Urustyla), *Exp. Cell Res.* **16**:648–656.

Poduslo, S.E., and Norton, W.T., 1972, The bulk separation of neuroglia and neuronperikarya, in: *Research Methods in Neurochemistry*, Vol. 1 (N. Marks and R. Rodnight, eds.), pp. 19–93, Plenum Press, New York.

Pun, J.Y., and Lombrozo, K., 1964, Microelectrophoresis of brain and pineal protein in polyacrylamide gel, *Anal. Biochem.* **9**:9–20.

Quentin, C.-D., and Neuhoff, V., 1972, Micro-isoelectric focusing for the detection of LDH isoenzymes in different brain regions of rabbit, *Int. J. Neurosci.* **44**:17–24.

Ramsay, J.A., Falloon, S.W.H., and Machin, K.E., 1951, An integrating flame photometer for small quantities, *J. Sci. Instrum.* **28**:75–80.

Rentzhog, L., 1970, Double isotope derivative assay of catecholamines, *Acta Pharmacol.* **28**(Suppl. 1):1–74.

Riley, R.F., and Coleman, M.K., 1968, Isoelectric fractionation of proteins on a micro-scale in polyacrylamide and agarose matrices, *J. Lab. Clin. Med.* **72**:714–720.

Rose, S.P.R., 1967, Preparation of enriched fractions from cerebral cortex containing isolated metabolically active neuronal and glial cells, *Biochem. J.* **102**:33–43.

Rosmus, J., and Deyl, Z., 1971, Chromatographic methods in the analysis of protein structure, *Chromatogr. Rev.* **13**:163–302.

Rüchel, R., Mesecke, S., Wolfrum, D.I., and Neuhoff, V., 1973, Mikroelektrophorese an kontinuierlichen Polyacrylamid Gradienten Gelen. I. Herstellung und Eigenschaften von Gelgradienten in Kapillaren: ihre Anwendung zur Proteinfraktionierung und Molgewichtsbestimmung, *Hoppe-Seyler's Z. Physiol. Chem.* **354**:1351–1368.

Rüchel, R., Mesecke, S., Wolfrum, D.I., and Neuhoff, V., 1974, Mikroelektrophorese an kontinuierlichen Polyacrylamid Gradienten Gelen. II. Mikroelektrophorese und elektrophoretische Zerlegung von SDS-protein-Komplexen in Polyacrylamidgel-Komplexen in Polyacrylamidgel-Gradienten, *Hoppe-Seyler's Z. Physiol. Chem.* **355**:997–1020.

Rude, S., Coggeshall, R.E., and van Orden, L.S., III, 1969, Chemical and ultrastructural identification of 5-hydroxy-tryptamine in an identified neuron, *J. Cell Biol.* **41**:832–854.

Saavedra, J.M., 1974, Enzymatic–isotopic assay for the presence of β-phenylethylamine in brain, *J. Neurochem.* **22**:211–216.

Saavedra, J.M., and Axelrod, J., 1972, A specific and sensitive assay for tryptamine in tissues, *J. Pharmacol. Exp. Ther.* **182**:363–369.

Saavedra, J.M., and Axelrod, J., 1973, The demonstration and distribution of phenylethanolamine in the brain and other tissues, *Proc. Natl. Acad. Sci. U.S.A.* **70**:769–772.

Saavedra, J.M., Brownstein, M., and Axelrod, J., 1973, A specific and sensitive enzymatic–isotopic microassay for serotonin in tissues, *J. Pharmacol. Exp. Ther.* **186**:508–515.

Saavedra, J.M., Palkovits, M., Brownstein, M.J., and Axelrod, J., 1974, Serotonin distribution in the nuclei of the rat hypotholamus and preoptic region, *Brain Res.* **77**:157–165.

Schiefer, H.G., and Neuhoff, V., 1971, Fluorometric microdetermination of phospholipids on the cellular level, *Hoppe-Seyler's Z. Physiol. Chem.* **352**:913–926.

Schlumpf, M., Lichtensteiger, W., Langemann, H., Waser, P.G., and Hefti, F., 1974, A fluorometric micromethod for the simultaneous determination of serotonin, noradrenaline and dopamine in milligram amounts of brain tissue, *Biochem. Pharmacol.* **23**:2337–2446.

Segundo, J.P., 1970, Functional possibilities of nerve cells for communication and for coding, *Acta Neurol. Latinoam.* **14**:340–344.

Seiler, N., 1970, Use of the dansyl reaction in Biochemical analysis, in: *Methods in Biochemical Analysis*, Vol. 18 (D. Glick, ed.), pp. 259–337, Interscience, New York.

Seiler, N., and Knödgen, B., 1973, Quantitative mass spectrometry by internal standardisation using a single focusing mass spectrometer and the peak switching facilities of a peak matching device, *Org. Mass Spectrom.* **7**:97–105.

Seiler, N., and Knödgen, B., 1974, Identification of amino acids in picomole amounts as their 5-dibutylamino-naphthalene-1-sulphonyl derivatives, *J. Chromatogr.* **97**:286–288.

Seiler, N., and Wiechmann, M., 1970, TLC analysis of amines as their dans-derivatives, in: *Progress in Thin-layer Chromatography and Related Methods*, Vol. 1 (A. Niederwieser and G. Pataki, eds.), pp. 94–144, Ann Arbor-Humphrey, Ann Arbor, Michigan.

Sharman, D.F., 1971, Methods of determination of catecholamines and their metabolites, in: *Methods of Neurochemistry*, Vol. 1 (R. Fried, ed.), pp. 83–128, Marcel Dekker, New York.

Shellenberger, M.K., and Gordon, J.H., 1971, A rapid, simplified procedure for simultaneous assay of norepinephrine, dopamine and 5-hydroxytryptamine from discrete brain areas, *Anal. Biochem.* **39**:356–372.

Snodgrass, S.R., and Iversen, L.L., 1973, A sensitive double isotope derivative assay to measure release of amino acids from brain *in vitro*, *Nature (London)* **241**:154–156.

Snyder, S.H., and Taylor, K.M., 1972, Assay of amines and their deaminating enzymes, in: *Research Methods in Neurochemistry*, Vol. 1 (N. Marks and R. Rodnight, eds.), pp. 287–316, Plenum Press, New York.

Snyder, S.H., Baldessarini, R., and Axelrod, J., 1966, A specific and sensitive enzymatic isotopic assay for tissue histamine, *J. Pharmacol. Exp. Ther.* **153**:544–549.

Stein, S., Böhler, P., Stone, J., Dairman, W., and Undenfriend, S., 1973, Amino acid analysis with fluorescamine at the picomole level, *Arch. Biochem. Biophys.* **155**:203–212.

Strumwasser, F., 1967, Types of information stored in single neurons, in: *Invertebrate Nervous Systems* (C.A.G. Wiersma, ed.), pp. 291–319, University of Chicago Press.

Svetashev, V.I., and Vaskovsky, V.E., 1972, A simplified technique for thin-layer microchromatography of lipids, *J. Chromatogr.* **67**:376–378.

Taylor, K.N., and Snyder, S.H., 1972, Isotopic microassay of histamine, histidine, histidine-decarboxylase and histamine methyltransferase in brain tissue, *J. Neurochem.* **19**:1343–1358.

Undenfriend, S., Stein, S., and Böhlen, p., 1972, A new fluorometric procedure for assay of amino acids, peptides and proteins in the picomole range, in: *Chemistry and Biology of Peptides*, Proceedings of the 3rd American Peptide Symposium (J. Meienhofer, ed.), pp. 655–663, Ann Arbor Science, Ann Arbor, Michigan.

Vurek, G.C., 1967, Emission photometry of picomolar amounts of calcium, magnesium and other metals, *Anal. Chem.* **39**:1599–1601.

Vurek, G.C., and Bowman, R.L., 1965, Helium-glow photometer for picomole analysis of alkali metals, *Science* **149**:448–450.

Weinreich, D., Weiner, C., and McCaman, R., 1975, Endogenous levels of histamine in single neurons isolated from CNS of *Aplysia california*, *Brain Res.* **84**:341–345.

Whittaker, V.P., 1968, The morphology of fractions of rat fore brain synaptosomes by continuous sucrose density gradients, *Biochem. J.* **106**:412–417.

Whittaker, V.P., 1973, The biochemistry of synaptic transmission, *Naturwissenschaften* **60**:281–289.

Willows, A.O.D., 1967, Behavioural acts elicited by the stimulation of single identifiable brain cells, *Science* **157**:570–574.

Willows, A.O.D., 1968, Behavioural acts elicited by stimulation of single identifiable nerve cells, in: *Physiological and Biochemical Aspects of Nervous Integration* (F.D. Carlson, ed.), pp. 217–244. Prentice-Hall, Englewood Cliffs, New Jersey.

Wilson, D.L., 1971, Molecular weight distribution of proteins synthesised in single, identified neurons of *Aplysia*, *J. Gen. Physiol.* **57**:26–40.

Wolfrum, D.I., Rüchel, R., Mesecke, S., and Neuhoff, V., 1974, Mikroelektrophorese in kontinuierlichen Polyacrylamid-Gradientengelen. III. Extraktion und Fraktionierung von Ribonucleinsäuren im Mikromassstab, *Hoppe-Seyler's Z. Physiol. Chem.* **355**:1415–1435.

Wrigley, C.W., 1968, Analytical fractionation of plant and animal proteins by gel electrofocusing, *J. Chromatogr.* **36**:362–365.

Zeuthen, E., 1961, The Cartesian diver balance, in: *General Cytochemical Methods*, Vol. 2 (J. Danielli, ed.), pp. 61–90, Academic Press, New York.

3

Tissue and Organ Culture

Werner T. Schlapfer

1. Introduction

The term "tissue culture" has been used to denote the *in vitro* cultivation and maintenance of living biological materials from multicellular organisms. The techniques that have been developed for this purpose fall into four broad classes: (1) methods to dissociate cells, possibly isolating individual cell types, and to maintain and propagate the dispersed cells as monolayers or suspensions; (2) methods to reassociate previously dissociated cells into aggregates and to maintain these reaggregates in an *in vitro* environment; (3) methods to cultivate tissue fragments over prolonged periods of time under conditions that encourage cellular interactions as well as differentiation and maturation of the cellular elements in a way resembling the *in vivo* situation; and (4) methods for maintenance of organ fragments under conditions that discourage the migration of cells from the explant, and that maintain the differentiated cells as a group in a viable, organized, and functioning state. Short-term incubation methods (24 hr or less) of tissue slices or fragments, as are commonly used in biochemical experiments, short-term isolations, or *in vitro* procedures used in neurophysiology (e.g., isolated ganglia or nerve–muscle preparations), do not qualify as culture systems in the context of this discussion and are not included in this review.

The first class of techniques is more precisely called "cell culture" and the second class is often referred to as "reaggregate cultures." Their application to cells of the nervous system is treated in Chapter 4. The third and fourth categories of methods share as a characteristic the encouragement of cell differentiation and cellular interactions; the third class is often referred to as "organotypic tissue cultures," sometimes as "intermediate tissue cul-

Werner T. Schlapfer • Western Research and Development Office, Veterans Administration Medical Center, Livermore, California 94550.

tures," "organized tissue cultures," "explant cultures," or simply "tissue culture," while the fourth class is called "organ culture." The distinction between these latter two classes is not always sharp, and some authors refer to both as "organ culture" or as "organotypic cultures." For the purposes of this chapter, however, it is valuable to distinguish between tissue culture and organ culture, since the methods, experimental justifications, and uses of the two approaches are quite distinct. In general, culture methods that encourage cellular outgrowth and the consequent flattening of the explanted tissue fragment will be considered under tissue-culture methods, while organ-culture methods will be reserved for techniques that discourage migration of cells from the explant.

In addition to the distinction between the tissue-culture and organ-culture methodologies, it may also be useful to distinguish between long-term tissue cultures and short-term tissue cultures. The techniques for the maintenance of long-term tissue cultures make more stringent demands as to the lack of toxicity, the appropriate nutrients, the proper ionic and physical environment, and the absence of contamination than do cultures maintained for only a relatively short incubation time. Under the proper conditions, cultures of nervous tissue have been maintained in an apparently healthy state for longer than a year. In this review, culture methods that are conducive to maturation and survival of the cellular components of the explant for 2–3 weeks or longer may be called long-term cultures, while methods used in experiments that do not require cultivation for longer than several days to about 1 or 2 weeks may be called short-term cultures. Most organ cultures would also fall into the category of short-term cultures. Short-term cultures can often utilize simplified techniques depending on the requirements of the particular experiment. For these reasons, the methods of long-term tissue cultures, short-term tissue cultures, and organ cultures will be treated separately, although the reader will notice that the distinctions are often quite arbitrary and are maintained here solely for the purpose of clarity of presentation. Methods for the cultivation of cold-blooded vertebrate and of invertebrate nervous tissues will be considered separately.

It is the purpose of this chapter to provide the reader with an overview of the various techniques that have been used to maintain nervous tissue in culture. Tissue and organ cultures are used for a large variety of experimental situations with a wide range of tissues of many species. It is therefore not surprising that a large number of different methods, culture vessels, tissue substrates, and media have been used. It is impossible in the space allotted for this chapter to cover all the variations of the tissue- and organ-culture methodologies in detail, and the main purpose of this chapter is therefore to guide the reader to the pertinent and vast original literature in this field.

1.1. Advantages and Disadvantages of Tissue Culture

Long-term isolation of tissues has both experimental disadvantages and advantages. It implies a considerable disruption of the normal environment of the cells, as well as a distortion of the *in vivo* topographic and physiological relationships of the tissue. If the environment is changed too drastically, cells may dedifferentiate and lose their characteristic morphology and function (Harris, 1964). For explant tissue cultures, or organ cultures in which an attempt is made to allow the cells to be integrated into a tissue, dedifferentiation is usually not a widespread phenomenon. Nevertheless, one has to be constantly on guard for such undesirable effects when dealing with cultured tissue. On the other hand, isolation and the ensuing simplification of the tissue in culture under suitable conditions provides the following features:

1. Control of the physical and chemical environment (e.g., nutrients, ions, substrates and cofactors for enzymatic synthesis, hormones, drugs, temperature, radiation).
2. Isolation from the controlling and modifying influences (neural, humoral, hormonal) of other tissues in the body, while allowing such influences to be experimentally controlled, e.g., by cocultivation of selected tissues.
3. Direct accessibility of the cells to labeled precursors, pharmacological agents, metabolic inhibitors, and tracers, thereby minimizing the problems of selective barriers, nonuniform diffusions, and general toxicities.
4. Recovery of the incubation medium for biochemical analysis.

In addition, the techniques of tissue culture often provide these further features:

5. Continuous observation of the cells in their living state under a suitable microscope (e.g., phase-contrast, polarizing, interference, fluorescent), permitting study of morphology and cell dynamics.
6. Accessibility of individual cells and even parts of cells to exploration with microelectrodes, microbeams, and other microinstruments.

Compared to tissue slices and other similar short-term isolation techniques, long-term isolation in cultures often permits recovery from the dissection trauma and from the transient changes occurring in the tissue in its adaptation to the new environment. The recovery may not be complete, but at least it can be hoped that some steady-state or slowly changing condition is reached that allows meaningful experimentation.

All these factors are of special significance in the study of nervous tissue, because the nervous system is an almost hopelessly complex organ consisting of various functionally and structurally different cell types. Simplification, coupled with the controllable environment, visualization, and accessibility made possible by the techniques of tissue and organ culture, therefore have been and will be one of the productive avenues to the study of some aspects of the nervous system.

Nervous tissue grown and maintained *in vitro* should be considered a model system for the study of some of the properties of neurons and their interactions under relatively simple and controllable conditions. Many investigations of cultured nervous tissue have been and are still concerned with demonstrating the integrity of its crucial cellular and organotypic parameters, such as its cytological, morphological, ultrastructural, bioelectrical, and biochemical properties. Most of the important structural and functional attributes of nerve cells are maintained in culture; thus, neurons differentiate and mature *in vitro*, they often form specific interactions with other cells (synaptic junctions, myelin formation), their electrical properties correspond well to the *in vivo* situation, and they seem to retain many of their biochemical characteristics. The vast recent literature dealing with the properties of nerve cells in culture has recently been reviewed by several authors (Murray, 1971; R.P. Bunge, 1975; Nelson, 1975; Crain, 1976; Fedoroff and Hertz, 1977; Giacolini *et al.*, 1980). The Tissue Culture Association (12111 Parklawn Drive, Rockville, Maryland 20852) publishes a yearly *Index of Tissue Culture*, a bibliography of tissue culture publications.

1.2. Historical Developments

Early attempts at culturing isolated tissue by von Recklinghausen (1866), Roux (1885), Ljunggren (1897), and many others may have resulted only in delayed death and decay of the tissues. But Jolly (1903), for instance, was able to maintain and observe amoeboid movement and cell divisions of leukocytes *in vitro* for about a month. To study the development of nerve fibers, Harrison (1907) developed a technique of culturing nerve tissue from embryonic frogs by embedding the explants in lymph clots on coverslips inverted and sealed over depression slides. In this way, he observed the formation of fibers by protoplasmic outflowing from the central perikarya and thus contributed substantially to the confirmation of the neuron doctrine.

In the early years of tissue culture, the main effort went into experiments on the regenerative and proliferative capacities of explanted cells, using tissues of many origins. Until better methods had been developed, the study of nervous tissue was limited by the apparent inability of the cultures to survive for longer than a week or two, after which the newly grown fibers usually started to degenerate (Ingebrigtsen, 1913). Nevertheless, significant contributions to many problems of neurology were made during this time. The review by Murray (1965) provides a detailed historical treatment of the earlier literature of nervous tissue in culture.

2. Development of Modern Methods of Nerve Tissue and Organ Culture

Any tissue- or organ-culture method must attempt to provide the tissue with an environment that is close to the *in vivo* situation while allowing for the simplification and accessibility demanded by the particular experiment. Successful long-term cultivation of nervous tissue seems to be possible with a number of techniques using various combinations of vessels, substrates for cell attachment, and compositions of nutrient media. However, several critical requirements are common to all techniques.

1. *Origin of the tissue.* Long-term organotypic tissue cultures have been most fruitful with embryonic or newborn tissues, and the vast majority of cultures are prepared with explants from embryos or neonates. There seems to be an optimum age, depending on the tissue and species, for the preparation of successful cultures (Murray, 1971; J.F. Schneider, 1973). Several laboratories have, however, succeeded in culturing adult nervous tissue (e.g., Murray and Stout, 1947; Costero and Pomerat, 1951; Hogue, 1953; Kiernan and Pettit, 1971; Spoerri and Glees, 1974; Tsiquaye and Zuckerman, 1974; Messing and Kim, 1979).

2. *An optimum size of the explant.* One dimension of the explant should not exceed about 0.5 mm, and the total volume of the tissue fragment is usually about 1 mm^3. The optimum size of the explant may depend on the origin tissue. The explant should be small enough to allow diffusion of nutrients and oxygen to all parts of the tissue, yet large enough to leave a maximum number of cells undamaged or with recoverable damage (Lumsden, 1968).

3. *Preparation of tissue explants with a minimum of trauma.* Extreme care should be taken to avoid tearing or crushing of the tissue fragments during the dissection. For example, the use of sharp knives, such as new razor blades or cataract knives, to cut the explants can make the difference between a viable culture and a mass of decaying tissue.

4. *A suitable substrate for cell attachment.* Since cultures of nervous tissue usually do not adhere well to uncoated glass, several methods are employed to hold the explants in place. Embedding the tissue in a plasma clot (chicken plasma clotted with chicken embryo extract) (Burrows, 1910) is a convenient method. Most neural tissues attach well to cover glasses coated with a film of reconstituted rat-tail collagen (Bornstein, 1958, 1973a; Masurovsky and Peterson, 1973), while commercial collagens have generally been found unsuitable for nervous tissue. Perforated cellophane has been used by several investigators (Breen and deVellis, 1975; Grunnet, 1973). Holding the explant under a strip of dialysis membrane has also been successfully employed (Pomerat, 1959; Chamley *et al.*, 1972). Organ cultures are commonly maintained on lens or filter paper (e.g., Kiernan and Pettit, 1971; Daniels and Moore, 1972), on cellulose acetate (e.g., Silverman *et al.*, 1973), on cellulose sponges (Cunningham, 1962), or on gelatin sponge foam (Rubinstein *et al.*, 1973; Sipe, 1976).

5. *An appropriate nutrient medium to provide an environment that approaches the in vivo environment as closely as possible.* In the design of appropriate media for tissue culture, the composition of the *in vivo* environment has been the guiding principle. However, the ideal environment for an explant may be different from the normal *in vivo* environment, and the medium formulations are usually devised empirically. The nutritional requirements of mammalian and avian nerve tissue cultures have been reviewed by Silberberg (1972).

The medium for nervous tissue cultures usually consists of a balanced salt solution (BSS), often containing a mixture of amino acids, vitamins, and other nutrients. These synthetic media are usually supplemented with sera and possibly other natural fluids (human placental serum; various fetal, newborn, or adult animal sera; various sera ultrafiltrates; and embryo extracts) and with glucose at a fairly high concentration. The pH is maintained between 6.8 and 7.1 (Lumsden, 1968). The ratio of the volume of the medium to that of the tissue should be kept at a minimum and the intervals between medium changes as long as feasible, since cultured nervous tissue seems to benefit from a self-conditioning of its medium and may suffer from the disturbance created by the medium change. Defined synthetic media would be desirable, since they allow precise control and standardization; however, cultures of nervous tissue have generally been unsuccessful in media lacking serum proteins (see Kim and Pleasure, 1978) (exceptions are cold-blooded vertebrate and some invertebrate cultures as discussed in Sections 6 and 7).

6. *Sufficient oxygenation of all parts of the tissue.* This is usually achieved by allowing only a thin film of nutrient medium to cover the culture or, as is the case in organ cultures, by placing the explant at the liquid–air interface.

7. *An optimum incubation temperature.* Tissues of mammalian origin are usually incubated at 34–37°C (Peterson and Murray, 1960; Bornstein, 1973*a*).

8. *Lack of toxicity in all parts of the culture chamber,* especially in the substrate for cell attachment, such as the cover glass bearing the explant (Peterson *et al.*, 1959). The glass-washing procedure often introduces toxicities into the tissue-culture chamber. On the other hand, disposable sterile plastic ware may exert subtle effects on cellular growth and function (Mithen *et al.*, 1980).

9. *Maintenance of strict sterility throughout all manipulations.* A detailed treatment of the problems of contamination in tissue culture can be found in Fogh (1973).

Many of these basic requirements are shared by the culture of cells and tissues of a variety of origins and are not peculiar to nervous tissue. The general methods of cell and tissue culture that are also relevant to the culture of nervous tissue (e.g., laboratory equipment preparation and sterilization of glassware, instruments, media) are described in great detail in a number of texts (e.g., Merchant *et al.*, 1960; Parker, 1961; Penso and Balducci, 1963; Willmer, 1965; Wasley and May, 1970; Whitaker, 1972; Kruse and Patterson, 1973; J. Paul, 1975). Another valuable source of specific methods and procedures of cell, tissue, and organ culture is the *TCA Manual*, a serial publication of the Tissue Culture Association.

3. Techniques for Long-Term Organotypic Cultures of Mammalian and Avian Nervous Tissue

The individual methods of culture of nervous tissues of mammalian and avian origin can be divided into three categories based on the design of the culture chamber: (1) slide cultures in sealed chambers; (2) petri-dish cultures or modifications thereof, usually open to the atmosphere; and (3) test-tube cultures.

3.1. General Considerations

Before I embark on a discussion of the various techniques that are being employed to culture nervous tissue, it may be valuable to elaborate a few general methodological details. Regardless of the particular culture method chosen, successful long-term cultures depend to a large extent on an awareness of some important technical details.

3.1.1. Sterility

Sterility is an absolute prerequisite for successful long-term cultures, especially since many workers claim that better cultures are obtained if no antibiotics are used in the culture medium. Fortunately, bacterial or fungal contaminations are usually also the easiest to pinpoint and remedy. Infections by mycoplasma are more difficult to detect, and they present a constant threat to the cultures. General texts on tissue culture (cited in Section 2) should be consulted for detailed discussion of aseptic manipulations. A scrupulously clean small room equipped with UV lights and filtered air intakes with a positive-pressure difference to the exterior usually serves well as a culture room. Alternatively, a culture hood (e.g., Edgeguard Hood by The Baker Co., Sanford, Maine) set in a small room may be sufficient. Before an explanting session is started, all laboratory surfaces should be washed with a disinfectant (e.g., Roccal) and possibly swabbed with alcohol. The entire room should then be exposed to UV radiation. The dissecting instruments are sterilized before the operation and are usually resterilized as often as possible during the dissection by rinsing in sterile saline, distilled water, and two changes of alcohol, and finally flaming before reuse. Glassware is sterilized properly by either dry heat or steam autoclave. Sterile media components are mixed under aseptic conditions, and a sample is incubated in a standard thioglycollate broth to test for sterility several days before use. Face masks, as well as caps and clean gowns or laboratory coats, should be worn during the culture session.

3.1.2. Tissue

The optimum explant size has already been discussed, and the need for sharp cuts, avoiding tears, gouging, and crushing, has been mentioned. The tissue of origin is dissected quickly but carefully from the animal and transferred into a dish containing balanced salt solution (BSS) or nutrient medium. After additional dissection, possibly under a dissecting microscope, to remove meninges and blood vessels, the tissue should then be cut into explant pieces of the proper size (about 1 mm^3 with one dimension usually not exceeding 0.5 mm). In addition to an optimum size of the explant, it is important to cut the explants commensurate with the cellular architecture and normal histogenesis of the tissue, e.g., in such a way as to damage the least number of neuronal processes. Cutting of the explants without delay after the dissection is important to facilitate diffusion of nutrients and oxygen to the cells. Some laboratories use cataract knives, iridectomy scissors, or No. 11 scalpel blades for cutting explants from the tissue of origin. Other laboratories (including the author's) have found triangular knives cut from stainless steel razor blades more convenient. The blade fragments can be clamped into surgical hemostats the handles of which have been altered to allow more convenient handling, or the blades can be held in a small chuck attached to a handle (modified needle holders). Dissecting razor-blade breakers and holders (Tiemann, Long Island City, New York) are also suitable for this purpose. New blades are used for each culture session, and care is taken not to damage the tips or the cutting edges. Clean cuts are made by quickly moving two blades pressed against each other in a scissorlike fashion through the tissue. The bottoms of plastic or glass petri dishes serve well as cutting surfaces, but some workers prefer sterile and well-cleaned frosted glass blocks. The explants are then transferred to a new pool of BSS containing at least glucose or to a complete nutrient solution and kept there until the dissection has been completed. Large-bore pipettes can be used for transferring the explants, or the dissecting knives can be used by flattening one blade onto the floor of the petri dish containing the tissue fragments (knives made from stainless steel razor blades are very convenient for this, since the blades are flexible) and gently pushing the explants onto the blade with a second knife.

3.1.3. Lack of Toxicity

By far the most common reason for unsuccessful long-term cultures of nervous tissue is the presence of toxic substances in the culture chamber, in the nutrient medium, on the dissecting instruments, or in any of the other materials that may come into contact with the tissue.

3.1.3a. Coverslips. Gold Seal (Clay Adams) coverslips have long been recommended (Peterson *et al.*, 1959; Bornstein, 1973*a*) for nerve-tissue culture. They should be washed and sterilized according to the procedure outlined by Bornstein (1973*a*) or J.F. Schneider (1973), which includes soaking in nitric acid, prolonged rinsing, and soaking in distilled water. Distilled water used in all operations should be either double-distilled or distilled deionized water. The distilled water should touch only glass, Teflon, Tygon, or high-grade stainless steel. Institutional distilled water is usually not sufficiently pure for tissue-culture use. Alternatively, coverslips can be cleaned in xylene, several rinses of acetone, alcohol, and finally distilled water, followed by boiling in distilled water and soaking overnight in ultrapure (redistilled) 95% ethyl alcohol. Sterilization should be by dry heat. Other materials used in lieu of glass coverslips (e.g., Aclar, cellophane) should be prepared as described in the original literature (cited below).

3.1.3b. Glassware. The problems involved in cleaning laboratory glassware are so great that it is often cheaper to use sterile disposable plastic wherever feasible. This includes petri dishes for dissection, cutting, and temporary storage of tissue fragments, pipettes, syringes, and media bottles (e.g., use 250-ml Falcon tissue-culture bottles to store media and media components). However, it has recently been reported that certain lots of polystyrene petri dishes are toxic to neurons in an air-tight tissue culture system (Mithen *et al.*, 1980). This raises the possibility that polystyrene containers release some agent(s) which may have subtle or sublethal effects on cultures even when the exposure to polystyrene is relatively short. The procedure for the preparation and sterilization of glassware is given in detail by Bornstein (1973*a*) and J.F. Schneider (1973). Dry-heat sterilization is usually preferable to autoclaving, since the steam of the autoclave may deposit toxic substances onto the glassware. A self-contained small autoclave filled with pure distilled water should be used if necessary, rather than an autoclave connected to a central steam supply as usually found in hospitals or research laboratories. Central steam supplies often contain anticorrosive agents that are added to the supply water. These agents may be good for the steam pipes, but cultured nervous tissue does better without them.

3.1.3c. Instruments. Dissecting instruments should be cleaned immediately after culturing by boiling for 15–20 min in tap water containing a pinch of Haemosol, rinsing several times with tap and distilled water, and then boiling in distilled water. Finally, they are rinsed in redistilled alcohol and dried in an oven. Knife blades cut from razor blades are first washed in xylene, acetone, and alcohol to remove oils.

3.1.3d. Nutrient Media. Some laboratories prepare standard BSSs or synthetic media from dried powder mixtures (cf. Bornstein, 1973*a*). In other cases, it may be necessary to prepare a special mixture of ingredients, in which case only the purest grade of chemical and distilled water of the highest purity should be used. Other laboratories use sterile BSSs (e.g., Simms', Hanks', Gey's, or Earle's BSS) and synthetic media [e.g., Eagle's Minimum Essential Medium (MEM), TC 1066, Medium 199, Eagle's Basal Medium, NCTC 109, F12, Medium BGJ[6]) prepared and marketed by a number of suppliers. The sera and other biological fluids included in the nutrient media can be purchased from various suppliers (for recent lists of sources of tissue-culture materials, including commercially available media, see J. Paul, 1975; J.F. Schneider, 1973). The quality of commercially available biological fluids, such as sera, is often variable, and poor cultures may result from inferior serum in the nutrient medium. It may therefore be necessary to test several lots of sera and then order a larger amount of a satisfactory batch. The serum used in the nutrient medium is usually heat-inactivated by heating to 56°C for 30 min before mixing with the other medium components. Some natural fluids may not be available commercially (e.g., mouse or rat embryo extract), or it may be preferable in some cases to use some natural fluids collected from a specific source. The special methods involved in preparation of various sera, embryo extracts, and ultrafiltrates are described in various texts on general tissue-culture methods (see Section 2).

3.2. Slide Cultures

3.2.1. Maximow Chamber

The Maximow double-coverslip assembly (Maximow, 1925) was first used for nervous tissue by Murray and Stout (1942) and subsequently developed by Peterson and Murray (1955). The Maximow technique favors maturation and differentiation of the tissue, including the appearance of complex bioelectrical activities (Crain, 1966, 1976) and *de novo* formation of myelin (Peterson and Murray, 1955; Ross *et al.*, 1962) and of functional synaptic connections (M.B. Bunge *et al.*, 1967; Crain and Peterson, 1967). This technique yields cultures that are often many cell layers thick (100–300 μm thick) and that are capable of differentiating in a way appropriate to the tissue of origin after isolation from the organism. Consequently, such cultures have been termed "organotypic" (Crain, 1966). The cultures obtained in this way are well suited to allow continuous or repeated observations, without disturbance, with a relatively high-power microscope.

The preparation and maintenance of Maximow cultures, including preparation of glassware, instruments, solutions, and culture media, have been described recently in great detail by Bornstein (1973*a*) and J.F. Schneider (1973).

Explants of nervous tissue are placed onto small circular glass or Aclar (chlorotrifluoroethylene, Type 33C, 1.5 mil, Allied Chemical Corp., Morristown, New Jersey) plastic coverslips (Masurovsky and Bunge, 1968) or at times on silicone rubber membranes (Sander Ltd., Harlow, Essex, England) (Shahar *et al.*, 1973) that have previously been coated with a film of reconstituted rat-tail collagen (for the method, see Bornstein, 1958, 1973*a*; Masurovsky and Peterson, 1973, 1975; or Price, 1975). The explant can also be embedded in a plasma clot on a coverslip (e.g., Peterson and Murray, 1955), but Maximow cultures are now made almost exclusively with collagen-coated coverslips. The coverslip bearing the tissue fragment is then attached to a larger cover glass with a drop of BSS. A single drop of nutrient medium is placed on the explant. The double coverslip is inverted over a depression slide, and the chamber is sealed with a paraffin–Vaseline mixture. These assemblies are usually incubated in a "lying-drop" position with the drop of medium forming a thin film over the culture. The cultures are fed with a fresh drop of nutrient medium twice a week. During the feeding procedure, the small round coverslip bearing the cultured tissue is removed from the assembly, washed in BSS, provided with a new drop of nutrient medium, and finally sealed over a new depression slide.

A large number of different parts of the nervous system have been successfully cultured using the Maximow double-coverslip assembly. The composition of the nutrient medium used with Maximow cultures depends on the origin of the cultured tissue. In addition, different workers seem to prefer different media, since the criteria for an optimum medium composition depend on the investigator's experimental aims.

1. *Spinal cord, dorsal root ganglia, and brainstem of neonate or embryonic rodents.* The nutrient medium typically consists of human placental cord serum, human serum, horse serum, calf serum, or fetal calf serum (25–40%), sometimes extract of chick, mouse, or rat embryos or mouse brains (up to 25%), or sometimes horse or bovine serum ultrafiltrates (up to 25%), Eagle's MEM, plus glutamate (up to 65%), and Simms', Earle's, or Hanks' BSS (up to 35%). The exact composition varies from laboratory to laboratory (Bornstein, 1973a; Dubois-Dalcq *et al.*, 1973; C.V. Paul and Powell, 1974; Roizin *et al.*, 1974; Sobkowicz *et al.*, 1973; Stern, 1973; Tischner and Thomas, 1973; Wolfgram and Myers, 1973).

2. *Cervical spinal ganglia of adult rats* have been cultured in a medium consisting of 33% fetal calf serum, 33% Eagle's MEM plus glutamine, and 33% chick embryo extract (Spoerri and Glees, 1974).

3. *Cerebellum of fetal or newborn rodents* is usually cultured in a nutrient medium with a composition similar to the one used for spinal cord (Allerand, 1972; Calvet and Lepault, 1975; Kim, 1971a; Privat *et al.*, 1974; M.K. Wolf and Holden, 1969). Some laboratories add 0.1–0.2 U/ml low-zinc, glucagon-free insulin (Squibb, New Brunswick, New Jersey) to the medium used for cerebellar cultures (Bornstein, 1973a; Seil, 1972; Silberberg *et al.*, 1972). Mouse-brain extract (7%) is sometimes included in the medium after the appearance of myelin (Seil, 1972). Some laboratories include 17% nonessential amino acid solution to enhance neuronal differentiation (Toran-Allerand, 1978).

4. *Cerebral cortex and hippocampus of fetal or newborn rodents.* The medium for cerebral and hippocampal cultures is composed of human placental cord serum (17–40%), Eagle's MEM plus glutamine (25–53%), sometimes Simms' BSS (up to 34%) or TC 1066 (33%), chick embryo extract (10–17%), and often 0.1 U/ml low-zinc, glucagon-free insulin (Bornstein, 1973a; Crain *et al.*, 1975; Goworek, 1974; Grosse and Lindner, 1974). Other workers prefer to culture hippocampal tissue in a medium containing 33% horse serum, 33% Medium 199, and 33% Hanks' BSS (Kim, 1973) or 25% human serum (adult male), 25% bovine serum ultrafiltrate, 25% chick embryo extract, and 25% BSS (J.H. LaVail and Wolf, 1973). Enhanced differentiation and maturation of cultures of embryonic rat hippocampus has been reported with the inclusion of 10^{-3} mg/ml "cerebrolysin," a pig brain hydrolysate (Ebewe, Unterach a/Attersee, Austria) in the nutrient medium (Wenzel *et al.*, 1977).

5. *Olfactory bulb* of neonatal mouse has been cultured in 33% horse serum, 33% Medium 199, and 33% Hanks' BSS (Kim, 1972b), and of fetal mouse in a medium consisting of human placental serum (34%), Eagle's MEM (55%), and chick embryo extract (11%) (Corrigall *et al.*, 1976).

6. *Hypothalamus of newborn mouse* has been successfully cultured in a medium containing up to 40% human adult serum, sometimes mixed in various proportions with horse serum, 50% Eagle's MEM plus glutamine, and 10–25% bovine serum ultrafiltrate (Sobkowicz *et al.*, 1974), or in 23% horse serum, 50% Eagle's MEM, 17% nonessential amino acid solution and

10% BSS (Toran-Allerand, 1978). Hypothalamic tissue of fetal guinea pig has been cultured in 30% human placental serum, 40% Eagle's MEM, and 30% Hanks' BSS (Marson and Privat, 1979).

7. *Human fetal spinal cord, brainstem, cerebellum, and cerebral cortex* can be cultured in a medium consisting of Eagle's MEM plus glutamine (75%), fetal calf serum (15%), and bovine serum (10%) (L. Hösli *et al.*, 1973*b*). Human fetal cerebellar cortex has also been cultured in a medium containing 33% horse serum, 33% Medium 199, and 33% Hanks' BSS (Kim, 1976*a*). For human fetal dorsal root ganglia the medium consists of 33% human placental serum, 50% Eagle's MEM, 10% chick embryo extract, and 7% BSS (Crain *et al.*, 1980).

8. *Sympathetic ganglia of embryonic rats or mice* can be cultured in a medium containing 27% human placental cord serum, 17% chick embryo extract, 21% bovine serum ultrafiltrate, 20% Eagle's MEM plus glutamine, 17% Simms' BSS, and 0.0015 U/ml low-zinc, glucagon-free insulin (Benitez *et al.*, 1974). Sympathetic ganglia of embryonic mouse have been grown in 33% human placental serum, 53% Eagle's MEM, 10% chick embryo extract, and 4% Simms' BSS (Cook and Peterson, 1974) or alternatively in 25% Gey's BSS, 25% bovine serum ultrafiltrate, 25% horse serum, and 25% chick embryo extract (Sano *et al.*, 1967).

9. *Sympathetic ganglia of 10- to 16-day chick embryos.* For chick sympathetic ganglia, the same medium as for rat superior cervical ganglion (Benitez *et al.*, 1974) can be used. Alternatively, a medium containing horse serum (25%), Medium 199 (25%), Hanks' BSS (25%), and chick embryo extract (25%), (Kim and Munkacsi, 1972), or human placental cord serum (25–33%), chick embryo extract (16–33%), Eagle's MEM plus glutamine (33–36%), and sometimes bovine serum ultrafiltrate (23%) and low-zinc insulin (0.0015 U/ml) has been used (Benitez *et al.*, 1973; Wenzel *et al.*, 1973).

10. *Embryonic chick neural tube* has been cultured in horse serum (33%), Medium 199 (33%), and Hanks' BSS (33%) (Kim and Wenger, 1972).

11. *Guinea pig myenteric plexus* (attached to explants of muscularis externa of small intestine) can be cultured in a growth medium containing 33% human placental serum, 53% Eagle's MEM, and 10% chick embryo extract (Dreyfus *et al.*, 1977).

12. *Retinas of fetal mice* have been co-cultured with tectal (superior colliculus) explants in 33% human placental serum, 53% Eagle's MEM, 10% chick embryo extract, and 4% Hanks' BSS (Smalheiser and Crain, 1978).

In all the media described above, the glucose concentration is increased to about 5–10 mg/ml, and the sera are heat-inactivated at 56°C for 30 min. Some laboratories add 10^{-2} M *N*-2-hydroxyethylpiperazine-*N*'-2-ethanesulfonic acid (HEPES) to their culture media for improved buffering (Bornstein, 1973*a*; Sobkowicz *et al.*, 1974; Toran-Allerand, 1976*a*). Achromycin at a concentration of 1.3–1.6 μg/ml is used as an antibiotic by some workers (e.g., Bornstein, 1973*a*; Corrigall *et al.*, 1976; Dreyfus *et al.*, 1977), while most laboratories use no antibiotics.

Cultures obtained by the Maximow technique have been studied with a variety of standard bioelectrical, biochemical, histological, and ultrastructural techniques. Some of the special methods are discussed separately in Section 8.

Such cultures have been used in a large number of studies (see review by Herschman, 1974). Some of the more recent applications include toxicological studies such as the effect of chlorpromazine, mescaline (Roubein *et al.*, 1973), and puromycin (Meller *et al.*, 1974) on mouse cerebellum; of chloroquine (Tischner, 1975) and LSD-25 (Roizin *et al.*, 1974) on rat spinal cord and ganglia; of iron-containing substances (Kristensson and Bornstein, 1974), aminopiperazines (Dubois-Dalcq *et al.*, 1973), and experimental hypocholesterolemic drugs (Zagoren *et al.*, 1975) on spinal cord and ganglia. Maximow cultures have been used in many neuropathological investigations (reviewed by Murray, 1971; Silberberg, 1972). Some of the more recent neuropathological reports include studies on experimental allergic encephalomyelitis (e.g., Bornstein, 1973*b*; Ulrich and Bornstein, 1973; Dorfman *et al.*, 1976), amyotrophic lateral sclerosis (Wolfgram and Myers, 1973), metachromatic leukodystrophy (Stern, 1973), muscular dystrophy (C.V. Paul and Powell, 1974), fetal anoxia (Privat *et al.*, 1972), and Refsum's disease (Dubois-Dalcq *et al.*, 1972*b*). A large number of neurophysiological studies (reviewed by Crain, 1976) and pharmacological studies (e.g., Crain, 1974) have been made with Maximow cultures. Several putative neurotransmitters have been examined using such cultures (e.g., L. Hösli *et al.*, 1976). These organotypic cultures have found recent applications in bacteriology (Fildes, 1974) and virology (Dubois-Dalcq *et al.*, 1972*a*; Kristensson *et al.*, 1974; Raine *et al.*, 1973; Feldman *et al.*, 1972; Willson *et al.*, 1974; Whetsell *et al.*, 1977; Ecob-Johnston, *et al.*, 1979), in nutrition research (Allerand, 1972), and in gerontology (Spoerri and Glees, 1973). In the field of developmental neurobiology, cultures have provided valuable information about the development of central neural connections (reviewed by Crain and Peterson, 1974; Crain, 1976).

The Maximow double-coverslip technique provides highly differentiated "organotypic" cultures, and is probably the best method yet developed to promote differentiation within CNS explants (Crain, 1976). However, this technique has several disadvantages: (1) the preparation of the glass depression slides and other associated glassware is very laborious; (2) the feeding of the cultures is a time-consuming task that impedes the rapid introduction of experimental solutions and increases the risk of contamination during medium changes; (3) while the small amount of medium used for each culture (about 0.1 ml) is very economical, only a very small volume of neural tissue can be supported and frequent feeding is necessary; and (4) the concave well impedes the use of phase optics for microscopic examination. If necessary, the optics can easily be improved by the use of phase slides (depression slides with a flat depression) instead of the concave Maximow depression slides, or, alternatively, the cultures can be transferred under sterile conditions for microscopic observation to a sandwich chamber with good optical properties (Billings-Gagliardi and Wolf, 1977).

To overcome some of these disadvantages, a variety of other culture chambers have been developed.

3.2.2. Rose Perfusion Chamber

The Rose perfusion chamber (Rose *et al.*, 1958) consists of two cover glasses on either side of an inert rubber gasket clamped between two suitable metal plates. The design of the chamber has been modified to fit particular circumstances (Sykes and Moore, 1959). The cultures are usually covered with a strip of dialysis membrane (to encourage spreading), or a film of collagen can be combined with a dialysis membrane cover (Hendelman and Booher, 1966). Feeding is accomplished by inserting sterile syringe needles through the rubber gasket. Permanent ports can be provided for perfusion or for rapid changes of the medium for pharmacological or toxicological studies during undisturbed microscopic observation (Pomerat, 1962).

This method eliminates several of the disadvantages of the Maximow technique. The optical properties of the Rose chamber allow the use of high-magnification phase-contrast optics, and the Rose chamber is therefore a favorite chamber for microcinematographic studies of cellular dynamics.

The degree of maturation of cultures maintained in Rose chambers has not been investigated as thoroughly as that in the Maximow cultures. Nevertheless, Rose-chamber cultures survive for prolonged periods of time, and myelin (e.g., Kitano and Iwai, 1974) as well as functional synaptic contacts are formed in culture (Purves *et al.*, 1974).

To prepare Rose-chamber cultures, the tissue fragments are placed on properly washed and sterilized No. 1, 45 × 50 mm, cover glasses that have usually been coated with collagen as used for the Maximow technique. They are then often covered with a strip of Visking dialysis tubing (Visking Co., Chicago, average pore radius 24 Å). The dialysis membrane is stretched only enough to remove wrinkles. A rubber gasket (usually about 3 mm thick and with an inside diameter of about $\frac{1}{2}$ inch) and another cover glass are placed on top of the cover glass bearing the tissue fragments. The resulting chamber is held together with two metal plates fastened with four screws or with two special clamps (Sight Instruments, Irvine, California). The procedure for the preparation of the cover glasses is the same as employed in the Maximow technique (see Bornstein, 1973a). The dialysis membrane is prepared by immersing the properly cut strips in 95% pure ethanol (redistilled) for a few minutes and then aseptically rinsing them through several changes of sterile culture medium.

About 1.8 ml medium is introduced through a sterile syringe needle pierced through the rubber gasket. A second needle is provided as vent. The medium is changed similarly twice weekly by insertion of syringe needles through the rubber gasket.

A variety of tissues have been grown in Rose chambers:

1. *Dorsal root ganglia from newborn rats* can be cultured with or without collagen-coated glass in a medium containing ascitic fluid (45%), Gey's BSS (50%), chick embryo extract (5%), and 12 mg/ml glucose (Quastel and Pomerat, 1963).

2. *Embryonic guinea pig cerebellum* has been cultured on collagen-coated coverslips, but without dialysis membranes, in a medium consisting of 75% Eagle's MEM, 20% calf serum, 5% chick embryo extract, and 0.5 mg/ml glucose (Hauw *et al.*, 1974).

3. *Rodent and chick sympathetic ganglia* have been grown on collagen and under dialysis membranes, in a medium containing 80% Medium 199, 20% fetal calf serum, 0.05 U/ml insulin, 5 mg/ml glucose, and 100 U/ml penicillin and sometimes 1 U/ml nerve growth factor (Chamley *et al.*, 1972; Heath *et al.*, 1974).

4. *Chick embryo acoustic ganglia* have been cultured directly on glass, under dialysis membranes, in a nutrient medium containing heat-inactivated newborn calf serum (30%), chick embryo extract (10%), Eagle's MEM (60%), 100 mg/ml penicillin, and 100 μg/ml streptomycin (Orr, 1965).

5. *Chick embryo dorsal root ganglia* seem to prefer collagen-coated cover glasses. They are successfully grown in a medium consisting of Eagle's MEM (60%), bovine serum (20%), whole-egg ultrafiltrate (70%), 12 mg/ml glucose, 100 U/ml penicillin, and 100 μg/ml streptomycin (Hendelman and Booher, 1966).

6. *Newborn guinea pig enteric ganglia* have been cultured on polylysine- and collagen-coated coverslips in Medium 199 supplemented with 10% fetal calf serum and 5 mg/ml glucose (Jessen *et al.*, 1978).

The Rose-chamber technique is a favorable method to investigate cellular dynamics of nerve tissue under a variety of experimental conditions such as studies on the specificity of nerve–effector organ contacts (Chamley *et al.*, 1973a,b; Mark *et al.*, 1973; Hill *et al.*, 1976). Rose chambers have also been used for pathological studies (e.g., Kitano and Iwai, 1974) and pharmacological studies (e.g., Hill *et al.*, 1973; Heath *et al.*, 1973).

The major disadvantage of the Rose chambers and other similar chambers with machined components is their high cost. In an active laboratory, several hundred cultures are often maintained simultaneously, and the purchase, maintenance, and preparation of the various parts is both expensive and time-consuming. In addition, the quality of the cultures of nervous tissue may not be as high as that obtained with other culture systems; in particular, the Maximow technique usually gives superior results (e.g., see Hauw *et al.*, 1974).

In recent years, several laboratories have attempted to devise culture chambers that give well-differentiated long-term cultures similar to the ones obtained with the Maximow technique, but that are less time-consuming to prepare and maintain. These newer methods take advantage of the availability of relatively cheap sterile plastic petri dishes that have been designed specifically for tissue culture.

3.3. Long-Term Cultures in Petri Dishes

3.3.1. Aclar Minidishes

R.P. Bunge and Wood (1973) have developed a culture dish that retains the advantages of the Maximow technique, namely, the use of a small amount of medium forming a shallow layer over the tissue and the capability for serial observation with high-resolution light microscopy. At the same time, these workers attempted to increase the amount of medium slightly so that more tissue could be cultured in one dish.

A shallow dish (25-mm diameter) is pressure-molded with an aluminum punch and die from sheets of 5-mil Aclar 33C (Allied Chemical Co.). These dishes are cleaned in 90% nitric acid, rinsed well in tap water and distilled water, and finally sterilized in 80% ethanol, dried, and then coated with collagen (Bornstein 1958, 1973a; Masurovsky and Peterson 1973). These Aclar minidishes are then placed into modified polystyrene petri dishes. The modification consists of cutting a 30- to 35-mm-diameter hole in the bottom of the petri dish and gluing a No. 3 Gold Seal (Clay Adams) coverslip over the opening with Silastic adhesive (medical grade, Type A, Dow Corning). This modification reduces the length of the optical path for high-magnification observation of the living cultures on an inverted microscope, and can be omitted if the experiment does not require high-power microscopic observation. The cultured tissue fragment is placed onto the collagen coat as in the Maximow technique and fed twice weekly with about 0.25 ml nutrient medium. The medium stays entirely within the minidish. Although these plastic petri dishes have tight-fitting lids, polystyrene is permeable to gases (including water vapor), and therefore the cultures have to be incubated in a humidified (98%) atmosphere containing 5% CO_2 to maintain the bicarbonate buffer at the proper pH.

Rat spinal cord, dorsal root ganglia, and olfactory bulb have been cultured in this fashion in a medium consisting of 65% Eagle's MEM, 25% human placental cord serum, 10% chick embryo extract, 6 mg/ml glucose, and sometimes 1.6 µg/ml achromycin (R.P. Bunge and Wood, 1973; Blank *et al.*, 1974; Pfenninger and Bunge, 1974). Embryonic rat superior cervical ganglia, spinal cord, and cerebral cortex have been cultured by this method in a similar medium except for the addition of crude extract of mouse salivary gland (Olson and Bunge, 1973).

This technique has the advantage that the tissue can be prepared for electron microscope directly in the Aclar minidish, since Aclar does not adhere to Epon (see below). The major disadvantage of this method is the considerable effort required for the manufacture and preparation of the minidishes. The dishes can be recycled, but the washing and sterilizing procedure again is time-consuming and carries the risk of introduction of toxicities into the culture system, a risk that is inherent whenever glassware is recycled.

3.3.2. Coverslip Cultures in Petri Dishes

A simplified version having essentially all the advantages of the minidish technique is to simply place a collagen-coated No. 1 coverslip into a plastic petri dish (J.F. Schneider and Rue, 1974). Displacement of the coverslip can be prevented by making a shallow groove along the sides of the coverslip with a warm sterile metal instrument (e.g., tip of forceps). The explant is placed on the collagen and a drop of medium (0.05–0.10 ml) is added as in the Maximow technique; care is taken to confine the medium to the glass coverslip. The incubation is carried out in a humidified atmosphere containing 5% CO_2. Rat embryo spinal ganglia, newborn rat motor cortex (J.F. Schneider and Rue, 1974), and mouse spinal cord primordia (Juurlink and Fedoroff, 1979) have been cultured in this fashion.

Cerebellar cultures of newborn rat that develop abundant myelin have been grown on collagen-coated coverslips in 35 × 10 mm plastic tissue-culture petri dishes in 0.4 ml medium containing 40% Earle's BSS, 20% Earle's MEM, 40% fetal calf serum, 6 mg/ml glucose, 0.4 U/ml zinc-free insulin, 0.002% phenol red, and 1.7 µg/ml tetracycline and 10 mM HEPES buffer. As in all petri-dish culture systems, incubation has to be in humidified air containing 5% CO_2 (Benjamins *et al.*, 1976). Dorsal root ganglia of chick embryos have been cultured for up to 7 weeks on collagen-coated glass or Melinex (ICI Ltd.) coverslips in petri dishes in a medium containing 10% horse serum and 90% Eagle's basal medium. Such cultures contain well-matured neurons and develop abundant myelin (Bird and Lieberman, 1976).

Some workers have used gelatin-coated coverslips in 35 × 10 mm plastic petri dishes for cultures of embryonic mouse spinal cord and dorsal root ganglia. In a medium consisting of 10% horse serum and 2.5% chick embryo extract in Medium 199, with 6 mg/ml glucose, 50 U/ml penicillin, 50 µg/ml streptomycin, and 10 mM HEPES buffer, such cultures did not develop myelin, but did form neuromuscular junctions with co-cultured muscle (Witkowski and Dubowitz, 1975).

Some nervous tissues can be successfully grown directly on the plastic surface of Falcon plastic dishes or on uncoated glass coverslips simply placed into Falcon plastic dishes (Lapham and Markesbery, 1971). In a later report from the same laboratory (Choi and Lapham, 1974), the coverslips were coated with commercial rat-tail collagen (Gibco). Explants from the cerebellum and cerebral cortex of human fetuses have been maintained in this way for up to 5 months in a medium containing 78% Medium F12, 20% fetal calf serum, 1% fresh L-glutamine (200 mM), 1% nonessential-amino-acid solution (Gibco), and 6.8 mg/ml glucose. Such cultures did not develop myelin, but the organization and maturation seem to be organotypic.

3.3.3. Sandwich Cultures

Sympathetic ganglia of chick embryos have been grown for up to 2 months sandwiched between the bottom of a 35-mm plastic tissue-culture dish and a small coverslip placed on top of the explanted ganglia and anchored with silicone high-vacuum grease (Hervonen and Rechardt, 1974). The "micromilieu" thus formed around the ganglia seems to encourage abundant fiber outgrowth, especially during the first few days of culture. The nutrient medium consisted of 90% Medium 199 and 10% fetal calf serum, was buffered with 0.05 M HEPES or TRICINE [*N*-tris(hydroxymethyl)methyl-glycine (Sigma, St. Louis, Missouri)] at a pH of 7.1–7.4, and contained 1–2 U/ml nerve growth factor (Burroughs-Wellcome) and 100 U/ml penicillin. The glucose concentration for optimum growth was found to be 3–6 mg/ml.

For biochemical experiments, a large amount of cultured tissue is often a necessity. A very simple and fast method for culturing large amounts of tissue has been developed by Breen and deVellis (1975) and described in detail by Cole and deVellis (1976). In this method, 8–10 explants (of embryonic rat cerebrum, cerebellum, midbrain, and brainstem) are sandwiched between two 3 × 4 cm sheets of perforated cellophane (Microbiological Associates)—properly prepared and sterilized (F.M. Price and Handleman, 1975)—and placed into 60-mm plastic tissue-culture dishes. Then, 3 ml medium is added and exchanged every 2 days. The medium consists of 50% Eagle's basal medium with Hanks' salts, 25% Gey's BSS, 25% heat-inactivated fetal calf serum, 0.1% glutamine, and 6 mg/ml glucose. Myelin develops in such cultures during the second week *in vitro*, and the maturation and development of the culture are described as comparing favorably with those achieved with the Maximow technique.

Fulton (1960) has developed a method of squashing organ fragments under polythene discs (Alkathine brand of polythene film, ICI). This method has been adapted for the long-term culture (up to 84 days) of adult rhesus monkey motor neurons in a chemically defined medium (Tsiquaye and Zuckerman, 1974). However, this method has not been sufficiently evaluated to allow a conclusion as to its general usefulness.

3.3.4. Summary

The preparation of petri-dish cultures is considerably simpler than that of Maximow- or Rose-chamber cultures. Except for the substrate bearing the cultures (coverslips or cellophane) and the dissecting instruments, disposable sterile plastic can be used for all the supplies needed for the preparation of the cultures (e.g., culture chambers, petri dishes for dissection and temporary tissue storage, media bottles, pipettes). The degree of maturation and differentiation of such cultures may in some instances attain nearly that obtained with the Maximow technique (e.g., Benjamins *et al.*, 1976). This is particularly true of the petri-dish methods employing small quantities of medium (the medium just forming a liquid film over the explant), methods that were essentially designed to simplify yet at the same time simulate the Maximow technique. On the other hand, the petri-dish methods that use large amounts of medium have, in general, not been so successful in attaining an optimum maturation, possibly because a deep layer of medium prevents sufficient oxygen from reaching the tissue. In addition, there exists the possibility that polystyrene may have subtle neurotoxic effects (Mithen *et al.*, 1980). In view of the widespread use of polystyrene in tissue culture work, this report should be carefully replicated and expanded to check under which conditions the use of polystyrene is not advisable.

3.4. Tube Cultures

3.4.1. Flying-Coverslip Cultures in Roller Tubes

For many experiments, in particular biochemical or virological studies, a large volume of medium readily accessible to the culture would be an advantage. To date, this is probably best achieved by the use of test tubes in roller drums as culture vessels. The larger amount of medium used permits a relatively longer interval between successive feedings, thus minimizing chances for contamination and sparing the cultures from the trauma of frequent disturbances. The most popular test-tube method is the technique of flying coverslips in roller tubes. Test-tube cultures have the additional advantage of relative simplicity of preparation.

Roller tubes were used by Hogue (1947), and Costero and Pomerat (1951) added flying coverslips to carry the explants. The use of a relatively large amount of nutrient medium (0.7–2.0 ml) allows lower concentrations of serum and embryo extract in the medium and requires a less frequent feeding schedule. The rolling action results in repeated draining, aeration, and refeeding of the cultures. Under these conditions, glial and mesenchymal migration seems to be encouraged. After approximately 2 weeks in culture, the three-dimensional arrangement of neurons becomes transformed into a nearly two-dimensional array of nerve cells embedded in a matrix of dendrites, axons, glial cells, and glial processes thin enough to allow good visualization with the phase-contrast microscope. While such cultures appear less complex than the thick Maximow-chamber cultures, the degree of maturation and differentiation seems similar to that obtained in cultures with the Maximow technique (Kim, 1973). Favorable visibility, accessibility of the neurons, and the relative ease of preparation of the cultures are the major advantages of the roller-tube technique.

Either collagen-coated coverslips (Bornstein 1958, 1973*a*) or coverslips bearing a plasma clot can be used to carry the cultures.

3.4.1a. Plasma-Clot Cultures. Two or three explants are placed into a drop of heparinized chicken plasma on a 11 × 22 or 12 × 50 mm No. 1 glass (Gold Seal) or fluoroplastic (Aclar) coverslip. The plasma is then clotted by adding a drop of chicken embryo extract (sometimes diluted with BSS to as little as 40%) or 0.2 mg/ml thrombin (Gähwiler and Dreifuss, 1979) and thoroughly mixing the two drops by stirring with a fine instrument (such as the tip of the dissecting knife used to transfer the explants). After the clot has solidified, the coverslips are inserted into test tubes, either 16 × 150 mm Pyrex tubes or Falcon plastic culture tubes (e.g., Falcon No. 3026 or 3033). Plastic tubes are preferable because of the lack of risk of introducing toxic substances from the washing procedure. In addition, the sharp corners of the glass coverslip get anchored at the bottom of the plastic culture tubes and do not slide as the roller drum turns. This ensures uniform draining, aeration, and refeeding. Tubes with screw caps are less likely to become contaminated than stoppered tubes with unprotected rims. If regular glass test tubes are used, they should be stoppered with silicone rubber stoppers. When small (11 × 22 mm) coverslips are used, the cultures are fed once or twice weekly with 0.7–1.0 ml nutrient medium. With the larger coverslips (12 × 50 mm), 2.0–2.5 ml medium is added to the test tubes and exchanged once every 5–7 days. The feeding is most easily accomplished by removing the spent medium with a small sterile disposable plastic pipette (e.g., 1-ml serological pipettes) connected to a gentle vacuum, taking care not to touch the culture with the tip of the pipette. The cultures are incubated in roller drums at 36°C. If plastic culture tubes are used, the atmosphere in the incubator should be humidified and contain 5% CO_2. The roller drums are titlted about 4–5° with respect to the horizontal axis to confine the culture medium to the bottom of the tube. The drums rotate at 8–12 revolutions/hr.

A variety of mammalian and avian nervous tissues have been grown in plasma clot cultures in roller tubes:

1. *Cerebellum and midbrain of newborn rats* have been cultured in a medium consisting of 25% fetal calf serum or horse serum, and either 75% Gey's BSS or 25% Hanks' BSS and 50% Eagle's basal medium with glucose increased to 5.5–6.5 mg/ml (Schlapfer *et al.*, 1972*b*; Gähwiler, 1975*b*, 1976) or 14% fetal calf serum and 2% chick embryo extract in Eagle's basal medium (Cechner *et al.*, 1970). Sometimes higher concentrations of sera are used, such as 45% calf serum, 5% chick embryo extract, and 50% Gey's BSS (Hild and Tasaki, 1962).

2. *Hypothalamus of neonatal mice* has been cultured in 84% Eagle's basal medium, 14% calf serum, and 2% chick embryo extract (Geller *et al.*, 1972), and hypothalamic explants of newborn rats have been cultured in 25% horse serum, 50% Eagle's basal medium, and 25% Hanks' BSS (Gähwiler and Dreifuss, 1979).

3. *Midbrain (mesencephalic root of fifth nucleus) and cerebellar folia of newborn kittens* have been grown in a medium composed of 50% Gey's BSS, 45% human ascitic fluid, and 5% chick embryo extract (Hild, 1957*a,b*).

4. *Retinal tissue of newborn rat* has been grown in plasma clots on coverslips in roller tubes in a medium containing 45% inactivated calf serum, 5% chick embryo extract, and 50% Gey's BSS (Hild and Callas, 1967; M.M. LaVail and Hild, 1971).

As usual, all the sera used in these media are heat-inactivated, and the glucose concentration is raised to 3–6 mg/ml.

The unknown composition of the plasma clot may interfere with some experimental procedures, and a collagen substrate may therefore be preferable. On the other hand, collagen-coated coverslips have the disadvantage of requiring a more time-consuming preparation.

3.4.1b. Roller-Tube Cultures on Collagen-Coated Coverslips. The procedure used to prepare roller-tube cultures on collagen-coated coverslips is essentially similar to the one used for plasma-clot cultures. The coverslips are prepared as in the Maximow technique according to the method of Bornstein (1958,

1973*a*). After the explants have been placed onto the collagen-coated coverslips, the coverslips are gently inserted into the culture tubes containing nutrient medium. The tubes are then left in a static position at 36°C for up to 24 hr prior to the commencement of the rolling action. The position of the coverslip in the culture tube should be such that the explant is near the liquid–air interface and covered with a thin film of medium. This gives the explants a chance to attach themselves firmly to the collagen substrate.

1. *Spinal cord, cerebral, and retinal cultures of chick embryos* have been grown on collagen-coated coverslips in roller tubes in a medium composed of 33% horse serum, 33% Medium 199, and 33% Hanks' BSS (Cho *et al.*, 1973, 1974; Kim, 1971*b*; Tunnicliff *et al.*, 1974; Kim and Tunnicliff, 1974), in 62% Medium 199, 33% horse serum, and 5% chick embryo extract (Fisher *et al.*, 1975), or in 30% bovine serum, 20% chick embryo extract, 30% Medium 199, and 20% Hanks' BSS (Ciani *et al.*, 1973). In all cases, the glucose concentration is increased to 3–6 mg/ml.

2. *Cultures of hippocampus, olfactory bulk, cerebral neocortex, and cerebellum of newborn mouse* have been grown in 33% horse serum, 33% Medium 199, 33% Hanks' BSS, and 6 mg/ml glucose (Kim, 1971*a*, 1972*a,b*, 1973). Cultures of newborn mouse cerebellum have been successful in 66% Earle's MEM, 34% horse serum, and 10 mg/ml glucose (Kim and Rizzuto, 1975). Neonate rat cerebellar cultures have been grown in 50% Earle's BSS and 40% fetal calf serum, supplemented with 6.5 mg/ml glucose, 0.3 U/ml zinc-free insulin, and 0.002% phenol red (Latovitzki and Silberberg, 1977).

3. *Hypothalamic cultures of rat embryos* have been grown in a medium consisting of Eagle's basal medium containing 15% fetal bovine serum, 1.6% chick embryo extract, and 4.7 mg/ml glucose (Geller, 1976).

4. *Human fetal cerebellar cortex* has been cultured in a medium composed of 33% horse serum, 33% Medium 199, 33% Hanks' BSS, and 10 mg/ml glucose (Kim, 1976*a*).

5. *Organs of Corti* of newborn mice have been cultured in a medium consisting of 40% fresh horse serum, 32% Simms' BSS, 25% Eagle's MEM plus glutamine, 20 mM HEPES buffer, and 3% glucose solution in Simms' BSS (20 g/100 ml) (Sobkowicz *et al.*, 1975).

6. *Retinal tissue from adult dogs* has been cultured in 80% Eagle's MEM, 10% horse serum, 10% fetal calf serum, 9 mg/ml glucose, 100 units/ml penicillin and 100 μg/ml streptomycin (Messing and Kim, 1979).

Flying-coverslip cultures in roller tubes have been used for a variety of pathological and toxicological studies (Kim and Rizzuto, 1975; Kim, 1975; Graham *et al.*, 1975; Kim and Wenger, 1973; Lumsden *et al.*, 1975), and for various neurochemical (e.g., Oh *et al.*, 1972; Kim *et al.*, 1972; Tunnicliff *et al.*, 1974; Kim and Tunnicliff, 1974; Cho *et al.*, 1974, Pleasure and Kim, 1976), neurophysiological (Hild and Tasaki, 1962; Schlapfer *et al.*, 1972*b*), and pharmacological investigations (Gähwiler, 1975*a,b*, 1976; Geller, 1976).

3.4.1c. Roller-Tube Cultures under a Dialyzing Membrane. It has also been possible to obtain myelinating and apparently well-maturing cultures of embryonic chick spinal cord by growing explants on uncoated glass coverslips in roller tubes. The explants are covered with a piece of Visking dialyzing membrane (Van Waters and Rogers, Inc., No. 25225) secured to the coverslip with a small plasma clot placed at each end. The medium consists of 60% Eagle's basal medium, 20% fetal bovine serum, and 20% whole-egg ultrafiltrate with a final glucose concentration of 6 mg/ml (Raiborn and Massey, 1965).

3.4.1d. Cultures on Cellulose Acetate Substrate in Roller Tubes. Nervous tissues (cerebellar cortex, caudate nucleus, hippocampus, mesencephalic tegmentum, and spinal gray matter) from young adult rats have also been cultured on cellulose acetate strips (Oxoid electrophoretic strips, 2.5 × 1.0 cm) in plastic screw-capped 140 × 14 mm roller tubes (Kiernan and Pettit, 1971). The cultures were fed every other day with 1.0 ml HEPES-buffered medium consisting of 20% calf serum in Medium 199 and containing 5 mg/ml glucose, 100 μg/ml ampicillin, and 10 μg/ml amphotericin B (Fungizone). Cultures prepared in this fashion often showed considerable degeneration during the 2-week culture period, but it is not clear whether this is a consequence of the culture method or is due to the use of nervous tissue from adult rather than from embryonic or newborn animals. The survival can be enhanced by incorporation of 154 mg/ml lyophilized aprotinin (Trasylol, Bayer, West Germany). This presumably suppresses neuronal lysosomal proteinases activated by the trauma of explantation (Davis *et al.*, 1975).

The roller drums are the only special equipment required for the preparation of roller-tube cultures. The test tubes are preferably 16 × 150 mm polystyrene culture tubes, but Pyrex test tubes with silicone rubber stoppers can also be used. The development and maturation of roller tube cultures is very similar whether polystyrene or Pyrex culture tubes are used. In view of the recent report about the possible neurotoxic effects of polystyrene (Mithen *et al.*, 1980), carefully controlled comparative experiments between the two culture vessels might be appropriate. Flying-coverslip cultures in roller tubes are easy to prepare in large quantities and yield well-differentiated cultures. The plasma-clot method in particular is probably one of the simplest of the long-term culture methods for nervous tissue. The main disadvantage of the "flying-coverslip" method is the inaccessibility of the culture to continuous undisturbed microscopic observation. For high-power microscopic observation, the coverslip has to be removed from the test tube and sterilely transferred to either a Maximow slide, a Rose chamber, or some other special sterile chamber. After observation, the culture can be returned to the roller tube. Nonsterile short-term microscopic observation is most easily carried out on an open bridge slide manufactured by cementing small glass blocks (cut from a microscope slide) at an appropriate distance onto a 1 × 3 inch microscope slide so that the rectangular coverslip carrying the culture can be inverted between the glass blocks. The space between the coverslip and the microscope slide is then filled with BSS or medium. Cultures examined in this fashion are of course no longer sterile and cannot be used for experiments requiring continued sterility.

The use of Leighton tubes (special test tubes that have an optically flat area at the bottom) partially overcome this disadvantage. Leighton tubes can be placed onto an inverted microscope. However, high-power observation, in particular with phase-contrast optics, is still not feasible in Leighton tubes.

3.4.2. Leighton-Tube Cultures

Leighton tubes have been used as culture chambers for nerve-tissue cultures by a number of investigators. The tissue explants are usually placed on rectangular (e.g., 10 × 22 mm or 12 × 32 mm) collagen-coated coverslips (Hansson, 1966; Hauw *et al.*, 1972a, 1974) or into plasma clots on coverslips (Hansson, 1966). Bare glass coverslips (Hansson, 1966; Storts and Koestner, 1968) or glass coverslips with a cellophane strip covering the explant (Spoerri *et al.*, 1973) have been used in some cases in conjunction with Leighton tubes. Perforated cellophane [Microbiological Associates (for preparation, see F.M. Price and Handleman, 1975)] can also be used as a substrate for cell attachment (Grunnet, 1973). The quality of cultures grwon in Leighton tubes often compares favorably with that of cultures obtained by the Maximow technique (e.g., see Hauw *et al.*, 1974).

Explants are prepared in the usual fashion and placed onto the substrate (e.g., the coverslips). The coverslips are then placed into the flat depression of the Leighton tube and a small amount of medium (as little as 0.2 ml) is added. The tubes are usually gassed with a moist air 5% CO_2 mixture, stoppered with silicone rubber stoppers and placed in a horizontal position into a tube rack in the incubator. Each tube is carefully rotated to make sure that the medium just contacts the explant and forms a thin film over the explant. It is sometimes necessary (especially during the first few days of culture) to rotate the cultures several times per day to ensure that the medium adequately covers the explant.

1. *Retinal tissue of embryonic rabbits* has been cultured on glass coverslips, sometimes coated with collagen, sometimes in plasma clots, inserted into Leighton tubes (Hansson, 1966). The nutrient medium consisted of 80% Hanks' BSS, 20% heat-inactivated calf serum, and 5–6 mg/ml glucose.

2. *Dorsal root ganglia of rat embryos and embryonic guinea pig cerebellum* have been cultured on collagen-coated coverslips in 0.4 ml medium consisting of Eagle's MEM (75%), calf serum (20%), chick embryo extract (5%), 0.5 mg/ml glucose, and sometimes 50 μg/ml streptomycin and 200 U/ml penicillin (Hauw *et al.*, 1972a).

3. *Cultures of the locus ceruleus of newborn mice* have been grown on collagen-coated coverslips in Eagle's MEM (60%), and fetal calf serum (40%) with 6 mg/ml glucose added (Victorov *et al.*, 1978).

4. *Chick sympathetic ganglia* have been grown on uncoated Melinex 0 strips (ICI) in 66% Medium 199, 17% chick embryo extract, 17% fetal bovine serum, and containing 133 U/ml penicillin and 66 μg/ml streptomycin (Lever and Presley, 1971).

5. *Explants of newborn canine cerebellum* have been grown directly on glass coverslips (without collagen) in Leighton tubes. The nutrient solution (0.4 ml) consisted of 35% Gey's BSS, 40% fetal calf serum, 25% bovine serum

ultrafiltrate, and 6 mg/ml glucose (Storts and Koestner, 1968). Such cultures survive in apparently healthy condition for several months and show maturation including myelin formation.

6. *Neonatal canine hypothalamus* has been cultured under similar conditions in a medium containing 90% Eagle's MEM and 10% fetal calf serum (Sakai *et al.*, 1974).

7. *Embryonic mouse cerebral cortex* has been cultured directly on glass coverslips covered with perforated cellophane and inserted into Leighton tubes containing 1 ml nutrient solution consisting of 60% Ham-F10 medium, 20% fetal calf serum, 4% NCTC 109 medium, 5% lactalbumin (0.5 g in 100 ml Hanks' solution), 10% bovine amniotic fluid, 6 mg/ml glucose, and buffered with HEPES to adjust the pH to 6.7 (Spoerri *et al.*, 1973).

8. *Explants of newborn mouse cerebellum* have been grown on perforated cellophane strips (1 × 4 cm) (Microbiological Association, Bethesda, Maryland) previously inserted into Leighton tubes. Nutrient medium, 0.6 ml, consisting of 68% Eagle's basal medium with glutamine, 30% fetal calf serum, 2% chick embryo extract, and 5 mg/ml glucose, is added to the tubes about 5 min after the explants have been placed onto the cellophane strips. The medium is changed three times a week (Grunnet, 1973). The cultures develop myelin and survive for more than a month under these conditions. This method is extremely simple to use, but the degree of maturation and differentiation of these cultures has not yet been sufficiently evaluated.

3.5. Summary

A number of techniques for long-term organotypic cultures of mammalian and avian nervous tissue have been reviewed. The method of choice for any particular experiment may depend on such factors as time consumption and cost in addition to the suitability of the culture technique to the experimental design. While cultures on collagen-coated coverslips in Maximow assemblies probably represent the most sophisticated long-term system, the cost and effort involved in establishing and maintaining a tissue-culture laboratory to produce large numbers of Maximow cultures can often be prohibitive. Similar considerations are true for techniques involving Rose chambers. The use of disposable plastic for all supplies (except the coverslips), as is possible with cultures in petri dishes or in roller tubes, makes elaborate dishwashing facilities unnecessary. Furthermore, several laboratories have found that culture techniques can be simplified considerably while still obtaining adequate differentiation, maturation, and survival for meaningful experimentation. The degree of long-term maturation and development achieved by the Maximow technique is often not necessary for a particular experiment. This is particularly true in that an evaluation of the degree of organotypic maturation and development in cultures is usually based on observations that are difficult to quantify accurately. Healthy cultures are judged by morphological observations such as myelin formation, lack of gross necrosis in the explant, relative absence of macrophages in the culture, "normal"-looking neurons, and abundance of synaptic profiles; on neuro-physiological observations such as the presence of complex bioelectric activity; and on biochemical measurements such as the detection of some critical enzymatic activity.

These criteria are amply achieved by many of the culture methods discussed. For example, myelin formation has been observed in nervous tissue cultured in Maximow assemblies, Rose chambers, petri dishes, roller tubes, and Leighton tubes. Complex bioelectrical activity suggestive of synaptic interactions is found in Maximow and roller-tube cultures, and a variety of biochemical parameters have been measured in nearly all culture systems mentioned.

While all the methods discussed so far attempt to achieve some degree of development and maturation comparable to that in the *in vivo* counterparts, there are many experimental situations wherein this goal need not be achieved or may indeed be an undesirable attribute. For example, the question of the mechanism of nerve-fiber growth and elongation or the formation of neuromuscular junctions allows the use of culture systems specifically designed for optimum axon growth. For this question, maturation of the culture is not necessary nor is long-term survival a critical factor. Therefore, simplified culture systems may be quite adequate for short-term culutres of up to 7–10 days of incubation.

4. Short-Term Tissue Cultures of Mammalian and Avian Nervous Tissue

Short-term cultures of mammalian or avian spinal cord, dorsal root ganglia, or sympathetic ganglia are often used to investigate such problems as fiber outgrowth or the formation of neuromuscular junctions.

4.1. Plasma-Clot Cultures

Chick dorsal root and sympathetic ganglia have been grown in plasma clots (1 or 2 drops of chicken plasma, 1 drop of chick embryo extract) on cover glasses sealed as hanging drops over depression slides. For short-term cultures (a few days up to about a week), no medium other than the plasma clot is required (Lumsden, 1951; Levi-Montalcini *et al.*, 1954; Dunn, 1971). The cover glass bearing the plasma clot with the dorsal root ganglia explants can also be placed in Sykes and Moore (1959) chambers or in Falcon plastic petri dishes. Chicken plasma can also be clotted by the addition of an equal volume of thrombin containing Eagle's basal medium (Handel, 1971). Instead of plasma clots, a semisolid medium consisting of one drop 1% agar in Hanks' BSS and two drops of Medium 199 containing 10% chick embryo extract and 20% fetal calf serum as well as antibiotics and nerve growth factor has been used for chick dorsal root ganglion cultures (Yamada *et al.*, 1970).

4.2. Maximow-Chamber Cultures

Maximow tissue-culture chambers with collagen-coated coverslips have been used for measurements of neurite outgrowth from chick dorsal root ganglia under experimental conditions (e.g., Roisen *et al.*, 1972). For short-term cultures, media relatively low in sera (e.g., 10% fetal calf serum in MEM or Medium 199) or even without serum (Blood, 1975) can be used. The collagen coat can be omitted if the cultures are grown on Mellinex 0 (300-gauge, ICI) coverslips (Mottram *et al.*, 1972).

4.3. Petri-Dish Cultures

In recent years, polystyrene tissue culture dishes have become the preferred culture vessels for short-term cultures. Collagen-coated glass coverslips in petri dishes have been used to culture chick ciliary ganglia (Hooisma *et al.*, 1975) and embryonic mouse spinal cord (Jockusch *et al.*, 1979). The glass coverslip can be omitted by coating the floor of the plastic petri dish with collagen according to the method Bornstein (1958), Elsdale and Bard (1972), or P.J. Price (1975) [e.g., Kasuya and Okada (1974) for newborn rat cerebellum, Rubin *et al.* (1976) for chick spinal cord; Schultzberg *et al.* (1978) for chick embryo spinal ganglia].

Collagen-coated plastic multitrays (e.g., Linbro 76-033-05) have been used as culture vessels for short-term cultures of newborn rat superior cervical ganglia. Such cultures have been used to study the modulation of the neurotransmitter choice by glucocorticosteroids (McLennan *et al.*, 1980). No assessment of the morphological or physiological integrity of cultures prepared by this method has been published yet.

For some experiments, it is often an advantage to use a chemically well-defined substrate. In addition, it is often desirable to be able to cleanly and easily remove the ganglia from the culture plate (e.g., for biochemical analysis) with the least amount of substrate contamination. Both plasma-clot and collagen-coat techniques are cumbersome in this regard. The plasma clots have the additional disadvantage of preventing the rapid addition or removal of experimental agents to the culture. For this reason, Formvar-coated glass coverslips in plastic petri dishes have been used as a culture substrate [Koda and Partlow (1976) for chick embryo sympathetic ganglia], and some authors have reported the successful culture directly on the plastic surface of culture dishes [Maxwell (1976) for chick embryo neural tube, Letourneau and Wessells (1974) for chick dorsal root ganglia, Hervonen and Rechardt (1974) for chick sympathetic ganglia; Chen and Chen (1975) for mouse embryo superior cervical ganglia; Coughlin and Rathbone (1977) for mouse parasympathetic submandibular ganglia] or on "Melinex" (ICI) plastic [James and Tresman (1968) for chick embryo muscle and spinal cord, Mottram *et al.* (1972) for chick embryo sympathetic ganglia]. Attachment of the tissue to the surface of the petri dish sometimes presents a problem. For chick and embryonic mouse dorsal root and sympathetic ganglia, tissue attachment is considerably improved if the culture surface of Falcon tissue-culture dishes is covered for about 1 min with F12 medium containing 5%

ovalbumin (Mizel and Bamburg, 1976a). Ganglia are then added, excess fluid is removed, and each ganglion is covered with a thin film of the 5% ovalbumin solution in medium F12. After 15 min incubation, 3–5 ml medium F12 with 0.5% ovalbumin is added, and the dishes are then incubated at 37°C in humidified air containing 5% CO_2. Successful attachment of the ganglia to the culture plates is also achieved by coating the plates with fetal calf serum (Mizel and Bamburg, 1976a; Coughlin et al., 1977) or with poly-L-lysine (Mizel and Bamburg, 1976b; Yavin and Yavin, 1974).

Short-term tissue cultures have recently been used to assess the effect of drugs, hormones, or toxins (e.g., Kasuya, 1974; Kasuya et al., 1974; Hervonen and Eränkö, 1975; Kormano and Hervonen, 1976), of nerve growth factor (reviewed by Levi-Montalcini and Angeletti, 1968; Mizel and Bamburg, 1976a), and of tissue explants (Ebendal and Jacobson, 1977; Jockusch et al., 1979) on neurite outgrowth, and to investigate the role of RNA and protein synthesis in fiber outgrowth (Mizel and Bamburg, 1976b). Several authors have studied the formation of neuromuscular junctions with such short-term cultures (e.g., James and Tresman, 1968; Hooisma et al., 1975; Rubin et al., 1976).

5. Organ Cultures of Mammalian and Avian Nervous Tissue

The methods for organ culture are quite distinct from the techniques commonly employed for organotypic tissue culture. Whereas tissue-culture methods emphasize long-term differentiation and maturation of nervous tissue under conditions that usually encourage favorable visualization of the cellular components of the tissue, e.g., outgrowth and flattening of the explants for optimum microscopic observation of the living tissue, organ-culture methods discourage cellular outgrowth and thereby encourage maintenance of the tissue organization. Cellular migration from the explant contributes to tissue disorganization and alteration of the normal morphological relationships among the cells of the tissue explant. Tissue environments are therefore sought that limit cell proliferation and outgrowth. This is usually accomplished by providing the explant with a medium that is less rich in nutrients than used in organotypic tissue cultures. In addition, the substrate used for organ cultures is chosen such that outgrowth is discouraged. Transparent culture substrates or optically favorable culture dishes are not critical requirements, since the thickness of the explant and the lack of outgrowth preclude high-power microscopic observation of the living culture. The major advantage of the techniques of organ culture is the degree of control of the tissue environment and accessibility of the tissue to exogenous agents (e.g., drug exposures) not usually feasible in the intact animal.

Organ cultures are usually relatively short-term cultures (up to one or two weeks) and are typically used for biochemical experiments that require incubation periods of longer than a day or that require a recovery from the dissection trauma before the biochemical assay. Developmental studies that are analyzed with histological, histochemical, or electron-microscopic techniques are often conducted with advantage on organ cultures.

The general methods of avian and mammalian organ culture have been discussed in considerable detail by Wessells (1967). Many of these techniques also apply to the organ culture of nervous tissue. The methods of organ culture can be grouped according to the culture vessel used in the procedure. The most common method uses specially designed organ-culture dishes, but embryo watchglasses, petri dishes, Erlenmeyer flasks, scintillation vials, or other chambers have been used. The explant is usually placed at the liquid–air interface so that there is easy access of oxygen to the organ fragment. This also seems to encourage the rapid development of a layer of epitheliumlike cells around the explant, which helps to preserve its structure and integrity.

5.1. Raft Cultures

The most popular organ-culture method uses the so-called "raft" technique (Trowell, 1954, 1959). The explant is placed onto a sterile Millipore filter disk (e.g., Daniels and Moore, 1972; Goodman *et al.*, 1974; Mackay, 1974; Lyser, 1971; Kiernan and Pettit, 1971; Halgren and Varon, 1972; Farbman, 1974), a piece of lens paper (e.g., Daniels and Moore, 1972; Kim *et al.*, 1975*b*; McKelvy *et al.*, 1975; Kiernan and Pettit, 1971), or a piece of fiberglass filter (Whatman Glass Fibre Paper, type GF/A) (Coyle *et al.*, 1973) or cellulose acetate (Silverman *et al.*, 1973; Kiernan and Pettit, 1971) that rests on a grid made of stainless steel mesh (e.g., 60-mesh stainless steel cloth, wire diameter 0.0075 inch, or Falcon No. 3010, or minimesh expanded metal, 1.5-mm mesh size, The Expanded Metal Co., London, England) or surgical titanium gauze. The explants of some tissues can be placed directly on the metal grid [e.g., pineal glands of adult rats (Klein and Weller, 1970), superior cervical ganglia of adult rats (Silberstein *et al.*, 1971; Silberstein, 1976; Brown *et al.*, 1977)], or sometimes on a nylon mesh (ASTM-400-37, Tetco, New York) (Parfitt *et al.*, 1976) or on a stainless steel grid coated with a thin layer of 2% agar (Ito *et al.*, 1977). Explants have also been cultured in a fold of vitelline membrane from an unincubated egg (Wolff, 1961), which is then placed on a filter membrane on the supporting grid (Lyser, 1971, 1975). Some workers have used gelatin sponge foam (Spongostan, Ferrosan International, Copenhagen, Denmark, or Gelfoam, Upjohn Company, Kalamazoo, Michigan) as support for the tissue fragments (Rubinstein *et al.*, 1973; Sipe, 1976). The foam rests on a stainless steel grid and is moistened with medium several minutes prior to the addition of the freshly explanted cultures to the surface of the foam.

The supporting grids can span the center wells of organ-culture dishes (sterile disposable plastic, e.g., Falcon or glass) (e.g., Goodman *et al.*, 1974; Kim *et al.*, 1975*b*; McKelvy *et al.*, 1975; Lyser, 1971; Halgren and Varon, 1972; Farbman, 1974) or the wells of plastic tissue-culture trays (e.g., Model FB-16-24-TC, Linbro, New Haven, Connecticut) (Parfitt *et al.*, 1976); rest on the bottom of an embryo watchglass (Coyle *et al.*, 1973; Silverman *et al.*, 1973); or form a bridge (made by bending down the two ends of a rectangular metal grid), a triangular platform (a triangular metal grid with the corners bent down as short legs), or a circular platform with a curled edge (Klein and Weller, 1970) on the floor of a glass or plastic petri dish (Daniels and Moore, 1972; Kiernan and Pettit, 1971), a sterile glass scintillation vial (Bensinger *et al.*, 1974), or a small (5-ml) borosilicate glass pot (Mackay, 1974). Instead of metal grids as filter supports, some investigators have used Plexiglas strips containing small holes (usually two holes of 3-mm diameter). Millipore filter membranes are placed over each hole and attached with chloroform (for type TH Millipore filters) or acetone (for type HA Millipore filters) (Norr, 1973; Rubinstein *et al.*, 1973).

Culture medium is added to the level of the metal grid so that the explant is just bathed in the nutrient solution, with the top surface of the explant exposed to the atmosphere for gas exchange. The atmosphere surrounding the tissue is kept very humid by incubation of the organ cultures in a humidified gas mixture of 95% air and 5% CO_2. Often, moistened cotton or filter paper is arranged around individual culture vessels or around the center well of the organ-culture dish to increase the humidity of the atmosphere surrounding the explant. The culture medium usually consists of a synthetic medium (e.g., Medium 199, MEM, Duebecco's modified Eagle medium, medium BGJ[b]) supplemented by 5–30% fetal or neonatal calf serum or sometimes chick embryo extract, or both. Glucose (5–6 mg/ml) and antibiotics are often added to the culture medium. Nerve growth factor (100 U/ml, Burroughs Wellcome Co., Research Triangle Park, North Carolina) is sometimes included in the medium for organ cultures of sympathetic ganglia.

A variety of nervous and neurosecretory tissues have been successfully cultured with the raft technique, among them rodent sympathetic ganglia (Mackay, 1974; Goodman *et al.*, 1974; Silberstein *et al.*, 1971; Silberstein, 1976; Brown *et al.*, 1977); chick embryo cerebral cortex and retina (Daniels and Moore, 1972); chick embryo spinal cord (Lyser, 1971); cerebellar cortex, caudate nucleus, hippocampus, mesencephalic tegmentum, and spinal cord of young adult rats (Kiernan and Pettit, 1971); raphe nucleus (Halgren and Varon, 1972) and substantia nigra and corpus striatum of newborn rat (Coyle *et al.*, 1973); rat cranial ganglia (Farbman, 1974); hypothalamo-neurohypophyseal complex of adult guinea pigs (McKelvy, 1974; Kim *et al.*, 1975b; Pearson *et al.*, 1975) median eminence (McKelvy *et al.*, 1975; Silverman *et al.*, 1973); fetal mouse otocysts (van deWater and Ruben, 1971); and pineal glands (e.g., Klein and Weller, 1970; Bensinger *et al.*, 1974; Parfitt *et al.*, 1976).

The essential characteristics of the raft technique have been adapted to a perfusion chamber basically similar to the Rose chamber described in Section 3.2.2. Instead of supporting the Millipore filter disk by a metal raft, the filter is held between two rubber gaskets that are in turn pressed together by glass coverslips held by two metal plates as in the Rose-chamber assembly. This results in a perfusion chamber with two compartments separated by the Millipore filter. The explants are placed onto the Millipore filter in the top compartment, and medium is introduced with a syringe (using a second syringe needle as vent) through the rubber gasket until the lower chamber is completely filled (Goube de LaForest *et al.*, 1967). Permanent ports could be installed, making such a chamber valuable for perfusion studies.

Relatively long-term organ culture of a variety of nervous tissues has been achieved in a slight modification of the Goube de LaForest chamber. The Millipore filter is replaced by a silk screen on which several drops of 0.5% agar in nutrient medium have been placed. After the agar has solidified, the explants are placed onto the agar substrate, and nutrient medium (20–30% calf or fetal calf serum, and sometimes 5% chick embryo extract in Eagle's MEM with a final glucose concentration of 5–6 mg/ml) is introduced into the lower compartment. Apparently healthy organ cultures of cerebellum of newborn mice, of chick and human embryo spinal ganglia, and of human fetal cerebellum and medulla have been maintained for up to 6 weeks in this fashion (Hauw *et al.*, 1969, 1972*b*; Hauw and Escourolle, 1975).

5.2. Transfilter Cultures

Another important variation of the raft technique is the transfilter organ-culture method, first introduced by Grobstein (1956) and subsequently used to study embryonic induction of neural tissue (Lash *et al.*, 1957). This method allows tissues of two different origins to be placed on opposing sides of Millipore filter membranes to study their mutual interactions by diffusible substances.

The filter supports consist of Plexiglas strips ($25 \times 6 \times 1$ mm) into which two or more 3-mm holes have been drilled (Auerbach, 1960). Small disks of Millipore filter membranes are arranged over the holes and attached by touching either chloroform (for type TH filters) or acetone (for HA type filters) to the edge of the filter. These filter assemblies are then sterilized in several changes of 70% ethanol and well rinsed in sterile balanced salt solution. Explants of one tissue type are placed on the top surface of the filter membrane and are incubated in an organ-culture dish as raft cultures for 4–10 hr. This allows the attachment of the tissue to the filter membrane. The filter supports are then inverted, and the transfilter tissue is placed in the well of the filter support. The explant is stranded against the filter membrane by withdrawing the medium remaining in the well. A drop of medium is then quickly added to the well so that the tissue is not pushed away from the membrane by medium seeping back through the filter. After a period of incubation in the inverted position, the filter supports are turned face-up again for continued incubation (Norr, 1973).

Millipore filter disks can also be glued to one side of a Plexiglas ring. This forms a shallow dish with a Plexiglas side. Explants of one tissue type (e.g., dorsal root ganglia) can then be cultured on the floor of this Millipore filter disk placed filter side down on a stainless steel grid such that the explant is at the air–medium interface just as in raft cultures. After the tissue has attached itself firmly (this may take several days), the Plexiglas ring with its filter is inverted and used as a raft for the culture of the second tissue type. Medium is again introduced to the level of the explant (Globus and Vethamany-Globus, 1977).

5.3. Agar Technique

An organ-culture technique using a semisolid agar medium has been developed by Wolff and Haffen (1952). Outgrowth of glial cells and neurites seems to be inhibited by the agar substrate (Chida and Shimizu, 1973), and the explants therefore retain their organotypic organization.

Explants are placed on the solidified agar base (1% purified agar in nutrient medium) in a petri dish, a depression slide, an embryo watchglass, or the center well of an organ-culture dish. The explants are then covered with a plasma clot and with a small amount of liquid medium. Newborn rat cerebellar and cerebral cortical cultures prepared in this fashion produced abundant myelin during the third week in culture (Chida and Shimizu, 1973). Instead of embedding the explant in a plasma clot, the explant can be placed in a fold of vitelline membrane of chicken egg (Wolff, 1961). This method has been used for organ cultures of chick embryo spinal cord (Lyser, 1971). An agar substrate has also been used in a perfusion chamber (Hauw and Escourolle, 1975), as discussed in Section 5.1.

5.4. Suspension Cultures

For some experimental purposes, it is possible to culture tissue fragments in flasks, as suspension cultures on rotary shakers. For example, Piddington (1973) has maintained chick diencephalon in suspension cultures for 4 days for a study of the effect of hydrocortisone on glutamine synthetase.

5.5. Petri-Dish Cultures

Short-term cultures from adult mouse brains have been maintained on scratched grids in plastic petri dishes (ESCO AA) in a medium of Eagle's MEM with 0.2% bovine serum albumin (Armour). Such cultures have been used for virus propagation (Woodward and Smith, 1975).

Fetal mammalian otocysts with their associated statoacoustic ganglion can be cultured directly on the plastic surface of the center well of plastic organ culture dishes in 0.15 ml of culture medium (Van DeWater, 1976).

5.6. Flask Cultures

It may be desirable for some experiments (in particular, for biochemical investigations) to culture relatively large amounts of tissue. This can be done in petri dishes as discussed above, but an alternative yet not widely used method for nervous tissue is the cultivation of explants in tissue-culture flasks.

Explants are spread over the culture surface of 30-ml disposable plastic flask tissue culture with the tip of a capillary pipette and moistened with a small amount of medium. After incubation for up to 12 hr, which allows the explants to adhere to the plastic surface, several milliliters of growth medium is added to the flask. Incubation is in 95% air and 5% CO_2. Such cultures have been used to carry viruses (Rogers *et al.*, 1967) and to investigate lipid metabolism of nervous tissue (Menkes, 1972).

Better adhesion of the explants is achieved and cellular outgrowth is encouraged if the growth surface of the tissue-culture flasks is pretreated with 10% fetal calf serum in distilled water at 34°C for 48 hr. Embryonic rat cerebral cortex has been grown for up to 40 days in 250-ml Falcon tissue-culture flasks (No. 3024) (J.F. Schneider, 1973). Myelination of these cultures has not yet been achieved, and the method, though potentially valuable, has not yet been critically tested.

6. Tissue and Organ Cultures of Nervous Tissue of Cold-Blooded Vertebrates

Vertebrate experimental embryology has long been dominated by work on amphibian embryos, and in fact, some of the first experiments on the culture of nervous tissue were performed with amphibian material (Harrison, 1907).

The most important advantages of the use of amphibian embryos for tissue and organ culture include the plentiful availability of amphibian embryos, their hardiness, and the fact that amphibian embryonic cells contain an endogenous food supply that makes it possible to culture embryonic tissues or organ rudiments in simple defined salt solutions. In addition, the incubation of tissues from poikilotherms at room temperature reduces the growth of contaminating microorganisms.

While adult nervous tissue from mammalian CNS has been, with a few exceptions, relatively refractory to long-term culture, CNS tissues from adult amphibians and fish can be cultured for several months. This may be due to the remarkable regenerative capacities of the CNS of some lower vertebrates.

The culture methods previously discussed for warm-blooded vertebrates can be used with little adaptation, except the incubation temperature and the medium composition, for amphibians and other cold-blooded vertebrates.

The special techniques of amphibian experimental embryology have been treated in detail by Rugh (1967), and the general tissue- and organ-culture methods including rearing of embryos, dissection techniques, and media compositions have been described by Jacobson (1967) and more recently by Monnickendam and Balls (1973). Furthermore, several articles concerned with the culture of nonnervous tissues of various cold-blooded vertebrates, including fish and reptiles, can be found in Kruse and Patterson (1973). In general, these methods apply to nervous tissues as well.

Some recent examples of cultures of nervous tissue from amphibian embryos include segments of neural tube of *Ambystoma tigrinum*, *Rana pipiens*, and *Xenopus laevis* that were explanted on uncoated glass coverslips in Niu–Twitty balanced salt solution and maintained as column drop cultures at 17°C (Berlinrood *et al.*, 1972); frog's heart interatrial septum in 80% Leibovitz medium (Gibco) to which 1 mmol/liter $CaCl_2$ had been added (McMahan and Kuffler, 1971); frog embryo retinal tissue in hanging-drop cultures over depression slides (Hollyfield and Witkovsky, 1974); and retinal cultures from embryonic *Xenopus laevis* on collagen in 30-mm plastic petri tissue-culture dishes in an amphibian medium containing about 28% fetal calf serum (Agranoff *et al.*, 1976).

Adult amphibian tissue has been cultured by several investigators. A detailed description of the method used for long-term cultivation of larval frog spinal cord explants has been provided by Pollack and Koves (1975). This method is an adaptation of the Maximow technique for vertebrate tissue described in Section 3.2.1. Sympathetic ganglia from adult frogs have been cultured on collagen-coated coverslips under strips of dialyzing cellophane in Rose chambers (as discussed in Section 3.2.2) in a medium having an ionic composition similar to that of frog plasma and containing amino acids and vitamins as in Eagle's Minimum Essential Medium and 20% fetal calf serum (Hill and Burnstock, 1975). Explants from adult goldfish brains (DeBoni *et al.*, 1976) and retina (Landreth and Agranoff, 1976) have been cultured in plastic petri dishes coated with either collagen or poly-L-lysine (Heacock and Agranoff, 1977).

Organ cultures of amphibian nervous tissues are prepared in a fashion similar to mammalian organ cultures. For example, sensory ganglia from adult newts have been maintained as organ cultures together with muscle tissue on Dacron cloth supported by stainless steel grids in plastic organ-culture dishes in K. Wolf and Quimby (1964) amphibian medium (Gibco) (Lentz, 1971).

7. Tissue and Organ Cultures of Invertebrate Nervous Tissue

General introductions to the culture of invertebrate tissues can be found in recent collections (Lutz, 1970; Vago, 1971; Maramorosch, 1976). Nutrient media culture methods, and applications of invertebrate cell, tissue, and organ culture, are discussed in great detail. Invertebrate tissue culture has made many significant contributions in genetics, pathology, and physiology (see Vago, 1971). The tissue and cell culture of insect tissues, in particular, has been developed by many laboratories due to the importance of insects as vectors of many plant and animal diseases (e.g., see Chao, 1973; Maramorosch, 1976). Hence, advances in insect virology were preceded by advances in insect-cell culture. However, the use of invertebrate tissue culture in neurobiology has thus far been scarce. This is rather surprising, since a significant part of our knowledge of the functioning of the nervous system derives from experiments on invertebrate systems. Indeed, the tissue culture of invertebrate nervous tissue may offer some particular advantage over the culture of vertebrate nervous tissue. Invertebrates generally have segmented nervous systems consisting of many rather small ganglionic masses that can be explanted separately or as a whole with the connective nerves intact. The dissection trauma and the tissue disorganization brought about by the cutting of organ fragments are therefore minimal. Furthermore, invertebrates seem to have evolved rapid diffusion processes for molecules and ions (Treherne, 1967). This is an important advantage over vertebrate tissue culture, in which necrosis often occurs due to limited access of nutrients to the central part of the explant. The nerve-cell bodies of invertebrate ganglia are situated peripherally, which not only facilitates the access of oxygen and metabolites, but also allows easy experimental access to the nerve-cell bodies. In addition, the neuronal somas are often large (e.g., in some molluscs such as *Aplysia californica* or *Helix aspersa*), which is a great advantage in neurophysiological and biochemical investigations. Invertebrate neurons are often morphologically and functionally identifiable, and neuronal circuits can often be mapped (e.g., see Kandel, 1976). The culture of such ganglia or groups of ganglia would provide experimental advantages that cannot be realized in vertebrate tissue and organ cultures.

A variety of culture methods, usually derived from vertebrate culture techniques, have been used for invertebrate tissues. Petri dishes, depression slides, Rose chambers, organ-culture dishes, roller tubes, and flasks have been employed for invertebrate tissues.

7.1. Insects

The general methods for insect-tissue culture, including culture media and operative techniques, have been reviewed by I. Schneider (1967). The preparation of several media for culture of insect tissue was described in detail by Vaughn and Goodwin (1977). One of the major difficulties of preparing primary explants from insects is the problem of adequate sterilization of their organs. This is the result of the small size and the irregular surface of insects, as well as the presence of internal contaminants such as cryptograms and bacteria in the digestive tract and the trachea. The external surface can be partially sterilized by immersion of the adult insect, the larvae, or eggs in a disinfectant solution such as potassium chloride, mercury chloride, or 70% ethyl alcohol. Internal microorganisms, however, can often be eliminated only by aseptic rearing of insects. The relevant techniques have been described by Meynadier (1971). Brains or brain fragments of a variety of insect species have been cultured, among them *Calliphora erythrocephala* (Diptera) pupae (Demal, 1956; Leloup, 1970), tsetse-fly *Glossina palpalis* (Trager, 1959), *Drosophila melanogaster* larvae (Demal, 1956; I. Schneider, 1966), and *Manduca sexta* (Borg and Marks, 1973) in hanging-drop cultures, usually over depression slides; *Glossina palpalis* in Porter flasks containing 0.3 ml medium (Trager, 1959); *Calliphora erythrocephala* (Courgeon, 1969) and *Aeschna cyanea* (Schaller and Meunier, 1967) as organ cultures on agar;

Leucophaea maderae in Rose chambers under strips of dialysis membrane (Marks and Reinecke, 1965); and ventral nerve cords (thoracic and abdominal ganglia) of *Galleria mellonella* (Lepidoptera) in small culture chambers made from two 25-mm-diameter circular cover glasses held with Vaseline on either side of a silicone O-ring (Robertson and Pipa, 1973). In none of the cases described above have the surviving neurons in these cultures been analyzed in detail. Rather, in most cases, the neural tissue was used as a source of hormones released into the culture medium to stimulate the survival or growth of the co-cultured tissues.

Chen and Levi-Montalcini (1969) have described a culture system for brains and ganglia from 16- to 18-day-old embryos of *Periplaneta americana*. Such cultures survive and mature for many weeks in a chemically defined medium consisting of 5 parts Schneider's insect solution and 4 parts Eagle's basal medium, or in a specially designed cockroach tissue culture medium (Seshan, 1976). This culture system has been analyzed in considerable detail and therefore warrants a more complete description.

The egg cases are sterilized by thoroughly wiping them with alcohol. Brains, subesophageal, thoracic, and abdominal ganglia are aseptically dissected out under a dissecting microscope. The ganglia are transferred to glass microculture dishes (13 × 16 mm) containing a circular coverslip and 0.5–0.8 ml medium. The explants are gently pressed onto the coverslip, and they usually adhere quickly to its surface. The culture dishes are incubated in a humidified atmosphere of 95% and 5% CO_2 at 29°C. Such cultures show vigorous outgrowth of nerve fibers, glial cells, and tracheolar cells. Nerve fibers develop connections to nearby ganglia (reviewed by Levi-Montalcini *et al.*, 1973), and specific functional connections seem to be established among ganglia (Provine *et al.*, 1976). Neurons in such cultures exhibit spontaneous bioelectrical activity (Provine *et al.*, 1973) and show at least some biochemical maturation (Schlapfer *et al.*, 1972a). This method has been adapted to the culture of nervous tissue from nymphal and adult cockroaches (Seshan and Levi-Montalcini, 1971; Seshan *et al.*, 1974; Seshan, 1975).

7.2. Molluscs

The culture of molluscan tissue and organs has been reviewed by Bayne (1976) and by Gomot (1972). The culture methods have generally been adaptations of either the agar organ-culture technique (see Section 5.3) or a liquid culture system in flasks or petri dishes. Few studies have been made of the properties of molluscan nervous tissue in organ culture. *Helix aspersa* cerebral ganglia survived for less than 2 weeks (Guyard, 1969) when cultured on an agar substrate according to the method of Wolff and Haffen (1952) as discussed in Section 5.3. Cerebral ganglia of *Viviparus viviparus* L. were cultured under similar conditions by Griffond (1969). A liquid medium was used by Benex (1964) to culture tentacles (containing ganglion cells at the base of the tentacle) of the planorb *Austoalorbis glabratus*. In none of the aforementioned studies was the functional or even the detailed structural integrity of the culture nervous tissue evaluated in any way. Rather, the nervous tissue was simply observed to "survive" for a number of weeks, as evidenced by the lack of necrosis. In addition, the presence of nervous tissue in cultures of other organs (such as ovotestis, gonads, penis) lengthened the life span of these organs (Houteville and Lubet, 1974; LeGall *et al.*, 1974), presumably by releasing some beneficial factors into the medium.

The effects of long-term organ culture (up to 45 days) on the neuronal function of the ganglia of *Aplysia californica* were recently evaluated (Strumwasser and Bahr, 1966; Strumwasser, 1971; Dewhurst and Weinreich, 1974). Isolated ganglia were incubated in the dark at 14°C in screw-capped 50-ml Erlenmeyer flasks containing 25 ml medium consisting of 1 part Millipore-filtered *Aplysia* hemolymph and 4 parts sterile synthetic seawater (Instant Ocean, Wickliffe, Ohio) and containing 4% 50× Minimum Essential Medium (MEM) essential amino acids, 1% 100× MEM nonessential amino acids, 1% 100× L-glutamine, 1% 100× Eagle's basal medium vitamin mixture, 10 mg/ml glucose, and 100 U/ml streptomycin-penicillin. Neurophysiological parameters (synaptic activity, action potentials, resting potentials) remained stable over at least 3 weeks in culture, and the activities of some of the enzymes involved in neurotransmitter synthesis were not markedly affected by the culture for the first 2 weeks *in vitro* (Dewhurst and Weinreich, 1974).

The electrical properties of neurons in the circumesophogeal ganglia of *Limnaea stagnalis* have been investigated during long-term (up to 50 days) culture in a synthetic liquid medium supplemented with 2% bovine serum (Geletiuk and Veprintsev, 1972; Kostenko and Veprintsev, 1972). Evoked and spontaneous action potentials, resting potentials, membrane resistance, and capacitance were found to be similar to *in vivo* measurements. Such cultures exhibited vigorous glial cells and neurite outgrowth.

7.3. Other Invertebrates

Successful organ culture of the nervous tissue of the horseshoe crab [(*Xiphosura polyphemus*) on agar (Wolff, 1962)], arachnids [*Leiobunnum longipes* ganglia in Falcon plastic tissue-culture flasks (Fowler and Goodnight, 1966a,b)], and Annelids [*Hirudo medicinalis* on an agar medium (Malecha, 1967)] has been reported. In all these studies, the condition of the cultured tissues was assessed only by light-microscopic observation. A recent promising preparation is the long-term cultivation of leech (*Hirudo medicinalis*) ganglia (Miyazaki and Nicholls, 1976; Wallace *et al.*, 1977). The ganglia were maintained in petri dishes in L-15 medium containing 5% fetal calf serum, 6 mg/ml glucose, 100 U/ml each of penicillin and streptomycin, and 50 U/ml mycostatin. The medium was changed every 3 days, and the cultures were incubated at 16°C. The morphological and bioelectrical properties of the ganglion cells are maintained for at least 6 weeks in culture.

8. Special Methodologies Associated with Mammalian and Avian Tissue Culture

In general, standard morphological, biochemical, and physiological techniques can also be employed with cultured nervous tissue. There are, however, a few special techniques or special adaptations that have been devised to facilitate the structural and functional analysis of nerve-tissue cultures.

8.1. Histological Techniques

Tissue cultures are most often stained as whole mounts without embedding and sectioning. Organ cultures, on the other hand, are usually treated exactly like fresh tissue fragments using the appropriate histological methods.

8.1.1. Methods for Cell Localization

It is often desirable to compare a certain area of the culture as observed in the living state with high-power optics to the identical area after fixation and staining. This is also an important preparatory step if certain specific areas of a culture are to be examined with the electron microscope. Several methods have been described to mark a region of interest in a living culture.

Cultures on glass coverslips that are removed from their culture chamber for microscopic observation (such as roller-tube cultures) are most easily marked by using a diamond scribe mounted in a special holder in the turret of the microscope objective to inscribe on the coverslip a small circle around the center of the microscopic field. Such objective diamond markers are available from microscope manufacturers (e.g., Carl Zeiss No. CZ 462960). The coverslips of Rose chambers can be similarly marked, and hence one can quickly and easily return to the same location for repeated observation if desired. The location of the inscribed circle can be highlighted by a mark made with Labink, which is not removed by normal solvents. After fixation and staining of the culture, care should be taken that the mounting medium (e.g., Permount) does not flow over the back of the coverslip and obscure the engraved mark.

Cultures grown on collagen in Maximow double-coverslip assemblies or any other culture chamber (e.g., petri dishes) that does not allow access for such a diamond scribe to the reverse side of the coverslip bearing the living culture can be provided with a coordinate system for repeated observation of the same cells at the light- and electron-microscopic level. This is accomplished by inserting an Aclar film (Aclar 33c, 1.5 mil) inscribed with reticles under the collagen layer (Masurovsky *et al.*, 1971; Masurovsky and Peterson, 1975). Aclar discs are dipped into dialyzed collagen solution and placed engraved side up onto the cover glass that will bear the culture. Gelation of the collagen and the subsequent culturing procedure are then carried out as usual. Similarly, collagen-coated "cell finder" object glasses, i.e., glass coverslips provided with a grid marked with numbers and letters to facilitate the localization of cells (e.g., Micropure, Driebergen, The Netherlands) can be used as culture substrate (Slaaf *et al.*, 1979). For comparative observation of the same field in the living culture and after staining, a special alignment holder for histological slides can be used (Hauw *et al.*, 1974). The holder is of the same dimensions as the culture chamber used (e.g., Maximow slide or petri dish). The culture-bearing coverslip is provided beforehand with two marks made with a diamond knife. When observing the living culture in its culture chamber, the coordinates of the marks with respect to some designated edges of the culture chambers are measured on the horizontal vernier of the microscope. After staining, the coverslip is mounted on a histological slide held in the special holder and the coverslip marks are aligned under the microscope so that they lie on the previously recorded coordinates.

8.1.2. *Silver and Mercury Stains for Nerve Cells*

The Holmes stain, the Bodian Protargol stain, and the Golgi technique are of particular value for the demonstration of nerve cells and axons in culture. These staining methods have been specifically altered for use with tissue culture.

a. *The Holmes silver stain* (Holmes, 1947) has been modified by M.K. Wolf (1964) for whole-mount staining of cultures. The coverslips bearing the cultures are treated as follows:

1. Fix in 10% formalin in balanced salt solution (BSS) at 4°C for 1 week or more (up to 4 months). Young cultures are preferably fixed in Lillie's formalin (Sobkowicz *et al.*, 1973).
2. Immerse in 70% ethanol at 4°C for at least 1 week.
3. Rinse in 50% ethanol.
4. Wash in distilled water for 10 min.
5. Place in 20% aqueous silver nitrate in the dark at room temperature for 1 hr.
6. Wash for 10 min in three changes of distilled water.
7. Impregnate at 37°C overnight in about 10 ml of solution containing the following: 55 ml boric acid solution (12.4 g boric acid/liter water, 45 ml borax buffer solution (19 g borax/liter water), 394 ml distilled water, 1 ml 1% aqueous silver nitrate, and 5 ml 10% aqueous pyridine.
8. Reduce for 2 min in a solution of 1 g hydroquinone and 10 g sodium sulfite (crystals) in 100 ml distilled water (this solution should not be more than a few days old).
9. Rinse well in several changes of distilled water.
10. Place in 0.2% aqueous gold chloride for 5 min.
11. Rinse in distilled water.

12. Place in 2% oxalic acid in H_2O for 10 min.
13. Rinse in distilled water.
14. Place in 5% sodium thiosulfate.
15. Wash in water for 10 min.
16. Dehydrate in 95% ethanol and two changes of 100% ethanol.
17. Clear in three changes of xylol and mount.

b. *The Bodian Protargol method* (Bodian, 1936) has long been used on whole-mount nerve-tissue cultures (e.g., Murray and Stout, 1947). A recent adaptation of the technique has been published by Kim (1970, 1971*b*):

1. Fix in formol–ammonium bromide solution for at least 24 hr at room temperature (14 g ammonium bromide, 70 ml formalin, distilled water to 750 ml).
2. Place in 95% ethanol for 2–3 days at 36°C.
3. Place into freshly prepared 0.7% Protargol* solution (silver proteinate) containing 1 drop 20% aqueous pyridine solution and about 0.2 g copper shots or copper fragments per 10 ml Protargol solution for 20–24 hr at 36°C.
4. Transfer to a fresh 0.7% Protargol solution with pyridine, but without the copper shot, for 6–18 hr at 36°C.
5. Reduce in 1% hydroquinone in 5% formalin for 10 min.
6. Place in 1% gold chloride solution for 5 min.
7. Transfer to 2% oxalic acid for 2–3 min.
8. Transfer to 5% sodium thiosulfate for 2–3 min.
9. Dehydrate, clear, and mount by inverting the coverslip over a microscope slide.

A modification of the Bodian technique for cultured tissue has recently been devised that employs much shorter impregnation and development times at higher temperatures (Zagon and Lasher, 1977). The impregnation is carried out in a 1% Protargol solution in the presence of polished copper sheets at 50–60°C for 1 hr, followed by development for 10 min at 50–60°C in 1% hydroquinone in 5% sodium sulfite.

c. *The Golgi–Cox method* has been modified for thick whole-mount preparations of organotypic cultures by Toran-Allerand (1976*b*). The cultures are treated as follows:

1. Fix in 10% acrolein in 0.9% NaCl for 30 min at room temperature.
2. Rinse in 0.9% NaCl for 15 min.
3. Postfix in gluteraldehyde solution [25 ml 25% gluteraldehyde (Taab Labs, Great Britain), 17.5 ml 0.2 M NaOH, 25 ml 0.2 M KH_2PO_4, 32.5 ml distilled water], pH 7.3–7.4, for 3 hr (one coverslip/dish).
4. Rinse with 0.9% NaCl for 15–20 min.

---

* *Note:* Not all commercial Protargols are of satisfactory quality. Kim (1971*b*) recommends the following manufacturers: Rogue (France), Winthrop (United States), and Chroma (Germany). In the author's laboratory, Winthrop Protargol has been used with satisfactory results.

5. Impregnate, with the cultures in a vertical position, for 18–19 days *in the dark* at room temperature in 1 part 5% potassium dichromate, 1 part 5% mercuric chloride, 1 part 4% potassium chromate, and 2 parts distilled water. Alternatively, 1 part distilled H_2O can be replaced by 2% sodium tungstate; in this case, incubation should be for 22–26 days. Not more than three coverslips are placed in porcelain staining racks (Chen, A.H. Thomas) and immersed in at least 300 ml impregnating solution in a staining dish (Wheaton, Fisher Scientific). A Teflon-coated stirring bar is added, and the dishes are sealed with Parafilm, covered, and placed on a magnetic stirrer to provide continuous slow circulation during the entire period of impregnation.

6. Rinse in distilled water until no more precipitate leaves the tissue.

7. Transfer to clean racks and dishes (three coverslips/rack) and alkalinize in the dark 1.5–6.5 hr in 300 ml 0.5% (wt./vol.) lithium hydroxide and 15% (wt./vol.) potassium nitrate in distilled water.

8. Rinse for at least 18 hr in several changes of distilled water containing 1 ml glacial acetic acid/500 ml water.

9. Wash well (2 hr) in distilled water.

10. Dehydrate, clear, and mount in Technicon (Technicon Instrument Corp.).

8.1.3. Myelin Stains

Identification of myelinated axons in living cultures can be made by using polarized light in the optical microscope. The histological staining method most often employed in whole mounts of nerve tissue cultures to demonstrate myelin is Sudan black B (Peterson and Murray, 1955).

1. The cultures are rinsed in BSS at 36–37°C to remove lipids and other ingredients of the culture medium.

2. Fix in 10% formalin in Locke's solution or BSS.

3. Rinse in 35% ethanol for 2 min.

4. Transfer to 70% ethanol for 2 min.

5. Stain for 60 min in a saturated Sudan black B solution (about 1.5 g/100 ml 70% ethanol prepared several days in advance, well stirred, and filtered before use).

6. Differentiate in 35% alcohol for 10 min.

7. Rinse in distilled H_2O for 5 min.

8. Mount in aqueous glycerine jelly. (Dissolve 10 g Knox gelatin in 60 ml distilled water in a 70°C water bath. Add 70 ml glycerine and 1 ml melted phenol. Keep at 60°C for use and apply to warm histological slides.)

An alternative, simple, and satisfactory myelin stain is a modified version of Heidenhain's iron hematoxylin stain (Suyeoka, 1969):

1. Rinse in BSS.
2. Fix in 10% formalin in BSS for several days or more.
3. Rinse in several changes of distilled water.
4. Mordant in 5% iron (ferric) alum solution for 30 min at room temperature.
5. Rinse in distilled water.
6. Stain in 0.5% hematoxylin solution for 2 hr at 37°C (prepare solution by diluting ripened 10% alcoholic hematoxylin solution with distilled water).
7. Rinse in distilled water.
8. Differentiate in 5% iron alum solution until the thinnest areas of the culture lose their color.
9. Wash in tap water.
10. Dehydrate, clear, and mount in Permount.

8.1.4. Basophilic Stains

Nissle stains can be employed to demonstrate nerve-cell bodies according to standard histological procedures. Cresyl violet, toluidine blue, Einarson's gallocyanine-chromalum, thionine, and methylene blue-azure blue are most often used.

8.2. Special Biochemical Techniques

The use of tissue cultures as neurochemical tools has been discussed in a recent review (Herschman, 1973). Routine biochemical methods can be used for analyses. The cultures are usually scraped from the substrate with as little collagen as possible and pooled with several other cultures before homogenization. Collagen contaminants and entrapped medium can be removed from the tissue by repeated rinses of the tissue fragments in BSS containing 1.7×10^{-4} M acetic acid (Giesing and Zilliken, 1975). Alternatively, thin cultures can be frozen rapidly in liquid-nitrogen-chilled Freon and then dried under vacuum without thawing. The dry culture can then be easily separated from underlying collagen (Lehrer, 1973). If contamination by substrate collagen is suspected, the amount of tissue protein, exclusive of substrate collagen included in the homogenate, can be measured by the microchemical method of Bonting and Jones (1957) as adapted to nerve-tissue cultures by Oh *et al.* (1975).

Long-term cultures have been used for a variety of biochemical investigations. Some recent examples include the study of lipid metabolism during synaptogenesis (Giesing *et al.*, 1975; Giesing and Zilliken, 1975), the correlation of the development of acetylcholinesterase activity with morphological development (Kim *et al.*, 1972, 1975*a*), and a number of other studies concerning the patterns of enzymatic development (Lehrer, 1973).

Standard enzyme and fluorescence histochemical procedures can be used on cultured tissue. Tissue cultures are usually treated as whole mounts after rinsing with buffer directly on the coverslip without sectioning, while organ cultures would be treated similarly to noncultured tissue. Recent examples of the use of histochemical techniques include the examination of the activity of acetylcholinesterase and other enzymes in long-term cultures of a variety of nervous tissues (e.g., E. Hösli and L. Hösli, 1970, 1971; Minelli *et al.*, 1971; Kim, 1976*b*) and of catecholamine fluorescence in sympathetic ganglia (Kim and Munkacsi, 1972; Hervonen and Eränkö, 1975).

An immunochemical method for quantifying myelin basic protein has been adapted for use in tissue culture by Sheffield and Kim (1977). This radioimmunoassay is a sensitive measure of myelination in tissue culture and lends itself well to quantitative studies of demyelination disorders.

8.3. Electron Microscopy

8.3.1. Transmission Electron Microscopy

Preparation of cultured tissue for electron microscopy involves the special problems of removing the tissue from the substrate without loss of biological material. Under some experimental circumstances, it is possible, usually after fixation of the tissue, to simply scrape the cultured tissue from its substrate with a razor blade and then to process it in a routine fashion for electron microscopy. This is of course particularly feasible for thick organ cultures, but it is also routinely done for cultures grown in Maximow chambers (e.g., Raine, 1973) or in roller tubes (e.g., Kim, 1973). For cultures grown on collagen-coated coverslips in Maximow assemblies, tissue is fixed on the coverslip and then removed from the coverslip with a razor blade on a "window" of collagen while in 70% ethanol, followed by dehydration and embedding in Epon (Raine, 1973). Other workers prefer to infiltrate the culture with a small amount of Epon on the coverslip, separate the culture from the coverslip with a razor blade, and then embed it in a silicone rubber mold or other suitable capsule (Kim, 1973). It is also possible to carry out the complete fixing, dehydrating, and embedding procedure on the coverslip (R.P. Bunge *et al.*, 1965; M.M. LaVail, 1968). After dehydration and infiltration with Epon or Araldite monomer, No. 00 BEEM (LKB) capsules with the tips cut off (preferably on a lathe to accomplish a clean, flat cut) and the caps removed are inverted over the culture or over a specific area of interest previously marked. The capsules are held in place with a small rubber band that is secured through a small notch cut at the top of the capsule. The capsules are then filled with Epon or Araldite and usually left overnight before polymerization in a 60°C oven. Removal of the coverslip is then easily accomplished by dipping the coverslip with its capsule into liquid nitrogen (Privat *et al.*, 1974) or, if embedded in Araldite, by setting the coverslip on a block of dry ice for 30–60 sec. This allows the coverslip to be snapped off cleanly due to the differential temperature coefficients of expansion of the glass and the plastic.

Several laboratories have experimented with methods that are at the same time simpler and lend themselves to the removal of extremely thin areas of the culture into the embedding medium without any loss of tissue.

These methods involve culturing either on a substrate that does not stick to Epon, such as Aclar plastic (e.g., Masurovsky and Bunge, 1968; Bunge and Wood, 1973), Melinex (e.g., Lever and Presley, 1971; Bird and Lieberman, 1976), or silicone rubber membranes (Shahar *et al.*, 1973), or on a substrate that can be cut with glass or diamond knives, such as perforated cellophane tape (Grunnet, 1973). Alternatively, the coverslips can be coated with a film of carbon by conventional vacuum evaporation, in which case the Epon-embedded culture separates cleanly from the coverslip (Shimada *et al.*, 1967; M.M. LaVail, 1968; M.M. LaVail and Hild, 1971; Chamley *et al.*, 1972).

All these latter methods that involve substrates other than the conventional collagen-coated Gold Seal coverslips have been criticized for producing cultures with poorer outgrowth and a less healthy appearance (Raine, 1973). On the other hand, cultures grown on collagen-coated Aclar plastic (Masurovsky and Bunge, 1968) or silicone rubber membranes (Shahar *et al.*, 1973) have been claimed to be indistinguishable from the conventional Maximow cultures on collagen-coated glass. Since the concept of a healthy culture is often based on a subjective judgment and, in any case, may depend on the experimental needs, these latter and possibly simpler methods might be of advantage in some experimental situations.

8.3.2. Scanning Electron Microscopy

Cultured nerve tissue easily lends itself to three-dimensional examination with the scanning electron microscope. In particular, areas of neurite and glial outgrowth have been examined in this fashion. No unique techniques are employed except that the cultures are dehydrated and dried directly on the coverslip (Chamley *et al.*, 1973a; Privat *et al.*, 1974; Hill *et al.*, 1974; Blood, 1975; Silberberg, 1975).

8.4. Autoradiography

Tissue culture offers particular advantages for autoradiographic studies. The problems of uneven substrate distribution are minimized. Thin tissue cultures are usually processed for autoradiography as whole mounts (e.g., L. Hösli and E. Hösli, 1972; L. Hösli *et al.*, 1973a; Choi and Lapham, 1974) by attaching the coverslip culture side up to a microscope slide and then treating the dehydrated culture like a tissue section. Thicker cultures and organ cultures are embedded, sectioned, and processed like other tissues (e.g., Ljungdahl and Hökfelt, 1973). Some recent applications of autoradiographic techniques on cultured nervous tissue include uptake studies of putative neurotransmitters (E. Hösli *et al.*, 1972; Ljungdahl and Hökfelt, 1973; L. Hösli *et al.*, 1973a; Hösli and Hösli, 1978) and studies on neuronal differentiation (Choi and Lapham, 1974).

8.5. Electrophysiological Techniques

A significant part of electrophysiological research has always been carried out on preparations isolated from the organism (e.g., ganglia, nerve–muscle preparations, tissue slices). Such short-term isolated preparations share with cultures the abiltity to control the tissue environment (e.g., ions, drugs, temperature). Long-term cultures might be required for experiments wherein prolonged treatments with drugs are desired. In addition, long-term tissue cultures often become thin enough to allow favorable visualization of individual cells and parts of cells, which then allows selective exploration with stimulating or recording electrodes (e.g., Hild and Tasaki, 1962) or with precisely aimed microiontophoretic application of pharmacological agents (e.g., Geller, 1976).

Standard electrophysiological recording and stimulating methods can easily be adapted to bioelectrical investigations of cultured nervous tissue. The special electrophysiological methods used in conjunction with tissue cultures have been reviewed recently by Crain (1973). For acute experiments, the cultures are transferred to a recording chamber on the microscope stage, and the recording and stimulating electrodes are placed with micromanipulators in the desired location under visual control.

8.5.1. Microscopes

Under some circumstances, the magnification of a dissecting microscope is sufficient for the placement of extracellular recording and stimulating electrodes. Often, however, more powerful optics are needed to position electrodes accurately into or near individual cells or cellular processes. High-power phase or interference optics are usually employed for this purpose.

An inverted microscope with a fixed stage (i.e., one with which focusing is achieved by vertical movement of the optics rather than by movement of the stage, e.g., a Tiyoda inverted microscope) is usually most convenient. A fixed stage is necessary so that the electrode positions are not disturbed during focusing. Similar considerations apply when a conventional upright microscope is used.

The tips of the recording or stimulating electrodes are brought into the optical field with suitable micromanipulators. Desirable areas of the culture are then selected by scanning the culture with the horizontal movements of the microscope stage.

This technique restricts the distance between accurately placed stimulating and recording electrodes to the dimensions of the optical field of view, but this limitation has been overcome by a number of investigators by the use of a rigidly fixed recording stage and a movable optical system that slides horizontally to scan the culture under investigation. In this way, electrodes can be placed in one area of the culture, the optical system (the microscope) can then be moved to another site, and additional electrodes can be positioned there. This allows stimulation and recording from distant, yet accurately selected, sites. Horizontal movement of the optical system is accomplished by sliding the entire microscope (without the stage) either on a large greased glass plate (Cechner *et al.*, 1970; Crain, 1970) or on specially designed horizontally movable bearings (Gähwiler, personal communication).

A variety of chambers, some of them temperature-controlled and some with provisions for perfusion of test solutions, have been used for electrophysiological experiments on cultured nervous tissue.

8.5.2. Open Chambers

The simplest chamber consists of an open 50-mm-diameter Falcon plastic petri dish placed on the stage of an inverted microscope. At lower magnifications, the coverslip bearing the culture is simply placed into the petri dish, while for higher magnification, it may be necessary to cut a hole in the bottom of the recording chamber and seal the coverslip over the hole in the floor of the dish. The dish is filled with BSS or nutrient solution, and the pH is maintained by gently blowing moist 5% CO_2–95% air over the surface of the solution. The stage of an inverted microscope may be suitably altered to allow temperature control of the dish, or a special temperature-controlled microchamber may be used (Gähwiler and Bauer, 1975). Electrodes are then introduced from above. Intracellular penetration of cultured cells has been found to be more successful if the electrode tip approaches the cell in a nearly vertical direction. This can be accomplished by bending the recording glass microelectrodes 3–4 mm from the tip at an angle of about 50° and positioning the micromanipulators carrying these electrodes at an angle so that the tips of the electrodes approach the culture more or less perpendicular to the coverslip (L. Hösli *et al.*, 1971; Ko *et al.*, 1976). A perfusion system can be added, in which case the perfusate would be warmed as it enters the petri dish by gravity flow (Gähwiler and Bauer, 1975). The fluid level can be kept constant by suction tube providing the outflow (L. Hösli *et al.*, 1976).

Specially designed open chambers have been used to suit certain experimental requirements (e.g., Crain, 1970, 1973; Lumsden *et al.*, 1975).

An upright (conventional) microscope can also be used with an open chamber, but the small space between the objective lens and the culture makes the introduction of electrodes at higher magnifications difficult.

8.5.3. Bridge Chambers

The stage of a standard upright microscope can be adapted to hold a bridge chamber, where the coverslip bearing the culture is mounted, culture down, to span the gap between two 4- to 6-mm-high glass or Teflon blocks attached to a glass baseplate. The coverslip with the culture then forms the roof of a chamber that has two open sides (Hild and Tasaki, 1963; Schlapfer *et al.*, 1972*b*). BSS or the desired medium is introduced with a syringe through the sides and is held in the chamber by surface tension. Ports for perfusion and exchange of the bathing medium can be provided through the bridge mounts. Since the sides of the chamber are open, perfusion has to be regulated by use of a syringe pump with two back-to-back syringes, one for introduction of the medium, the other for removal of an equal amount of medium. Provision for temperature control can be added, either by using a heating coil or by using small thermoelectric units under the chamber (Gähwiler *et al.*, 1972). This arrangement has the advantage that a conventional microscope can be used, which allows, with a long-working-distance condenser, the use of high-power phase optics.

8.5.4. Hanging-Drop Chambers

A variation of the bridge chamber is the use of a moist chamber with open sides, in which the coverslip bearing the culture again forms the roof of the chamber, but only a drop of medium ("hanging drop") is used. Evaporation from the small amount of liquid is cut down by keeping the environment at a high humidity (Crain, 1970). Rapid exchange of medium during recording is difficult in this kind of chamber, and the rounded drop surface introduces vibrations into the recording system and amplifies them. In addition, the use of phase optics is hindered by the curvature of the drop surface.

8.5.5. Chambers for Long-Term Stimulation or Recording

Under some circumstances, it may be desirable to record from or stimulate certain areas of a culture for prolonged periods of time. The recording chambers discussed above are all open to the atmosphere and are therefore subject to contamination.

For prolonged or repetitive recording or stimulating sessions, sterile conditions would be desirable. This can be accomplished by incorporating relatively large and fixed metal electrodes into the culture substrate inside the culture chamber (Cunningham, 1962).

Long-term recordings from nerve trunks of the abdominal ganglion of the mollusc *Aplysia californica* have been obtained by culturing the ganglion in a special sterilizable culture chamber manufactured from Silastic rubber and Kel-F plastic. The chamber is equipped with several nerve cuffs that contain circular platinum–iridium recording electrodes through which the ganglionic nerves can be pulled (Strumwasser, 1971). Spike activity has been recorded in this fashion for as long as 42 days in culture. Such a chamber could also be used for prolonged repetitive stimulation of selected nerves.

Another useful approach has been taken by Baer and Crain (1971) (see also Crain, 1973), who have developed a hermetically sealed chamber that encloses not only the culture but also a set of small micromanipulators and microelectrodes. The coverslip bearing the cultures is introduced into this special closed chamber under sterile conditions. The chamber is then placed on the microscope stage, and the micromanipulators with their microelectrodes are manipulated from the exterior of the chamber with small magnets, under visual control, to the desired locations in the culture. The stimulating and recording system is connected to the chamber via a special plug. Since the chamber containing the culture can be returned to the incubator without endangering the sterility or necessarily altering the electrode positions, this system allows repeated electrical stimulation and recording from defined areas of a culture over many days during development or prolonged experimental treatments.

9. Conclusion

This chapter has attempted to review the various methods used to culture nervous tissues of many origins. It is in no way intended to be a substitute for a thorough study of the original literature in a particular area. Rather, it is hoped that this chapter provides a guide to the vast literature in this field.

The culture of nervous tissue provides many unique opportunities to address some of the important problems in neurobiology. Since the techniques can be very complex and often involve considerable cost and long-term commitment, the appropriateness of the use of tissue-culture methods for the solution of a particular biological question should be considered carefully. If tissue culture is chosen, the particular culture method selected should be the one best suited to the specific problem under investigation. In this way, tissue culture can be a means to the end, and not an end in itself.

ACKNOWLEDGMENTS. The author wishes to thank Drs. Beat H. Gähwiler, Abdel-Megid Mamoon, and Paul B.J. Woodson for their careful reading of the manuscript and their helpful suggestions. Support for this work was provided by the Medical Research Service of the Veterans Administration and by a grant from the National Institute on Alcohol Abuse and Alcoholism.

References

Agranoff, B.W., Field, P., and Gaze, R.M., 1976, Neurite outgrowth from explanted *Xenopus* retina: An effect of prior optic nerve section, *Brain Res.* **113**:225–234.

Allerand, C.D., 1972, Effect of pre-natal protein deprivation on neonatal cerebellar development *in vitro*, *Nature (London) New Biol.* **239**:157–158.

Auerbach, R., 1960, Morphogenetic interactions in the development of the mouse thymus gland, *Dev. Biol.* **2**:271–284.

Baer, S.C., and Crain, S.M., 1971, Magnetically coupled micromanipulator for use within a sealed chamber, *J. Appl. Physiol.* **31**:926–929.

Bayne, C.J., 1976, Culture of molluscan organs: A review, in: *Invertebrate Tissue Culture: Research Applications* (K. Maramorosch, ed.), pp. 61–74, Academic Press, New York.

Benex, J., 1965, Sur la dédifférenciation des tentacules de planorbes en survie: Rôle de la présence d'éléments nerveux sur le retard de cette dédifférenciation, *C. R. Acad. Sci.* **258**:2193–2196.

Benitez, H.H., Murray, M.R., and Côté, L.J., 1973, Responses of sympathetic chain-ganglia isolated in organotypic culture to agents affecting adrenergic neurons: Fluorescence histochemistry, *Exp. Neurol.* **39**:424–428.

Benitez, H.H., Masurovsky, E.B., and Murray, M.R., 1974, Interneurons of the sympathetic ganglia in organotypic culture: A suggestion as to their function, based on three types of study, *J. Neurocytol.* **3**:363–384.

Benjamins, J.A., Fitch, J., and Radin, N.S., 1976, Effects of ceramide analogs on myelinating organ cultures, *Brain Res.* **102**:267–281.

Bensinger, R.E., Klein, D.C., Weller, J.L., and Lovenberg, W., 1974, Radiometric assay of total tryptophan hydroxylation by intact cultured pineal glands, *J. Neurochem.* **23**:111–117.

Berlinrood, M., McGee-Russel, S.M., and Allen, R.D., 1972, Patterns of particle movement in nerve fibres *in vitro*: An analysis by photokymography and microscopy, *J. Cell Sci.* **11**:875–886.

Billings-Gagliardi, S., and Wolf, M.K., 1977, A simple method for examining organotypic CNS cultures with Nomarski optics, *In Vitro* **13**:371–377.

Bird, M.M., and Lieberman, A.B., 1976, Microtubule fascicles in the stem processes of cultured sensory ganglion cells, *Cell Tissue Res.* **169**:41–47.

Blank, W.F., Bunge, M.B., and Bunge, R.P., 1974, The sensitivity of the myelin sheath, particularly the Schwann cell–axolemmal junction, to lowered calcium levels in cultured sensory ganglia, *Brain Res.* **67**:503–518.

Blood, L.A., 1975, Scanning electron microscope observations of the outgrowth from embryonic chick dorsal root ganglia in culture, *Neurobiology* **5**:75–83.

Bodian, D., 1936, A new method for staining nerve fibers and nerve endings in mounted paraffin sections, *Anat. Rec.* **65**:89–97.

Bonting, S.L., and Jones, M., 1957, Determination of microgram quantities of deoxyribonucleic acid and protein in tissues grown *in vitro, Arch. Biochem.* **66**:340–353.

Borg, T.K., and Marks, E.P., 1973, Ultrastructure of the median neurosecretory cells of *Manduca sexta in vivo* and *in vitro, J. Insect Physiol.* **19**:1913–1920.

Bornstein, M.B., 1958, Reconstituted rat-tail collagen used as substrate for tissue cultures on coverslips in Maximow slides and roller tubes, *Lab. Invest.* **7**:134–137.

Bornstein, M.B., 1973*a*, Organotypic mammalian central and peripheral nerve tissue, in: *Tissue Culture: Methods and Applications* (P.F. Kruse and M.K. Patterson, eds.), pp. 86–92, Academic Press, New York.

Bornstein, M.B., 1973*b*, The immunopathology of demyelinative disorders examined in organotypic cultures of mammalian central nerve tissue, in: *Progress in Neuropathology*, Vol. 2 (H.M. Zimmerman, ed.), pp. 69–90, Grune and Stratton, New York.

Breen, G.A.M., and deVellis, J., 1975, Regulation of glycerol phosphate dehydrogenase by hydrocortisone in rat brain explants, *Exp. Cell Res.* **91**:159–169.

Brown, J.H., Nelson, D.L., and Molinoff, P.B., 1977, Organ culture of rat superior cervical ganglia, *J. Pharmacol. Exp. Ther.* **201**:298–311.

Bunge, M.B., Bunge, R.P., and Peterson, E.R., 1967, The onset of synapse formation in spinal cord cultures as studied by electron microscopy, *Brain Res.* **6**:728–749.

Bunge, R.P., 1975, Changing use of nerve tissue culture 1950–1975, in: *The Nervous System*, Vol. 1, *The Basic Neurosciences* (D.B. Tower, ed.), pp. 31–42, Raven Press, New York.

Bunge, R.P., and Wood, P., 1973, Studies on the transplantation of spinal cord tissue in the rat. I. The development of a culture system for hemisections of embryonic spinal cord, *Brain Res.* **57**:261–276.

Bunge, R.P., Bunge, M.B., and Peterson, E.R., 1965, An electron microscope study of cultured rat spinal cord, *J. Cell Biol.* **24**:163–191.

Burrows, M.T., 1910, The cultivation of tissue of the chick-embryo outside the body, *J. Am. Med. Assoc.* **55**:2057–2058.

Calvet, M.C., and Lepault, A.M., 1975, *In vitro* Purkinje cell electrical behavior related to tissular environment, *Exp. Brain Res.* **23**:249–258.

Cechner, R.L., Fleming, D.G., and Geller, H.M., 1970, Neurons *in vitro*: A tool for basic and applied research in neural electrodynamics, in: *Biomedical Engineering Systems* (M. Clynes and J.H. Milsum, eds.), pp. 595–653, McGraw-Hill, New York.

Chamley, J.H., Mark, G.E., Campbell, G.R., and Burnstock, G., 1972, Sympathetic ganglia in culture. I. Neurons, *Z. Zellforsch.* **135**:287–314.

Chamley, J.H., Campbell, G.R., and Burnstock, G., 1973*a*, An analysis of the interactions between sympathetic nerve fibers and smooth muscle cells in tissue culture, *Dev. Biol.* **33**:344–361.

Chamley, J.H., Goller, I., and Burnstock, G., 1973*b*, Selective growth of sympathetic nerve fibers to explants of normally densely innervated autonomic effector organs in tissue culture, *Dev. Biol.* **31**:362–379.

Chao, J., 1973, The application of invertebrate tissue culture to the *in vitro* study of animal parasites, *Curr. Top. Comp. Pathobiol.* **2**:107–144.

Chen, J.S., and Chen, M.G.M., 1975, A method to grow pieces of superior cervical ganglion *in vitro*, *TCA Man.* **1**:5–6.

Chen, J.S., and Levi-Montalcini, R., 1969, Axonal outgrowth and cell migration *in vitro* from nervous system of cockroach embryos, *Science* **166**:631–632.

Chida, N., and Shimizu, Y., 1973, Biosynthesis of myelin lipids of cultured nervous tissues—incorporation of choline and CDP-choline into myelin phospholipids, *Tohoku J. Exp. Med.* **111**:41–49.

Cho, Y.D., Martin, R.O., and Tunnicliff, G., 1973, Uptake of (^{3}H) glycine and (^{14}C) glutamate by cultures of chick spinal cord, *J. Physiol. (London)* **235**:437–446.

Cho, Y.D., Tunnicliff, G., and Martin, R.O., 1974, The uptake process for γ-aminobutyric acid in cultures of developing chick cerebrum, *Exp. Neurol.* **44**:306–312.

Choi, B.H., and Lapham, L.W., 1974, Autoradiographic studies of migrating neurons and astrocytes of human fetal cerebral cortex *in vitro*, *Exp. Mol. Pathol.* **21**:204–217.

Ciani, F., Contestabile, A., Minelli, G., and Quaglia, A., 1973, Ultrastructural localization of alkaline phosphatase in cultures of nervous tissue *in vitro*, *J. Neurocytol.* **2**:105–116.

Cole, R., and deVellis, J., 1976, The cellophane sandwich explant system, *TCA Man.* **2**:261–263.

Cook, R.D., and Peterson, E.R., 1974, The growth of smooth muscle and sympathetic ganglia in organotypic tissue cultures: Light and electron microscopy, *J. Neurol. Sci.* **22**:25–38.

Corrigall, W.A., Crain, S.M., and Bornstein, M.B., 1976, Electrophysiological studies of fetal mouse olfactory bulb explants during development of synaptic function in culture, *J. Neurobiol.* **7**:521–536.

Costero, I., and Pomerat, C.M., 1951, Cultivation of neurons from the adult human cerebral and cerebellar cortex, *Am. J. Anat.* **89**:405–467.

Coughlin, M.D., and Rathbone, M.P., 1977, Factors involved in the stimulation of parasympathetic nerve outgrowth, *Dev. Biol.* **61**:131–139.

Coughlin, M.D., Boyer, D.M., and Black, I.B., 1977, Embryonic development of a mouse sympathetic ganglion *in vivo* and *in vitro*, *Proc. Natl. Acad. Sci. U.S.A.* **74**:3438–3442.

Courgeon, A.-M., 1969, L'activité mitotique, en culture organotypique, dans les disques oculo-antennaires de larves de *Calliphora erythrocephala* Meig. (*insecte diptère*), *C. R. Acad. Sci. Ser. D.* **268**:950–952.

Coyle, J.T., Jacobowitz, D., Klein, D., and Axelrod, J., 1973, Dopaminergic neurons in explants of substantia nigra in culture, *J. Neurobiol.* **4**:461–470.

Crain, S.M., 1966, Development of "organotypic" bioelectric activites in central nervous tissues during maturation in culture, *Int. Rev. Neurobiol.* **9**:1–43.

Crain, S.M., 1970, Bioelectric interactions between cultured fetal rodent spinal cord and skeletal muscle after innervation *in vitro*, *J. Exp. Zool.* **173**:353–370.

Crain, S.M., 1973, Microelectrode recording in brain tissue cultures, in: *Methods in Physiological Psychology: Bioelectric Recording Techniques: Cellular Processes and Brain Potentials*, Vol. 1 (R.F. Thompson and M.M. Patterson, eds.), pp. 39–75, Academic Press, New York.

Crain, S.M., 1974, Selective depression of organotypic bioelectric activities of CNS tissue cultures by pharmacologic and metabolic agents, in: *Drugs and the Developing Brain* (A. Vernadakis and N. Weiner, eds.), pp. 29–57, Plenum Press, New York.

Crain, S.M., 1976, *Neurophysiologic Studies in Tissue Culture*, Raven Press, New York.

Crain, S.M., and Peterson, E.R., 1967, Onset and development of functional interneuronal connections in explants of rat spinal cord ganglia during maturation in culture, *Brain Res.* **6**:750–762.

Crain, S.M., and Peterson, E.R., 1974, Development of neural connections in culture, *Ann. N. Y. Acad. Sci.* **228**:6–34.

Crain, S.M., Raine, C.S., and Bornstein, M.B., 1975, Early formation of synaptic networks in culture of fetal mouse cerebral neocortex and hippocampus, *J. Neurobiol.* **6**:329–336.

Crain, S.M., Peterson, E.R., Leibman, M., and Schulman, H., 1980, Dependence on nerve growth factor of early human fetal dorsal root ganglion neurons in organotypic cultures, *Exp. Neurol.* **67:**205–214.

Cunningham, A.W.B., 1962, Qualitative behavior of spontaneous potentials from explant of 15-day chick embryo telencephalon *in vitro, J. Gen. Physiol.* **45:**1065–1076.

Daniels, E., and Moore, K.L., 1972, A direct analysis of early chick embryonic neuroepithelial responses following exposure to EDTA, *Teratology* **6:**215–226.

Davis, H., Gascho, C., and Kiernan, J.A., 1975, Action of aprotinin on the survival of adult cerebellar neurons in organ culture, *Acta Neuropathol. (Berlin)* **32:**359–362.

DeBoni, U., Seger, M., Scott, J.W., and Crapper, D.P., 1976, Neuron culture from adult goldfish, *J. Neurobiol.* **7:**495–512.

Demal, J., 1956, Culture *in vitro* d'ébauches imaginales de Diptères, *Ann. Sci. Nat. Zool.* **18:**155–161.

Dewhurst, S.A., and Weinreich, D., 1974, Effects of long-term organ culture on neurotransmitter metabolism in the ganglia of *Aplysia californica, J. Neurobiol.* **5:**21–31.

Dorfman, S.H., Holtzer, H., and Silberberg, D.H., 1976, Effect of 5-bromo-2'-deoxyuridine or cytosine-β-D-arabinofuranoside hydrochloride on myelination in newborn rat cerebellum cultures following removal of myelination inhibiting antiserum to whole cord or cerebroside, *Brain Res.* **104:**283–294.

Dreyfus, F., Sherman, D.L., and Gershon, M.D., 1977, Uptake of serotonin by intrinsic neurons of the myenteric plexus grown in organotypic tissue culture, *Brain Res.* **128:**109–123.

Dubois-Dalcq, M., Buyse, M., Lefebvre, N., and Sprecher-Goldberger, S., 1972a, Herpes virus hominis type 2 and intranuclear tubular structures in organized nervous tissue cultures, *Acta Neuropathol. (Berlin)* **22:**170–179.

Dubois-Dalcq, M., Menu, R., and Buyse, M., 1972b, Influence of fatty acids on fine structure of cultured neurons: An experimental approach to Refsum's disease, *J. Neuropathol. Exp. Neurol.* **31:**645–667.

Dubois-Dalcq, M., Buyse, M., DeBruyne, J., Van Tieghem, N., and Thiry, L., 1973, Ultrastructural study of the action of aminopiperazines on transformed cell lines and on nervous system tissue culture, *Exp. Cell Res.* **77:**303–311.

Dunn, G.A., 1971, Mutual contact inhibition of extension of chick sensory nerve fibres *in vitro, J. Comp. Neurol.* **143:**491–507.

Ebendal, T., and Jacobson, C.-O., 1977, Tissue explant affecting extension and orientation of axons in cultured chick embryo ganglia, *Exp. Cell Res.* **105:**379–387.

Ecob-Johnston, M.S., Elizan, T., Schwartz, J., and Whetsell, W.O., Jr., 1979, Herpes simplex virus types 1 and 2 in organotypic cultures of mouse central and peripheral nervous system. 2. Electron microscopic observations of myelin degeneration, *J. Neuropathol. Exp. Neurol.* **38:**10–18.

Elsdale, T., and Bard, J., 1972, Collagen substrata for studies on cell behavior, *J. Cell. Biol.* **54:**626–637.

Farbman, A.I., 1974, Trophic functions of the neuron, VI. Other trophic systems: Taste bud regeneration in organ culture, *Ann. N. Y. Acad. Sci.* **228:**350–354.

Fedoroff, S., and Hertz, L. (eds.), 1977, *Cell, Tissue and Organ Cultures in Neurobiology*, Academic Press, New York.

Feldman, L.A., Raine, C.S., Sheppard, R.D., and Bornstein, M.B., 1972, Virus–host cell relationships in measles-infected cultures of central nervous tissue, *J. Neuropathol. Exp. Neurol.* **31:**624–638.

Fildes, C., 1974, Organized nerve tissue cultures infected with *Mycobacterium leprae* and *Mycobacterium lepraemurium, Int. J. Lepr.* **42:**154–161.

Fisher, K.R.S., Fedoroff, S., and Wenger, E.L., 1975, Effect of osmotic pressure on neurogenesis in cultures of chick embryo spinal cords, *In Vitro* **11:**329–337.

Fogh, J. (ed.), 1973, *Contamination in Tissue Culture*, Academic Press, New York.

Fowler, D.J., and Goodnight, C.J., 1966a, The maintenance of adult opilionid tissues *in vitro, Trans. Am. Microsc. Soc.* **85:**378–389.

Fowler, D.J., and Goodnight, C.J., 1966b, Neurosecretory cells: Daily rhythmicity in *Leiobunum longipes, Science* **152**:1078–1080.

Fulton, F., 1960, Tissue culture on polythene, *J. Gen. Microbiol.* **22**:416–422.

Gähwiler, B.H., 1975a, The effects of GABA, picrotoxin and bicuculline on the spontaneous bioelectric activity of cultured cerebellar Purkinje cells, *Brain Res.* **99**:85–95.

Gähwiler, B.H., 1975b, Bioelectric effects of isoproterenol and propranolol on nerve cells in explants of rat cerebellum, *Brain Res.* **99**:393–399.

Gähwiler, B.H., 1976, Spontaneous bioelectric activity of cultured Purkinje cells during exposure to glutamate, glycine and strychnine, *J. Neurobiol.* **7**:97–107.

Gähwiler, B.H., and Bauer, W., 1975, Design of a microchamber for electrophysiological experiments *in vitro, Experientia* **31**:868–869.

Gähwiler, B.H., and Dreifuss, J.J., 1979, Phasically firing neurons in long-term cultures of the rat hypothalamic supraoptic area: Pacemaker and follower cells, *Brain Res.* **177**:95–103.

Gähwiler, B.H., Mamoon, A.M., Schlapfer, W.T., and Tobias, C.A., 1972, Effects of temperature on spontaneous bioelectric activity of cultured nerve cells, *Brain Res.* **40**:527–533.

Geletiuk, V.I., and Veprintsev, B.N., 1972, Electrical properties of *Limnaea stagnalis* neurons in tissue culture, *Tsitologiya* **14**:1133–1139.

Geller, H.M., 1976, Effects of some putative neurotransmitters on unit activity of tuberal hypothalamic neurons *in vitro, Brain Res.* **108**:423–430.

Geller, H.M., Cechner, R.L., and Fleming, D.G., 1972, Effect of goldthioglucose on long-term cultures of mouse hypothalamus, *Neurobiology* **2**:154–161.

Giacobini, E., Vernadakis, A., and Shahar, A. (eds.), 1980, *Tissue Culture in Neurobiology*, Raven Press, New York.

Giesing, M., and Zilliken, F., 1975, Lipid metabolism of developing central nervous tissues in organotypic cultures. II. A comparative study of medium and tissue fatty acids in developing rat cerebral cortex and cerebellum, *Nutr. Meta.* **19**:251–262.

Giesing, M., Neumann, G., Egge, H., and Zilliken, F., 1975, Lipid metabolism of developing central nervous tissues in organotypic cultures. I. Lipid distribution and fatty acid profiles of the medium for rat brain cortex *in vitro, Nutr. Metab.*.**19**:242–250.

Globus, M., and Vethamany-Globus, S., 1977, Transfilter mitogenic effect of dorsal root ganglia on cultured regeneration blastemata, in the newt, *Notophthalmus viridescens, Dev. Biol.* **56**:316–328.

Gomot, L., 1972, The organotypic culture of invertebrates other than insects, in: *Invertebrate Tissue Culture*, Vol. 2 (C. Vago, ed.), pp. 42–136, Academic Press, New York.

Goodman, R., Oesch, F., and Thoenen, H., 1974, Changes in enzyme patterns produced by high potassium concentration and dibutyryl cyclic AMP in organ cultures of sympathetic ganglia, *J. Neurochem.* **23**:369–378.

Goube de LaForest, P., Robineaux, R., and Voisin, J., 1967, Perfectionnement et modalités d'utilisation d'une chambre à perfusion pour la culture d'organes *in vitro, Ann. Inst. Pasteur* **113**:449–454.

Goworek, K., 1974, Differenzierung der Neurone in Explantkulturen des Cerebrocortex neonataler Ratten, *Z. Mikrosk. Anat. Forsch.* **88**:1125–1136.

Graham, D.I., Kim, S.U., Gonatas, N.K., and Guyotte, L., 1975, The neurotoxic effects of triethyltin (TET) sulfate on myelinating cultures of mouse spinal cord, *J. Neuropathol. Exp. Neurol.* **34**:401–412.

Griffond, B., 1969, Survie et évolution, en culture *in vitro*, des testicules de *Viviparus viviparus* L., gastéropode prosobranche à sexes séparés, *C. R. Acad. Sci. Ser. D.* **268**:963–965.

Grobstein, C., 1956, Transfilter induction of tubules in mouse metanephrogenic mesenchyme, *Exp. Cell Res.* **10**:424–440.

Grosse, G., and Lindner, G., 1974, Zytodifferenzierung des Hippocampus fetaler Ratten in der Zellkultur, *Z. Mikrosk. Anat. Forsch.* **88**:705–712.

Grunnet, M.L., 1973, A simple method for maintaining neurons *in vitro, Stain Technol.* **48**:207–211.

Guyard, A., 1969, Elaboration d'un milieu synthétique enrichi destiné à la culture d'organes de mollusques, *C. R. Acad. Sci. Ser. D.* **268**:162–164.

Halgren, E., and Varon, S., 1972, Serotonin turnover in cultured raphe nuclei from newborn rat: *In vitro* development and drug effects, *Brain Res.* **48:**438–442.

Handel, M.A., 1971, Effects of experimental degradation of microtubules on the growth of cultured nerve fibers, *J. Exp. Zool.* **178:**523–532.

Hansson, H.-A., 1966, Selective effects of metabolic inhibitors on retinal cultures, *Exp. Eye Res.* **5:**335–354.

Harris, M., 1964, *Cell Culture and Somatic Variation*, Holt, Rinehart and Winston, New York.

Harrison, R.G., 1907, Observations on the living developing nerve fiber, *Anat. Rec.* **1:**116–118.

Hauw, J.J., and Escourolle, R., 1975, Organ culture of the developing human cerebellum, *Brain Res.* **99:**117–123.

Hauw, J.J., Goube de LaForest, P., Anteunis, A., Cathala, F., and Robineaux, R., 1969, Culture organotypique prolongée de tissu nerveux en chambre perfusable, *C. R. Acad. Sci. Ser. D* **269:**1205–1208.

Hauw, J.J., Novikoff, A.B., Novikoff, P.M., Boutry, J.M., and Robineaux, R., 1972*a*, Culture of nervous tissue on collagen in Leighton tubes, *Brain Res.* **37:**301–309.

Hauw, J.J., Berger, B., and Escourolle, R., 1972*b*, Présence de synapse en culture organotypique *in vitro* de cervelet humain, *C. R. Acad. Sci. Ser. D* **274:**264–266.

Hauw, J.J., Boutry, J.M., Crosnier-Suttin, N., and Robineaux, R., 1974, Morphology of cultured guinea-pig cerebellum. I. Pattern of development: Comparison of phase contrast cinematography and silver impregnations of various cell types, *Cell Tissue Res.* **152:**141–164.

Heacock, H.M., and Agranoff, B.W., 1977, Clockwise growth of neurites from retinal explants, *Science* **198:**64–66.

Heath, J., Eränkö, O., and Eränkö, L., 1973, Effect of guanethidine on the ultrastructure of the small, granule-containing cells in cultures of rat sympathetic ganglia, *Acta Pharmacol. Toxicol. (Copenhagen)* **33:**209–218.

Heath, J.W., Hill, C.E., and Burnstock, G., 1974, Axon retraction following guanethidine treatment: Studies of sympathetic neurons in tissue culture, *J. Neurocytol.* **3:**263–276.

Hendelman, W., and Booher, J., 1966, Factors involved in the culturing of chick embryo dorsal root ganglia in the Rose chamber, *Tex. Rep. Biol. Med.* **24:**83–89.

Herschman, H.R., 1973, Tissue and cell culture as a tool in neurochemistry, in: *Proteins of the Nervous System* (D.J. Schneider *et al.*, eds.), pp. 95–115, Raven Press, New York.

Herschman, H.R., 1974, Culture of neural tissue and cells, in: *Research Methods in Neurochemistry*, Vol. 2 (N. Marks and R. Rodnight, eds.), pp. 101–160, Plenum Press, New York.

Hervonen, H., and Eränkö, O., 1975, Fluorescence histochemical and electron microscopical observations on sympathetic ganglia of the chick embryo cultured with and without hydrocortisone, *Cell Tissue Res.* **156:**145–166.

Hervonen, H., and Rechardt, L., 1974, Observations on closed tissue cultures of sympathetic ganglia of chick embryos in media buffered with *N*-Tris-(Hydroxymethyl)methyl-glycine or *N*-2-hydroxy-ethylpiperazine-*N*-2-ethanesulfonic acid, *Acta Physiol. Scand.* **90:**267–277.

Hild, W., 1957*a*, Myelogenesis in cultures of mammalian central nervous tissue, *Z. Zellforsch.* **46:**71–95.

Hild, W., 1957*b*, Observations on neurons and neuroglia from the area of the mesencephalic fifth nucleus of the cat *in vitro*, *Z. Zellforsch.* **47:**127–146.

Hild, W., and Callas, G., 1967, The behavior of retinal tissue *in vitro*: Light and electron microscopic observations, *Z. Zellforsch.* **80:**1–21.

Hild, W., and Tasaki, I., 1962, Morphological and physiological properties of neurons and glial cells in tissue culture, *J. Neurophysiol.* **25:**277–304.

Hill, C.E., and Burnstock, G., 1975, Amphibian sympathetic ganglia in tissue culture, *Cell Tissue Res.* **162:**209–233.

Hill, C.E., Mark, G.E., Eränkö, O., Eränkö, L., and Burnstock, G., 1973, Use of tissue culture to examine the actions of guanethidine and 6-hydroxydopamine, *Eur. J. Pharmacol.* **23:**162–174.

Hill, C.E., Chamley, J.H., and Burnstock, G., 1974, Cell surfaces and fibre relationships in sympathetic ganglion cultures: A scanning electron-microscopic study, *J. Cell Sci.* **14:**657–669.

Hill, C.E., Purves, R.D., Watanabe, H., and Burnstock, G., 1976, Specificity of innervation of iris musculature by sympathetic nerve fibres in tissue culture, *Pfluegers Arch.* **361:**127–134.

Hogue, M.J., 1947, Human fetal brain cells in tissue culture: Their identification and motility, *J. Exp. Zool.* **106:**85–107.

Hogue, M.J., 1953, A study of adult human brain cells grown in tissue culture. *Am. J. Anat.* **93:**397–427.

Hollyfield, J.G., and Witkovsky, P., 1974, Pigmented retinal epithelium involvement in photo-receptor development and function, *J. Exp. Zool.* **189:**357–378.

Holmes, W., 1947, The peripheral nerve biopsy, in: *Recent Advances in Clinical Pathology* (S.C. Dyke, ed.), pp. 402–417, Brakiston, Philadelphia.

Hooisma, J., Slaaf, D.W., Meeter, E., and Stevens, W.F., 1975, The innervation of chick striated muscle fibers by the chick ciliary ganglion in tissue culture, *Brain Res.* **85:**79–85.

Hösli, E., and Hösli, L., 1970, The presence of acetylcholinesterase in cultures of cerebellum and brain stem, *Brain Res.* **19:**494–496.

Hösli, E., and Hösli, L., 1971, Acetylcholinesterase in cultured rat spinal cord, *Brain Res.* **30:**193–197.

Hösli, E., and Hösli, L., 1978, Autoradiographic localization of the uptake of ^{3}H-β-alanine in rat nervous tissue cultures, *Experientia* **34:**1519–1521.

Hösli, E., Ljungdahl, Å., Hökfelt, T., and Hösli, L., 1972, Spinal cord tissue cultures—a model for autoradiographic studies on uptake of putative neurotransmitters such as glycine and GABA, *Experientia* **28:**1342–1344.

Hösli, L, and Hösli, E., 1972, Autoradiographic localization of the uptake of glycine in cultures of rat medulla oblongata, *Brain Res.* **45:**612–616.

Hösli, L., Andrès, P.F., and Hösli, E., 1971, Effects of glycine on spinal neurons grown in tissue culture, *Brain Res.* **34:**399–402.

Hösli, L., Hösli, E., and Andrès, P.F., 1973a, Nervous tissue culture—a model to study action and uptake of putative neurotransmitters such as amino acids, *Brain Res.* **62:**597–602.

Hösli, L., Hösli, E., and Andrès, P.F., 1973b, Light microscopic and electrophysiological studies of cultured human central nervous tissue, *Eur. Neurol.* **9:**121–130.

Hösli, L., Andrès, P.F., and Hösli, E., 1976, Ionic mechanism associated with depolarization by glutamate and aspartate on human and rat spinal neurones in tissue culture, *Pfluegers Arch.* **363:**43–48.

Houteville, P., and Lubet, P., 1974, Analyse expérimentale, en culture organotypique, de l'action des ganglions cérébio-pleureux et viscéraux sur le manteau de la moule mâle, *Mytilus edulis* L. (Mollusque Pélécypode), *C. R. Acad. Sci. Ser. D* **278:**2469–2472.

Ingebrigtsen, R., 1973, Studies of the degeneration and regeneration of the axis cylinders *in vitro, J. Exp. Med.* **17:**182–191.

Ito, Y., Donahoe, P.K., and Hendren, W.H., 1977, Differentiation of intramural ganglia in the dissociated rectosigmoid of the rat: An organ culture study, *J. Pediatr. Surg.* **12:**969–975.

Jacobson, A.G., 1967, Amphibian cell culture, organ culture, and tissue dissociation, in: *Methods in Developmental Biology* (F.H. Wilt and N.K. Wessells, eds.), pp. 531–542, Thomas Y. Crowell, New York.

James, D.W., and Tresman, R.L., 1968, *De novo* formation of neuro-muscular junctions in tissue culture, *Nature (London)* **220:**384–385.

Jessen, K.R., McConnell, J.D., Purves, R.D., Burnstock, G., and Chamley-Campbell, J., 1978, Tissue culture of mammalian enteric neurons, *Brain Res.* **152:**573–579.

Jockusch, H., Jockusch, B.M., and Burger, M.M., 1979, Nerve fibers in culture and their interaction with non-neural cells visualized by immunofluorescence, *J. Cell Biol.* **80:**629–641.

Jolly, M.J., 1903, Sur la durée de la vie et de la multiplication des cellules animales en dehors de l'organisme, *C. R. Soc. Biol.* **55:**1266–1268.

Juurlink, B.H., and Fedoroff, S., 1979, The development of mouse spinal cord in tissue culture. I. Cultures of whole mouse embryos and spinal-cord primordia, *In Vitro* **15:**86–94.

Kandel, E.R., 1976, *The Cellular Basis of Behavior: An Introduction to Behavioral Neurobiology*, Freeman, San Francisco.

Kasuya, M., 1974, Toxicity of phthalate esters to nervous tissue in culture, *Bull. Environ. Contam. Toxicol.* **12:**167–172.

Kasuya, M., and Okada, A., 1974, A simple tissue culture method of nervous tissue, *Jpn. J. Exp. Med.* **44:**219–222.

Kasuya, M., Sugawara, N., and Okada, A., 1974, Toxic effect of cadmium stearate on rat cerebellum in culture, *Bull. Environ. Contam. Toxicol.* **12:**535–540.

Kiernan, J.A., and Pettit, D.R., 1971, Organ culture of the central nervous system of the adult rat, *Exp. Neurol.* **32:**111–120.

Kim, S.U., 1970, Observation on cerebellar granule cells in tissue culture: A silver and electron microscopic study, *Z. Zellforsch.* **107:**454–465.

Kim, S.U., 1971*a*, Electron microscope study of mouse cerebellum in tissue culture, *Exp. Neurol.* **33:**30–44.

Kim, S.U., 1971*b*, Neuronal types in long-term culture of avian retina, *Experientia* **27:**1319–1321.

Kim, S.U., 1972*a*, Light and electron microscope study of mouse cerebral neocortex in tissue culture, *Exp. Neurol.* **35:**305–321.

Kim, S.U., 1972*b*, Light and electron microscope study of neurons and synapses in neonatal mouse olfactory bulb cultured *in vitro*, *Exp. Neurol.* **36:**336–349.

Kim, S.U.., 1973, Morphological development of neonatal mouse hippocampus cultured *in vitro*, *Exp. Neurol.* **41:**150–162.

Kim, S.U., 1975, Brain hypoxia studied in mouse central nervous system cultures. I. Sequential cellular changes, *Lab. Invest.* **33:**658–669.

Kim, S.U., 1976*a*, Tissue culture of human fetal cerebellum: A light and electron microscopic study, *Exp. Neurol.* **50:**226–239.

Kim, S.U., 1976*b*, Acetylcholinesterase distribution in chick spinal cord cultures: A light and electron microscope study, *Histochemistry* **48:**205–217.

Kim, S.U., and Munkacsi, I., 1972, Cytochemical demonstration of catecholamines and acetylcholinesterase in cultures of chick sympathetic ganglia, *Experientia* **28:**824–825.

Kim, S.U., and Rizzuto, 1975, Neuroaxonal degeneration induced by sodium diethyldithiocarbamate in cultures of central nervous tissue, *J. Neuropathol. Exp. Neurol.* **34:**531–541.

Kim, S.U., and Tunnicliff, G., 1974, Morphological and biochemical development of chick cerebrum cultured *in vitro*, *Exp. Neurol.* **43:**515–526.

Kim, S.U., and Wenger, E.L., 1972, De novo formation of synapse in cultures of chick neural tube, *Nature (London) New Biol.* **236:**152–153.

Kim, S.U., and Wenger, B.S., 1973, Neurotoxic effects of 6-aminonicotinamide on cultures of central nervous tissue, *Acta Neuropathol. (Berlin)* **26:**259–264.

Kim, S.U., and Pleasure, D.E., 1978, Tissue culture analysis of neurogenesis: Myelination and synapse formation are retarded by serum deprivation, *Brain Res.* **145:**15–25.

Kim, S.U., Oh, T.H., and Johnson, D.D., 1972, Developmental changes of acetylcholinesterase and pseudocholinesterase in organotypic cultures of spinal cord, *Exp. Neurol.* **35:**274–281.

Kim, S.U., Oh, T.H., and Johnson, D.D., 1975*a*, Increased activity of choline acetyltransferase and acetylcholinesterase in developing cultures of chick spinal cord: A correlation with morphological development, *Neurobiology* **5:**119–127.

Kim, S.U., Pearson, D., and Paik, W.K., 1975*b*, Studies on S-adenosylmethionine:protein-carboxyl methyltransferase in the hypothalamoneurohypophysial complex in organ culture, *Biochem. Biophys. Res. Commun.* **67:**448–454.

Kitano, S., and Iwai, H., 1974, Demyelination of the spiral ganglion cultured under anoxic conditions, *Arch. Otorhinolaryngol. (N. Y.)* **208:**157–162.

Klein, D.C., and Weller, J., 1970, Input and output signals in a model neural system: The regulation of melatonin production in the pineal gland, *In Vitro* **6:**197–204.

Ko, C.-P., Burton, H., and Bunge, R.P., 1976, Synaptic transmission between rat spinal cord explant and dissociated superior cervical ganglion neurons in tissue culture, *Brain Res.* **117:**437–460.

Koda, L.Y., and Partlow, L.M., 1976, Membrane marker movement on sympathetic axons in tissue culture, *J. Neurobiol.* **7:**157–172.

Kormano, M., and Hervonen, H., 1976, Use of tissue culture to examine neurotoxicity of contrast media, *Radiology* **120:**727–729.

Kostenko, M.A., and Veprintsev, B.N., 1972, Cultivation of nerve tissue of an adult mollusc *Limnaea stagnalis* in organ cultures *in vitro*, *Tsitologiya* **14**:1392–1397.

Kristensson, K., and Bornstein, M.B., 1974, Effects of iron-containing substances on nervous tissue *in vitro*, *Acta Neuropathol.* (*Berlin*) **28**:281–292.

Kristensson, K., Sheppard, R.D., and Bornstein, M.B., 1974, Observations on uptake of herpes simplex virus in organized cultures of mammalian nervous tissue, *Acta Neuropathol.* (*Berlin*) **28**:37–44.

Kruse, P.F., Jr., and Patterson, M.K., Jr. (eds.), 1973, *Tissue Culture: Methods and Applications*, Academic Press, New York.

Landreth, G.E., and Agranoff, B.W., 1976, Explant culture of adult goldfish retina: Effect of prior optic nerve crush, *Brain Res.* **118**:299–303.

Lapham, L.W., and Markesbery, W.R., 1971, Human fetal cerebellar cortex: Organization and maturation of cells *in vitro*, *Science* **173**:829–832.

Lash, J., Holtzer, S., and Holtzer, H., 1957, An experimental analysis of the development of the spinal column, *Exp. Cell Res.* **13**:292–303.

Latovitzki, N., and Silberberg, D.H., 1977, UDP-galactose:ceramide galactosyltransferase and 2′,3′-cyclic nucleotide 3′-phosphohydrolase activities in cultured newborn rat cerebellum: Association with myelination and concurrent susceptibility to 5-bromodeoxyuridine, *J. Neurochem.* **29**:611–614.

LaVail, J.H., and Wolf, M.K., 1973, Postnatal development of the mouse dentate gyrus in organotypic cultures of the hippocampal formation, *Am. J. Anat.* **137**:47–66.

LaVail, M.M., 1968, A method of embedding selected areas of tissue cultures for electron microscopy, *Tex. Rep. Biol. Med.* **26**:215–222.

LaVail, M.M., and Hild, W., 1971, Histotypic organization of the rat retina *in vitro*, *Z. Zellforsch.* **114**:557–579.

LeGall, S., Griffond, B., and Streiff, W., 1974, Existance de facteurs endocriniens de la morphogenèse et de la régression du pénis chez *Buccinum undatum* et *Viviparus viviparus*, mollusques, gastéropodes, gonochoriques, *C. R. Acad. Sci. Ser. D* **278**:773–776.

Lehrer, G.M., 1973, The tissue culture as a model for the biochemistry of brain development, *Prog. Brain Res.* **40**:219–230.

Leloup, A.-M., 1970, Culture d'organs d'insecte: Influence de la glande en anneau et du complexe cérébral sur la différenciation de l'ovaire prépupal de *Calliphora erythrocephala*, *Ann. Biol.* **9**:447–453.

Lentz, T.L., 1971, Nerve trophic function: *In vitro* assay of effects of nerve tissue on muscle cholinesterase activity, *Science* **171**:187–189.

Letourneau, P.C., and Wessells, N.K., 1974, Migratory cell locomotion *versus* nerve axon elongation: Differences based on the effects of lanthanum ion, *J. Cell Biol.* **61**:56–69.

Lever, J.D., and Presley, R., 1971, Studies on the sympathetic neurone *in vitro*, *Prog. Brain Res.* **34**:499–512.

Levi-Montalcini, R., and Angeletti, P.U., 1968, Nerve growth factor, *Physiol. Rev.* **48**:534–569.

Levi-Montalcini, R., Meyer, H., and Hamburger, V., 1954, *In vitro* experiments on the effects of mouse sarcomas 180 and 37 on spinal and sympathetic ganglia of the chick embryo, *Cancer Res.* **14**:49–57.

Levi-Montalcini, R., Chen, J.S., Seshan, K.R., and Aloe, L., 1973, An *in vitro* approach to the insect nervous system, in: *Developmental Neurobiology of Arthropods* (D. Young, ed.), pp. 5–36, Cambridge University Press, Cambridge.

Ljungdal, A., and Hökfelt, T., 1973, Autoradiographic uptake patterns of (^{3}H) GABA and (^{3}H) glycine in central nervous tissues with special reference to the cat spinal cord, *Brain Res.* **62**:587–595.

Ljunggren, C.A., 1897, Von der Fähigheit des Hautepithels, ausserhalb des Organismus sein Leben zu behalten, mit Berücksichtigung der Transplantation, *Dtsch. Z. Chir.* **47**:608–615.

Lumsden, C.E., 1951, Aspects of neurite outgrowth in tissue culture, *Anat. Rec.* **110**:145–179.

Lumsden, C.E., 1968, Nervous tissue in culture, in: *The Structure and Function of Nervous Tissue*, Vol. 1 (G.H. Bourne, ed.), pp. 67–140, Academic Press, New York.

Lumsden, C.E., Howard, L., Aparicio, S.R., and Bradbury, M., 1975, Antisynaptic antibody in allergic encephalomyelitis. II. The synapse-blocking effects in tissue culture of demyelinating sera from experimental allergic encephalomyelitis, *Brain Res.* **93**:283–299.

Lutz, H. (ed.), 1970, *Invertebrate Organ Cultures*, Gordon and Breach, London.

Lyser, K.M., 1971, Early differentiation of the chick embryo spinal cord in organ culture: Light and electron microscopy, *Anat. Rec.* **169**:45–64.

Lyser, K.M., 1975, Organotypic culture of early embryonic nervous system, *TCA Man.* **1**:121–126.

Mackay, A.V., 1974, The long-term regulation of tyrosine hydroxylase activity in cultured sympathetic ganglia: Role of ganglionic noradrenaline content, *Br. J. Pharmacol.* **51**:509–520.

Malecha, J., 1967, Etude en culture organotypique de l'influence endocrine de la masse nerveuse péripharyngienne sur la maturation testiculaire chez *Hirundo medicinalis* L., *C. R. Acad. Sci. Ser. D* **265**:1806–1808.

Maramorosch, K. (ed.), 1976, *Invertebrate Tissue Culture: Research Applications*, Academic Press, New York.

Mark, G.E., Chamley, J.H., and Burnstock, G., 1973, Interactions between autonomic nerves and smooth and cardiac muscle cells in tissue culture, *Dev. Biol.* **32**:194–200.

Marks, E.P., and Reinecke, J.P., 1965, Regenerating tissues from the cockroach *Leucophaea maderae:* Effect of endocrine glands *in vitro*, *Gen. Comp. Endocrinol.* **5**:241–247.

Marson, A.M., and Privat, A., 1979, *In vitro* differentiation of hypothalamic magnocellular neurons of guinea pigs, *Cell Tissue Res.* **203**:393–401.

Masurovsky, E.B., and Bunge, R.P., 1968, Fluoroplastic coverslips for long-term nerve tissue culture, *Stain Technol.* **43**:161–165.

Masurovsky, E.B., and Peterson, E.R., 1973, Photo-reconstituted collagen gel for tissue culture substrates, *Exp. Cell Res.* **76**:447–448.

Masurovsky, E.B., and Peterson, E.R., 1975, A procedure for preparing photogelled collagen substrates containing Aclar reticles for tissue culture, *TCA Man.* **1**:107–109.

Masurovsky, E.B., Peterson, E.R., and Crain, S.M., 1971, Aclar film reticles for precise cell location in nerve tissue culture, *In Vitro* **6**:379.

Maximow, A., 1925, Tissue cultures of young mammalian embryos, *Contrib. Embryol. Carnegie Inst.* **16**:47–113.

Maxwell, G.D., 1976, Cell cycle changes during neural crest cell differentiation *in vitro*, *Dev. Biol.* **49**:66–79.

McKelvy, J.F., 1974, Biochemical neuroendocrinology. I. Biosynthesis of thyrotropin releasing hormone (TRH) by organ cultures of mammalian hypothalamus, *Brain Res.* **65**:489–502.

McKelvy, J.F., Sheridan, M., Joseph, S., Phelps, C.H., and Perrie, S., 1975, Biosynthesis of thyrotropin-releasing hormone in organ cultures of the guinea pig median eminence, *Endocrinology* **97**:908–918.

McLennan, I.S., Hill, C.E., and Hendry, I.A., 1980, Glucocorticosteroids modulate transmitter choice in developing superior cervical ganglion, *Nature (London)* **283**:206–207.

McMahan, U.J., and Kuffler, S.W., 1971, Visual identification of synaptic boutons on living ganglion cells and of varicosities in postganglionic axons in the heart of the frog, *Proc. R. Soc. London Ser. B* **177**:485–508.

Meller, K., Mestres, P., Breipohl, W., and Waelsch, M., 1974, Time-lapse cinematographic studies on the inhibition of cell motility by puromycin in cultured nervous tissue, *Cell Tissue Res.* **148**:227–235.

Menkes, J.H., 1972, Lipid metabolism of brain tissue in culture, *Lipids* **7**:135–141.

Merchant, D.J., Kahn, R.H., and Murphy, W.H., Jr., 1960, *Handbook of Cell and Organ Culture*, Burgess, Minneapolis.

Messing, A., and Kim, S.U., 1979, Long-term culture of adult mammalian central nervous system neurons, *Exp. Neurol.* **65**:293–300.

Meynadier, R., 1971, Aseptic rearing of invertebrates for tissue culture, in: *Invertebrate Tissue Culture*, Vol. 1 (C. Vago, ed.), pp. 141–167, Academic Press, New York.

Minelli, G., Ciani, F., and Contestabile, A., 1971, The occurrence of some enzymatic activities in the differentiation of nerve tissue of *Gallus* embryos in cultures *in vitro*, *Histochemie* **28**:160–169.

Mithen, F.A., Cochran, M., Johnson, M.I., and Bunge, R.P., 1980, Neurotoxicity of polystyrene containers detected in a closed tissue culture system, *Neurosci. Lett.* **17**:107–111.

Miyazaki, S., and Nicholls, J.G., 1976, The properties and connections of nerve cells in leech ganglia maintained in culture, *Proc. R. Soc. London Ser. B* **194**:295–311.

Mizel, S.B., and Bamburg, J.R., 1976a, Studies on the action of nerve growth factor. I. Characterization of a simplified *in vitro* culture system for dorsal root and sympathetic ganglia, *Dev. Biol.* **49**:11–19.

Mizel, S.B., and Bamburg, J.R., 1976b, Studies on the action of nerve growth factor. III. Role of RNA and protein synthesis in the process of neurite outgrowth, *Dev. Biol.* **49**:20–28.

Monnickendam, M.A., and Balls, M., 1973, Amphibian organ culture, *Experientia* **29**:1–17.

Mottram, D.R., Presley, R., Lever, J.D., and Ivens, C., 1972, Some characteristics of chick embryo sypmathetic ganglion fragments in a simple culture system with and without nerve growth factor, *Z. Anat. Entwicklungsgesch.* **133**:127–133.

Murray, M.R., 1965, Nervous tissue *in vitro*, in: *Cells and Tissues in Culture*, Vol. 2 (E.N. Willmer, ed.), pp. 373–455, Academic Press, London.

Murray, M.R., 1971, Nervous tissues isolated in culture, in: *Handbook of Neurochemistry*, Vol. 5A (A. Lajtha, ed.), pp. 373–424, Plenum Press, New York.

Murray, M.R., and Stout, A.P., 1942, Characteristics of human Schwann cells *in vitro*, *Anat. Rec.* **84**:275–293.

Murray, M.R., and Stout, A.P., 1947, Adult human sympathetic ganglion cells cultivated *in vitro*, *Am. J. Anat.* **80**:225–273.

Nelson, P.G., 1975, Nerve and muscle cells in culture, *Physiol. Rev.* **55**:1–61.

Norr, S.C., 1973, *In vitro* analysis of sympathetic neuron differentiation from chick neural crest cells, *Dev. Biol.* **34**:16–38.

Oh, T.H., Johnson, D.D., and Kim, S.U., 1972, Neurotrophic effect on isolated chick embryo muscle in culture, *Science* **178**:1298–1300.

Oh, T.H., Kim, S.U., and Johnson, D.D., 1975, Measurement of protein in cultures of nervous tissue grown on collagen substrate, *Neurobiology* **5**:188–191.

Olson, M.I., and Bunge, R.P., 1973, Anatomical observations on the specificity of synapse formation in tissue culture, *Brain Res.* **59**:19–33.

Orr, M.F., 1965, Development of acoustic ganglia in tissue cultures of embryonic chick otocysts, *Exp. Cell Res.* **40**:68–77.

Parfitt, A., Weller, J.L., and Klein, D.C., 1976, Beta adrenergic blockers decrease adrenergically stimulated N-acetyltransferase activity in pineal glands in organ culture, *Neuropharmacology* **15**:353–358.

Parker, R.C., 1961, *Methods of Tissue Culture*, 3rd ed., Paul B. Hoeber, New York.

Paul, C.V., and Powell, J.A., 1974, Organ culture studies of coupled fetal cord and adult muscle from normal and dystrophic mice, *J. Neurol. Sci.* **21**:365–379.

Paul, J., 1975, *Cell and Tissue Culture*, 5th ed., Churchill Livingstone, Edinburgh and London.

Pearson, D., Shainberg, A., Osinchak, J., and Sachs, H., 1975, The hypothalamo-neurohypophysial complex in organ culture: Morphologic and biochemical characteristics, *Endocrinology* **96**:982–995.

Penso, G., and Balducci, D., 1963, *Tissue Culture in Biological Research*, Elsevier, Amsterdam, London, and New York.

Peterson, E.R., and Murray, M.R., 1955, Myelin sheath formation in cultures of avian spinal ganglia, *Am. J. Anat.* **96**:319–355.

Peterson, E.R., and Murray, M.R., 1960, Modification of development in isolated dorsal root ganglia by nutritional and physical factors, *Dev. Biol.* **2**:461–476.

Peterson, E.R., Deitch, A.D., and Murray, M.R., 1959, Type of glass as a factor in maintenance of coverslip cultures, *Lab. Invest.* **8**:1507–1512.

Pfenninger, K.H., and Bunge, R.P., 1974, Freeze–fracturing of nerve growth cones and young fibers: A study of developing plasma membrane, *J. Cell Biol.* **63**:180–196.

Piddington, R., 1973, Glutamine synthetase in the avian diencephalon, *J. Exp. Zool.* **184**:167–175.

Pleasure, D., and Kim, S.U., 1976, Enzyme markers for myelination of mouse cerebellum *in vivo* and in tissue culture, *Brain Res.* **104**:193–196.

Pollack, E.D., and Koves J., 1975, *In vitro* cultivation of larval frog spinal cord explants, *TCA Man.* **1**:193–197.

Pomerat, C.M., 1959, Rhythmic contractions of Schwann cells, *Science* **130**:1759–1760.

Pomerat, C.M., 1962, Cinematography in the service of neuropathology, in: *Fourth International Congress of Neuropathology*, Vol. 2 (H. Jacob, ed.) pp. 207–216, Georg Thieme, Stuttgart.

Price, F.M., and Handleman, S.L., 1975, The preparation and use of perforated cellophane for tissue culture, *TCA Man.* **1**:59–60.

Price, P.J., 1975, Preparation and use of rat-tail collagen, *TCA Man.* **1**:43–44.

Privat, A., Drian, M.J., and Gruner, J.E., 1972, Retardation in the outgrowth of dysmature rat cerebellum cultivated *in vitro*, *Biol. Neonate* **20**:414–424.

Privat, A., Drian, M.J., and Mandon, P., 1974, Synaptogenesis in the outgrowth of rat cerebellum in organized culture, *J. Comp. Neurol.* **153**:291–307.

Provine, R.R., Aloe, L., and Seshan, K.R., 1973, Spontaneous bioelectric activity in long-term cultures of the embryonic insect central nervous system, *Brain Res.* **56**:364–370.

Provine, R.R., Seshan, K.R., and Aloe, L., 1976, Formation of cockroach interganglionic connectives: An *in vitro* analysis, *J. Comp. Neurol.* **165**:17–30.

Purves, R.D., Hill, C.E., Chamley, J.H., Mark, G.E., Fry, D.M., and Burnstock, G., 1974, Functional autonomic neuromuscular junctions in tissue culture, *Pfluegers Arch.* **350**:1–7.

Quastel, M.R., and Pomerat, C.M., 1963, Irradiation of cells in tissue culture. VIII. Morphological observations of neurons following cobalt-60 gamma irradiation of dorsal root ganglia from the newborn rat, *Z. Zellforsch.* **59**:214–223.

Raiborn, C.W., and Massey, J.R., 1965, Dialyzing membrane for improved spinal cord maintenance in roller tube cultures, *Stain Technol.* **40**:293–294.

Raine, C.S., 1973, Ultrastructural applications of cultures of organized nervous tissue to neuropathology, in: *Progress in Neuropathology*, Vol. 2 (H.M. Zimmerman, ed.), pp. 27–68, Grune and Stratton, New York.

Raine, C.S., Feldman, L.A., Sheppard, R.D., and Bornstein, M.B., 1973, Subacute sclerosing panencephalitis virus in cultures of organized central nervous tissue, *Lab. Invest.* **28**:627–640.

Robertson, J., and Pipa, R., 1973, Metamorphic shortening of interganglionic connectives of *Galleria mellonella* (Lepidoptera) *in vitro*: Stimulation by ecdysone analogues, *J. Insect Physiol.* **19**:673–679.

Rogers, N.G., Basnight, M., Gibbs, C.J., Jr., and Gajdusek, D.C., 1967, Latent viruses in chimpanzees with experimental kuru, *Nature (London)* **216**:446–449.

Roisen, F.J., Murphy, R.A., and Braden, W.G., 1972, Neurite development *in vitro*. I. The effects of adenosine 3′5′-cyclic monophosphate (cyclic AMP), *J. Neurobiol.* **3**:347–368.

Roizin, L., Schneider, J., Willson, N., Liu, J.C., and Mullen, C., 1974, Effects of prolonged LSD-25 administration upon neurons of spinal cord ganglia tissue cultures, *J. Neuropathol. Exp. Neurol.* **33**:212–225.

Rose, G.G., Pomerat, C.M., Shindler, T.O., and Trunnell, J.B., 1958, A cellophane-strip technique for culturing tissue in multipurpose culture chambers, *J. Biophys. Biochem. Cytol.* **4**:761–764.

Ross, L.L., Bornstein, M.B., and Lehrer, G.M., 1962, Electron microscope observations of rat and mouse cerebellum in tissue culture, *J. Cell Biol.* **14**:19–30.

Roubein, I.F., Samuelly, M., and Keup, W., 1973, The toxicity of chlorpromazine and mescaline on mouse cerebellum and fibroblast cells in culture, *Acta Pharmacol. Toxicol. (Copenhagen)* **33**:326–329.

Roux, W., 1885, Beiträge zur Entwicklungsmechanik des Embryo, *Z. Biol.* **21**:411–526.

Rubin, L.L., Gorio, A., and Mauro, A., 1976, Effect of cytochalasin B on neuromuscular transmission in tissue culture, *Brain Res.* **104**:171–175.

Rubinstein, L.J., Herman, M.M., and Foley, V.l., 1973, *In vitro* characteristics of human glioblastomas maintained in organ culture systems, *Am. J. Pathol.* **71**:61–80.

Rugh, R., 1967, *Experimental Embryology*, 3rd ed., Burgess, Minneapolis.

Sakai, K.K., Marks, B.H., George, J.M., and Koestner, A., 1974, The isolated organ-cultured supraoptic nucleus as a neuropharmacological test system, *J. Pharmacol. Exp. Ther.* **190**:482–491.

Sano, Y., Odake, G., and Yonezawa, T., 1967, Fluorescence microscopic observations of catecholamines in cultures of sympathetic chains, *Z. Zellforsch.* **80:**345–352.

Schaller, F., and Meunier, J., 1967, Resultats de culture organotypiques du cerveu et du ganglion sous-oesophagieu d'*Aeschna cyanea* Müll. (insecte obonate): Survie des organs et évolution des éléments neurosécréteurs, *C. R. Acad. Sci. Ser. D* **264:**1441–1444.

Schlapfer, W.T., Haywood, P., and Barondes, S.H., 1972a, Cholinesterase and choline acetyltransferase activities develop in whole explant but not in dissociated cell cultures of cockroach brain, *Brain Res.* **39:**540–544.

Schlapfer, W.T., Mamoon, A.M., and Tobias, C.A., 1972b, Spontaneous bioelectric activity of neurons in cerebellar cultures: Evidence for synaptic interactions, *Brain Res.* **45:**345–363.

Schneider, I., 1966, Histology of larval eye-antennal disks and cephalic ganglia of *Drosophila* cultured *in vitro*, *J. Embryol. Exp. Morphol.* **15:**271–279.

Schneider, I., 1967, Insect tissue culture, in: *Methods in Developmental Biology* (F.H. Wilt and N.K. Wessells, eds.), pp. 543–554, Thomas Y. Crowell, New York.

Schneider, J.F., 1973, Culture of nerve tissue, in: *Methods of Neurochemistry*, Vol. 4 (R. Fried, ed.) pp. 1–68, Marcel Dekker, New York.

Schneider, J.F., and Rue, C.E., 1974, A simplified method of nerve organ culture, *Experientia* **30:**829–830.

Schultzberg, M., Ebendal, T., Hökfelt, T., Nilsson, G., and Pfenninger, K., 1978, Substance P-like immunoreactivity in cultured spinal ganglia from chick embryos, *J. Neurocytol.* **7:**107–117.

Seil, F.J., 1972, Neuronal groups and fiber patterns in cerebellar tissue cultures, *Brain Res.* **42:**33–51.

Seshan, K.R., 1975, Apparatus and techniques for the preparation of cockroach neuroendocrine organs for long-term maintenance *in vitro*, *TCA Man.* **1:**215–220.

Seshan, K.R., 1976, Tissue culture medium and cockroach ringer for the cockroach *Periplaneta americana*, *TCA Man.* **2:**319–322.

Seshan, K.R., and Levi-Montalcini, R., 1971, *In Vitro* analysis of corpora cardiaca and corpora allata from nymphal and adult specimems of *Periplaneta americana*, *Arch. Ital. Biol.* **108:**81–109.

Seshan, K.R., Provine, R.R., and Levi-Montalcini, R., 1974, Structural and electrophysiological properties of nymphal and adult insect medial neurosecretory cells: An *in vitro* analysis, *Brain Res.* **78:**359–376.

Shahar, A., Monzain, R., and Straussman, Y., 1973, Silicone rubber membrane as a support for long-term cultivation and electron microscopic processing of nervous tissue, *Tissue Cell* **5:**691–696.

Sheffield, W.D., and Kim, S.U., 1977, Basic protein radioimmunoassay as a monitor of myelin in tissue culture, *Brain Res.* **120:**193–196.

Shimada, Y., Fischman, D.A., and Moscona, A.A., 1967, The fine structure of embryonic chick skeletal muscle cells differentiated *in vitro*, *J. Cell Biol.* **35:**445–453.

Silberberg, D.H., 1972, Cultivation of nerve tissue, in: *Growth, Nutrition and Metabolism of Cells in Culture*, Vol. 2 (G.H. Rothblat and V.J. Cristofalo, eds.), pp. 131–167, Academic Press, New York.

Silberberg, D.H., 1975, Scanning electron microscopy of organotypic rat cerebellum cultures, *J. Neuropathol. Exp. Neurol.* **34:**189–199.

Silberberg, D., Benjamins, J., Herschkowitz, N., and McKhann, G.M., 1972, Incorporation of radioactive sulphate into sulphatide during myelination in cultures of rat cerebellum, *J. Neurochem.* **19:**11–18.

Silberstein, S.D., 1973, Sympathetic ganglia in organ culture, in: *Neurosciences Research*, Vol. 5, *Chemical Approaches to Brain Function* (S. Ehrenpreis and I.J. Kopin, eds.), pp. 1–34, Academic Press, New York.

Silberstein, S.D., Johnson, D.G., Jacobowitz, D.M., and Kopin, I.J., 1971, Sympathetic reinnervation of the rat iris in organ culture, *Proc. Nat. Acad. Sci. U.S.A.* **68:**1121–1124.

Silverman, A.J., Knigge, K.M., Ribas, J.L., and Sheridan, M.N., 1973, Transport capacity of median eminence. 3. Amino acid and thyroxine transport of organ-cultured median eminence, *Neuroendocrinology* **11:**107–118.

Sipe, J.C., 1976, Gap junctions between astrocytes during growth and differentiation in organ culture systems, *Cell Tissue Res.* **170**:485–490.

Slaaf, D.W., Hooisma, J., Meeter, E., and Stevens, W.F., 1979, Effects of innervation by ciliary ganglia on developing muscle *in vitro*, *Brain Res.* **175**:87–107.

Smalheiser, N.R., and Crain, S.M., 1978, Formation of functional retinotectal connections in co-cultures of fetal mouse explants, *Brain Res.* **148**:484–492.

Sobkowicz, H.M., Hartmann, H.A., Monzain, R., and Desnoyers, P., 1973, Growth, differentiation and ribonucleic acid content of the fetal rat spinal ganglion cells in culture, *J. Comp. Neurol.* **148**:249–283.

Sobkowicz, H.M., Bleier, R., and Monzain, R., 1974, Cell survival and architectonic differentiation of the hypothalamic mamillary region of the newborn mouse in culture, *J. Comp. Neurol.* **155**:355–375.

Sobkowicz, H.M., Bereman, B., and Rose, J.E., 1975, Organotypic development of the organ of Corti in culture, *J. Neurocytol.* **4**:543–572.

Spoerri, P.E., and Glees, P., 1973, Neuronal aging in cultures: An electron-microscopic study, *Exp. Gerontol.* **8**:259–263.

Spoerri, P.E., and Glees, P., 1974, The effects of dimethylaminoethyl *p*-chlorophenoxyacetate on spinal ganglia neurons and satellite cells in culture: Mitochondrial changes in the aging neurons: An electron microscope study, *Mech. Ageing Dev.* **3**:131–155.

Spoerri, P.E., Jentsch, J., and Glees, P., 1973, Apamin from bee venom: Effects of the neurotoxin on cultures of the embryonic mouse cortex, *Neurobiology* **3**:207–214.

Stern, J., 1973, The formation of sulfatide inclusions in organized nervous tissue culture, *Lab. Invest.* **28**:87–95.

Storts, R.W., and Koestner, A., 1968, General cultural characteristics of canine cerebellar explants, *Am. J. Vet. Res.* **29**:2351–2364.

Strumwasser, F., 1971, The cellular basis of behavior in *Aplysia*, *J. Psychiatr. Res.* **8**:237–257.

Strumwasser, F., and Bahr, R., 1966, Prolonged *in vitro* culture and autoradiographic studies of neurons in *Aplysia*, *Fed. Proc. Fed. Am. Soc. Exp. Biol.* **25**:512.

Suyeoka, O., 1969, Myelin stain for the cultured brain tissue, *Acta Neuropathol. (Berlin)* **13**:369–371.

Sykes, J.A., and Moore, E.B., 1959, A new chamber for tissue culture, *Proc. Soc. Exp. Biol. Med.* **100**:125–127.

Tischner, K., 1975, Changes caused by chloroquine in cultured nervous tissues of the rat, *Acta Neuropathol. (Berlin)* **33**:23–33.

Tischner, K., and Thomas, E., 1973, Development and differentiation of fetal rat sensory ganglia and spinal cord segments *in vitro*: An enzyme histochemical study, *Z. Zellforsch. Mikrosk. Anat.* **144**:339–351.

Toran-Allerand, C.D., 1976a, Sex steroids and the development of the newborn mouse hypothalamus and preoptic area *in vitro:* Implications for sexual differentiation, *Brain Res.* **106**:407–412.

Toran-Allerand, C.D., 1976b, Golgi–Cox modifications for the impregnation of whole-mount preparations of organotypic cultures of the CNS, *Brain Res.* **118**:293–298.

Toran-Allerand, C.D., 1978, The luteinizing hormone-releasing hormone (LH-RH) neuron in cultures of the newborn mouse hypothalamus preoptic area: Ontogenetic aspects and responses to steroid, *Brain Res.* **149**:257–265.

Trager, W., 1959, Tsetse-fly tissue culture and the development of trypanosomes to the infective stage, *Ann. Trop. Med. Parasitol.* **53**:473–491.

Treherne, J.E., 1967, Axonal function and ionic regulation in insect central nervous tissue, in: *Insects and Physiology* (J.W.L. Beament and J.E. Treherne, eds.), pp. 175–188, American Elsevier, New York.

Trowell, O.A., 1954, A modified technique for organ culture *in vitro*, *Exp. Cell Res.* **6**:246–248.

Trowell, O.A., 1959, The culture of mature organs in synthetic medium, *Exp. Cell. Res.* **16**:118–147.

Tsiquaye, K.N., and Zuckerman, A.J., 1974, Maintenance of adult rhesus monkey motor neurons in tissue culture, *Cytobios* **9**:207–215.

Tunnicliff, G., Cho, Y.D., and Martin, R.O., 1974, Kinetic properties of the GABA uptake system in cultures of chick retina, *Neurobiology* **4**:38–42.

Ulrich, J., and Bornstein, M.B., 1973, Experimental allergic encephalomyelitis (EAE): Delayed myelination-inhibition *in vitro* with EAE-serum: Changes in vulnerability of oligodendroglia and newly formed myelin sheaths, *Acta Neuropathol. (Berlin)* **25**:138–148.

Vago, C. (ed.), 1971, *Invertebrate Tissue Culture*, Vols. 1 and 2, Academic Press, New York.

Van DeWater, T.R., 1976, Effects of removal of the statoacoustic ganglion complex upon the growing otocyst, *Ann. Otol. Rhinol. Laryngol.* **85**(Suppl.) 33:2–32.

Van DeWater, T.R., and Ruben, R.J., 1971, Organ culture of the mammalian inner ear, *Acta Otolaryngol.* **71**:303–312.

Vaughn, J.L., and Goodwin, R.H., 1977, Preparation of several media for the culture of tissues and cells from invertebrates, *TCA Man.* **3**:527–537.

Victorov, I., Nguyen-Legros, J., Boutry, J.M., Gay, M., Berger, B., and Hauw, J.J., 1978, Technique simple de culture du noyau de locus coeruleus du souriceau nouveau-né, *C. R. Acad. Sci. Ser. D (Paris)* **286**:1893–1894.

von Recklinghausen, F.D., 1866, Über die Erzeugung von rothen Blutkörperchen, *Arch. Mikrosk. Anat.* **2**:137–139.

Wallace, B.G., Adal, M.N., and Nicholls, J.G., 1977, Regeneration of synaptic connections of sensory neurons in leech ganglia maintained in culture, *Proc. R. Soc. London Ser. B* **199**:567–585.

Wasley, G.D., and May, J.W., 1970, *Animal Cell Culture Methods*, Blackwell, Oxford and Edinburgh.

Wenzel, M., Grosse, G., Tapp, R., and Wenzel, J., 1973, Electron microscopic studies of the development of neurons in explant cultures of the trigeminal ganglion of the chick embryo, *Z. Mikrosk. Anat. Forsch.* **87**:379–409.

Wenzel, M., Wenzel, J., Grosse, G., Lindner, G., and Matthies, H., 1977, Morphometrische Untersuchungen zur *in-vitro*-Differenzierung von Hippocampusneuronen unter pharmakologischer Beeinflussung, *Verh. Anat. Ges.* **71**:121–126.

Wessells, N.K., 1967, Avian and mammalian organ culture, in: *Methods in Developmental Biology* (F.H. Wilt and N.K. Wessells, eds.), pp. 445–456, Thomas Y. Crowell, New York.

Whetsell, W.O., Schwartz, J., and Elizan, T.S., 1977, Comparative effects of herpes simplex virus types 1 and 2 in organotypic cultures of mouse dorsal root ganglion, *J. Neuropathol. Exp. Neurol.* **36**:547–560.

Whitaker, A.M., 1972, *Tissue and Cell Culture*, Baillière Tindall, London.

Willmer, E.M. (ed.), 1965, *Cells and Tissues in Culture*, Vol. 1, Academic Press, New York.

Willson, N.J., Schneider, J.F., Rosen, M., and Belisle, E.H., 1974, Ultrastructural pathology of murine cytomegalovirus infection in cultured mouse nervous system tissue, *Am. J. Pathol.* **74**:467–480.

Witkowski, J.A., and Dubowitz, V., 1975, Growth of diseased human muscle in combined cultures with normal mouse embryonic spinal cord, *J. Neurol. Sci.* **26**:203–220.

Wolf, K., and Quimby, M.C., 1964, Amphibian cell culture: Permanent cell line from bullfrog (*Rana catesbeiana*), *Science* **144**:1578–1580.

Wolf, M.K., 1964, Differentiation of neuronal types and synapses in myelinating cultures of mouse cerebellum, *J. Cell Biol.* **22**:259–279.

Wolf, M.K., and Holden, A.B., 1969, Tissue culture analysis of the inherited defect of central nervous system myelination in jumpy mice, *J. Neuropathol. Exp. Neurol.* **28**:195–213.

Wolff, E., 1961, Utilisation de la membrane vitelline de l'oeuf de poule en culture organotypique. I. Technique et possibilités, *Dev. Biol.* **3**:767–786.

Wolff, E., 1962, Culture organotypique de tissue de jeunes limules (*Xiphosura polyphemus*), *Bull. Soc. Zool. Fr.* **87**:120–126.

Wolff, E., and Haffen, K., 1952, Sur une méthode de culture d'organes embryonnaires "*in vitro*," *Tex. Rep. Biol. Med.* **10**:463–472.

Wolfgram, F., and Myers, L., 1973, Amyotrophic lateral sclerosis: Effect of serum on anterior horn cells in tissue culture, *Science* **179**:579–580.

Woodward, C.G., and Smith, H., 1975, Production of defective interfering virus in the brains of mice by an avirulent, in contrast with a virulent, strain of Semliki Forest virus, *Br. J. Exp. Pathol.* **56**:363–372.

Yamada, K.M., Spooner, B.S., and Wessells, N.K., 1970, Axon growth: Roles of microfilaments and microtubules, *Proc. Natl. Acad. Sci. U.S.A.* **66:**1206–1212.

Yavin, E., and Yavin, Z., 1974, Attachment and culture of dissociated cells from rat embryo cerebral hemispheres on polylysine-coated surface, *J. Cell Biol.* **62:**540–546.

Zagon, I.S. and Lasher, R.S., 1977, A modification of the Bodian technique for embedded, frozen, and cultured nervous tissue, *Trans. Am. Microsc. Soc.* **96:**91–96.

Zagoren, J.C., Suzuki, K., Bornstein, M.B., Chen, S.M., and Suzuki, K., 1975, Effect of hypocholesterolemic drug AY9944 on cultured nervous tissue: Morphologic and biochemical studies, *J. Neuropathol. Exp. Neurol.* **34:**375–387.

4

Cell Culture

Kedar N. Prasad

1. Introduction

The cells of mammalian nervous tissue divide, differentiate, migrate, and mature in a highly ordered pattern. Each aspect of neuronal development is expressed at a precise time, for a defined purpose. A defect in the regulation of any of the aforenamed steps can result in abnormal nervous tissue. Depending on the type of defective regulation, the stage of development during which it occurs, and the types of cells (neuronal vs. glial) that are involved, an abnormality of nervous tissue could be expressed in the form of malignancy or of various other neurological disorders. Therefore, understanding of the regulation of growth rate, differentiation, migration, maturation, and the expression of individual differentiated functions in nerve cells would be helpful in developing new approaches to the therapy of neurological diseases. Obviously, these problems cannot be systematically investigated in a complex experimental system such as the intact mammalian organism. Techniques of cell culture that allow mammalian nervous tissue to be "simplifed" prior to experimentation provide the possibility of increasing our understanding of these problems.

During the last 70 years, attempts have been made to investigate various aspects of neuronal development and functioning by growing nervous tissue in a well-defined chemical environment outside the body (i.e., *in vitro*). By using the *in vitro* growth condition, remarkable progress has been made with respect to understanding some aspects of the regulation of expression of differentiated functions. Although cell-culture techniques have great advantages over whole-animal techniques for some experimental designs, the cells that are generally studied are already preprogrammed and have expressed

Kedar N. Prasad • Department of Radiology, University of Colorado Medical Center, Denver, Colorado 80220.

functioning to varying degrees. All the experimenter can do is to increase or decrease the differentiated functions in nerve cells by manipulating a well-defined chemical environment. This has led to the identification of some molecular constituents that play important roles in the regulation of a given differentiated function. However, methods that can detect the molecular events that induce neural differentation have yet to be developed. Therefore, many basic questions remain to be answered. For example: What are factors that trigger differentation of neuroectoderm from presumptive undetermined epidermis? What factors direct neuroectoderm to form neuronl or glial elements? What factors regulate neuronal specificity and the building or wiring of circuits between nerve-cell populations and between nerve cells and their end organs? What factors control the formation of synapses?

2. Outline of Cell Culture Techniques

Five major types of *in vitro* techniques have been used to investigate the regulation of differentiated functions and malignancy in nerve cells: (1) explant culture; (2) aggregate culture; (3) dissociated cell culture; (4) monolayer cell culture; and (5) suspension cell culture.

None of these experimental systems is suitable for the study of all aspects of neurobiology. Rather, each system provides some unique advantages (as well as disadvantages) for explorations of a given aspect of nerve-cell function. The purpose of this chapter is to describe the techniques of each type of culture, and discuss the usefulness and limitations of each system in the study of the regulation of growth rate, malignancy, and differentation of nerve cells.

2.1. Explant Culture

In 1907, Harrison (1907) was first to use a neural-explant culture. He explanted a small portion of neural tube from a frog embryo to clotted lymph and demonstrated the development of an axis cylinder from the perikaryon. Since then, this technique has been used extensively and has provided better understanding with respect to problems such as cellular morphology and movements, neurite outgrowth, neurophysiological properties, neurosecretion, effects of drugs and nerve growth factors, the ultrastructures of long-term-cultured neurons, maintenance and formation of new synapses, myelination, axoplasmic flow, and experimental allergic encephalomyelitis. The explanted neural tissue can achieve full cytotypic, organotypic, and functional maturation *in vitro* (Levi-Montalcini, 1971; Murray, 1965; Bornstein and Model, 1972; Crain *et al.*, 1968; Pomerat *et al.*, 1967; Weiss, 1934). Although explant culture is discussed by Schlapfer in Chapter 3, in some ways it does prove to be a transitional technique between tissue and cell culture and will therefore be briefly discussed here. More details are available in Chapter 3. Various investigators have used different techniques of explant culture, and they are described below.

Explant-culture techniques of neural tissues are now well refined. The most commonly utilized tissues for such studies are cerebellum, cerebrum, dorsal root ganglia, and autonomic ganglia from neonatal rats, kittens, or chick embryos. The most commonly used procedure is the double-coverslip modification (Murray and Stout, 1947) of the Maximow assembly. Small pieces of neonatal tissue (approximately 1 mm square) are placed on collagen-coated (Bornstein and Murray, 1958) coverslips, and the round coverslip is fixed by capillarity to a larger square coverslip. The explant is fed with a drop or two of nutrient medium containing a balanced salt solution (BSS) serum, embryo extract, and glucose. When the explants are firmly in place and fed properly, the assembly is completed by placing the double coverslip onto a depression slide. The chamber is then sealed with paraffin and incubated at 35°C. To feed such cultures or to change the consitution of their medium, the inner round coverslip containing the culture is removed, washed, placed on a new square coverslip, refilled with the appropriate medium, and sealed in a new depression slide.

The nutrient media used in explant cultures vary from one investigator to another and from one tissue to another. For 18-day mouse hippocampus or neocortex (Crain and Bornstein, 1974), the nutrient medium contains human placental serum (33%), Eagle's Minimum Essential Medium (MEM) (54%), chick embryo extract (10%), glucose (600 mg%), and achromycin with ascorbic acid (1.2 μg/ml). Explants (Bunge *et al.*, 1974) of rat spinal cord (15.5 days old) and rat superior cervical ganglion neurons (19–21 days old) are cultured in 88% Eagle's MEM, 2% 9-day chick embryo extract, 10% human placental serum, 600 mg% glucose, 20 U/ml nerve growth factor, 2-deoxy-5-fluororidine (10^{-5} M), uridine (10^{-5} M), and 15 mM added KCl. Increased potassium levels are beneficial for neuronal survival in culture (Scott, 1971). The addition of 2-deoxy-5-fluororidine prevents the growth of dividing nonneural elements, which then allows one to observe the neurite growth in neurons.

A different explant technique has been used for chick embryo sympathetic ganglia (Phillipson and Sandler, 1975). Chick embryos of 13–15 days are suitable for culturing explants of sympathetic ganglia. Embryos are removed aseptically from their shells and placed in a sterile petri dish. Those showing developmental abnormalities are discarded. The lumbar sympathetic chains are removed under a dissecting microscope fitted with a 20 × objective. Only the first ganglion of the chain and of its contralateral partner are used.

The culture system used is a modification of that of Basrur *et al.* (1962) using a Leighton tube. In brief, aseptically obtained tissue fragments are rinsed in Hanks' BSS before they are placed in a petri dish containing a few drops of growth medium. The tissue is chopped into fine cubes of 1 mm or less, and 6 or 8 such pieces are drawn into a curved Pasteur pipette along with a little medium. These tissue fragments are spaced out over the surface of an 11 × 35 mm coverslip on the floor of a Leighton tube, and the excess medium is drawn off. A second coverslip is placed over the explant in such a way that its edge overlaps that of the lower coverslip by a few millimeters. Each Leighton tube with explant is then placed, flat surface up, on a slightly sloped rack, and 2 ml growth medium is pipetted down its round surface. Five or six such tubes, sealed with siliconized rubber stoppers, are placed on their flat surface in the rack and incubated at 38°C. The only point of difference from the original technique is the use of a single- rather than a double-coverslip system. The ganglia are pipetted onto the glass coverslip and covered with 1.5 ml culture medium. The Leighton tube is then gassed with air containing 5% CO_2, and incubated in an oven at 37°C. The coverslips are prepared by washing for 1 hr in 2 N HCl followed by thorough washing with water and a final rinse with distilled water. Coating the coverslips with collagen is unnecessary for culture periods up to about 3–4 days. The culture medium contains 90% Medium 199 supplemented with 500 mg/100 ml D-glucose and 10% fetal calf serum. Nerve growth factor is added in varying concentrations (1–20 U/ml), and embryo extract is not used. The method is highly effective, producing a success rate of about 100%. It is simple and rapid, since 10 ganglia can be grown on a single coverslip and 20 tubes can be set up in a morning's work. Oxygen tensions in the medium under these conditions are approximately 6-fold greater than those present in the egg *in vivo*, but the CO_2 tension and pH values are within the physiological range.

Another different method has been used for culturing explants of sensory ganglia (Levi-Montalcini and Angeletti, 1963). Sensory ganglia are dissected out from 7- to 10-day chick embryos and collected in physiological solution. They are then transferred in 0.5% trypsin solution and incubated at 37°C for 20 min. On careful washing in several changes of physiological solution, the ganglia are gently dissociated by suction with a micropipette and dispersed in 0.5 ml Eagle's medium containing 10% horse serum. A homogeneous suspension of nerve and supporting cells is prepared. Four drops of this suspension are placed in plastic dishes containing medium. These dishes are maintained at 4% CO_2 in air. Cells of sympathetic ganglia from 9- to 14-day chick embryos are prepared in a similar manner except the trypsin exposure time is longer (35 min). The volume of the medium in each dish is 1.5 ml. Nerve growth factor is added in a final concentration of 0.05 µg/ml every day.

2.1.1. Advantages of Explant Culture

The neural tissue can fully achieve cytotypic, organotypic, and functional maturation *in vitro*. Since these explants are maintained in a well-defined chemical environment, the effect of pharmacological agents, hormones, and nerve growth factor can be evaluated without the indirect effects of other agents that are present *in vivo*. Environmental variables can be manipulated in either numerical or temporal combinations that are not possible *in vivo*. The precise cell-to-cell contact is maintained similar to that closure *in vivo*.

Cellular architecture and assembly more closely resemble the *in vivo* condition with explant cultures than with other techniques discussed below.

2.1.2. Disadvantages of Explant Culture

There are several disadvantages of explant culture. Although explants are very small, centrally located cells frequently receive less nutrient and may become necrotic. The expression of biochemical properties cannot be studied in separate neuronal and nonneuronal populations. Cells within the explants cannot be visualized, and this limits the study of the morphology of inner cellular components. Some cells within explants die and many others migrate out along the attachment surface until the explant becomes flattened and resolved. Neurons generally are passive, but may be pulled around by other cells. This causes a dynamic change in the explant culture that is unique to the *in vitro* growth condition. Neuron populations in explants generally do not divide, although glial cells do, and therefore the regulation of growth rate of nerve cells cannot be studied. The sample size for biochemical analysis is small.

2.2. Aggregate Culture

The culture technique of aggregation allows dissociated cells to maintain cell-to-cell contacts and thus produces histotypic patterns characteristic of the original tissue. This technique has been extensively used by Moscona (1965, 1974) for various neurobiological studies. Many biochemical and morphological features of differentiation are expressed in aggregate culture in a fashion similar to that observed *in vivo* (Moscona, 1946, 1965, 1974; DeLong, 1970; Sidman and Wessells, 1975; Seeds, 1973; Trinkaus and Groves, 1955; Garber and Moscona, 1972; Morris and Moscona, 1971; Varon and Raiborn, 1969, 1972; Lowry, 1963). The tendency to aggregate is a property of all embryonic cells, and in general, the more undifferentiated the tissue the better the aggregation. There are several other factors that influence cellular aggregation. These include tissue of origin, state of differentiation of cells, rotation speed *in vitro*, temperature, calcium, serum protein, and cell-specific aggregation factors (Seeds, 1973). Aggregation formation is enhanced by gently rotating suspensions of trypsin-dissociated cells in Erlenmeyer flasks on a gyrotary shaker. A speed (about 70 rpm) is selected such that the cells are brought into a vortex, thereby greatly increasing the number of collisions between cells. At higher speeds, the shearing forces increase; therefore, the aggregate size decreases, but the number of aggregates per flask increases. The volume and composition of the cell culture medium are also important variables. Aggregation is also directly dependent on temperature. The maximal aggregation occurs at 37–38°C. Lower temperatures reduce aggregate size, and temperatures below 15°C completely block aggregation. New protein synthesis is necessary for reaggregation, since puromycin, an inhibitor of protein synthesis, inhibits this phenomenon. The aggregation of cells from a specific tissue decreases with increasing differentiation. The technique of reaggregation has been described by several investigators, but the methods described below have been taken from a recent study of Seeds (1973).

2.2.1. Cell Dissociation

C56BL/6J mice, 17 days pregnant, are the source of fetuses for the cell culture. The whole fetal brain is removed and finely minced. The tissue is dissociated in a solution of 0.25% trypsin (Difco 1:250) in saline (0.138 M NaCl, 5.4 mM KCl, 1.1 mM Na_2PO_4, 1.1 mM KH_2PO_4, 0.4% glucose, and 0.01% calcium chloride) and incubated at 37°C with constant rotation for 15 min. The tissue is dispersed by gently pipetting (three times). Large, undispersed pieces are allowed to settle out, and the suspended cells are removed and passed through a nylon screen to complete the dissociation into single cells. The large pieces are swirled in fresh trypsin solution for an additional 10 min, then pipetted and screened as before. After the addition of fetal serum to a final concentration of 10%, the cells are collected by centrifugation (200g) for 5 min.

2.2.2. Cell Culture

Aggregate cultures are prepared from approximately 1×10^7 cells in 25-ml Erlenmeyer screw-cap flasks containing 3.0 ml Eagle's Basal Medium (EBM) with 0.4% glucose and 10% fetal calf serum. The fasks are equilibrated with 5% CO_2 in air and incubated at 37°C with constant rotation (40 rpm). Aggregation is complete within 12–24 hr, and these initial aggregates are composed of loosely packed and randomly dispersed cells. These early aggregates have small cellular extensions. After several days, there is an increase in cellular outgrowth, forming a more stable complex. During this time, the cells undergo extensive migration and reaggregate within the aggregate into clusters of similar cell types. The ability of cells derived from different tissues to sort out in aggregates of mixed cell populations has been known for many years (Trinkaus and Groves, 1955; Moscona, 1946). Since there are many cell types in the brain, and the various regions develop at different rates, it is not surprising that cells of specific brain regions segregate themselves from cells of other regions (Garber and Moscona, 1972).

The technique of aggregate culture has also been used for studying the biochemical properties of developing chick retina (Morris and Moscona, 1971).

2.2.3. Advantages of Aggregate Culture

Some of the advantages of this culture technique are similar to those described for explant culture (Section 2.1.1). There are some additional advantages. Large amounts of nerve tissue can be obtained for the analysis of many biochemical differentiated functions. The method also permits the study of cell-surface properties relevant to histogenesis.

2.2.4. Disadvantages of Aggregate Culture

The disadvantages of aggregate culture are similar to those described for dissociated cell culture in Section 2.3.6 and are treated fully there. There may be some additional disadvantages. The individual cells may be altered to varying extents during the dissociation process. Furthermore, since cells of similar kinds tend to aggregate themselves, the influence of different cell types on one another, if any, is lost.

2.3. Dissociated Cell Culture

The cellular heterogenity characteristic of explant cultures as well as problems of neuronal viability and the complex interrelationships among various cell types have emphasized the need for developing a method of isolating at least neuronal from nonneuronal components to allow for studies of the function of neurons and glial cells separately. These cell populations can then be brought together selectively for studies of the interaction effects of these cell types *in vitro* on the expression of various cellular functions, which then can be compared with the properties of the original undissociated nerve tissue.

2.3.1. Chick Embryo Cerebrum

Varon (Varon and Raiborn, 1969) has developed an elegant technique of isolating individual neuronal and nonneuronal cell types that involves dissociation, fractionation, and culture. The procedure described below is carried out under sterile conditions and at room temperature.

1. The cerebrum of an 11-day-old chick embryo is selected as a starting material for the isolation of various cell types because by 12 days, neural differentiation is practically complete. From day 12 on, neuroglia increase, fiber tracts thicken, important metabolic changes including initiation of sphingomyelin synthesis take place, and bioelectrical activity appears.

2. The tissue, dissected free of its meningeal membranes, is placed in a nylon bag (200-mesh), which is then immersed in EBM and stroked on the outside for a few minutes until it no longer releases cellular materials into the outer fluids. The tissues are dissociated mechanically suspended in EBM (4 ml/cerebrum), and then passed through nylon cloth (200-mesh). The outer medium contains a large number of dispersed cells, various cell aggregates, and minute cell debris.

3. The outer cell suspension is further filtered on nylon cloth (30-mesh) to remove large aggregates, centrifuged 2 min at 700 rpm, and suspended in the same volume of buffer.

4. Aliquots (4 ml) of the suspension described above are layered in round-bottom polyethylene tubes over 5 ml of a 1 : 1 mixture of 15% bovine serum albumin (Fraction IV) and 20% sucrose, both in EBM. The tubes are centrifuged for 10 min at 700 rpm. The upper phase is discarded, and the interface and underlying bovine serum albumin—sucrose phase (S_1) are collected. The pellet (P_1) is resuspended in 7 ml fresh EBM and layered over bovine serum albumin–sucrose and centrifuged as above. The second S_2 and second P_2 are obtained. S_1 and S_2 are pooled, diluted with an equal volume of EBM, and centrifuged for 5 min at 1000 rpm. The sediment is resuspended in 4 ml EBM (S fraction). The P_2 pellet is also suspended in 4 ml EBM, P_2 and S fractions being pooled, when so desired.

5. An equal volume of Ca- and Mg-free BSS containing 0.002% trypsin solution is added to the solution containing the P_2 or S fraction or both. After 3 min of incubation, the suspensions are centrifuged for 5 min at 1000 rpm; the pellets are resuspended in small volumes of EBM and are referred to as P_2T or ST (for trypsin-treated P_2 or S fraction).

6. A 3-ml aliquot of P_2T containing 20% calf serum is seeded in T-15 flasks and incubated at 38°C for 24 hr. The media, with their contents of unattached cells, are withdrawn from the flask and centrifuged 5 min at 700 rpm, and the pellet is resuspended in EBM. This fraction contains primarily (about 90%) A cells, as defined below. The cells left attached in each flask (P_2TA) are harvested in 3 ml Ca- and Mg-free BSS containing 0.1% trypsin and centrifuged for 5 min at 1000 rpm. The suspension of this fraction contains two distinct types of cells, i.e., B and C cells, as well as small cells. However, these cells are not easily distinguishable from each other. The PT_2A suspension is seeded on a glass surface containing EBM and 20% calf serum. After 24 hr only C cells are attached. The pure population of cells has been termed CC. It has not been possible to isolate B cells and small cells in a purified form.

The three major cells types that are isolated by this technique can be described as follows.

A cells. The large, round cells when grown in a monolayer culture develop a number of processes. Most A cells are multipolar. Their cell bodies are 15 μm in diameter. Their nuclei are distinct and usually located eccentrically. They contain two nucleoli. The A cells do not attach to glass even in a serum-supplemented medium. There has been no evidence of mitotic activity. These cells probably represent well-differentiated neurons.

B cells. The B cells are small (7–10 μm in diameter) and dense, and have smaller nucleoli. They also develop processes, but unlike A elements, B cells are mostly bipolar. B-cell fibers, unlike A-cell fibers, do not take up silver stain; however, like A cells, B cells do not attach to glass and do not show proliferation. They also probably represent neuronal elements.

C cells. The C cells are in many respects similar to B cells, but they behave differently in culture. C cells spread out into very large, extremely thin, multiform elements with ragged contours and gross processes. The nuclei are oval and usually centralized. These cells attach to uncoated glass surfaces in the presence of serum-supplemented medium. These are nonneuronal cells.

This type of culture develops very rapidly. A and B cells begin developing processes within the first few hours, and by 4 hr, many of them are long and branched. By 24 hr, long fibers criss-cross the field in complex lattices. Fibers make numerous contacts with one another, and can make intricate arrangements or appear to merge with one another, or bundle into tight nervelike fascicles. Fiber-to-cell contacts are also frequent between A cells, or between A and B cells. Most A and B cells start dying after 4–7 days of culture, although a few such elements have been known to survive without apparent change for over 1 month. In most cultures, proliferation of C cells becomes the dominant feature by day 4, and by day 7 these cells have taken over.

2.3.2. Chick Embryo Sympathetic Ganglia

Different methods have been used for preparing dissociated cell cultures from chick embryo sympathetic ganglia (Varon and Raiborn, 1972). The lumbar halves (10 ganglia each) of the paravertebral sympathetic chains are dissected from 7- to 15-day-old chick embryos and collected in either Hank's BSS or Ca- and Mg-free saline [(CMF) 1 liter of solution containing NaCl, 8 g; KCl, 0.2 g; $Na_2HPO_4 \cdot 2H_2O$, 0.05 g; $NaHCO_3$, 2.2 g; and glucose, 6 g] at pH 7.5. A ratio of 1 ml collecting solution per four half-chains (40 ganglia) is used. Tissue from up to 10 embryos is pooled, with a maximal collection time of 30 min. At the end of the collection step, the collection fluid is aspirated and replaced with an equal volume of BSS containing 0.25% trypsin, and the material is incubated for 30 min at 37°C. After brief centrifugation, the trypsin solution is replaced with an equal volume of low-bicarbonate culture medium (LMS), consisting of EBM supplemented with glucose (increased to 6 g/liter), penicillin (200,000 U/liter), fetal calf serum (5%), and 7 S nerve growth factor at 100 BU/ml. Full dispersion of tissue on LMS fluid is achieved by 10 aspirations through a Pasteur pipette (tip diameter ≈500 μm). The dissociated cell suspension (harvest) is differentially counted for neuronal and nonneuronal elements in a hematocytometer chamber under phase contrast. The neuronal populations can be differentiated by size and shape, namely, larger (≥20 μm diameter) polygonal elements and smaller (≈8 μm diameter) round ones. The nonneuronal cells are easily distinguished from the neurons because of their considerably smaller size (≤5 μm diameter), irregular contour, and cytoplasmic granulations.

For culture purposes, the harvest suspension is diluted approximately 8-fold to 50,000 neurons/ml using high-bicarbonate culture medium (HMS) with the same composition as the LMS except for a 10-fold increase in bicarbonate level to 2.2 g/liter. Aliquots 2 ml; 100,000 neurons plus about equal numbers of accompanying nonneurons) are seeded into 35-mm plastic Falcon dishes, precoated with reconstituted rat-tail collagen, and the dishes incubated at 37°C in a water-saturated, 5% CO_2–95% air mixture. For long-term culture, the medium is changed every other day.

The collection of ganglia directly in CMF rather than Hank's BSS or LMS results in greater yield of both neuronal and nonneuronal cells, with no noticeable impairment of their subsequent viability in culture (Varon and Raiborn, 1972). The advantage of CMF collection is greater for younger ganglia (almost 3-fold) than for older ones (about 1.5-fold). The duration of the tissue exposure to either CMF or BSS in the collection step can be varied from 2 to 60 min with no noticeable influence on the final dissociation yield. Lowering the trypsin concentration or shortening the time of treatment progressively reduces the final cell yields, but increasing either or both modalities provides no significant further improvement. Omission of the 7 S nerve growth factor from the dispersion medium does not result in lower yields. Prolonged exposure to alkaline (pH 9) medium at any one stage of the dissociation procedure decreases the final cell yield. The ratio of medium to tissues is an important factor. Ratios lower than 25 μl or greater than 75 μl per ganglion lead to significantly poorer yield of both neuronal and nonneuronal elements.

2.3.3. Comparison of Chick, Rat, and Human Embryonic Brain

Booher and Sensenbrenner (1972) compared the growth and cultivation of dissociated neuronal and glial cells from embryonic chick, rat, and human brain grown in flask cultures with those grown in reconstituted rat-tail-collagen-coated flasks. Various ages of embryonic brain tissue were studied in each species: in the chick, 5-, 7-, 8-, 11-, 14-, and 18-day-old embryos were used; in the rat, 7-, 9-, and 16-day-old embryos were used; in humans, 13- and 19-week-old embryos were used. Prior to the dissociation procedure, the meningeal tissue and any connective tissue were removed from the brain with the aid of the stereodissecting microscope, to ensure that only cerebral and subcortical tissue cells were used for culture. Two standard techniques were employed in the dissociation of the cerebral tissue: trypsin (0.25%) dissociation and the sieve technique, in which the brain tissue was gently passed through a nylon mesh (48-μm pore) into a small reservoir of nutrient medium. The inoculum of cell suspension was between 2 and 2.5 $\times$ 10^5 cells/ml in all culture preparations. The nutrient medium consisted of EBM supplemented with 20% fetal bovine serum, 20% (vol./vol.) whole-egg ul-trafiltrate, and 600 mg% (wt./vol.) glucose, penicillin (100 U/ml), and strep-tomycin (100 μg/ml). The pH range of the medium was maintained between 6.9 and 7.3. During the growth period, the fresh medium was changed periodically. The cell cultures were incubated at 37°C in an atmosphere of 95% oxygen and 5% CO_2. Relative humidity was maintained between 80 and 90%. Subcultures were made by trypsinizing the cells and reseeding into collagen- or non-collagen-coated flasks at one half the original cell concentration. The basic difference between the cultures made on the collagen-coated substance and the plastic alone was evident during the first week of culture, in which the cells grown on collagen remained well isolated, while cellular clumping was observed in the cultures grown on plastic. Following the establishment of a monolayer of mesenchymal cells, on which the neuronal elements appeared to develop and differentiate, these aspects of the two cultures were nearly identical.

The best result with respect to neuronal survival and maturation was obtained with 8-day-old chick and 7-day-old rat embryos. Cultures made from older chick and rat brains resulted in a high yield of mesenchymal cells along with glial-like elements. Both 13- and 19-week human fetal dissociated brain cells grew and differentiated into mature neurons on both collagen and plastic. As in the rat and chick, these two brain specimens were derived from near midterm of gestation, which is prior to the onset of complete morphological neuronal differentiation in the human brain.

2.3.4. Other Methods of Isolation

Several other methods of isolating neuronal and nonneuronal cells have been used. The freehand dissection of individual neurons from frozen sections (Lowry, 1963) or of individual neurons and glial clumps from fresh sections (Hydèn, 1962) has been used. Larger numbers of individual neurons have been isolated by mechanical disruption of tissue and then pouring the suspension through nylon sieves (Roots and Johnston, 1964). Preparation of dissociated neuronal and glial cells on an even larger scale (Cohen *et al.*, 1964; Nakai, 1956; Varon *et al.*, 1963) has been achieved by following the procedure for other tissue (Rinaldini, 1958). Others have tried to fractionate the cell suspensions by differential centrifugation in sucrose (Korey, 1957; LeBaron, 1966), sucrose–serum mixture (Chu, 1954), Ficoll (Rose, 1967), and acetone–glycerol–water (Satake and Abe, 1966).

Another study (Dezerga *et al.*, 1970) has reported the isolation of viable neurons from chick embryonic spinal ganglia by centrifugation through albumin gradients. Thirty to forty dorsal root ganglia were removed from 7-day chick embryos and immersed in a petri dish for 5 min at 37°C with 5 ml 0.25% trypsin in Hank's BSS. The ganglia were gently pipetted back and forth for 10 min with a small-bore Pasteur pipette to remove the capsule and intact nonneuronal tissue. The cells were then incubated in the trypsin solution at 37°C for 15 min. The ganglia became dissociated by continuous pipetting and incubating for about 1.5 hr. The mixed cell suspension was then centrifuged at 1000g for 2 min, and then 1 ml Medium 199 supplemented with 700 mg% glucose was added to the pellet and a suspension was prepared.

The neuronal fraction was then separated from the other cells in the suspension by ultracentrifugation in the Spinco Model L with the SW 39 rotor using discontinous bovine serum albumin gradients with the following specific gravities: 1.100, 1.079, and 1.065. Layering of the gradient fractions was done at 4°C, and the tubes were kept cold at all times. A 1-ml sample of each of the different albumin fractions was added to sterile 5-ml centrifuge tubes with the highest density on the bottom. The tubes were then centrifuged at 30,000 rpm for 1 hr at 4°C, and enriched neuronal fractions were collected from the 1.079 and 1.100 interfaces. Aliquots (1 ml) of the isolated cell suspension were cultured in Rose chambers after they were mixed with 1 ml Medium 199 containing 400 mg% glucose and 1 ml fresh rooster plasma and nerve growth factor. Fiber outgrowths from neurons reached 12 times the length of the original soma within 48 hr.

2.3.5. Advantages of Dissociated Cell Culture

This technique provides an opportunity to compare the behavior and biochemical properties of neuronal and nonneuronal cells both separately and in conjunction with one or more cell types. With this advantage, it may help to clarify which neuronal functions depend on perfect tissue organization and cell-to-cell contact for maximal expression.

2.3.6. Disadvantages of Dissociated Cell Culture

One cannot study the induction and regulation of expression of differentiated functions, since these cells are already fully differentiated. Furthermore, the regulation of growth rate in nerve cells cannot be studied, since the nerve cells obtained after dissociation do not divide in culture.

Other disadvantages may be mentioned here that apply not only to dissociation techniques but also to cell-culture techniques in general. The original organization of the tissue no longer exists and cell–cell contacts are lost. Intercellular distances are increased depending on the volume of fluid in which the cells are dispersed, and the numerical balance among different cell types may be modified. Individual cells may be altered to varying degrees depending on their type, source, and the procedures employed. Although trypan blue has been used to estimate the viability of isolated cells, this is a highly unreliable method.

2.4. Monolayer Cell Culture

None of the techniques described aboved is suitable for the study of regulation of growth rate, differentiation, and gene expression in homogeneous populations of nerve cells. The availability of monolayer cell cultures provides a unique opportunity to study these problems. Unfortunately, cultures of normal, dividing nerve cells are not available. However, monolayer cultures of nerve cells that have been derived from tumors are available. Monolayer cultures of mouse and human neuroblastoma are now being used in several laboratories to investigate the problems of growth-rate regulation, differentiation, malignancy, and other aspects of neurobiological interest. Many of the responses of neuroblastoma cells to various agents are similar to those observed in explants of embryonic nervous tissue (as reviewed in Prasad, 1975). Techniques for culturing mouse and human neuroblastoma will be described separately.

2.4.1. Mouse Neuroblastoma

Mouse C1300 tumor spontaneously arose in the body cavity of a mouse of strain A/J in 1940, and has been maintained by subcutaneous transplantation in A/J mice at the Jackson Laboratory, Bar Harbor, Maine. This tumor was initially diagnosed as a round-cell tumor, possibly neuroblastoma (Dunham and Stewart, 1953). However, its neuronal nature was not confirmed until 1969, when two groups of investigators (Schubert *et al.*, 1969; Augusti-Tocco and Sato, 1969) independently established the cell line in monolayer culture. Schubert *et al.* (1969) used the following technique: The solid tumor was adapted to tissue-culture conditions by dispersing the cells in modified Eagle's medium containing 20% fetal calf serum. The cell cultures were maintained at 37°C in an 85% air–15% CO_2 incubator. Cells were cloned twice by spreading dilute cell suspensions on solid agar (0.5% agar, 4.5 mg/ml tryptic soy broth in Eagle's modified medium plus 20% fetal calf serum) and picking visible colonies with a platinum loop after a 2-week incubation.

Augusti-Tocco and Sato (1969) used a different technique but obtained a similar result. The technique of alternate-passage animal culture (Buonassisi *et al.*, 1962) was followed to adapt the neuroblastoma cells to the monolayer growth condition. The tissue was dissociated by treatment with a Viokase (0.25%) solution, and the single cells were plated in Falcon plastic flasks pretreated with 5% gelatin solution. Ham's F-10 medium supplemented with 15% horse serum and 2.5% fetal calf serum was used. Clonal lines were isolated from cultures at the second or third passage *in vitro* following the single-cell plating.

Another method has been followed in the author's laboratory. Solid tumor, free of necrotic tissue, is dissected from the animal. Solid tumor of relatively small (3- to 4-pea) size has very few if any necrotic cells. Solid tumor tissue, after being washed three times in Ca-free MEM, is placed on an organ-culture grid that is immersed in Ca-free MEM. The tissue is grated on the organ-culture grid, and clumps of tissue that float into the medium are collected. The remaining, fibrous tissue is excluded from further treatment. The collected tissue clumps are washed twice with Ca-free MEM and then incubated in the presence of a Viokase solution (0.25% in Ca-free medium) for 20 min. A single-cell suspension is prepared, and cells are washed twice with F12 medium. Cells are then plated in Falcon plastic flasks containing F12 medium with 10% α,γ-globulin, newborn calf serum, penicillin (100 μg/ml), and streptomycin (100 U/ml), and are maintained at 36°C in a humidified atmosphere of 5% CO_2 in air. Clones are isolated as follows (Prasad *et al.*, 1973): Two hundred cells are placed in Falcon plastic dishes (60 mm) and are allowed to form clones. A stainless steel ring coated with nontoxic Vaseline is placed on a well-isolated clone, and the ring is half filled with 0.25% Viokase. After 5 min of incubation, the cells are transferred to another dish. The procedure is repeated at least twice before the clone is selected for the experiment.

Despite variations in techniques used for monolayer cultures by different laboratories, the morphology and biochemical features of neuroblastoma cells are expressed in a similar fashion. In our experience, we have found the following factors important for healthy growth of neuroblastoma cells in culture: Cells should not be allowed to reach confluency before splitting, because confluent cells do not grow well after replating; however, they may recover eventually. Commencing 2 days after plating, the medium must be changed daily because neuroblastoma cells produce lactic acid at a relatively high rate. The medium, the Viokase, and the centrifuge tubes should be warmed at 37°C, and should be used immediately for splitting. In addition, the medium should be prewarmed in the flask in which new cells will be plated.

2.4.2. Human Neuroblastoma Cells

Several laboratories have developed monolayer cultures of human neuroblastoma. Like mouse neuroblastoma cells, human neuroblastoma cells may be adapted to monolayer growth by using different methods and different growth conditions, but the cells behave similarly under all these conditions. Tumilowicz *et al.* (1970) used the following procedures: Neuroblastoma tissue was obtained from an abdominal mass of a 13-month-old Caucasian boy during exploratory surgery. This tumor mass had rare areas of organoid differentation. Cells (IMR-32) were explanted by mincing tissue into fine fragments with a sharp blade and placing the resulting suspension in milk-dilution bottles containing 10 ml growth medium. To facilitate fragment attachment, the bottles were inverted so that adhering fragments were not in contact with medium introduced on the opposite side. After about 1 hr of incubation at 37°C when the fragments became more firmly adhered, the bottle was turned so that the medium covered the fragments. When abundant outgrowth had occurred, remaining fragments were removed with 0.25% trypsin in Hanks' BSS (trypsin and medium components, except antibiotics). Cells were grown in Eagle's MEM plus nonessential amino acids in Hanks' BSS and 30% heat-inactivated (30 min at 58°C) fetal bovine serum containing penicillin (100 U/ml) and streptomycin (100 μg/ml). Beginning with the 37th subcultivation, the composition of the growth medium was altered to consist of 80% Medium 199 and 20% fetal bovine serum (not heat-inactivated) containing antibiotic concentrations as described above. The growth medium was replaced twice weekly. Temperatures for cultivation ranged between 35.5 and 37°C. When the cell layer became confluent, cells were detached from glass for subcultivation with trypsin.

The doubling time of IMR-32 was about 48 hr, and the chromosome number varied from 46 to 48 (Tumilowicz *et al.*, 1970). This cell line is available from the American Tissue Culture Collection. We have used this cell line for several studies and have made the following technical observations: These cells are extremely sensitive to acidic pH; therefore, the medium is changed frequently. Cells are replated when the culture has reached the subconfluent stage. The growth rate is better and cells are healthier when the initial number of cells plated in Falcon flasks is at least 0.25×10^6.

Biedler *et al.* (1973) have followed a different procedure. A bone marrow sample from a patient with neuroblastoma was cultured in Eagle's MEM supplemented with 20% fetal bovine serum, penicillin (100 U/ml), streptomycin (100 μg/ml), and fungizone (2.5 μg/ml) in plastic flasks. Erythrocytes were washed off after 2 days when attachment and some growth of cells was noted. Cultures were transferred three or four times at irregular intervals. After 3 months, SK-N-SH cells were routinely transferred every 2 weeks with medium replacement on days 5, 9, and 12 of the 2-week cycle.

Monolayer cultures have also been established from neuroblastoma tumor tissue obtained from different patients. Small pieces of tumor tissue were minced in the medium described, and tumor pieces were treated with 0.125% trypsin and 0.02% EDTA in Ca- and Mg-free phosphate-buffered salt solution to obtain a cell suspension. Clumps of cells attached to plastic culture flasks after 2–3 days, and by about the 2nd month, SK-N-MC cells were transferred at weekly intervals with a medium replacement on day 5. In plastic flasks, cells were attached after approximately 24 hr, while in glass vessels, attachment occurred only after 2 days. Both cell lines were routinely maintained in Eagle's medium containing 15% fetal bovine serum, penicillin, streptomycin, and nonessential amino acids (Eagle's formulation). For culture transfer, cells were exposed to the EDTA–trypsin solution for 3–5 min.

The chromosome number per cell varies from 44 to 50 with a modal chromosome number of 46 (Biedler *et al.*, 1973). These cell lines are available only from Dr. Biedler's laboratory.

Another line of cells designated as NJB has been maintained since 1964 (Goldstein *et al.*, 1964). Small fragments of neuroblastoma tissue were explanted under perforated cellophane in fluid media or in chicken plasma-clot cultures on perforated cellophane, in D-35 Carrel flasks. Cultures were also prepared on coverslips in plasma clots in 50-mm petri dishes and gassed with 5% CO_2 and 95% O_2. The fluid medium consisted of either 60% Medium 199 or a modified Eagle's medium, containing higher concentrations of the essential amino acids, nonessential amino acids, and either 40% human serum or 40% calf serum. Glucose was added at a concentration of 5 mg/ml in some cultures. Penicillin was initially added at a concentration of 200 U/ml.

2.4.3. Advantages of Monolayer Cell Culture

There are several unique advantages of monolayer cultures of nerve cells. It would be ideal if such a culture could be established for normal dividing neuroblasts; however, all monolayer cultures of nerve cells at present have been derived from tumors. Nevertheless, such cultures provide for the first time an opportunity to study the regulation of growth rate, differentiation, and maturation in a homogeneous clonal population of nerve cells.

Furthermore, the expression of individual gene products can be measured in dividing nerve cells. The toxicity of pharmacological agents on nerve cells can also be studied. The mechanism of neurotransmitter effects can be investigated at the molecular level. The effect of a given agent on nerve and glial cells can be separately studied. Recent success in synapse formation with hybrid neuroblastoma cells (neuroblastoma × glial) and muscle cells (Nelson *et al.*, 1976) provides a unique opportunity to study the regulation of synapse formation.

2.4.4. Disadvantages of Monolayer Cell Culture

A monolayer cell culture is completely removed from *in vivo* tissue architecture. Therefore, the information obtained from such cultures may not be pertinent to the *in vivo* condition. The expression of certain genes is lost, whereas the expression of others is increased. For example, the activity of glucose-6-phosphate dehydrogenase *in vivo* is less than that observed in culture, whereas the reverse is true for 6-phosphogluconate dehydrogenase (Prasad *et al.*, 1976). Hydrocortisone induces glutamine synthetase activity in explants of neural retina. However, it fails to do so when dissociated retinal cells are cultured in monolayers (Morris and Moscona, 1970). All monolayer cultures of nerve cells are derived from tumor cells and may not have pertinence for normal cells. The mouse neuroblastoma cells are highly aneuploid; the chromosome number per cell varies from 40 to 200 (Amano *et al.*, 1972), whereas human neuroblastoma cells are nearer to the diploid condition, with chromosome numbers varying from 44 to 48 (Tumilowicz *et al.*, 1970; Biedler *et al.*, 1973).

2.5. Suspension Cell Culture

Neuroblastoma cells derived from C1300 tumor have also been grown in suspension cultures. These cells have also been grown as monolayer cultures. Cells were grown in Eagle's MEM for suspension culture. The culture medium was supplemented with 5% calf serum, 5% fetal calf serum, 2 mM glutamine, 25 U/ml penicillin G, and 10 μg/ml streptomycin sulfate. These cells were maintained at 37°C in suspension (spinner culture) in Erlenmeyer flasks by continuous stirring with a Teflon-coated magnetic bar. The flasks were flushed with 5% CO_2 in air and were then closed with a rubber stopper through which a small cotton-plugged glass tube was inserted. The density was adjusted daily to 3×10^5 cells/ml by dilution with fresh suspension-culture medium.

2.5.1. Advantages of Suspension Culture

Suspension culture has all the advantages of monolayer culture (Section 2.4.3). However, it provides an additional tool in dissecting out those neuronal functions that require interaction with solid substrates from those that do not. This has been demonstrated in a recent study (Truding *et al.*, 1974). For example, dibutyryl cyclic AMP induces the synthesis of glycoprotein of molecular weight 105,000 in both suspension cultures and monolayer cells, whereas it induces the synthesis of a protein of molecular weight 78,000 only in monolayer cells.

2.5.2. Disadvantages of Suspension Culture

Suspension culture also has the disadvantages that were described for monolayer cell culture (Section 2.4.4). It has the additional disadvantage that many biochemical functions may not be induced if cells are grown in suspension cultures. For example, the regulation of neurite formation and synapse formation cannot be studied in suspension culture.

3. Concluding Remarks

At least five techniques for growing nerve cells in culture are available to study the regulation of growth rate, differentiation, and malignancy. Each technique has its own advantages and disadvantages. Therefore, it is essential to use more than one technique before making a general conclusion with respect to any one aspect of regulation of neuronal features or cancer. What molecule or molecules trigger neural differentiation? This important question in neural development still remains to be answered despite extensive work. The various techniques described in this section have used cells that are relatively well differentiated. It would be advantageous to have monolayer cultures of dividing nerve cells as well as of presumptive undetermined epidermis. The latter would provide an opportunity to study the molecular mechanisms for inducing neural differentiation.

References

Amano, A., Richelson, E., and Nirenberg, M., 1972, Neurotransmitter synthesis by neuroblastoma clones, *Proc. Natl. Acad. Sci. U.S.A.* **69**:258.

Augusti-Tocco, G., and Sato, G., 1969, Establishment of functional clonal lines of neurons from mouse neoroblastoma, *Proc. Natl. Acad. Sci. U.S.A.* **64**:311.

Basrur, P.K., Basrur, V.R. and Gilman, J.P.W., 1962, A simple method for short term cultures from small biopsies, *Exp. Cell Res.* **30**:229

Bornstein, M.B., and Model, P.G., 1972, Development of synapse and myelin in cultures of dissociated embryonic mouse spinal cord, medulla and cerebrum, *Brain Res.* **37**:287.

Bornstein, M.B., and Murray, M.R., 1958, Serial observation on patterns of growth, myelin formation, maintenance and degeneration in cultures of newborn rat and kitten cerebellum, *J. Biophys. Biochem. Cytol.* **4**:499.

Bunge, R.P., Rees, R., Wood, P., Burton, H., and Ko, C.P., 1974, Anatomical and physiological observation on synapses formed on isolated autonomic neurons in tissue culture, *Brain Res.* **66**:401.

Buonassisi, V., Sato, G., and Cohen, A.I., 1962, Hormone-producing cultures of adrenal and pituitary tumor origin, *Proc. Natl. Acad. Sci. U.S.A.* **48**:1184.

Chu, L.W., 1954, A cytological study of anterior horn cells isolated from human spinal cord, *J. Comp. Neurol.* **100:**381.

Cohen, A.J., Noncl, E.C., and Richter, W., 1964, Nerve growth requirement for development of dissociated embryonic sensory and sympathetic ganglia in culture, *Proc. Soc. Exp. Biol. Med.* **116:**784.

Crain, S.M., and Bornstein, M.B., 1974, Early onset in inhibitory functions during synaptogenesis in fetal brain cultures, *Brain Res.* **68:**351.

Crain, S.M., Peterson, E.R., and Bornstein, M.B., 1968, Formation of functional interneuronal connections between explants of various mammalian central nervous systems during development *in vitro*, in: *Growth of the Central Nervous System* (G.E.W. Wolstenholme, ed.), pp. 13–40, Churchill, London.

DeLong, G.R., 1970, Histogenesis of fetal mouse isocortex and hippocampus in reaggregating cell cultures, *Dev. Biol.* **22:**563.

Dezerga, G., Johnson, L., Morrow, J., Kasten, F.H., 1970, Isolation of viable neurons from embryonic spinal ganglia by centrifugation through albumin gradient, *Exp. Cell Res.* **63:**189.

Dunham, L.C., and Stewart, H.L., 1953, A survey of transplantable and transmissible animal tumors, *J. Natl. Cancer Inst.* **13:**1299.

Garber, B.B., and Moscona, A.A., 1972, Reconstruction of brain tissue from cell suspension. I. Aggregation patterns of cells dissociated from different regions of developing brian, *Dev. Biol.* **24:**217.

Goldstein, M.N., Burdman, J.A., and Journey, L.J., 1964, Long-term tissue culture of neuroblastomas: Morphologic evidence for differentiation and maturation, *J. Natl. Cancer Inst.* **32:**165.

Harrison, R.G., 1907, Observation on the living developing nerve fibre, *Proc. Soc. Exp. Biol. Med.* **4:**140.

Hydèn, H., 1962, Cytophysiological aspects of the nucleic acids and protein of nervous tissue, in: *Neurochemistry* (K.A.C. Elliott, J.H. Page, and J.H. Quartel, eds.), pp. 331–335, Charles C. Thomas, Springfield, Illinois.

Korey, S., 1957, Some characteristics of a neuroglia fraction, in: *Metabolism of the Nervous System* (D. Richter, ed.), pp. 87–90, Pergamon Press, New York.

LeBaron, F.N., 1966, Determination of protein bound phosphoinositides in glial cell concentrates of brain white matter, in: *Variation in Chemical Composition of the Nervous System* (G.B. Ansell, ed.), p. 65, Pergamon Press, New York.

Levi-Montalcini, R., 1971, Two control mechanisms of growth and differentiation of the sympathetic nervous system, in: *Cellular Aspects of Neural Growth and Differentiation* (D. Pease, ed.), pp. 253–268, University of California Press, Los Angeles.

Levi-Montalcini, R., and Angeletti, P.U., 1963, Essential role of the nerve growth factor in survival and maintenance of dissociated sensory and sympathetic embryonic nerve cells *in vitro*, *Dev. Biol.* **7:**653.

Lowry, O., 1963, The chemical study of single neurons, *Harvey Lect.* **58:**1019.

Morris, J.E., and Moscona, A.A., 1970, Induction of glutamine synthetase in embryonic retina: Its dependence on cell interaction, *Science* **167:**1736.

Morris, J.E., and Moscona, A.A., 1971, The induction of glutamine synthetase in cell aggregates of embryonic neural retina: Correlations with differentiation and multicellular organization, *Dev. Biol.* **25:**420.

Moscona, A.A., 1946, Development of heterotypic combinations of dissociated embryonic chick cells, *Proc. Soc. Exp. Biol. Med.* **92:**410.

Moscona, A.A., 1965, Recombination of dissociated cells and the development of cell aggregates, in: *Cell and Tissue Culture* (E.N. Willmer, ed.), pp. 489–529, Academic Press, New York.

Moscona, A.A., 1974, Surface specification of embryonic cells: Lectin receptors, cell recognition and specific cell ligands, in: *The Cell Surface in Development* (A.A. Moscona, ed.), pp. 67–99, Wiley, New York.

Murray, M.R., 1965, Nervous tissue *in vitro*, in: *Cells and Tissue in Culture*, Vol. 2 (E.N. Willmer, ed.), pp. 373–455, Academic Press, New York.

Murray, M.R., and Stout, A.P. 1947, Distinctive characteristics of the sympatheticoblastoma cultivated *in vitro*: A method for prompt diagnosis, *Am. J. Anat.* **23**:429.

Nakai, J., 1956, Dissociated dorsal root ganglia in tissue culture, *Am. J. Anat.* **99**:81.

Nelson, P., Christian, C., and Nirenberg, M., 1976, Synapse formation between clonal neuroblastoma × glioma hybrid cells and striated muscle cells, *Proc. Natl. Acad. Sci. U.S.A.* **73**:123.

Phillipson, O.T., and Sandler, M., 1975, The influence of nerve growth factor, potassium depolorization and dibutyryl (cyclic) adenosine 3′, 5′-monophosphate on explant culture of chick embryo sympathetic ganglia, *Brain Res.* **90**:273.

Pomerat, C.M., Hendelman, W.J., Raiborn, C.W., Jr., and Massey, J.F., 1967, Dynamic activities of nervous tissue *in vitro*, in: *The Neuron* (H. Hyden, ed.), pp. 119–178, Elsevier, New York.

Prasad, K.N., 1975, Differentiation of neuroblastoma cells in culture, *Biol. Rev.* **50**:129.

Prasad, K.N., Mandal, B., Waymire, J.C., Lees, G.J., Vernadakis, A., and Weiner, N., 1973, Basal level of neurotransmitter synthesizing enzymes and effect of cyclic AMP agents on the morphological differentiation of isolated neuroblastoma clones, *Nature (London) New Biol.* **241**:117.

Prasad, N., Prasad, R., and Prasad, K.N., 1977, Electrophoretic patterns of glucose metabolizing enzymes and acid phosphatase in mouse and human neuroblastoma cells, *Exp. Cell Res.* **104**:273–277.

Rinaldini, L.M.J., 1958, The isolation of living cells from animal tissue, *Int. Rev. Cytol.* **7**:587.

Roots, B., and Johnston, P.U., 1964, Neurons of ox brain nuclei: Their isolation and appearance by light and electron microscopy, *J. Ultrastruct. Res.* **10**:350.

Rose, S.P.R., 1967, Preparation of enriched fractions from cerebral cortex containing isolated, metabolically active neuronal and glial cells, *Biochem. J.* **102**:33.

Satake, M., and Abe, S., 1966, Preparation and characterization of nerve cell perikaryon from rat cerebral cortex, *J. Biochem. (Tokyo)* **79**:72.

Schubert, D., Humphreys, S., Baronia, C., and Cohn, M., 1969, *In vitro* differentiation of a mouse neuroblastoma. *Proc. Natl. Acad. Sci. U.S.A.* **64**:316.

Scott, B.S., 1971, Effect of potassium on neuron survival in cultures of dissociated human nervous tissue, *Exp. Neurol.* **30**:297.

Seeds, N.W., 1973, Differentiation of aggregating brain cell cultures, in: *Tissue Culture of the Nervous System* (G. Sato, ed.), pp. 35–53, Plenum Press, New York.

Sidman, R.L., and Wessells, N.K., 1975, Control of direction of growth during the elongation of neurites, *Exp. Neurol.* **43**:237.

Trinkaus, J.P., and Groves, P.W., 1955, Differentiation in culture of mixed aggregates of dissociated tissue cells, *Proc. Natl. Acad. Sci. U.S.A.* **41**:787.

Truding, R., Shelanski, M.L., Daniels, M.P., and Morrel, P., 1974, Comparison of surface membranes isolated from cultured murine neuroblastoma cells in the differentiated or undifferentiated cells, *J. Biol. Chem.* **249**:3973.

Tumilowicz, J.J., Nichols, W.W., Cholon, J.J., and Greene, A.E., 1970, Definition of a continuous human cell line derived from neuroblastoma, *Cancer Res.* **30**:2110.

Varon, S., and Raiborn, C.W., 1969, Dissociation, fractionation and culture of embryonic brain cells, *Brain Res.* **12**:180.

Varon, S., and Raiborn, C., 1972, Dissociation, fractionation and culture of chick embryo sympathetic ganglion cells, *J. Neurocytol.* **1**:211.

Varon, S., Raiborn, C.W., Jr., Seto, T., and Pomerat, C.M., 1963, A cell line from trypsinized adult rabbit brain tissue, *Z. Zellforsch.* **59**:35.

Weiss, P., 1934, *In vitro* experiments on the factors determining the course of nerve fibre, *J. Exp. Zool.* **68**:393.

Enzyme Kinetics

P. Kontro and S. S. Oja

1. Introduction

Biological research has benefited greatly by studies on reaction kinetics; in particular, those carried out with enzyme-catalyzed reactions have been rewarding. Thorough kinetic analyses constitute an essential part in the characterization of any enzyme. In neurobiology, the general principles of enzyme kinetics have also been applied to phenomena other than metabolic reactions proper, e.g., carrier-mediated membrane transport, drug–receptor interactions, or transmitter-binding to postsynaptic membranes. Neurobiologists thus often need a basic knowledge of enzyme kinetics even if they are not directly involved in problems of enzymology and intermediary metabolism. It is to be regretted that at present, enzyme kinetic methods are frequently misused in neurobiology when sound consideration of their fundamental principles and inherent limitations is neglected. Furthermore, oversimplified treatment of data may be deceptive, leading to erroneous conclusions and wasting good experimental work. On the other hand, very elaborate treatment is difficult to accomplish successfully. Moreover, primary observations in biological experiments, in which an exact control of all contributing factors is not within the bounds of possibility, are seldom accurate enough for extremely sophisticated analyses.

Many biologists might be intimidated by the apparent complexity of enzyme kinetics and the abundance of mathematical expressions and graphic plots therein. Despite experimental, and especially analytical, difficulties, we hope to encourage the reader to undertake appropriate enzyme-kinetic studies. Even the most important enzymes have not yet been thoroughly characterized kinetically, not to mention other newer applications of enzyme-

P. Kontro and S. S. Oja • Department of Biomedical Sciences, University of Tampere, Tampere, Finland.

kinetic principles within neurobiology. Much of this unknown continent remains to be conquered. In this chapter, we shall discuss some basic concepts of enzyme kinetics, particularly those that are pertinent to neurobiological applications, and give the practical framework to help conduct meaningful experiments on the mechanisms of enzyme action. We shall first describe the kinetic principles of uncatalyzed chemical reactions, and then those of simple unireactant enzyme reactions; their treatments constitute the background for all enzyme kinetics. The other subjects will be inhibition, allosteric effects, and cooperativity, as well as the effects of pH and temperature. Of multireactant enzyme systems, only the two-substrate mechanisms will be described. In particular, this last section contains examples of neurobiological applications of enzyme-kinetic methods. For the most part, we have concentrated on the kinetics of reactions in steady-state conditions.

A number of monographs on enzyme kinetics have been published. We can recommend *Principles of Enzyme Kinetics* by Cornish-Bowden (1976) to all neurobiologists who wish to get basic kinetic information in a modern—but yet easily understood—manner. Also, a reader not mathematically orientated can readily follow the text. The book contains all the important areas of enzyme kinetics relevant in neurobiology, including a clear presentation of allosteric effects and cooperativity, which are often neglected in other textbooks. *The Chemical Kinetics of Enzyme Action* by Laidler and Bunting (1973) is meant for both biochemists and physical chemists. Its approach is rather mathematical, as is that of *Enzyme Kinetics* by Plowman (1972). *Enzyme Kinetics* by Segel (1975) is a very thorough book, covering all elementary aspects and also many new subjects in enzyme kinetics. A neurobiologist with a wider interest in kinetic investigation should acquaint himself with this book. *Kinetics of Enzyme Mechanisms* by Wong (1975) has a modern approach to enzyme kinetics, different from the others. The reader of this sophisticated book should preferably have some previous knowledge of the behavior of enzymes. The monographs by Cleland (1970) and Koshland (1970) in *The Enzymes* edited by Boyer are also recommended to neurobiologists involved in kinetic problems. We have ourselves greatly benefited from the aforementioned texts on enzyme kinetics in preparing this chapter.

2. Basic Kinetic Concepts

2.1. Kinetics of Uncatalyzed Reactions

The kinetic work done on nonenzymic systems gives the basis for knowledge of the kinetics of enzyme reactions. Fundamental mechanisms and terminology of uncatalyzed reactions are therefore discussed first.

If a chemical reaction occurs in one phase, it is called a *homogeneous reaction*. When substances in different phases react with each other, the reaction is *heterogeneous*. A substance involved in a chemical reaction is called a *reactant*. The *reaction velocity* is defined as changes of concentrations of the reactants per unit of time. This is dependent on various environmental factors, such as the initial concentrations of the reacting molecules, pH of the reaction medium, temperature, pressure, and the presence or absence of catalysts. The velocity of a reaction is usually measured by following the rate of disappearance of a reactant or the rate of appearance of a product as a function of time. The latter method is generally preferred, because it is easier to measure the increasing concentration of a newly formed product than to follow small decrements occurring in the concentration of a reactant. The velocity that is measured at the initial stages of the reaction is termed the *initial velocity*. The reaction velocity is a function of the initial concentrations of the reactants.

Chemical reactions are classified according to their *molecularity* and according to *order*. The molecularity of a reaction refers to the number of molecules transformed in the reaction. Thus, the reaction in which a substance A^* is irreversibly converted to a product P, e.g., $A \rightarrow P$, is *unimolecular* or *monomolecular*. The reactions $A + B \rightarrow P$ and $2\,A \rightarrow P$ are *bimolecular*, and $A + B + C \rightarrow P$ is *trimolecular* or *termolecular*. The order of the reaction is determined by the number of the reactants the concentrations of which change proportionally to the reaction velocity. Thus, in a *first-order reaction*, the reaction velocity is proportional to one changing concentration. The velocity of the simple reaction $A \rightarrow P$ is the rate of disappearance of A or the appearance of P. According to the law of mass action, the reaction velocity v at any time t is given by

$$v = -\frac{da}{dt} = \frac{dp}{dt} = ka \qquad (1)$$

in which a and p denote the concentrations of A and P, respectively.

Equation (1) is the *rate equation* or *kinetic equation (law)* characteristic for the first-order type of reaction. The constant k is the *first-order rate constant* or *specific reaction rate*. Its dimension is time^{-1}. Important first-order reactions include, for example, the radioactive decay processes that are discussed in Volume 2, Chapter 10.

In a *second-order reaction*, such as $A + B \rightarrow P$, the velocity is proportional to two concentrations (a and b) of reactants:

$$v = \frac{dp}{dt} = kab \qquad (2)$$

If two similar molecules react, $2\,A \rightarrow P$, the expression becomes

$$v = \frac{dp}{dt} = ka^2 \qquad (3)$$

The constant k is now the *second-order rate constant* and has the dimensions of concentration^{-1} time^{-1}.

---

* Definitions of the symbols used throughout this chapter will be found in Section 9.

For simple one-step reactions, the order is usually the same as the molecularity. Many complex reactions consist of unimolecular and bimolecular steps, however, and then the molecularity of the whole reaction is not necessarily the same as its order. Also, in reversible reactions, neither the molecularity nor the order of the reverse reaction need be the same as in the forward reaction. True unimolecular reactions are very rare. Reactions of order greater than two are rare as well, although reactions of a molecularity greater than two are common. Trimolecular reactions do not usually occur as a single trimolecular step, however, and they are thus not of the third order. The overall reaction may consist of two or more elementary steps, e.g.:

$$A + B \rightarrow X$$

$$X + C \rightarrow P$$

If one of these reactions is much slower than the others, the whole reaction rate equals the velocity of the slow step. This is called the *rate-limiting step*.

A chemical reaction may also be of *zero order*, which means that the reaction velocity is independent of the concentrations of the reactants, e.g., $v = ka^0$, i.e., the rate is constant. Catalyzed reactions are of zero order when every reactant is present in such great excess that the reaction proceeds at its maximum velocity. If the concentration of one reactant is much larger than that of all others, it remains virtually constant. The constant-concentration term can then be combined with the rate constant in the kinetic equation. Thus, the kinetic order of the reaction is reduced by one. A bimolecular reaction becomes kinetically an *apparent-* or *pseudo-first-order* reaction. This has very important consequences for enzyme kinetics.

In reversible reactions, the back reaction is also included in the rate equation, and the kinetics of the mechanism become rather complicated. For example:

$$A \underset{k_{-1}}{\overset{k_{+1}}{\rightleftharpoons}} P$$

in which k_{+1} and k_{-1} are the rate constants for the forward and reverse reactions, respectively. The kinetic equations are as follows:

$$-\frac{da}{dt} = k_{-1}p - k_{+1}a \tag{4}$$

$$\frac{dp}{dt} = k_{+1}a - k_{-1}p \tag{5}$$

The concentration of A at zero time is a_0 and that of P: $p_0 = 0$. Then $a_0 =$

$a + p$ at any time. The integral solutions of differential equations (4) and (5) for the concentrations of a and p at time t are

$$a = a_0 \left[\frac{k_{-1} + k_{+1} \exp[-(k_{-1} + k_{+1})t]}{k_{-1} + k_{+1}} \right] \qquad (6)$$

$$p = a_0 \left[\frac{k_{+1} - k_{+1} \exp[-(k_{-1} + k_{+1})t]}{k_{-1} + k_{+1}} \right] \qquad (7)$$

Reversible bimolecular reactions of the type $A + B \rightleftharpoons P$ can be treated similarly, but the expressions for the concentrations of the reactants become very complicated (e.g., Segel, 1975; Cornish-Bowden, 1976).

In a sequence involving two or more reaction steps, separate kinetic equations must be written for each step, e.g., in

$$A + B \underset{k_{-1}}{\overset{k_{+1}}{\rightleftharpoons}} X \overset{k_{+2}}{\longrightarrow} P$$

This system can be characterized by the following expressions:

$$-\frac{da}{dt} = -\frac{db}{dt} = k_{+1}ab - k_{-1}x \qquad (8)$$

$$\frac{dx}{dt} = k_{+1}ab - (k_{-1} + k_{+2})x \qquad (9)$$

$$\frac{dp}{dt} = k_{+2}x \qquad (10)$$

These three differential equations can also be numerically integrated with the aid of a computer for the most general case when there exist no restrictions in the magnitude of the rate constants, the initial concentrations of the reactants, and the particular time for which the rate law is established. Simple explicit solutions are obtainable, however, for certain restricted cases.

2.2. Determination of Order of Reaction

After initial mixing of the reactants, the concentration of any reactant can be followed as a function of time. The graphic representation of the results is called the *progress curve* (Fig. 1). Such a curve can be used for determination of the order of the reaction, which is also one of the most important properties characterizing enzyme reactions. For instance, the rate of a reaction of the nth order involving one reacting substance is proportional to the nth power of its concentration:

$$-\frac{da}{dt} = ka^n \tag{11}$$

$$\log\left(-\frac{da}{dt}\right) = \log k + n \log a \tag{12}$$

In the plot log v vs. log a, a straight line is obtained. Its slope indicates the order of the reaction. This method is called the *differential method*, since in it a number of tangents are drawn to the concentration–time curves. This differential method can be practiced in two different ways: (1) The progress of the reaction is followed as a function of time and the diminishing concentration of A is determined at different intervals. Tangents are drawn for the curve of a vs. t at different concentrations (Fig. 1A). Their slopes are used as estimates for da/dt and inserted in equation (12). In this case, the slope of the plot of log $(-da/dt)$ vs. log a is called the *order with respect to time*, because in this procedure time is the variable. Possible interference from the products of the reaction is the main disadvantage of this method. This order is easy to determine, however, since only a single initial concentration of substrate is needed. (2) In the other application, only the initial reaction rates are estimated, but the experiments are carried out at various initial concentrations of A (Fig. 1B). The order of the reaction is again determined according to equation (12) by plotting log $(-da/dt)$ against log a. This order is known as the *order with respect to concentration* or the *true order*, since the concentration is the variable. This application using initial rates and initial

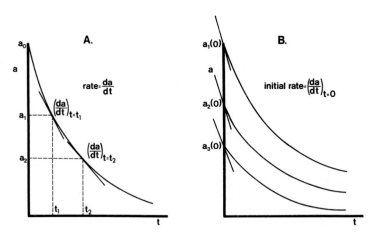

Fig. 1. Progress curves of a reaction showing the concentration of the reactant A as a function of time. (A) Use of the differential method for determination of the order with respect to time for the reaction. The rate da/dt is estimated at varying intervals, at t_1 and t_2, for instance, when the concentration a gradually diminishes. (B) Use of the differential method for determination of the order with respect to concentration. The initial rate $(da/dt)_{t=0}$ is reckoned at different initial concentrations, e.g., at $a_1(0)$, $a_2(0)$, and $a_3(0)$, of the reactant A.

concentrations is recommended for enzymic reactions in which products may seriously interfere with the reaction. Initial velocities can often be obtained very accurately when the reaction is adjusted to occur slowly (e.g., in reduced enzyme concentrations). However, the tangents for the initial points of the rate curves are not easy to draw precisely.

The order of the reaction can also be determined with an entirely different method. In this procedure, one measures a time interval (τ) required for a given extent (x) of the reaction to proceed, i.e., for A to have reached a value equal to $a_0(1-x)$. This time interval is called *fractional time*. It is *half-time* when $x = 0.5$. In general, τ is proportional to $1/a_0^{n-1}$, where n is the order of the reaction. Thus, n can be determined by measuring two fractional times τ_1 and τ_2 corresponding to two different initial concentrations $a_1(0)$ and $a_2(0)$. Then

$$\frac{\tau_1}{\tau_2} = \left(\frac{a_2(0)}{a_1(0)}\right)^{n-1} \tag{13}$$

$$\log \frac{\tau_1}{\tau_2} = (n-1) \log \frac{a_2(0)}{a_1(0)} \tag{14}$$

$$n = 1 + \frac{\log \tau_1 - \log \tau_2}{\log a_2(0) - \log a_1(0)} \tag{15}$$

Sometimes in multireactant reactions, the order with respect to each reactant must be known separately. This can be achieved with the differential methods by altering the concentration of each reactant individually in turn and keeping the other concentrations constant. Then the slope of the straight line drawn according to equation (12) will equal the order with respect to the variable reactant. It is not always possible to determine the order of the reaction. This is often the case with complicated enzyme reactions.

2.3. Determination of First-Order Rate Constants

The first-order rate constant can be determined fairly easily for simple unimolecular reactions. Integration of the rate equation in the logarithmic form (1) gives:

$$\ln \frac{a_0}{a} = kt \tag{16}$$

or

$$\log \frac{a_0}{a} = \frac{kt}{2.303} \tag{17}$$

where a_0 is the initial concentration and a the concentration after time t. The plot of log (a_0/a) against t yields a straight line with a slope of $k/2.303$. In equations (16) and (17), the concentration a is often replaced by the difference $(a_0 - p)$, in which p is the concentration of the product. Because many reactions are of the first order with respect to each individual reactant, many cases can be reduced in practice to the problem of determining the rate constant for a first-order reaction. The reaction is then carried out under so-called pseudo-first-order conditions, in which every other reactant except one is added in turn in large excess.

The determination of the first-order rate constant for a reversible reaction is more complicated. In equation (7), the equilibrium value of the product p_∞ equals $k_{+1}a_0/(k_{+1} + k_{-1})$, because the exponential term disappears when t is large. Then

$$p_\infty - p = p_\infty \exp[-(k_{+1} + k_{-1})t] \tag{18}$$

and

$$\log (p_\infty - p) = \log p_\infty - (k_{+1} + k_{-1})t/2.303 \tag{19}$$

The plot of $\log (p_\infty - p)$ vs. t is a straight line with a slope of $-(k_{+1} + k_{-1})/2.303$. The reliability of this method depends largely on the accurate determination of p_∞. If this is not possible, an application according to Guggenheim (1926) should be used. In this procedure, two sets of concentrations of the product, (p_i and p'_i), are determined at times t_i and t'_i, respectively, so that every $t'_i = t_i + \tau$, where τ is a constant. From equation (18), we then obtain

$$p_\infty - p_i = p_\infty \exp[-(k_{+1} + k_{-1})t_i] \tag{20}$$

$$p_\infty - p'_i = p_\infty \exp[-(k_{+1} + k_{-1})(t_i + \tau)] \tag{21}$$

Subtracting equation (21) from equation (20) gives

$$p'_i - p_i = p_\infty\{1 - \exp[-(k_{+1} + k_{-1})\tau]\} \exp[-(k_{+1} + k_{-1})t_i] \tag{22}$$

which signifies that

$$\ln (p'_i - p_i) + (k_{+1} + k_{-1})t_i = \text{constant} \tag{23}$$

The plot of $\ln (p'_i - p_i)$ against t_i is a straight line with a slope of $-(k_{+1} + k_{-1})$. It is called the Guggenheim plot.

Equations (20) and (21) can also be combined by dividing (20) by (21) as follows:

$$\frac{p_\infty - p_i}{p_\infty - p'_i} = \exp[(k_{+1} + k_{-1})\tau] \tag{24}$$

Equation (24) can be rearranged to

$$p'_i = p_\infty\{1 - \exp[-(k_{+1} + k_{-1})\tau]\} + p_i \exp[-(k_{+1} + k_{-1})\tau] \quad (25)$$

The plot of p'_i vs. p_i is also a straight line with a slope of $\exp[-(k_{+1} + k_{-1})\tau]$. This plot is known as the Kézdy–Swinbourne plot (Kézdy *et al.*, 1958; Swinbourne, 1960). It is less widely used than the Guggenheim plot. Both these plots have the disadvantage that they do not show any deviations from first-order kinetics. They yield rather good straight lines even if the reaction is not of the first order at all. Therefore, the order of the reaction should be known before either of these plots is used. Another disadvantage is that the individual rate constants, k_{+1} and k_{-1}, cannot be calculated separately without some additional knowledge. It is usually convenient also to measure the equilibrium constant of the reaction. Since it equals k_{+1}/k_{-1}, both rate constants can then be evaluated separately.

2.4. Determination of Second-Order Rate Constants

The second-order rate constants for simple one-way reactions can be calculated by methods similar to those applied above for determination of the first-order rate constants. Equation (2) can be transformed after integration to

$$\frac{1}{a_0 - b_0} \log \frac{ab_0}{a_0 b} = \frac{kt}{2.303} \quad (26)$$

in which a_0 and b_0 are the initial concentrations at $t = 0$ and a and b the concentrations at t. The plot of $\log ab_0/a_0 b$ vs. t has a slope of $k(a_0 - b_0)/2.303$, from which k can be calculated. For the special case of $2 A \rightarrow P$ or when $a_0 = b_0$, the rate equation (3) can be transformed after integration to

$$\frac{1}{a} = kt + \text{constant} \quad (27)$$

Between the limits a_0 at $t = 0$ and a at any time t, the following relationship obtains:

$$\frac{1}{a} - \frac{1}{a_0} = kt \quad (28)$$

Equation (2) can be written in an alternative way using the initial concentrations a_0 and b_0 and denoting with p the amount of the product formed after time t:

$$\frac{dp}{dt} = k(a_0 - p)(b_0 - p) \quad (29)$$

Integrating and inserting $p = 0$ at $t = 0$, we obtain

$$\ln \frac{a_0(b_0 - p)}{b_0(a_0 - p)} = (b_0 - a_0)kt \tag{30}$$

or

$$t = \frac{2.303}{k(a_0 - b_0)} \log \frac{b_0(a_0 - p)}{a_0(b_0 - p)} \tag{31}$$

According to equation (31), for a second-order reaction the plot of log $[b_0(a_0 - p)/a_0(b_0 - p)]$ against t is linear with a slope of $k(a_0 - b_0)/2.303$. For the special case when $a_0 = b_0$, we get

$$\frac{dp}{dt} = k(a_0 - p)^2 \tag{32}$$

and

$$\frac{p}{a_0(a_0 - p)} = kt \tag{33}$$

It should be borne in mind that the dimensions of the second-order rate constants are concentration^{-1} time^{-1}, and the first-order rate constants have the dimension time^{-1}. Thus, a direct comparison between first- and second-order rate constants is not possible. This fact is often forgotten.

2.5. Effect of Temperature on Rate Constants

Temperature has a profound effect on the velocity of chemical reactions. Therefore, the temperature of the reaction medium must always be controlled in all kinetic work. Studies on the temperature dependence of kinetic constants may give valuable information.

Van't Hoff (1884) demonstrated the dependence of any equilibrium constant K on the absolute temperature T:

$$\frac{d \ln K}{dT} = \frac{\Delta H^\circ}{RT^2} \tag{34}$$

in which R is the gas constant and ΔH° the standard enthalpy change in the reaction. Arrhenius (1889) further proposed a similar relationship between the rate constant k and temperature:

$$\frac{d \ln k}{dT} = \frac{E}{RT^2} \tag{35}$$

in which E is the activation energy of the reaction.

Integration of equation (35) with respect to T gives

$$\ln k = \ln C - \frac{E}{RT} \tag{36}$$

where C is the integration constant. The magnitude of the activation energy can be obtained from the slope of the straight line when $\ln k$ is plotted against $1/T$. In practice, it is usually simpler to plot $\log k$ against $1/T$. The slope of the resulting straight line is $-E/2.303R$ (Fig. 2). This plot, often used in enzyme kinetics, is known as the Arrhenius plot. Its use and applications are discussed further in Section 7.2.2.

3. Kinetics of Unireactant Enzymes

Enzymes may be defined as protein catalysts for chemical reactions in biological systems. Virtually all chemical reactions occurring in living cells are subject to catalysis and control by enzymes. The investigation of the mechanisms of enzyme reactions differs somewhat from that of ordinary chemical reactions. The structure of the reacting molecules is known in most chemical reactions, whereas the detailed three-dimensional conformation of most enzymes is not yet completely known. The kinetics of enzyme-catalyzed reactions were an object of study long before the complex structure of these catalysts themselves. Until recently, kinetic studies have also been the main source of information on the nature of the functionally active substructures of enzymes. The structures of the enzymes and the enzyme–substrate complexes have now also been investigated with X-ray analyses. Wider prospects for understanding the behavior of enzymes have been opened by correlating structural and kinetic information. Kinetic studies in the context of chemical and structural information on the enzyme complete the picture of the whole catalytic process.

Every enzyme facilitates only one or a very limited number of reactions,

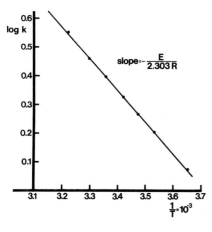

Fig. 2. Arrhenius plot: the logarithm of the rate constant (k) of a reaction as a function of $1/T$. The magnitude of the activation energy (E) can be assessed from the slope of the resulting straight line as indicated in the figure.

which are usually of the same reaction type occurring with structurally related compounds. The reactant the transformation of which is catalyzed by an enzyme is called a *substrate*. The high catalytic efficiency and specificity exhibited by enzymes imply the participation of several distinct functional groups of the enzyme in the catalytic process. Such a conception has been the impetus for acceptance of the concept of an *active site*, consisting of those side chains and peptide bonds in the enzyme molecule that come into direct physical contact with the substrate and coenzyme and directly take part in the catalytic process. The term can also be extended to comprise regulatory sites to which some regulatory substances may attach.

The amount of an enzyme present is usually stated by measuring the rate of the enzyme-catalyzed reaction under suitable conditions. This rate is proportional to the quantity of the enzyme. *Enzymic activity* is defined as the rate of reaction of substrate that may be attributed to catalysis by an enzyme. *Specific activity* is the phenomenological coefficient that relates the activity of the enzyme, under specified conditions, to its mass, and *molar activity* is the phenomenological coefficient that relates the activity under specified conditions to the amount of enzyme substance (moles). The *enzyme unit* (U) was formerly defined as the amount of enzyme that catalyzes the transformation of 1 μmol of the substrate per minute under standard conditions. Enzyme units are still most commonly used in expressing the activity of an enzyme. However, the Enzyme Commission (1972) of the International Union of Biochemistry discourages the use of the enzyme unit and recommends instead a new unit of enzymic activity named the *katal* (symbol: *kat*). One katal is the amount of enzymic activity that converts 1 mol of substrate per second. Thus, 1 kat = 1 mol/sec = 60 mol/min = 60×10^6 μmol/min = 6×10^7 U. The other quantities of enzymic activity should also be expressed by using katals. The specific activity of an enzyme will then be katals per kilogram of protein and the molar activity katals per mole of enzyme, or suitable multiples thereof, e.g., microkatals per kilogram (μkat/kg) and nanokatals per mole (nkat/mol).

3.1. Basic Theory of Enzyme Kinetics

3.1.1. Steady-State Rate Law

Enzyme-catalyzed reactions were studied for the first time in the latter part of the 19th century. It is a remarkable achievement that the pioneers of enzyme kinetics made progress in their work, even though no enzyme had been purified and buffer solutions to adjust pH in the reaction medium had not yet been introduced. The fundamental idea of the enzyme–substrate complex emerged early. Wurtz (1880) had shown that papain appeared to form an insoluble compound with fibrin previous to hydrolysis. O'Sullivan and Thompson (1890) studied the enzyme invertase, which catalyzes the hydrolysis of sucrose to glucose and fructose. They noticed that the reaction was dependent on the acidity and had an optimum temperature above which the rate fell to zero. They considered that the augmented thermal stability of the enzyme in the presence of its substrate, sucrose, indicated the existence of some kind of combination between the enzyme and sucrose molecules.

Brown (1902) suggested that the formation of an enzyme–substrate complex was the limiting factor for the reaction rate in yeast fermentation, because the complex had a fixed lifetime before breaking down into products. The maximum velocity of the reaction would be reached when there was a sufficient amount of substrate to convert all the enzyme to the complex form. The idea of the fixed lifetime of the enzyme–substrate complex was not accepted by Henri (1902, 1903). He proposed instead an equilibrium between the free enzyme and the enzyme–substrate and enzyme–product complexes and characterized reaction mechanisms with precise mathematical equations. Michaelis and Menten (1913) confirmed these earlier observations in carefully controlled kinetic experiments. They measured initial rates of the reaction at different substrate concentrations. They assumed that the reversible formation of the enzyme–substrate complex was sufficiently rapid—compared to its breakdown—to be represented by an equilibrium constant K. For their invaluable contributions, Michaelis and Menten are regarded as the founders of modern enzymology. At the same time, Van Slyke and Cullen (1914) formulated the same mathematical equation, assuming that the complex formation was irreversible. Later, Briggs and Haldane (1925) presented the rate equation for an enzymic reaction in a general form that includes all the aforementioned assumptions as special cases. They assumed that there exists a steady state in which the amount of the intermediate complex is constant:

$$E + S \underset{k_{-1}}{\overset{k_{+1}}{\rightleftharpoons}} ES \overset{k_{+2}}{\longrightarrow} E + P$$

E, S, ES, and P denote the enzyme, substrate, enzyme–substrate complex, and product, respectively.

There are three elementary steps in the mechanism represented above: (1) The enzyme and substrate interact in a bimolecular step to form the enzyme–substrate complex ES, which reaction has the rate constant k_{+1}. The complex may then (2) dissociate back to the free enzyme and substrate before catalysis has occurred or (3) yield the free enzyme and product after catalysis. These latter monomolecular reactions have the rate constants k_{-1} and k_{+2}, respectively. The rate law for this kind of reaction is conveniently expressed with the rate constants and the measurable concentrations of the total enzyme, e, and total substrate, s. The concentrations of the enzyme–substrate complex, x, and the free enzyme, $(e - x)$, are usually eliminated because they are not easily amenable to measurement. The formation of the enzyme–substrate complex is very rapid when the enzyme is added to the solution containing the substrate. This phase during which ES accumulates is called the *transient phase*. The transient phase eventually becomes a steady state when most of the enzyme is bound to ES. The velocity of product formation is proportional to the amount of the enzyme–substrate complex:

$$v = \frac{dp}{dt} = k_{+2}x \tag{37}$$

The rate of alteration of ES must equal its formation via the k_{+1} step minus its breakdown via the k_{-1} and k_{+2} steps:

$$\frac{dx}{dt} = k_{+1}s(e - x) - k_{-1}x - k_{+2}x \tag{38}$$

Rearranging, we get

$$x = \frac{k_{+1}se - (dx/dt)}{k_{+1}s + k_{-1} + k_{+2}} \tag{39}$$

At the start of the reaction, the rate of ES formation is $k_{+1}se$, but when the steady state is reached, it falls with the decreasing concentration of S. If s is much larger than e, this fall is very slow in the steady-state phase, and then dx/dt is much smaller than $k_{+1}se$. We may write

$$x = \frac{k_{+1}se}{k_{+1}s + k_{-1} + k_{+2}} \tag{40}$$

Inserting the expression for x from equation (40) into equation (37), we have the steady-state rate law first presented by Briggs and Haldane in 1925:

$$v = \frac{k_{+2}se}{s + (k_{-1} + k_{+2})/k_{+1}} \tag{41}$$

and in more general form

$$v = \frac{Vs}{s + K_m} \tag{42}$$

where the maximum velocity or saturation velocity of the reaction $V = k_{+1}e$ and the Michaelis constant $K_m = (k_{-1} + k_{+2})/k_{+1}$. Equation (42) is generally known as the Michaelis–Menten equation.

Michaelis and Menten assumed that the complex formation from the enzyme and substrate reaches a thermodynamic equilibrium; i.e., k_{+2} is much smaller than k_{-1}. In their treatment, K_m equaled k_{-1}/k_{+1}. The irreversible enzyme–substrate complex proposed by Van Slyke and Cullen (1914) means that k_{-1} should be much smaller than k_{+2} and then K_m is k_{+2}/k_{+1}. The general form of the steady-state rate law of Briggs and Haldane covers both these earlier assumptions as limiting cases. They ensue from extreme relative values of the rate constants. The common adaptation of the steady-state method proposes that dx/dt is actually zero, but this stipulation is clearly unjustified. The steady-state approximation causes a biphasic error in calculations of experimental data compared to more precise numerical methods, as has been shown by Swoboda (1957). The difference between the steady-state approximation and numerical calculations is larger in the transient phase and smaller in the steady state. The discrepancy can be suppressed by a high substrate/enzyme ratio, which prolongs the steady-state phase in relation to the transient phase. The steady-state approximation is then sufficiently accurate for most kinetic work, but its limitations should not be forgotten.

The Michaelis–Menten equation is well suited to most experimental data, not only to one-substrate mechanisms, but also to many other enzyme mechanisms. Equations of the same type have also been formulated outside the sphere of enzyme kinetics. Such analogously treated processes are adsorption of gases by solids (Langmuir, 1916, 1918) and chain-reaction mechanisms. Langmuir's treatment of simple adsorption resembles the binding of enzymes to substrates described by Michaelis and Menten. He also emphasized similarities between solid surfaces and enzymes. Later, Hitchcock (1926) was clearly aware of the similarity between equations that describe the binding of ligands to solid surfaces on one hand, and to proteins on the other. The basic equation of Michaelis and Menten has been widely applicable for many biological purposes, including transport phenomena and the binding of various substances to receptor molecules at membranes, as is discussed in Section 8.

3.1.2. Kinetic Parameters K_m and V

The graphic representation of equation (42) is a rectangular hyperbola with asymptotes $s = -K_m$ and $v = V$ (Fig. 3). When s is very small, v is directly proportional to s: $v = Vs/K_m$. The reaction is now of the first order with respect to s. Thus, at very low substrate concentrations, the relationship V/K_m has the meaning of a first-order rate constant. When the substrate concentration equals K_m, the velocity is half-maximal; i.e., $v = 0.5V$. Thence, K_m is most often defined as the substrate concentration at which the velocity

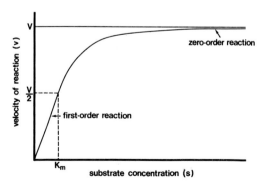

Fig. 3. Dependence of velocity (v) of an enzyme-catalyzed unireactant reaction on the concentration of the substrate (s). K_m (Michaelis constant) is the substrate concentration at which the reaction velocity is half the maximum velocity V.

of the reaction is one half the maximum velocity. The dimensions of K_m are naturally those of the substrate concentration. At very high s, v approaches V and the reaction is of zero order with respect to s. The enzyme is then thought to be saturated by substrate molecules.

The maximum velocity V depends on the enzyme concentration. The reaction velocity usually approaches the saturation velocity rather slowly, and for this reason V cannot be directly measured experimentally. It is often estimated by a graphic extrapolation, as will be dicussed later. If the enzyme concentration is known, the *catalytic constant* or *turnover number* can be calculated as V/e.

K_m does not equal the equilibrium (dissociation) constant, $K = k_{-1}/k_{+1}$ (often termed the substrate constant, K_s), of the reaction $ES \rightleftharpoons E + S$ unless k_{+2} is negligible in comparison to k_{-1} (the equilibrium approximation). Since K_m is not the dissociation constant of the ES complex, its reciprocal is not simply the affinity or binding constant (k_{+1}/k_{-1}) of an enzyme for its substrate either, even if in the literature such a misinterpretation commonly persists. K_m represents the apparent dynamic dissociation constant under steady-state conditions because it relates to the dynamic equilibrium between at least three reactions the individual rates of which are characteristic of the mode of action of the enzyme. K_m may thus be larger or smaller than or the same as the dissociation constant, depending on the mechanisms and the relative sizes of rate constants. However, K_m is still the most useful constant in enzyme chemistry despite discrepancies and inaccuracies concerning its real meaning. K_m is the only characteristic parameter used to check the validity of enzyme assays, for example. Furthermore, it is often necessary to describe complex reactions in simple terms, in which case the basic kinetic parameters K_m and V and their changes with experimental conditions are evaluated.

3.1.3. Reversible One-Substrate Mechanisms

The Michaelis–Menten equation decribes the simplest mechanism, in which the enzyme–substrate complex breaks down virtually irreversibly to yield the reaction product and the regenerated free enzyme. There also remains the possibility that the product may be formed from the product side*:

$$A + E \underset{k_{-1}}{\overset{k_{+1}}{\rightleftharpoons}} AE \underset{k_{-2}}{\overset{k_{+2}}{\rightleftharpoons}} E + P$$

The rate equation of this reversible reaction is as follows (the reversible form of the Michaelis–Menten equation):

$$v = \frac{V^f a/K_m^A - V^r p/K_m^P}{1 + a/K_m^A + p/K_m^P} \tag{43}$$

---

* The substrate is now designated A, because S is reserved for substrates in general. $A, B. . .$ are used for particular substrates in the forward reaction and $P, Q, . . .$ for substrates in the reverse reaction.

where V^f and V^r are the maximum velocities of the forward and reverse reactions, respectively, and K_m^A and K_m^P are the respective Michaelis constants of the substrate A and the product P. In equilibrium, when the net reaction velocity is zero, we get the so-called Haldane relationship:

$$K = \frac{V^f K_m^P}{V^r K_m^A} \tag{44}$$

where K is the equilibrium constant of the reaction. This Haldane relationship states that the kinetic parameters of a reversible enzyme-catalyzed reaction are not independent of each other and that they are limited by the thermodynamic equilibrium constant for the whole reaction.

3.2. Time Course of Enzyme Reactions

The enzyme-catalyzed reaction usually proceeds through three definite phases separated on the time scale (Fig. 4). The very rapid period before the steady state is reached is the transient or pre-steady-state phase. The middle phase represents the initial-velocity phase in which the reaction velocity is almost constant. This is the period most closely investigated since the work of Michaelis and Menten. The final phase is called the progress-curve phase, in which the concentrations of the substrate and product gradually change and the velocity begins progressively to decrease.

Experimental methods for studying the very fast reactions of the pre-steady state with half-times much less than 1 sec differ considerably from those applied for slower reactions. It is also obvious that the kinetic equations for fast reactions will be different from steady-state equations. The major disadvantage of steady-state measurements is that the steady-state velocity of a multistep reaction is the rate of the slowest step and provides no information about any of the faster steps. Nor do steady-state experiments give any indication of the number of steps involved in the reaction. In transient-state kinetics, there are no such limitations. The slowest reactions of the pre-steady state can be measured by rapid-flow or stopped-flow techniques, usually applied to absorption or fluorescence spectrometry. The transient phase of

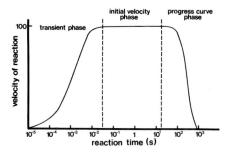

Fig. 4. Time course of an enzyme reaction. The reaction velocity is plotted as a function of the logarithm of the reaction time.

the Michaelis–Menten mechanism can be analyzed by so-called "burst" kinetics using very high concentrations of the enzyme. This method has later led to an important application of titrating enzymes.

The most useful method for the study of exceedingly rapid reactions is to observe the relaxation of a system back to equilibrium after a small perturbation (relaxation kinetics). A number of physical perturbations have been used in relaxation studies, e.g., pressure shock, sound absorption, and electrical-field change. The technique of temperature-jump has, however, been that most widely applied to enzyme systems. In the temperature-jump method, perturbation of the equilibrium results from a sudden increase in temperature. The individual rate constants of enzyme reactions can be determined in the transient-state phase with the methods of relaxation kinetics. Despite these modern methods, it is likely that the steady-state approximation will continue to predominate for many years because of its much simpler theory and the less expensive equipment needed in experiments. Another advantage of the steady-state methods is the very small amount of enzyme needed for measurements. Enzymologists and also neurobiologists should, however, be aware of the usefulness of the transient-state methods.

Another new approach for enzymologists is to follow the enzyme-catalyzed reaction over a prolonged period and to observe the effects of progressive accumulation of the product and depletion of the substrate in the medium. Accurate values for kinetic parameters can be obtained from a relatively small number of experiments by analyzing this progress curve. An integrated Michaelis–Menten equation is used for data analysis. It is likely that in the future the analysis of progress curves will prove a very valuable tool for enzyme kineticists.

3.3. Determination of Kinetic Parameters

3.3.1. Practical Considerations

For reliable results, all glassware and incubation vessels used in enzyme-kinetic work must be very clean, since even small amounts of foreign substances may denature the enzyme protein, inactivate the enzyme, or otherwise affect the rate of reaction. The enzyme protein should preferably be exhaustively purified. However, kinetic experiments can be made with impure preparations if no side reactions interfere with the assay procedure. Many enzymes are bound to a membrane or constitute a part of a multienzyme complex and in such natural form may exhibit kinetic properties different from those of the purified enzyme. Kinetic experiments can be carried out on systems in which the enzyme molecules are attached to a support. This is, in fact, rather seldom done, although in many cases it is more important to study these more natural forms than purified enzymes in free solution. The techniques for the supported enzyme systems have been discussed, for example, by Laidler and Bunting (1973).

Kinetic experiments should be conducted under conditions in which the enzyme is reasonably stable. Conditions in the reaction mixture must also be maintained as constant as possible, especially temperature and pH. It is always advisable to carry out the experiments in a thermostat with good temperature control. All reactants must be brought to the correct temperature before being added to the reaction mixture. Correct choice of buffer solution is very important. The buffering capacity should be adequate to resist any pH changes as the reaction proceeds, especially if there is production of acid or base. The optimum pH sometimes depends on the buffer used. The buffer must not interact in any way with the enzyme reaction itself. The reaction is started by mixing all the reactants together. This mixing should be done properly so that all the reactants come into immediate contact with each other. Care must be taken that all reactants, particularly all the enzyme, are added to the reaction vessels and not left in the tips of pipettes, for example. Carelessness in such basic matters may cause considerable experimental error because the amounts of the reactants are often very small. A precise time-check should also be kept on every reaction step.

Only a few enzymes can be assayed directly by spectrophotometric or other methods. Therefore, the presence of enzymes is evidenced by the occurrence of the specific reactions that they catalyze. The amount of an enzyme present and its activity are estimated from the reaction velocity. In most kinetic work, the initial velocity of an enzyme reaction is measured during the steady state, in which the accumulation of reaction products or the progressive fall in the substrate concentration does not diminish the reaction velocity. The contribution of the reverse reaction is also assumed to be negligible as compared to that of the forward reaction. Estimation of initial velocity often involves an extrapolation of the progress curve back to zero time. Extrapolation is facilitated by the use of a computer. The slope of the tangent of the progress curve in the origin corresponds to the intitial steady-state velocity. The methods for measuring the initial velocity of an enzyme reaction include spectrophotometric, fluorometric, manometric, polarimetric, chromatographic, and electrode methods, and various chemical estimations. We shall not discuss these methods; the reader is referred for details to the available enzyme handbooks.

3.3.2. Graphic Estimation of K_m and V

Estimation of these kinetic parameters is based on a series of measurements of initial reaction velocities at different substrate concentrations. Graphic analysis methods are still most common. The data are presented as a graph from which K_m and V are evaluated either visually or by applying statistical principles. There are several ways to plot the experimental results for this purpose. The different modes of graphic parameter estimation are by no means equal, however, and the investigator must be aware of their relative merits and shortcomings.

1. The Michaelis–Menten equation (42) yields a rectangular hyperbola if

the initial velocity is plotted as a function of the substrate concentration. The maximum velocity V is the horizontal asymptote of the hyperbola (see Fig. 3). The substrate concentration at half-maximum velocity reveals K_m. The parameter estimation here is not accurate. It is difficult to draw correct rectangular hyperbolas. There is also a tendency to draw the asymptotes too close to the curve. Precise evaluation of the asymptotic value of the saturation velocity is difficult, and errors in the estimation of V also cause a corresponding bias in K_m values. The possible deviations of the data from the Michaelis–Menten equation often remain undetected in these plots.

2. A relatively uncommon method for estimating the kinetic parameters is to plot v against $-\log s$ (ps). This corresponds to the logarithmic form of the basic Michaelis–Menten equation (42):

$$\mathrm{p}s = \mathrm{p}K_m + \log \frac{V - v}{v} \tag{45}$$

The resulting curve is rather similar to those obtained for dissociations (Fig. 5). If $v = V/2$, the last term on the right-hand side in equation (45) becomes zero and ps = pK_m. The main difficulty with this graph lies in getting experimental points at substrate concentrations high enough to reach the maximum velocity.

3. For the aforestated reasons, linear transformations of the original Michaelis–Menten equation are in common use. The most straightforward and common of these linearizations of equation (42) is the *double-reciprocal* or *Lineweaver–Burk* plot (Lineweaver and Burk, 1934):

$$\frac{1}{v} = \frac{K_m}{V} \frac{1}{s} + \frac{1}{V} \tag{46}$$

The plot of $1/v$ against $1/s$ is linear, and K_m/V represents the slope of the

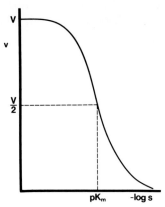

Fig. 5. Plot of the logarithmic transformation of the Michaelis–Menten equation. Reaction velocity (v) is presented as a function of -log s.

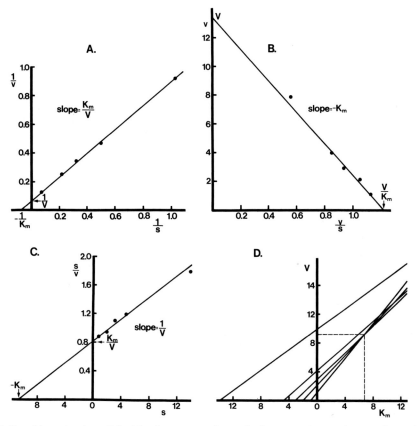

Fig. 6. Graphic evaluation of the kinetic constants for rat brain tryptophan aminotransferase (L-tryptophan:2-oxogluzarate aminotransferase, EC 2.6.1.27) with four different linear plots: (A) Lineweaver–Burk plot ($1/v$ vs. $1/s$), apparent K_m = 12.5 mmol/liter and V = 16.7 μkat/kg tissue. (B) Eadie plot (v vs. v/s), apparent K_m = 11.0 mmol/liter and V = 13.4 μkat/kg tissue. (C) Hanes plot (s/v vs. s), apparent K_m = 10.8 mmol/liter and V = 13.3 μkat/kg tissue. (D) Direct linear plot (V vs. K_m), apparent K_m = 6.8 mmol/liter and V = 9.2 μkat/kg tissue. Dimensions: v = μkat/kg fresh tissue and s = mmol/liter. Straight lines have been fitted to unweighted data by the method of least squares. Recalculated and redrawn from Oja (1968).

straight line (Fig. 6A). The intercept on the ordinate is $1/V$ and on the abscissa $-1/K_m$.

4. The second linear transformation of equation (42) yields the so-called *Eadie* (Eadie, 1942) or *Hofstee* plot (Hofstee, 1959) of v vs. v/s:

$$v = V - K_m \frac{v}{s} \tag{47}$$

This graphic representation is a straight line that intersects the ordinate at V (Fig. 6B). The slope of the line is $-K_m$. The so-called *Scatchard* plot of v/s vs. v (Scatchard, 1949), which is most widely used in binding studies, is simply the inverse of the Eadie plot.

5. The third linear plot is s/v against s, which is known as the *Hanes* (Hanes, 1932) or *Woolf* plot (Woolf, 1932):

$$\frac{s}{v} = \frac{1}{V} s + \frac{K_m}{V} \tag{48}$$

Here, it is the slope that gives $1/V$, the intersection with the vertical axis K_m/V and with the horizontal axis $-K_m$ (Fig. 6C).

6. A completely different approach to present kinetic data by using initial velocities has recently been suggested by Eisenthal and Cornish-Bowden (1974). Their method is called the direct linear plot. The Michaelis–Menten equation (42) is rearranged to show the dependence of V on K_m:

$$V = v + \frac{v}{s} K_m \tag{49}$$

In the graph V against K_m, a straight line is drawn through the points of $-s$ on the abscissa and v on the ordinate, which points represent one single observation (Fig. 6D). The slope of the line is then v/s. This straight line relates all values of K_m and V that satisfy the given values of s and v exactly. If such lines are drawn for many observations, they should all intersect at the same point. The coordinates of this point indicate the only values of V and K_m that satisfy all the observations and are thus the parameters sought. It is fairly easy to reveal a poor observation in this plot, since it gives rise to a line that clearly disagrees. This direct linear plot has the advantage that no calculations are required.

The relative merits of the linear transformations are not the same. If the experimental data were absolutely accurate, all plots would give identical estimates for K_m and V. However, experimental errors and biological variation affect the determination of v. For this reason, the plots are not equally suited for graphic estimation of V and K_m. The double-reciprocal plot is the most widely used, although there have been serious objections to it. Dowd and Riggs (1965) have even claimed that it ought to be abandoned unless the experimental points are properly weighted. If the increments of the substrate concentration are equally spaced, the points will tend to concentrate near the ordinate in the double-reciprocal plot, and the points corresponding to the lowest concentrations (and thus often most susceptible to error) will chiefly direct the drawing of the "best-fit" straight line. The plot also tends to give a deceptively good "fit" (i.e., a straight line) even with results not obeying Michaelis–Menten kinetics. In the Hanes plot (s/v vs. s), the points are generally more evenly spaced, and it is more sensitive to any departure from linearity. The Eadie plot (v vs. v/s) gives the "worst fit" (the points do not neatly fall in a straight line), because the errors in measurements of the reaction velocity manifest themselves in both the ordinate and the abscissa. The points tend to be relatively well spaced. This plot is superior in the graphic detection of nonlinearity. We must unfortunately omit the aforementioned new Eisenthal–Cornish-Bowden plot (V vs. K_m) from this discussion, since its applicability has not yet been thoroughly evaluated by other authors. For the sake of comparison, the same data on brain tryptophan aminotransferase have been plotted in Fig. 6 by applying the four linear transformations discussed above.

Dowd and Riggs (1965) compared the estimates of K_m and V obtained with the aid of a computer from the three classic linear plots. The Hanes transform was found to be slightly better than the Eadie plot when the errors in the determination of v were small, but the Eadie plot was the most reliable when the errors were large. The Lineweaver–Burk plot was definitively inferior in all cases. Zydowo *et al.* (1971) drew similar inferences from their analyses. Endrenyi and Kwong (1973) also preferred the Hanes transform, particularly when there were constant errors or constant relative errors in v and the substrate concentrations tested were spaced harmonically or geometrically. When using any of these linearized plots, it should be kept in mind that they cannot tell more than the original equation; they only do it more illustratively. If the original experimental results are unsatisfactory, these mathematical manipulations will not improve them. On the contrary, these plots may be deceptive, because it is rather easy to draw on a graph straight lines that please the eye and fit almost any series of values, especially if the points are widely scattered or few in number and the coordinate axes are cunningly selected. The linear transformations are also likely to distort or reduce awareness of experimental errors.

If K_m and V are estimated solely by graphic means, it is advisable to examine the data by all linear transformations and compare them. After the linearity is established, the values of K_m and V should be estimated statistically. It is recommended that the least-squares method be used to fit a straight line to experimental values with proper weighting of the individual points. For technical reasons, measurements of the reaction velocity in high substrate ·concentrations can have a quite different scatter from those in low substrate concentrations. If the velocity measurements obtained in different substrate concentrations have a different variance—as they usually have—more weight should be assigned to the more reliable measurements and less to the less reliable ones according to their relative invariances.

3.3.3. Algebraic Determination of K_m and V

For accurate work, it is unwise to use any graphic plot for parameter estimation. Only for illustrative purposes is a linear plot, preferably either v vs. v/s or s/v vs. s, appropriate. Enzymic rate laws are mostly nonlinear functions. Their parameters must thus be estimated by nonlinear regression methods. A number of approaches have been published. In the method of Wilkinson (1961), provisional estimates for K_m and V—found by one of the linear transformations of equation (42)—were adjusted with the aid of Taylor's theorem by fitting a bilinear regression of v on the hyperbolic Michaelis–Menten equation (42) and its first derivative and then iteratively improving the estimation. The basic principles in the method of Bliss and James (1966) are similar. The calculations involved in these nonlinear regression methods are tedious. They can be made decisively easier by digital computers. Computer programs following the principles of Wilkinson have been devised by Cleland (1963). The method of Bliss and James has been programmed by Hanson *et al.* (1967). A logarithmic transform of the Michaelis–Menten equation was iteratively used for parameter estimation by Barber *et al.* (1967). Computer programs following the method of Barber *et al.* have since been used by Paumgartner *et al.* (1969). Hoare (1972) and Atkins (1971*a,b*) have introduced multipurpose digital-computer programs for fitting many types of nonlinear functions, among them enzyme-kinetic equations, to experimental data. All these iterative approaches are based on the general principles of regression analyses. With the calculation procedures named above, the estimates of kinetic parameters with their confidence limits can be assessed objectively and with precision.

Neame and Richards (1972) have presented a simple algebraic method for the estimation of K_m and V. The calculations are so easy that no computers are needed; a desk calculator suffices. The reaction velocities measured at two different substrate concentrations are inserted into two equations (42). Now we have a pair of Michaelis–Menten equations with two unknowns, K_m and V. The equation pair is easily solvable. The solutions for a number of v and s pairs allow evaluation of mean values of K_m and V and of their variance. To avoid a systematic bias in each pair of s values, one should be above K_m and one below. The authors are of the opinion that this simple calculation method yields estimates that match well in reliability those obtained by sophisticated computing procedures.

3.3.4. Integrated-Rate Method for the Estimation of K_m and V

Both the reaction velocity and the substrate concentration change continuously during the enzymic reaction. It is possible, then, to determine the kinetic parameters from a single progress curve if a sufficient number of simultaneous measurements of the reaction velocity and substrate concentration have been made at varying intervals. The time course of a reaction that obeys equation (42) for an extended period is defined by the integrated form of the Michaelis–Menten equation:

$$Vt = s_0 - s + K_m \ln (s_0/s) \tag{50}$$

in which s_0 is the substrate concentration at the start of the reaction.

A linear transformation of equation (50) is possible in different ways:

$$\frac{t}{\ln (s_0/s)} = \frac{1}{V} \left[\frac{s_0 - s}{\ln (s_0/s)} \right] + \frac{K_m}{V} \tag{51}$$

$$\frac{s_0 - s}{t} = V - \frac{K_m}{t} \ln (s_0/s) \tag{52}$$

$$\frac{t}{s_0 - s} = \frac{K_m}{V} \left[\frac{1}{(s_0 - s)} \right] \ln (s_0/s) + \frac{1}{V} \tag{53}$$

A plot of $t/\ln (s_0/s)$ vs. $(s_0 - s)/\ln (s_0/s)$ [equation (51)] is a straight line with a slope of $1/V$ and an intercept of K_m/V on the $t/\ln (s_0/s)$ axis. It resembles the plot of s/v vs. s derived from equation (48), in which the initial velocities were used. Similarly, equations (52) and (53) yield straight lines in the plots of $(s_0-s)/t$ vs. $\ln (s_0/s)/t$ and $t/(s_0-s)$ vs. $\ln (s_0/s)/(s_0-s)$ analogous to the plots of v vs. v/s and $1/v$ vs. $1/s$, respectively. The integrated-rate methods should be used only when the reaction is irreversible or there is no inhibition caused by accumulating reaction products. It is almost impossible, however, to differentiate between the effects of the product inhibition and the depletion of the substrate in a single run. One must be very careful, therefore, when using an integrated-rate method to seek an accurate determination of the kinetic parameters.

4. Inhibition Studies

4.1. Basic Nomenclature

A chemical substance may influence the rate of enzyme-catalyzed reactions. The reduction of the rate is *inhibition* and the effector substance is an *inhibitor*. When the reaction velocity is increased, the phenomenon is called *activation* and the effector substance is then an *activator*. Substances that are either inhibitors or activators may be called *modifiers, moderators,* or *effectors,* because the distinction between inhibition and activation often becomes blurred. A modifier can be either an inhibitor or an activator under different conditions for the same enzyme.

Inhibitors can be either enzyme products or substrates, other molecules that resemble these, or completely foreign substances. Inhibition is classified as *reversible* or *irreversible*. Irreversible inhibitors are often catalytic poisons causing denaturation of the enzyme protein. This kind of inhibition does not aid in elucidating kinetic mechanisms and therefore will not be further considered in this chapter. In contrast, reversible inhibitors are major tools in the study of enzyme kinetics. Reversible inhibitors usually form complexes with various enzyme forms and thus lower the amount of enzyme available for participation in the normal reaction sequence. The activity of the enzyme returns on removal of the inhibitor by dialysis or by other means. Thus, there is an equilibrium among the enzyme–inhibitor complex, the free inhibitor, and the enzyme. Inhibition may further be *partial* or *total*. A total inhibitor reduces the reaction velocity to zero when the inhibitor concentration is infinite, whereas a partial inhibitor diminishes the reaction velocity to a certain degree. *Product inhibition* results from the formation of the same enzyme–reactant complexes that form when the product is a substrate in the reverse reaction. Thus, the product of a reaction inhibits by recombining with the enzyme form that is liberated on dissociation of this particular product. An *alternative product* is a product that would be created if a different (alternative) substrate were used. It causes *alternative product inhibition*. The alternative substrate itself may also be an inhibitor. A *dead-end inhibitor* does not act as a substrate or product, but forms complexes with one or more forms of the enzyme. In this case, the formation of an enzyme–inhibitor complex constitutes a dead-end step in the reaction sequence. The special case in which a substance can act both as a substrate and as a dead-end inhibitor is called *substrate inhibition*.

If there is an increase in the K_m value, the inhibition is called *competitive*. When maximum velocity is reduced, the inhibition is *noncompetitive* or *pure noncompetitive* or *classic noncompetitive* according to the most widely used classic nomenclature. Such an inhibition does not frequently occur in practice, and therefore, some investigators regard noncompetitive inhibition as only a special case of *mixed inhibition*. In mixed inhibition, there is an increase in K_m and a decrement in V. When both K_m and V are reduced by a constant ratio, we have *uncompetitive inhibition*, which is also called *anticompetitive* or *coupling inhibition*.

4.2. Inhibition Mechanisms

We shall use the letter I to designate an inhibitor and the letter i to designate its concentration. An inhibitor may form two kinds of dynamic complexes with the enzyme, EIS and EI, involving or not involving the substrate as well. These complexes have catalytic properties different from those of the free enzyme. The most common type of inhibition is *competitive inhibition*. A competitive inhibitor gives valuable information on the specificity of the reaction mechanism. The simplest explanation for the action of a competitive inhibitor is that the inhibitor resembles the substrate enough structurally to compete with it in the active center of the enzyme and thereby reduces the velocity of reaction. This leads to the formation of a nonproductive enzyme–inhibitor complex, EI, that can only revert to E and I. In this case, we can measure the true equilibrium constant of the dissociation reaction: $EI \rightleftarrows E + I$. It is termed the *inhibition constant*, K_i.

The steady-state rate equation for competitive inhibition is

$$v = \frac{Vs}{K_m(1 + i/K_i) + s} \tag{54}$$

Equation (54) can be reduced to the simple Michaelis–Menten equation as follows:

$$v = \frac{V^{app}s}{K_m^{app} + s} \tag{55}$$

where V^{app} and K_m^{app} are the apparent (observed) values for these constants in the presence of inhibition. It emerges that $V^{app} = V$ and $K_m^{app} = K_m(1 + i/K_i)$. The competitive inhibitor thus increases the value of K_m by the factor $(1 + i/K_i)$ and does not change V. The degree of inhibition diminishes with the increasing substrate concentration.

In *mixed inhibition*, the complex EIS does not break down, being a dead-end product. K_i' is the inhibitor dissociation constant of *EIS* and K_i the dissociation constant of *EI*. K_m (or more correctly K_s) is the dissociation constant of *ES*. The following equations apply:

$$V^{app} = \frac{V}{1 + i/K_i'} \tag{56}$$

$$K_m^{app} = \frac{K_m(1 + i/K_i)}{(1 + i/K_i')} \tag{57}$$

Both K_m and V are altered, but by different factors. Mixed inhibition occurs commonly in cases of product inhibition. In pure noncompetitive inhibition, $K_m^{app} = K_m$; i.e., K_i must equal K_i'. It must be assumed that here the inhibitor interferes with the catalytic properties of the enzyme but has no effect on the binding of substrate. The degree of inhibition is unaffected by the substrate concentration. Such a situation is in practice very rare; possibly it obtains only for very small inhibitors such as protons and metal ions.

Uncompetitive inhibition means that the inhibitor combines only with those species of the enzyme that are themselves unable to combine with the substrate. This implies that the inhibitor-binding site in the enzyme becomes available only after the substrate has been bound. Thus, *EI* is not formed and *EIS* is a dead-end product. The binding of the inhibitor is unaffected by an increasing substrate concentration. An increment in the concentration of the substrate does not relieve the inhibition, but may instead enhance it. Uncompetitive inhibitors exert equal effects on V and K_m:

$$V^{app} = \frac{V}{1 + i/K_i'} \tag{58}$$

$$K_m^{app} = \frac{K_m}{1 + i/K_i'} \tag{59}$$

Uncompetitive inhibition often occurs as a special case of product inhibition.

4.3. Analysis of Inhibition

Two experimental strategies are customary when the nature of the inhibition of an enzyme reaction is studied by measuring the initial reaction velocities: either the concentration of the inhibitor is kept constant and the concentration of the substrate varied or vice versa. The choice depends on the nature of the investigation.

4.3.1. Determination of the Type of Inhibition

The pure types of inhibition described above can be readily discriminated by various graphic methods. For this purpose, usually only one plot, a *primary plot*, is needed.

1. The double-reciprocal plot, $1/v$ against $1/s$, is the most common despite its shortcomings. For this plot, the concentration of the substrate is varied in the presence of fixed concentrations of the inhibitor. For a competitive inhibitor, the rate equation (54) can be arranged as follows:

$$\frac{1}{v} = \frac{1}{V} + \frac{K_m}{V}\left(1 + \frac{i}{K_i}\right)\frac{1}{s} \tag{60}$$

Equation (60) shows that the intercept on the ordinate of the double-reciprocal plot is not altered by a competitive inhibitor. It does, however, increase the slope of the straight line by the factor of $(1 + i/K_i)$. Since the line intersects the abscissa at $-1/K_m$ and $-1/K_m(1 + i/K_i)$ in the absence and presence of the inhibitor, respectively, these intercepts can be used for evaluation of the inhibition constant, K_i, as well as the slopes (Fig. 7A).

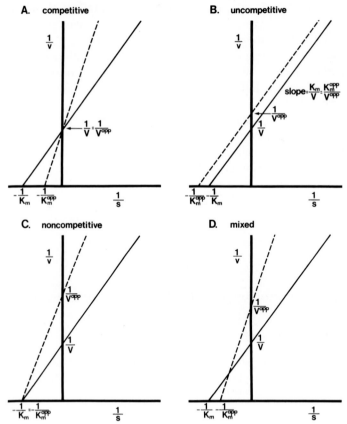

Fig. 7. Effects of competitive, uncompetitive, noncompetitive, and mixed inhibitors on the velocity of an enzyme reaction as shown by four linear plots of the Michaelis–Menten equation. (———) Uninhibited reactions; (– – – –) inhibited reactions. V^{app} and K_m^{app} are the kinetic constants in the presence of the inhibitors and V and K_m the constants in their absence. (A–D) Lineweaver–Burk plots of $1/v$ vs. $1/s$. (E–H) Eadie plots of v vs. v/s. (I–L) Hanes plots of s/v vs. s. (M–P) Direct linear plots of V^{app} vs. K_m^{app} in which $\hat{V}^{app}$ and $\hat{K}_m^{app}$ are the final estimates for the apparent kinetic constants in the presence of the inhibitors.

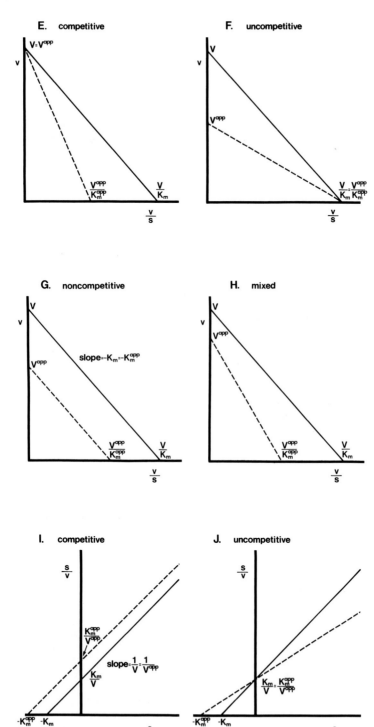

Fig. 7 (*Continued*)

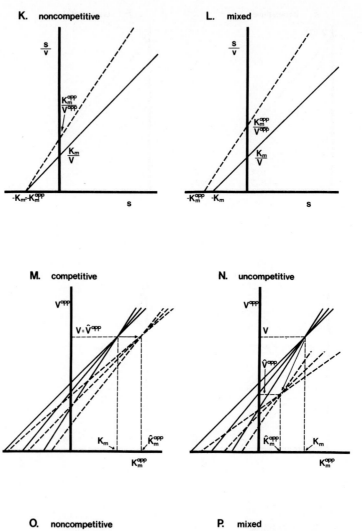

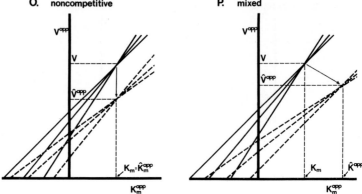

Fig. 7 (*Continued*)

The rate equation for uncompetitive inhibition can be written in the form of

$$\frac{1}{v} = \frac{K_m}{V}\frac{1}{s} + \frac{1}{V}\left(1 + \frac{i}{K_i'}\right) \tag{61}$$

In this case, the slope of the line in the plot $1/v$ vs. $1/s$ remains constant, but the intercepts on both the ordinate and the abscissa will be multiplied by the factor $(1 + i/K_i')$ in the presence of the inhibitor (Fig. 7B).

The rate equation in the case of pure noncompetitive inhibition is transformable to

$$\frac{1}{v} = \frac{K_m}{V}\left(1 + \frac{i}{K_i}\right)\frac{1}{s} + \frac{1}{V}\left(1 + \frac{i}{K_i}\right) \tag{62}$$

In the plot $1/v$ vs. $1/s$, the intercept on the abscissa is $-1/K_m$ in both the presence and the absence of the inhibitor. The slope of the line is $(1 + i/K_i)K_m/V$ in the presence of the inhibitor and the intercept on the ordinate is $(1 + i/K_i)/V$. They can thus be used for evaluation of K_i (Fig. 7C).

Finally, Fig. 7D is the double-reciprocal plot depicting mixed inhibition. All three properties of the straight line, the intercepts on both axes and the slope, are affected by the inhibition.

2, 3. The two further classic linear transforms of the Michaelis–Menten equation can also be used for assessing the type of inhibition and the inhibition constants as shown in Fig. 7E–L.

4. A new method for revealing the inhibition type appears in the book by Cornish-Bowden (1976). In this method, the reaction velocity at different substrate concentrations must first be measured both in the presence and in the absence of the inhibitor. When straight lines are drawn in the plot V^{app} against K_m^{app} through the points representing each pair of v and $-s$ on the ordinate and the abscissa, respectively, the common intersection point shifts in a direction that will indicate the type of inhibition: in competitive inhibition, the change is to the right; in uncompetitive inhibition, toward the origin; in mixed inhibition, the shift is intermediate; and in the special case of pure noncompetitive inhibition, downward (Fig. 7M–P).

4.3.2. Determination of the Inhibition Constants

There are several procedures for evaluating the inhibition constants. The type of inhibition must first be known to ensure that the constants are evaluated in the correct way. For this purpose, the aforementioned primary plots should be consulted. If the ratios $1/V^{\mathrm{app}}$ and $K_m^{\mathrm{app}}/V^{\mathrm{app}}$ are linear functions of the inhibitor concentrations when a *secondary plot* is constructed, the inhibition is said to be linear. In more complex cases, the replots are hyperbolas or parabolas. These different types are thus referred to as *linear, hyperbolic,* or *parabolic inhibition* according to the characteristics of the secondary plots. In this section, we shall concentrate only on linear inhibition. Hyperbolic inhibition is, however, very common, although not often reported. This is probably due to difficulties in diagnosing hyperbolic inhibition safely, since the nature of the inhibition must be established over a wide range of inhibitor concentrations. Cornish-Bowden (1976) states that linear inhibition is likely to be provoked only by products of the reaction and by close substrate analogues. *Allosteric* inhibition or activation resembles hyperbolic inhibition or activation. An allosteric effector bears no structural similarity to the substrates or products of the reaction and binds to a separate site in the enzyme, thus affecting the velocity of the reaction. Allostery is discussed in Section 5.

Of the graphic methods for evaluating inhibition constants, probably the most noteworthy are the following.

1. All the linear transformations of the Michaelis–Menten equation can be used for estimation of the inhibition constants as discussed in the previous section (cf. Fig. 7).

2. The inhibition constant K_i can also be determined by a simple graphic method without any calculations according to Dixon (1953). The initial velocity is measured at a series of varying inhibitor concentrations while the substrate concentration is kept constant. A plot of $1/v$ against i yields a straight line. If two (or more) such lines are drawn corresponding to different concentrations of S, the point of intersection is at $i = -K_i$ (Fig. 8). It can be calculated by equaling the expressions for $1/v$. This applies for any of the linear types of inhibition and can be used for competitive, noncompetitive, and mixed inhibition. In uncompetitive inhibition, K_i is infinite and the straight lines are parallel.

3. The aforementioned Dixon plot does not provide the "uncompetitive inhibition constant," K_i'. This can be estimated by plotting s/v against i at several concentrations of S. A set of straight lines is obtained that intersect at $i = -K_i'$. This applies for uncompetitive and mixed inhibition. In competitive inhibition, the lines are again parallel.

4. In another method, the apparent kinetic constants, K_m^{app} and V^{app}, are first determined at several concentrations of I. If then, for example, K_m^{app}/V^{app} and $1/V^{app}$ are plotted against i, straight lines are obtained. The point of intersection with the abscissa is $-K_i$ if the ordinate is K_m^{app}/V^{app} and $-K_i'$ if the ordinate is $1/V^{app}$.

5. The inhibition constant K_i can also be determined by finding the concentration of a competitive inhibitor that reduces the reaction velocity by one half. The experiments must be done using two different substrate concentrations (s_1 and s_2). The inhibitor concentration sought, i_1, is easily found from the plot of v against i. The reaction velocity, v_i, at that inhibitor concentration is

$$v_i = \tfrac{1}{2}v = \tfrac{1}{2}\frac{Vs_1}{s_1 + K_m} \tag{62}$$

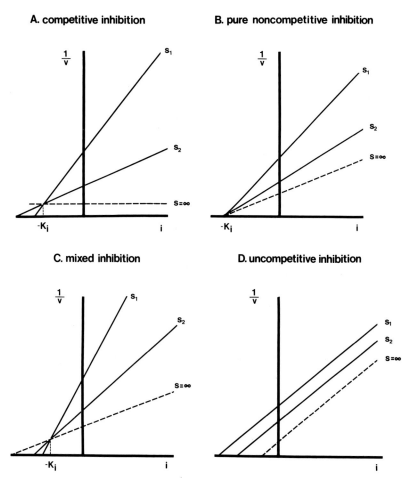

Fig. 8. Dixon plots of the four types of inhibition. The initial reaction velocity is measured at varying inhibitor concentrations (i) at two different fixed substrate concentrations (s_1 and s_2). The inhibition constant K_i is obtained from the plots as indicated.

or

$$v_i = \frac{V s_1}{s_1 + K_m(1 + i_1/K_i)} \tag{63}$$

Expressions for K_m and K_i can be derived from equations (62) and (63) as follows:

$$K_m = \frac{s_1 K_i}{i_1 - K_i} \qquad K_i = \frac{i_1 K_m}{s_1 + K_m} \tag{64}$$

The same experiments and calculations are then repeated at another substrate concentration s_2. Now we have two sets of experimental values (s_1, i_1 and s_2, i_2) and two expressions for K_m. They must equal:

$$\frac{s_1 K_i}{i_1 - K_i} = \frac{s_2 K_i}{i_2 - K_i} \tag{65}$$

The solution for K_i from equation (65) is

$$K_i = \frac{s_1 i_2 - s_2 i_1}{s_1 - s_2} \tag{66}$$

Similarly, K_m can be estimated from two separate expressions for K_i according to equation (64):

$$K_m = \frac{s_1 i_2 - s_2 i_1}{i_1 - i_2} \tag{67}$$

In this procedure, both K_m and K_i are evaluated simultaneously, but it should be kept in mind that equations (63)–(67) apply only for competitive inhibition.

6. K_i and K_i' can also be determined from a single plot introduced by Hunter and Downs (1945). In the case of mixed inhibition, the difference between the uninhibited reaction velocity and the inhibited velocity at each substrate concentration ($v - v_i$) is given by

$$v - v_i = \frac{Vs}{K_m + s} - \frac{Vs/(1 + i/K_i')}{K_m(1 + i/K_i)/(1 + i/K_i') + s} \tag{68}$$

Equation (68) can be rearranged to

$$\frac{iv_i}{v - v_i} = \frac{K_m + s}{K_m/K_i + s/K_i'} \tag{69}$$

Then $iv_i/(v - v_i)$ is plotted against s. We get a rectangular hyperbola with an intercept K_i on the ordinate. When s increases, the value of $iv_i/(v - v_i)$ approaches a limiting value of K_i'. This K_i' is usually difficult to locate graphically but easy to calculate if K_m is known, because $iv_i/(v - v_i)$ is equal to $2/(1/K_i + 1/K_i')$ when $s = K_m$. In the limiting case of pure noncompetitive inhibition when $K_i' = K_i$, the right-hand side of equation (69) is simply K_i. In competitive inhibition, $K_i' \rightarrow \infty$ and equation (69) simplifies to

$$\frac{iv_i}{v - v_i} = K_i(1 + s/K_m) \tag{70}$$

which describes a straight line with an intercept K_i on the ordinate. In uncompetitive inhibition when $K_i \to \infty$, equation (69) takes the form

$$\frac{iv_i}{v - v_i} = K_i'(1 + K_m/s) \tag{71}$$

When $s \to 0$, the curve approaches the ordinate asymptotically, and when s increases, it similarly approaches a limiting value of K_i'. The Hunter–Downs plot is useful when only a few rate determinations can be carried out over a wide range of s and i values and in particular when inhibition is either competitive or pure noncompetitive.

Finally, we must reiterate the statement made in discussion of the various graphic representations of the basic Michaelis–Menten equation (42): all graphic plots visualize the situation informatively and give valuable hints as to the magnitude of inhibition constants, but the eventual estimation of the constants must be done algebraically if any accuracy is desired. Computing procedures mentioned above or their subsequent modifications may decisively assist in evaluation of the constants and their fiducial limits.

4.4. Substrate Inhibition

In a number of single-substrate reactions, the reaction velocity passes through a maximum and then falls, approaching zero at very high substrate concentrations. The simplest explanation for this kind of behavior is the formation of a dead-end complex that contains two molecules of the substrate attached to one enzyme molecule, i.e., SES or ES_2. The following rate equation applies for the initial reaction velocity:

$$v = \frac{Vs}{K_m + s + s^2/K_{SI}} \tag{72}$$

where K_{SI} is the dissociation constant of the complex SES. Equation (72) differs from the simple Michaelis–Menten equation (42) by the term s^2/K_{SI}, which, however, becomes significant only at high substrate concentrations. Thus, the rate follows the Michaelis–Menten equation (42) at low substrate concentrations but approaches zero, instead of V, at high values of s. Figure 9A depicts v as a function of s. The Lineweaver–Burk plot of $1/v$ against $1/s$ is linear at large values of $1/s$ but deviates from linearity at low $1/s$ values (Fig. 9B) according to the rearranged rate equation:

$$\frac{1}{v} = \frac{K_m}{V} \frac{1}{s} + \frac{1}{V} + \frac{1}{VK_{SI}} s \tag{73}$$

From the slope of the linear part of the curve, K_m/V can be estimated, and $1/V$ can be calculated from the intercept on the $1/v$ axis of the corresponding extrapolated straight line. The linear transform of s/v vs. s is a parabola (Fig.

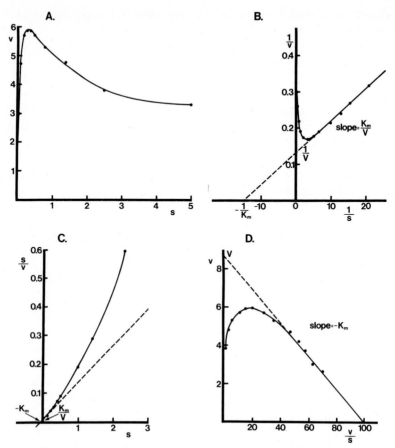

Fig. 9. Inhibition by the substrate presented in four different plots. The kinetic constants (K_m and V) can be obtained from the extrapolated straight lines as indicated. (A) Hyperbolic curve of v vs. s. (B) Lineweaver–Burk plot of $1/v$ vs. $1/s$. (C) Hanes plot of s/v vs. s. (D) Eadie plot of v vs. v/s. The points in different plots have been calculated from the same hypothetical primary data to visualize their displacement from the straight lines in each plot.

9C). If K_{SI} is much larger than K_m, the latter plot is also almost linear at low values of s, and thus K_m and V can be estimated as usual. Also, the plot v vs. v/s can be employed in the same way (Fig. 9D). The dissociation constant K_{SI} must be determined by using data obtained at high substrate concentrations. A plot of $1/v$ against s is linear at high values of s and the slope gives $1/VK_{SI}$.

Of the many enzymes inhibited by their own substrates, succinate semialdehyde dehydrogenase (succinate semialdehyde:NAD^+ oxidoreductase, EC 1.2.1.16) is a good example. The enzyme isolated from the rat brain was inhibited by succinate semialdehyde at concentrations above 10^{-4} mol/liter (Kammeraat and Veldstra, 1968). The authors assumed that the aldehyde group of the succinate semialdehyde molecule is involved in the formation of the inactive complex SES. A K_m value (4.7×10^{-6} mol/liter) was calculated from the data obtained at low succinate semialdehyde concentrations. The dissociation constant of the complex *SES* (2.2×10^{-3} mol/liter) was determined at high substrate concentrations. Hall and Kravitz (1967*b*) have obtained rather similar results with the same enzyme purified from the lobster central nervous system.

Substrate inhibition does not normally occur at physiological substrate concentrations. It generally sets only the upper limits to experiments done by the neurobiologist.

The use of substrate inhibition in analyzing two-substrate reaction mechanisms is discussed in Section 6.3.

4.5. Product Inhibition

A study of inhibition caused by reaction products should be a part of all thorough kinetic investigations. The products are the substrates for the reverse reaction and presumably also form complexes with the enzyme. These complexes tie up enzyme molecules, so that the effects of the products are always inhibitory. The type of inhibition depends on the substrate tested and on the product used as inhibitor. A complete product-inhibition study involves the use of each product in turn as an inhibitor vs. each substrate as a variable substrate. Thus, inhibition by the product may be competitive, noncompetitive, mixed, or uncompetitive in nature, depending on the experimental combination. Enzyme mechanisms may be predicted accurately according to the observed product-inhibition behavior, as discussed briefly previously. In certain cases, alternative products may also be used for a demonstration of distinctive kinetic patterns and hence to confirm or identify kinetic mechanisms.

As an example of product inhibition, the enzyme choline acetyltransferase (acetyl-CoA : choline *O*-acetyltransferase, EC 2.3.1.6) is competitively inhibited by its reaction product, CoA. CoA competes with the substrate acetyl-CoA for the active site of the enzyme. The inhibition constant has been determined to be about 16×10^{-6} mol/liter (Potter, 1972). Product inhibition may constitute an important factor in the control of enzymatic reactions under normal physiological conditions in the nervous system.

5. Cooperative Effects

5.1. Sigmoid Kinetics

Ordinary enzyme-kinetic laws do not provide an explanation for every kind of enzyme behavior. The studies on allosteric effects and cooperativity have developed apart from the mainstream of enzyme kinetics and are often neglected in the textbooks on enzyme kinetics. However, a neurobiologist should be aware of certain main concepts concerning the regulation of enzyme activity, because these same ideas are often applied to receptor binding and transport phenomena within neurochemistry.

There are many enzyme-catalyzed reactions that do not exhibit a hyperbolic relationship between the reaction velocity and the substrate concentration as the Michaelis–Menten equation predicts. The relationship is often sigmoid, or S-shaped (Fig. 10), giving rise to the concept of sigmoid kinetics. Sigmoid kinetics can be observed in the interactions between the substrate and the enzyme or between inhibitors or activators and the enzyme. The simplest equation consistent with sigmoid behavior relates the reaction velocity v to the substrate concentration s as follows:

$$v = \frac{js^2}{k + ms^2} \tag{74}$$

where j, k, and m are constants. An inversion of equation (74) gives

$$\frac{1}{v} = \frac{k}{j}\frac{1}{s^2} + \frac{m}{j} \tag{75}$$

According to equation (75), a plot of $1/v$ against $1/s$ is not a straight line. Instead, a straight line is obtained when $1/v$ is plotted against $1/s^2$ (Fig. 11).

5.2. Allostery and Cooperativity

It should be kept well in mind that not all nonhyperbolic results in enzyme kinetics are due to sigmoid kinetics. Laidler and Bunting (1973) have collected those mechanisms that are likely to lead to sigmoid kinetics. These include the following cases: (1) the enzyme contains an impurity that interacts with the substrate; (2) the enzyme can exist in two forms with different activities; (3) two molecules of the substrate bind to one molecule of the enzyme (in this case, the complex SES must react more rapidly than ES, thus giving rise to a quadratic dependence of the reaction velocity on s at low substrate concentrations); (4) the substrate acts as a modifier forming an active complex ES and an inactive complex SE; (5) two substrates react by a random-order ternary-complex mechanism (see Section 6.1.2), or they can

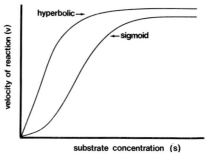

Fig. 10. Sigmoid dependence of the reaction velocity (v) on the concentration of the substrate (s) when compared to hyperbolic (Michaelis–Menten) kinetics.

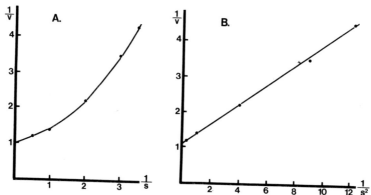

Fig. 11. Sigmoid kinetics presented in the Lineweaver–Burk plot of $1/v$ vs. $1/s$ (A) and in a plot of $1/v$ vs. $1/s^2$ (B).

react along two or more alternative pathways; (6) enzyme molecules consist of several subunits that can interact with each other. The last case comprises the concept of *allostery*, which is the most common explanation for sigmoid kinetics. Very often, allostery and sigmoid kinetics are wrongly used synonymously. This is not always true, as discussed above.

The term allostery means that a substrate or a modifier binds to the enzyme at a site other than the catalytic site, causing conformational changes in the enzyme protein, which changes in turn affect the velocity of the reaction. The allosteric effectors can cause inhibition or activation. Allosteric effectors are often structurally dissimilar to the substrate molecule. The enzymes that possess allosteric binding sites for allosteric effectors are called *allosteric enzymes*. Allostery itself does not provide an explanation for sigmoid kinetics. It should be considered only as a special type of inhibition or activation. It is also clear that sigmoid kinetics alone give no proof for allostery, since other mechanisms may be involved as well. The most direct evidence for allostery occurs when an enzyme can be made insensitive to a modifier without losing its catalytic activity.

Allostery may be regarded as a part of a still wider concept termed *cooperativity*. Many enzymes possess a propensity to respond very sensitively to changes in metabolite concentrations. This unusual property is thought to arise from "cooperation" between the active sites or ligands of polymeric enzymes. Binding of one molecule of the ligand enhances the binding or reaction rates of the others (positive cooperativity). The biological significance of this phenomenon is easily conceived. Figure 10 shows that at low concentrations of the substrate, the activity of the enzyme is lower in the presence of an allosteric modifier, when sigmoid kinetics prevails, than in the absence of any modifier. The degree of inhibition becomes less severe when the concentration of the substrate increases. Cooperativity in this case signifies that at a low substrate concentration, the allosteric effector is a powerful inhibitor, most effectively, then, regulating enzyme activity. On the other hand, when more substrate becomes available, relatively small changes in the substrate concentration result in large changes in enzyme activity. Many enzymes are both allosteric and cooperative, but this is not a rule.

5.3. Hill Equation

Hill (1910) formulated a purely empirical expression for ligand-binding. This equation was first applied to the binding of oxygen to hemoglobin, but has since found a large number of further applications. The binding of a ligand A to a polymeric protein E can be written as

$$E + hA \underset{}{\overset{K_h}{\rightleftharpoons}} EA_h$$

and characterized by the expression:

$$\bar{Y}_s = \frac{K_h a^h}{1 + K_h a^h} \tag{76}$$

in which a is the concentration of the ligand; $\bar{Y}_s$ is the fractional saturation, i.e., number of occupied binding sites/total number of binding sites; K_h is the association constant; and h is the number of ligands that the polymeric enzyme can bind (interaction coefficient). Equation (76) is known as the *Hill equation*. It can be rearranged as follows:

$$\log \frac{\bar{Y}_s}{1 - \bar{Y}_s} = \log K_h + h \log a \tag{77}$$

If it is assumed that the reaction velocity v is proportional to the fraction of the sites occupied

$$\log \frac{v}{V - v} = h \log s - \log K' \tag{78}$$

in which K' is a constant comprising interaction factors and the dissociation constant K_s. The Hill equation frequently appears in this form in neurobiology. The plot of $\log [v/(V - v)]$ against $\log s$ is a straight line with the slope equal to h (cf. Fig. 23). The *Hill plot* fits a variety of binding data for values of $\bar{Y}_s$ within the range of 0.1 to 0.9. The exponent h is known as the *Hill coefficient*. The Hill coefficient is widely used as an index for cooperativity: a high h means a high degree of cooperativity. Qualitatively, h below 1.0 means negatively cooperative effects, $h = 1.0$ a noncooperative effect, and h over 1.0 (positively) cooperative effects. (Negative cooperativity means that the binding of the first molecule of the ligand makes it more difficult for the next one to bind.) The Hill coefficient also provides the lower limit for the number of subunits in the fully associated protein.

Taketa and Pogell (1965) have used a different parameter for expressing the degree of cooperativity, the *cooperativity index*, R_x. It is defined as the

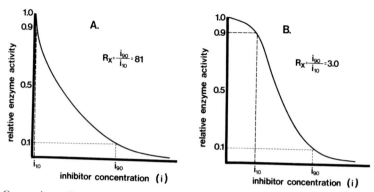

Fig. 12. Comparison of hyperbolic (A) and sigmoid (B) inhibition curves. Assay of the cooperativity index R_x. R_x equals 81 in linear inhibition of the Michaelis–Menten type and is less than 81 in sigmoid inhibition.

ratio of the *x*-values coinciding with $\bar{Y}_s = 0.9$ and $\bar{Y}_s = 0.1$. It is thus the ligand concentration ratio at 90 and 10% saturation (or inhibition as demonstrated in Fig. 12). R_x is 81 for all Michaelis–Menten kinetics, less than 81 in positive cooperativity, and larger than 81 in negative cooperativity. R_x is a purely empirical measure not based on any models. The relationship between the Hill coefficient and the cooperativity index can be evaluated by substituting $\bar{Y}_s = 0.1$ and $\bar{Y}_s = 0.9$ into equation (76) and solving for *a*. R_x is seen to equal $81^{1/h}$.

5.4. Graphic Presentation of Cooperative Effects

The kinetics of allosteric enzymes are often illustrated by so-called saturation curves. In a saturation curve, some function of sites occupied (the saturation function $\bar{Y}_s$) is plotted against some function of the concentration (*a*) of the ligand. If it is assumed that the reaction velocity is proportional to the fraction of the sites occupied, $\bar{Y}_s$ can be replaced by v/V. This assumption leads to the Michaelis–Menten equation. This substitution for $\bar{Y}_s$ is very useful in practice, although in complex systems it is not applicable. Saturation curves can be made in a number of different ways, each giving some diagnostic information regarding the ligand-binding effects. A plot of $\bar{Y}_s$ vs. *a* is a hyperbolic curve when there are no cooperative effects. Positive cooperative effects make the plot sigmoid, while negative effects manifest themselves with a slow tailing-off of the curve as compared to a hyperbola.

The estimation of kinetic parameters and cooperative effects from a sigmoid curve is difficult, but, for instance, the double-reciprocal form does prove useful. Deviations from linearity are revealed by this plot if there are subunit interactions. Positively cooperative effects render the double-reciprocal plot concave upward and negative effects, concave downward. Also, the other linearized plots of the Michaelis–Menten equation deviate from linearity. The plot $\bar{Y}_s$ vs. $\bar{Y}_s/s$ is curved upward when cooperativity is negative and downward when it is positive. This is the most common plot with kinetic data. The Hanes plot $s/\bar{Y}_s$ vs. s, on the other hand, is curved upward in positive cooperative cases and downward in negative ones.

The inverse of the Eadie plot, the Scatchard (or Klotz) plot, is used for determination of the total number of binding sites in allosteric proteins. This plot is especially useful when there are no cooperative effects. The slope is proportional to the binding constant, and the limiting intercept equals the number of binding sites. The plot is not linear if there occurs negative or positive cooperativity. In a positively cooperative case, the curve is concave downward and in a negatively cooperative case, upward. The Scatchard plot is commonly used when receptor binding data are analyzed.

The previously mentioned Hill equation has been widely applied in studies not only on enzyme kinetics but also on membrane transport and receptor binding. It is suitable for inhibitors that give a sigmoid response as well, in this case generally in the form of

$$\log \frac{v_i}{v - v_i} = \log K_i' - h \log i \tag{79}$$

where v_i is the inhibited velocity, v the uninhibited velocity, and $K_i' = s_{50}^h$, in which s_{50} is the substrate concentration that maintains 50% of V.

A good example of an enzyme with allosteric binding sites is (Na^+,K^+)-activated adenosinetriphosphatase (EC 3.6.1.3). The enzyme purified from crustacean axonal membranes has recently been very thoroughly characterized by Gache *et al.* (1976). The cooperativity of the binding of Na^+, K^+ and ATP to their respective sites in the enzyme was determined by using classic titration curves of the Michaelis–Menten type (enzyme activity vs. logarithm of ligand concentration) and Hill plots. The normal steady-state kinetic behavior of ATP binding changed at very low concentrations of K^+ and negative cooperativity was observed (Hill coefficient below 1). The binding of Na^+ to its internal stimulatory sites had a positive cooperative effect (Hill coefficient about 2). Alterations in the K^+ concentration did not affect the homotropic effects between Na^+ sites. No cooperative actions were observed in the binding of K^+ to its external stimulatory sites. Mg^{2+} had no heterotropic effects. The temperature dependence of some kinetic parameters of the (Na^+,K^+)-ATPase was also analyzed by using the Arrhenius plot of $\log v$ vs. $1/T$ as described in Section 2.5.

5.5. Models of Cooperativity

All modern theories of cooperativity are based on the theory of *induced fit* by Koshland (1958, 1959). This concept, in simple terms, means that enzymes possess a flexible active site. A conformational change occurs in the enzyme on substrate-binding and causes the proper alignment of the catalytic groups of the enzyme with the site of reaction in the substrate. The most widely accepted models for explaining cooperativity are the *symmetry* or *allosteric model* by Monod *et al.* (1965) and the *sequential model* by Koshland *et al.* (1966). In the allosteric model, a distinction is made between the different effects that a ligand exerts on another ligand. A *homotropic effect* means that the binding of one molecule of the ligand affects the binding of subsequent ligands of the same sort. A *heterotropic effect* means that the binding of one type of the ligand influences the binding of another type. Homotropic effects are always positively cooperative, but heterotropic effects can be either (positively) cooperative or antagonistic (negatively cooperative).

A detailed discussion of the models lies outside the scope of this article, and the reader is referred to the monographs in the reference list. The potential of the cooperative and allosteric enzymes should be kept in mind when considering, for instance, the control mechanisms of neuronal activity. An allosteric enzyme could be a powerful modifier of the biosynthesis and catabolism of neurotransmitters. The binding of neurotransmitters and drugs to membrane receptor sites could be modified by allosteric effects as well.

6. Kinetics of Two-Substrate Reactions

6.1. Reaction Mechanisms

Most enzymes catalyze the conversion of two or more substrates to two or more products. The simple unireactant mechanism described earlier is in fact very rare, being confined only to some isomerization reactions. Most enzyme reactions in biochemistry involve more than one substrate, the most common type being

$$A + B \rightleftharpoons P + Q$$

This is a typical two-substrate, two-product reaction representing 60% of all known enzyme-catalyzed reactions. There exist more complicated reactions as well, involving three, four, or even more substrates. They can, however, be characterized by an extension of the principles of the two-substrate, two-product type. A general knowledge of the main types of two-substrate reactions is essential.

Cleland (1963) introduced in 1963 a system for the schematic representation of reaction mechanisms. He has also proposed a general nomenclature for enzyme mechanisms that is important to understand. All mechanisms that require the binding of every substrate before any product can be released are called *sequential mechanisms*. The existence of ternary complexes of the type *EAB* has thus been proposed in these cases. The mechanisms in which some products are released before every substrate has been bound are called *nonsequential* or *Ping-Pong mechanisms*. The formation of binary complexes *EA* and *EB* has been assumed in these latter cases. Mechanisms that imply isomerization of the enzyme are called *iso mechanisms*. The sequential mechanisms include two common alternatives: the *random-order* and the *ordered* or *compulsory-order mechanisms*.

Substrates are usually designated by the letters *A, B, C* in the order in which they bind to the enzyme. Products are written as *P, Q, R* following the order in which they leave the enzyme. The enzyme–substrate complexes are generally called *transitory complexes*, and there exist two types of them. If the active site of the enzyme is not completely filled so that it can still bind another reactant, the complex may be called a *noncentral transitory complex.* When the active site is completely filled with reactants, it is a *central complex.* In the case of the central complex, only the dissociation of the substrate or products is possible. The central complex corresponds to the complex of Michaelis in classic enzyme theory. The complexes are normally enclosed within parentheses.

The number of kinetically important substrates or products is designated by the terms *Uni, Bi, Ter, Quad,* etc. Reactance in a given direction is the number of reactants involved in the reaction for that direction. A Uni Bi reaction is thus unireactant in the forward direction and bireactant in the reverse direction. A reaction with two substrates and two products is then Bi Bi, being bireactant in both directions. At constant pH in an aqueous medium, hydrogen ions and water are usually not regarded as substrates.

The steady-state rate equation for any enzyme mechanism can in principle be derived in the same way as the simple Michaelis–Menten equation. In practice, however, this is very laborious and liable to error. The rate equations are usually very complicated except for a few of the simplest cases. King and Altman (1956) have developed a schematic method that can be applied to any mechanism that consists of a series of reactions between different forms of an enzyme. Kinetic theories for two-substrate reactions have been developed by Alberty, Dalziel, Cleland, Hanes, and Wong. The analysis methods for two-substrate reactions are quite analogous to those for one-substrate reactions: changes in the initial velocity as a function of the concentration of one of the two substrates are expressed by a Michaelis–Menten equation, since most two-substrate enzymes behave much like one-substrate enzymes if only one substrate concentration is varied. This substrate is then called the *variable substrate,* and the second substrate, the concentration of which is held constant, is the *fixed substrate.*

Schematic diagrams for the description of enzyme reactions are in common use. The whole reaction sequence is written from left to right with a horizontal line representing the enzyme in its various forms. Substrate additions and product dissociations are marked by vertical arrows. For example,

is a so-called ordered Bi Bi mechanism.

The three main types of two-substrate reactions are discussed below.

6.1.1. Ordered or Compulsory-Order Ternary-Complex Mechanism

This mechanism implies that no binding site exists in the enzyme for one of the two substrates until the other is bound. There thus prevails a compulsory order of binding, and the reaction mechanism is ordered. The example above, an ordered Bi Bi mechanism, is this kind of reaction. The same sequence can also be expressed as follows:

$$A + E \rightleftarrows AE$$

$$AE + B \rightleftarrows (AEB \rightleftarrows QEP) \rightleftarrows QE + P$$

$$QE \rightleftarrows E + Q$$

A and its product Q are usually called *leading* or *obligatory reactants* and are often coenzymes of the reaction. A variety of NAD^+- or $NADP^+$-requiring dehydrogenases obey this mechanism.

A simple example of the ordered mechanisms is phosphoserine phosphohydrolase (EC 3.1.3.3) purified from mouse brain (Bridgers, 1969). It catalyzes the reaction:

$$Phosphoserine + H_2O \rightleftarrows serine + orthophosphate$$

The mechanism is consistent with an ordered Uni Bi mechanism:

Phosphoserine		Serine		Phosphate	
↓		↑		↑	
E	E(phosphoserine)		E(phosphate)		E

There exists a special case of the compulsory-order mechanism that is called the Theorell–Chance mechanism, originally proposed for alcohol dehydrogenase. In this case, the steady-state concentrations of the ternary complexes are very low. The first product P appears to be formed directly from the substrate B by interaction with an EA complex. Kinetic analyses of the brain choline acetyltransferase show the enzyme to behave in a manner consistent with this Theorell–Chance mechanism (Potter, 1972). Choline acetyltransferase catalyzes the reaction:

$$Choline + acetyl\text{-}CoA \ (AcCoA) \rightleftarrows acetylcholine \ (ACh) + CoA$$

The proposed mechanism of this reaction is

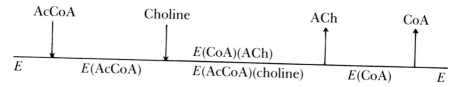

6.1.2. Random-Order or Random-Order Ternary-Complex Mechanism

In this case, the enzyme is assumed to have two binding sites, one for each substrate or product. The substrates can bind themselves to the enzyme in a random order, independently of each other. This mechanism is also called *random Bi Bi*. All possible binary enzyme–substrate complexes are formed rapidly and reversibly. The slow step is the interconversion of the two ternary complexes. The reaction mechanism may be expressed as follows:

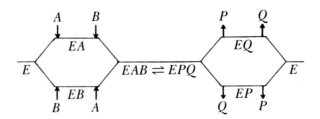

or

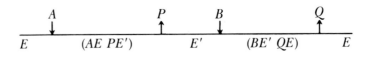

This kind of mechanism is common for many kinases, e.g., for phosphotransferases.

6.1.3. Ping-Pong or Substituted-Enzyme Mechanism

In this nonsequential mechanism, the reaction proceeds through a substituted-enzyme intermediate and not via any ternary complex:

$$E + A \rightleftarrows (AE \rightleftarrows PE') \rightleftarrows P + E'$$

$$B + E' \rightleftarrows (BE' \rightleftarrows QE) \rightleftarrows Q + E$$

or

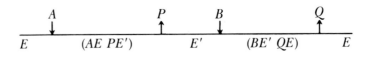

Thus, products are formed prior to the reaction with the second substrate at the expense of conversion of the enzyme from a form E to E'. Some important reactions appear to evince this kind of behavior, for example, those catalyzed by the pyridoxal-phosphate-requiring aminotransferases and dehydrogenation reactions catalyzed by some flavoproteins. Among the aminotransferases, GABA-T (4-aminobutyrate:2-oxoglutarate aminotransferase, EC 2.6.1.19)—the enzyme catabolizing γ-aminobutyric acid (GABA)—catalyzes the reversible transamination of GABA with 2-oxoglutarate. The reaction mechanism of GABA-T is consistent with the Ping-Pong mechanism (Hall and Kravitz, 1967a). Another neurobiologically important enzyme that obeys this mechanism is acetylcholinesterase (acetylcholine hydrolase, EC 3.1.1.7) (Potter, 1972):

$$ACh + H_2O \rightleftarrows choline + acetic\ acid$$

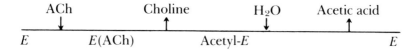

The last step in the reaction, deacetylation, is rate-limiting.

6.2. Determination of the Kinetic Parameters: Initial-Velocity Measurements

Steady-state kinetic measurements help decisively in differentiating the reaction mechanisms discussed above. The complete rate equations derived by the King–Altman method constitute the basis of analyses. These equations, written in terms of rate constants, are rather formidable, however, and we limit the discussion of the rate laws to initial-velocity studies. Here, all terms involving concentrations of products can be eliminated and the equations look simpler. Also, more informative terms are used instead of rate constants. However, more constants than simply one Michaelis constant per reactant must now be defined. The mechanisms themselves may be quite complex, but usually in a consideration of one individual reactant, they can be reduced to the form of the Michaelis–Menten equation, in which K_m and V are functions of the other reactants.

The initial velocity for a two-substrate reaction obeying the sequential mechanism can be expressed by the following equation:

$$v = \frac{Vab}{K_i^A K_m^B + K_m^B a + K_m^A b + ab} \tag{80}$$

and that for a reaction obeying the Ping-Pong or substituted-enzyme mechanism by the following equation:

$$v = \frac{Vab}{K_m^B a + K_m^A b + ab} \tag{81}$$

Equations (80) and (81) contain four kinetic parameters. V represents the maximum velocity when both substrates are saturating. K_m^A is defined as the limiting Michaelis constant for the substrate A when the other substrate B is saturating. Similarly, K_m^B is the limiting Michaelis constant for B when A is saturating. K_i^A is the limiting value of the Michaelis constant of A when the concentration of B becomes zero. It is also the true equilibrium dissociation constant of the complex EA.

The type of kinetic mechanism is rather easy to distinguish, but the determination of all the kinetic constants requires the fitting of kinetic data in both primary and secondary plots. The most common experimental procedure is to keep the concentration of one of the substrates (B) constant as a changing fixed substrate and vary the concentration of the other (A). Equation (80) for the sequential mechanisms can be rewritten as follows:

$$v = \frac{(Vb/K_m^B + b)^a}{(K_i^A K_m^B + K_m^A b)/(K_m^B + b) + a} \tag{82}$$

Equation (82) resembles the simple Michaelis–Menten equation (42), in which the apparent V and K_m appear as

$$K_m^{\text{app}} = \frac{K_i^A K_m^B + K_m^A b}{K_m^B + b} \tag{83}$$

$$V^{\text{app}} = \frac{Vb}{K_m^B + b} \tag{84}$$

Thus, both V^{app} and K_m^{app} are functions of the concentration of B. The corresponding apparent constants for the nonsequential Ping-Pong mechanism are

$$K_m^{\text{app}} = \frac{K_m^A b}{K_m^B + b} \tag{85}$$

$$V^{\text{app}} = \frac{Vb}{K_m^B + b} \tag{86}$$

V^{app} and K_m^{app} can be graphically determined in exactly the same way as in the one-substrate case. We prefer a plot in which a/v is plotted against a, although double-reciprocal plots are the most common in the literature. In the plot a/v vs. a for the sequential mechanisms, straight lines representing various concentrations of B intersect at a common point $a = -K_i^A$ and $a/v = (K_m^A - K_i^A)/V$ (Fig. 13A). This primary plot does not yet yield all kinetic constants, and a secondary plot must be consulted. Equation (84) can be rearranged to

$$\frac{b}{V^{\text{app}}} = \frac{K_m^B}{V} + \frac{1}{V} b \tag{87}$$

Equation (87) shows that a plot of b/V^{app} vs. b gives a straight line with a slope of $1/V$ and an intercept on the b/V^{app} axis of K_m^B/V (Fig. 14A). From

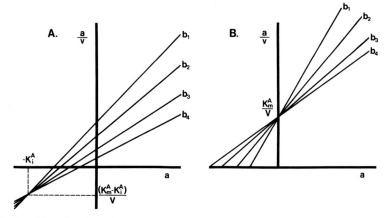

Fig. 13. Graphic estimation of the kinetic constants for a bireactant enzyme with substrates A and B. Primary plots of a/v vs. a at various fixed values of b. (A) Sequential mechanism; (B) nonsequential mechanism.

equations (83) and (84) we get

$$\frac{bK_m^{\text{app}}}{V^{\text{app}}} = \frac{K_i^A K_m^B}{V} + \frac{K_m^A}{V} b \tag{88}$$

A plot of $bK_m^{\text{app}}/V^{\text{app}}$ against b is a straight line with a slope of K_m^A/V and the intercept on the bK_m^{app}/V axis of $K_i^A K_m^B/V$ (Fig. 14B). With these three graphic representations, all four kinetic parameters for the sequential mechanisms can be evaluated.

A plot of a/v vs. a for a Ping-Pong mechanism consists of a series of straight lines intersecting on the ordinate (Fig. 13B). This pattern is easy to distinguish from the primary plot of a sequential mechanism. Only one secondary plot is required for the Ping-Pong mechanism, viz., that of b/V^{app} against b, which is the same plot used in the sequential mechanism (Fig. 14A).

Another method to plot the kinetic data for two-substrate reactions resembles the double-reciprocal treatment of one-substrate reactions. Equations (80) and (81) can be rewritten in the form

$$\frac{1}{v} = \left[\frac{K_m^A}{V} \left(1 + \frac{K_i^A K_m^B}{K_m^A b} \right) \right] \frac{1}{a} + \frac{1}{V} \left(1 + \frac{K_m^B}{b} \right) \tag{89}$$

(Sequential mechanisms)

$$\frac{1}{b} = \frac{K_m^A}{V} \frac{1}{a} + \frac{1}{V} \left(1 + \frac{K_m^B}{b} \right) \tag{90}$$

(Ping-Pong mechanism)

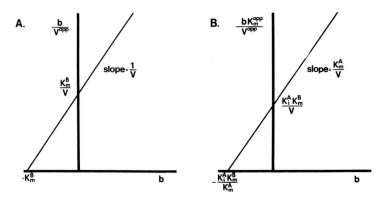

Fig. 14. Secondary plots for estimation of the kinetic constants for bireactant enzymes. Both plots (A) and (B) are needed in the case of a sequential mechanism. Only (A) is necessary in nonsequential mechanisms.

In a plot of $1/v$ against $1/a$, a family of lines appears, corresponding to the different concentrations of B. If the lines intersect (Fig. 15A), the mechanism is sequential, but if they are parallel (Fig. 15B), the mechanism is Ping-Pong. The slopes and intercepts on the ordinate of these lines are indicated in Fig. 15. Two secondary plots are needed, i.e., those of the intercepts and slopes obtained from the primary plots plotted against the reciprocal of the concentrations of the variable substrate. The slope of the slope-plot is $K_i^A K_m^B/V$ and the intercept K_m^A/V (Fig. 15C). The slope of the intercept-plot is K_m^B/V and the intercept $1/V$ (Fig. 15D). With this information, the four kinetic parameters can again be evaluated.

The main disadvantage of both treatments using initial-velocity measurements is that one cannot discriminate between the random- and compulsory-order ternary-complex mechanisms. Only a separation into either sequential or nonsequential mechanisms can be achieved. Also, initial-velocity studies do not distinguish the substrates A and B in a bireactant sequential mechanism, since the equations retain their forms if A and B are interchanged. Product-inhibition or isotope-exchange studies are necessary for such a distinction. Finally, we emphasize here once again: graphic presentations should be used only for preliminary work and illustrating results. Accurate estimation of the kinetic parameters must be done by some computational procedure.

As a neurobiological example, we may choose the reaction

$$\text{2-Oxoglutaramate + phenylalanine} \rightleftarrows \text{glutamine + phenylpyruvate}$$

Its kinetics are consistent with a Ping-Pong reaction mechanism. The enzyme glutamine transaminase (glutamine: 2-oxo-acid aminotransferase, EC 2.6.1.15) has been purified from rat brain by Van Leuven (1976). Reaction kinetics and kinetic constants were determined by using the double-reciprocal treatment. The reaction velocity was measured as a function of the phenylalanine concentration at fixed concentrations of 2-oxoglutaramate. The primary plot was a set of parallel lines. Comparable lines were obtained when phenylalanine was the fixed substrate. The secondary plots of the intercepts against the reciprocals of the concentration of the second substrate were linear and intersected on the ordinate. With these plots, the nonsequential features of the reaction were confirmed and the kinetic constants estimated. The K_m values were 21.0×10^{-3} mol/liter for 2-oxoglutaramate (K_m^A) and 0.092×10^{-3} mol/liter for phenylalanine (K_m^B).

Also, the kinetic constants of GABA-T, which obeys a nonsequential mechanism, have been determined using the double-reciprocal treatment (Hall and Kravitz, 1967a). The K_m value of GABA (K_m^A) was 0.61×10^{-3} mol/liter and that of 2-oxoglutarate (K_m^B) 0.11×10^{-3} mol/liter. 2-Oxoglutarate substrate inhibits the enzyme at higher concentrations, having a dissociation constant (K_i^B) of about 0.86×10^{-3} mol/liter (see below).

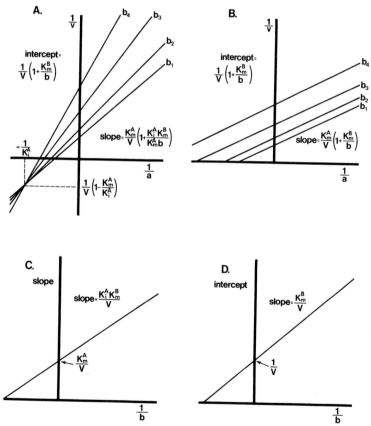

Fig. 15. Double-reciprocal treatment for graphic estimation of the kinetic constants for bireactant enzymes. Primary plots of $1/v$ vs. $1/a$ at various fixed values of b for sequential mechanism (A) and non-sequential mechanism (B). Secondary plots of the slopes (C) and intercepts (D) vs. $1/b$ obtained from the primary plots (A) and (B).

6.3. Substrate-Inhibition Studies

It has been assumed in the foregoing account of initial-velocity studies that the concentrations of the substrates are low. In higher substrate concentrations, at least one of the four reactants may bind itself to a wrong enzyme form. The result is so-called substrate inhibition (as discussed in Section 4.4). Substrate inhibition is not a nuisance, however, since the reaction mechanism of a two-substrate reaction can often be inferred from substrate-inhibition studies.

In the compulsory-order ternary-complex mechanism, substrate inhibition is effective at high concentrations of the substrate A. B is understood as the substrate that inhibits via substrate inhibition. Uncompetitive substrate inhibition is characteristic of ordered systems and occurs when B reacts with EQ to form a dead-end complex BEQ. Primary plots of b/v against b are parabolic with a common point of intersection at $b = -K_i^A K_m^B / K_m^A$. Primary plots of a/v against a are linear without a common intersection.

In the substituted-enzyme mechanism, the dead-end complex is EB. Thus, the inhibition by substrate is most effective when the concentration of A is small. Competitive inhibition is common in these Ping-Pong systems. In the primary plots of a/v vs. a, straight lines without a common intersection point are obtained, but every pair of lines intersects individually at a positive value of a. Primary plots of b/v against b are parabolic, and they all intersect on the b/v axis.

In general, inhibition by a substrate is not harmful in low substrate concentrations, because substrates normally bind themselves better to the correct enzyme form than to the wrong ones. The phenomenon is sometimes a useful additional tool for distinguishing between ordered and Ping-Pong mechanisms because it adds to the observed differences of these mechanisms. It is often very useful to confirm kinetic interpretations by elevating a substrate concentration to an inhibitory level and analyzing the results. Competitive substrate inhibition can be regarded as fairly strong positive evidence for a Ping-Pong mechanism. When substrate inhibition occurs in the ordered mechanism, the identification of the second substrate is possible.

6.4. Product-Inhibition Studies

Product-inhibition studies are the most useful and informative technique for elucidating reaction pathways. A reasonable interpretation of their results requires a great deal of basic kinetic knowledge. The product of the reaction can act as a competitive, uncompetitive, or mixed inhibitor, depending on which substrate is variable. The inhibition is linear if the product can combine with only one form of the enzyme. If the product can also bind to "wrong" enzyme forms producing dead-end complexes, nonlinear inhibition is possible. Competitive inhibition can occur in two ways: (1) the product inhibitor is able on binding to displace the variable substrate or (2) the product and the variable substrate exclude each other. Uncompetitive inhibition ensues from the absence of any reversible pathway between the binding of the substrate and the binding of the product.

As a simple example, we may mention the previously discussed classic Ping-Pong Bi Bi mechanism, which can be elucidated by product-inhibition studies. In this mechanism, A and Q combine with E, while B and P combine with E'. Both these pairs thus provoke competitive product inhibition. The pairs A and P or P and Q must cause pure noncompetitive inhibition. In each case, the inhibition is eliminated by saturating with the nonvariable substrate. This kind of product-inhibition pattern is observed, for instance, with a number of aminotransferases.

6.5. Isotope-Exchange Studies

In the study of two-substrate or, more often, multiple-substrate reactions, the method of isotope exchange is a very valuable tool alongside initial-velocity and product-inhibition studies. This technique is the most powerful in detecting minor alternative pathways undiscovered by other methods and measuring their relative contribution to the overall mechanism. The method of isotope exchange can also be used for confirming a predicted mechanism and checking its kinetic constants that have been obtained by other means.

When the isotope-exchange method is used, one must stipulate that labeled radioactive compounds react in the same manner as their nonradioactive counterparts. Isotope effects are thus assumed to be minimal (cf. Volume 2, Chapter 10). Also, the concentrations of the labeled compounds must be so low as not to affect the concentrations of unlabeled reactants. This is usually easy to arrange through the use of labeled compounds of high specific radioactivity. Very highly purified enzyme preparations are required in isotope-exchange experiments, because this method is much more sensitive to the presence of impurities than conventional kinetic experiments.

In this technique, an atom in the reactant molecule is isotopically labeled and the isotopic exchanges between reactant molecules are measured. Isotope-exchange studies are always carried out in the presence of one or more products and often at chemical equilibrium. The transfer of the label is followed from the substrate to products or from products back to the substrate. The details of the method are rather complicated, and the reader is referred to the textbooks available. One simple example illustrates the possibilities of this technique. The following Ping-Pong Bi Bi mechanism is easy to identify by isotopic exchange:

$$\begin{array}{ccccccc} & A & & P & & B & & Q \\ & \downarrow & & \uparrow & & \downarrow & & \uparrow \\ \hline E & & (AE\ PE') & & E' & & (BE'\ QE) & & E \end{array}$$

The substrate A interacts with the enzyme, forming P even in the absence of the other substrate B. The label in A will appear in P, and the rate of the transfer depends on the concentration of *A* but is independent of *B*. If *B* is labeled, the label cannot appear in *Q* in the absence of *A*. If *B* is permitted to react with the free enzyme to form *Q*, there will be an isotope exchange between *B* and *Q*. By such deductions, the kinetic pattern can be identified.

The isotope-exchange method is used most widely to distinguish the compulsory-order ternary-complex mechanism from the random-order Bi Bi mechanisms. Initial-velocity studies do not differentiate these. The details of Ping-Pong mechanisms have also been investigated by this means.

7. Effects of pH and Temperature on Enzyme Kinetics

7.1. Effects of pH

7.1.1. General Effects

All enzyme-kinetic experiments must be conducted at a carefully controlled pH, because enzymes are extremely sensitive to variations in the environmental pH. Almost all enzymes are active in aqueous solutions at pH values within the range of 5 to 9. Only a few exceptions such as pepsin and alkaline phosphatase are known. Enzymes contain several chemical groups, e.g., the imidazole group of histidine, the sulfhydryl group of cysteine, and the N-terminal amino groups, that are either deprotonated or protonated in the aforementioned pH range. Important information on the nature of ionizable groups in the active center or in its close vicinity can be obtained by investigating the pH dependence of the enzyme reaction and measuring kinetic parameters as a function of pH. Very seldom in neurochemistry has a thorough analysis of pH effects been done; usually, only the optimum pH of the studied enzyme at one substrate concentration has been determined. The analysis of pH effects is the major bridge between kinetic mechanisms and the molecular structure of enzymes. It should be kept in mind, however, that a kinetic study can reveal the presence in the enzyme protein of only those groups the ionization state of which affects the reaction velocity.

The hydrogen ion or proton, H^+, can be considered as a hyperbolic inhibitor or activator of the enzyme in the simplest cases in which only a single acidic or basic group is involved. The proton may often be regarded also as a substrate that displays substrate inhibition in certain concentrations. The type of effect generally depends on the enzyme form combining the proton. In its modifier properties, the proton differs considerably from other modifiers. It is smaller in size than any other chemical substance and therefore exerts no steric effects. Protons may thus cause pure noncompetitive inhibition, which otherwise seldom occurs. Protons are also bound by many sites of the enzyme. At the extremes of pH, the protein structure changes irreversibly and the reaction velocity falls to zero. The proton is the most important modifier because it influences all enzymes. pH also has a profound effect on the ionization of the substrate and its binding to the enzyme. Thus, the whole reactivity in catalysis is influenced by the hydrogen ion. The initial velocity of an enzyme reaction exhibits three phases as a function of pH: at low pH values there is an increase in velocity, at high pH values a decrease is observed, and within an intermediate range, usually around neutrality, the enzymic activity is maximal in the pH optimum. A characteristically bell-shaped curve illustrates this behavior (Fig. 16).

7.1.2. Effects on Kinetic Constants

A general theory for one-substrate reactions implies that the free enzyme and the enzyme–substrate complex have three protonation states in a quasi-equilibrium relative to each other. Only the middle ionization forms are considered functionally active:

$$
\begin{array}{ccc}
H_2S & & H_2ES \\
\updownarrow K_1 & & \updownarrow K_1' \\
S + HE^- & \underset{k_{-1}}{\overset{k_{+1}}{\rightleftharpoons}} HES^- & \overset{k_{+2}}{\longrightarrow} HE^- + P \\
\updownarrow K_2 & & \updownarrow K_2' \\
E^{2-} & & ES^{2-}
\end{array}
$$

At any hydrogen ion concentration h^+, the observed kinetic constants are

$$
V^{\mathrm{app}} = V \Big/ \left(\frac{h^+}{K_1'} + 1 + \frac{K_2'}{h^+} \right) \tag{91}
$$

$$
K_m^{\mathrm{app}} = \left(\frac{h^+}{K_1} + 1 + \frac{K_2}{h^+} \right) K_m \Big/ \left(\frac{h^+}{K_1'} + 1 + \frac{K_2'}{h^+} \right) \tag{92}
$$

$$
V^{\mathrm{app}}/K_m^{\mathrm{app}} = V/K_m \Big/ \left(\frac{h^+}{K_1} + 1 + \frac{K_2}{h^+} \right) \tag{93}
$$

In these expressions, K_1, K_2, K_1', and K_2' are the respective dissociation

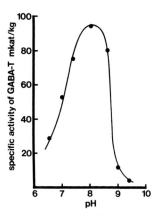

Fig. 16. pH-dependence curve of GABA transaminase (GABA-T) purified from the mouse brain. Redrawn by permission from Wu (1976).

constants indicated in the ionization scheme above. The observed kinetic parameters of an enzyme reaction are thus functions of pH. V^{app} reflects the ionization of the enzyme and K_m^{app} is influenced by both ionizations; V^{app} and K_m^{app} are therefore of fundamental importance in studies of pH effects. In low substrate concentrations, V^{app}/K_m^{app} is the apparent first-order rate constant of the reaction and V^{app} is the apparent zero-order rate constant in high substrate concentrations.

The maximal velocity is proportional to the relative amounts of the complex HES^- [equation (91)]. The dissociation constants K_1' and K_2' can be determined from the bell-shaped curve when V^{app} is plotted as a function of pH (Fig. 17A). The variation of V^{app} reflects the pH-induced changes in the rate-limiting steps, in the protonation or deprotonation of a group in the transitory complex. The effects are usually due to the ionization of the groups in the enzyme molecule responsible for the actual catalysis. Such groups are the imidazole group, the carboxyl group, the hydroxyl group, and the α- and ε-amino groups.

V^{app}/K_m^{app} is proportional to the complex HE^- [equation (93)]. The dissociation constants K_1 and K_2 can be determined from the bell-shaped curve of V^{app}/K_m^{app} vs. pH (Fig. 17B). V^{app}/K_m^{app} reflects the ability of the enzyme and substrate to combine. Thus, the ionizations occurring in the substrate or in the enzyme form that combines with the substrate can be detected on this plot.

The effect of pH on K_m^{app} is usually not profitable to analyze because its behavior is very complicated. K_m^{app} is proportional to both complexes HES^- and HE^- [equation (92)].

As mentioned above, the plots of V^{app} vs. pH and V^{app}/K_m^{app} vs. pH are

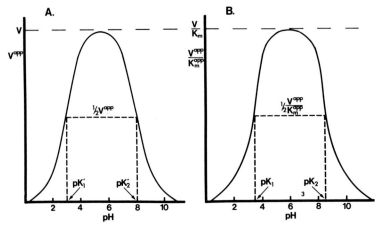

Fig. 17. Dependence of the kinetic constants on pH. Evaluation of pK values for (A) the enzyme–substrate complex (pK_1' and pK_2') and (B) the free enzyme (pK_1 and pK_2) from the V^{app} vs. pH and V^{app}/K_m^{app} vs. pH plots, respectively. The horizontal lines drawn at $V^{app}/2$ and $(V^{app}/K_m^{app})/2$ intersect the curves at those points the abscissae of which indicate the pKs sought.

both bell-shaped (Fig. 17). The reason is that only the complexes HES^- and HE^- are assumed to be active. If ES^{2-} or H_2ES were as active as HES^-, the pH profile for V^{app} would be left or right sigmoid, respectively. If E^{2-} or H_2E were as active as HE^-, the pH profile for V^{app}/K_m^{app} would also be left or right sigmoid. When pK values are separated by at least 3.5 units, the bell-shaped plots of V^{app} vs. pH and V^{app}/K_m^{app} vs. pH will have maxima very close to V and V/K_m, respectively. The pK values can be read directly from the plots as the pH values at $V^{app}/2$ and $(V^{app}/K_m^{app})/2$ (Fig. 17). If pK values are separated by less than 3.5 pH units, the maxima of the plots will be significantly lower than the theoretical maxima, and consequently the pH values of the half-maximum points will not correspond to the correct pK values.

There is yet another way to depict the pH dependence of kinetic constants. This is the so-called Dixon–Webb plot (Dixon and Webb, 1964) of $\log V^{app}$ vs. pH and $\log V^{app}/K_m^{app}$ vs. pH. $\log V^{app}$ is given by the equation:

$$\log V^{app} = \log V - \log\left(1 + \frac{h^+}{K_1'} + \frac{K_2'}{h^+}\right) \tag{94}$$

At low pH values, the term K_2'/h^+ may be ignored, and since the ratio of h^+/K_1' is much larger than 1, the 1 in equation (94) can be ignored as well. The equation is then

$$\log V^{app} = \log V - \log h^+ + \log K_1' \tag{95}$$

or

$$\log V^{app} = \log V + pH - pK_1' \tag{96}$$

or

$$\log \frac{V^{app}}{V} = pH - pK_1' \tag{97}$$

The Dixon–Webb plot is illustrated by Fig. 18. At low pH values, the plot approaches a straight line with a slope of $+1$. At that point on this line where $V^{app} = V$, $\log(V^{app}/V) = 0$ and $pH = pK_1'$. To obtain this pK_1', a horizontal line is drawn at $\log V$ and another line with a slope of $+1$ as a tangent to the curve at low pH. The coordinate of the intersection point on the abscissa is the sought pK_1'. At high pH, the logarithmic form of equation (94) reduces to

$$\log V^{app} = \log V + pK_2' - pH \tag{98}$$

A tangent that has a slope of -1 is then drawn to the curve at high pH. The point at which this line intersects the horizontal line of $\log V$ indicates pK_2'

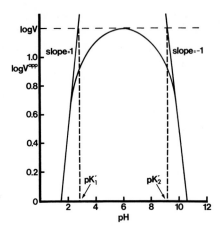

Fig. 18. Dixon–Webb plot of V^{app} vs. pH for estimation of pK_1' and pK_2' for the enzyme–substrate complex. Tangents drawn for the curve within low and high pH ranges intersect the horizontal log V line at those points the abscissae of which indicate pK_1' and pK_2', respectively.

(see Fig. 18). A similar plot of log (V^{app}/K_m^{app}) vs. pH can be used for the detection of pK_1 and pK_2.

The apparent pK values obtained by such pH studies are identical with the true pKs in very simple enzyme reactions, which are assumed to be at either rapid-equilibrium or steady-state conditions. These assumptions do not hold for many enzymes. Thus, the pKs determined from the linear or logarithmic plots must be considered tentative values only, giving no more than an indication of the nature of the reacting protonic groups. Additional information is needed to confirm the results. One should be very careful in drawing conclusions from the studies of pH effects. A correct interpretation of pH effects is the most difficult thing in kinetic deductions, requiring a very thorough knowledge of the kinetic mechanisms and the various steps in the reaction sequence in question.

As an example of recent neurobiological studies of pH effects on enzyme activity, we may mention the paper by Gabay *et al.* (1976). The activity of monoamine oxidase (MAO, amine:oxygen oxidoreductase, deaminating, flavin-containing, EC 1.4.3.4) in the mitochondrial fraction of the bovine cerebral cortex was shown to be markedly affected by pH. The kinetic parameters at three different pH values were measured for various substrates. The results were discussed on the assumption that the effect of pH on the activity of MAO is related to the ionization of the substrate.

7.2. Effects of Temperature

7.2.1. General Effects

The effects of temperature on enzyme reactions are extremely complex. Temperature may affect reaction velocity by influencing, for example, the stability of the enzyme, the rate of breakdown of the enzyme–substrate complex, the affinity of the enzyme for the substrate, or the pH functions of one or all of the reaction components. Temperature may also alter the affinity of the enzyme for various effectors. It is understandable that temperature has scarcely been used as a variable in steady-state enzyme studies on reaction mechanisms. Temperature effects may, however, give valuable information on the energetics of various steps in the reaction and help in identifying protonic groups in the active centers and in determining the specificity of the reaction.

The effects of temperature on reaction velocity are mainly due to two factors: denaturation of the enzyme protein and kinetic effects on reaction rates. Denaturation of the enzyme protein at elevated temperatures reduces the concentration of the active enzyme and thus slows down the reaction. The kinetic effects facilitate the reaction velocity as temperature increases. As an overall effect, the initial velocity usually first increases with temperature and then decreases beyond an apparent optimum temperature. This optimum temperature is not a characteristic feature for an enzyme, however, since it varies greatly, first of all with time, but also with pH and other reaction conditions (Fig. 19).

The action of temperature on reactions is often characterized by the temperature coefficient, Q_{10}. Q_{10} is the factor by which the reaction velocity increases when temperature increases by 10°C. The temperature coefficient for most enzyme reactions lies between 1 and 2. In general, the coefficient is lower for a catalyzed reaction than for the same uncatalyzed reaction.

The temperature instability of a new enzyme is always tested. This is done by exposing the enzyme to various temperatures for a fixed period before measuring its activity at a temperature where the enzyme is stable. The effects on affinity can be eliminated by using concentrations of the substrate high enough to saturate the enzyme. Since the pK values of the buffer components also change with temperature, the pH of each buffer should be measured at each temperature tested.

7.2.2. Thermodynamic Considerations

The theoretical treatment of the temperature dependence of simple chemical reactions (discussed previously) applies equally well to enzyme

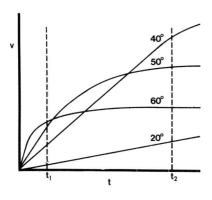

Fig. 19. Progress curves of an enzyme reaction at various temperatures. They show that the optimum temperature varies with the duration of incubation.

reactions. The equilibrium constant K for any chemical reaction depends on temperature according to the van't Hoff equation (34). If, in equation (34), $\Delta H°$ is positive and K increases with temperature, the reaction is endothermic. In an exothermic reaction, $\Delta H°$ is negative and K decreases when temperature increases. The equilibrium constant is also given by the equation:

$$\Delta G° = -RT \ln K \qquad (99)$$

where $\Delta G°$ is the free-energy change between products and reactants in their standard states. $\Delta G°$ consists of two components:

$$\Delta G° = \Delta H° - T\Delta S° \qquad (100)$$

where $\Delta S°$ is the change in entropy, which is the measure of molecular chaos.

The previously discussed Arrhenius equations (35) and (36) characterized temperature effects on the rate constants. Equations (34) and (35) were strikingly similar. Analogously, E in equation (35) may thus be considered an energy difference between the reactants of a reaction step and an activated complex that is an essential intermediate in the reaction. The activated-complex theory of reaction rates states that the reactants and the activated complex are in equilibrium. The rate constant k is proportional to this equilibrium constant $K^{\ddagger}$:

$$k = \frac{RT}{Nh} \times K^{\ddagger} \times \text{transmission coefficient } (\sim 1) \qquad (101)$$

where N is Avogadro's number and h is Planck's constant.

On the basis of equations (99) and (100), $K^{\ddagger}$ can be expressed in terms of $\Delta G^{\ddagger}$, $\Delta H^{\ddagger}$, and $\Delta S^{\ddagger}$, which are now the free-energy, enthalpy, and entropy differences between the reactants and the activated complex:

$$K^{\ddagger} = \exp(-\Delta G^{\ddagger}/RT) = \exp(-\Delta H^{\ddagger}/RT) \exp(\Delta S^{\ddagger}/R) \qquad (102)$$

Inserting equation (102) into equation (101) gives

$$k = \frac{RT}{Nh} \exp(-\Delta G^{\ddagger}/RT) = \frac{RT}{Nh} \exp(-\Delta H^{\ddagger}/RT) \exp(\Delta S^{\ddagger}/R) \qquad (103)$$

and

$$\ln k = \ln \frac{RT}{Nh} - \frac{\Delta G^{\ddagger}}{RT} \qquad (104)$$

and

$$\frac{d \ln k}{dT} = \frac{1}{T} + \frac{\Delta H^{\ddagger}}{RT^2} \tag{105}$$

Furthermore, equations (35) and (105) show that

$$E = RT + \Delta H^{\ddagger} \tag{106}$$

The integration constant in equation (36), the so-called frequency factor C, is, according to equations (100), (104), and (106)

$$C = \frac{RT}{Nh} \exp (\Delta S^{\ddagger}/R) \tag{107}$$

if $\Delta H^{\ddagger}$ is close to E.

Thus, when k is estimated at any temperature, $\Delta G^{\ddagger}$ can be calculated [equation (104)]. If k is measured at a series of different temperatures, it is possible to estimate E and C from the Arrhenius plot (see Fig. 2) of $\ln k$ against $1/T$. Furthermore, $\Delta H^{\ddagger}$ can be calculated by equation (106) and $\Delta S^{\ddagger}$ by equation (100) or (107). The $\Delta G^{\ddagger}$ profile for an enzyme mechanism is obtained by calculating $\Delta G^{\ddagger}$ for all its steps. Equation (104) can be rearranged for this purpose:

$$\Delta G^{\ddagger} = RT \left(\ln \frac{RT}{Nh} - \ln k \right) \tag{108}$$

Studies made of the dependence of the kinetic parameters on temperature have been of little practical importance. If K_m is the true dissociation constant, its temperature dependence can give useful thermodynamic information on the enzyme. The temperature dependence of V is useful only if it is known which step in the mechanism it refers to.

The change in the equilibrium constant K with temperature can be used in connection with pH studies to identify protonic groups in the active center of the enzyme. For this purpose, equations (99) and (100) are combined:

$$-\log K = \frac{\Delta H^{\circ}}{2.303R} \frac{1}{T} - \frac{\Delta S^{\circ}}{2.303R} \tag{109}$$

When $-\log K$ (pK) is plotted against $1/T$, a straight line with a slope of $\Delta H^{\circ}/2.303R$ is obtained. The experimental procedure consists of measurements of the apparent V at various pH at different temperatures. The plots of V^{app} vs. pH are drawn for several temperatures. From such plots (see Fig. 17A), the pK values can be estimated, as discussed in Section 7.1.2. The observed pK values are then replotted against $1/T$ according to equation (109). The standard enthalpy of ionization is obtained from the slope. This value can then be compared to the tabulated enthalpy values.

8. Certain Neurobiological Applications

8.1. Transport of Organic Solutes across Nervous Membranes

Principles of enzyme kinetics have also been widely applied to describe mediated transfer of solutes across cell membranes. The current concepts of transport mechanisms have been recently reviewed, for instance, by Christensen (1975). Analysis of experimental transport data has been critically discussed by Neame and Richards (1972) in general, and by Oja and Vahvelainen (1975) with special emphasis on neurobiologal aspects. Only a brief account is given in this chapter of the kinetics of transport phenomena. The reader is referred to the aforementioned reviews for a more thorough discussion of transport kinetics and for further illustrative neurobiological examples.

The carrier-mediated transport of solutes has many features common with enzyme action. Within low ranges of concentration, the rate of transport is directly proportional to the concentration of the solute. At higher concentrations, the transport becomes saturable until the rate no longer increases with the solute concentration. The carrier-mediated transport thus phenomenologically conforms to Michaelis–Menten kinetics. This behavior is commonly explained by assuming a limited number of specialized sites on the cell membranes, called *carriers*. The carrier corresponds to the enzyme and the solute transported to the substrate transformed. The formation of an enzyme–substrate complex is equivalent to the combination of the carrier with the solute. The conversion of the enzyme–substrate complex to reaction products and free enzyme is analogous to the transfer of the solute across the membrane with consequent unloading of the carrier. Thus, the velocity of unidirectional transport of a solute mediated by one single species of carrier is also given by the Michaelis–Menten equation (42).

In this case, s is the concentration of the solute, V the maximum velocity of the transport, and K_m the transport constant equivalent to the Michaelis constant. The same kinetic equation for carrier-mediated transport can be derived theoretically from the adsorption equilibrium of carrier, solute, and solute-loaded carrier and from the forward transfer of the loaded carrier across the cell membrane (Neame and Richards, 1972). It should be borne in mind, however, that the kinetic equation is also based on certain simplifying assumptions in mediated transport; e.g., the solution should initially be on only one side of the membrane, or transport should be in one direction only; equal numbers of identical carrier sites move across the membrane in each direction at equal rates; the carrier sites come into equilibrium with the solute before moving to the opposite surface; and other assumptions.

If the carrier-mediated transport occurs against a concentration gradient requiring metabolic energy, it is called *"active transport"* or *"concentrative transport"*. Carrier-mediated transport may also operate down the concentration gradient, resulting in equalization of the concentrations of the solute on both sides of the membrane. This phenomenon is called *"equalizing transport"* or *"passive mediated transport."* It also has the common misnomer *"facilitated diffusion,"* although it is a mediated phenomenon.

Transport operates naturally two-way across membranes. In most cases, only the unidirectional initial influx into tissue spaces has been considered; the reverse efflux processes—probably at least partially carrier-mediated as well—are less characterized. The majority of studies have been made *in vitro*, although analogous kinetic formalizations are applicable *in vivo* as well (e.g., Neame and Richards, 1972; Oja, 1974). There probably also occurs some nonsaturable transmembrane penetration of solutes, i.e., diffusion, concomitantly with carrier-mediated saturable transport. The total unidirectional transfer is then given by the following expression:

$$v = \frac{Vs}{K_m + s} + K_d s \tag{110}$$

in which K_d is the transmembrane diffusion constant. Due to the diffusion term in equation (110), the rate of total transport increases indefinitely with increasing solute concentration (Fig. 20). After subtraction of the contribution of the nonsaturable component from the total transport (Oja and Vahvelainen, 1975), the transport constant and the maximum velocity of transport are amenable for estimation in the same manner as the corresponding constants in enzyme kinetics.

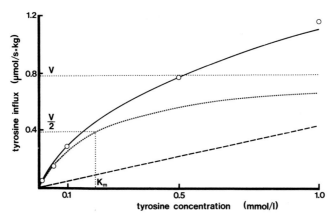

Fig. 20. Unidirectional influx of tyrosine into cerebral cortex slices of 7-day-old rats as a function of the tyrosine concentration in the medium. The total influx (○—○) was divided into two components; one saturable conforming to Michaelis–Menten kinetics (······) and another nonsaturable (---). The maximum velocity of transport $V = 0.78$ μmol $\times$ sec^{-1} $\times$ kg^{-1} incubated tissue, the transport constant $K_m = 0.20$ mmol/liter, and —supposing the volume of 1 kg tissue to be 1 liter—the diffusion constant $K_d = 0.43 \times 10^{-3}$ sec^{-1} [equation (110)]. Redrawn from Vahvelainen and Oja (1972).

A number of solutes penetrate into nervous tissue preparations with the aid of at least two separate carrier systems. If it is assumed that the two carriers act in parallel for one solute, the observed velocity of the transport is the sum of the action of both saturable components and possible nonsaturable diffusion (Fig. 21A,B):

$$v = \frac{V_1 s}{K_{m_1} + s} + \frac{V_2 s}{K_{m_2} + s} + K_d s \tag{111}$$

in which the subsubscripts "1" and "2" refer to the two separate carrier

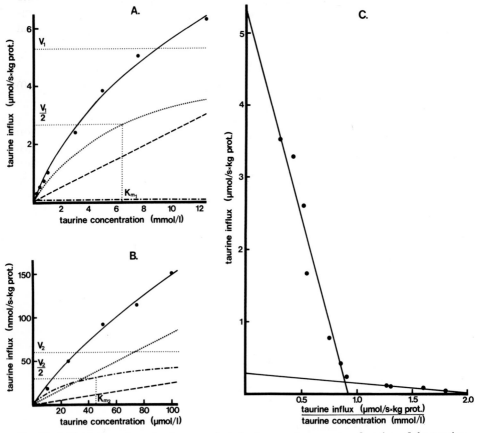

Fig. 21. Initial influx of taurine into rat whole brain synaptosomes as a function of the taurine concentration in the incubation medium. (A) Total influx (●—●) divided into three components: diffusion (– – –) and high-affinity (–·–·–) and low-affinity (· · · · · · ·) saturable influxes, according to Lähdesmäki *et al.* (1975). (B) Lower concentration range of (A) shown on an expanded scale. (C) Saturable influx of taurine depicted in the plot v vs. v/s suggesting the existence of two separate transport systems. The estimated transport parameters [equation (111)] are $V_1 = 5.25$ μmol × sec^{-1} × kg^{-1} synaptosomal protein, $V_2 = 60$ nmol × sec^{-1} × kg^{-1}, $K_{m_1} = 6.3$ mmol/liter, $K_{m_2} = 46$ μmol/liter, and $K_d = 0.23 \times 10^{-3}$ sec^{-1}. Redrawn from Kontro and Oja (1978).

systems. One saturable component is often characterized by very small K_m and V (*"high-affinity transport"*) and the other by large K_m and V (*"low-affinity transport"*). In particular, the influx of neurotransmitters is assumed to exhibit such dualism. The existence of more than one carrier-mediated transport is to be suspected when the transport rate observed does not fit the single-carrier hypothesis, although such behavior is subject to a number of other interpretations as well. A fitting of the transport data to linear transformations of the Michaelis–Menten equation is illustrative. The graphic presentation is a kinked line when there is more than one carrier involved (Fig. 21C). The best plot for showing deviations from linearity is the plot of v vs. v/s, as discussed in Section 3.3.2.

In a two-carrier case, the estimates for transport parameters must not be read directly from the linear plots. The literature is full of erroneous estimates obtained in this way. The contribution of the high-affinity transport must first be arbitrarily estimated and subtracted from the total transport (any possible diffusion has been eliminated beforehand). The remaining transport is assumed to represent the low-affinity component. Its parameters can now be estimated. The parameters for both components can be iteratively improved by repeating the estimation after the contribution of each transport component in turn has been subtracted from the total transport by using its arbitrarily estimated parameters. This kind of division of the saturable transport components is successful only if the kinetic constants differ widely enough in magnitude (Neame and Richards, 1972; Oja and Vahvelainen, 1975).

The specificity of carriers may not be sharply demarcated. If several solutes are transported by the same carrier, they inhibit the transport of each other in a manner comparable to competitive inhibition. Such competitive inhibition gives valuable information as to the specificity of the transport system. Analysis of inhibition in transport systems is carried out following the principles of enzyme inhibition. In transport studies, K_i is a constant characterizing the ratio between the rate constants of the adsorption and desorption reactions of the inhibiting solute with the carrier. Activation also occurs in carrier-mediated transport. In the presence of an activator, the transport velocity of the solute increases due to an increment either in the affinity between the solute and the carrier or in the number of carrier sites available. The most common activators are metal ions, e.g., Na^+, K^+, Ca^{2+}, and Mg^{2+}. Activation of transport is analyzed similarly to inhibition, including the possibility that the activator may be an allosteric modifier. For instance, the ion dependence of neurotransmitter uptake has been elucidated in this fashion (e.g., Martin and Smith, 1972; Martin, 1973).

Inhibition and activation analyses of transport become very complicated if the effects on both inward and outward fluxes must be taken into account. The effector may also be itself transported. In multiple-carrier systems, only one of the carriers or again all of them may be affected. The velocity of transport can be affected by another solute present on either the *cis* or the *trans* side of the membrane. *Cis-inhibition* is in question when an effector present on the same membrane side inhibits the transport. *Cis-stimulation* signifies an enhancement of transport under the same premises. In *trans-inhibition*, unidirectional transport is impeded by an effector on the other side of the membrane, while an enhancement of the transport in this case signifies *trans-stimulation*. The use of the misnomer "*exchange diffusion*" for this last-mentioned variant of mediated transport should be discontinued.

8.2. Binding of Ligands to Receptors

The progress of enzyme kinetics has profoundly influenced the concept of ligand–receptor or drug–receptor interaction. The ligand (agonist or antagonist) is assumed to become bound to the membrane receptor to form a *ligand–receptor complex* in a fashion analogous to the formation of an enzyme–substrate complex. The binding of the agonist or antagonist and its physiological effect are discrete but sequential processes independently amenable to study. The association of the ligand to the receptor is a bimolecular event:

$$R + A \underset{k_{-1}}{\overset{k_{+1}}{\rightleftharpoons}} RA$$

Here, R denotes the free receptor, A the free ligand, RA the ligand–receptor complex, and k_{+1} and k_{-1} are the rate constants of association and dissociation, respectively.

The total receptor concentration is r, the concentration of the complex x, and that of the free receptor $(r - x)$. At equilibrium

$$k_{+1}a(r - x) = k_{-1}x \tag{112}$$

Analogously to the Michaelis—Menten treatment, the concentration of the ligand–receptor complex at equilibrium can be expressed by the equation:

$$x = \frac{ar}{a + K_D} \tag{113}$$

in which the *dissociation constant*, K_D, is defined as the ratio k_{-1}/k_{+1}. The ratio k_{+1}/k_{-1} is often called the *affinity constant*, K_{aff}. The ratio x/r is designated $\bar{Y}_s$, and it represents the fractional saturation of receptors by the ligand:

$$\bar{Y}_s = \frac{a}{a + K_D} \tag{114}$$

The plot of $\bar{Y}_s$ against a is a hyperbolic curve.

The biochemical properties of membrane receptors for peptide hormones and neurotransmitters, for instance, have been studied by applying this simple concept. A general policy has been to measure the binding of a radioactive compound (agonist or antagonist) to intact cells or isolated membrane fragments. The interaction between the ligand and the receptor must be "specific," i.e., "true" binding to the receptor. The *specific binding* is considered strictly structurally specific, tissue-specific, reversible, and saturable. The saturability of the process indicates a finite number of binding sites. The ligand should also have a high affinity for the receptor.

In practice, the kinetics of the binding of neurotransmitters, for example, to synaptic membrane fractions have been investigated as follows:

1. The specific binding has been measured as a function of the ligand concentration to indicate the saturability of the process. The data have been further analyzed with Scatchard plots often corresponding to the equation

$$\frac{a_b}{a_f} = -\frac{1}{K_D} a_b + \frac{nr}{K_D} \tag{115}$$

in which a_b is the concentration of the bound ligand per mole of protein, a_f the concentration of the free ligand, K_D the dissociation constant of a receptor site, n the number of independent ligand binding sites per protein molecule, and thence nr is the total concentration of ligand-binding sites. In most studies on initial velocities *in vitro*, it is safe to assume the concentration of the free ligand to be the same as the total ligand concentration, even if this is not true in the equilibrium-binding studies, when a relatively large proportion of the added ligand is bound. A plot of the ratio of the bound ligand to the free ligand against the concentration of the bound ligand is thus linear in the majority of cases. The slope of the straight line is $-1/K_D$, and the intercept on the vertical axis gives nr/K_D and on the horizontal axis nr, which indicates the total receptor amount. A linear Scatchard plot signifies a single population of receptor sites, while concave inflection indicates "second-order" binding sites, negative cooperativity between receptors, self-association of the ligands, and so on.

2. The association and dissociation of the ligand can also be measured as functions of time. From these experiments, the rate constants and the dissociation constant can be estimated.

3. The properties of the ligand-binding can be further analyzed by using the previously mentioned Hill equation (76).

The postsynaptic receptor site for glycine has been characterized in the manner described above by Snyder and his collaborators (Young and Snyder, 1974*a,b*). Glycine is a postulated inhibitory neurotransmitter in spinal cord synapses specifically antagonized by the convulsant drug strychnine. The properties of the glycine receptor have been studied by measuring the binding of [³H]strychnine to isolated spinal cord membranes. Strychnine is used instead of glycine because strychnine has a high affinity for postsynaptic glycine receptors but none for presynaptic glycine carriers. The total binding of strychnine consists of two components: specific and nonspecific binding. The specific binding is obtained by subtracting, from the total, bound radioactivity that is not displaceable by high glycine concentrations. The specific binding of strychnine is saturable, obeying Michaelis–Menten kinetics with a half-maximal binding at a strychnine concentration of 2.7×10^{-9} mol/liter. In these experiments, the data were further analyzed with Scatchard plots (Fig. 22). A straight line indicated a single population of receptor sites. The total concentration of strychnine-binding sites obtained from the intercept on the abscissa was about 1.7 μmol/kg synaptic membrane protein. The dissociation constant calculated from the slope was 2.6×10^{-9} mol/liter.

The reaction kinetics of the specific strychnine binding were also assayed. Half-maximal binding was attained in about 34 sec and maximal binding in 5 min. The association was assumed to obey bimolecular second-order rate kinetics. The second-order rate constant of association, 1.0×10^{-7} mol^{-1}sec^{-1}, was worked out from the initial association velocity, which was assumed to equal $k_{+1}ar$ [cf. equation (112)].

Dissociation of the strychnine–receptor complex obeys first-order kinetics. The rate constant of dissociation of the complex was 1.54×10^{-2} sec^{-1} [equation (17)] and thence its half-life about 45 sec. The dissociation constant, K_D, of the strychnine–receptor complex is then $k_{-1}/k_{+1} = 1.54 \times 10^{-9}$ mol/liter, matching relatively well with the estimate obtained from the Scatchard plot. Nonradioactive strychnine and glycine displace the specifically bound [³H]strychnine in a saturable manner. Straight lines intersecting on the ordinate in the double-reciprocal plot (the reciprocal of specific [³H]strychnine binding against the reciprocal of [³H]strychnine concentration) indicated competitive displacement by nonradioactive strychnine, whereas the plot in the presence of glycine was not linear but concave upward, suggesting that the effect of glycine is cooperative with strychnine. This assumption is supported by the Hill plots of log $[v/(V - v)]$ against the

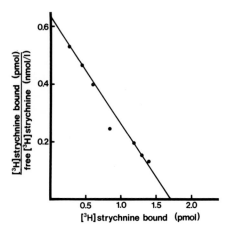

Fig. 22. Specific binding of strychnine to isolated rat spinal cord membranes with increasing concentration of [³H]strychnine. Synaptic membrane suspensions (1 mg membrane protein per tube) were incubated with [³H]strychnine for 10 min at 4°C and the amount of [³H]strychnine specifically bound to membrane protein was assayed. The results are plotted according to the method of Scatchard for evaluation of the dissociation constant of the ligand–receptor complex, $K_D = 2.6 \times 10^{-9}$ sec^{-1}. Redrawn by permission from Young and Snyder (1974a).

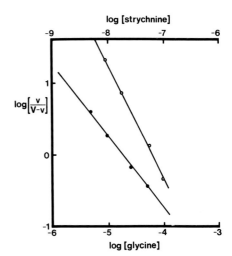

Fig. 23. Hill plot of displacement of specific [³H]strychnine binding by nonradioactive glycine and strychnine. The Hill coefficient of strychnine displacement by glycine (○—○) is 1.7 and by strychnine (●—●) 1.0. Concentrations are given in mol/liter. Redrawn by permission from Young and Snyder (1974a).

logarithm of the concentration of glycine (V represents the specific strychnine binding and v the remaining binding in the presence of nonradioactive strychnine or glycine). Both plots are linear (Fig. 23). The Hill coefficient (the slope of the line) for strychnine is about 1.0 and for glycine 1.7. The latter indicates that glycine displaces the strychnine-binding in a positive cooperative fashion. The interaction of some anions (Cl^-, Br^-, I^-) with the strychnine-binding sites was similarly analyzed with Hill plots, which indicated that these ions inhibit strychnine-binding in a positive cooperative fashion.

9. Symbols and Their Definitions

A, B, S	Free reactants, substrates, or ligands
a, b, s	Concentrations of A, B, or S
$a_0, a_1(0),$ $a_2(0), b_0, s_0$	Initial concentrations of A, B, or S
a_b	Concentration of the bound ligand
a_f	Concentration of the free ligand
C	Integration constant (Arrhenius frequency factor)
E	Free enzyme, activation energy
e	Total concentration of the enzyme
EI	Enzyme–inhibitor complex
EIS	Enzyme–inhibitor–substrate complex
ES	Enzyme–substrate complex
ΔG°	Free-energy change
$\Delta G^\ddagger$	Free-energy difference between the reactants and the activated complex

ΔH°	Standard enthalpy change
$\Delta H^\ddagger$	Enthalpy difference between the reactants and the activated complex
h	Planck's constant, Hill coefficient, number of ligands that a polymeric protein can bind
h^+	Hydrogen ion concentration
I	Free inhibitor
i	Inhibitor concentration
K	Equilibrium constant
K_1, K_2	Dissociation constants of the free enzyme
K_1', K_2'	Dissociation constants of the enzyme–substrate complex
K_{aff}	Affinity constant
K_D	Dissociation constant of the ligand–receptor complex
K_d	Transmembrane diffusion constant
K_h	Association constant
K_i	Dissociation constant of the enzyme–inhibitor complex EI (inhibitor constant)
K_i'	Dissociation constant of the enzyme–inhibitor–substrate complex EIS
K_i^A	Limiting Michaelis constant of the substrate A when the concentration of the other substrate B is zero
K_m	Michaelis constant, transport constant
K_m^A, K_m^B	Limiting Michaelis constant of the substrate A or B, when the other substrate B or A, respectively, is saturating
K_m^{app}	Apparent Michaelis constant
K_m^P	Michaelis constant of the product P
K_s	Dissociation constant of the enzyme–substrate complex ES (substrate constant)
K_{SI}	Dissociation constant of the enzyme–substrate complex SES
k	Rate constant of the reaction
k_{+1}, k_{+2}	Rate constants of forward reactions
k_{-1}, k_{-2}	Rate constants of reverse reactions
N	Avogadro's number
n	Order of the reaction, number of ligand-binding sites per molecule
P, Q	Products of the reaction
p, p_i, p_i'	Concentrations of the product of the reaction
p_∞	Equilibrium concentration of the product
Q_{10}	Temperature coefficient
R	Gas constant, free receptor
R_x	Cooperativity index
r	Total concentration of the receptor
RA	Ligand–receptor complex
$\Delta S^\ddagger$	Entropy difference between the reactants and the activated complex
s_{50}	Substrate concentration that maintains 50% of the maximum reaction velocity

T	Absolute temperature
t	Time
τ	Time interval
U	Enzyme unit
V	Maximum velocity of the enzyme reaction, maximum velocity of transport
V^{app}	Apparent maximum velocity
V^f	Maximum velocity of the forward reaction
V^r	Maximum velocity of the reverse reaction
v	Reaction velocity
v_i	Reaction velocity in the presence of the inhibitor
X	Reaction intermediate
x	Concentration of the enzyme–substrate complex, the ligand–receptor complex, or the reaction intermediate
$\bar{Y}_s$	Fractional saturation (saturation function)

References

Arrhenius, S., 1889, Über die Reaktionsgeschwindigkeit bei der Inversion von Rohr-Zucker durch Säuren, *Z. Phys. Chem.* **4**:226.

Atkins, G.L., 1971*a*, A versatile digital computer program for non-linear regression analysis, *Biochim. Biophys. Acta* **252**:405.

Atkins, G.L., 1971*b*, Some applications of a digital computer program to estimate biological parameters by non-linear regression analysis, *Biochim. Biophys. Acta* **252**:421.

Barber, H.E., Welch, B.L., and Mackay, D., 1967, The use of the logarithmic transformation in the calculation of the transport parameters of a system that obeys Michaelis–Menten kinetics, *Biochem. J.* **103**:251.

Bliss, C.J., and James, A.T., 1966, Fitting the rectangular hyperbola, *Biometrics* **22**:573.

Bridgers, W.F., 1969, Purification of mouse brain phosphoserine phosphohydrolase and phosphotransferase, *Arch. Biochem. Biophys.* **133**:201.

Briggs, G.E., and Haldane, J.B.S., 1925, A note on the kinetics of enzyme action, *Biochem. J.* **19**:338.

Brown, A.J., 1902, Enzyme action, *J. Chem. Soc.* **81**:373.

Christensen, H.N., 1975, *Biological Transport*, Benjamin, New York.

Cleland, W.W., 1963, The kinetics of enzyme-catalyzed reactions with two or more substrates or products. I. Nomenclature and rate equations, *Biochim. Biophys. Acta* **67**:104.

Cleland, W.W., 1970, Steady state kinetics, in: *The Enzymes*, Vol. II (P.D. Boyer, ed.), pp. 1–65, Academic Press, New York.

Cornish-Bowden, A., 1976, *Principles of Enzyme Kinetics*, Butterworths, London.

Dixon, M., 1953, The determination of enzyme inhibitor constants, *Biochem. J.* **55**:170.

Dixon, M., and Webb, E.C., 1964, *Enzymes*, Longmans, London.

Dowd, J.E., and Riggs, D.S., 1965, A comparison of estimates of Michaelis–Menten kinetic constants from various linear transformations, *J. Biol. Chem.* **240**:863.

Eadie, G.S., 1942, The inhibition of cholinesterase by physostigmine and prostigmine, *J. Biol. Chem.* **146**:85.

Eisenthal, R., and Cornish-Bowden, A., 1974, The direct linear plot: A new graphical procedure for estimating enzyme kinetic parameters, *Biochem. J.* **139**:715.

Endrenyi, L., and Kwong, F.H.F., 1973, Some problems of estimating macromolecule-ligand binding parameters, *Acta Biol. Med. Ger.* **31**:495.

Enzyme Committee, 1972, *Enzyme Nomenclature*, Elsevier, Amsterdam.

Gabay, S., Achee, F.M., and Mentes, G., 1976, Some parameters affecting the activity of monoamine oxidase in purified bovine brain mitochondria, *J. Neurochem.* **27**:415.

Gache, C., Rossi, B., and Lazdunski, M., 1976, (Na$^+$, K$^+$)-activated adenosinetriphosphatase of axonal membranes: Cooperativity and control. Steady-state analysis, *Eur. J. Biochem.* **65**:293.

Guggenheim, E.A., 1926, On the determination of the velocity constant of a unimolecular reaction, *Philos. Mag. Ser. VII* **2**:538.

Haldane, J.B.S., and Stern, K.G., 1932, *Allgemeine Chemie der Enzyme*, pp. 119–120, Steinkopf, Dresden and Leipzig.

Hall, Z.W., and Kravitz, E.A., 1967*a*, The metabolism of γ-aminobutyric acid (GABA) in the lobster nervous system—I. GABA-glutamate transaminase, *J. Neurochem.* **14**:45.

Hall, Z.W., and Kravitz, E.A., 1967*b*, The metabolism of γ-aminobutyric acid (GABA) in the lobster nervous system—II. Succinic semialdehyde dehydrogenase, *J. Neurochem.* **14**:55.

Hanes, C.S., 1932, Studies of plant amylases. 1. The effect of starch concentration upon the velocity of hydrolysis by the amylase of germinated barley, *Biochem. J.* **26**:1406.

Hanson, K.R., Ling, R., and Havir, E., 1967, A computer program for fitting data to the Michaelis–Menten equation, *Biochem. Biophys. Res. Commun.* **29**:194.

Henri, V., 1902, Théorié générale de l'action de quelques diastases, *C. R. Acad. Sci.* **135**:916.

Henri, V., 1903, *Lois Générales de l'Action des Diastases*, Hermann, Paris.

Hill, A.V., 1910, The possible effects of the aggregation of the molecules of haemoglobin on its dissociation curves, *J. Physiol. (London)* **40**:IV.

Hitchcock, D.J., 1926, The formal identity of Langmuir's adsorption equation with the law of mass action, *J. Am. Chem. Soc.* **48**:2870.

Hoare, D.G., 1972, The temperature dependence of transport of L-leucine in human erythrocytes, *J. Physiol. (London)* **221**:311.

Hofstee, B.H.J., 1959, Non-inverted versus inverted plots in enzyme kinetics, *Nature (London)* **184**:1296.

Hunter, A., and Downs, C.E., 1945, The inhibition of arginase by amino acids, *J. Biol. Chem.* **157**:427.

Kammeraat, C., and Veldstra, H., 1968, Characterization of succinate semialdehyde dehydrogenase from rat brain, *Biochim. Biophys. Acta* **151**:1.

Kézdy, F.J., Jaz, J., and Bruylants, A., 1958, Cinétique de l'action de l'acide nitreux sur les amides. I. Méthode générale, *Bull. Soc. Chim. Belg.* **67**:687.

King, E.L., and Altman, C., 1956, A schematic method of deriving the rate laws for enzyme-catalyzed reactions, *J. Phys. Chem.* **60**:1375.

Kontro, P., and Oja, S. S., 1978, Taurine uptake by rat brain synaptosomes, *J. Neurochem.* **30**:1297.

Koshland, D.E., Jr., 1958, Application of a theory of enzyme specificity to protein synthesis, *Proc. Natl. Acad. Sci. U.S.A.* **44**:98.

Koshland, D.E., Jr., 1959, Enzyme flexibility and enzyme action, *J. Cell. Comp. Physiol.* **54**(Suppl.):245.

Koshland, D.E., Jr., 1970, Molecular basis for enzyme regulation, in: *The Enzymes*, Vol. I (P.D. Boyer, ed.), pp. 342–369, Academic Press, New York.

Koshland, D.E., Jr., Némethy, G., and Filmec, D., 1966, Comparison of experimental finding data and theoretical models in proteins containing subunits, *Biochemistry* **5**:365.

Lähdesmäki, P., Pasula, M., and Oja, S.S., 1975, Effect of electrical stimulation and chlorpromazine on the uptake and release of taurine, γ-aminobutyric acid (GABA) and glutamic acid in mouse brain synaptosomes, *J. Neurochem.* **25**:675.

Laidler, K.J., and Bunting, P.S., 1973, *The Chemical Kinetics of Enzyme Action*, Clarendon Press, Oxford.

Langmuir, J., 1916, The constitution and fundamental properties of solids and liquids. Part I. Solids, *J. Am. Chem. Soc.* **38**:2221.

Langmuir, J., 1918, The adsorption of gases on plane surfaces of glass, mica and platinum, *J. Am. Chem. Soc.* **40**:427.

Lineweaver, H., and Burk, D., 1934, The determination of enzyme dissociation constants, *J. Am. Chem. Soc.* **56**:658.

Martin, D.L., 1973, Kinetics of the sodium-dependent transport of gamma-aminobutyric acid by synaptosomes, *J. Neurochem.* **21**:345.

Martin, D.L., and Smith, A.A., III, 1972, Ions and the transport of gamma-aminobutyric acid by synaptosomes, *J. Neurochem.* **19**:841.

Michaelis, L., and Menten, M.L., 1913, Die Kinetik der Invertinwirkung, *Biochem. Z.* **49**:333.

Monod, J., Wyman, J., and Changeux, J.-P., 1965, On the nature of allosteric transitions: A plausible model, *J. Mol. Biol.* **12**:88.

Neame, K.D., and Richards, T.G., 1972, *Elementary Kinetics of Membrane Carrier Transport,* Blackwell, Oxford.

Oja, S.S., 1968, Activity of aromatic aminotransferases in rat brain, *Ann. Med. Exp. Fenn.* **46**:541.

Oja, S.S., 1974, Determination of transport rates *in vivo,* in: *Research Methods in Neurochemistry,* Vol. 2 (N. Marks and R. Rodnight, eds.) pp. 183–216, Plenum Press, New York.

Oja, S.S., and Vahvelainen, M.-L., 1975, Transport of amino acids in brain slices, in: *Research Methods in Neurochemistry,* Vol. 3 (N. Marks and R. Rodnight, eds.), pp. 67–137, Plenum Press, New York.

O'Sullivan, C., and Thompson, F.W., 1890, Method of estimating and recording the activity of preparation of invertase, *J. Chem. Soc.* **57**:865.

Paumgartner, G., Huber, J., and Grabner, G., 1969, Kinetik der hepatischen Farbstoffaufnahme von Indocyaningrün: Einfluss von Bilirubin und Natriumglykocholat, *Experientia* **25**:1219.

Plowman, K.M., 1972, *Enzyme Kinetics,* McGraw-Hill, New York.

Potter, L.T., 1972, Synthesis, storage and release of acetylcholine from nerve terminals, in: *The Structure and Function of Nervous Tissue,* Vol. IV (G.H. Bourne, ed.), pp. 105–128, Academic Press, New York.

Scatchard, G., 1949, The attractions of proteins for small molecules and ions, *Ann. N. Y. Acad. Sci.* **51**:660.

Segel, I.H., 1975, *Enzyme Kinetics: Behavior and Analysis of Rapid Equilibrium and Steady-State Enzyme Systems,* Wiley, New York.

Swinbourne, E.S., 1960, Method for obtaining the rate coefficient and final concentration of a first order reaction, *J. Chem. Soc.* 2371.

Swoboda, P.A.T., 1957, The kinetics of enzyme action, *Biochim. Biophys. Acta* **23**:70.

Taketa, K., and Pogell, B.M., 1965, Allosteric inhibition of rat liver fructose 1,6-diphosphatase by adenosine 5'-monophosphate, *J. Biol. Chem.* **240**:651.

Vahvelainen, M.-L., and Oja, S.S., 1972, Kinetics of influx of phenylalanine, tyrosine, tryptophan, histidine and leucine into slices of brain cortex from adult and 7-day-old rats, *Brain Res.* **40**:477.

Van Leuven, F., 1976, Glutamine transaminase from brain tissue: Further studies on kinetic properties and specificity of the enzyme, *Eur. J. Biochem.* **65**:271.

Van Slyke, D.D., and Cullen, G.E., 1914, The mode of action of urease and of enzymes in general, *J. Biol. Chem.* **19**:141.

Van't Hoff, J.H., 1884, *Études de Dynamique Chimique,* pp. 114–118, Muller, Amsterdam.

Wilkinson, G.N., 1961, Statistical estimations in enzyme kinetics, *Biochem. J.* **80**:324.

Wong, J.T.-F., 1975, *Kinetics of Enzyme Mechanisms,* Academic Press, London.

Woolf, B., 1932, cited by Haldane and Stern (1932).

Wu, J.-Y., 1976, Purification, characterization and kinetic studies of GAD and GABA-T from mouse brain, in: *GABA in Nervous System Function* (E. Roberts, T.N. Chase, and D.B. Tower, eds.), pp. 7–55, Raven Press, New York.

Wurtz, A., 1880, Sur la papaine: Nouvelle contribution à l'histoire des ferments solubles, *C. R. Acad. Sci.* **91**:787.

Young, A.B., and Snyder, S.H., 1974*a,* Strychnine binding in rat spinal cord membranes associated with the synaptic glycine receptor: Cooperativity of glycine interactions, *Mol. Pharmacol.* **10**:790.

Young, A.B., and Snyder, S.H., 1974*b,* Glycine synaptic receptor: Evidence that strychnine binding is associated with the ionic conductance mechanism, *Proc. Natl. Acad. Sci. U.S.A.* **71**:4002.

Zydowo, M., Kaletha, K., and Dudek, A., 1971, Computer statistical analysis of the Michaelis constant estimations, *Acta Biochim. Pol.* **18**:367.

6

Spectrophotometry and Fluorometry

Robert Lahue

1. Introduction

Before going into detail, it may be advisable to provide an overview of the principles that underlie the techniques to be discussed in this chapter. The most important physical principles involve the relationships between the wave and particle properties of both light and matter and the interaction of light with matter. Some understanding of certain concepts of quantum physics is fundamental to an understanding of the theory of the techniques described below, although it clearly is not necessary for the routine use of these techniques.

Basically, both light and matter can be conceived as consisting of energy states that are discretely, rather than continuously, distributed. These energy levels can be directly correlated with the frequency of either light particles (photons) or material particles (e.g., electrons). Absorption of radiant energy by matter is qualitatively dependent on the energy states of the matter (i.e., its atomic and molecular constitution) and the energy states, or frequency, of the incident radiation. When conditions are appropriate, matter will absorb light energy, and this absorption results in an alteration in the energy states of the material particles. The selectivity of matter in absorbing light of particular wavelengths provides a powerful tool for the qualitative identification of an unknown material. Quantitative measures of the concentration of a particular constituent of a sample are also possible, and this forms the basis for the techniques to be described below. The altered energy states of matter that has absorbed radiant energy are extremely unstable and therefore transient. There is a strong tendency for the energy states to revert to their original or ground distribution. In this process, some of the energy of the

Robert Lahue • Department of Psychology, Renison College, University of Waterloo, Waterloo, Ontario, Canada.

material particles may be emitted in the form of radiant energy. Analysis of the emitted energy also provides a qualitative tool for atomic and molecular identification. However, in general, such emission techniques will not be discussed. The exception to this will be a discussion of those instances in which the emitted energy arises from fluorescence, because analysis based on this phenomenon can provide one of the most sensitive quantitative measures.

Both theoretical and practical aspects of the various techniques will be discussed. However, attention will first be directed toward the theoretical background for such techniques.

1.1. Spectrum of Electromagnetic Radiation

The electromagnetic spectrum is the series of various forms of electromagnetic radiation ordered according to wavelength. As such, it spans a range of wavelength from 10^5 m and longer (radio waves) to those as short as 10^{-13} m (gamma rays) and 10^{-14} m (cosmic rays). This chapter is primarily concerned with the visible portion of the spectrum, which ranges from 4×10^{-7} to 7.5×10^{-7} m, as well as the ultraviolet and infrared portions of the spectrum, which are immediately adjacent to the visible portion at shorter and longer wavelengths, respectively.

As an electronic charge accelerates, it radiates an electromagnetic wave. This wave consists of two components, a component of electrical force that parallels the oscillation of the accelerated particle and a magnetic force that lies perpendicular to the electrical force. Both the electrical and magnetic components, while perpendicular to one another, are also perpendicular to the direction of propagation of the wave that radiates away from the oscillating charge. The propagated electromagnetic wave is sinusoidal in form. Actually, both components take on the form of a sine wave, but since the magnetic component is of negligible importance for the following discussion, only the electrical component will be considered. The points of maximum displacement of the wave (i.e., the amplitude or ψ) are called antinodes, and the distance between two antinodes of like sign is called the wavelength (λ). The units used to designate wavelength vary depending on the magnitude of the wavelengths so designated. In the ultraviolet and visible regions of the spectrum, the designation is usually made in angstroms, Å (10^{-10} m), or in millimicrons, mμm (10^{-9} m). The millimicron, which is equal to 10 Å, is now more properly called the nanon or nanometer (nm). Since wavelengths are longer in the infrared, it is more convenient to express wavelengths in this region in microns, μm (10^{-6} m).

An alternative measure of electromagnetic radiation is the frequency (ν) or number of wave maxima (antinodes of like sign) passing a fixed point in a unit of time. The most commonly used interval of time is the second, and thus frequency is expressed in cycles (maxima) per second (cps). A third measure, wave number (σ or $\bar{\nu}$), is defined as the number of waves per unit of length. The most commonly used length unit is cm, and thus wave numbers are expressed as reciprocal centimeters. A wavelength can be converted to the corresponding wave number by simply taking its reciprocal and making the proper adjustment in the decimal place to produce cm^{-1}. If the wavelength is measured in μm, as in the infrared, then $\sigma = (1/\lambda) \times 10^4$. For example, for a wavelength of 5 μm, $\sigma = 2000$ cm^{-1}, i.e., has a wave number of 2000 or 2000 waves/cm.

The frequency of electromagnetic radiation and its wavelength are related by the formula $\lambda = c/\nu$, in which c is the velocity of light in a vacuum and is approximately equal to 3×10^{-10} cm/sec. The value of c is a constant for all forms of electromagnetic radiation. However, the velocity of radiation is not constant or equal to c when it travels through some medium and, in fact, varies dependent on the medium. The frequency of radiation, on the other hand, is constant and independent of the medium. The previous equation, then, may be rewritten $\lambda\nu = v$, in which the velocity of the radiation (v) is equal to the product of the wavelength and the frequency. In a vacuum, $v = c$, while $v < c$ in a medium. The ratio of the speed of light through a vacuum to the speed of light through a medium is known as the refractive index (n). Thus, $n = c/v$. While both the velocity and the wavelength of light of a given frequency vary dependent on the medium, the magnitude of this variation is dependent on the frequency in question. Since the refractive index of a medium is dependent on frequency (or wavelength corrected to its vacuum value), it is always reported in relation to frequency. Refractive index is inversely proportional to frequency (i.e., higher frequencies corresponding to smaller n's) but directly proportional to wavelength (i.e., shorter wavelengths corresponding to smaller n's). The variations in refractive index dependent on frequency (or wavelength) are generally small, but not insignificant, particularly for those wavelengths of most interest in relation to the techniques to be discussed here. Since, usually, $n > 1$, both the velocity (v) and wavelength (λ) will be less in a medium than in a vacuum.

1.1.1. Wave–Particle Duality of Electromagnetic Radiation

The properties of electromagnetic radiation have been described by both wave models and particle models. In situations concerned only with the propagation of light, the radiation follows the laws of wave motion. This does not necessarily mean that light is a wave, but only that certain aspects of its behavior such as reflection, refraction, diffraction, and other properties can be understood through the mathematical laws of wave motion. However, in all instances in which light interacts with individual atoms, molecules, or subatomic systems, it appears to be of a discrete nature, behaving as though it consisted of indivisible particles (i.e., photons). Again, it is not that light need really consist of such photons, but only that certain aspects of the behavior of light when interacting with matter can be understood through the mathematical description of particles.

Attempts to produce a unified description of light phenomena, and of electromagnetic radiation in general, with concepts derived from classic physics have failed because these concepts resulted from observation of macroscopic phenomena. Measurements that are relevant to the macroworld and have led to concepts pertaining to it are impossible to carry out in microscopic systems, and this limits the applicability of macroworld, or classic physical, concepts to such microsystems. The general forms of quantum theory, which are relevant here, resulted from a redefinition of the concepts of classic physics in terms of the new methods of measurement appropriate to the microworld. The principles of classic physics are retained not because they appropriately describe processes in the microworld, but because they allow microphenomena to be visualized in familiar terms (Freeman, 1968).

In contrast to mechanical waves (e.g., waves in water), electromagnetic waves do not produce any displacement of the medium through which they travel and, of course, can even be freely propagated through a pure vacuum. The wave amplitude, ψ, then, has no direct physical meaning. Its conceptual value is based on the fact that quantum-mechanical problems are solved mathematically in the same way that water-wave or other classic wave problems are solved. Classic waves and the particle waves of electromagnetic radiation both obey the same kind of mathematical wave equation. However, in the case of classic waves, the wave amplitude is directly observable, whereas ψ generally is not. Following from these mathematical equations, the intensity of a particle wave is the square of the absolute value of the amplitude or $|\psi|^2$. The intensity of a particle wave at a given point, $|\psi(x)|^2$, is proportional to the probability of finding the particles at position x.

Quantum theory assumes that waves of electromagnetic radiation are not continuous in regard to their energy distribution. Rather, this energy is distributed discontinuously, in discrete quanta or photons. As components of electromagnetic radiation, photons can be described in terms of their frequency, with the energy of each photon being directly proportional to its frequency. The equation describing this relationship is $E = h\nu$, in which the constant, h, is called Planck's constant and is approximately equal to 6.6×10^{-27} erg sec. Thus, not only do quanta of differing frequencies have different energies, but also the differences in energy between two frequencies are discontinuous and always in multiples of h. Since the frequency of radiation is inversely proportional to its wavelength, higher frequencies that are correlated with higher energies are also correlated with shorter wavelengths and vice versa.

Thus, only two properties are necessary to characterize electromagnetic radiation whether it is conceptualized in terms of the classic wave model for light or in terms of the photon model. The first property, the quality, is defined by the frequency (ν) or the energy per photon (E), which is directly related, or the wavelength (λ) in a vacuum. The second property is the quantity or intensity of the radiation ($|\psi|^2$).

1.1.2. Wave Nature of Matter

Although the necessity for both wave and particle models of electromagnetic radiation as conceptualized in quantum physics is fundamental to

the understanding of interactions between light and matter, this conceptualization may be broadened. The key to the generality of quantum physics
concepts is that they not only explain the particle behavior of light but also
incorporate wave characteristics of matter. Thus, the nature of both energy
and matter and their interaction may be understood through a single system
of mathematical concepts. The comparability between light and matter that
has been empirically demonstrated will be presented.

First, one other mathematical concept relating to the behavior of photons
must be presented. To fully conceptualize photon behavior as observed in
various experiments, it was necessary to show that both the classic physical
laws of conservation of energy and those of conservation of momentum were
applicable. The relevant algebraic expression is derived as follows: A quantity
of electromagnetic radiation with an energy E has a momentum E/c. Since,
for a photon, $E = h\nu$, it must have a momentum $h\nu/c$ that could also be
expressed as h/λ, since $c = \nu\lambda$. The momentum of a photon (ρ), then, is
described by the following equation: $\rho = h\nu/c = h/\lambda$.

The second step is to relate the momentum of a photon to that of a
particle of matter. The mass (m) of a particle, such as an electron, traveling
with a velocity, v, is defined as $m = m_0/\sqrt{1 - (v^2/c^2)}$, in which m_0 represents
the mass of the particle at rest, i.e., when $v = 0$. Further, the momentum of
the particle is given by $\rho = mv$ and the total energy by $E = mc^2$. To
incorporate a wave description of the particle, the principal characteristics of
waves, i.e., frequency (ν) and wavelength (λ), must be considered. The total
energy of a particle considered in terms of either its particle properties or
its wave properties is $E = mc^2 = h\nu$, and similarly, the momentum of the
particle is $\rho = mv = h/\lambda$. The latter equation could be rewritten $\lambda = h/mv$,
thus combining both wave and particle properties.

The equations for the momentum of both photons and particles are
identical, i.e., $\rho = h/\lambda$. However, if the alternate expression for the momentum
of a photon ($\rho = h\nu/c$) is considered, it becomes clear that it cannot also
represent the momentum of a particle. The reason for this is that the velocity
of a wave associated with a photon is c, and therefore, for photons, $\nu\lambda = c$
and $1/\lambda = \nu/c$. For any particle, the general formula $\rho = h/\lambda$ holds, but only
when the wave velocity is c is it permissible to use the identity $1/\lambda = \nu/c$ and
transform the general momentum equation into $\rho = h\nu/c$. The wave velocity
of material particles is not c, and therefore this equation does not apply to
them.

Another essential difference between a photon and a material particle
becomes apparent when the relativistic expression for the total energy of a
material particle is considered. Since $E = mc^2$ and $m = m_0/\sqrt{1 - (v^2/c^2)}$, E
$= m_0c^2/\sqrt{1 - (v^2/c^2)}$. If this equation were to apply to a photon, clearly, in
the denominator it would be true that $v = c$. The denominator would then
become zero, yielding an energy expression of infinite magnitude. This is
why a material particle can never reach the velocity of light. However, if it
is recalled that the rest mass of a photon is zero, the numerator as well as the
denominator of the expression becomes zero (or $E = 0/0$), which is indeter-

minate, which is to say that E can take on any value. The energy of a photon, in fact, has a finite value of $h\nu$. On the other hand, if the resting mass of a particle is zero and its speed is less than c, then the numerator in the expression is zero while the denominator takes on a nonzero value and the total energy, then, is zero. Thus, a particle with a rest mass of zero can possess energy only if it travels at the speed of light, in which case, as mentioned above, both the numerator and the denominator are equal to zero (Atkins, 1972).

Since material particles follow the same mathematical wave functions as photons, the previous discussion of the amplitude and intensity of photon waves also applies to particle waves. To restate briefly, particle waves do not displace the medium in which they occur. Furthermore, the intensity of a particle wave at a point (x) that is given by the expression $|\psi(x)|^2$ is proportionate to the probability of actually finding a particle at that point x. The amplitude of the particle wave, ψ, has no physical meaning.

Quantum theory, as briefly outlined above, has made several contributions to the understanding of the phenomena of interest in this chapter. These may be briefly summarized at this point. In the first place, the classic laws of conservation of energy and conservation of mass are now subsumed under a single principle of conservation of mass–energy. This follows based on the fact that a photon of energy E has a linear momentum of $\rho = E/c$. If a photon could be regarded as having a dynamic mass m, it would be necessary to assign it a momentum mc, since this entity travels with the speed c. By equating these two expressions, $E/c = mc$ or $E = mc^2$ is obtained. The special theory of relativity is able to prove that this relationship has a significance much broader than its application to photons. The principle of mass–energy equivalence states that to every quantity of energy E there corresponds an equivalent mass m, and with every mass there is associated a certain intrinsic energy.

A related contribution lies in the clear equivalence in many aspects of photons and material particles regarding the wave and particle properties. From this equivalence, it becomes clear that the use of Planck's constant to describe the discontinuous nature of the energy distribution in photon waves is also applicable to a description of the energy states of material particles. As mentioned previously, it is important to keep in mind the distinction between macro- and microsystems. The discontinuous or quantal distribution of energy as observed in the microworld is not apparent in the familiar macroworld of classic physics. This is not to say that the principle does not hold for these more observable phenomena, but rather that the discontinuous nature of the microworld averages out and tends not to be observed when many microsystems combine to form macrophenomena (Freeman, 1968).

2. Energy States of Matter

The total energy of a molecule, excluding nuclear energies, is equal to the sum of the energy due to the electrons of the constituent atoms, or the electronic energy; the vibrational energies of the molecular bonds; and the rotational energy of the molecule as a whole. Thus, $E_T = E_E + E_V + E_R$. The absorption of light results in an increase in the total energy of a molecular system due to one or a combination of electronic, vibrational, and rotational transitions from lower to higher energy states. Energy levels for all three types of transitions are quantized. However, rotational energy levels are relatively closely spaced, and transitions among them are produced by absorption of relatively low-energy radiation, which is to say radiation of relatively low frequency, i.e., microwave and far-infrared radiation. The energy of radiation just able to produce rotational transitions is not of sufficient magnitude to produce vibrational or electronic transitions. Light of infrared and visible wavelengths has sufficient energy to produce transitions among the less closely spaced vibrational energy levels. Generally, if such radiation is well controlled in terms of frequency, all the photons incident upon the radiated material will be too high in energy to interact with quantized rotational levels. However, as higher vibrational states return to the ground state, some of the energy released by a molecule may be of appropriate magnitude to elicit rotational transitions. Thus, vibrational transitions, as a rule, do not occur in the absence of concomitant rotational transitions. Similarly, the still higher photon energies of higher-frequency visible and ultraviolet radiation, while capable of producing electronic transitions, usually also result in both vibrational and rotational transitions.

2.1. Rotational Energy

As with ordinary-sized objects, a molecule can rotate or revolve about an axis passing through its center of gravity. Three axes, mutually perpendicular, pass through the center of gravity of molecules. Since rotation may take place about any of these axes, molecules are said to possess three rotational degrees of freedom. Diatomic, and more generally all linear molecules, can rotate only about two axes perpendicular to the length of the molecule and, consequently, are characterized by only two degrees of freedom.

The simplest explanations of many molecular characteristics are based on examples using diatomic molecules. In general, only such relatively uncomplicated examples will be used here. The energy of rotation of molecules can be treated initially with the familiar methods of classic mechanics, which can then be qualified in light of quantum-mechanical restrictions. Thus, the kinetic energy of a rotating molecule is given as $kE = \frac{1}{2}I\omega^2$, where ω is the angular velocity, I is the moment of inertia, and $I = \mu r^2$. The formula for the inertial moment contains a term that is frequently useful in consideration of both rotational and vibrational states of molecules. This term, μ, is called the reduced mass of the molecule and combines the masses of the constitutent atoms of the molecule. In the case of diatomic molecules, $\mu = m_1 m_2/(m_1 + m_2)$. The term r in the inertial moment equation is simply the internuclear distance. The product of the angular velocity and the moment of inertia in such a classically described system is the angular momentum, $I\omega$.

Since, in such a molecular system, the region in which the particle (molecule) is free to move is quite limited, the rotational energies allowed will be subject to quantum-mechanical restrictions and therefore quantized. Such restrictions are more easily discussed in terms of angular momentum rather than energy. The allowed amounts of angular momentum are multiples of $h/2\pi$. Diatomic molecules can rotate only with the angular momentum given by the following expression: $I\omega = \sqrt{J(J + 1)}\,(h/2\pi)$, where J can be equal to zero or any positive integer. Thus, J is the rotational quantum number.

Substitution of the last equation for the angular-momentum component of the kinetic-energy equation yields a formula that defines the energies (E_R) with which a diatomic molecule is allowed to rotate according to quantum mechanics. Thus, $E_R = (h^2/8\pi^2 I)\,J(J + 1)$.

Since the energy of molecules is to some extent dependent on temperature, the rotational energies actually possessed by a collection of molecules will vary with temperature. Furthermore, all molecules held at a given temperature need not be characterized by identical rotational energies. In fact, although a rather large proportion of the molecules may be characterized by a relatively small number of rotational energies, some molecules will be in higher energy states. The ratio between the number of molecules (n_i) occupying some energy level and the number of molecules (n_0) occupying the lowest energy level is defined as $(n_i/n_0) = e^{-(\epsilon_i - \epsilon_0)/kT}$, where $\epsilon_i - \epsilon_0$ is the difference between the two energy levels. The value of the Boltzmann constant, k, is equal to 1.38×10^{-16} erg/degree, and T is the temperature in degrees Kelvin. This equation defines the Boltzmann distribution of the number (n_i) of molecules at various energy levels as this relates to temperature. This distribution suggests that for energy-level differentials, $\epsilon_i - \epsilon_0$, that are not much larger than kT, the ratio n_i/n_0 is not much smaller than unity, and therefore such energy levels, ϵ_i, will be appreciably populated. When the energy differential is considerably larger than kT, the ratio is very small and the higher energy levels will not be populated to a great extent. Examples given in Barrow (1963) demonstrate that at room temperature (i.e., 25°C), the first ten allowable rotational energies of CO are less than kT, and therefore many rotational-energy levels of CO will be appreciably populated at room temperature. Only the first five allowed rotational energy levels of HBr fall below kT at this temperature, and although these levels will be well populated, it is clear that many fewer levels will be populated with HBr as compared to CO.

This latter comparison suggests an inevitable outcome of the quantum-mechanical formula defining the allowable rotational energies (E_R) of molecules as mentioned above. As the moment of inertia (I) increases, the spacing between the allowed energy levels decreases. Thus, the number of allowed energy levels below kT for CO in the example above was ten, while there were only five allowed levels below this value for HBr. Obviously, the energy levels for CO must be more closely spaced than those for HBr. This difference is consequent on the fact that I for CO is more than four times as large as I for HBr.

The absorption of electromagnetic energy by a rotating molecule leads to an increase in the rotational energy of the molecule. The rotational energy of molecules can be influenced by electromagnetic radiation only if they possess a dipole moment, that is, only if charge is differentially distributed in the molecule such that one end is more positively charged than the other, which is more negatively charged. Molecules such as H_2 have identical ends and possess no dipole moment, and thus their rotational energies cannot be affected by electromagnetic radiation.

The wave nature of radiation presents an oscillating electrical field to molecules in its path. The differential charge distribution on the molecule will tend to oscillate with the positive and negative aspects of the electrical field of the radiation. If the frequency of the radiation, i.e., its rate of oscillation, and the rate at which the molecule is rotating are equal, the electrical field of the radiation can interact with the dipole of the molecule and increase its rotational energy. When the frequencies of rotation and radiation are not equal, the oscillating electrical field will tend to increase the rate of rotation during one half of its cycle and decrease it during the next half, having little or no net effect.

Closer examination than is possible here of the mathematical wave functions that give rise to quantized rotational energy levels reveals that there is a restriction on the energy levels that a molecule may enter after having absorbed radiation. This restriction, or selection rule, states that on absorption of radiation, a molecule can move only to the next higher energy level. In other words, J may increase by only 1, or $\Delta J = 1$. Similarly, if a molecule loses energy, it can move only to the next lower energy level, i.e., $\Delta J = -1$. The general selection rule describing changes in rotational energy levels is $\Delta J = \pm 1$.

The rotational movement of molecules can be affected by factors other than electromagnetic radiation. It is clear from the discussion above that according to the Boltzmann distribution, there will be a considerable range of rotational energies exhibited by a collection of molecules at a given temperature, and alterations in temperature will affect the range of states exhibited. Very simply, it is well known that increases in temperature increase the random movements of molecules and can easily be imagined, therefore, to affect their rotational energies. On the other hand, at room temperature, or even lower temperatures, the considerable movement of molecules results in the inevitability of frequent collisions with other molecules in the liquid state. Such collisions are likely to result in an alteration of the rotational states of the colliding molecules. Furthermore, collisions may occur at a rate high enough to produce a significant probability that molecules undergoing a rotational transition due to absorption of electromagnetic radiation will collide with other molecules before the transition is complete. Because molecular collisions are likely to alter the absorption-induced transition either before it is completed or soon after, the lifetime of excited states will be relatively short, and analysis of the rotational states of molecules in the liquid phase is severely restricted. Rotational spectra are best studied when the molecules are in the gaseous phase, in which, if the pressure is not too high, the volume of molecules is small relative to the volume of the gas. Gaseous molecules do collide, but with relatively low frequency. Thus, gaseous molecules to a great extent are free to move independently of one another and can remain in excited rotational states for extended periods.

2.2. Vibrational Energy

Molecules also possess energy due to vibrations of the bonds between pairs of atoms. The forces that bond atoms together to form molecules are not entirely rigid. Rather, they can be thought of as being somewhat like springs that are able to be compressed or extended along their main axis, thus allowing bonded atoms to move closer to or farther away from one another. To continue the spring analogy, such bonds, like springs, are capable of movement along their other axes, e.g., bending, which results in distortions of the angles between the atoms in a molecule. This intramolecular movement about bonds is termed vibration, and the energy required for such movement is vibrational energy.

Vibrations occur between two atoms within a molecule, and there are more vibrations possible for a molecule as the number of atoms constituting the molecule increases. A single atom can be considered to have three degrees of freedom of movement related to three axes that pass through the center of gravity of the atom and are perpendicular to one another. In a molecule composed of n atoms, then, $3n$ degrees of freedom can be conceptualized. Thus, movement of a given molecule will exhibit $3n$ degrees of freedom. As mentioned previously, rotational movement of molecules implies three degrees of freedom. Additionally, the movement of a molecule through a medium, as a whole and regardless of its rotational movement, is called its translational movement. Since the translational movement of a molecule can also be described with reference to three mutually perpendicular directions of motion, there are three degrees of freedom of translational movement. There are, then, three degrees of freedom of translational movement and three degrees of freedom of rotational movement (or two degrees of freedom in the case of linear molecules) for each molecule. If each molecule has a total of $3n$ degrees of freedom, subtraction of the translational and rotational degrees of freedom will yield a number describing the degrees of vibrational freedom. Thus, molecules have $3n - 6$ (or $3n - 5$ for linear molecules) degrees of freedom of vibrational movement. The number of vibrational degrees of freedom corresponds to the number of basic ways that a molecule can vibrate.

When a mass (m) attached to one end of a spring that has a force constant k and that in turn is fixed at its opposite end is considered, classic physics has shown that vibration in a given system (i.e., a system defined by particular values of m and k) will occur with a characteristic frequency, v_c, according to the following equation: $v_c = (1/2\pi) \sqrt{k/m}$. (The symbol v_c is used to denote the vibrational frequency of this classic system in contrast to the quantum number v to be used subsequently when discussing molecular systems.) Similarly, if the system is composed of a spring with a mass at each end, rather than with one end fixed, the characteristic frequency of vibration is given by $v_c = (1/2\pi) \sqrt{k/\mu}$, where μ is the reduced mass as described previously. The energy stored in such a system for a particular spring and two specific masses is dependent on the amplitude of the vibrations. A diatomic molecule can be conceptualized as such a system of two masses (the two atoms) and a spring (the molecular bond). However, quantum mechanics has demonstrated that certain restrictions must be added to the mathematical description of the classic system before it can be applied to the microworld of molecules.

The primary restriction is that molecular systems cannot assume a continuous range of vibrational energies as classic systems can, but rather only discrete energy states are allowed. The equation describing the allowed energy levels is $E_v = hv_c (v + \frac{1}{2})$ and includes the vibrational quantum number v, which may assume the value of zero or of a positive integer. Substitution for v_c from the equation defining it above shows that the allowable energy states of a diatomic molecule can be described by $E_V = (h/2\pi) \sqrt{(k/\mu)} (v + \frac{1}{2})$. This equation clearly suggests two factors. In the first place, vibrational energies, for a given diatomic system as defined by k and μ, may assume only certain discrete values in accordance with the quantum function of v. Second, even in the lowest vibrational state, i.e., $v = 0$ or v_0, the energy of vibration does not go to zero, but rather is equal to $(h/2\pi) \sqrt{(k/\mu)} (\frac{1}{2})$. In contrast to classic systems, molecular systems are not allowed to have zero vibrational energy. Although at the lowest of temperatures all molecules will assume their lowest vibrational energy level, the molecules will not be in a state of zero energy. In this lowest state of vibrational energy, molecules are said to possess zero-point energy, which, of course, does imply that there is some energy in that state.

Calculations of the energies of various vibrational states for a given molecule show that they are considerably larger than the rotational energies for the same molecule. Furthermore, the magnitude of the energy difference between adjacent states is also much larger than the spacing between rotational energies. Consideration of the Boltzmann distribution has demonstrated that the spacing between vibrational energy states is large compared to the value of kT. Thus, at room temperature, most molecules will be at zero-point energy, their lowest energy state, v_0.

As was the case with rotational transitions, only those molecules that possess a dipole moment can interact with electromagnetic radiation. Because molecules vibrate, the magnitude of their dipole moment constantly oscillates. The interaction of electromagnetic radiation with the vibrational-energy states of molecules is analogous to the interaction with rotational states. If the oscillation of the charged ends of the diatomic system is in phase with the electrical field of the electromagnetic radiation—that is, if the electromagnetic radiation has the same frequency as the vibration of the molecule—during each vibration, the electrical field will enhance the vibration of the molecule. Changes from one vibrational state to another, i.e., vibrational transitions, follow the selection rule that $\Delta v = \pm 1$. The interaction of a molecule with electromagnetic radiation can result in a change of vibrational energy only to an adjacent energy level. When molecules are originally in the v_0 state, which is to be expected at room temperature, or when the absorption of electromagnetic energy is of interest, then an increase in vibrational energy is all that is allowed, and the relevant aspect of the selection rule is $\Delta v = +1$. Relatively weak exceptions to this rule are not uncommon, but need not be considered here.

The number of degrees of vibrational freedom for a diatomic molecule is 1 [i.e., $3(2) - 5 = 1$], and it follows that diatomic molecules will absorb radiation of a single wavelength as determined by the equation discussed above. It should be apparent that a transition from v_0 to v_1, for example, will involve an increase in vibrational energy, ΔE_V, equal to the energy differential between the v_0 and v_1 states or $\Delta E_V = E_{V_{(v_1)}} - E_{V_{(v_0)}}$. This required energy can be related to radiation of a particular wavelength, as previously discussed, and only that wavelength of radiation can be absorbed and effect the vibrational transition.

The vibrational characteristics of polyatomic molecules are more complex, but a simplistic description will be presented here. A triatomic molecule will possess three degrees of vibrational freedom [i.e., $3(3) - 6 = 3$] and can be said to have three basic, or natural, vibrations. Each of these vibrations can be considered as equivalent to the single vibration of a diatomic molecule. Each of these basic vibrations can, then, be described in terms of its allowable energy levels. Furthermore, the characteristic energy levels of each vibration lead to the possibility that different wavelengths of electromagnetic radiation will be absorbed by each basic vibration, resulting in a transition from v_0 to v_1. Two of the natural vibrations in a triatomic molecule are associated with the stretching (and compressing) vibration of the two molecular bonds. The third natural vibration corresponds to bending, or distortion, of the angle formed by the three atoms. The spacing between the energy levels for bond-stretching vibrations is much larger than the spacing for bond-bending vibrations. Thus, absorption affecting bond-stretching will occur at higher frequencies or shorter wavelengths than for bond-bending. Diatomic molecules and other linear molecules, of course, do not allow the possibility of bond-bending vibrations. More complex, polyatomic molecules are characterized by even more complex vibrational states.

2.3. Combination of Rotational and Vibrational Energies

Infrared radiation, which, when absorbed, results in vibrational transitions in molecules, provides more than enough energy to elicit rotational transitions, as well. In the liquid state, which is most common in neurobiological work, intermolecular interactions (i.e., collisions) effectively limit the occurrence of rotational transitions, as discussed previously. However, in the gas phase, rotational transitions do occur in conjunction with vibrational transitions. Although this chapter will not be concerned specifically with information that can be obtained by detailed examination of rotational and vibrational transitions, either combined or independently, it will be instructive to briefly examine such combined transitions.

The energy of diatomic gas molecules due to rotation and vibration can be summarized fairly well by a simple summation of the two separate energies, or $E_{R-V} = (h^2/8\pi^2 I) J(J + 1) + (h/2\pi) \sqrt{(k/\mu)} (v + \frac{1}{2})$, in which both J and v are quantum numbers as previously discussed. The previous discussion of rotational states can be assumed to apply to the lowest, v_0, vibrational state, since most molecules would have been in this state at room temperature and in the absence of infrared absorption. The main focus on vibrational energies centered around the v_0 and v_1 states. At room temperature, then, most molecules are in the v_0 state, but at the same time a wide range of rotational states are inhabited. It follows that at room temperature, the vibrational state, usually v_0, of a molecule and its rotational state are essentially independent, with the rotational states assumed by a collection of molecules determined by the Boltzmann distribution.

Molecules that are characterized by higher vibrational states such as v_1, after absorption for example, still rotate. The allowable rotational energies of a molecule in the v_1 vibrational state are virtually the same as those for a molecule in v_0. The allowed-energy-level pattern for a rotating and vibrating molecule consists of a series of relatively widely spaced vibrational energies as defined by $E_V = (h/2\pi) \sqrt{(k/\mu)} (v + \frac{1}{2})$, with each vibrational-energy level subdivided into a series of much more closely spaced rotational energies. The spacing of the rotational energies is essentially identical for each of the allowed vibrational states.

When both rotational and vibrational energies are considered, the interaction of molecules, even diatomic molecules, with electromagnetic radiation becomes quite complex. Infrared absorption leads to vibrational transitions that follow the vibrational selection rule. Of course, the most common transition is from v_0 to v_1. The selection rule for rotational transitions must also be followed, with both increases and decreases in rotational energy requiring consideration. Consider, for example, those molecules of a diatomic gas that make a $v_0 \rightarrow v_1$ transition due to infrared absorption. Since these molecules when they initially occupied v_0 were in a variety of rotational states—J_0, J_1, J_2, etc.—a similar variety of rotational transitions will accompany the vibrational transition dependent on the initial rotational state and following the rotational selection rule, i.e., $\Delta J = \pm 1$. Molecules in J_0 will move to J_1 when $v_0 \rightarrow v_1$. No other transition is possible, since $J_0 \rightarrow J_0$ is a forbidden transition (it violates the selection rule) and there are no rotational states lower than J_0. Molecules initially in J_1 can move to J_2 or J_0. Those in J_2 can move to J_3 or J_1, and so on. Thus, depending on the number of rotational states initially occupied by the molecules, as determined by the Boltzmann distribution, a fairly large number of combined rotational–vibrational transitions may occur when the molecule makes a transition from v_0 to v_1.

For a given molecule in the state characterized by v_0 and J_n, a combined energy, E_{R-V}, can be calculated as indicated by the formula given above. A similar calculation can be made for the energy of the molecule following a transition when its state is v_1 and $J_{n\pm1}$. The energy required for the transition can then be determined, $\Delta E_{R-V} = E_{R-V(V_1,J_{n\pm1})} - E_{R-V(V_0,J_n)}$. The value of ΔE_{R-V} can then be related to the particular frequency of infrared radiation that will elicit the transition according to $E = hv$, or more specifically $v = h/\Delta E_{R-V}$. Each of the allowed transitions will be produced by absorption of radiation of the appropriate frequency. Conversely, if the molecules are subjected to heterogeneous infrared radiation, each molecule will be able to absorb that wavelength that is necessary to achieve a transition, and therefore a variety of transitions will occur in the gas as a whole. The frequency of occurrence of any particular transition and, consequently, the amount of radiation of the appropriate frequency that is absorbed is related to the initial distribution of the molecules in the various allowed rotational states when $v = 0$.

As might be expected, this brief description does not fully detail all aspects of combined rotational–vibrational transitions. However, it is sufficient for present purposes. As can also be imagined, more complex molecules exhibit a wider range of allowable transitions. Polyatomic molecules, as already discussed, have in contrast to diatomic molecules more than one basic mode of vibration. The description of the possible combined transitions for a diatomic molecule that has but one natural mode of vibration must be extended to each of the natural vibrational modes of the polyatomic molecule. This, of course, introduces considerable complexity into the situation.

2.4. Electronic Energy

The final type of energy that contributes to the total energy of a molecule is electronic energy. Each of the electrons of the atoms constituting a molecule has energy, and each of these energies as well as the total energy of a molecule due to the energy of its electrons can be fairly accurately determined. When the energy of one of the electrons changes, as can happen when electromagnetic radiation is absorbed, the energy of the molecule also changes. Although it would have been impossible to discuss rotational and vibrational energies except with reference to molecules, since rotation and vibration are characteristics of molecules, electronic energies can be discussed with reference both to molecules and to single atoms. Since a knowledge of the electronic states of atoms underlies the description of molecular electronic states, atomic electronic states will be considered first.

2.4.1. Electronic States of Atoms

The electron-orbital model of the atom is basic to an understanding of electronic states and their interactions with electromagnetic radiation. According to this model, electrons are conceptualized, in part, as occupying orbits at more or less fixed distances from the nucleus. However, this is really a conceptualization from the older, planetary model of atomic structure and is primarily concerned with the particle properties of electrons. To be more precise, the quantum-mechanical description of orbiting electrons must be considered. The quantum-mechanical model also takes into account the wave properties of electrons. This model describes the properties of electron waves in atoms with a mathematical wave function equation, the Schroedinger equation. Solutions to this equation are called wave functions, or orbitals in the case of electrons. They measure the probability of finding an electron in a given region in space. Rather than traveling in definite planar orbits of fixed radius, electrons are seen to travel in certain volumes in space. The orbital of a 1s electron (this designation will be considered below), for example, can be fairly accurately represented by a sphere. The wave function specifies that the probability of finding this electron is at a maximum at a certain distance from the nucleus (which, in fact, corresponds to the radius of the 1s electron orbit in the planetary model). This probability decreases with increasing radius but never reaches zero. Ultimately, the 1s orbital is infinite in size, since, in fact, it is possible to find that electron anywhere. For practical purposes, a radius is chosen that defines a volume within which there is a 90% probability of finding the electron.

The volumes defined by orbitals for all s electrons are spherical. However, the orbitals of other electrons vary in shape. For example, p electrons are commonly described as being shaped somewhat like dumbbells. A given shell can have three p orbitals, and they are oriented perpendicularly to one another. The d and f orbitals take on more complex shapes.

Each uncharged atom has a number of electrons that is equal to its atomic number, i.e., the number of protons in the nucleus. The electrons are distributed about the nucleus in orbitals, each of which can be characterized as requiring a specific energy for an electron occupying them. All the orbitals differ systematically with regard to these energy levels. Each orbital may be unoccupied or occupied by either one or two electrons. When a pair of electrons occupies a single orbital, they will be of opposite or antiparallel spin. (The spin of an electron describes its behavior in a magnetic field and is discussed further below.) Simplistically, each electron can be conceived as spinning on its axis in either a clockwise or a counterclockwise direction. Differences in the spin of paired electrons within a single orbital result in a small energy differential between the two electrons. Because of this spin-related energy differential as well as the fact that no two orbitals possess the same energy, it follows that no two electrons in a single atom can be in the same energy state, since each orbital can be occupied by a maximum of only two electrons that differ in their spin. As will be seen subsequently, the energy of each electron can be defined by four quantum numbers. Three of these specify the energy characteristics of the orbital and the fourth describes the spin of the electron. Another way of stating the previous principle is that no two electrons in the same atom can possess the same values of these four quantum numbers. This is called the Pauli Exclusion Principle. In the unexcited or ground state, orbitals will always be occupied by electrons in order of the energy level of the orbital, with lower energy levels filled first. This is termed the Aufbau Principle and can be stated in another manner. In determining the electronic structure of the ground state of an atom, all electrons may be thought of as removed from the nucleus, and then fed back into the atom, filling up the available orbitals in the order of their energies.

In a manner reminiscent of the planetary model, the orbitals are still commonly viewed as being arranged as shells. The shells are numbered with integers beginning with 1 to designate the lowest-energy orbital. Each shell is composed of one or more subshells that are labeled s, p, d, or f in increasing order of energy. An s subshell contains but one orbital and therefore can be occupied by a maximum of 2 electrons. Since the first shell has only an s subshell, it can be occupied by a maximum of 2 electrons. There are a maximum of three p orbitals in a particular shell, each of which can be occupied by 2 electrons. Therefore, the maximum number of p electrons in a given shell is 6. In similar fashion, d subshells can have five orbitals and, consequently, 10 electrons, while f subshells have at most seven orbitals and 14 electrons. Some subshells in higher-numbered shells are of lower energy than some subshells of lower-numbered shells. Since the orbitals must be occupied in order of their energies, some elements are characterized by having more than one incomplete shell.

Each electron in an atom can be characterized by a series of four quantum numbers. These quantum numbers, when substituted in the appropriate formula, provide a means of calculating the energy of an electron in a particular orbital and thus defining the orbital. Although the principal quantum number alone is the main factor in determining the energy of an orbital, the other quantum numbers serve to specify other features of the orbital.

The principal quantum number, n, determines the most probable distance (r) of an electron from the nucleus according to the equation $r = n^2h^2/4\pi^2k_0me^2$, in which k_0 is the Coulomb-force constant, m is the mass of the electron, and e is its charge. The principal quantum number can take all integer values from 1 upward and is equivalent to the number of the shell.

The orbital quantum number l determines the permitted values of the orbital angular momentum L, the magnitude of which is given by $L = \sqrt{l(l + 1)}\,(h/2\pi)$. Angular momentum is also quantized, and therefore, l takes on only integer values that range from 0 to $n - 1$. Thus, the lowest energy state for an electron that is defined by a principal quantum number of $n = 1$ is associated only with a single orbital quantum number of 0. When $n = 2$, two orbital-angular-momentum states are possible, with l taking on the values of 0 and 1. The orbital quantum number, l, corresponds to the subshell of the electron. When $l = 0$, the orbital is an s orbital. When $l = 1$, p orbitals are defined, and so on.

Since L, the orbital angular momentum, is a vector, its direction can be specified in addition to its magnitude. The electron can be conceived as having the properties of a small magnet that always finds itself in a magnetic field. Thus, the total energy of the atom must depend slightly on the orientation of this magnet with respect to the external field. The orientation is specified by the component of L along the line of the magnetic field β. In its normal position, L sets itself as close as possible to the direction of β, and this corresponds to the lowest state of magnetic energy. An increase in magnetic energy is required for L to assume any other orientation. The energy required is at a maximum when L is oriented as nearly as possible in a direction opposite to β. Quantum mechanics specifies that only certain values of L_β, the component of L along the direction of the magnetic field, are possible. The magnetic quantum number m_l determines the permitted values of L_β according to the formula $L_\beta = m_l(h/2\pi)$. The magnetic quantum number, m_l, can take any integer value from $-l$ through zero to $+l$. Since when $n = 1$, the only possible value of l is $l = 0$, m_l must also be equal to 0. When $n = 2$, $l = 1$, which defines the p subshell of the second shell, and m_l can take on the value of -1, 0, or $+1$, which defines the mutually perpendicular orientations of the three p orbitals.

One further quantum number is based on electron spin, as already mentioned, and is due to the angular-momentum contribution that the electron itself makes as it spins on its axis. The magnitude of the electron-spin angular momentum, L_s, is defined as $L_s = m_s (h/2\pi)$. The spin magnetic quantum number, m_s, can take on one of two values, either $m_s = -\frac{1}{2}$ or $m_s = +\frac{1}{2}$. Thus, according to m_s, an electron may be oriented in space in one of two ways with the spin-angular-momentum vector either parallel or antiparallel to the magnetic field. This results in there being slightly different energies for an electron dependent on which state it is in.

Experimental observations have shown that no two electrons in an atom may be in the same energy state. The generalization from these observations has been called the Pauli Exclusion Principle, which has already been mentioned but which can be restated. According to this principle, no two electrons in a given atom can be described as having the same values for all four quantum numbers. Two electrons in the same orbital would have the same values of n, l, and m_l, but would differ in their values of m_s. The quantum-number description of the electrons of an atom may readily be compared with the planetary model. All electrons occupying a single shell in the planetary model can be described by a single principal quantum number, or n. Electrons occupying a single subshell possess the same orbital quantum number, l, and where a subshell contains more than one orbital, the electrons in a single orbital will also possess the same magnetic quantum number, m_l. The third shell, for example, contains three subshells according to the planetary model. There can be a maximum of one pair of electrons in an s subshell, 6 electrons in the p subshell, and 10 electrons in the d subshell. The orbital or quantum-number description of the atom defines the same structure, but more explicitly. The third shell would contain electrons for which $n = 3$. When $n = 3$, l, which ranges from 0 to $n - 1$, can take on the value of 0, 1, or 2, which define the three subshells. The orbitals within each subshell are defined by m_l, which can take on values from $-l$ through zero to $+l$ for each value of l. Thus, when $l = 0$, only one orbital is possible; i.e., $m_l = 0$, as in the s subshell. When $l = 1$, m_l may equal -1, 0, or 1, which yields the three orbitals of the p subshell. When $l = 2$, m_l may equal -2, -1, 0, 1, or 2, which yields the five orbitals of the d subshell. Quantum mechanics, then, specifies that there may be a maximum of nine orbitals in which $n = 3$. Since, of course, each orbital may be occupied by 2 electrons that differ in m_s, the third shell (i.e., $n = 3$) may contain 18 electrons. Any shell may be treated in a similar fashion. This rather qualitative description does not fully specify the order in which the orbitals will be filled, but a detailed analysis of the energies associated with each unique set of four quantum numbers will, of course, allow a ranking of energy states that will then specify the order in which orbitals are filled, following the Aufbau Principle that lower-energy orbitals are always filled before higher-energy orbitals.

In a rather general way, the electronic state of an atom can be viewed as a summation of the states of the electrons of that atom or, more specifically, the resultant of the angular-momentum vectors of the electrons. For most atoms of neurobiological importance, i.e., those of relatively low atomic numbers, the orbital angular momenta of the individual electrons interact to form a total orbital angular momentum. As is the case with individual electrons, the orbital angular momentum of the atom is quantized. The effects of electron spin also couple to form a resultant spin for the atom. The total orbital angular momentum and the spin resultant couple to yield a quantized total angular momentum that describes the ground state of the atom. Excited states, of course, are also possible and can be described in a similar fashion. It is the energy differential between the ground and excited states of the atom that determines the energy, and hence the wavelength, of light that that atom will absorb.

All electrons are paired in completed shells or subshells of atoms, and the resultant orbital and spin momenta are therefore zero. Atoms that are characterized by having all electrons in completed shells and subshells, then, have resultant orbital and spin momenta of zero and are classified as being in an s-state. The total state of atoms that have incomplete subshells can be derived by consideration of the electrons in the incomplete subshells, the valence electrons, only. The resultant orbital and spin momenta of a monovalent atom have the same value as those of the single valence electron. The resultant momenta of polyvalent atoms are based on an algebraic summation of the orbital angular momenta and spin of each of the valence electrons. A detailed treatment of the term symbols and energy levels for the electronic states of atoms is not necessary for present purposes and will be omitted. However, the subject is discussed more thoroughly in some of the references given in Section 3.2. Although not entirely accurate, it is certainly sufficient for the purposes of this discussion to understand the interaction of light with matter with reference to the energy states of individual outer-shell electrons rather than to the electronic state of the whole atom.

2.4.2. Electronic States of Molecules

The classification of the electronic states of atoms, i.e., the classification of electronic orbitals, briefly described above, is possible because the electrical field of an atom is, essentially, spherically symmetrical. Thus, the angular-momentum vectors for any state are constants of the motion. The force fields about molecules are not spherically symmetrical. In diatomic molecules, for example, a strong electrical field holds the atoms together, and this force field is cylindrically symmetrical about the internuclear axis. The state of the molecule is related to that of the individual atoms in a strong electrical field. As with the determination of the electronic states of atoms, molecular electronic states are related to the total orbital and spin momenta of electrons. However, because of the molecule's cylindrical rather than spherical sym-metry, it is the components of the orbital and spin momenta about the direction of the field that are important for the classification of molecular electronic states. Similar considerations apply to polyatomic molecules, al-though their description is more complicated.

The creation of most chemical bonds implies the formation of a new orbital into which one electron from each of the two combining atoms is fed. Since both electrons occupy the same orbital, they must have opposite spins. It is easiest to imagine the combination of atomic orbitals to form molecular orbitals if only valence electrons are considered. These are the most important electrons to consider, and in any case, the orbitals of nonvalence electrons remain basically unchanged in molecules. A system of terminology and the mathematics necessary to determine the energy states of electrons in molec-ular orbitals are available, but, as was the case with atomic orbitals, a detailed description will not be given here. Again, an understanding of light–matter interactions with reference to the states of single electrons will provide sufficient background for discussion of electronic transitions in molecules. However, some of the more general principles and related terminology of electronic configurations (as opposed to energy states) of molecules will be mentioned.

As indicated earlier, the Schroedinger equation allows orbitals, or probability distributions, to be constructed for electrons in different subshells. Electrons in the lowest-energy, s, subshell could be described by a spherical orbital. Electrons in the p subshell were described by dumbbell-shaped orbitals with the further clarification that a three-dimensional orientation factor was also necessary for a complete description. Higher-energy subshells also exhibit distinctive orbitals, although these were not discussed in detail. The orbitals of valence electrons in molecules can be categorized basically in two ways.

The first type of categorization is rather analogous to the shape of the orbitals of individual atoms. While s orbitals of atoms are spherically symmetrical, σ orbitals of molecules are cylindrically symmetrical about the internuclear axis. There is no nodal plane, i.e., region in which the probability distribution for the electrons is zero, containing the axis. Electrons from any subshells may jointly form a σ molecular orbital. The combination of two s electrons into a σ orbital is most easily conceptualized. The two spherical orbitals simply overlap between the two nuclei, yielding a cylindrical orbital. Because of the overlap, the greatest density for the probability distribution of the electrons in the orbital lies between the two nuclei. That is, the greatest chance of finding either one electron or the other is in the internuclear space. Theoretically, it is not possible to ascribe a probability distribution to either electron alone, since the probability of finding one specific electron cannot be determined. Since both electrons must be considered together, it should be clear, at least intuitively, that the greatest density of the probability distribution will occur where the two initial s orbitals overlap.

Two p orbitals may also overlap to form σ orbitals if the atomic orbitals happen to be oriented in such a way initially that they lie along the internuclear axis in the molecular state. An s orbital and a properly oriented p orbital can also overlap to form a σ orbital. Indeed, electrons from any subshells may overlap with one another to form σ molecular orbitals. Molecular orbitals are generally characterized by a lower energy state than the corresponding atomic orbitals, which results in the stability of the molecular bonds since the transition from the relatively low energy state of the molecular orbital to the high energy state of the corresponding atomic orbitals requires a considerable addition of energy to the system.

The only other molecular orbitals of interest here are termed π orbitals. Electrons in s atomic orbitals cannot partake in π orbitals. However, p electrons or those of higher-energy subshells may combine to form π orbitals. Only the combination of p electrons will be considered here. Although π orbitals are similar to σ orbitals in that they are roughly cylindrical, π orbitals are characterized by having a nodal plane containing the nuclear axis that is analogous to the nodal plane through the nucleus for p atomic orbitals. Only p orbitals that are perpendicular rather than parallel to the internuclear axis combine to form π orbitals. Individual p orbitals of two atoms each of which has the same orientation to the internuclear axes (i.e., perpendicular to the internuclear axis but parallel to one another), then, combine to form a π bond. As previously mentioned, the p orbitals are dumbbell-shaped with a node, or region of extremely low electron density, at the nucleus. In π bonding, the ends of the p dumbbells that are on the same side of the internuclear axis merge to form two separate areas of the π orbital, one on each side of the internuclear axis nodal plane. The nodal plane containing the internuclear axis is a region of absolutely minimal electron density. Although any physical representation or conceptualization of the π orbital would suggest that electrons would be confined in their movement to one half of the orbital or the other with the nodal plane presenting a barrier to exchange between the halves of the orbital, this is an artifact that results from a more or less classic description of a quantum-physical phenomenon. In fact, no such barrier exists and electrons may move anywhere within the orbital.

These two types of molecular orbitals, σ and π, can be further characterized in terms of bonding or antibonding configurations. The descriptions of σ and π molecular orbitals already given are actually for their bonding configuration. As the name implies, bonding configurations are those that promote bonding or maintenance of bonds because the energy states of the electrons and of the molecule as a whole are more favorable. Bonding configurations may be considered the norm for molecular orbitals, while antibonding orbitals occur more often as a result of energy added to a molecular system due, for example, to the absorption of electromagnetic radiation. Antibonding orbitals in one sense can be described as quite similar to bonding orbitals except for the introduction of an additional nodal plane. Thus, the antibonding σ orbital has a nodal plane perpendicular to the internuclear axis, which results in a division of the orbital into two parts separated by a region of very low electron density. The π orbital already has a nodal plane containing the internuclear axis, and the antibonding π orbital introduces another nodal plane perpendicular to the first but parallel to the original p orbitals. The two regions of electron density in the π orbital are thus further subdivided into two regions each. Both σ and π antibonding orbitals are higher energy states than the corresponding bonding orbitals. Furthermore, in comparison to the bonding configuration, that tends to hold the molecule together, antibonding configurations represent an increase in repulsive interactions between the constituent atoms and decrease molecular stability. A further type of orbital is called the nonbonding orbital. This refers to the orbitals of valence electrons that do not take part in molecular bonds in a given molecule. The "free" electron pairs of oxygen and nitrogen are examples of electrons that are often nonbonding.

While the absorption of electromagnetic radiation can be conceptualized in terms of the energy states of individual electrons, and this less than satisfactory view will be followed here for the sake of simplicity, from the molecular point of view, the most common phenomenon associated with electronic excitation is the introduction of electrons into antibonding orbitals. These are frequently π electrons or nonbonding electrons that make a transition to antibonding π orbitals. The symbols b, n, and * are often used to designate bonding, nonbonding, and antibonding orbitals, respectively, although bonding is usually implied in the absence of b. Thus, the excitation transitions that are of most frequent interest for the neurobiologist may be symbolized as $\pi \rightarrow \pi^*$ and $n \rightarrow \pi^*$.

3. Quantum Absorption of Radiant Energy

Quantum mechanics allows, at least in theory, a precise determination of the various energy states of matter. As indicated in the previous sections, the total energy of a molecule, excluding nuclear energies and translational energy, is composed of rotational, vibrational, and electronic energy. Although, in practice, such energy determinations are most easily and accurately made only for diatomic molecules, larger molecules, while greatly complicating the situation, can nonetheless be considered at least in an analogous manner. While of necessity rotational and vibrational energies have been discussed in molecular systems in the foregoing sections, more attention was directed toward individual electrons when considering electronic energies. Many general principles readily transfer from atomic to molecular systems; for example, the Pauli Exclusion Principle and the Aufbau Principle are equally important for molecular orbitals, although they have been presented only for the atomic case. For the sake of brevity, and simplicity, the orbitals of individual electrons have been and will continue to be emphasized at the expense of discussion of molecular orbitals. In any case, for the purposes of this chapter and the usual applications of the neurobiologist, an understanding of the basic principles involved is much more relevant than the ability to generate mathematically accurate and detailed descriptions of the phenomena of interest. Such descriptions are more in the realm of organic and quantum chemists. From the biochemical, or neurobiological, point of view, the relative distribution of energy states is more important than their absolute distribution. And, as will be discussed subsequently, the necessary information concerning the relative distributions of energy states is usually obtained empirically.

As already described, the energy distribution of the electrons in an atom or molecule is discontinuous. Only certain energy values are allowed, and these are defined by various combinations of the four quantum numbers, for atoms, or other characteristics, for molecules. If the absorption of radiant energy by an electron has the effect of raising that electron to a higher energy state, then it should be clear that the higher energy state must also be a discrete one, definable by the same procedures as are relevant to a definition of the ground state. It should also be clear that the energy differential between the initial, usually ground, state and the excited state in any particular case will be of a readily definable magnitude. Thus, the energy differential (ΔE) is equal to the difference between the energies of the initial (E_I) and excited (E_E) states, or $\Delta E = E_E - E_I$. Since the energy of both states can be accurately determined by quantum mechanics, at least in theory, ΔE can also be defined. This energy differential can be viewed as the energy that must be added to a particular electron to move it from a specific initial state to a specific excited state. Addition of more or less energy than ΔE would not exactly move E_I to E_E. Since this is the only transition allowed by the discrete, quantum nature of the distribution of energy states, no transition would occur.

If ranges of possible ΔE values for electrons in various ground states are calculated, it becomes apparent that these energies are of the same magnitude as photons of light from visible and ultraviolet frequencies. Absorption of a photon of visible or ultraviolet light of precisely the correct energy, as defined by ΔE, will raise an electron from one specific state to another. Since $E = hv$, or $v = E/h$, photons of the correct energy (substituting ΔE for E) will be specifically defined by their frequency. Thus, matter and energy, both of which are quantized with reference to h, interact in terms of their quantum characteristics. If the relationship $E = hv$ (or $\Delta E = hv$) cannot be met for light of a particular frequency striking specific electrons, then no absorption of radiant energy is possible. On the other hand, if the energy of the incident photons precisely matches an allowable ΔE, then absorption will readily occur.

The last sentence suggested that only certain values of ΔE are allowable, which is to say that although an electron in a particular ground state could potentially be raised to any one of a number of higher energy states, in fact, only certain transitions occur with more than a negligible probability. As was the case with rotational and vibrational transitions, a series of selection rules defines which transitions are allowable. All other transitions are said to be forbidden. Forbidden transitions do occur, but only in a limited range of circumstances and with a small probability.

Without selection rules, it would seem that any of the electrons of a particular atom or molecule could make a transition to any one of a number of excited states. Heterogeneous light incident upon a sample could produce several possible transitions in each electron in a given ground state because photons equivalent to all relevant values of ΔE would be available. Since there would be electrons in a variety of ground states, a great number of transitions would be possible. If the incident radiation were limited to one frequency or a small range of frequencies, an electron in a given ground state would be subject to either one transition or no transition dependent on the frequency of the incident radiation. However, since there would be electrons in various ground states (dependent on the composition of the sample), the sample as a whole might still exhibit a variety of transitions. The effects of the interaction of light with matter would have been, in the absence of selection rules, most complex.

One selection rule, of several, serves to greatly simplify the situation while at the same time describing the relatively accurately the transitions that actually do occur. It should be noted that this rule will apply to electrons considered individually. Analogous rules exist for descriptions of atomic and molecular states, but this simplest example will be used here. The rule states that transitions that would result in a change of the orbital quantum number (l) of more than 1 are forbidden. That is, transitions from the orbital quantum of the initial state (l_I) to that of the excited state (l_E) such that $l_E = l_I \pm 1$ are permitted. All other transitions are forbidden. When the absorption of energy is considered, only increases in l are relevant.

Although other selection rules further reduce the number of allowable transitions, this orbital-quantum-number-selection rule alone greatly limits the interaction of electrons (and atoms and molecules) with incident radiation. Considering this one rule, only those transitions that are compatible with an increase in l of 1 are allowed. When all the electrons of an atom are considered, the possibilities become more restricted. This orbital-quantum-number-selection rule allows transitions to the next-higher-energy subshell only. Electrons in the outermost subshell can move to a higher, but previously unoccupied, energy state. However, because of the effects of the Pauli Exclusion Principle and the Aufbau Principle, electrons closer to the nucleus will not be able to move to a higher state, since those higher states allowed by the selection rule will already be occupied.

This description of the interactions that occur between light and matter, though providing sufficient background for the purposes of this chapter, has been rather sketchy. However, the points that have been made and their implications are of value. To summarize, both energy and matter are quantized, and their interactions with one another are described in quantum terms. Furthermore, as described by the selection rules of quantum chemistry (not only for electronic but also for rotational and vibrational transitions), the interactions that do occur (i.e., are allowed) can be precisely described and are relatively limited in number.

3.1. After Absorption

As will be discussed below, after absorption of electromagnetic radiation and the consequent transition of a molecule to a rotational, vibrational, or electronic state that is higher in energy than its original state, most molecules lose energy and return to their initial state, or one close to it, in a relatively short time. That is, the high-energy states that are the result of absorbed radiation are quite transient. It follows that there must be a process whereby molecules can effectively lose energy to return to their ground states. There are actually several such processes, which can be separated into two categories dependent on the manner in which the energy is lost. Radiative processes are those in which some or all of the excess energy is emitted in the form of electromagnetic radiation. In nonradiative processes, the excess energy is transferred to nearby molecules, often by collision.

3.1.1. Nonradiative Processes

Although nonradiative losses of energy are certainly deserving of discussion, they apparently have not as yet been treated very rigorously at a theoretical level. Consequently, this discussion will have to proceed in a rather qualitative manner, though this is quite sufficient for present purposes. Three factors seem to be important in determining the course of nonradiative processes. The frequency with which molecules collide is the first factor. The type of excited state the molecule is in, i.e., rotational, vibrational, or electronic, and the amount of excess energy that the molecule possesses are the other factors, although they are obviously related.

The rate at which molecules collide when in the liquid state has been estimated at 10^{13} times per second. Thus, a molecule will experience a collision with a neighboring molecule about once every 10^{-13} sec. This contrasts with a collision rate in the gaseous state of about 10^{10} times per second. On the basis of these figures, one may conclude that any molecular behavior that occurred in less than 10^{-10} sec for gaseous molecules and 10^{-13} sec for molecules in the liquid state might proceed between collisions. For example, molecules rotate about 10^{11} to 10^{12} times per second, the time for a complete rotation, then, being in the range 10^{-12} to 10^{-11} sec. It should be apparent that a molecule in the gaseous state could rotate hundreds of times before colliding with another molecule. On the other hand, the time between collisions for liquid molecules, approximately 10^{-13} sec, does not provide sufficient time for a complete rotation and, for all practical purposes, molecules in the liquid state are not free to rotate. As has been previously discussed, the amount of energy added to a molecule in the course of a rotational transition, and conversely the amount that must be released for a return to the initial state, is relatively small. It appears that this relatively small amount of energy is readily released by molecules. Collisions are quite effective means of energy transfer for rotational energies. The rotational energy is transferred to one or both of the colliding molecules as translational energy, perhaps subsequently resulting in heat.

Vibrational transitions from excited states to lower energy states are not as readily induced by collisions as are rotational transitions. Molecular vibrations seem typically to have a period of about 10^{-13} sec, and thus a molecule must experience collisions at about the same rate as it vibrates. However, it appears that as many as 10^4 collisions on the average may occur before one vibrational transition will be effected. With 10^4 collisions required, occurring once every 10^{-13} sec, an excited vibrational state may be expected to last 10^{-9} sec, which will allow even a liquid molecule to vibrate thousands of times in the excited state. The excess vibrational energy eventually lost probably results in alteration of the translational and rotational energies of both colliding molecules. It is not clear why transfer of vibrational energy is so much less efficient than the transfer of rotational energy. It might be recalled that vibrational transitions do involve somewhat larger amounts of energy than do rotational transitions.

Electronic transitions require even larger energy differentials, and the excess energy is correspondingly more difficult to transfer. When a molecule undergoes an electronic transition, it moves from one of the vibrational levels of the ground electronic state (often the lowest vibrational level for that electronic state) to one of the vibrational levels of the excited electronic state. The new vibrational level is usually not the lowest vibrational level of that electronic state, but rather an elevated level. The molecule, then, not only is in an excited electronic configuration but also is in an elevated vibrational level within this configuration. The magnitude of the energy differential between the ground and excited states requires a substantial transfer of energy before the ground state can be attained.

It seems that it is virtually impossible for such a large amount of energy to be lost at one time. Rather, the energy is lost in successive stages. Collisions between molecules result in a stepwise loss of vibrational energy, with the lowest vibrational level of the excited electronic configuration being achieved quite rapidly, in about 10^{-9} sec, as described above. For the molecule to now move from the lowest vibrational level of the excited electronic configuration to some vibrational level of the ground electronic configuration would require the removal of a large amount of energy in a single collision as well as the rearrangement of the electronic structure of the molecule. It is apparently not very likely for this to occur. Such an end state can be achieved by emission of electromagnetic radiation, as will be discussed below, but it can also occur through a nonradiative process called internal conversion.

Although it has not been discussed previously, between the ground and excited electronic states there will normally be a number of possible, though perhaps not very probable, electronic arrangements. Each of these will have a number of vibrational levels. In many cases, the energy of the lowest vibrational level of the excited state is equal to the energy of one of the upper vibrational levels of the intermediate electronic configuration. At the point where the energy levels of the two electronic arrangements are equivalent, a particular molecular geometry may allow the intermediate state to be entered. Energy may be dissipated through downward vibrational transitions in this intermediate state until a point is reached at which its energy is equivalent to one of the higher vibrational levels of the ground electronic state. At that point, a crossover may again occur, this time from the intermediate state to the ground state. Additional downward vibrational transitions through the levels of the ground state can eventually lead to the lowest vibrational level of the ground state, which is usually the initial state prior to absorption of electromagnetic radiation. Thus, by this combination of vibrational deactivation and internal conversion, a molecule can move from an excited electronic state to the initial ground state by nonradiative processes alone.

This process does not always occur with consistency from one molecular species to another. It would seem that the most difficulty with this form of energy dissipation arises from excited states characterized by electrons with unpaired spins. Most molecules in the ground state have a net spin of zero. That is, there are an even number of electrons, and all of these are paired in orbitals with antiparallel spins. Absorption of electromagnetic radiation usually results in a transition by one electron to a higher-energy orbital. This leaves, of course, two orbitals that contain only one electron. Since these are in different orbitals, there is no longer a requirement that their spins be paired, although the spins of the two electrons most often remain paired. This state in which the spins are paired is termed a singlet state. When the excited electron changes its spin, resulting in unpaired electrons, i.e., electrons having parallel spins, the configuration is called a triplet state. A return to the initial energy state of a molecule in the triplet excited state requires that a singlet state be resumed. The internal conversion path by which a molecule returns from a triplet excited state to a singlet ground state does not seem to be very efficient. Nonetheless, triplet-to-singlet conversions do occur by nonradiative processes.

3.1.2. Radiative Processes

Molecules can also dissipate excess energy by a process in which electromagnetic energy is emitted by the molecule. This radiative dissipation takes, perhaps, as much as an order of magnitude longer to occur than do nonradiative processes. Thus, nonradiative processes compete, under most conditions, quite effectively with radiative processes, and little radiative transfer occurs. However, sufficient reductions in the kinetic translational energies of molecules, for example by cooling, can limit the number of molecular collisions resulting in nonradiative energy transfer sufficiently to allow a significant increase in the radiative process. Radiative transfer is an important process for many molecules even at room temperature, and it will be considered further since it is the basis of fluorescence.

The previous description indicated that molecules in excited electronic configurations move quite rapidly to the lowest vibrational level of the excited electronic configuration. After this point, a relatively large amount of energy must be dissipated to return to the ground electronic state. This can occur due to internal conversion or the emission of electromagnetic radiation, which is known as fluorescence. Since a certain amount of the absorbed energy is dissipated by vibrational deactivation through the vibrational levels of the excited electronic configuration, the amount of energy that must be lost in the return from the lowest vibrational level of the excited configuration to the ground electronic configuration is less than the amount originally absorbed. Thus, the emitted radiation will be of a lower energy, and thus of longer wavelength, than the absorbed radiation. This particular relationship between the frequencies of the absorbed and emitted radiation is almost always true and, as will be seen, is one of the keys to the sensitivity and power of fluorescence as an analytical tool.

In addition to a direct jump from the excited to ground states with the emission of radiation, fluorescence can occur between the lowest vibrational level of an intermediate electronic configuration reached, as described above, by internal conversion and the ground state. However, fluorescence emission can occur only between two singlet states. If the process of internal conversion results in the formation of a triplet configuration, fluorescence can occur only after the triplet has returned to a singlet state. The conversion from singlet to triplet is rather difficult to achieve, and to distinguish it from the more readily occurring internal conversion, it is often called intersystem crossing. When a triplet state is achieved, it is followed by a stepwise vibrational deactivation to the lowest vibrational level of the triplet state. With the molecule having reached this energy level, further dissipation of energy must await a further intersystem crossing back to a singlet configuration. This occurs at a very slow rate. Thus, the emission of radiation in molecules that pass through a triplet state is greatly delayed. While the eventual radiative process is equivalent to fluorescence, it is termed phosphorescence to make a distinction between the time course of the nonradiative processes leading to radiation in each case. The delay between absorption and emission of radiation for fluorescence is in the range 10^{-9} to 10^{-4} sec, while phosphorescence delay ranges from the longer fluorescence times up to several seconds or longer. At present, phosphorescence is of little value as a neurobiological tool and will not be discussed further. However, fluorescence will be treated further below.

3.2. Selectively Indexed References

The discussions in the many preceding sections revolved around what might be considered as fairly general knowledge, and there did not seem to be a need for detailed referencing of the material. However, the discussion drew on many sources ranging from very general ones to some quite advanced treatises. For those interested in a further treatment of the various topics, some references are given here. The references are grouped under several general headings that suggest their content but make no attempt to fully categorize it.

3.2.1. Quantum Theory

Thorough treatment of quantum theory, including more detailed discussions: Adam (1956), Atkins (1972), Barrow (1963), Bauman (1962), Freeman (1968), Grum (1972), Orear (1967), West (1956a), White (1934).

More abbreviated treatment of quantum theory: Alford (1962), Duncan (1956a,b), Jaffe and Orchin (1962), Jorgensen (1962), Lothian (1969), Pesce *et al.* (1971).

3.2.2. Atomic or Molecular Orbital Theory or Both

Thorough treatment of atomic or molecular orbital theory or both: Adam (1956), Atkins (1972), Bauman (1962), Brode (1943), Duncan (1956b), Freeman (1968), Jaffe and Orchin (1962), Jorgensen (1962), Matsen (1956), Orear (1967), Pesce *et al.* (1971), West (1956a), White (1934).

More abbreviated treatment of atomic or molecular orbital theory or both: Barrow (1963), Bladen (1964), Corwin and Bursey (1966), Jaffe (1962), Lothian (1969), Roberts and Caserio (1967), White (1964).

3.2.3. Absorption or Molecular Transitions or Both

Thorough treatment of absorption of electromagnetic energy or molecular transitions or both: Adam (1956), Barrow (1963), Bauman (1962), Brode

(1943), Cheng (1971), Duncan (1956*b*), Hiskey (1955), Jaffe and Orchin (1962), Matsen (1956), Pesce *et al.* (1971).

More abbreviated treatment of absorption of electromagnetic energy or molecular transitions or both: Bladen (1964), Bladen and Eglinton (1964), Grum (1972), Jorgensen (1962), Mellon (1950), Morton (1962), Newman (1964), Roberts and Caserio (1967), Scott (1955).

3.2.4. Fluorescence

Thorough treatment of fluorescence: Pesce *et al.* (1971), Udenfriend (1969), Van Duuren and Chan (1971), West (1956*b*).

More abbreviated treatment of fluorescence: Barrow (1963), Ehrenberg and Theorell (1962), Jaffe and Orchin (1962), Rosenberg (1955), Udenfriend (1962), White and Argauer (1970), Wotherspoon *et al.* (1972).

3.2.5. Infrared

Both *detailed* and *general* treatments of infrared techniques, which are not covered in this chapter but are of relevance regarding vibrational motions: Anderson and Woodall (1972), Clark (1955), Duncan (1956*a*), Eglinton (1964), Jones and Sandorfy (1956), Smith (1964).

3.2.6. Very General Treatments of the Entire Topic

Blum (1950), Burris (1972), Rosenberg (1955), Skoog and West (1963), Stearns (1969), Van Holde (1971), West (1949*a,b*).

3.3. Absorption Spectra

The major role of visible and ultraviolet spectrophotometric and fluorometric techniques in neurobiological work is quantitative, that is, estimation of the concentration of particular neurochemicals. These techniques have also been used qualitatively, although not very frequently by neurobiologists. Qualitative work emphasizes the relationship between molecular structure and the absorbing characteristics of the molecules. A great deal of information has been revealed about the energy states of molecules through analysis of their interaction with electromagnetic radiation. Much of the basic information in the preceding sections derived from such qualitative analysis, and many of the references given in the last section more or less fully detail this as well as the emission photometric techniques that provide complementary data on molecular energy states and structure. Quantitative work relies on previous qualitative analysis to specify the optimum conditions under which the concentration of a particular neurochemical can be assessed. Although qualitative analysis is not of primary importance here, the basic principles that have been derived from it will be briefly discussed as a prelude to a more detailed treatment of quantitative techniques.

Perhaps the easiest way to gain an understanding of absorption spectra is to limit the discussion at first to visible light and leave details of instrumentation for later treatment. Assume that a sample of some known substance is placed in a beam of white light, i.e., light containing photons of all frequencies. Some of the photons will be absorbed by the sample dependent on both the frequency of the photons and the energy states of the sample molecules. The principles presented in the previous sections showed not only that photons of specific frequencies would be absorbed but also that it is possible, at least theoretically, to predict which frequencies will be absorbed. Even the simplest molecules will absorb light at many frequencies corresponding to the many vibrational and electronic transitions that are allowed for the molecule in question. The beam of light as it leaves the sample will be reduced in intensity as a function of the number of frequencies at which absorption occurred and the number of photons absorbed at each frequency. Depending on the concentration and properties of the sample, it may be possible for the human eye to detect the decrease in the intensity of the beam and, perhaps, a shift from white to some color. However, such a method of detection would provide a very small amount of the information that could possibly be derived about the interaction of the sample with the beam of light.

On the other hand, the light leaving the sample could be passed through a prism to disperse it according to wavelength. The dispersed light produces a spectrum with light of different wavelengths arranged according to wavelength. If the light from the prism is allowed to fall upon a photographic plate, a permanent record of the spectrum may be obtained. If a spectrum of the incident radiation were also produced without passing the light through the sample, all frequencies would be represented, and variations in the intensity of the developed image across the spectrum would be a function of the intensity of the incident light at each frequency. This variation in intensity also depends on the sensitivity of the photographic emulsion to light of different wavelengths, but this source of variation can be easily calibrated and eliminated from further consideration. Color emulsions can be used, and the distribution according to frequency in the resulting spectrum is easily discerned. Black-and-white emulsions are, in fact, more commonly used, in part because of their greater sensitivity to light. The distribution of frequencies in a black-and-white spectrum can be easily calibrated, thus allowing the relationship between intensity and frequency to be determined. A similar spectrum produced by the light after it has passed through the sample will appear discontinuous. Rather than all frequencies being represented, as in the spectrum of the incident light, the intensity at some frequencies will be greatly reduced, resulting in dark or black bands or lines across the spectrum at the points where those frequencies would normally have affected the emulsion.

The absorption spectrum of a molecule, then, consists of a series of discontinuities in the spectrum of the incident radiation. These discontinuities correspond to the absorption by the sample of particular frequencies of light.

The frequencies absorbed in turn correspond to certain vibrational and electronic transitions of the molecule that can, as previously discussed, be related to the energy states and other characteristics of the molecule. As already mentioned, such absorption spectra typically consist of a great number of discontinuities. Furthermore, a variety of factors can influence the characteristics of the spectrum. Although it is theoretically possible to account for every feature of an absorption spectrum, this is a very complicated procedure, often impossible with larger molecules, and of little relevance as it applies to quantitative methods. The references mentioned above should be consulted for a more detailed discussion of the interpretation of absorption spectra. What is of interest for this discussion is the fact that absorption spectra can, and in fact often must, be characterized by a relatively small number of more prominent features that can be related to classes of molecular structures and that provide the basis for quantitative methods. Some of these prominent characteristics of the absorption spectra of visible and ultraviolet light will be discussed.

3.3.1. Relevant Features of Absorption Spectra

For any but the simplest molecules, it is generally impossible to determine all the allowed vibrational energies because of the many vibrational degrees of freedom of polyatomic molecules with their correspondingly complicated pattern of allowed energies. Obviously, the energies of various electronic arrangements of such molecules as a function of the various possible vibrational modes are also difficult to determine because of the complexity of the latter. In any case, the spacing between the lines in an absorption spectrum corresponding to the frequencies absorbed in the various vibrational states of a particular electronic configuration is relatively small due to the relatively close spacing of vibrational energy levels and instrumental limitations in the production of an absorption spectrum, which make it difficult or impossible to resolve the vibrational fine structure. What is obtained, then, is a rather broad band of absorption corresponding to the various unresolved frequencies of absorption due to the vibrational components of the particular electronic transition. The quantitative work of interest to the neurobiologist usually involves samples studied in solution, which, as previously discussed, further blurs the vibrational fine structure. What can be more easily done is to relate an observed absorption band to some electronic transition in the molecule without attempting to understand the vibrational structure of the excited state. This, in fact, is all that is necessary for an understanding of quantitative methods. Four different classes of molecules, or groups within molecules, are recognized as being responsible for most of the electronic transitions of interest. These will be mentioned briefly, and then the first two, which are of most practical significance for the neurobiologist, will be explored in greater detail.

Compounds containing hydrogen and carbon atoms only provide the model for the first class of molecules, or groups within molecules, that absorb visible or ultraviolet radiation. Saturated hydrocarbons, in which each carbon atom is bonded to four other atoms by four single bonds, or four separate pairs of shared electrons, do not absorb radiation in the visible or ultraviolet regions of the spectrum. In saturated hydrocarbons, all the valence electrons of the molecule are in single bonds. They cannot be rearranged to an excited state by absorbing radiation without a disruption of the molecular bonds. Radiation from the far-ultraviolet region of the spectrum provides sufficient energy for such disruption, but discussion of ultraviolet radiation in this chapter excludes radiation from the far-ultraviolet.

In unsaturated hydrocarbons, more than a single pair of electrons is shared between two adjacent carbon atoms; e.g., in a double bond, two pairs of electrons are shared. Rearrangement of the electronic configuration to an excited state is possible in unsaturated hydrocarbons without a resultant disruption of bonds. A double bond between two carbon atoms is formed from two components. One bond involves an electron from each carbon atom, each of which occupies an orbital along the internuclear axis. The overlapping of these two orbitals along the internuclear axis results in a σ bond, which was discussed in a previous section and which is also the typical single bond between carbon and another atom. The second component of the double bond involves one electron from each carbon atom, each of which occupies an orbital oriented perpendicularly to the internuclear axis. These are called π electrons, and the merging of their orbitals both above and below the axis of the molecule, as has been previously discussed, results in a π-bonding orbital.

Absorption of radiation by a double bond between two carbon atoms can result in the rearrangement of the π orbital to an excited state while leaving the σ bond, and thus the linkage between the two carbon atoms, intact. The rearrangement results in each of the π electrons occupying its own orbital about the appropriate carbon atom. The electrons are said to be in anti-bonding π orbitals, designated as π^* orbitals. The transition is then $\pi \rightarrow \pi^*$. The π bond does not actually exist any longer; the σ bond is all that holds the two carbon atoms together, and therefore the atoms are less strongly bound.

A second class of absorbing structures involves valence electrons that do not actually partake in a bond in the molecule. An example of such nonbonding (n) electrons are the four free outer electrons of the oxygen atom in the carbonyl group. The two other valence electrons of the oxygen atom in this group form a double bond with the carbon atom consisting of both a σ- and a π-bonding orbital as in the double bond between two carbon atoms. Both the n and the π electrons can be raised to an excited state on absorption of electromagnetic radiation. In both cases, the electron occupies a π^* orbital in the excited state. However, the n $\rightarrow \pi^*$ transition requires considerably less absorbed energy than the $\pi \rightarrow \pi^*$ transition.

The ions of the transition metals and their complexes form another class of absorbing structures. Unlike the previous two classes, in which atoms that are from the first row of the periodic table and therefore contain s and p electrons provided the primary constituents of the absorbing structures, the metal ions of this class have valence electrons in d-subshell orbitals. There are five d orbitals, each of which can be occupied by two electrons at most. The structures of interest are those in which the ion of a transition metal has other chemical groups attached, or coordinated, to it. Many metal ions form coordinated ions, sometimes with two or four but most often with six coordinating groups, or ligands. Thus, the ion of nickel, Ni^{2+}, may complex with six ammonium, NH_3, ligands to form the coordinated ion $[Ni(NH_3)_6]$. Although the energies of the electrons in the d orbitals of the uncoordinated ion are equivalent, the arrangement of the ligands about the metal ion in the coordinated ion changes the energies of some d orbitals in relation to other d orbitals because of the electrical field produced by the electrons of the ligands. The basic transition that occurs on absorption of electromagnetic radiation by such coordinated ions is that of an electron in a low-energy d orbital moving to one of the higher-energy d orbitals. The absorption of electromagnetic radiation by ions of transition metals is dependent on the d orbitals being partially filled. The complexing does not occur if all the orbitals are filled. Absorption is also dependent on the presence of electron-repelling groups, or ligands, located appropriately around the metal ion to split the energies of the d orbitals. This phenomenon belongs, of course, in the realm of inorganic chemistry. However, although it will not be discussed further, it is of some interest in the present context, since the actual structures quantitatively assayed in some reactions of interest to the neurobiologist are, in fact, metallic ion complexes as end-products of those reactions.

3.3.2. Chromophores and Absorption

The realization that certain molecular structures led to characteristic colors of molecules resulted in these structures' being termed chromophores, or color-bearers. The relationship between color and a particular chromophore, however, is not absolute. Rather, the color produced is dependent on the position of the chromophoric group within the molecule, the presence of other chromophoric groups within the molecule, and the presence, also within the molecule, of other groups that do not produce color themselves but that affect the color produced by chromophoric groups. These latter groups have been termed auxochromes, or color aids. Ultraviolet and visible absorption is usually a function of chromophores, rather than of the whole molecule. The more common chromophores are usually covalently bonded, unsaturated groups.

Chromophores have functional groups that absorb in the visible and near-ultraviolet when they are bonded to a saturated residue possessing no unshared, nonbonding valence electrons that do not absorb in this region. An example of such a residue would be a hydrocarbon chain. Auxochromes contain functional groups that have nonbonding valence electrons and absorb only in the ultraviolet or far-ultraviolet, exhibiting a $n \rightarrow \pi^*$ transition. When an auxochrome is combined with a chromophore, the absorption band of the chromophore will frequently be shifted to a longer wavelength and the intensity of the absorption band will be increased. Alterations in the characteristic absorption of a chromophore, whether due to its combination with auxochromes or with other chromophores, may be resolved into two effects, changes in the intensity of the absorption band and shifts in the frequency of the radiation absorbed. A move to longer wavelengths of absorption is called a bathochromic shift, while the move to shorter wavelengths is called a hypsochromic shift. The hyperchromic effect is an increase in the intensity of an absorption band, while a decrease in intensity is called a hypochromic effect.

The effects on absorption of the presence of more than one chromophoric group within a molecule is dependent on their positions in the molecule. When the chromophores are isolated, which is to say that there are at least two saturated bonds between each pair of chromophores, each chromophore tends to absorb independently, and the absorption of the molecule as a whole approximates the sum of the characteristic absorptions of the individual chromophores. When two chromophores are separated by only one single bond, they are said to be conjugated. Conjugated systems usually result in a bathochromic shift with hyperchromic effects; i.e., the wavelength of the absorbed radiation is increased, as is the intensity of the absorption band. The greater the degree of conjugation, i.e., the number of conjugated chromophores, the greater are the bathochromic and hyperchromic effects. In general, then, conjugation results in a new absorption spectrum. When two or more chromophores are bonded to one another without intervening single bonds, the system is said to be continuous. Continuous chromophoric systems also result in an absorption pattern that differs from that of the original chromophore. In comparison with a conjugated molecule containing the same chromophores, the continuous chromophoric system exhibits a greater bathochromic shift but less of a hyperchromic effect.

In conjugated or continuous systems, bonding is not restricted to the carbon atoms of the individual chromophores, but can be shared with adjacent carbon atoms as well. Thus, a π electron distribution over all the chromophores can be visualized. The π electrons are said to be delocalized. The energy levels of such delocalized molecular orbitals are more closely spaced than orbitals that are not delocalized. Thus, the energies of absorbed quanta necessary to produce a transition to an excited state are smaller for conjugated and continuous systems, and so absorption occurs at longer wavelengths.

The basic principles just outlined are applicable whether the multiple chromophores of a molecule are the same or whether they differ in structure. Of course, when the chromophores differ from one another, the resulting absorption spectrum can become even more complex because of the different initial spectra of the individual chromophores. Other factors can also influence the appearance of the spectra. In particular, different solvents for the absorbing species may alter the spectrum of that species because of the ability of some solvents to affect the electronic distribution within solute molecules. More advanced treatment of chromophore absorption as well as a consideration of the properties of a wide variety of chromophores will not be presented here, but it is available from other sources (Bauman, 1962; Brode, 1943; Calder, 1969; Cheng, 1971; Lothian, 1969; Morton, 1962).

A knowledge of the basic empirical findings of qualitative absorption spectroscopy in the visible and ultraviolet regions of the electromagnetic spectrum, as discussed above, is valuable for the understanding of quantitative methods in several senses. Of course, the same theoretical and practical principles transfer from the use of one technique to the other. Second, many important molecules of neurobiological interest contain chromophoric groups and can be quantitatively assessed using absorption techniques. Qualitative methods are important in themselves as aids to the identification of the compounds of interest and as a check that the quantitative method, as it is being developed, is actually assaying the compound of interest. This methodology is also invaluable in establishing the most effective conditions under which a quantitative assay should be run, including such variables as solvent composition.

Many compounds of interest to the neurobiologist cannot be measured directly by absorption techniques. These include molecules that do not absorb in the visible or ultraviolet region, as well as enzymes, which are basically too complex to deal with using such techniques. The nonabsorbing molecules can often be treated chemically to form a chromophore-bearing derivative, while the activity, though not the absolute amount, of various enzymes can be assessed by following, spectrophotometrically, the formation of chromophore-bearing enzyme products. These products are usually analogues of the naturally occurring product of the enzyme's action. In addition to the usefulness of qualitative techniques as briefly outlined above, the extensive empirical work that has been done with various chromophoric groups has provided a vast array of molecules that can be applied in quantitative assays.

4. Quantitative Absorption Spectrophotometry

While many of the procedures and much of the instrumentation for qualitative work are applied directly to quantitative techniques, with in fact some simplification, additional theoretical considerations are necessary before the quantitative method can be fully understood.

4.1. Laws of Absorption

Consider again the simple experimental setup used as an example in the previous discussion. Rather than white light as the source of radiation, monochromatic or homogeneous light (i.e., light restricted to only one wavelength) is used. Although the sample is contained in a sample cell, for the time being the effects of the cell material itself on the light passing through it may well be ignored. The first law of absorption, known variously as Bouguer's Law, Lambert's Law, or the Bouguer–Lambert Law, expresses the relationship between the proportion of light absorbed by a transparent medium and the thickness of the medium. If the medium is imagined to be divided by a number of cross sections of equal thickness and lying perpendicular to the incident radiation, every cross section will absorb the same proportion of the light incident upon it. For example, if the intensity of the initial beam of light is taken to be 1, and each cross section absorbs 1/20th of the radiation, then the amount absorbed by the first cross section will be 0.05, and the intensity of the beam leaving that cross section will be 0.95 (i.e., $1 - 0.05$). The intensity of the beam incident upon the second cross section is, then, 0.95, and the amount absorbed is $0.95 \times 0.05 = 0.0475$. The intensity of the beam leaving the second cross-section is $0.95 - 0.0475 = 0.9025$. The same process is repeated for each cross section or layer of the medium.

Though a solution to this problem of how much light will be absorbed by a sample can obviously be derived with a differential equation, it can also be stated, generally, with a logarithmic equation. This statement of the law is, then, $\log_e (I_0/I) = \alpha b$, where I_0 is the intensity of the incident light beam, I is the intensity of the light beam after passing through the absorbing medium, b is the thickness of the absorbing medium in centimeters, and α is an absorption coefficient characteristic of the medium. As denoted by e, natural logarithms are used. If logarithms to the base 10 are used, the Bunsen–Roscoe extinction coefficient, K, replaces α. The constant K depends on the medium examined. Since it is equal to $[\log (I_0/I)]/b$, it is the reciprocal of the medium thickness required to weaken the intensity of the incident light by one tenth. Because of the conversion from logarithms to the base e to base 10, $\alpha = 2.303K$. The absorption coefficients α and K are applicable only to pure materials, since they contain no concentration factor.

A second law, Beer's Law, is necessary before solutions of differing concentrations may be considered. Beer's Law takes into account the fact that a photon of light can be absorbed by a molecule only if it actually strikes the molecule. The number of photons that will collide with molecules depends on the number of molecules in the path of the incident light beam, and Beer's Law states that the amount absorbed by a sample solution is proportional to the number of absorbing molecules through which the light passes. If a sample absorbing at a particular wavelength is dissolved in a solvent that does not absorb at the same wavelength, the amount of light absorbed is proportional to the concentration of the solute. As has been implied, this proportionality applies only to parallel, monochromatic light. The effects of deviations from this really unattainable ideal will be considered later. Incorporation of the absorption-proportional-to-concentration aspect of Beer's Law into the first law of absorption yields $\log (I_0/I) = abc$, in which a is the absorptivity, a property characteristic of the molecules of the sample. The absorptivity is independent of the concentration, denoted by c. The thickness of the absorbing medium is given by b. This fully summarizes the laws of absorption. Since the effect of the concentration of the absorbing substance has been the most important practical aspect of these absorption laws, the equation just presented is usually referred to simply as Beer's Law. Most spectrophotometers actually measure a quantity called absorbance, A (sometimes called optical density), which can be put in the form of the previous equation $\log (I_0/I) = abc = A$. When the concentration of the sample (c) is expressed in moles per liter and the thickness of the sample solution (b) is expressed in centimeters, the absorptivity a is called the molar absorptivity of ϵ. The molar absorptivity, ϵ, is sometimes called the molar extinction coefficient. The absorbance, A, is a dimensionless quantity determined experimentally. Since $abc = A$, $a = A/b$ (cm) c(mol/liter) $= A/b$(cm) c(mol/1000 cm^3) $= A/bc \times 1000$ cm^2/mol. Therefore, $\epsilon = A/bc \times 1000$ cm^2/mol. Molar absorptivity, then, has units of 1000 cm^2/mol, although ϵ is rarely expressed in units. Another expression sometimes encountered is the transmittance, T, defined by the equation $\log (1/T) = abc$. As this equation implies, the transmittance refers to the percentage of the incident light beam transmitted by the absorbing medium, or $T = (I/I_0)$.

According to the relationship described by Beer's Law, the amount of light of the appropriate frequency absorbed by a molecular species is a direct function of the concentration of the solution being examined or of the length of the light path through the solution, or of both. The absorbance, A, as shown above, is equal to abc. Since a is a constant characteristic for a given molecular species, i.e., independent of concentration and path length, A will vary linearly as a function of b or c or both. If the path length, b, is held constant while A is determined for solutions of differing concentrations, a plot of A by c, i.e., absorbance by concentration, will be linear. Similarly, if c is held constant and A is determined for different path lengths, then a plot of A by b will also be linear.

When these conditions of linearity obtain for a particular procedure, then Beer's Law is said to hold. However, these conditions are frequently not obtained, and in such a case Beer's Law is not applicable. Although deviations from the relationship described by the law might suggest it to be invalid, there apparently has never been an instance in which Beer's Law has been shown to be invalid. The explanation for this seeming paradox is that it has been possible to demonstrate a variety of factors that, though they limit the practical applicability of the law, do not render it theoretically invalid. These factors will be discussed below, as will the techniques available to either increase the likelihood of Beer's Law holding or allow quantitation of samples under conditions in which the law does not apply. Before proceeding with that discussion, an overview of the structure and function of the appropriate instrumentation will be presented.

4.2. Instrumentation

4.2.1. Spectrophotometric Instrumentation

A spectrophotometer is actually a combination of two instruments. The first, a spectrometer, is a unit capable of producing light of any particular wavelength. A photometer, the second component, is a unit that measures the intensity of light. Other types of instruments can be used for many of the same tasks and will be briefly mentioned. However, the modern spectrophotometer is not only perhaps the most versatile type of instrument available but also, because of its wide commercial availability, probably the most commonly employed instrument. The major portion of the following discussion will be devoted to spectrophotometers. Further discussion of other forms of instrumentation can be found elsewhere (e.g., Calder, 1969; Mellon, 1950).

4.2.2. Nonspectrophotometric Instrumentation

The two basic types of instruments that are not classified as spectrophotometers are colorimeters and filter (or abridged) photometers. As the name suggests, colorimeters rely on variation in the color of a solution with changes in the concentration of the solute that produces the color. In contrast to spectrophotometric instruments, natural or white light is used in the incident light beam, and the human eye is used in place of a photometer to assess absorption. Obviously, such methods relying on human vision are applicable only for absorption occurring in the visual region of the electromagnetic spectrum. Quantitative colorimetric determinations require a visual matching between the color density produced by the sample solution and the colors of standard solutions in which the concentration of the substance being analyzed is known. It is also possible with certain instruments to compare the color density of the sample solution with standardized color disks. With either method, of course, the human element is critical, and a considerable amount of variability must therefore be expected. In addition to the term colorimeter, such instruments are often called color comparitors.

Filter photometers eliminate the variability due to human judgments and allow not only a greater experimental precision but also a decrease in the time needed to complete analysis. This advance is made possible by the use of photoelectric instrumentation, a photometer, rather than the eye to measure absorption. Additionally, the original white light is passed through a filter or filters prior to being beamed upon the sample. The filters function, in general, to transmit only a limited range of wavelengths from the white source of light. Thus, the sample may be irradiated largely by light of the frequency that is absorbed by the sample molecules. Although filters are not capable of isolating a single wavelength of light, the range of wavelengths passed by filters is considerably smaller than that of the source light. Since light of relatively few wavelengths is very strongly absorbed by any particular molecular species, the reduction in the intensity of a beam of white light due to such limited absorption will itself be limited. As pointed out above in the discussion of Beer's Law, the empirical datum of interest is the ratio between incident and transmitted light. As a first approximation, the amount of light of absorbable frequencies actually absorbed can be taken as a constant factor determined by the intensity of the incident light at that frequency. Thus, decreasing the amount of nonabsorbable light—i.e., light of frequencies that are not absorbed by the system under investigation—while allowing the intensity of the absorbable frequencies to remain constant will increase the sensitivity of the assay method. There is another advantage to restricting the number of wavelengths included in the incident beam. Although a molecule of interest will exhibit strong absorption at a relatively limited number of frequencies, other constituents of the solution are also likely to absorb, though at different frequencies. If the incident light is heterogeneous, then, virtually all components of the sample solution will be able to absorb, and the amount of light absorbed will not depend only on the concentration of the molecules of interest. When the range of wavelengths included in the incident beam is restricted so that the only molecules that can absorb are those of interest, the increased sensitivity of the method is obvious.

As indicated above, other sources should be consulted for a further discussion of colorimetric and filter photometric instrumentation and methodology. The manufacturers of the various instruments can provide even more detailed information about the particular models they produce. Grum (1972) has provided a fairly comprehensive listing of the photometers and also spectrophotometers that were commercially available at that time. This listing also provides numerous fundamental characteristics of each of the more than 35 models covered.

4.3. Spectrophotometer Components

Although the sophistication of the instruments available today provides for extremely simple operation readily managed by persons with very little training, there are many occasions on which the researcher needs a working knowledge of spectrophotometer operation. This will be discussed here, as will some of the fundamental aspects of instrument design. Since the optics and electronics employed in both the visible and ultraviolet regions of the spectrum are basically the same, principles of operation for both regions will be discussed together.

The essential components of a spectrophotometer include a source of radiant energy, a monochromator to allow isolation of the separate frequencies emitted by the radiant source, a component capable of detecting radiant energy, and a component that indicates the intensity of the energy detected. Additionally, a means of placing the sample into the light beam must be included either before or after the monochromator. In virtually all units that function in the ultraviolet as well as most of those that can be employed only in the visible region, the sample is usually placed after the monochromator. Each component will now be treated separately.

4.3.1. Radiation Sources

The ideal spectrophotometer would allow the selection of specific frequencies of radiation for passage through the sample. Although it is possible to select radiation sources that emit radiation at selected frequencies only, such a strategy has two shortcomings. In the first place, every time a different wavelength of light was needed for the incident beam, a new light source would have to be employed. Second, it would be impossible to have different light sources capable, in total, of producing all frequencies of radiation. Since monochromators produce the same result, namely, isolating radiation by frequency, the best source of radiation has different character-istics. However, even with the use of monochromators, light sources that have limited emission frequencies are sometimes of value.

In general, the best radiation source would be one that produces light at all frequencies within the visible and ultraviolet regions. Furthermore, the intensity of the radiant energy should be constant over all frequencies radiated. Stray radiation becomes problematic whenever the intensity of the radiation is very much greater at one frequency or group of frequencies than at another. This results in radiation of one frequency appearing in the incident beam even though the monochromator is not set to pass that frequency. If the source is uniform with regard to intensity, it is easy to design monochromators that trap the great majority of radiation at frequen-cies that are not desired. As might be expected, this ideal radiation source has not been attained, although the types to be mentioned can closely approximate the ideal.

Perhaps the most common sources of radiation are tungsten-filament lamps, in which a coiled tungsten filament is placed in a gas-filled enclosure, or bulb. These lamps provide a high-intensity output, and when the filament is straight, as in the ordinary automobile headlamp, the radiation is easily imaged upon the entrance to the monochromator. Since glass absorbs in the ultraviolet, the normal tungsten-filament lamp emits little energy in this region. However, if the bulb is made of a material that does not absorb in this region, or has a window made of such a material (e.g., silica), the tungsten lamp can also serve in the ultraviolet. Running the lamp above its normal voltage increases its output, especially at shorter wavelengths because of increased color temperature. This enhances their usefulness in the ultraviolet, where there normally is a drop-off in intensity. A modification of this lamp, the quartz-iodine lamp, places the tungsten filament inside a fused silica bulb containing iodine vapor. This allows the filament to be heated to higher temperatures without deterioration and results in an appreciable increase in ultraviolet energy.

Another type of lamp utilizes the radiant energy produced when an electrical arc is passed through hydrogen under reduced pressure. A similar lamp employing deuterium produces higher-intensity radiation while also having a longer working life. The tungsten-filament bulb and its variations and the hydrogen and deuterium arc lamps are the most common sources of radiation. They are all characterized by a relatively continuous emission spectrum of fairly constant intensity.

A discharge, or arc, lamp filled with xenon gas under high pressure provides much greater intensity in the ultraviolet than the lamps mentioned above. However, the light is produced as a small, albeit very intense, spot that is somewhat difficult to stabilize spatially. The xenon arc also produces much greater intensities in the visible region, which can lead to problems with stray radiation. High-pressure mercury discharge lamps emit a series of intense lines of radiation that are broadened under atmospheric pressure to provide a degree of continuity between one another. However, the radiation emitted is not really continuous and is not of much value for absorption work. On the other hand, both mercury and xenon discharge lamps are of considerable value in fluorometric work.

Most of the lamps mentioned above can be operated on normal alternating current, although some require direct current. Since the intensity of the radiation emitted at different frequencies varies with the voltage applied to the lamp, it is essential that the power driving the lamp be stable. Most instruments provide a satisfactory degree of voltage regulation to eliminate this source of variability.

4.3.2. Monochromators

Whether, as is most common, a source of continuous radiant energy is used or a discontinuous source, an essential requirement for spectrophotometric work is that the number of different wavelengths of light included in the incident beam be as restricted as possible at any given time. Monochromators serve this function. Although there may be many variations in the detailed structure of different monochromators, they all function in a similar fashion. The description to be given here is aimed at the properties of monochromators in general.

Basically, a monochromator is an enclosed device that is light-tight except for two openings, or slits. Light from the radiant source enters the monochromator through one of the slits, and light of the selected frequency exits through the other slit. The inside of the device is painted black to absorb light of all frequencies except those that are being passed. There is a very large reduction in the intensity of the light leaving the monochromator compared to that entering, since the majority of the wavelengths of which the initial light is composed are trapped by the device. It is important, then, that the size of the entrance slit and the spatial orientation of the light source coincide to allow the maximum amount of initial radiation to enter the monochromator. The exit slit of the monochromator can also be adjusted to vary the intensity of the light leaving it. However, as will be mentioned below, such an adjustment has an effect not only on the intensity of the beam of light but also on the wavelengths that are included in the beam. Practically speaking, adjustments of the exit slit must always derive from a compromise between beam intensity and spectral purity.

Of course, the key function of the monochromator is the isolation of light of specific wavelengths from the heterogeneous source light, thus allowing transmission of the isolated light through the exit slit and into the rest of the spectrophotometric system. There are two fundamental methods for isolation of light according to wavelength that are used in monochro-mators. The first method makes use of the refraction of light and its subsequent spectral dispersion when the refracting object is a prism. In the second method, light is dispersed by diffraction on an optical grating. A good discussion of both types of monochromators is included in Bauman (1962), from which much of the following description is taken.

4.3.2a. Prism Monochromators. As was mentioned previously, the direction of travel of a light beam is altered as it obliquely passes from one transparent medium into another. The initial displacement of the beam from an orientation perpendicular to the interface between the two media is called the angle of incidence. The displacement of the beam from the perpendicular after the interface has been crossed is the angle of refraction. As the beam passes from a less dense to a more dense medium (e.g., from vacuum or air to water or glass), it is bent toward the perpendicular, which is to say that the angle of refraction is less than the angle of incidence. Conversely, if the beam passes from a more dense to a less dense medium, the reverse effect on the beam is seen. The ratio between the sine of the angle of incidence and the sine of the angle of refraction is a constant for any two media. Additionally, the ratio is constant only for a particular wavelength of radiation. The value will vary with other wavelengths. When a vacuum is used as a standard, less dense medium and light is passed from it into other media, the values for the ratios of the sines of incident and refracted angles are called the refractive indexes of the various more dense media. This, of course, is still dependent on the wavelength of the light used. Since all refractive indexes are based on the same less dense medium (i.e., a vacuum) for the incident angle, they may readily be compared. Furthermore, this makes it an easy matter to determine the amount of refraction that takes place between any two media based on the ratio of the refractive indexes of the two media. The following equation summarizes the determination of refractive indexes (i.e., between a vacuum and any other medium): $(\sin i)/(\sin r) = n$, with i representing the incident angle, r the refracted angle, and n the index of refraction. Now, if the refractive indexes of two media are known (i.e., n_1 and n_2), the angle of refraction of a beam passing from one medium (with a refractive index of n_1) to the other for a particular incident angle can be determined as follows: $(\sin i)_{n_1}/(\sin r)_{n_2} = n_2/n_1$. This relationship is most useful, since, of course, the neurobiologist is rarely interested in the special case in which one of the media is a vacuum.

When a beam of light passes from one medium into another and then back into the first medium (e.g., from air to glass and back into air), the direction of the beam when in the first medium (air) will be unchanged on passing out of the second medium (glass) if both the interfaces between the media are parallel (e.g., when the glass plate has parallel faces). The beam will, however, be displaced to the side. Basically, even though many fre-quencies of heterogeneous light are refracted differentially on entering the glass and therefore take separate paths through it, they will be reunited on returning to air. On the other hand, if the two interfaces are not parallel, the paths taken by the light of different frequencies will not be reunited. Rather, light of each frequency will be refracted out of the second medium at a different angle. This is what occurs in a prism. Heterogeneous light that is dispersed on entering the prism is further dispersed on leaving it. Furthermore, since the degree of dispersion, actually refraction, at each interface is directly dependent on wavelength, the light leaves the prism in the form of a spectrum. Any material absorbs light at some wavelength, which can occur at any portion of the electromagnetic spectrum. In general, the materials used in the construction of prisms for visible and ultraviolet instruments absorb only at shorter wavelengths. In practice, the absorption range sets the upper limit for transmitted frequencies. All lower frequencies of radiation are transmitted. However, in addition to the fact that the refractive index varies systematically with wavelength, the change in refractive index with wavelength is not a linear function. Rather, the shorter wavelengths immediately below the absorption band of the medium are refracted, or bent, more strongly than the longer wavelengths further below the absorption band. The function relating refraction to wavelength is more exponential in nature, and the end result for all materials of interest here is that the degree of dispersion produced by a prism is much greater for the shorter wavelengths of light. The importance of this relationship, stated another way, is that the light of differing frequencies in a spectrum is much more widely spaced at higher frequencies, and thus discrimination of one frequency of light from another is potentially easier in that portion of the spectrum.

4.3.2b. Materials for Prisms. Due to dual considerations of absorption by the prism material and variations in dispersion dependent on wavelength, the choice of a prism material must be based on the region of the electromagnetic spectrum for which an instrument is intended. Furthermore, since ideally the wavelengths of interest must be transmitted by all portions of the instrument with absorption taking place due to the sample material alone, discussion of the absorption properties of prism materials applies, as well, to other components such as lenses, mirrors, and sample containers (cells). Although the ideal may be difficult to achieve for a wide range of frequencies, and such deficiencies can often be readily managed, it is necessary even when some absorption is due to the instrument itself, as opposed to the sample, that a sufficient amount of light be transmitted to allow an accurate measurement of the absorption by the sample.

Materials can be categorized according to the region of the spectrum in which they are most efficient and, since it is more than a minor consideration, most cost-effective. The cost and complexity of an instrument generally increase with its capability at shorter wavelengths. Thus, although it is possible to obtain an instrument that functions in both the visible and ultraviolet regions, such an instrument would not be practical if used only for work in the visible region. In any case, other factors that limit versatility must be considered. Materials can be categorized as most valuable in the visible and near-ultraviolet, in the ultraviolet, and in the far- or vacuum-ultraviolet. The latter presents special problems, since light in this region is absorbed by air. For instruments to function in this region of the spectrum, it must be possible to evacuate them. The increased design and cost problems this presents as well as the limited versatility of such instruments renders them of little interest to the neurobiologist. This discussion will focus primarily on instrumentation for the visible and ultraviolet regions.

The least expensive, yet nearly ideal, prism material for the visible and near-ultraviolet is glass. It is possible to vary the amount of dispersion for various regions within this part of the spectrum by manipulating the composition of the glass. Flint glass, which is a compound containing lead, is used most commonly. When the lead content of the glass is relatively low, the prism is most efficient at the higher frequencies within this range. A denser flint, i.e., with a higher lead content, is more practical for the lower frequencies of the visible region and into the very near-infrared. Of course, it is not necessary to emphasize one region of the spectrum over another, and more versatility can be achieved with flint glasses that function through-out the visible and near-ultraviolet.

Glass prisms are not suitable for work in the ultraviolet region. Silicon dioxide, or silica, is an economical though efficient material for work in the ultraviolet region. Quartz, which is a crystalline form of silica, transmits slightly further into the ultraviolet than does silica. Absorption apparently due to trace impurities limits the frequency range of transmission for both materials. Silicon tetrachloride is a much purer form of silica that is transparent into the near-vacuum-ultraviolet. Though silica prisms obviously extend the working range of the monochromator into the ultraviolet, silica exhibits very little dispersion in the visible region of the spectrum when compared to glass. Glass prisms are preferred in this region. In fact, for best results, it is preferable to exchange a glass for a silica prism, or vice versa, dependent on which region of the spectrum is being examined. When this is possible, it is nonetheless a rather inconvenient procedure that often leads to an acceptance of the relatively poorer performance of silica prisms in the visible region.

Special materials that are generally of limited value in the visible and near-ultraviolet are necessary for both the vacuum-ultraviolet and infrared regions. Just a few examples will be mentioned. Calcium difluoride and lithium fluoride prisms may be used in the vacuum-ultraviolet, though grating monochromators are perhaps preferable in this region. Sodium chloride, potassium bromide, thallium bromide-iodide, cesium bromide, and cesium iodide are more or less common materials for use in the infrared. In general, any material for use in the vacuum-ultraviolet or the infrared will be found to absorb somewhere in that region and therefore limit slightly the range of frequencies that can be examined.

There are obviously many prism materials that are useful for one or more regions of the spectrum. While the absorption band(s) of a material has a clear influence on its applicability in one region or another, the relationship between dispersion and wavelength is also important. This relationship becomes slightly complicated when the absorption band occurs in the middle of an otherwise useful region of the spectrum. This is particularly important with materials for the infrared. However, as has been mentioned, the materials commonly used for the visible and near-ultraviolet do not absorb in these regions, and so the relationship is less complex. As discussed above, as the wavelength decreases toward the absorption band, dispersion increases. However, as the absorption band is approached at shorter wavelengths, the amount of light transmitted by the prism decreases. To maintain sufficient intensity for the light to be passed through the sample, it becomes necessary to increase the width of the exit slit on the monochromator (see below). This allows more light to leave the device but at the expense of also increasing the heterogeneity of that light as frequencies adjacent to the desired frequency are included in the beam. Thus, the resolution, in terms of frequency, is decreased as the absorption band is approached despite increased dispersion. The optimum region of performance for a prism, then, is near but not too near the absorption band of the prism material. For prisms that transmit over a large portion of the spectrum, as is desirable for a visible and near-ultraviolet instrument, the region of optimum performance is relatively small compared to the large range of frequencies that are subject to considerably less dispersion. To increase performance over a range of frequencies, prisms are often made of a combination of materials each of which provides optimum performance for a different range of frequencies.

The use of gratings instead of prisms will be discussed next, followed by a discussion of features that are common to both types of monochromators.

4.3.2c. Grating Monochromators. The phenomena of interference and diffraction are the most important aspects of the behavior of light that indicate the need for a wave theory of light. Interference is explained by the principle of the superposition of wave amplitudes. This principle expresses the fact that two or more waves that may happen to coincide in space and time will continue to act independently. In other words, the intersecting waves show no interaction. However, the intensity of the wave pattern at a specific point is proportional to the square of the sum of the amplitudes of all waves present taking into account the signs of the individual amplitudes. In general, the square of the sum of the amplitudes is not equal to the sum of the squares of the amplitudes, and therefore the intensity at a given point is rarely equal to the sum of the intensities of the waves. The waves are said to interfere. Interference may be either constructive interference, in which the intensity at a point is greater than the intensities of any of the individual waves, or destructive interference, in which the intensity at a point is less than any of the intensities of the individual waves. Interference can be observed by combining two waves of the same frequency and some known phase difference. Radiation such as this that has a fixed phase relationship between any two parts of the wave front is called coherent radiation. When this fixed phase relationship is not present, the radiation is said to be incoherent.

If two point sources of coherent light are imagined, each will be surrounded by a series of concentric circles each representing a wave front and with the separation between each front equal to the wavelength of the radiation. As the wave fronts move away from one of the point sources, they encounter fronts moving from the other radiant source. An interference pattern is produced in this situation that can be observed as an array of systematically varying brighter and darker regions. The brightest regions occur at those points where waves from each light source interact and when each wave is at either a maximum or a minimum intensity. At such an intersection of two maxima or minima, the intensity due to constructive interference is 4 times as great as the intensity arising from either wave front alone. At points where one wave maximum and one wave minimum intersect, the intensity is zero due to destructive interference. The positions of the maxima or minima of individual wave fronts continually move away from the light sources, but the points at which constructive or destructive interference take place remain fixed as wave fronts move through them.

Diffraction is another familiar phenomenon that occurs when waves pass a fixed object and are seen to bend. When a light beam is passed through a small aperture in an otherwise opaque screen placed in its path, a purely corpuscular or particle theory of light would predict a straight beam of light, equal in size to the aperture, to proceed beyond the screen. What in fact happens, and what necessitates a wave theory of light, is that the aperture appears to serve as another, or secondary, source of radiation, and waves proceed away from it. On the far side of the screen, the waves, of course, are seen as semicircular. The amount of diffraction, or bending, of the light to form a semicircular wave front is dependent on both the size of the aperture and the wavelength of the light, with greater diffraction occurring with smaller apertures for a given wavelength of radiation or with longer wavelengths for a given size of aperture.

If light from a single source is allowed to pass simultaneously through two or more relatively closely spaced apertures in the same opaque screen, a series of wave fronts will proceed from each aperture due to diffraction. Furthermore, since the wave fronts from all the secondary sources (i.e., apertures) will move through the same region in space, interference will occur as discussed above. If the light from the secondary sources is focused upon some detector, e.g., a photographic plate, a series of light and dark regions are observed that correspond to the pattern of constructive and destructive interference produced. Light rays originating from areas of maximum constructive interference that are parallel to the optical axis of the focusing device, e.g., lens, may be focused on the center of the photographic plate. If the plate is flat, these will be the only rays that can be focused for maximum sharpness. Rays originating from other areas of constructive interference will not be parallel to the optical axis and will not focus as sharply. The pattern that appears upon the plate is one with a sharply focused region of maximal brightness in the center with pairs of images of decreasing sharpness and brightness on either side. Each of the more or less bright images will be separated from every other image by a dark band corresponding to an area of destructive interference. The exact positions of the images on the plate are dependent on the wavelength of the light, since this determines the points at which the wave trains are in phase, which is a prerequisite for constructive interference. For similar reasons, the size of the aperture and the distance of the plate from it also affect the pattern produced. The total energy incident upon the plate remains essentially constant as these factors are varied. The only thing that varies with these factors is the distribution of the energy among the various components of the diffraction pattern.

While the interference pattern produced by two secondary sources and detected by the photographic plate consists of a central image of peak intensity (the maxima) with images of gradually decreasing intensity (secondary maxima) adjacent to it, a somewhat different pattern is observed when more sources are used. As the number of sources is increased, most of the secondary maxima decrease in intensity, while some of them increase in intensity and become nearly as intense as the central maximum. Furthermore, the width of each maximum decreases and the spacing between the more intense of the secondary maxima also increases as the number of secondary sources is increased. The central maximum is called the zero-order maximum, and each pair of intense secondary maxima is called an order. The pair of secondary maxima immediately adjacent is the first order, the next pair is the second order, and so on.

The principle of the diffraction grating is based on this phenomenon and the additional fact that if the radiation contains two or more frequencies of light, each frequency will produce its own diffraction pattern independent of the others. The central maxima of the diffraction pattern for each frequency will fall in the same position, but in general the secondary maxima will not coincide. This results in the production of a complete spectrum (of the available frequencies) for each order. The zero order is of no value, since it contains all the frequencies present in the incident light and is undispersed. The frequencies are well dispersed in each of the other orders, though this depends on the design of the monochromator. One problem does arise, especially when the incident light covers a broad range of frequencies, in that the spectra produced in the various orders may overlap. This is less of a problem when the frequencies of interest are at the higher end of the spectrum, as is the case here. When necessary, additional design elements such as filters or prisms may be incorporated to alleviate this problem by reducing unwanted frequencies.

Some diffraction gratings are quite similar to the basic model described here. Such gratings consist of a transparent screen with slits that allow light to be transmitted through it. The slits are formed by opaque lines that are ruled on the transparent screen. The main difference is that rather than there being only a few slits, there are very many. The number of slits per centimeter is roughly equal to the number of waves per centimeter for the higher frequencies to be diffracted. Thus, when working with the visible and near-ultraviolet, as many as 6000–18,000 slits (or grooves) per centimeter are necessary. One difficulty with constructing such a grating is that all the slits must be equal in size and the spacing between the slits must also be precise and constant. Since the light must pass through the transparent screen in this type of grating, a question arises as to the material used, since it is likely to absorb at some frequencies that may be of interest and therefore produce a distorted spectrum. Another type of design circumvents this problem.

The other type of grating, which is, in fact, the more common type, functions by reflecting light from an aluminized surface. This eliminates absorption or dispersion effects due to passage of the radiation through the glass plate on which the other type of grating is ruled. Such a reflection grating consists of a highly polished surface, usually aluminum, in which equally spaced, parallel grooves are cut. Anywhere from 6000 to 18,000 grooves per centimeter are cut for instruments of interest here. The precision required in the construction of a reflection grating requires the use of a diamond cutting tool and rather sophisticated engines to drive the tool, as well as ancillary techniques for plating and other steps of manufacture. The grooves, as cut, take on the shape of a right angle with unequal sides. The same side of each groove is cut at a steep angle relative to the surface of the plate, while the other side is cut at a much more gradual angle. The surface of the grating as presented to the light consists of a series of sharp peaks corresponding to the points between each groove and with alternating steep- and shallow-angled surfaces between each peak. Diffraction occurs as light is reflected from the surface of the grating. As light is diffracted from each groove in the grating, it sets up interference patterns with light diffracted from the other grooves. The shape of the grooves, as well as their spacing, both of which may be varied during the initial cutting of the grooves, interact with the angle at which the incident light strikes the grating (as well as the frequency of the light) to determine the degree of dispersion as well as the angle at which the various orders will be reflected from the grating and the intensity of each of the orders. The total intensity of the light striking the grating is distributed among the various orders. Since, as a rule, only one order is used while the others are absorbed by the inside surface of the monochromator, it is desirable to concentrate as much of the original intensity as possible into the single order that is to be used. The greatest intensity occurs in that order for which the ordinary law of reflection from the surface of the groove is obeyed, that is, when the angle of incidence is equal to the angle of reflection. The intensity also depends on the total reflecting area, or on the ratio of the area occupied by the reflecting surface to that occupied by the groove edge. The angle that the shallow-shaped, and therefore larger, surface of each groove makes with a surface through the bottom of the grating is called the blaze angle because the grating, if viewed from this angle, appears as a blaze of light. Appropriate manipulation of the blaze angle and the other variables mentioned can produce a grating, in this case called an echelette, in which 75% or more of the reflected intensity is concentrated in a single order.

Another interesting aspect of the functioning of a grating, whether transmission or reflection, and an aspect of great practical importance, is the geometry of the spectrum produced. As discussed above, the dispersion produced by a prism is basically an exponential function of wavelength, with shorter wavelengths being more greatly dispersed. If the angle of incidence is held constant, as it is in practice, the angular dispersion of the light is proportional to wavelength. This results in the linear dispersion of the spectrum produced being normal, which is to say that the dispersion produced is a constant function of wavelength. The spacing between wavelengths is constant throughout the spectrum. This feature of gratings is important because it provides the possibility of determining the wavelength at a particular point in the spectrum by simple interpolation between two known wavelengths. As can be imagined, this greatly simplifies the task of selecting specific wavelengths for transmission through the exit slit of the monochromator.

Before the overall functioning of monochromators can be summarized, it is necessary to first consider a couple of other components commonly incorporated into one monochromator design or another.

4.3.2d. Focusing Devices. Both prisms and gratings function most effectively and reliably when light is incident upon them at some fixed angle. Since light shining directly from a source of radiation will contain rays of various spatial orientations that would strike the dispersive device at differing angles of incidence, it is necessary to collimate the light from the source. In collimated light, each ray is parallel to every other ray, and so all light in a collimated beam can strike the dispersive element at the same angle. In addition, after dispersion, whether by a prism or a grating device, it is generally necessary to focus the spectrum, or at least selected portions of it, upon the exit slit so that it may be passed through the sample and detection portions of the instrument.

Lenses can serve either of these necessary functions. Nonparallel light from a point source, or an approximation of a point source such as the entrance slit of a monochromator, can be collimated by passing it through a lens. Conversely, divergent rays striking a lens can be brought to a sharp focus at a particular location such as the exit slit of the monochromator. Furthermore, lenses can be used to either magnify or reduce the size of the image being focused.

The action of a lens depends on the difference between the refractive index of the lens material and the refractive index of the surrounding medium, as was the case with the prism. Other important factors that influence the functioning of a lens are the curvatures of the two lens surfaces. The function of lenses constructed of glass, silica, or other materials discussed above in relation to prisms is also affected by the dispersive power of these materials. Since the refractive index of the lens actually varies with frequency, the focal length of the lens, which is in part a function of the refractive index, also varies with frequency. With light in the visible region of the spectrum, a simple or cheap lens produces a colored image, since it can provide a sharp focus for only one color at a time. As the magnifying power of the lens increases, the disparity in focal lengths for different wavelengths also increases and the focusing problem becomes even more serious. The relationship between focal length and wavelength as displayed by such a lens is called chromatic aberration. Simple lenses made of differing materials can be combined to form a compound lens with the aim of greatly reducing chromatic aberration. Such lenses are said to be achromatic. The strategy followed in the construction of a compound lens is to combine lenses of varying refractive indexes and shapes so that the dispersion produced by one lens will be canceled out by another lens. Each component lens compensates for chromatic aberration at only one frequency; thus, to achieve a perfect achromatic lens would be an incredibly complex task. However, a lens that provides compensation at a few spaced frequencies generally reduces aberration at intermediate frequencies to an acceptable level. As was the case with prisms, it is relatively easy to vary the composition of the glass in such a way as to produce a sufficiently achromatic lens for the visible region of the spectrum. Other materials may be combined to form achromatic lenses for other regions of the spectrum; e.g., a combination of silica and fluorite is suitable for the near-ultraviolet region, but technical as well as economic considerations severely restrict the feasability of lenses for such applications. Because of the difficulty in producing good achromatic lenses for regions other than the visible, other components are often used rather than lenses whenever good focusing ability is necessary. When focusing is less critical, lenses are used more frequently.

The most common means of eliminating lenses, and thereby the problem of chromatic aberration, is by substituting mirrors in place of lenses. Mirrors can be constructed such that their optical properties are independent of refractive index and, hence, frequency. However, the aluminized surface of mirrors is much more delicate than the surface of most lenses, and even with proper care, they deteriorate and need refinishing after a number of years. Of course, while the use of mirrors in place of lenses when good focusing is critical does eliminate the problem of chromatic aberration, the composition of other absorbing materials in the light path such as lenses, windows, and prisms can still affect the spectral performance of an instrument.

Another problem, spherical aberration, arises with both lenses and mirrors and is due to the fact that while spherical lenses or mirrors can sharply focus light rays that are close to the optical axis, rays farther from the axis are not focused as sharply. One way of resolving this is to use lenses or mirrors in which the diameter is small compared to the focal length. A workable combination of diameters large enough to function effectively and the correspondingly large focal length required tend to conflict with other practical limitations on instrument design such as overall size. Other solutions to the problem are often more feasible. One approach is the use of an aspheric lens to deviate the path of the incoming light in such a fashion that the spherical distortion of a mirror is compensated. Of course, the dispersive properties of such a lens may present problems in themselves. A mirror in the shape of a paraboloid can either collimate light rays coming from a point source or focus a beam of collimated light upon a single point. An aspheric lens is not needed in front of a paraboloidal mirror, and so the frequency of the incoming light has no effect on the function of the mirror. The fact that a paraboloidal mirror can provide a sharp focus at only one point at a time makes it unsuitable in some applications, such as when a large spectral region is to be photographed. However, the techniques of interest here require only one region of a spectrum to be in focus at a given time and are, then, quite compatible with such mirrors.

   4.3.2e. General Monochromator Design. Although the prism or grating used to disperse light is obviously the most critical component of a monochromator, the roles of other components are not unimportant. As suggested previously, all the optics of the monochromator are contained in a light-tight container that also serves as a light trap for stray radiation within the monochromator. Source light enters the monochromator and after passing through the optics leaves through an exit slit. If monochromatic radiation were incident upon the entrance slit of a monochromator, an image of the entrance slit would appear at the exit slit. If the focusing optics before and after the dispersive element have the same focal lengths, then the width of the image at the exit slit is the same as the width of the entrance slit. Because of diffraction effects around the edges of the entrance slit and whatever imperfections may be inherent in the optics, the corners of the image will not be perfectly sharp. As the width of the entrance slit is decreased, the image also becomes narrower, until diffraction and optical imperfections reduce the sharpness of the image to the extent that further entrance-slit-narrowing is seen to have little effect on the image.

   The width of the image produced at the exit slit for each wavelength of light in heterogeneous radiation would, under ideal circumstances, be equal to the width of the entrance slit. The ability of the monochromator to distinguish between two frequencies that differ only slightly depends on the width of the images produced for each frequency relative to the separation between the images at the two frequencies. If the images overlap significantly, only one image will be observed, rather than two. The spread of the image relative to the wavelength scale of the spectrum of dispersed light is called the spectral slit width. The width of the entrance slit (the mechanical slit width) and the spectral slit width are approximately proportional except for very narrow slits. However, the dispersing element and the optics of the monochromator can also influence the spectral slit width.

   The resolving power of a monochromator is a theoretical limit to the ability of an instrument to separate adjacent frequencies and is determined solely by the optical design of the instrument, especially the prism or gratings. Although the theoretical resolving power is a convenient measure of an instrument's capabilities, in a practical sense there is no way in which it can be empirically verified for any instrument used in absorption measurements. However, the resolution, or resolution limit, of a monochromator, which is not to be confused with its theoretical resolving power, is of more practical importance. The resolution, the smallest frequency or wavelength interval that can be separated, depends on the resolving power and the quality of the optics as well as on the mechanical slit widths. Of course, when the entire spectrophotometer is considered, its ability to detect whatever separation the monochromator achieves also becomes a factor.

The light-gathering power of a monochromator can be measured in terms of its speed, or f-number, to borrow an optical term. The speed sets a limit on the slit widths that normally can be employed. The f-number is defined as the ratio between the focal length of the focusing device in front of the dispersing element and the diameter of the aperture. Since apertures are not circular, the diameter is taken to be the diameter of a circle equal in area to the area of the entrance slit. A low value of f represents a fast instrument or one of superior light-gathering power. In absorption work, the importance of resolution is not really due to an interest in how closely adjacent frequencies can be separated, since the absorption bands encountered are typically broader than the resolution limit of the instrument. A more critical consideration that is related to resolution is the effect of variations in mechanical slit width on the shapes and intensities of absorption bands as observed. Spectral slit width is usually the term considered in this regard.

Since monochromators never attain the ideal of producing truly monochromatic light, the light leaving the exit slit of such an instrument contains a range of frequencies. If it were possible to obtain a plot of intensity vs. frequency for the radiation at the exit slit, such a plot would define the exit-slit function. It is usually practical to consider that the exit-slit function is approximately triangular in shape. That is, the intensity is at a maximum in the middle of the spectral band and drops off sharply and linearly on either side of this maximum to a value of zero intensity at the extreme frequencies. This is only an approximation, since it would not be unreasonable for the intensity distribution about a central frequency to be assymmetrical because the energy available at the entrance slit would rarely be equal at all frequencies. However, the assumption concerning symmetry is workable and simplifies the problem. The spectral slit width is defined as the width, in units of frequency or wavelength, of a band the limits of which are defined as those frequencies or wavelengths at which the intensity is one half the maximum intensity. In absorption work, this is also called the half band width. As defined here, the spectral slit width clearly depends on the true exit-slit function. Since it is not possible to determine the intensity of the radiation at each frequency contained in a band at the exit slit, and therefore the true exit-slit function cannot be established, the value of the spectral slit width remains only an approximation. Taking into account its nature as an approximation, the spectral slit width still remains an important descriptor for the operation of a monochromator. As may have been noted, this definition of spectral slit width differs somewhat from that previously given, which was more similar to the resolution limit of the instrument. In rather simplified terms, the spectral slit width or the resolution limit of a monochromator may be considered as a sum of the effects due to mechanical slit width, diffraction, and imperfections or aberrations in the optical system. Knowledge of the size and dispersive properties of the prism or grating as well as the mechanical slit width and an estimate of the optical aberration allow some approximation of the spectral slit widths of an instrument. Such approximations are usually obtainable from the manufacturer of the instrument.

Most instruments are designed in such a way that the sizes of the entrance and exit slits are varied simultaneously. The energy passing through the entrance slit is a direct function of the size of the slit, since the intensity of the radiant source itself is usually fixed. The intensity of the spectral image at the exit slit depends directly on the energy passing through the entrance slit. The proportion of the energy focused in the plane of the exit slit that actually passes out of the instrument is a function of the width of the exit slit. The energy that passes completely through the monochromator is proportional to the product of the entrance and exit slit widths.

In consideration of the fact that the spectrophotometer as a whole can do no better than provide a measure proportional to the total energy passing through the exit slit and measured intensities or absorptivities are thus basically averages over some small range of frequencies, some of the effects caused by changes in resolution, or spectral slit width, can be understood. As slit widths are increased, the bands become broader, and some of their characteristic features may be diminished. Furthermore, the peak intensity diminishes, resulting in an apparent decrease in peak absorptivity. Asymmetric bands shift position slightly. The area under an absorption band is less sensitive to spectral slit width than is peak absorptivity. Again, these effects are predictable in light of the fact that increasing the range of frequencies in the band tends to average out variations.

To summarize the organization of a simple monochromator, the following points may be made. Radiant energy enters the instrument through an adjustable entrance slit and leaves through an exit slit. The widths of these slits play an important role in controlling both the intensity and the spectral purity of the light transmitted by the monochromator. While the range of frequencies passed by the instrument may be restricted by reducing slit widths, this occurs at the expense of the total intensity transmitted and consequently the energy available to the rest of the spectrophotometric system. Maintaining sufficient intensity to produce valid and reliable absorption measurements usually requires a considerable compromise regarding spectral purity.

Light entering the monochromator is focused by mirrors or lenses and thrown onto a prism or diffraction grating to produce dispersion. The dispersed light is again focused in the plane of the exit slit, although only a portion of the spectrum is focused directly upon the exit slit. While varying from one instrument to another, the portion of the spectrum actually passing through the exit slit is controlled by a dial, usually calibrated in wavelength or frequency, that alters the geometry of the optical components of the monochromator. Either the focusing element or the dispersive element or both in unison are rotated when the dial is turned such that the spectrum can be displaced back and forth across the exit slit. Numerous designs have been used to accomplish this displacement, and an even greater number of designs have been employed for the monochromator as a whole. The simple design presented here certainly exposes the reader to the functioning of such devices. The wide variety of designs that are commercially available and that may include any number of variations on this simple design have been reviewed by Bauman (1962) (see also Grum, 1972), from which much of this discussion has been adapted and which should be consulted for more detailed treatment.

4.3.3. Sample Cells

Although the containers within which a sample material may be confined in a spectrophotometer may vary quite widely in terms of their size, shape, and composition dependent not only on the nature and physical state of the sample but also on the spectral region of interest, the discussion here is concerned with cells that are used for visible and near-ultraviolet work with liquid samples. Two criteria for satisfactory cells become immediately apparent. In the first place, it is important that the material of which the cell is constructed be chemically inert in relation to any solutions with which it will come in contact. Just as important, the cell material should not, itself, absorb in the region of the spectrum for which its use is intended. With regard to this criterion, the previous discussion of the absorptive properties and problems of various prisms and lens materials provides a good framework within which choices of sample-cell materials can be made. For the visible and ultraviolet regions, the choice of materials is fairly uncomplicated, and, as with the optical components of the monochromator, glass, silica, and quartz are the most common cell materials.

The laws of absorption indicate a relationship between absorption and concentration of the absorbing material and the path length of the radiation through the absorbing material. When the absorbing substance is present in extremely small amounts, as in gaseous samples, it may be necessary to use cells with relatively long dimensions along the optical axis to achieve measurable amounts of absorption. Samples of solid materials present the opposite problem as well as difficulties with achieving sample homogeneity. Liquid samples are generally much easier to work with. The homogeneity of the sample is usually quite readily achieved, as is an optimal concentration of the absorptive material in relation to the instrument's capabilities, as will be discussed subsequently. Thus, most spectrophotometers use rather standard sample cells. When necessary to obtain satisfactory absorption determinations, it is much more common to alter either the sample solution or other aspects of the instrument than it is to vary the dimensions of the sample cell. Because of the central importance of the length of the path that light takes through the sample in the determination of absorption, it is also necessary that this dimension be accurately established. This is usually based on the data supplied by manufacturers of quality cells. The standard inside diameter of absorption cells for liquids is 1 cm, which is obviously quite convenient in light of the standard dimensions used for the laws of absorption. Although the nominal dimensions of liquid absorption cells are standard, the cells are available in two shapes. Rectangular cells, several centimeters in height and 1 cm in width, are constructed to high tolerances with special care that the faces of the cell are parallel. Use of such rectangular cells is certainly preferable in many situations, although their cost is relatively high. Alternative cells are cylindrical in shape, though generally of similar dimensions. They are less expensive, in part because of less precision in production. They also represent a compromise in terms of their suitability for precise absorption determinations when compared with rectangular cells. Two reasons for this are readily apparent. The first has to do with the angle at which the incident light strikes the cell and consequently, because of refraction, the path the light follows through the sample. Since the incident light is collimated as it leaves the monochromator, ignoring slit diffraction, essentially parallel rays of light strike the sample cell. If the face of the cell is everywhere perpendicular to the incident light, the incident angle for all light entering the cell will be constant. Thus, refraction will be constant and at a minimum because of the dependence of refraction angle on incidence angle. Essentially the same consideration applies to the effects observed at the parallel rear face of the cell as the light passes out of it. In contrast to this, the curved surface presented by a cylindrical cell to the incident light results in a whole range of angles of incidence and correspondingly complicated refraction effects and light paths through the sample. Furthermore, only a limited proportion of the incident light rays will strike the cell at a perpendicular such that their path through the cell will equal its nominal dimension of 1 cm. The greater proportion of the incident light will travel shorter distances through the sample solution. Both these facts suggest the fundamental inadequacy of cylindrical cells for some absorption determinations. Establishment of the molar extinction coefficient, for example, would be most difficult unless rectangular cells are used. On the other hand, as will be seen later, cylindrical cells are often adequate and in such cases have the added benefit of relatively low cost.

While the characteristics of the sample cells are of undeniable importance, optimum results, regardless of the cell used, can be obtained only if the cell can be accurately and repeatedly positioned in the light path. Although some cells are integral components of the instrument, the neurobiologist is more likely to use cells that are removed from the instrument for washing and refilling after each measurement. This is the general type of cell being discussed here. The sample-cell holder of the spectrophotometer serves the function of accurately aligning the cell each time it is placed in the instrument. However, careless handling can partially defeat its function, and the human component of the alignment procedure must be made as reproducible as possible. Cylindrical cells seem to be more troublesome in this regard than are rectangular cells, though both types should be handled carefully. Either inherent in their construction or in the typical manner in which they are handled, rectangular cells will have one set of parallel faces that are intended to be positioned perpendicular to the light path, and of course it is important to see that they are so positioned.

Inhomogeneity in the cell surfaces exposed to the light beam can have considerable effects on their refraction and absorption characteristics. Though such inhomogeneity is controlled as far as possible during manufacture and the quality achieved at this point becomes a fixed parameter, subsequent handling can result in less than optimal performance. Scratches on the surfaces of the cell are particularly troublesome, and since they are generally irreversible, lead eventually to a need for replacement of the scratched cells. In addition to the optical effects of scratches, they can make thorough cleaning of the cell difficult. Chemicals that are not completely removed from scratches can also have optical effects, including absorption. When the scratches are on the interior of the cell, the material incompletely washed from them may chemically contaminate the sample, depending obviously on the nature of both the contaminant and the sample solution. Optical inhomogeneities, which are fortunately reversible, can also result from careless handling. Fingerprints as well as various reagents that may be carried by the fingers must be avoided on, or alternatively removed from, the surfaces of the cell that are exposed to the light path. Similarly, residues on these surfaces resulting, for example, from spillage are a not uncommon source of inhomogeneity. Finally, when cleaning the surfaces of the cell, care must be taken that they are dust- and lint-free.

4.3.4. Radiation Detectors

After the light beam has passed through the sample cell, it is necessary to measure its intensity. For light in the visible region, one of the oldest and certainly a very sensitive measuring device is the human eye. However, human vision is a rather variable measure and not satisfactory any longer for quantitative work. Furthermore, its usefulness is obviously limited to the visible region. As with so many components of spectrophotometers, the best device to be used for measuring the intensity of radiation varies with the spectral region of interest. Since many authors have reviewed detectors suitable for various regions, only visible and ultraviolet detectors will be briefly discussed here. In fact, only the detector most frequently encountered in absorption-research instrumentation will be considered.

Nearly all visible and ultraviolet absorption instruments employ photo-electric detectors—more specifically, phototubes. A phototube consists of a semicylindrical cathode and a wire anode sealed within an evacuated glass envelope. The envelope, or at least a portion of it (a window), is made of a material that is transparent in the visible or ultraviolet regions or both, such as glass, silica, or quartz. The concave surface of the cathode, which faces the incoming light, supports a layer of photoemissive material, often an alkali-metal or alkaline-earth oxide. The composition of this layer determines the spectral region to which the tube is sensitive. To cover the whole visible and ultraviolet region, a combination of materials sensitive in different frequency ranges is employed. A potential difference is established with the anode positive in relation to the cathode. The phototube operates according to the photoelectric principle such that absorption of electromagnetic radiation by the cathode causes electrons to be ejected by the photoemissive layer and drawn to the anode by the applied potential. The number of electrons ejected and consequently the current produced are proportional to the intensity of the incident radiation. The fraction of the ejected electrons collected by the anode increases as the applied potential is increased. At some potential, all electrons are collected and further increases in the potential applied have no effect. This is called the saturation potential. The current produced in a phototube operated at or above saturation is linear with respect to the intensity of the light incident upon the cathode.

If the anode of a phototube is also covered with a photoemissive material, the electrons from the cathode that strike it will each, in turn, cause several electrons to be ejected. These electrons may, then, be accelerated toward another anode at a higher potential (with the original anode serving as a secondary cathode or dynode), causing further multiple emissions of electrons. By combining 10 or 12 such amplification stages, each with about 100V of potential difference, it is possible to achieve stable amplification of the original signal by a factor of one million or more. This device is called a photomultiplier tube.

Phototubes, and especially photomultiplier tubes, have become the preferred detectors for the visible and ultraviolet regions for a number of reasons. In the first place, the simple fact that such devices can be designed so as to function in the ultraviolet is not easily reproduced. The fact that the functional range can be extended throughout the visible and ultraviolet is of great practical value. In contrast to some alternative types of detectors, the design of phototubes and photomultiplier tubes allows for easy signal amplification, which in conjunction with their inherent sensitivity gives them a large working range.

Some characteristics of this type of detector determine performance limits. Since the current produced by a given intensity of light is proportional to the potential applied to the anode, as well as each dynode, instability in the power source will result in fluctuations in output. Thus, a very stable source of power is required, which, in the case of photomultiplier tubes, must be capable of voltages as high as 1000V. With all other aspects of system design controlled, the stability of the power source provides a fundamental limitation on the sensitivity, although it is possible to use very stable sources that minimize this limitation. Whatever variations there may be in the voltage applied to the anode and dynodes, no matter how small, it represents noise. Levels of radiation that do not produce currents somewhat in excess of the magnitude of the current fluctuations due to power instability cannot be discriminated from that instability.

Another source of noise is called dark current. In the absence of illumination, the photoemissive surfaces of the detector spontaneously eject electrons, albeit at a very low rate. Thus, even when the device is illuminated, some of the electrons ejected from the cathodes are not due to the absorption of radiation. This dark current, being a property of the photoemissive coating, is proportional to the area of the active cathode coating. Again, light intensities that produce the same relative amount of electron ejection as does the dark current will be difficult to discriminate from it. At higher levels of illumination, the dark-current effect, representing something of a constant factor, would be readily accommodated except that the dark current is also affected by temperature. When it becomes a serious problem, it is often possible to deal with the temperature relationship by controlling the temperature of the detector. In any case, it is certainly advisable to allow the temperature of the spectrophotometer to stabilize before any measurements are attempted.

A further source of noise and, under ideal conditions, the truly limiting factor is called shot-noise. Although the number of electrons ejected by the photoemissive coating is proportional to the intensity of the incident light, which is to say that each photon of radiation has a specific probability of causing an electron to be ejected, in fact there is a certain amount of variability in the process. The output signal of the detector fluctuates because the process of photoemission is statistical in nature. Such fluctuations are proportional to the square root of the number of photons striking the detector. Thus, as the intensity of the radiation striking the detector increases, the absolute noise output also increases. However, since the total output is directly proportional to the intensity rather than its square root, as the intensity increases the level of absolute noise decreases relative to the total output of the detector. This may be termed the signal-to-noise ratio, and it increases as the intensity of the illumination increases. In general, when all sources of noise are considered, the signal-to-noise ratio still increases with intensity. Thus, optimum performance from the detector is obtained at high levels of illumination. Factors to be discussed later, though, suggest that there are some practical limitations that for other reasons must be placed on the level of illumination used in the spectrophotometric system.

A rather different sort of problem is characteristic of virtually all radiation detectors, though often for differing reasons. This problem is based on the fact that the detectors are differentially sensitive to frequency even within the spectral range appropriate to them. In the case of the photoemissive detectors being described here, the principle underlying the variation in spectral response is due simply to the frequency-related process of absorption as treated previously. Any material used as a cathode coating can be characterized by an absorption band, or limited range of frequencies, of radiant energy that is capable of removing an electron from its parent molecule. Of course, within this absorption band, the middle frequencies are more efficiently absorbed than the others. For a given coating compound, then, there is a limited range of effective frequencies, and the effectiveness varies even within this range. To broaden the functional range of the detector, a combination of photoemissive compounds is used to coat the cathode. Each compound is characterized by its own relatively narrow absorption band, which, ideally, overlaps slightly with the absorption bands of the other compounds that absorb at adjacent frequencies. While it is possible to cover the whole visible and ultraviolet region in this manner, the response of the detector in relation to frequency is not linear. The absorption band of the combined coating material contains peaks for the individual compounds, some of which are greater than others due to the differences in absorption for each compound. Between the peaks are the lower shoulders of absorption bands overlapping in part with one another, and so on. The net effect, as stated, is a deviation from a flat or linear frequency response. Any deviations from linearity in the detector itself can be problematical for some absorption techniques. The absorption spectrum of a sample, for example, will reflect in part the varying sensitivity of the detector to light of different frequencies. Furthermore, in addition to the effects of the absorption spectrum of the photoemissive coating, the envelope or window through which the light must pass before striking the cathode may also show absorption somewhere in the region. Since this material is likely to be similar to that used in other components, i.e., lenses, prisms, and sample cells, its absorption band is more a characteristic of the whole instrument rather than of just the detector. In any case, while the spectral sensitivity of the detector is an important consideration, it does not cause any insurmountable difficulties. Techniques for evaluation of this effect and compensation for it will be discussed later.

Since the signals produced by the detector are generally small in magnitude, it is necessary for them to be amplified prior to their being displayed. There are at least a few different amplification systems that can be used. There is also some variation in the methods used to display the output. Both these components, i.e., amplification and display, are dependent on the overall design and operating principles of the spectrophotometer. Before considering these two components any further, the basic design of the spectrophotometer will be discussed.

4.3.5. Single-Beam and Double-Beam Spectrophotometers

Two basic approaches have been taken to the design of spectrophotometers for absorption measurements. These can be fairly simply classified according to the geometry of the light path through the instrument, although the two types of instruments vary in several other aspects. Although the actual practice of quantitative-absorption measurements will be discussed later, certain features dictate spectrophotometer design and will be briefly mentioned here. In the type of measurements with which this chapter is concerned, the sample of unknown concentration is present in solution. Common sense suggests that the solvent should be transparent, or free of absorption bands, at the same frequencies at which the sample absorbs. If this were always the case, measurement of sample absorption would be a simple matter. In reality, it is rare that a solvent has no effect on the absorption of a sample solution. Consequently, it is normal practice to compare absorption by the sample solution with a control solution or blank. Most simply, the blank contains all the components of the sample solution in equivalent amounts with the exception of the sample material itself. All other things being equal, the difference in absorption between the sample solution and the blank solution can be attributed to the sample material.

In a single-beam instrument, which is basically the type described in previous sections, a single beam of light passes from the monochromator and through the sample cuvette (cell) and finally falls upon the detector. The sample solution and the blank must be placed in the beam at different times, and the absorption due to each is measured independently. If a spectrum of absorption by the sample over a wide range of frequencies is to be produced, absorption due to both the sample solution and the blank must be determined independently at each wavelength. In single-beam instruments, both the scanning of different frequencies and adjustments to monochromator slit widths are effected manually, as necessary. Similarly, the absorption obtained is read and recorded manually, in general.

A dark-current potentiometer determines the bias voltage applied to the phototube and amplifier. While the spectrophotometer is turned on and after it has warmed up, or stabilized electronically and thermally, a phototube switch is placed in the off position, which places a shutter in front of the phototube, excluding all light from striking the photoemissive cathode. The dark-current potentiometer is then adjusted so that the reading on the galvanometer display scale is at the zero point of the scale. For as long as the dark current remains stable, this adjustment has the effect of minimizing the influence of the dark current on the absorption measurements.

Adjustment for the dark current having been made, the phototube switch is turned on, allowing the light beam to strike the detector. However, the sample compartment is left empty at this point, and thus the full amount of available light for the particular wavelength and slit widths being used passes through the instrument to the detector. A second adjustment, the transmittance or absorption potentiometer, is then manipulated such that the deflection on the readout scale is equal to 100% transmittance or zero absorbance. The scale is usually calibrated both linearly from 0 to 100% transmittance and logarithmically for absorbance. The dark-current adjustment, then, is used to establish the zero transmittance or 100% absorbance end of the scale, while the second adjustment establishes the opposite end of the scale. The transmittance or absorbance of a sample cell now placed in the light beam can be directly read from the scale. While all single-beam instruments work on this basic principle, there are, of course, design variations, and the design presented here is meant only as a simple model. Other sources (e.g., Bauman, 1962) discuss the design details of specific commercial instruments.

The amplifier in this type of instrument, as long as it is stable, does not affect the accuracy of absorption measurements. The accuracy depends only on the assumption that any potential drop across the phototube is proportional to the intensity of the radiation incident upon the tube. Of course, the transmittance scale itself must also be linear and in proportion to the potential drop across the phototube. The total change in voltage that can be achieved by adjusting the transmittance potentiometer limits the range of intensities over which the instrument can operate. Any alterations in this range when necessary to accommodate particular sample materials or alternate light sources or detectors of different sensitivities must be accomplished by changing the slit widths. However, some additional adjustment is usually possible with a sensitivity control that varies the total voltage range controlled by the transmittance potentiometer.

Since, as was mentioned above, the signal tends to increase more rapidly than noise with increasing intensities, the precision of an intensity measurement is always improved when the total intensity of the radiation striking the detector increases as long as both the detector and amplifier are operated within their proper ranges. Precision refers to the repeatability of a measurement and is, in large part, a function of the signal-to-noise ratio. Wide slit widths allow more radiant energy to strike the detector and thus increase the precision of absorption measurements. However, as slit widths are increased, the resolution of a spectrophotometer decreases, approaching a minimum at the largest slit widths. In such a situation, because of the decrease in spectral purity, the absorption measured very often is not characteristic of absorption at the desired frequency, i.e., the frequency nominally designated for the spectral band produced by the wavelength setting of the instrument. The ability of the spectrophotometer to produce absorption measurements that correspond to the true absorption at the desired frequency can be termed its accuracy. The accuracy of the instrument, then, is at a minimum when its precision is at its greatest. Enhancing the accuracy of a measurement by reducing slit widths increases the variability of the measurement and thus decreases precision because of the corresponding reduction in the signal-to-noise ratio. Clearly, optimal absorption measurements will derive from a compromise made between the competing needs for both accuracy and precision. As will be mentioned below, there are some situations in which either accuracy or precision must obviously be sacrificed. For example, establishment of the entire absorption spectrum for an unknown compound is an exercise in accuracy. True absorption at each frequency is the objective, and all efforts should be made to maximize accuracy including the use of narrow slit widths. Procedures to increase precision without sacrificing accuracy will be considered later. Another example of the type of absorption measurement of more routine interest to neurobiologists demonstrates the possibility of trading off spectral purity for increased precision. Such a routine spectrophotometric application involves determination of the concentration of a sample compound of known identity with a previously established absorption spectrum. In this type of quantitative determination, absorption is frequently measured at the single wavelength at which absorption is maximum. For many compounds of interest, though not all, the absorption maximum is rather broad, which is to say that the relationship between absorption and wavelength is fairly constant over a range of wavelengths. Widening of the monochromator slits to increase radiant intensity, though at the same time broadening the range of frequencies passed, will not affect the accuracy of the determination if the frequency band being passed corresponds roughly to the band of frequencies over which absorption is constant. Thus, precision may be increased, especially with strongly absorbing samples, without a simultaneous loss of accuracy. The manual mode of operation typical of single-beam spectrophotometers also contributes to greater precision compared to double-beam instruments. On the other hand, the fact that the sample and blank determinations must be made sequentially renders precision vulnerable to many problems. Fluctuations in the intensity of the light source, in the voltage applied to the detector, and in the amplification circuitry all have the effect of causing measured absorption to vary over time. High-quality single-beam instruments must place a premium on minimizing such fluctuations. Additionally, the operator must take precautions to control for such fluctuations, which, for example, in the extreme may require a blank determination for each reading of sample absorption. This obviously involves a considerable amount of work. Nevertheless, for many researchers including neurobiologists who are primarily interested in routine quantitative assays, single-beam instruments are frequently quite satisfactory and have the added advantage of being less expensive than double-beam instruments.

Double-beam instruments, as the name suggests, are characterized by two, at least partially independent, light paths through the instrument. The aim is to combine the measurement of sample and blank solution in a single reading, rather than in separate measurements as with single-beam instruments. Such simultaneous measurements eliminate the problems of temporal fluctuations in various components of the instrument that are inherent in single-beam techniques. Although it would be possible to produce an instrument with entirely independent light paths, great difficulty is encountered in producing truly matched paths, i.e., paths that would produce identical measured light intensities if each were under identical sample conditions. While the various components of the spectrophotometer, including light sources, monochromators, detectors, and others, can be manufactured quite precisely, it is virtually impossible for any two like components, e.g., two detectors, to be perfectly identical. Of course, the more components that are duplicated within the instrument, the greater will be its cost. Thus, identical but independent light paths are a practical impossibility in this sense.

Nevertheless, this is clearly the basic design concept of the double-beam spectrophotometer. To maintain this concept while avoiding the difficulty of unmatched components, the only portion of the instrument that includes double light paths is the sample compartment. In other words, there is only one light source, one monochromator, one radiation detector, and so on. Various devices are used to pass light from the chosen spectral region alternately through the cuvettes containing either the sample or the blank. Although the incident light could be divided and continuously passed through both cuvettes, this would require, among other things, the use of two detectors. If, in fact, the beam of light is divided by a beam-splitter, which, essentially, produces two continuous beams, shutters are used to alternate the passage of light through the two cuvettes, or at a later point to alternate the light coming from the two cuvettes. Another device makes use of a rotation mirror, which again can send the light through one cuvette at a time. In either case, light striking the detector at a particular instant has passed through only one cuvette or the other. In a sense, this arrangement seems to be analogous to that of the single-beam instrument with manual alternation of the sample and blank cuvettes replaced by instrumental alternation. The problem with the single-beam arrangement is due to instrumental fluctuations, which generally take place over seconds or minutes and roughly coincide with the rate at which the cuvettes are alternated in the instrument. In double-beam instruments, the light paths are alternated many times a second, allowing very little possibility of instrumental fluctuations between particular sample and blank measurements. The independent sample and blank light paths converge upon a single detector that, of course, is subjected to a light flickering alternately from one cuvette to the other. The circuitry that amplifies and subsequently displays the absorbance or transmittance is phased with the alternation of the light paths. Very simply, in fact, separate circuits amplify the detector response for each cuvette. Furthermore, the measure actually displayed is one in which the absorbance of the sample cuvette is already adjusted for the absorbance of the blank. This results in one of the shortcomings of the double-beam instrument as compared with the single-beam. The double-beam instrument does not readily allow the absolute absorbance of the sample solution to be determined. The absorbance measure available is usually one that is relative to the absorbance of the blank.

4.4. Quantitative-Absorption Procedures

This chapter is concerned essentially with the use of absorption spectro-photometry as a quantitative method. Such usage is based on the principle, mentioned above, that the fraction of incident radiation absorbed by a material will increase, reproducibly, as the concentration of the absorbing material is increased. If the relationship between the concentration of the absorbing material and the fraction of incident radiation absorbed is expo-nential, i.e., if the absorbance is directly proportional to concentration, analysis is somewhat simplified. However, this is not the only situation to which quantitative procedures can be applied. The following sections will outline the simpler methodology in which proportionality holds. Subse-quently, the factors that can result in deviations from proportionality, i.e., Bee r's Law, will be discussed, as will alternative procedures.

4.4.1. Development of the Procedure

Although the actual procedure used for a given quantitative analysis may be quite simple, considerable work is usually necessary during the development of the procedure. The first step in such development is essentially qualitative, since the absorption spectrum of the material of interest is obtained. It certainly would help if it were possible to obtain the spectrum from a pure sample of the material, but this is generally impractical. Obviously, many compounds of interest are not liquids in their pure state (remember that neurobiological work almost always involves the use of liquid samples) and therefore must be dissolved before analysis. Furthermore, while even a pure sample will transmit considerable amounts of light in the spectral regions where absorption is low, it is likely that for frequencies at which any absorption does occur, so little light will be transmitted that it will be impossible to ascertain the true characteristics of the absorption spectrum.

In any case, it is not practical with the substances of interest to neurobiologists to obtain visible or ultraviolet spectra with pure samples. Granting that the sample material must be dissolved or diluted, an important factor becomes the nature of the solvent. As previously discussed, the solvent selected should not absorb, or should be transparent, in the region of the spectrum being investigated. Of course, this may not always be possible, since specific materials may require solvents that do not happen to be transparent at all wavelengths. One may use for different regions of the spectrum different solvents that in combination may provide transparency throughout the region of interest. Alternatively, and probably advisedly, an absorption spectrum should be obtained for the solvent that can be used to correct for solvent absorption peaks in the sample spectrum.

The solvent and appropriate sample concentration having been chosen, the absorbance of the sample material is determined at regularly spaced frequencies. Generally, the determinations are made approximately every 20 cycles over most of the spectrum, but the spacing between the determinations is reduced to 5- or 10- cycle intervals in the regions where the relationship between frequency and absorption changes rapidly (i.e., in the vicinity of the absorption peaks). The closer spacing reduces the likelihood of overlooking the detailed structure of absorption maxima, since they can often be rather irregular in shape.

Quantitative-absorption determinations, of the type of interest here, of course, are usually made at one wavelength only. The preliminary work with a qualitative-absorption spectrum allows selection of the best frequency for the quantitative determinations. Most compounds of interest are characterized by relatively complex absorption spectra even when in the liquid state, which yields basically electronic transitional information but not rotational or vibrational fine structure. Such spectra typically display absorption peaks in more than one region of the spectrum. The most advantageous peak region for quantitative work would be that in which absorption is maximum, since the signal-to-noise ratio of the sample component of the spectrophotometric system would be maximal at such places. However, other considerations may indicate that less than maximum peak absorption bands may be preferable. The design of the spectrophotometer itself, and concomitantly its functional capabilities, may rule out the use of maxima that occur too close to the extremes of the spectral region covered by the instrument, since it is likely to be less sensitive and more imprecise at these extremes. The shape of the absorption peak is also an important consideration. It may be desirable to make quantitative determinations at a frequency that corresponds to the center of a broad absorption peak of relatively lower intensity, rather than at a frequency at which absorption is much more intense but that also peaks very steeply. As discussed previously, very steep, narrow absorption bands present problems because of the difficulties involved in obtaining truly monochromatic radiation within the spectrophotometer compounded by the difficulty of repeatedly and precisely obtaining radiation of a certain nominal frequency. Thus, sharply peaked maxima seriously impede the applicability of Beer's Law. Of course, the same instrumental problems apply at all nominal frequencies, but when an absorption band is broad so that basically all the frequencies contained in the incident light correspond to frequencies also contained in the absorption peak, no deviations from Beer's Law are apparent.

Other factors must also be considered when selecting a nominal frequency for quantitative-absorption determinations. In the first place, it is advisable to consider the absorption characteristics of any other substances that are to be included in the sample solution. The solvent itself is of primary importance, as has already been mentioned. Buffers, centrifugation-gradient or extraction media, and trace contaminants are also important. It is possible, and sometimes necessary, to determine sample absorption for frequencies at which other components of the solutions also absorb. When the other components of the solution remain at a constant concentration and only the concentration of the sample material varies, the nonsample absorption is clearly a constant factor from which the sample absorption should be discriminable. However, for a given absorption peak, the ratio between sample and nonsample absorption may be so small that, in fact, the sample itself may contribute relatively little to the absorption peak. Such a small contribution is really synonymous with a poor signal-to-noise ratio and is therefore undesirable. Futhermore, even this is a more or less ideal situation, since it assumes that all components of the sample solution remain at constant concentration while the concentration of the sample material itself varies. In most cases, this assumption is quite unwarranted, since variation in the concentration of the sample almost invariably necessitates changing concentrations of other components of the solution. Clearly, then, it is important to have analyzed the absorption characteristics of all components of the solution beforehand. Whenever possible, the nominal frequency at which absorption of the sample is determined should be one at which other components of the solution are as transparent as is feasible.

Another strategy is also useful when dealing with this question. When it is difficult to select a frequency for absorption determinations or when the material of interest does not exhibit strong absorption at a convenient wavelength, the sample material may be chemically altered to yield a new compound with more convenient absorption characteristics. Similarly, when the compound of interest is a product of an enzyme reaction, an artificial substrate yielding a more suitable end-product may be useful. Examples of this strategy will be considered later. For now, the critical factor is that the material on which the absorption determinations are made be one that exhibits strong absorption at a wavelength that is convenient in terms of both instrumental limitations and the other components of the sample solution.

4.4.2. Applicability of Beer's Law

Having determined the most fundamental aspect of the procedure, namely, the wavelength at which the absorption is to be measured, one should perhaps undertake as the next step a check for the conditions under which Beer's Law holds, if in fact it does. To do this, a series of sample solutions of increasing sample concentration are prepared. The absorbance of each solution is then determined spectrophotometrically, in relation, of course, to the absorbance of the appropriate blank and dependent on whether a single- or double-beam instrument is used. A graph is then constructed of the absorbance values (plotted on the ordinate) vs. concentration (plotted on the abcissa). If the experimental points and the origin of the axes lie along a straight line, then the sample solution is said to obey Beer's Law over the range of concentrations studied. As will be mentioned below, increased sample concentrations, sharp absorption bands, and other factors can result in deviations from Beer's Law. Usually, such deviations result in the curve's being concave relative to the concentration axis. Dependent on the range of concentrations examined, the result may be a curve that is linear for some range of the lower concentrations though changing to an exponential function at higher concentrations. In such cases, Beer's Law still holds for that range of concentrations that exhibit linear increases in absorbance, but does not hold for the higher concentrations.

The decision as to whether a given set of experimental points is linear or not must take into account the degree of accuracy desired in the analysis. Repeated measurements on each sample solution as well as on duplicate or triplicate solutions for each concentration yield information about the amount of variability in the process of measuring absorbance, including both instrument and operator variability. Such information provides a means of assessing the extent to which observed deviations from linearity are expected due to measurement variability as opposed to true deviations from Beer's Law. Calder (1969) further discusses the manner in which simple statistical considerations can help in the interpretation of experimental results.

The linearity for the degree of accuracy required, having been determined, the slope of the absorbance, concentration plot can be made either visually or, for greater accuracy, by a least-squares method. If the plot is linear, i.e., if Beer's Law holds, the slope of the line represents the product of the absorptivity of the solute and the cell length. As mentioned in the previous discussion of the laws of absorption, the absorbance of a solution is equal to the product of the cell length, concentration of the absorbing species, and the absorptivity, which is a constant of the absorbing material. Furthermore, when the cell length is given in centimeters and the concentration is expressed in moles per liter, the constant is known as the molar extinction coefficient. An identity can then be constructed, if the proper units are used, between the slope of the experimental line and the absorptivity—or more precisely the molar extinction coefficient—that becomes, in fact, equivalent to the reduction in incident-light intensity due to the absorbance of 1 mol/liter of the absorbing substance. In summary, the construction of an empirical plot of concentration vs. absorbance yields, if Beer's Law holds, a constant, the molar extinction coefficient, of the material in question.

Knowledge of the molar extinction coefficient of a compound provides a simple means of determining the concentration of a sample solution of that compound. If the absorbance of the sample solution is obtained under the same conditions (including the composition of the blank) used in the determination of the coefficient, the concentration of the material in the sample solution is calculated by dividing the absorbance of the sample by the molar extinction coefficient. This is particularly convenient when the coefficient has been previously determined or is available in the literature, as is often the case. Of course, if the original graph is available, it is certainly just as easy to read the concentration that corresponds to a particular absorbance directly from the graph. As will be discussed next, this is the procedure that must be followed when Beer's Law does not hold.

If a sample solution of interest does not follow Beer's Law, or for higher concentrations of a solution that does follow Beer's Law over a lower range of concentrations, a plot of concentration vs. absorbance is constructed, as described above. However, it is advisable to include more experimental points so that a smooth curve of the function may be drawn. When Beer's Law holds, only enough points need be established to indicate that the function is linear and what its slope is. Even when the concentration of an unknown sample is read directly from the curve, interpolation between experimental points is perfectly feasible because of the linearity of the function. It should be noted, though, that extrapolation is an objectionable procedure, since the linearity of the function can be assured only over the actual range of concentrations examined. The situation is somewhat different when Beer's Law does not apply. The concentrations of unknown must be read directly from the calibration plot, and no assumptions about the nature of the absorbance function can be made that will warrant interpolation. Thus, as mentioned, a greater number of experimental points must be established so that any unknown concentration will be the same as or very close to one of the calibration concentrations. This suggests that an enormous number of experimental points must be established to cover all possible concentrations, but this can be simplified. Very often, an approximate range of concentrations is known within which the sample is expected to fall. The calibration curve can then contain relatively few points outside this range while being more detailed within it. When the concentration of a sample falls outside a certain range of concentrations that have been well calibrated, it is usually a relatively simple matter to either concentrate or dilute the sample so that its concentration falls within the range. This again, alleviates the need for extending the number of points in the calibration curve.

Whenever a new supply of reagents is put into routine laboratory use, or in fact whenever new reagent solutions are prepared, it is advisable to construct new calibration curves for quantitative measurements. Even the highest-grade reagents may vary from batch to batch, and this should be considered as sufficient cause for routine reexamination of the calibration data. Similarly, preparation of new reagent solutions may affect results due to unavoidable variations in the process of making the solutions in the lab. The need for frequent preparation of new reference data may also be a function of the particular quantitative technique in question. A procedure that embodies, for example, a broad, intense absorption peak and a very high signal-to-noise ratio between the component of interest and other components, such as solvent, buffers, or others, may be relatively immune to normal variations in reagent quality and the composition of reagent solutions. Other techniques may be more sensitive. Methods that require reagents to be prepared anew each day are, of course, very vulnerable to the type of problem being discussed here unless they also happen to be very robust techniques in the sense that they are relatively invulnerable to variations in reagent composition. In any case, previously published methods usually indicate the frequency with which calibration curves should be repeated. When in doubt, it certainly does no harm to produce reference data more frequently than is absolutely necessary. When a new method is being developed, calibration curves should be repeated very frequently and under a reasonable variety of normal experimental conditions to establish the frequency of calibration necessary when the method is put into routine use. Finally, one fairly common source of difficulty should be mentioned. Reagents produced by different manufacturers often vary considerably, presumably due to unreported or unidentified impurities. Methods should always report the exact source of reagents used as well as the grade (or purity) of the reagents. It is not always necessary to use materials from the same manufacturer, but when problems arise in the application of a technique, the source of the reagents should immediately be considered as a possible source of the problem. In some cases, it proves necessary to obtain reagents from exactly the same source as reported in the original method.

4.4.3. Deviations from the Laws of Absorbance

The linear relationship between absorbance and the concentration of the absorbing substance has been described above as a fundamental principle of absorption spectrophotometry. If monochromatic, or nearly monochromatic, light is used and the concentration figures used in the absorbance laws are equal to the actual concentration of the absorbing substance, the linear relationship will be observed. While this relationship does generally hold, it should be clear from the previous discussion that this is not always the case. Either serious deviation from monochromaticity or lack of coincidence between the true and apparent concentrations of the absorbing substance may be responsible for lack of linearity. Although the use of calibration curves, as described above, allows measurements to be made when a linear relationship is not present, an understanding of the possible causes of nonlinearity leads to an ability to predict in which situations nonlinearity may be expected and also to practical advice on modifying assays so as to achieve linearity.

Previous discussion of monochromator design and performance suggested that light leaving the exit slit of the device could, under some circumstances, be contaminated by stray white light. Such white-light impurity is a major source of deviation from the laws of absorbance. In the simplest situation, which will suffice to demonstrate the problem, light of the desired nominal frequency will be absorbed by the substance under investigation, while the majority of the stray radiation will not be absorbed. In essence, the sample is irradiated by both absorbable and unabsorbable light, both of which are capable of producing a response in the photodetector. A constant response is produced in the photodetector by the unabsorbable light even though the response produced by the absorbable light diminishes with increasing sample concentration. However, as the sample concentration increases, the ratio between absorbable and unabsorbable light striking the photodetector decreases exponentially such that the observed absorbance of increasing sample concentration is also exponentially diminished. At low concentrations, the absorbance relationship is linear, but rapidly loses linearity as concentration is increased until a point is reached at which all absorbable light is, in fact, absorbed and further increases in sample concentration do not affect absorbance. Of course, the greater the ratio of stray to absorbable light, the more serious are the deviations from linearity. Although calibration plots that account for such deviations may be used, it is clearly desirable to minimize sources of stray light. Many monochromators available today deal quite well with this problem. However, as discussed previously, even the best monochromators suffer a relative loss of intensity near the extreme ends of the spectrum, and at those extreme frequencies, stray white light can easily account for a larger proportion of the intensity of the exit beam. Manufacturer's suggestions for monochromator operation and maintenance should be followed closely to allow optimal functioning and minimum white-light problems.

A similar problem arises from the fact, also previously discussed, that no monochromator produces a completely pure, monochromatic beam of light. To the extent that some of the frequencies included in the beam are unabsorbable, deviations from linearity may be expected as in the case of stray white light. The deviations are not likely to be as severe as in the case of stray white light, since intensity at the contaminating wavelengths will usually be somewhat less relative to the absorbable wavelength. However, the deviation from linearity increases in severity as the concentration of the absorbing species increases. Since the range of frequencies included in a monochromatic beam is usually small—dependent on slit widths, of course— and symmetrical about the nominal wavelength with rather sharply diminishing intensity with distance from the nominal peak, substances with relatively flat absorption peaks may absorb all wavelengths in the beam equally, in which case no deviation from linearity will occur. When the absorption peak coinciding with the nominal exciting wavelength is sharp, or when measurements are made on the side of an absorption band, deviations are inevitable. This does not pose an insurmountable problem, but simply requires the use of calibration curves. Furthermore, dependent on the characteristics of the calibration curve, it may be possible to define a range of sample concentrations for which the linearity assumption is correct within the limits of error acceptable to the experimenter.

In combination with the optical sources of deviation from the laws of absorption, several chemical effects may also result in deviations. Such chemical effects, of course, result only in apparent deviations from the laws, since inevitably the absorbing molecule is actually altered in some way, thus altering the conditions of measurement as well.

Frequently, the concentration of an absorbing species can affect the nature of the absorption process in unexpected ways. One example is relatively simple and not too often problematic. It is possible that when the intensity of the exciting radiation is low, concentrations of the absorbing species that virtually deplete the entire exciting light beam may be prepared. At such concentrations, a sharp deviation from linearity is experienced. Such circumstances are not often encountered, and since they are likely to result at very high concentrations, all other things being equal there should be a relatively broad range of concentrations over which linearity is exhibited.

More commonly problematic are situations in which the concentration of the absorbing species actually alters the color, i.e., absorption spectrum, of the solution. The usual cause for such changes in absorption characteristics is actual alterations in the absorbing molecule. The most commonly cited example of this phenomenon, and one that will be familiar to any student of an introductory chemistry course, concerns cobaltous chloride. Dilute solutions of cobaltous chloride in 4 M HCl result in a pink $[Co(H_2O)_4]^{2+}$ ion. However, concentrated solutions yield a blue $CoCl_4^{2-}$ ion. As the concentration of a cobaltous chloride solution is gradually increased, the colors produced change from pink through shades of purple to blue as the equilibrium between the pink and blue ionic forms shifts. The deviation from linearity encountered, if the pink ionic form is being measured as concentration increases, should be obvious. Many other examples of this phenomenon have been encountered. Mixtures of colored substances do not always behave as would be expected from the laws of absorption, and in all cases of such mixtures, the assumption of additivity of absorbances should be empirically established (see e.g., Calder, 1969).

Another source of difficulty with the laws of absorbance can arise from absorbing species that possess acidic or basic groups or both. As the pH of the solution is varied, the concentrations of the various conjugates invariably differ from one another, and therefore the absorbance of such solutions is bound to change as a function of pH. In poorly buffered solutions, the addition of increasing quantities of an absorbing species may result in a progressive shift in pH, which in turn affects the distribution of conjugates and, ultimately, the absorption characteristics of the solution. Such difficulties may be reduced by adequate buffering, or by working at a pH sufficiently extreme to maintain a predominance of one conjugate form in the solution, or by both measures. An additional complication may arise when the absorbing species is an end-product of a reaction, e.g., an enzymic process, that is pH-dependent. Alterations in pH due, for example, to increasing concentrations of substrate or to by-products of the reactions, or to both, may affect the rate constants of the reaction and consequently may cause apparent deviations from linearity. Again, adequate buffering may be a first line of defense against such unwanted reactions.

The pH of a sample solution may contribute indirectly to another phenomenon that affects the linearity of concentration vs. absorption plots. Many organic absorbing substances may be easily oxidized by atmospheric oxygen, especially when the mixing of solutions results in considerable aeration. Alkaline pH tends to promote such oxidation reactions and, if possible, should be avoided. When this is not possible, the absorption characteristics of the solution are likely to vary as functions of both concentration and time. Rigorous standardization of procedures, especially timing of the absorption determination, although not capable of eliminating the oxidation effect, will normally allow for reproducible results.

Further sources of deviation from the absorption laws can be related to electrolyte concentrations as well as the presence of colloidal materials such as proteins or detergent micelles. In both cases, the nature of the interaction of electromagnetic radiation with the states of the absorbing material may be altered. Essentially, electrolytes may alter the dipole moment of the absorber, and, somewhat similarly, colloids may alter the electronic and vibrational states of the absorbing molecule. Of course, such effects are of little concern as long as they have been taken into account in the development of a quantitative procedure. Problems are more likely to arise when a previously standardized procedure is altered even slightly. For example, analysis of materials isolated on concentrated salt gradients will introduce a strong, potential electrolyte factor into an assay in which such a factor may not have been considered during development.

Temperature is known in many instances to significantly affect the equilibrium of chemical reactions. If the absorbing species is a product of such a temperature-sensitive equilibrium, obviously the temperature of the reaction, i.e., sample, must be controlled. Some substances also exhibit a temperature-lability. The basic resolution of this source of nonlinearity is to rigorously standardize both temperature and time in the procedure. It is worth remembering that in the many spectrophotometers that do not provide temperature control for the sample compartment, the sample may be subjected to an increase over ambient temperature when it is placed in the instrument. This temperature increase will obviously be affected by the length of time that the sample remains in the instrument.

Several comments in summary are appropriate at this point. In the first place, it should be clear that quantitative procedures that exhibit apparent deviations from the linear relationship between absorption and the concentration of the absorbing material as expressed by the laws of absorption are far from uncommon. Of course, this is more of a theoretical aside, but emphasis should be placed on the fact that it has always been possible to demonstrate that a lack of linearity in a given situation is due to an apparent deviation from the absorption laws rather than a violation of them. When a procedure is being developed, all the factors discussed above should be taken into consideration and their effects included in the discussion of the procedure. For the reader who is more likely to employ a previously developed assay rather than to develop a new procedure, certain warnings are in order. An established procedure is likely to have taken into account the variety of factors discussed above. Any deviation from proven methodology may introduce a variable that could significantly influence absorption. Particular care is warranted when an established procedure has demonstrated linearity and therefore provided the possibility of basing concentration calculations on a molar extinction coefficient. Such procedures should never be employed without a verification that the results are reproducible in the user's own laboratory. Variations in instrumentation, reagent sources, operator capabilities, and so on can all contribute to interlaboratory variation. Once the importance of such sources of variability has been discounted, the procedure may be routinely employed. Procedures in which deviations from linearity are expected and which require calibration curves are less likely to result in an oversight of interlaboratory variability. When discrepancies are discovered, consideration of the factors discussed above will often lead to an indentification of the source of variation.

4.5. Fluorescence Spectra

The discussion of absorption spectra in a previous section established a connection between the frequencies of light that molecules absorb and the actual structure, particularly the electronic configurations, of the molecules. Considerations of the states of molecules after the absorption of electromagnetic radiation indicated that there are occasions when some of the absorbed radiation is lost by the molecule through radiative emission. Although the emission of radiation can occur after a relatively long time in the excited state, which is then known as phosphorescence, the emissions of concern here occur more rapidly and are termed fluorescence. As has already been pointed out, the process of fluorescence occurs only when a molecule is in the lowest vibrational level of an excited electronic configuration. The loss of a photon of light results in the molecule's falling to one of the vibrational levels of a lower-energy electronic configuration. The energy of the photon and hence the frequency of the emitted radiation are dependent on the energy differential between the two states. Generally speaking, there are many lower states that a molecule can assume, and therefore there are a similar number of different frequencies at which radiative emission can occur. Without considering the instrumentation at this point, suffice it to say that it is possible to determine the frequencies at which a molecule emits radiation in a manner analogous to that in which the absorption spectrum of a molecule is obtained. Information obtained from such fluorescence spectra can be used to establish the various vibrational and electronic configurations of a molecule in much the same way as absorption spectra are used. While fluorescence spectra are of considerable value in the interpretation of molecular structure, the primary interest here is in the use of fluorescence as a quantitative technique. As was the case with absorption spectra, the primary value of fluorescence spectra in this context is their use in establishing optimum conditions for quantitative procedures.

Since the phenomenon of fluorescence is obviously and intimately linked with that of absorption, many of the considerations that affect the latter can have repercussions on fluorescence techniques. As with absorption measurements, an ideal quantitative-fluorescence procedure is one in which the fluorescence measured is proportional to the concentration of the sample. It should be clear that the amount of fluorescence, as an end product of absorption, must be proportional to the amount of radiation absorbed by the sample. Perhaps it is obvious, then, that interpretation of fluorescence data is simplified when the amount of electromagnetic radiation absorbed by the sample is proportional to the concentration of the sample. In other words, it is helpful if the laws of absorption hold. Of course, the previous sections have suggested numerous reasons for apparent deviation from the absorption laws. In fact, however, the major source of such deviation is sample concentration. It is truly rare for there not to be a linear relationship between absorption and concentration as long as the sample concentrations are kept low enough. Fortunately, fluorescence techniques are so sensitive that sample concentrations can usually be kept at such low levels that the linearity between absorption and concentration can be maintained. When sample solutions are maintained at very dilute concentrations so that only a small fraction of the incident light is absorbed, measured fluorescence is, in fact, directly proportional to the concentration of the fluorescent material. The linear relationship between concentration and measured fluorescence breaks down when the concentration is large enough to significantly diminish the incident light beam such that the energy available for excitation is not uniformly distributed throughout the sample solution. Actually, due to the sensitivity of fluorescence techniques and consequently the low concentrations that can be assayed, linearity can often be maintained over a 10,000-fold increase in concentration.

Quantitative techniques rarely require such an extended range of linearity, since sample concentrations can usually be limited to a smaller range.

The relationship between the amount of energy absorbed by a sample and the actual intensity of the energy emitted as fluorescence is determined by a proportionality constant termed the quantum yield of fluorescence. This is essentially equal to the ratio between the number of quanta emitted and the number of quanta absorbed. Under ideal conditions, the numbers of emitted and absorbed quanta are equal and the proportionality constant is unity. This rarely occurs, and therefore the value of the constant is usually some lower figure. In fact, under ordinary conditions, the quantum fluorescence efficiency of most absorbing molecules is quite low. As discussed previously, the virtual nonfluorescence of many strongly absorbing molecules can be explained by alternative, nonradiative processes that dissipate the excess energy of the excited states of molecules. Such processes, which tend to reduce the lifetime of the excited electronic state and therefore decrease the quantum yield of fluorescence, are termed quenching processes. Quenching processes include both those that result, through internal conversion, in the loss of energy directly by the excited molecule and those that result from an interaction between the excited molecule and another molecule in the sample solution. The latter may either reduce the lifetime of the excited state so as to decrease the probability of fluorescence or interact with a fluorescing species in such a way as to alter the characteristics of the emitted radiation. On the other hand, many biochemical structures do exhibit fluorescence. Furthermore, it is frequently possible to chemically modify essentially nonfluorescing molecules to yield strongly fluorescing products. The excellent reviews of the subject by Udenfriend (1962, 1969), from which, incidentally, much of this discussion is drawn, thoroughly cover the relationship between chemical structure and fluorescence.

4.5.1. Fluorescence Instrumentation

Most of the components of a fluorescence-measuring system are similar or identical to those of an absorption spectrophotometer, as are many of the theoretical and practical considerations regarding their use. This fact will serve to simplify the following presentation, since much of the relevant presentation and discussion has appeared above. The basic arrangement of the fluorometric instrument involves the irradiation of the sample by an incident light beam and, subsequently, measurement of the intensity of the fluorescent emission. While the sensitivity of absorption instruments is limited by their capacity to discriminate reductions in the intensity of the incident beam, a rather different situation is involved in fluorometric work. It will be recalled from the previous discussion that the wavelength of the light emitted by an excited molecule is virtually always longer than the wavelength of the absorbed radiation. The great sensitivity of fluorescence measurements lies in the ability of the appropriate instrumentation to screen the photodetector from the incident radiation. To the extent that the photodetector is exposed to fluorescent light only, it "sees" that light against a black background. In other words, under ideal conditions, the instrument responds to the intensity of the fluorescent radiation alone. The signal-to-noise ratio is greatly enhanced under such conditions, and therefore sensitivity is also increased.

As mentioned, fluorescence instrumentation bears a fundamental similarity to absorption instrumentation. The general features of quantitative instrumentation will be discussed here, but Udenfriend (1962, 1969) can be consulted for more detailed treatment. A quantitative fluorescence instrument includes a light source, a filter or monochromator for isolating the desired frequency of incident light, a sample container, another filter or monochromator arrangement for isolating the desired frequency of fluorescent emission, a photodetector, and, finally, the necessary amplification and readout devices. Thus, the major difference between most absorption and fluorescence instruments is the inclusion of a filter or monochromator between the sample and the photodetector. Incidentally, since the fluorescence emission occurs in all directions, the light path to the photodetector is usually off the axis of the incident radiation. This simplifies the work of the fluorescence filter or monochromator, since the intensity of the incident radiation leaving the sample container in such an off-axis direction is much lower than the intensity along the axis of the incident light beam.

4.5.2. Procedural Considerations

Most, or all, of the practical factors that require consideration in the development and utilization of absorption procedures also apply to fluorometric work. However, the nature of the fluorescence phenomenon, especially when coupled with the fact that most neurobiologists work at or near the limits of sensitivity of fluorometric assays, dictates further practical considerations. These considerations revolve essentially around the sample component of the instrumentation and, as such, deal with the cuvettes, solvents, solutes, impurities, and other factors.

As with absorption work, the composition of the sample cuvettes can influence quantitative determinations. Since virtually all fluorometers function in the single-beam mode, i.e., sample and blank determinations must be made alternately, care must be taken to match sample cuvettes with one another. Care must also be taken to avoid using cuvettes that have become too greatly scratched. Not only do the scratches cause excessive light scattering, but if they are present on the inside of the cuvette, they also become extremely difficult to clean of impurities. The light used to induce fluorescence in the sample material is usually high-frequency and often ultraviolet. As has been previously discussed, the material of which the cuvette is constructed often absorbs at such high frequencies. If the absorption is particularly strong at a given wavelength, it will be difficult to pass sufficient radiant energy into the sample solution. An even more complicated problem arises from the fact that the cuvette material after absorption may itself exhibit strong fluorescence, which can be quite misleading. In general, the suggestions previously discussed for avoiding particular materials when specific wavelengths are used can simply be followed in fluorescence work.

The choice of solvent for a particular assay can also be a critical variable. The same problems encountered with cuvette materials also arise with solvents, namely, both absorption and fluorescence emission at specific wavelengths. Solvents that absorb in a particular region of the spectrum cannot be used if fluorescent determinations are to be made in that region. Furthermore, because of the great potential sensitivity of fluorescence techniques, even traces of an absorbing solvent—remaining, for example, from a previous stage in an isolation procedure—may significantly interfere with a fluorescence assay. Perhaps even more troublesome are impurities in solvents. High-quality solvents suitable for spectrophotometric work are often unsuitable for fluorescence procedures because of trace impurities that are detectable due to the greater sensitivity of fluorescent techniques. Some solvents are now available in a fluorescence grade, but others may need to be purified by the user. Fortunately, it is not always necessary to use absolutely pure solvents. The purity necessary is a function of the level of sensitivity of the assay, the absorption and fluorescence characteristics of the impurities, and the ability of the instrument being used to discriminate fluorescence due to the sample from that due to the impurities. Though it may not be necessary to purify a solvent for a given assay, the solvent should always be run as a blank to determine its contribution to fluorescence. If the solvent produces an appreciable amount of fluorescence relative to the fluorescence of the sample material, it can usually be reduced by distillation or washing with acid, then alkali, then water, provided that the fluorescence is due to impurities and not to the solvent material itself. The water used for washing or when used as a solvent itself should be as free as possible from impurities. Ordinary tap water is highly fluorescent itself, and even distilled water contains a significant level of impurity. The most sensitive procedures require distilled water to be redistilled in a glass still, and often this double-distilled water must be passed over a deionizing resin immediately before use.

Although care with solvent purity is essential for most fluorescent procedures, other aspects of the procedure require equal attention lest the sensitivity gained with solvent purity be lost through other sources of impurities. One common source of contamination is stopcock grease, which can, of course, add impurities to a solvent even as the solvent is being distilled. Rubber and cork can also release a significant amount of fluorescent impurities, especially if they come into contact with organic solvents. Their use should be avoided. As suggested already in regard to cuvettes, the glassware used in a fluorescent procedure can contribute impurities to the sample solution. Synthetic detergents should be avoided because of fluorescent components that are most difficult to remove, especially from scratches. Chromic acid is quite absorbent in the ultraviolet region and must be thoroughly removed by rinsing if used to clean glassware. Chromic acid should not be used to clean cuvettes. Best results with cuvettes and all glassware when maximum purity is required are achieved by washing in boiling nitric acid followed by thorough rinsing with distilled water. In the sequence of glassware that samples, solvents, reagents, and other materials traverse on their way to the final sample cuvette, the closer in the sequence to the assay, the more important is cleanliness. As in other cases, the required level of cleanliness is determined in large part by the degree of sensitivity required for a particular assay. More often than not, the neurobiologist is working near the detection limits of an assay, and so greatest concern with purity is essential.

In addition to the points mentioned above, clearly, all aspects of a procedure must be monitored for fluorescent impurities. This certainly applies to all reagents and buffers that are used. Of particular interest to the neurobiologist who is working with tissue rather than chemicals alone is the level of impurities in the tissue itself. It is usually much more important for fluorescent procedures as compared to absorption procedures to extract the substance of interest, in a fairly pure state, from the tissue prior to an assay. Of course, care must be taken with all of the reagents, solvents, and other materials used in such extraction.

In summary, impurities can present two types of problems for fluorescence assays. As has just been emphasized, the impurities may absorb light of the wavelength used in the procedure. Additionally, after such absorption, fluorescence emission may occur at wavelengths that cannot be discriminated from those emitted by the sample. This, of course, does not necessarilly reduce the number of quanta emitted by the sample material. However, any fluorescence affecting the photodetector but not emitted by the sample can be considered as noise. Fluorescence due to the sample could, then, be termed signal. The sensitivity, accuracy, and precision of an assay are in a large sense a function of the signal-to-noise ratio. Obviously, the more noise relative to signal, the lower will be the quality of the measurement. The other obvious difficulty encountered with impurities occurs when they actually cause a quenching of the fluorescence of the sample material.

As mentioned in the previous discussion of radiative vs. nonradiative processes for the loss of excess energy following absorption, the amount of energy lost through nonradiative processes is proportional to the kinetic energy of the molecules of the solution and, hence, proportional to the temperature of the solution. Coversely, then, the intensity of fluorescence emission from a solution is inversely proportional to the temperature of the solution. Theoretically, if the temperature of the solution is lowered sufficiently, virtually all energy loss by the excited molecules will occur through radiative processes. A fair amount of work has been done with low-temperature fluorescence, but it is of little practical significance to the neurobiologist, who is more likely to work at room temperature. What is important, however, is the fact that fluorescence is extremely sensitive to temperature. It is necessary to be aware of this and to control temperature during fluorometric assays. As with most other practical considerations, more attention to temperature fluctuation is necessary as the sensitivity limits of the procedure are approached.

5. Spectrophotometric Application: Cholinesterase Lability

A portion of a series of studies that grew from investigations of cardioregulation in the invertebrate *Limulus polyphemus* involved spectrophotometric determinations of enzyme activity. These will be briefly described to illustrate the application of the methodology described in this chapter.

5.1. *Limulus polyphemus*—A Simple System

The aim of the comparative approach is to take advantage of physiological variation in order to develop fundamental concepts, and the wide range of invertebrate types provides such variation (Van der Kloot, 1967; Corning and Lahue, 1972). While morphology and organization vary widely in living systems, the comparative physiologist may take advantage of the fact that many biochemical and physiological features of most systems are quite similar. This allows various processes to be studied in organisms that offer special advantages for such study (Corning and Lahue, 1972). The giant axon of the squid, for example, offers several special advantages as a simplified neuronal system. These have been described by Rosenberg (Chapter 1) (also, P. Rosenberg, 1973). Osborne (Chapter 2) Schlapfer (Chapter 3), and Prasad (Chapter 4) have described other approaches that yield simplified systems for neurobiological investigations. The horseshoe crab, *Limulus polyphemus,* provides another example of such a simple system that offers special advantages to the neurobiologist.

The nervous system of *Limulus* is quite large and extremely easy to expose following a simple dissection. The central nervous system consists of a fusion of eight cephalothoracic ganglia into a single mass that lies in a circumesophageal ring. Complete segmentation of the nervous system is maintained in the abdominal portion of the animal. A pair of nerve cords arise from the posterior portions of the subesophageal ganglionic mass and course through the abdomen, ipsilaterally connecting a pair of ganglia (hemiganglia) in each abdominal segment. An intraganglionic commissure in each segment provides the only contralateral connection of the two cords. From each pair of abdominal hemiganglia (hereafter referred to as a ganglion) arise two pairs of mixed peripheral nerves.

Localization of recording and stimulating parameters is facilitated by the segmental morphology of the abdomen. Lesions may also be precisely located from one animal to the next. Furthermore, the entire central nervous system is surrounded by a tough arterial sheath that when cannulated, allows the entire system to be perfused with various agents.

Work by Corning and Von Burg (Corning and Von Burg, 1968, 1970;

Von Burg and Corning, 1969, 1970) revealed behavioral plasticity and response hierarchies as well as many features of functional central nervous system cardioregulatory organization. The effects of various pharmacological agents, including acetylcholine, D-tubocurarine, and eserine, on the abdominal ganglia and cardioregulation have been outlined (Von Burg and Corning, 1968, 1971; Corning *et al.*, 1971; Lahue and Corning, 1973*b*).

A series of investigations have demonstrated that the cardioregulatory output of the abdominal ganglia is modifiable by experience. That is, stimuli that are capable of exerting an inhibitory effect on the heart become less effective after repeated stimulation (Lahue, 1974; Lahue and Corning, 1971, 1973*a,b*; Corning and Lahue, 1972). These studies allowed an assessment of the role of system integrity in this habituation phenomenon. Generally, as a ganglion was successively isolated from other portions of the central nervous system, the rate of habituation declined. Various pharmacological agents, when perfused through the abdominal ganglia during stimulation, were also able to modify the course of habituation.

The pharmacological aspects of the organization of habituation in the abdominal ganglia of *Limulus* were studied in two ganglion systems. The abdominal nerve cord was cannulated posterior to the second ganglion, and in the process the connections with the lower cord were severed. The cord was also severed anterior to the first abdominal ganglion. Thus, the system examined consisted of two ganglia that were isolated from the rest of the system and that were also capable of being perfused. Spontaneous dorsal nerve activity remained quite constant when saline or acetylcholine (1×10^{-4} M) was perfused. Higher concentrations of acetylcholine (8×10^{-4} M) resulted in increased spontaneous activity. Spontaneous activity at first decreased, then increased, when the perfusate included D-tubocurarine (6×10^{-4} and 2×10^{-3} M). The rate of habituation was affected when the perfusate included saline only in comparison to that of preparations that were not perfused. The rate of habituation was greatly increased, relative to saline, when the perfusate included the lower acetylcholine concentration. The higher concentration of acetylcholine tended to produce sensitization rather than habituation. The two levels of D-tubocurarine exerted effects opposite to those of acetylcholine. The low level of D-tubocurarine tended to cause sensitization, while the higher level resulted in habituation.

5.2. Cholinesterase Activity

The activity of cholinesterase enzymes was assayed in various preparations according to the well-established procedure of Ellman *et al.* (1961). All experimental values were specific activities and were calculated using these authors' extinction coefficient of 13,600. The enzyme substrate was acetylthiocholine. Following enzyme hydrolysis, the reaction product was converted to a yellow-colored product. Enzyme activity was expressed as moles of acetylthiocholine hydrolyzed per liter per minute per milligram of protein. The reaction was run under linear conditions as demonstrated in Table 1.

Table 1. Linearity of the Cholinesterase Assay

Time (min)	Change in optical density per minute[a]		
	A	B	C
1.0	0.023	0.029	0.033
2.0	0.023	0.028	0.033
3.0	0.022	0.030	0.031
4.0	0.024	0.030	0.033
5.0	0.023	0.029	0.032
6.0	0.022	0.029	0.032
7.0	0.023	0.029	0.033

[a] Data as observed in three separate preparations.

The rate of change of optical density per minute is expressed for several minutes in individual replicates of three different experiments.

In all enzyme experiments, following whatever procedural variations may have been employed, liquid nitrogen was poured over the abdominal ganglia at some standardized time, and the tissue was virtually immediately frozen. The rapid freezing was due to the relatively small size of the tissue being frozen. A typical ganglion, for example, was only about 100 mm^3 in size. The ganglia were maintained in liquid nitrogen until they were to be homogenized prior to determination of enzyme activity.

Initial enzyme determinations were performed on abdominal ganglia

Table 2. Cholinesterase Activity in Untreated Tissues[a]

Ganglia	Connectives
First preparation	
1.200×10^{-7}	3.900×10^{-7}
1.060×10^{-7}	3.990×10^{-7}
1.290×10^{-7}	4.550×10^{-7}
Second preparation	
0.756×10^{-7}	1.630×10^{-7}
0.867×10^{-7}	1.747×10^{-7}
Third preparation	
0.690×10^{-7}	1.953×10^{-7}
0.690×10^{-7}	1.953×10^{-7}

[a] Data from pooled ganglia and pooled interconnective tissue in three separate preparations. Three replications were run on tissue from the first preparation, and two replications were run on tissues from the second and third preparations.

and nerve cords that had been perfused with saline but not stimulated. Each of the first three ganglia was isolated from the nerve cord interconnectives and pooled. The remaining interconnective tissue was pooled separately. When enzyme activity in the two samples was determined, the interconnective tissue was found to exhibit about 3 times the activity of the ganglionic tissue. The results of two or three replicate determinations on tissues from three animals are presented in Table 2. The general levels of specific activity lie within the range obtained in tissues from other organisms (Potter, 1969). When the cholinesterase inhibitor eserine was included in the assay at a final concentration of 4×10^{-4} M, activity in both groups of samples was completely inhibited. This tends to confirm that the hydrolysis of the substrate was due to the activity of cholinesterase and not some other nonspecific source. The inclusion of D-tubocurarine in concentrations ranging from 8×10^{-4} to 1.6×10^{-2} M resulted in inhibition of enzyme activity ranging from 25 to 57%.

For studies of habituation, the perfusion cannula was inserted through the nerve cord sheath to a point behind the first abdominal ganglion. The cannula was firmly tied between the first and second ganglia and between the second and third ganglia. The compression of the nerve cord between the tying material and the cannula was sufficient to completely disrupt the connections between ganglia. This was verified electrophysiologically and morphologically. The result was that the first and second abdominal ganglia were completely isolated from one another and from the rest of the central nervous system. The first ganglion could be perfused, while the second was not. Following perfusion and in many cases stimulation of the first ganglion, the two ganglia were rapidly frozen and subsequently assayed separately for cholinesterase activity. A ratio was obtained between the activities in the perfused and (often) stimulated first ganglion and the nonperfused, unstimulated second ganglion in each preparation. The results of these experiments are presented in Table 3.

When both ganglia were treated as controls, that is, prepared exactly as described above but without any perfusion or stimulation, the specific activity of the second ganglion was consistently higher than that in the first ganglion. Since both ganglia were treated identically in this case, this may be taken to be the normal or control situation. In all other comparisons, the second ganglion was treated in exactly the same fashion, while the first ganglion was subjected to various other treatments. When the first ganglion was perfused with saline, the specific enzyme activity increased significantly. The increase was sufficiently large to reverse the two ganglion ratios. The inclusion of 1×10^{-4} M acetylcholine to the perfusate had approximately the same effect on enzyme activity as did perfusion with saline alone. However, when 360 habituation stimuli were applied in conjunction with 1×10^{-4} M acetylcholine perfusion, specific enzyme activity in the first ganglion decreased, and the two ganglion ratios were again similar to that obtained in the control (untreated) preparations.

There is certainly a marked lability of the cholinesterase system in

Table 3. First Ganglion/Second Ganglion Cholinesterase-
Activity Ratios[a]

Group One	Group Two	Group Three	Group Four
0.911	1.330	1.130	0.890
0.877	1.110	1.030	0.904
0.847	0.973	1.220	0.910
0.749	1.230	0.780	0.849
0.950	1.170	0.960	0.950
0.945	1.630	0.980	0.862
0.724	0.894	1.080	
	1.350	0.970	
	1.140	1.380	
	0.970	1.070	
	1.500	1.270	

[a] Each data point represents the ratio between the specific activities of the first abdominal ganglion and the second abdominal ganglion in a single preparation treated under one of four conditions. Three replicate assays were performed on each ganglion, and the ratios reported are averages of the replicates. Group One: untreated control; Group Two: saline control; Group Three: acetylcholine (1×10^{-4} M); Group Four: acetylcholine (1×10^{-4} M) plus habituation. Please refer to the text for further details.

Limulus. Some of this lability can be related to the process of habituation. Although the procedures employed were not exactly comparable, the enzyme-activity findings seem compatible with the results of electrophysiological experiments.

References

Adam, N.K., 1956, *Physical Chemistry*, Clarendon Press, Oxford.

Alford, W.P., 1962, Atomic structure, in: *Comprehensive Biochemistry*, Vol. I, *Atomic and Molecular Structure* (M. Florkin and E.H. Stotz, eds.), pp. 1–33, Elsevier, Amsterdam.

Anderson, D.H., and Woodall, N.B., 1972, Infrared spectroscopy, in: *Techniques of Chemistry*, Vol. I, *Physical Methods of Chemistry: Part IIIB. Spectroscopy and Spectrometry in the Infrared, Visible, and Ultraviolet* (A. Weissberger and B.W. Rossiter, eds.), pp. 1–84, Wiley-Interscience, New York.

Atkins, K.R., 1972, *Physics-Once-Over-Lightly*, John Wiley and Sons, New York.

Barrow, G.M., 1963, *The Structure of Molecules*, W.A. Benjamin, New York.

Bauman, R.P., 1962, *Absorption Spectroscopy*, John Wiley and Sons, New York.

Bladen, P., 1964, Ultraviolet and visible spectroscopy, in: *Physical Methods in Organic Chemistry* (J.C.P. Schwarz, ed.), pp.126–167, Holden-Day, San Francisco.

Bladen, P., and Eglinton, G., 1964, Ultraviolet, visible and infrared spectroscopy, in: *Physical Methods in Organic Chemistry* (J.C.P. Schwarz, ed.), pp. 22–34, Holden-Day, San Francisco.

Blum, H.F., 1950, Action spectra and absorption spectra, in: *Biophysical Research Methods* (F.M. Uber, ed.), pp. 417–449, Interscience, New York.

Brode, W.R., 1943, *Chemical Spectroscopy*, John Wiley and Sons, New York.

Burris, R.H., 1972, Spectrophotometry, in: *Manometric and Biochemical Techniques* (W.W. Umbreit, R.H. Burris, and J.F. Stauffer, eds.), pp. 276–294, Burgess, Minneapolis.

Calder, A.B., 1969, *Photometric Methods of Analysis*, American Elsevier, New York.

Cheng, K.L., 1971, Absorptimetry, in: *Advances in Analytical Chemistry and Instrumentation*, Vol. 9, *Spectrochemical Methods of Analysis* (J.D. Wineforder, ed.), pp. 321–385, Wiley-Interscience, New York.

Clark, C. 1955, Infrared spectrophotometry, in: *Physical Techniques in Biological Research*, Vol. 1 (G. Oster and A.W. Pollister, eds.), pp. 205–323, Academic Press, New York.

Corning, W.C., and Lahue, R., 1972, Invertebrate strategies in comparative learning studies, *Am. Zool.* **12**:455–469.

Corning, W.C., and Von Burg, R., 1968, Behavioral and neurophysiological investigations of *Limulus polyphemus*, in: *Neurobiology of Invertebrates* (J. Salanki, ed.), pp. 463–477, Plenum Press, New York.

Corning, W.C., and Von Burg, R., 1970, Neural origin of cardioperiodicities in *Limulus, Can. J. Zool.* **48**:1450–1454.

Corning, W.C., Lahue, R., and Von Burg, R., 1971, Supraesophageal ganglia influences on *Limulus* heart rhythmm: Confirmatory evidence, *Can. J. Physiol. Pharmacol.* **49**:387–393.

Corwin, A.H., and Bursey, M.M., 1966, *Elements of Organic Chemistry*, Addison-Wesley, Reading, Massachusetts.

Duncan, A.B.F., 1956a, Theory of infrared and Raman spectra, in: *Technique of Organic Chemistry*, Vol. IX, *Chemical Applications of Spectroscopy* (W. West, ed.), pp. 187–245, Interscience, New York.

Duncan, A.B.F., 1956b, Electronic spectra in the visible and ultraviolet. Part 1. Theory of electronic spectra, in: *Technique of Organic Chemistry*, Vol. IX, *Chemical Applications of Spectroscopy* (W. West, ed.), pp. 581–628, Interscience, New York.

Eglinton, G., 1964, Infrared and Raman spectroscopy, in: *Physical Methods in Organic Chemistry* (J.C.P. Schwarz, ed.), pp. 35–125, Holden-Day, San Francisco.

Ehrenberg, A., and Theorell, H., 1962, Fluorescence, in: *Comprehensive Biochemistry*, Vol. 3, *Methods for the Study of Molecules* (M. Florkin and E.H. Stotz, eds.), pp. 169–188, Elsevier, Amsterdam.

Ellman, G.L., Courtney, K.D., Andres, V., and Featherstone, R.M., 1961, A new and rapid colorimetric determination of acetylcholinesterase activity, *Biochem. Pharmacol.* **7**:88–95.

Freeman, I.M., 1968, *Physics: Principles and Insights*, McGraw-Hill, New York.

Grum, F., 1972, Visible and ultraviolet spectrophotometry, in: *Techniques of Chemistry*, Vol. I, *Physical Methods of Chemistry: Part IIIB. Spectroscopy and Spectrometry in the Infrared, Visible, and Ultraviolet* (A. Weissberger and B.W. Rossiter, eds.),pp. 207–427, Wiley-Interscience, New York.

Hiskey, C.F., 1955, Absorption spectroscopy, in: *Physical Techniques in Biological Research* (G. Oster and A.W. Pollister, eds.), Vol. I, pp. 73–130, Academic Press, New York.

Jaffe, H.H., 1962, Electronic theory of organic molecules, in: *Comprehensive Biochemistry*, Vol. I, *Atomic and Molecular Structure* (M. Florkin and E.H. Stotz, eds.), pp. 34–112, Elsevier, Amsterdam.

Jaffe, H.H., and Orchin, M., 1962, *Theory and Applications of Ultraviolet Spectroscopy*, John Wiley and Sons, New York.

Jones, H.N., and Sandorfy, C., 1956, The application of infrared and Raman spectrometry to the elucidation of molecular structure, in: *Technique of Organic Chemistry*, Vol. IX, *Chemical Applications of Spectroscopy* (W. West, ed.), pp. 247–580, Interscience, New York.

Jorgensen, C.K., 1962, *Absorption Spectra and Chemical Bonding in Complexes*, Pergamon Press, Oxford.

Lahue, R., 1974, Habituation characteristics and mechanisms in the abdominal ganglia of *Limulus polyphemus*, Ph.D. disertation, University of Waterloo, Waterloo, Ontario, Canada.

Lahue, R., and Corning, W.C., 1971, Habituation in *Limulus* abdominal ganglia, *Biol. Bull.* **140**:427–439.

Lahue, R., and Corning, W., 1973a, Incremental and decremental processes in *Limulus* ganglia: Stimulus frequency and ganglion organization influences, *Behav. Biol.* **8**:637–653.

Lahue, R., and Corning, W., 1973b, The effects of curare in *Limulus* central nervous system, *Can. J. Physiol. Pharmacol.* **51**:366–370.

Lothian, G.F., 1969, *Absorption Spectrophotometry*, Adam Hilger, London.

Matsen, F.A., 1956, Electronic spectra in the visible and ultraviolet. Part 2. Applications, in: *Technique of Organic Chemistry*, Vol. IX, *Chemical Applications of Spectroscopy* (W. West, ed.), pp. 629–706, Interscience, New York.

Mellon, M.G., 1950, *Analytical Absorption Spectroscopy*, John Wiley and Sons, New York.

Morton, R.A., 1962, Spectrophotometry in ultraviolet and visible regions, in: *Comprehensive Biochemistry*, Vol. 3, *Methods for the Study of Molecules* (M. Florkin and E.H. Stotz, eds.),pp. 66–132, Elsevier, Amsterdam.

Newman, D.W., 1964, Ultraviolet and visible absorption spectroscopy, in: *Instrumental Methods of Experimental Biology* (D.W. Newman, ed.), pp. 324–359, Macmillan, New York.

Orear, J., 1967, *Fundamental Physics*, John Wiley and Sons, New York.

Pesce, A.J., Rosen, C.-G., and Pasby, T.L., 1971, *Fluorescence Spectroscopy*, Marcel Dekker, New York.

Potter, L.T., 1969, Acetylcholine, choline, acetyltransferase, and acetylcholinesterase, in: *Handbook of Neurochemistry*, Vol. IV (A. Lajtha, ed.), pp. 263–284, Plenum Press, New York.

Roberts, J.D., and Caserio, M.C., 1967, *Modern Organic Chemistry*, W.A. Benjamin, New York.

Rosenberg, J.L., 1955, Photochemistry and luminescence, in: *Physical Techniques In Biological Research*, Vol. I (G. Oster and A.W. Pollister, eds.), pp. 1–49, Academic Press, New York.

Rosenberg, P., 1973, The giant axon of the squid: A useful preparation for neurochemical and pharmacological studies, *Methods of Neurochem.* **4**:97–160.

Scott, J.F., 1955, Ultraviolet absorption spectrophotometry, in: *Physical Techniques in Biological Research*, Vol. I (G. Oster and A.W. Pollister, eds.), pp. 131–203, Academic Press, New York.

Skoog, D.A., and West, D.M., 1963, *Fundamentals of Analytical Chemistry*, Holt, Rinehart and Winston, New York.

Smith, C.D., 1964, Infrared absorption spectroscopy, in: *Instrumental Methods of Experimental Biology* (D.W. Newman, ed.), pp. 360–393, Macmillan, New York.

Stearns, E.I., 1969, *The Practice of Absorption Spectrophotometry*, Wiley-Interscience, New York.

Udenfriend, S., 1962, *Fluorescence Assay in Biology and Medicine*, Academic Press, New York.

Udenfriend, S., 1969, *Fluorescence Assay in Biology and Medicine*, Vol. II, Academic Press, New York.

Van der Kloot, W.G., 1967, Goals and strategy of comparative pharmacology, *Fed. Proc. Fed. Am. Soc. Exp. Biol.* **26**:975–980.

Van Duuren, B.L., and Chan, T.L., 1971, Fluorescence spectrometry, in: *Advances in Analytical Chemistry and Instrumentation*, Vol. 9, *Spectrochemical Methods of Analysis* (J.D. Wineforder, ed.), pp. 387–450, Wiley-Interscience, New York.

Van Holde, K.E., 1971, *Physical Biochemistry*, Prentice-Hall, Englewood Cliffs, New Jersey.

Von Burg, R., and Corning, W.C., 1968, Neurologically active drugs in *Limulus* central nervous system, *Am. Zool.* **8**:379.

Von Burg, R., and Corning, W.C., 1969, Cardioinhibitor nerves in *Limulus*, *Can. J. Zool.* **47**:735–737.

Von Burg, R., and Corning, W.C., 1970, Cardioregulatory properties of *Limulus* abdominal ganglia, *Can. J. Physiol. Pharmacol.* **48**:333–341.

Von Burg, R., and Corning, W.C., 1971, The effects of drugs on *Limulus* cardioregulators, *Can. J. Physiol. Pharmacol.* **49**:1044–1048.

West, W., 1949a, Spectroscopy and spectrophotometry, in: *Physical Methods of Organic Chemistry*, Vol. I, Part III (A. Weissberger, ed.), pp. 1241–1398, Interscience, New York.

West, W., 1949b, Colorimetry, photometric analysis, fluorimetry, and turbidimetry, in: *Physical Methods of Organic Chemistry*, Vol. I, Part II (A. Weissberger, ed.), pp. 1399–1490, Interscience, New York.

West, W. 1956a, Introductory survey of molecular spectra, in: *Technique of Organic Chemistry*, Vol. IX, *Chemical Applications of Spectroscopy* (W. West, ed.), pp. 1–70, Interscience, New York.

West, W.,1956b, Fluorescence and phosphorescence, in: *Technique of Organic Chemistry*, Vol. IX, *Chemical Applications of Spectroscopy* (W. West, ed.), pp. 707–758, Interscience, New York.

White, C.E., and Argauer, R.J., 1970, *Fluorescence Analysis: A Practical Approach*, Marcel Dekker, New York.

White, E.H., 1964, *Chemical Background for the Biological Sciences*, Prentice-Hall, Englewood Cliffs, New Jersey.

White, H.E., 1934, *Introduction to Atomic Spectra*, McGraw-Hill, New York.

Wotherspoon, N., Oster, G.K., and Oster, G., 1972, The determination of fluorescence and phosphorescence, in: *Techniques of Chemistry*, Vol. I, *Physical Methods of Chemistry: Part IIIB. Spectroscopy and Spectrometry in the Infrared, Visible, and Ultraviolet* (A. Weissberger and B.W. Rossiter, eds.), pp. 429–484, Wiley-Interscience, New York.

Immunological Techniques in Biochemical Investigation

J.S. Woodhead and R.J. Thompson

1. Introduction

While the mechanism of antibody production in response to antigenic challenge is poorly understood, the opposite is true of reactions between antigen and antibody. This is not surprising, since these reactions can be studied both *in vivo* and *in vitro* in a range of model systems. The understanding of these processes has proved of enormous benefit to the biochemist during recent years, and a wide range of immunological techniques for both the qualitative and quantitative estimation of antigens and antibodies have been developed. The usefulness of antibodies for the detection of biologically important molecules derives from two fundamental properties. First, their remarkable specificity for antigen can make possible the identification and quantitation of molecules in complex biological media such as plasma even in the presence of structurally related molecules. Second, the energy of reaction between antigen and antibody is frequently extremely high, thus allowing the reaction to be monitored at very low concentrations provided a suitable method of detection is used.

Reactions between antigen and antibody can be monitored either directly or by following secondary events that are dependent on the primary interaction. Direct methods for monitoring antigen-antibody interaction such as fluorometry, ammonium sulfate precipitation, or equilibrium dialysis have limited applications, so that measurement of the primary reaction requires that either the antigen or the antibody be labeled. Secondary events that can be used to follow the primary reaction include immunoprecipitation, hem-

J.S. Woodhead • Department of Medical Biochemistry, Welsh National School of Medicine, Cardiff CF4 4XN, Wales. *R.J. Thompson* • Department of Clinical Biochemistry, University of Cambridge, Cambridge, England.

agglutination, and complement fixation. In this review, we shall consider first the principles and applications of these latter techniques and then the development of quantitative assay methods using labeled antigens or antibodies. The discussion will be confined mainly to common techniques for the detection of soluble antigens. For discussion of some of the more specialized procedures (e.g., histochemical immunofluorescence techniques for fixed antigens), the reader is referred to comprehensive standard works (Nairn, 1969; Williams and Chase, 1971; Weir, 1973).

2. Basic Immunological Techniques

2.1. Immunoprecipitation

Addition of increasing amounts of antigen to a fixed amount of antiserum (containing antibody directed against that antigen) leads to the characteristic sequence of events depicted in Fig. 1. At first, increasing amounts of antigen produce increasing quantities of immunoprecipitate; this is the zone (or interval) of antibody excess, and essentially all the antigen added appears in the immunoprecipitate. Addition of larger amounts of antigen takes the system into the zone of equivalence, in which the ratio of the concentrations of antigen and antibody is such that all the antigen added and all the antibody present react and precipitate, the supernatant being devoid both of free antigen and of free antibody, and also of soluble antibody-antigen complexes. Increasing amounts of antigen added beyond the zone of equivalence result in a progressively decreasing amount of immunoprecipitate; this is the zone of antigen excess, and the supernatant now contains

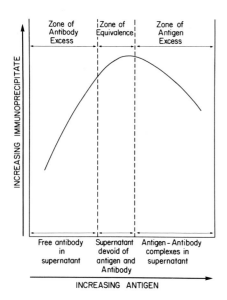

Fig. 1. Formation of immunoprecipitate between antigen and antibody. Increasing amounts of antigen added to a constant amount of antibody produce an immunoprecipitate according to the curve shown. Maximum immunoprecipitate is formed in the zone of equivalence; at antigen concentrations below this, there is free antibody left in the supernatant, while at antigen concentrations above this, there are free antigen and unprecipitated antibody–antigen complexes left in the supernatant.

soluble antigen–antibody complexes and also, possibly, free antigen. The terms "antigen excess" and "antibody excess" thus refer to the supernatant left after precipitation. The generally accepted explanation of the phenomenon described above is the "lattice" theory. Immunoglobulin G (IgG) molecules are bivalent, i.e., possess two available binding sites for antigen, while most antigens have several antigenic groupings (usually 10–50 per molecule, depending on size). A single antigen molecule can thus bind several antibody molecules, and each antibody molecule can combine with two antigen molecules. In this way, a large aggregate or "lattice" of antigen and antibody molecules can be built up, which, if large enough, will appear as an immunoprecipitate. No lattice formation is possible without sufficient antigen to "polymerize" the antibody molecules, while too high an antigen concentration occupies antibody sites with free antigen rather than with antigen bound to a second antibody (Humphrey and White, 1970). Lattice formation also means that the relative amounts of antigen and antibody in the immunoprecipitate will vary according to the initial reacting concentration.

In mass terms, the immunoprecipitate is formed largely from antibody rather than antigen. Excess antigen not only inhibits the formation of immunoprecipitate, but also dissolves immunoprecipitate already formed; i.e., the antigen-antibody interaction is reversible. Some antisera (horse-type antisera) produce immunoprecipitates that are soluble in antibody excess as well as in antigen excess. The shape of the precipitin curve varies with different antigens and different antisera, some antigen-antibody systems producing very narrow zones of equivalence, others producing very broad zones. Physical factors are also of importance in determining the production of an immunoprecipitate. While the precipitin reaction is relatively insensitive to pH variation between pH 6.6 and 8.5, the immunoprecipitate is soluble below pH 4.2 and above pH 9.5. Usually, more immunoprecipitate is formed at 0° than at 37°C, although this varies with different antigen–antibody systems. The effects of ionic strength, organic solvents, lipids, and other factors on the precipitation reaction have recently been discussed (Williams and Chase, 1971).

2.2. Immunodiffusion

The production of an immunoprecipitate depends on a correct concentration ratio between antigen and antibody. The actual physical amounts of antigen and antibody present are often unknown, and consequently the correct concentration ratio for precipitation has to be found by trial and error. This means adding increasing concentrations of antigen to a series of tubes containing a fixed amount of antiserum (or/ vice versa) until the zone of equivalence is reached. A simpler method of establishing a zone of equivalence is the technique of immunodiffusion. The techniques of immunoprecipitation in transparent gels have largely been established by Oudin (1946), Ouchterlony (1948), Elek (1948), and Oakley and Fulthorpe (1953). Immunodiffusion and immunoelectrophoresis have recently been discussed by Ouchterlony and Nilsson (1973), and a practical monograph is now available (Clausen, 1971). The diffusion of an antigen in free solution creates a concentration gradient varying from 100% at the source of diffusion to 0% at a sufficient distance from that source. Consequently, establishing a concentration gradient of antigen within a solution of antibody must lead to a zone of equivalence at some point along the gradient, providing the initial concentration of antigen is greater than the concentration of the antibody. The diffusion of antigen (or antibody) in one dimension in a fluid medium can be described mathematically (see Williams and Chase, 1971, Volume 3, pp. 108–118). The diffusion coefficient of a molecule in free solution varies with the size and degree of hydration of the molecule and is also related to temperature. The distance from the point of origin of a given amount of the diffusing substance is directly proportional to the square root of the time of diffusion, and the distance of diffusion is also proportional to the logarithm of the initial concentration of the substance. Diffusion in fluid media can be influenced by convection currents; these can be countered by performing the diffusion in a gelified medium that stabilizes the fluid and prevents convection disturbances and, furthermore, traps and stabilizes immunoprecipitates. Several media have been used for immunodiffusion; these include agar (a mixture of agarose and agaro-pectin), agarose, starch gel, cellulose acetate membranes, and polyacrylamide, but for most purposes agar or agarose gels are suitable. Methods of preparation and use of these gel media are described by Clausen (1971). The micellar structure of agar gel does not restrict the diffusion of substances with molecular weights up to about 200,000, and since the gel consists nearly entirely of water (99% in some cases), antigen–antibody reactions and diffusion characteristics are the same as those occurring in free solution. The main immunoglobulin species participating in immune reactions in gels is IgG, partly because this forms 83% of total serum immunoglobulins and partly because its molecular weight of 160,000 permits free diffusion within the gel; other immunoglobulin classes of higher molecular weight (IgM and some IgA molecules) are excluded from the gel (Clausen, 1971). The restriction property of gel media to very-high-molecular-weight substances is useful in that immunoprecipitates formed within the gel that are above this size limit remain trapped, while unreacted antigen and antibody can be washed out with saline. This facilitates detection of immunoprecipitates with protein stains, since it reduces the background staining to very low levels.

Immunodiffusion can be classified according to whether one or both reactants (i.e., antigen or antibody or both) are allowed to form a concentration gradient. In single immunodiffusion, the antigen (or antibody) is allowed to diffuse into a gel containing the antibody (or antigen) at a constant relatively low concentration. In double immunodiffusion, both antigen and antibody are allowed to diffuse independently into the same gel; at the start of the diffusion, the gel contains neither antigen nor antibody and is therefore serologically "neutral." Immunodiffusion is also described as "one-dimensional" if both antigen and antibody diffuse along the same axis, or as "two-dimensional" either if diffusion of one reactant is radial or if antigen and antibody diffuse along different axes (Ouchterlony and Nilsson, 1973). If the concentrations of antigen and antibody at the source of diffusion are equivalent (i.e., the correct ratio for immunoprecipitation), then the system is said to be "balanced"; if the relative concentrations are not equivalent, then the system is "unbalanced." Single-immunodiffusion techniques require an unbalanced system; double immunodiffusion is normally performed with balanced systems. Further terms that are sometimes encountered in immunodiffusion are "simple," "complex," and "multiple"; these refer not to the type of technique being used but to the nature of the serological system being examined. A simple system contains only one antigen–antibody system, (a complex system consists of more than one antigen–antibody system), a complex system consists of more than one serologically related antigen–antibody system, and a multiple system consists of several unrelated antigen–antibody systems (Ouchterlony and Nilsson, 1973).

2.2.1. Single Immunodiffusion

2.2.1a. Tube Technique. This technique was introduced by Oudin (1946) in 1946. The antibody is mixed with agar and allowed to set in the bottom of a glass tube (Fig. 2A). The antigen solution is layered on top of the set gel, and the system is incubated at room temperature for up to 3 days. The antigen diffuses into the gel, and at the zone of equivalence an immunoprecipitate forms. As antigen continues to diffuse into the gel, a region of antigen excess builds up behind the immunoprecipitate, which therefore

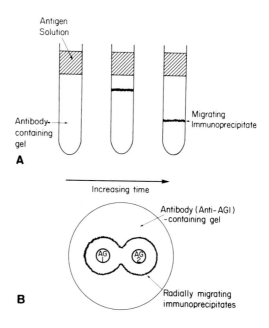

Fig. 2. (A) Oudin tube technique. Antigen diffuses into agar gel containing antibody to form a migrating immunoprecipitate. This is an unbalanced one-dimensional single-immunodiffusion system. (B) Petrie technique. Holes are punched in an agar plate containing antibody to antigen 1 (AG 1). AG 1 and antigen 2 (AG 2) are placed in the central wells and diffuse radially to form an unbalanced single-radial-immunodiffusion system with radially migrating immunoprecipitates. Fusion of the migrating immunoprecipitates as shown indicates that the antibody (anti-AG 1) cannot distinguish AG 1 from AG 2; hence, the two antigens are immunologically identical.

dissolves. As the antigen concentration builds up in front of the zone of equivalence, more immunoprecipitate forms, and the gel therefore appears to contain a migrating immunoprecipitate. The distance of migration is an exponential function of the antigen concentration; hence, the technique can be used quantitatively by standardization with known concentrations of the same antigen. The antigen must be in excess, i.e., the system must be unbalanced, and the technique requires accurate temperature control, since abrupt changes of temperature lead to the appearance of artifact bands that may be more or less dense than the immunoprecipitate. These bands, however, remain stationary. Different antigens diffusing into the gel react independently with their specific antibodies.

2.2.1b. Single Radial Immunodiffusion. This method was originally developed by Petrie (1932) and used to compare different antigens. Subsequently, the method was developed for quantitative use by Mancini *et al.* (1964). The antigen is placed in a well punched in a plate of agar gel containing the antibody and allowed to diffuse radially (Fig. 2B). Ring-shaped precipitates appear around the wells; these migrate centrifugally by the same mechanism as in the Oudin tube technique (see above). However, unlike the latter, the antigen is not used in such excess, and eventually the rings stop migrating, since insufficient antigen is present to continue dissolving the center of the precipitation ring. If two wells are used, then fusion of the rings indicates antigenic identity with the particular antibody used (Fig. 2B). The area covered by the rings at the end of diffusion is directly proportional to the initial antigen concentration (Mancini *et al.*, 1964); hence, by using a known amount of antigen, a standard curve can be constructed and the method used quantitatively.

2.2.2. Double Immunodiffusion

With this technique, both antigen and antibody diffuse toward each other in the same gel, forming a precipitate at the point of equivalence. Double immunodiffusion uses balanced systems; i.e., the initial concentrations of antigen and antibody are close to equivalence, so that the precipitate does not migrate. Double immunodiffusion can be performed in tubes, a one-dimensional method developed from the Oudin technique by Oakley and Fulthorpe (1953), but is more often performed two-dimensionally using wells punched out of agar plates (Ouchterlony, 1948; Elek, 1948). The arrangement of wells for antigens and antibodies varies; some of the more common patterns used are shown in Fig. 3A together with the usual patterns of immunoprecipitates. Since both antigen and antibody are diffusing radially, the zone of equivalence, and hence the immunoprecipitate, grows from the ends as shown in Fig. 3B. The shape of the immunoprecipitate is related to the molecular weight of the antigen (and hence to its diffusion rate) relative to the molecular weight of the antibody (about 160,000 for IgG). If the antigen has a molecular weight similar to that of the antibody, the precipitate appears as a straight line between the antigen and antibody wells; if the

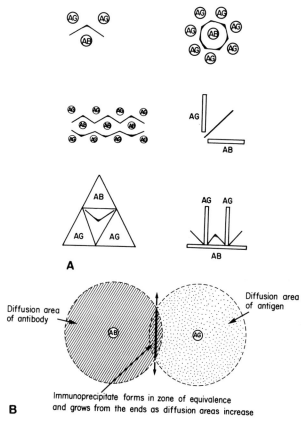

Fig. 3. Double immunodiffusion systems. (A) Common patterns of wells and troughs used in double-immunodiffusion systems in agar plates. Circular wells or rectangular troughs are cut in serologically neutral agar plates, and the antigen (AG) and respective antibody (AB) solutions are added as shown. The usual position of the immunoprecipitate after immunodiffusion is shown by the black lines. These represent two-dimensional balanced double-immunodiffusion systems. (B) Precipitate formation following double immunodiffusion in agar. The antibody (AB) and antigen (AG) diffuse radially from separate wells; a zone of equivalence is formed at some point within the overlap of the two diffusion areas. The immunoprecipitate forms at the zone of equivalence and grows from the ends as the diffusion areas increase.

antigen is smaller than the antibody, the precipitate appears as a curved arc with the concavity toward the antibody well; while if the antigen has a significantly higher molecular weight than the antibody, the precipitate appears as a curved arc with the concavity toward the antigen well (Williams and Chase, 1971, Volume 3). If the concentration of the antibody is kept constant, then progressive dilution of the antigen results in the immunoprecipitate's appearing closer to the antigen well. Conversely, at constant antigen concentration, diluting the antibody results in the immunoprecipitate's appearing closer to the antibody well. Each antigen–antibody system behaves independently in precipitating in the gel; one immunoprecipitate does not act as a barrier to unrelated antigens and antibodies, but only to the antigen and antibody that form that immunoprecipitate. The number of precipitin lines is therefore equal to, or less than, the number of antigens placed in the antigen well.

Double immunodiffusion is widely used to detect similarities between two antigens when tested against the same antibody. Figure 4 shows the types of precipitation lines that can be obtained. Figure 4a shows the pattern obtained with a "reaction of identity"; in this reaction, the two precipitation lines fuse completely. This means that the antibody being used cannot distinguish between the two antigens, which are therefore immunologically identicial. In view of the great specificity shown by antibodies, this may indicate that the two antigens are also chemically identical. Figure 4b shows the pattern obtained with a "reaction of nonidentity"; in this reaction, the two precipitating systems behave independently and cross with no interference of the precipitin lines. This means that the two antigens are immunologically distinct and each is reacting independently with its own separate antibody diffusing from the antibody well. Figure 4c illustrates the intermediate situation in which the two antigens have several antigenic groupings in common (which produces line fusion), but one of the antigens has additional groupings not present in the second antigen. These additional antigenic determinants form a spur opposite the well of the antigen carrying them; the spur is less dense than the rest of the precipitate, since it is formed from less antibody. Figure 4d shows a fourth possible pattern, which is found in cases in which both antigens have individually specific groupings but also share at least one common grouping and is usually seen with two cross-reacting antigens different from the one used to produce the antiserum (Williams and Chase, 1971, Volume 3). Further discussion of the form and significance of precipitation spurs can be found in Ouchterlony and Nilsson (1973).

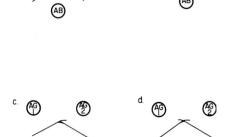

Fig. 4. Patterns of immunoprecipitates in double immunodiffusion. Two antigens, AG 1 and AG 2, are compared for cross-reactivity by immunodiffusion against a single antiserum (AB). The patterns shown are a reaction of identity (a), a reaction of nonidentity (b), a reaction of partial identity (c), and a more complex form of partial identity (d). See the text for further explanation.

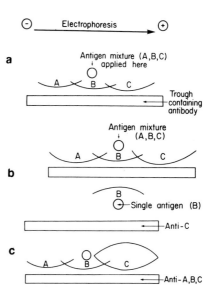

Fig. 5. Diagrammatic representation of immunoelectrophoresis. (a) A mixture containing antigens
A, B, and C is placed in a well in an agar slide and
electrophoresed as shown. After electrophoresis,
a trough is cut in the agar and filled with antisera
directed against all three antigens; curved immunoprecipitation arcs appear as shown. The
identity of individual antigens can be investigated
by comparison with a known single antigen (b) or
by use of a second antiserum monospecific toward
one of the antigens (c).

2.3. Immunoelectrophoresis

The ability of immunodiffusion techniques to distinguish a mixture of
antigen–antibody systems is limited by their ability to resolve a number of
distinct precipitin lines. The resolving power of immunodiffusion can be
increased if a preliminary separation procedure is applied first. This is
generally done by electrophoresis, although other procedures, e.g., thin-
layer gel chromatography (Hanson *et al.*, 1966), have also been used.
Electrophoresis as a preliminary to immunodiffusion was introduced by
Grabar and Williams (1953). The technique is illustrated in Fig. 5. The
mixture of antigens is placed in a well in a strip of agar and electrophoresed,
usually at pH 8.6 for 30–90 min. Following electrophoresis, a trough is cut
parallel to the axis of electrophoresis, an antiserum to the original mixture
of antigens is placed in the trough, and the system is allowed to diffuse as
with double immunodiffusion. A series of precipitation arcs form between
the electrophoretic strip and the trough. Fusion of arcs to give a double-
humped appearance indicates an immunologically identical pair of antigens
with the same electrophoretic mobility. Micromethods for immunoelectrophoresis were introduced by Scheidegger (1955). The procedure can also be
performed with cellulose acetate strips. Practical details are given in Clausen
(1971).

2.3.1. Electrophoresis into Antibody-Containing Gels

Diffusion of antigens into an antibody-containing gel can be accelerated
by the application of an electrical field (Ressler, 1960). This has been

exploited (Clarke and Freeman, 1968; Laurell, 1965, 1966) to produce two further immunoelectrophoretic techniques: "rocket" immunoelectrophoresis and antigen–antibody crossed electrophoresis.

2.3.2. Rocket Immunoelectrophoresis

This technique is primarily a quantitative one. Wells are cut into an agar gel containing antibody, and the relevent antigen is placed in the well and electrophoresed toward the anode. Flame-shaped immunoprecipitates appear that migrate toward the anode at a decreasing rate (Fig. 6A). Unbound antigen migrating within the flame moves into the precipitate, which redissolves. When antigen excess can no longer be achieved, the flame-shaped precipitate becomes stationary; the distance finally traveled by the peak is directly proportional to the initial antigen concentration. By constructing a standard curve with known antigen concentrations, the unknown antigen solution can be quantitated.

The rocket technique has been modified for the quantitation of circulating insulin antibodies in insulin-treated diabetics (Christiansen, 1970). In this technique, ^{125}I-labeled insulin is incubated overnight with serum samples. The mixture is then electrophoresed into a gel containing anti-human IgG with the resulting formation of rocket-shaped IgG–anti-IgG precipitates. Unreacted label is electrophoresed out of the gel. The height of the rocket

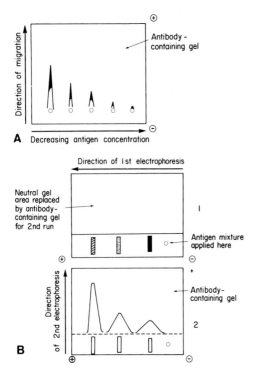

Fig. 6. Electrophoresis into antibody-containing gels. (A) Laurell "rocket" technique. Antigen is placed in wells cut in antibody-containing agar electrophoresed to produce migrating flame-shaped immunoprecipitates. The height of the "rocket" is proportional to the amount of antigen used. (B) Two-dimensional crossed immunoelectrophoresis. The antigen mixture is electrophoresed along one side of an agar plate to produce a first-dimension separation. The rest of the agar is then removed and replaced by antibody-containing agar. The second-dimension electrophoresis, at right angles to the first, drives the antigens into the antibody-containing agar and produces a series of peaks, each peak corresponding to a single antigen–antibody system.

relates to the total IgG content of the serum, and the amount of ^{125}I in the precipitate is proportional to the concentration of insulin-binding antibodies.

2.3.3. Two-Dimensional Crossed Immunoelectrophoresis

This system probably possesses a higher resolving power than any other form of immunodiffusion or immunoelectrophoresis. The first separation is an agarose gel electrophoretic run as with the standard immunoelectrophoresis described in Section 2.3 (Grabar and Williams, 1953). However, instead of using immunodiffusion in the second dimension, the antigens are electrophoresed into an antibody-containing gel (Fig. 6B). During the second run, both antibodies and antigens electrophorese, the former largely toward the cathode (Laurell, 1965) and the latter toward the anode, creating a series of flame-shaped precipitin arcs. Unlike standard immunoelectrophoresis, the precipitin arcs are considerably concentrated and the formation of precipitin lines is therefore facilitated. A prerequisite of this system is that antigenic electrophoretic mobility be different from that of immunoglobulin. Each individual peak represents an independent antigen–antibody system; asymmetry of single peaks indicates antigenic microheterogeneity. Various artifacts are described by Clarke and Freeman (1968). The uses of immunoelectrophoresis in identification and quantitation of antigens and antibodies have recently been reviewed (Axelsen and Bock, 1972).

2.4. General Considerations and Uses of Immunodiffusion and Immunoelectrophoresis

2.4.1. Limitations and Artifacts

Antigens can be proteins, carbohydrates, lipids, or nucleic acids, or any combination of these. While most antigens encountered giving rise to immunoprecipitates in immunodiffusion and immunoelectrophoresis are probably proteins, the formation of precipitate does not define the chemical nature of the antigen. A classic example of this has been the misidentification of cerebroside as a brain-specific "protein" (Tremblay *et al.*, 1974). Furthermore, the formation of a precipitate in an immunodiffusion or immunoelectrophoretic gel need not necessarily be an immunologically specific event, since albumin and β-lipoprotein, for example, can precipitate nonspecifically. Extraction of antigens with detergents followed by immunodiffusion in the absence of detergent can also lead to nonspecific precipitation (Clausen, 1971). The formation of a specific immunoprecipitate depends on both antigen's and antibody's being able to diffuse in the gel system. Some antigens will be too large to do this, and some, e.g., lysozyme, will react with the agar gel (Clausen, 1971). The formation of a visible precipitate depends on the antigen's and antibody's being present at the site of precipitation in optimal proportions, and also for that precipitate to be sufficiently dense to be visible, i.e., a minimum of 1 μg/ml. For these reasons, immunodiffusion can never be used as evidence of purity or homogeneity of an antigen solution. An apparently homogeneous antigen preparation should normally give a single precipitation line when tested against a specific antibody. Though the antigen contains several antigenic determinants and the antiserum may contain antibodies specific for individual determinants, the fact that these determinants are on the same molecule means that immunoprecipitation occurs at a single site in the gel. The production of more than one precipitin line with an apparently single antigen can be due to unsuspected impurities in the antigen preparation, breakdown of an originally single antigen by enzymic or other degradation in the gel diffusing system, temperature artifacts, or greatly imbalanced systems producing rapidly migrating precipitates (Williams and Chase, 1971). In practice, however, the presence of two precipitin lines usually means the presence of two precipitating antigen–antibody systems.

2.4.2. Qualitative Uses and Identification of Precipitates

Immunodiffusion and immunoelectrophoresis are widely used for the qualitative analysis of antigen mixtures to find the minimum number of antigens present or to compare two antigens for immunological cross-reactivity. Examples of this are the use of immunodiffusion to provide evidence for the identity of the brain-specific antigen-α and 14-3-2 protein (Bennett, 1974) and to compare choline acetyltransferase cross-reactivity among various species (Singh and McGeer, 1974). Immunoelectrophoresis has been used qualitatively to compare antigens from synaptosomes and synaptic vesicles with soluble brain antigens (Bock *et al.*, 1974).

The identification of antigens in immunoprecipitates can be approached in several ways. Several general or specific staining procedures exist that can be used directly on the immunoprecipitates after removal of unreacted (soluble) antigen and antibody by washing in saline. All precipitates will stain with protein stains (e.g., amido black, Ponceau red, nigrosine), since immunoprecipitates are largely composed of antibody. However, staining procedures also exist for lipoproteins (e.g., Oil Red O, Sudan Black), glycoproteins and polysaccharides (periodic acid procedures), nucleoproteins and nucleic acids (e.g., pyronine Y, Fuelgen-formazan), and metalloproteins (alizarin blue S, bathophenanthroline). Many enzymes still retain activity when combined with antibody in an immunoprecipitate, e.g., the enolase activity found in 14-3-2 protein (Bock and Dissing, 1975). A wide variety of enzyme-staining procedures, largely adapted from histochemistry, are available to detect specific enzyme activities in gel immunoprecipitates. A comprehensive list of these procedures can be found in Williams and Chase (1971). Further identification of the antigenic component of an immunoprecipitate requires a supply of the pure antigen that can be used to show immunological reaction of identity with the unknown antigen and to absorb out specific antibodies from the antiserum directed toward the antigen. Practical details are given in Williams and Chase (1971).

2.4.3. Quantitative Uses of Immunodiffusion and Immunoelectrophoresis

Each of the immunodiffusion and immunoelectrophoresis techniques described above may be used quantitatively or at least semiquantitatively to estimate the concentrations of either antigen or antibody. In practice, these techniques are mostly used for the quantitative estimation of antigen, and accurate absolute measurement requires that each system be standardized with varying amounts of the pure antigen. The measurements required are rates of diffusion with the Oudin tube technique, the area of the precipitation ring at the end of diffusion with the Mancini method, the distance from the antibody trough to the precipitation arc with immunoelectrophoresis (Grabar and Williams, 1953), the final height or rate of migration of Laurell "rockets," and the area of the immunoprecipitation peaks with antigen–antibody crossed electrophoresis (Clarke and Freeman, 1968). The Ouchterlony-type plate can also be used quantitatively by titrating known and unknown antigen concentrations against the same antibody and finding the concentration at which the precipitate is no longer visible. The limiting constraint on sensitivity with these techniques is the ability to detect the immunoprecipitate. This must be at approximately 1 µg/ml to be visible, although the detection limits for immunoprecipitates can be greatly increased by specialized techniques such as interferometry, autoradiography, and fluorescent labeling of antibodies (Williams and Chase, 1971; Clausen, 1971). Since about 10% of the antigen–antibody complex is normally antigen, about 0.1 µg/ml of antigen is usually necessary at the precipitation site to give a visible precipitate. The concentration in the original diffusion well must then be considerably higher than this. A further limit on sensitivity is the relatively small volumes that are normally applied in standard immunoelectrophoresis and immunodiffusion techniques. This may necessitate concentration of samples before analysis. Techniques involving a preliminary separation of the antigen mixture before a second-dimension immunodiffusion or immunoelectrophoresis are less sensitive than one-step diffusion techniques, since individual antigens are inevitably diluted. Normally, the minimal amount of antigen that can reliably be detected by the Oudin tube technique and the Laurell rocket technique is of the order of 0.5 µg (Clausen, 1971). However, under special conditions, the Mancini plate technique can detect 2 µl of a 1.25 µg/ml solution, i.e., 2.5 ng. The actual practical limit of sensitivity varies from one antigen–antibody system to another. A complete discussion of quantitative applications of immunodiffusion and immunoelectrophoresis is given in Williams and Chase (1971).

3. Further Immunological Techniques

3.1. Agglutination Reactions

Many immunological tests rely on the property of immune complexes to form lattices. The reaction between antigens present on cell surfaces and

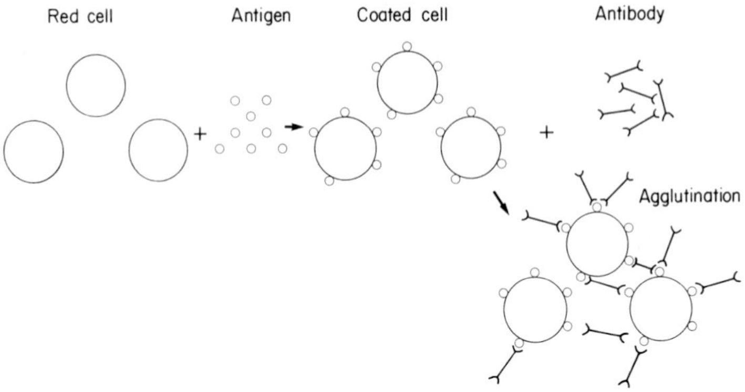

Fig. 7. Diagram representing hemagglutination. Antigen-coated red cells are agglutinated in the presence of antibody.

antibody may thus lead to agglutination of a cell suspension. Agglutination reactions may be used directly to detect cell-surface antigens, though indirect (passive) procedures can be developed using antigens coupled to the surface of cells or other suspended particles (Fig. 7). The direct procedure is used routinely in the identification of red-cell-surface antigens and their antibodies (Stratton and Renton, 1967).

Certain nonagglutinating (incomplete) antibodies will produce agglutination after the treatment of cell membranes with proteolytic enzymes.. Morton and Pickles (1947) first described a hemagglutination test for anti-Rh antibody following treatment of the cells with trypsin. This modification has extended the range of surface antigens that can be detected by agglutination (Stratton and Renton, 1967), though clearly the digestion procedure will itself alter the antigenic composition of the membrane (Morton, 1962).

A simple but ingenious way of dealing with the problem of incomplete antibodies was developed by Coombs *et al.* (1945) and involves the detection of the immune complex by means of addition of an appropriate anti-IgG antiserum. Though nonagglutinating, the initial complex is sufficiently stable to be washed in saline before addition of the agglutinating anti-IgG. The technique has found applications in the detection of blood-group antigens and also in the identification of nonagglutinating bacterial antibodies (e.g., Kerr *et al.*, 1966). It is interesting that the principle of this indirect technique for detecting immune complexes has been adopted in recent years for a wide range of diagnostic methods. These procedures involve the uptake of antibody already labeled with fluorescent compounds, enzymes, or radioactive isotopes. Thus, fluorescein- or peroxidase-labeled anti-IgG is used routinely in the histological identification of cellular antigens (Nairn, 1969). Anti-IgG antibodies labeled with radioiodine have been used for measuring antigen–antibody complexes in the immunoradiometric assay of polypeptide hormones (Beck and Hales, 1975). Wide *et al.* (1971) developed a means of detecting circulating reaginic IgE antibodies for the diagnosis of allergic conditions. The method involves extraction of plasma using the allergen coupled to Sephadex. The uptake of IgE onto the immunoadsorbent is detected by reacting the complex with ^{125}I-labeled anti-IgE antibodies. The so-called radioallergosorbent test (RAST) has been applied commercially to the detection of a wide range of common allergens. The applications of ^{125}I-labeled antibodies as assay reagents are described in detail below.

Since agglutinating antibodies are produced in response to many bacterial and parasitic infections, the ability of a patient's serum to agglutinate cell suspensions may be useful diagnostically. By the use of standardized antigen suspensions, it is possible to detect changes in antibody titer by measuring the changes in concentration required to produce agglutination. Frequently, however, the antibodies produced in response to infectious diseases may recognize antigenic determinants from widely differing species, thereby limiting the diagnostic value of such tests.

Passive hemagglutination tests use mainly suspensions of sheep cells to which antigens are coupled by physical adsorption or covalent bonding. Many bacterial lipolysaccharides absorb readily to sheep red cells, which are then agglutinated by antibody. The technique has been applied to the diagnosis of tubercular infections (Hall and Manion, 1951). The inhibition of agglutination of sensitized cells by added antigen has been used as a sensitive test for the detection of bacterial cells in body fluids (Neter, 1956) and recently for the detection of morphine (Adler and Lau, 1971).

The discovery that mild treatment of red cells with tannic acid would increase enormously the range of proteins adsorbed to the cells (Boyden, 1951) has extended the use of passive hemagglutination as a means of antibody detection (Stavitsky and Arquilla, 1958). Quantitation of antibodies has been achieved using tanned cells in the presence of a fixed amount of antigen (Fulthorpe *et al.*, 1961). A number of procedures involving the covalent coupling of antigens to cells for use in hemagglutination tests have been described, using such coupling agents as bis-diazotized benzidine (Stavitsky and Arquilla, 1955), 2,4-diisocyanate (Schick and Singer, 1961), water-soluble carbodiimide (Johnson *et al.*, 1966), and glutaraldehyde, cyanuric chloride, and tetrazotized dianisidine (Avrameas *et al.*, 1969).

A useful technique for coupling antigens indirectly to red cells was devised by Coombs *et al.* (1953). In this procedure, the antigen is coupled covalently to nonagglutinating anti-sheep red cell antibodies, the complex then being adsorbed onto the cell preparation. The treated cells are agglutinated by antibodies to the coupled antigen. The advantages of this technique over the direct-coupling methods are that damage to the red cells is avoided in the coupling procedure and that the use of antibody to link the antigen prevents any subsequent nonspecific uptake of IgG onto the cells.

The ability of proteins to adsorb to particles other than red cells has been used to devise alternative agglutination procedures. Thus, suspensions of bentonite particles have been used in the detection of rheumatoid factor (Bozicevich *et al.*, 1958). Alternatively, latex particles have been used for the detection of rheumatoid factor and also for the detection of high concentrations of human chorionic gonadotrophin, the latter forming the basis of a simple immunological pregnancy test (Wide and Gemzell, 1960).

3.2. Complement-Fixation Tests

The biochemsitry of the complement system has been studied extensively in recent years. As a result, considerable insight has been gained into the structural components of the system and their action at the molecular level (Fearon and Austin, 1976). The presence of a group of factors in serum and their involvement in immune reactions has been recognized for many years (Mayer, 1970). *In vivo*, complement is involved in facilitating phagocytosis and promoting inflammation. A widely recognized property of complement is its ability to lyse antibody-sensitized red cells *in vitro*. The hemolysis results from a series of reactions involving binding of the complement proteins to immune complexes on the red-cell surface. The observation that preincubation of complement with soluble immune complexes will inhibit its hemolytic activity has resulted in the development over the past 50 years of a number of techniques for the detection and quantitation of antigens and antibodies. The principle of assays based on inhibition of complement fixation is illustrated in Fig. 8.

As with hemagglutination tests, the favored reagent for complement-fixation assays is the sheep erythrocyte. Cell preparations are said to be stable for periods of 2 months (Levine, 1973), and after sensitization with rabbit anti-red cell antibody are uniformly susceptible to hemolysis by guinea pig complement. The presence of other immune complexes in the system results in competition for the available complement and a decrease in hemolysis. In practice, therefore, it is normal to use a concentration of complement that is marginally limiting in its ability to lyse the cell preparation. In this way, the presence of small quantities of immune complexes can be detected by monitoring changes in supernatant absorption at 413 nm.

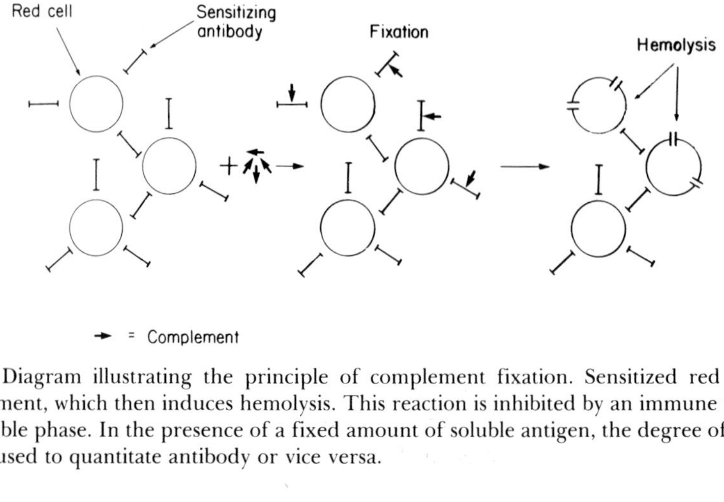

= Complement

Fig. 8. Diagram illustrating the principle of complement fixation. Sensitized red cells bind complement, which then induces hemolysis. This reaction is inhibited by an immune reaction in the soluble phase. In the presence of a fixed amount of soluble antigen, the degree of inhibition can be used to quantitate antibody or vice versa.

The most important property of complement-fixation tests is the ability to quantitate either antigen or antibody by measuring differences in optical density. Thus, in addition to providing useful diagnostic tests, e.g., HLA typing and the Wasserman reaction for the detection of antibodies in syphilis, a number of quantitative assays have been developed that offer much-improved sensitivity when compared with hemagglutination tests. It has been possible, for example, to develop assays for luteinizing hormone and growth hormone (Trenkle *et al.*, 1961), human chorionic gonadotropin (Brody and Carlstrom, 1960), and bovine parathyroid hormone (Tashjian *et al.*, 1964*a*). Though these assays were not sensitive enough to measure the hormones in serum, the assay of parathyroid hormone enabled structure–function studies to be carried out on relatively small quantities of peptide (Tashjian *et al.*, 1964*b*). Of neurochemical interest is the development by Moore *et al.* (1968) of complement-fixation assays for the brain proteins S 100 and 14-3-2 that were used to establish the specific localization of these proteins in nervous tissue.

The major problem in the use of complement-fixation assays on a reliable quantitative basis is that of standardization of reagents. Since the reaction involves a number of variables, it is necessary to standardize all reagents very carefully before use.

3.3. Radioimmunoassay

In the techniques described so far, the important property of antibodies has been the high degree of specificity that they impose on the analytical system. The discovery that antigen–antibody reactions could be monitored directly without the formation of precipitating complexes or the introduction of complicated secondary reactions has led to the development of a group of extremely sensitive immunological assays. Yalow and Berson (1960) described the first radioimmunoassay of insulin, in which the binding of insulin to antibody was monitored by the presence of ^{131}I-labeled insulin as a "tracer." At the same time, Ekins (1960) described an analogous assay technique for the measurement of thyroxine using endogenous thyroxine-binding globulin. Yalow and Berson used the name radioimmunoassay, since the technique employed a radioisotope label and an antibody, though as Ekins has consistently pointed out, assays dependent on the reaction between a ligand and a specific binding reagent are obligatory of neither radioactivity nor antibody. For this reason, he preferred the term "saturation analysis" to describe techniques that involve the progressive saturation by ligand of the binding reagent. While it is doubtful whether there will ever be agreement on the terminology applied to these techniques, it is worth considering the principles on which the techniques are based.

Under conditions in which the product of ligand–binding reagent interaction is not readily detectable, it is possible to monitor the reaction by the incorporation into the system of a trace amount of labeled ligand. If the concentration of binding reagent is constant, increasing the dose of ligand leads to progressive saturation of binding sites and consequently a reduction in the proportion of ligand actually bound. This concentration-dependent distribution of ligand between "bound" and "free" phases is monitored by the distribution of a fixed amount of labeled ligand, normally referred to as "tracer" (Fig. 9). Terms such as "displacement" of label by ligand and "competition" between ligand and label for binding sites have been variously introduced in attempts to provide a pictorial representation of the nature of the reaction involved. As such, both terms are misleading in that they do not describe accurately the reactions involved. Nevertheless, the term "competitive protein binding" is commonly used to describe those methods that employ endogenous binding proteins rather than antibodies. The name has no theoretical justification, but does illustrate the problem of providing meaningful terminology in a rapidly developing field.

It will be appreciated that to monitor the reaction between ligand and binding reagent, certain criteria must be met. First, it is necessary to produce a label of high specific activity so that only a minute amount is required in the reaction. Second, it is important that a highly specific binding reagent be produced with a suitably high affinity for the ligand, since this ultimately will determine the sensitivity of the assay system. Finally, it is necessary to develop a satisfactory means of efficiently separating the bound from the unreacted ligand.

3.3.1. Preparation of Labeled Antigens

One of the most significant steps in the development of radioimmunoassays during the last decade was the introduction by Greenwood *et al.* (1963) of a method for iodinating proteins to high specific activity. This classic technique involves the oxidation of $Na^{125}I$ by chloramine-T in the

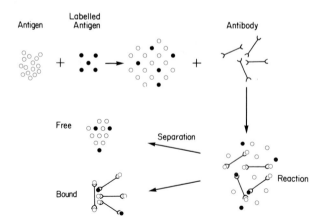

Fig. 9. Radioimmunoassay principle. A fixed amount of antibody is progressively saturated by increasing concentrations of antigen. The concentration-dependent distribution of antigen between bound and free phases is monitored using a trace amount of labeled antigen.

presence of peptide, the free iodine being readily incorporated into tyrosine residues. The reaction is then stopped by addition of sodium metabisulfite. The iodinated peptide may be separated from the reagents by gel filtration or physical adsorption onto particles of silica (Yalow and Berson, 1966) or powdered cellulose (Stevenson and Spalding, 1968), depending on the particular characteristics of the peptide being used. Purification of the iodinated protein is also necessary because of the inevitable "damage" that occurs during the iodination procedure as a result of exposure of the peptide either to radioiodine, to oxidizing agent, or to a combination of both. Thus, while the procedure outlined above remains in popular usage, a number of modifications have been developed to minimize damage during iodination and improve the quality in terms of immunoreactivity of the labeled product.

Thorell and Johansson (1971) developed a method devised by Marcholonis (1969) for the enzymic introduction of ^{125}I into tyrosyl residues using lactoperoxidase. This procedure avoids exposure of the proteins to powerful oxidizing or reducing agents and may be advantageous for labeling peptides (e.g., adrenocorticotrophic hormone, parathyroid hormone) that are particularly susceptible to oxidation. Recently, a technique has been described using immobilized lactoperoxidase that is capable of producing high-quality labeled proteins (Karonen *et al.*, 1975). An improved technique for labeling follicle-stimulating hormone has been described by Butt (1972) in which the oxidizing agent is chlorine gas.

An interesting method has been developed by Bolton and Hunter (1973) that involves reaction of the protein with an iodinated ester, thereby avoiding contact with potentially damaging oxidizing agents or the radioisotope itself. In this procedure, 3-(*p*-hyroxyphenyl)propionic acid *N*-hydroxysuccinimide ester is iodinated using chloramine-T. The labeled ester reacts spontaneously in aqueous solution with ε-amino groups of lysine residues of the NH_2-terminal ends of the protein to form a covalent complex. The iodinated ester is available commercially and would seem to be particularly useful in the labeling of peptides that are lacking in tyrosine (e.g., human C-peptide, porcine parathyroid hormone) or for which the conventional procedure has proved difficult.

The availability of steroids and other metabolites (e.g., cyclic nucleotides) that have been labeled to high specific activities with tritium has been largely responsible for the extension of the basic assay procedure into the small-molecule field. In recent years, however, attempts have been made to improve specific activities of steroid labels and also get away from the problems of β-counting by the preparation of iodinated steroid derivatives. Iodinated tyrosine methyl ester (Oliver *et al.*, 1968) was conjugated with digitoxin to produce a suitable label. This ester together with iodinated tyramine and histamine have also been used for steroid radioimmunoassay with some success (Hunter *et al.*, 1975), though the main advantage over tritiated derivatives is their convenience rather than any dramatic increase in assay sensitivity. This is due partly to the fact that the affinity of antibody for free hapten is invariably the limiting factor in steroid radioimmunoassay. It is also important to realize that any modifications of a steroid molecule are likely to produce profound effects on the reactivity of that molecule with either antibody or binding protein, resulting in widely discrepant behavior between label and hapten. For this reason, the use of methyl-[^{75}Se] derivatives (Chambers *et al.*, 1975) may offer advantages, since they represent relatively minor molecular modifications. It is significant, however, that certain antisera can apparently discriminate between estradiol and its tritiated forms (Jeffcoate *et al.*, 1975).

It is interesting that much of the recent development work in the field of radioimmunoassay has been aimed at the production of nonisotopically labeled antigens. Thus, tracers have been produced using a stable free radical (spin-labeling) (Leute *et al.*, 1971), fluorescent groups (Aalberse, 1973), bacteriophages (Haimovich *et al.*, 1970; Andrieu *et al.*, 1975), and enzymes (Engvall and Perlmann, 1971; Van Weemen and Schuurs, 1971). The motivation for such developments has been primarily the problem of shelf life of radioactive tracers, particularly that of iodinated antigens. It is well recognized that in addition to the problem of radioactive decay during storage, there is the apparently related phenomenon of loss of immunoreactivity of the labeled antigen. Of the labels listed above, only enzymes offer a combination of sensitivity (i.e., those enzymes with high catalytic activity) and convenience of detection (e.g., fluorometric or spectrophotometric) that approaches that of radioisotopes. The application of enzyme labels to immunological assay systems has recently been reviewed (Wisdom, 1976).

A method worthy of particular consideration is the homogeneous enzyme immunoassay system developed by Rubenstein *et al.* (1972). In this procedure, the enzyme is coupled to a hapten in such a way that its activity is lost when the hapten reacts with antibody. The loss of activity may be due to a conformational change following the immunological reaction, rather than to steric hindrance (Rowley *et al.*, 1975). Though the method seems applicable only to hapten assays, it has been developed widely for the assay of drugs (Bastiani *et al.*, 1973). A unique property of the system is that the nature of the assay precludes the requirement of a separation stage.

3.3.2. Separation Procedures

The separation of reacted from unreacted antigen relies on the physical characteristics of the antibody-bound fraction differing from those of the free antigen. In the case of low-molecular-weight antigens, the reaction with antibody produces a large change in molecular size and possibly charge. In such cases, satisfactory separation can be achieved by adsorption of unreacted antigen by charcoal (Herbert *et al.*, 1965) or powdered silica (Rosselin *et al.*, 1966), or alternatively by precipitation of the bound fraction using ethanol (Odell *et al.*, 1965) or polyethylene glycol (Desbuquois and Aurbach, 1971). Where the properties of bound and free fractions are essentially similar, alternative procedures must be sought, the most useful probably being the immunological precipitation of the bound fraction by a second anti-IgG antibody (Morgan and Lazarow, 1963; Hales and Randle, 1963), or alternatively using antibody coupled covalently to a solid phase such as Sephadex (Wide and Porath, 1966) or physically adsorbed onto plastic tubes (Catt and Tregear, 1967). Because of their universal application, both the second-antibody separation and the solid-phase techniques are now widely used. The range of separation techniques in radioimmunoassay and their particular merits have been reviewed recently (Ratcliffe, 1974).

3.3.3. Assay Design

While it has been stressed that the potential sensitivity of an assay is determined ultimately by the affinity of the antibody for the antigen, several other factors, such as concentration of reagents, may well be limiting in a nonoptimized system. A traditional approach to assay design has been to select a concentration of antiserum that binds approximately 50% of a fixed amount of tracer. Standard curves are then produced by addition of increasing amounts of unlabeled antigen and relating the quantity of label bound to the concentration of added antigen. While this totally empirical approach is no doubt capable of producing an assay, many factors will govern its performance. For example, for a tracer to be ideal, not only must it be of high specific activity, but also it should be of high immunoreactivity, and furthermore, this immunoreactivity should be identical to that of the antigen. The consequence of "damage" to a tracer as a result of either iodination or incubation in plasma will detract from assay performance by increasing the apparent ("nonspecific") binding in that assay. The production and assessment of iodinated antigens for use in radioimmunoassay have been reviewed by Hunter (1971, 1974).

A theoretical approach to assay optimization has been developed over the years by Ekins and his co-workers (Ekins *et al.*, 1968; Ekins and Newman, 1970; Ekins, 1974). While the complex mathematics of this approach acts as a deterrent to some workers, there is no doubt that a number of important concepts have appeared. The most important of these is the realization that optimal reagent concentrations are influenced by the experimental error in the system. The elimination of certain experimental errors has in fact resulted in marked improvement. For example, Wide (1969) has developed an assay for thyrotropin in which the antibodies are linked covalently to Sephadex beads, thereby allowing extensive washing of the bound fraction following incubation. This washing results in a reduced and consistent error at all concentrations of hormone, with the result that assay sensitivity is maximal when binding in the absence of added antigen is less than 10% and also when delayed addition of large amounts of tracer is employed.

It is important to note that optimization does not necessarily mean obtaining maximum sensitivity. For example, the monitoring of growth-hormone levels in treated acromegalics requires the ability to detect changes against high background levels. Thus, the development of precise measurement of large amounts of antigen involves quite different reagent concentrations from those used when maximal sensitivity is required (Ekins, 1974).

3.3.4. Applications

Radioimmunoassay represents one of the most sensitive quantitative techniques yet developed. While both protein and steroid endocrinology have reaped enormous benefit from the development of radioimmunoassays, there are few areas of biochemical investigation in which the techniques are not used. It would be foolish, therefore, to contemplate a list of all the compounds of interest for which radioimmunoassays exist. However, some idea of the diversity of the potential applications can be gained from recent reviews covering such areas as pharmacology (Marks *et al.*, 1974, 1975), oncology (Bagshawe, 1974), hematology (Newmark and Gordon, 1974), and virology (Cameron and Dane, 1974).

A number of radioimmunoassays have been developed in recent years for the quantitation of peptides of neurochemical interest. It has been possible to determine the tissue content of the acidic proteins S 100 (Uozomi and Ryan, 1973) and 14-3-2 (Rovoltella *et al.*, 1976) and also myelin basic protein (Cohen *et al.*, 1975). The full potential of such assays in neurochemical investigations has yet to be realized.

3.4. Immunoradiometric Assay

The immunoradiometric technique was introduced by Miles and Hales (1968) as an alternative to radioimmunoassay. In principle, the method

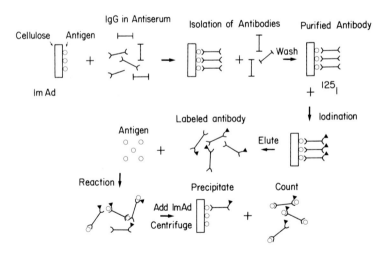

Fig. 10. Stages in the immunoradiometric assay. Antibody is purified by means of an immunoadsorbent (ImAd). After being labeled with ^{125}I, it is eluted from the ImAd and is added in excess to solutions of antigen. Unreacted antibody is then removed following incubation of the equilibrated reaction mixture with ImAd. Supernatant radioactivity is a function of antigen concentration.

involves the conversion of antigen to a measurable derivative by reaction with purified ^{125}I-labeled antibodies. Though the technique involves the radioisotopic monitoring of an *in vitro* immune reaction, it differs quite funamentally from radioimmunoassay in principle and in practice (Woodhead *et al.*, 1974).

The steps involved in the immunoradiometric technique are summarized in Fig. 10. Antibody protein is extracted from antiserum using an immunoadsorbent that consists of insolubilized antigen. The most satisfactory solid phase for immunoadsorbent production is a diazonium salt of powdered cellulose (Gurevich *et al.*, 1961). Immunoadsorbents produced from this base combine the advantages of a high antigen content with very low nonspecific-protein uptakes. The extracted antibody can be iodinated by the conventional chloramine-T method (Greenwood *et al.*, 1963) while still bound to immunoadsorbent, and then eluted by lowering the pH. In the assay, solutions of antigen are incubated with an excess of labeled antibody. Immunoadsorbent is added to bind unreacted label and then removed by centrifugation. The quantity of label remaining in the supernatant is then a function of the initial antigen concentration.

Though the immunoradiometric assay procedure is simple to perform, the preparation of reagents is generally more demanding than that required for conventional radioimmunoassay, though several applications of the technique have been reported (Woodhead *et al.*, 1974). However, there are a number of situations in which labeled-antibody techniques offer distinct advantages over the conventional methods. The most obvious advantage relates to methods in which the antigen has proved difficult to iodinate or is unstable in assay incubations. Thus, the immunoradiometric technique has been particularly useful in the assay of parathyroid hormone (Addison *et al.*, 1971), where conventional assays have been subject to the problems of iodination and incubation damage. Furthermore, the stability of the label that can be stored bound to immunoadsorbent means that a single iodinated preparation can be used over a considerable period of time (Woodhead *et al.*, 1974).

The immunoradiometric technique has proved extremely successful in the assay of ferritin in human serum (Addison *et al.*, 1972). This large molecule is difficult to iodinate, and conventional radioimmunoassays lack the sensitivity to measure normal serum levels. The high sensitivity of the immunoradiometric technique relates to the ability of each ferritin molecule to bind several molecules of labeled antibody. It is likely that the immunoradiometric technique will be of value generally in the assay of high-molecular-weight proteins.

An extension of the basic immunoradiometric technique has been the development of a universal label (Beck and Hales, 1975). In this procedure, the antibody is not labeled directly with ^{125}I but indirectly with [^{125}I]anti-IgG. In addition to avoiding the problems of iodinating precious antibodies, the method has the advantage that a single preparation of [^{125}I]anti-rabbit (or guinea pig) IgG can be used for a wide range of polypeptide assays.

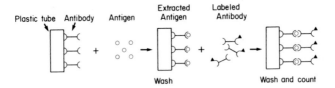

Fig. 11. Stages involved in a two-site assay using antibody-coated plastic tubes.

3.4.1. Two-Site Assay

This method has extended the basic immunoradiometric technique to provide assays of increased sensitivity and specificity. The principle is outlined in Fig. 11. An immunoadsorbent is prepared by adsorbing antibodies from immune sera onto the walls of plastic tubes (Catt and Tregear, 1967). After a washing procedure, the tubes are used to extract antigen from solution. After further washing, the uptake of antigen is measured by addition of labeled antibody. The amount of label bound during the second incubation is a function of the concentration of antigen in the first incubation.

This technique offers three major advantages over the assay systems discussed above. First, the assay procedure is simple. While the extensive washing of plastic tubes is tedious, it is possible to store antibody-coated tubes for long periods at $-20°C$ without loss of immunological reactivity (Readhead

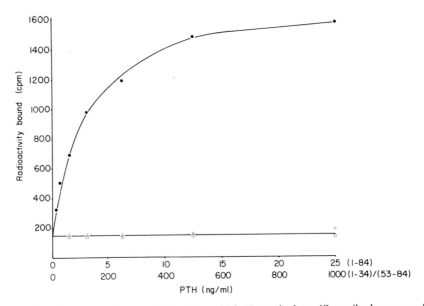

Fig. 12. Two-site assay of bovine PTH (●) in which N-terminal-specific antibody was used to extract the hormone and uptake was measured with $[^{125}I]$-C-terminal antibody. Note the absence of reactivity with high concentrations of either N-terminal (1–34) PTH (△) or C-terminal (53–84) PTH (○). Reprinted from Woodhead *et al.* (1977) with permission from the *Journal of Endocrinology.*

et al., 1973). Thus, large batches can be prepared at one time. Second, the technique is very sensitive, a feature that results primarily from the very low nonspecific uptake of labeled antibody. Because of the high capacity of antibody-coated tubes to bind antigen, the system can be used to measure accurately over a wide range of antigen concentrations. Third, the fact that each antigen molecule is reacted with at least two antibody molecules results in an increase in assay specificity. By the use of antibody populations selected for their reactivity with defined molecular sequences, it is possible to exploit this property further. It has thus been possible to develop an assay for parathyroid hormone (PTH) in which the tubes are coated with antibody specific for the N-terminal (1–34) region of the molecule and the hormone uptake is measured with ^{125}I-labeled antibodies specific for the C-terminal (53–84) region (Woodhead *et al.*, 1977). This system is thus capable of measuring a large sequence of the peptide (containing both N- and C-terminal components) in the presence of molecular fragments (Fig. 12). This type of assay specificity is thus of value in investigating the peripheral metabolism of secreted molecules such as PTH (Habener *et al.*, 1972) and also in the assay of molecules with closely related structures present in blood (e.g., the pituitary glycoprotein hormones). It is likely that with the development of automated washing procedures, this technique will find wide application in the quantitation of proteins in biological fluids.

4. Antibody Production

The potential of any immunological assay system is ultimately determined by the characteristics of the antibody. If the aim is to develop an assay of maximum sensitivity, then the affinity of the antibody for antigen is the critical factor. Another equally important objective is the measurement of molecules in biological fluids that may contain related or unrelated compounds potentially capable of affecting the reaction being monitored. It is common practice, for example, to raise antisera to partially purified antigens and rely on the preparation of highly purified tracers to give the assay system its required specificity. However, the ability of an assay to distinguish among structurally related compounds present in the same sample (e.g., the pituitary glycoprotein hormones, estrogens) depends on the inherent specificity of the antibody molecules present, or the ability to introduce into the system steps that can overcome the problems of cross-reactivity (e.g., extraction and chromatography).

The majority of peptide and protein molecules are antigenic when administered to animal species different from those of their origin, though molecular size may be an important factor in determining the ultimate response (Playfair *et al.*, 1974). The ability of small molecules, which are inherently nonantigenic, to function as haptens when conjugated to macromolecular carriers has been well known for many years (Landsteiner, 1945). This potential of endogenous small molecules to behave as antigens has extended the scope of radioimmunoassay into the field of steroid endocrinology and other areas of metabolism wherein high sensitivity of measurement is required. The chemistry of the preparation of steroid antigens has been reviewed by Kellie *et al.* (1975).

Though there are no simple practical rules that can be guaranteed to produce suitable antisera, some general points are worth noting. The amount of immunogen required to elicit a response is small, doses of 50–100 µg per animal being commonly used. In the case of haptens, both the type of attachment to the carrier and the degree of substitution may be critical in the production of satisfactory antisera (Kellie *et al.*, 1975; Niswender *et al.*, 1975). Antigen is normally administered in a stable emulsion using Freund's adjuvant, which has the effect of delayed release into the circulation and possible rapid degradation and also enhanced stimulation of the reticuloendothelial system. Various routes of administration and immunization schedules have been developed, probably the most successful being multiple-site intradermal injections (Vaitukaitis *et al.*, 1971). This technique appears to combine maximum economy in use of antigen (as little as 20 µg being effective) with a rapid response. Peak responses seem to occur within a few weeks of a primary injection and within 10–14 days of any further booster injections (Playfair *et al.*, 1974).

References

Aalberse, R.C., 1973, Quantitative fluoroimmunoassay, *Clin. Chim. Acta* **48**:109.

Addison, G.M., Hales, C.N., Woodhead, J.S., and O'Riordan, J.L.H., 1971, Immunoradiometric assay of parathyroid hormone, *J. Endocrinol.* **49**:521.

Addison, G.M., Beamish, M.R., Hales, C.N., Hodgkins, M., Jacobs, A., and Llewellin, P., 1972, An immunoradiometric assay for ferritin in the serum of normal subjects and patients with iron deficiency and iron overload, *J. Clin. Pathol.* **25**:326.

Adler, F.L., and Lau, C.T., 1971, Detection of morphine by haemagglutination inhibition, *J. Immunol.* **106**:1684.

Andrieu, J.M., Mamas, S., and Dray, F., 1975, Viroimmunoassay of steroids: Method and principles, in: *Steroid Radioimmunoassay* (E.H.D. Cameron, S.G. Hillier, and K. Griffiths, eds.) p. 189, Alpha Omega, Cardiff.

Avrameas, S., Tandou, B., and Chuilon, J., 1969, Glutaraldehyde, cyanuric chloride and tetrazotized *O*-dianisidine as coupling agents in the passive haemagglutination test, *Immunochemistry* **6**:67.

Axelsen, N.H., and Bock, E., 1972, Identification and quantitation of antigens and antibodies by means of quantitative immunoelectrophoresis—a survey of methods, *J. Immunol. Methods* **1**:109.

Bagshawe, K.D., 1974, Tumour associated antigens, in: *Radioimmunoassay and Saturation Analysis*, *Br. Med. Bull.* **30**:68.

Bastiani, R.J., Phillips, R.C., Schneider, R.S., and Ullman, E.F., 1973, Homogeneous immunochemical drug assays, *Am. J. Med. Technol.* **39**:211.

Beck, P., and Hales, C.N., 1975, Immunoassay of serum polypeptide hormones by using ^{125}I-labelled anti-(immunoglobulin G) antibodies, *Biochem. J.* **145**:607.

Bennett, G.S., 1974, Immunologic and electrophoretic identity between nervous system specific proteins antigen alpha and 14-3-2, *Brain Res.* **63**:365.

Bock, E., and Dissing, J., 1975, Demonstration of enolase activity connected to the brain specific protein 14-3-2, *Scand. J. Immunol.* **4**(Suppl. 2):31.

Bock, E., Jørgensen, O.S., and Morris, S.J., 1974, Antigen–antibody crossed electrophoresis of rat brain synaptosomes and synaptic vesicles: Correlation to water-soluble antigens from rat brain, *J. Neurochem.* **22**:1013.

Bolton, A.E., and Hunter, W.M., 1973, The labelling of proteins to high specific radioactivities by conjugation to a ^{125}I-containing acylating agent, *Biochem. J.* **133**:529.

Boyden, S.V., 1951, Adsorption of proteins on erythrocytes treated with tannic acid and subsequent haemagglutination by antiprotein sera, *J. Exp. Med.* **93**:107.

Bozicevich, J., Bunim, J.J., Freund, J., and Ward, S.B., 1958, Bentonite flocculation test for rheumatoid arthritis, *Proc. Soc. Exp. Biol. Med.* **97**:180.

Brody, S., and Carlstrom, G., 1960, Estimation of human chorionic gonadotrophin in biological fluids by complement fixation, *Lancet* **2**:99.

Butt, W.R., 1972, The iodination of follicle-stimulating hormone and other hormones for radioimmunoassay, *J. Endocrinol.* **55**:453.

Cameron, C.H., and Dane, D.S., 1974, Viruses, in: *Radioimmunoassay and Saturation Analysis, Br. Med. Bull.* **30**:90.

Catt, K.J., and Tregear, G.W., 1967, Solid phase radioimmunoassay in antibody-coated tubes, *Science* **158**:1570.

Chambers, V.E.M., Glover, J.S., and Tudor, R., 1975, ^{75}Se-radioligands in steroid immunoassay, in: *Steroid Radioimmunoassay* (E.H.D. Cameron, S.G. Hillier, and K. Griffiths, eds.), p. 177, Alpha Omega, Cardiff.

Christiansen, A.H., 1970, A new method for determination of insulin-binding immunoglobulins in insulin-treated diabetic patients, *Horm. Metab. Res.* **2**:187.

Clarke, H.G.M., and Freeman, T., 1968, Quantitative immunoelectrophoresis of human serum proteins, *Clin. Sci.* **35**:403.

Clausen, J., 1971, Immunochemical techniques for the identification and estimation of macromolecules, in: *Laboratory Techniques in Biochemistry and Molecular Biology*, Vol. 1, Part 3 (T.S. Work and E. Work, eds.), p. 399, North-Holland, Amsterdam.

Cohen, S.R., McKhann, G.M., and Guanieri, M., 1975, A radioimmunoassay for myelin basic protein and its use for quantitative measurements, *J. Neurochem.* **25**:371.

Coombs, R.R.A., Mourant, A.E., and Race, R.R., 1945, A new test for the detection of weak and "incomplete" Rh agglutinins, *Br. J. Exp. Pathol.* **26**:255.

Coombs, R.R.A., Howard, A.N., and Mynors, L.S., 1953, A serological procedure theoretically capable of detecting incomplete and non precipitating antibodies to soluble protein antigens, *Br. J. Exp. Pathol.* **34**:525.

Desbuquois, B., and Aurbach, G.D., 1971, Use of polyethylene glycol to separate free and antibody-bound peptide hormones in radioimmunoassays, *J. Clin. Endocrinol.* **33**:732.

Ekins, R.P., 1960, The estimation of thyroxine in human plasma by an electrophoretic technique, *Clin. Chim. Acta* **5**:453–459.

Ekins, R.P., 1974, Basic principles, in: *Radioimmunoassay and Saturation Analysis, Br. Med. Bull.* **30**:3.

Ekins, R.P., and Newman, G.B., 1970, Theoretical aspects of saturation analysis, *Acta Endocrinol. (Copenhagen)*, Suppl. 147, p. 11.

Ekins, R.P., Newman, G.B., and O'Riordan, J.L.H., 1968, Theoretical aspects of "saturation" and radioimmunoassay, in: *Radioisotopes in Medicine: In vitro studies* (R.L. Hayes, F.A. Goswitz, and B.E.P. Murphy, eds.), p. 59, U.S. Atomic Energy Commission Publication, Conf. 671111.

Elek, S.D., 1948, The recognition of toxicogenic bacterial stains *in vitro*, *Br. Med. J.* **1**:493.

Engvall, E., and Perlmann, P., 1971, Enzyme-linked immunosorbent assay (ELISA): Quantitative assay of immunoglobulin G, *Immunochemistry* **8**:871.

Fearon, D.T., and Austin, K.F., 1976, The human complement system: Biochemistry, biology and pathobiology, *Essays Med. Biochem.* **2**:1.

Fulthorpe, A.J., Roitt, I.M., Doniach, D., and Couchman, K.G., 1961, A stable sheep cell preparation for detecting thyroglobulin autoantibodies and its clinical application, *J. Clin. Pathol.* **14**:654.

Grabar, P., and Williams, C.A., 1953, Méthode permettant l'étude conjugée des propriétés electrophorétiques et immunochimique d'un mélange de protéines: Application au sérum sanguin, *Biochem. Biophys. Acta* **10**:193.

Greenwood, F.C., Hunter, W.M., and Glover, J.S., 1963, The preparation of ^{131}I-labelled human growth hormone of high specific radioactivity, *Biochem. J.* **89**:114.

Gurevich, A.E., Kuzovleva, O.B., and Tumanova, A.E., 1961, Production of protein–cellulose complexes (immunoadsorbents) in the form of suspensions able to bind great amounts of antibodies, *Biokhimiya* **26**:803.

Habener, J.F., Segre, G.V., Powell, D., Murray, T.M., and Potts, J.T., 1972, Immunoreactive parathyroid hormone in circulation of man, *Nature (London) New Biol.* **238**:152.

Haimovich, J., Hurwitz, E., Novik, N., and Sela, M., 1970, Use of protein–bacteriophage conjugates for detection and quantification of proteins, *Biochim. Biophys. Acta* **207**:125.

Hales, C.N., and Randle, P.J., 1963, Immunoassay of insulin with insulin–antibody precipitate, *Biochem. J.* **88**:137.

Hall, W.H., and Manion, R.E., 1951, Haemagglutinins and haemolysins for erythrocytes sensitized with tuberculin in pulmonary tuberculosis, *J. Clin. Invest.* **30**:1542.

Hanson, L.A., Johansson, B.G., and Rymo, L., 1966, Further applications of thin-layer gel filtration: Two dimensional gel filtration–electrophoresis and immuno-gel filtration, *Clin. Chim. Acta* **14**:391.

Herbert, V., Lau, K.S., Gottlieb, C.W., and Bleicher, S.J., 1965, Coated charcoal immunoassay of insulin, *J. Clin. Endocrinol. Metab.* **25**:1375.

Humphrey, J.H., and White, R.G., 1970, *Immunology for Students of Medicine*, 3rd ed., Blackwell, Oxford.

Hunter, W.M., 1971, The preparation and assessment of iodinated antigens, in: *Radioimmunoassay Methods* (K.E. Kirkham and W.M. Hunter, eds.), p. 3, Churchill Livingstone, Edinburgh.

Hunter, W.M., 1974, Preparation and assessment of radioactive tracers, in: *Radioimmunoassay and Saturation Analysis, Br. Med. Bull.* **30**:18.

Hunter, W.M., Nars, P.W., and Rutherford, F.J., 1975, Preparation and behaviour of ^{125}I-labelled radioligands for phenolic and neutral steroids, in: *Steroid Radioimmunoassay* (E.H.D. Cameron, S.G. Hillier, and K. Griffiths, eds.), p. 141, Alpha Omega, Cardiff.

Jeffcoate, S.L., Edwards, R., Gilby, E.D., and White, N., 1975, The use of ^{3}H-labelled ligands in steroid radioimmunoassays, in: *Steroid Radioimmunoassay* (E.H.D. Cameron, S.G. Hillier, and K. Griffiths, eds.), p. 133, Alpha Omega, Cardiff.

Johnson, H.M., Brenner, K., and Hall, H.E., 1966, The use of a water-soluble carbodiimide as a coupling reagent in the passive haemagglutination test, *J. Immunol.* **97**:791.

Karonen, S.L., Morsky, P., Siren, M., and Seuderling, U., 1975, An enzymatic solid-phase method for trace iodination of proteins and peptides with ^{125}iodine, *Anal. Biochem.* **67**:1.

Kellie, A.E., Lichman, K.V., and Samarajeewa, P., 1975, Chemistry of steroid–protein conjugate formation, in: *Steroid Radioimmunoassay* (E.H.D. Cameron, S.G. Hillier, and K. Griffiths, eds.), p. 33, Alpha Omega, Cardiff.

Kerr, W.R., Coughlan, J.D., Payne, D.J.H., and Robertson, L., 1966, Laboratory diagnosis of chronic brucellosis, *Lancet* **2**:1181.

Landsteiner, K., 1945, *The Specificity of Serological Reactions*, Harvard University Press, Cambridge.

Laurell, C.B., 1965, Antigen–antibody crossed electrophoresis, *Anal. Biochem.* **10**:358.

Laurell, C.B., 1966, Quantitative estimation of proteins by electrophoresis in agarose gel containing antibodies, *Anal. Biochem.* **15**:45.

Leute, R.K., Ullman, E.F., Goldstein, A., and Herzenberg, L.A., 1971, Spin immunoassay technique for determination of morphine, *Nature (London)* **236**:253.

Levine, L., 1973, Micro-complement fixation, in: *Handbook of Experimental Immunology*, Vol. I (D.M. Weir, ed.), p. 221, Blackwell, Oxford.

Mancini, G., Vaerman, J.P., Carbonara, A.J., and Heremans, J.F., 1964, A single radial immunodiffusion method for the immunological quantitation of proteins, in: *Protides of the Biological Fluids: 11th Colloquium* (H. Peeters, ed.), p. 370, Pergamon Press, Oxford.

Marcholonis, J.J., 1969, An enzymic method for the trace iodination of immunoglobulins and other proteins, *Biochem. J.* **113**:229.

Marks, V., Morris, B.A., and Teale, J.D., 1974, Pharmacology, in: *Radioimmunoassay and Saturation Analysis, Br. Med. Bull.* **30:**80.

Marks, V., Morris, B.A., Teale, J.D., Robinson, J.D., and Aherne, G.W., 1975, The radioimmunoassay of drugs, in: *Radioimmunoassay in Clinical Biochemistry* (C.A. Pasternak, ed.), p. 37, Heyden, London.

Mayer, M.M., 1970, Highlights of complement research during the past 25 years, *Immunochemistry* **7:**485.

Miles, L.E.M., and Hales, C.N., 1968, Labelled antibodies and immunological assay systems, *Nature (London)* **219:**186.

Moore, B.W., Perez, V.J., and Gehring, M., 1968, Assay and distribution of a soluble protein characteristic of the nervous system, *J. Neurochem.* **15:**265.

Morgan, C.R., and Lazarow, A., 1963, Immunoassay of insulin: Two antibody systems: Plasma insulin levels of normal, subdiabetic and diabetic rats, *Diabetes* **12:**115.

Morton, J.A., 1962, Some observations on the action of blood-group antibodies on red cells treated with proteolytic enzymes, *Br. J. Haematol.* **8:**134.

Morton, J.A., and Pickles, M.M., 1947, Use of trypsin in the detection of incomplete anti-Rh antibodies, *Nature (London)* **159:**779.

Nairn, R.C., 1969, *Fluorescent Protein Tracing*, 3rd ed., Livingstone, Edinburgh.

Neter, E., 1956, Bacterial haemagglutination and haemolysis, *Bacteriol. Rev.* **20:**166.

Newmark, P.A., and Gordon, Y.B., 1974, Haematology, in: *Radioimmunoassay and Saturation Analysis, Br. Med. Bull.* **30:**86.

Niswender, G.D., Nett, T.M., Meyer, D.L., and Hagerman, D.D., 1975, Factors influencing the specificity of antibodies to steroid hormones, in: *Steroid Radioimmunoassay* (E.H.D. Cameron, S.G. Hillier, and K. Griffiths, eds.), p. 61, Alpha Omega, Cardiff.

Oakley, C.L., and Fulthorpe, A.J., 1953, Antigenic analysis by diffusion, *J. Pathol. Bacteriol.* **65:**49.

Odell, W.D., WIlber, J.F., and Paul, W.E., 1965, Radioimmunoassay of thyrotropin in human serum, *J. Clin. Endocrinol. Metab.* **25:**1179.

Oliver, G.C., Parker, B.M., Brasfield, D.L., and Parker, C.W., 1968, The measurement of digitoxin in human serum by radioimmunoassay, *J. Clin. Invest.* **47:**1035.

Ouchterlony, O., 1948, *In vitro* method for testing the toxin-producing capacity of diphtheria bacteria, *Acta Pathol. Microbiol. Scand.* **25:**186.

Ouchterlony, O., and Nilsson, L.A., 1973, Immunodiffusion and immunoelectrophoresis, in: *Handbook of Experimental Immunology*, Vol. I (D.M. Weir, ed.), p. 191, Blackwell, Oxford.

Oudin, J., 1946, Méthode d'analyse immunochimique par précipitation spécifique en milieu gélifié, *C. R. Acad. Sci.* **222:**115.

Petrie, G.F., 1932, A specific precipitin reaction associated with the growth on agar plates of *Meningococcus, Pneumococcus* and *B. dysenteriae* (Shiga), *Br. J. Exp. Pathol.* **13:**380.

Playfair, J.H.L., Hurn, B.A.L., and Schulster, D., 1974, Production of antibodies and binding reagents, in: *Radioimmunoassay and Saturation Analysis, Br. Med. Bull.* **30:**24.

Ratcliffe, J.G., 1974, Separation techniques in saturation analysis, in: *Radioimmunoassay and Saturation Analysis, Br. Med. Bull.* **30:**32.

Readhead, C., Addison, G.M., Hales, C.N., and Lehmann, H., 1973, Immunoradiometric and two-site assay of human follicle-stimulating hormone, *J. Endocrinol.* **59:**313.

Ressler, N., 1960, Electrophoresis of serum protein antigens in an antibody containing buffer, *Clin. Chim. Acta* **5:**359.

Revoltella, R., Bertolini, L., Diamond, L., Vigneti, E., and Grasso, A., 1976, A radioimmunoassay for measuring 14-3-2 protein in cell extracts, *J. Neurochem.* **26:**831.

Rosselin, G., Assan, R., Yalow, R.S., and Berson, S.A., 1966, Separation of antibody-bound and unbound peptide hormones labelled with iodine-131 by talcum powder and precipitated silica, *Nature (London)* **212:**355.

Rowley, G.L., Rubenstein, K.E., Huisjen, J., and Ullman, E.F., 1975, Mechanism by which antibodies inhibit hapten–malate dehydrogenase conjugates, *J. Biol. Chem.* **250:**3759.

Rubenstein, K.E., Schneider, R.S., and Ullman, E.F., 1972, "Homogeneous" enzyme immunoassay: A new immunochemical technique, *Biochem. Biophys. Res. Commun.* **47:**846.

Scheidegger, J.J., 1955, Une micro-méthode de l'immuno-électrophorèse, *Int. Arch. Allergy* **7**:103.

Schick, A.F., and Singer, S.J., 1961, On the formation of covalent linkages between two protein molecules, *J. Biol. Chem.* **236**:2477.

Singh, V.K., and McGeer, P.L., 1974, Cross-immunity of antibodies to human choline acetyltransferase in various vertebrate species, *Brain Res.* **82**:356.

Stavitsky, A.B., and Arquilla, E.R., 1955, Micromethods for the study of proteins and antibodies. III. Procedure and applications of haemagglutination and haemagglutination-inhibition reactions with bis-diazotized benzidine and protein conjugated red blood cells, *J. Immunol.* **74**:306.

Stavitsky, A.B., and Arquilla, E.R., 1958, Studies of proteins and antibodies by specific haemagglutination and haemolysis of protein conjugated erythrocytes, *Int. Arch. Allergy* **13**:1.

Stevenson, P.M., and Spalding, A.C., 1968, Experiences with a radioimmunoassay for luteinising hormone, in: *Protein and Polypeptide Hormones*, Part II (M. Magoulies, ed.), p. 401, *Exerpta Medica Int. Congr. Ser.*, No. 161.

Stratton, F., and Renton, P.H., 1967, *Practical Blood Grouping*, 2nd ed., Blackwell, Oxford.

Tashjian, A.H., Levine, L., and Munson, P.L., 1964a, Immunoassay of parathyroid hormone by quantitative complement fixation, *Endocrinology* **74**:244.

Tashjian, A.H., Ontjes, D.A., and Munson, P.L., 1964b, Alkylation and oxidation of methionine in bovine parathyroid hormone: Effects on hormonal activity and antigenicity, *Biochemistry* **3**:1175.

Thorell, J.I., and Johansson, B.G., 1971, Enzymic iodination of polypeptides with ^{125}I to high specific activity, *Biochem. Biophys. Acta* **251**:363.

Tremblay, J., Simon, M., and Barondes, S.H., 1974, Cerebroside may be falsely identified as a soluble "brain specific protein," *J. Neurochem.* **23**:315.

Trenkle, A., Mougdal, N.R., Sadri, K.K., and Li, C.H., 1961, Complement-fixing antibodies to human growth hormone and sheep interstitial cell stimulating hormone, *Nature (London)* **192**:260.

Uozomi, T., and Ryan, R.J., 1973, Isolation, amino acid composition and radioimmunoassay of human brain S 100 protein, *Mayo Clin. Proc.* **48**:50.

Vaitukaitis, J., Robbins, J.B., Nieschlag, E., and Ross, G.T., 1971, A method for producing antisera with small doses of immunogen, *J. Clin Endocrinol. Metab.* **33**:988.

Van Weemen, B.K., and Schuurs, A.H.W.M., 1971, Immunoassay using hapten–enzyme conjugates, *FEBS Lett.* **15**:232.

Weir, D.M. (ed.), 1973, *Handbook of Experimental Immunology*, 2nd ed., Blackwell, Oxford.

Wide, L., 1969, Radioimmunoassay employing immunosorbents, *Acta Endocrinol. (Copenhagen)*, Suppl. 142, p. 207.

Wide, L., and Gemzell, C.A., 1960, An immunological pregnancy test, *Acta Endocrinol.* **35**:261.

Wide, L., and Porath, J., 1966, Radioimmunoassay of proteins with the use of Sephadex-coupled antibodies, *Biochim. Biophys. Acta* **130**:257.

Wide, L., Bennich, J., and Johansson, S.G.O., 1971, Diagnosis of allergy by an *in vitro* test for allergen antibodies, *Lancet* **2**:1105.

Williams, C.A., and Chase, M.W. (eds.), 1971, *Methods in Immunology and Immunochemistry*, Academic Press, New York.

Wisdom, G.B., 1976, Enzyme-immunoassay, *Clin. Chem.* **22**:1243.

Woodhead, J.S., Addison, G.M., and Hales, C.N., 1974, The immunoradiometric assay and related techniques, in: *Radioimmunoassay and Saturation Analysis*, *Br. Med. Bull.* **30**:44.

Woodhead, J.S., Davies, S.J., and Lister, D., 1977, Two-site assay of bovine parathyroid hormone, *J. Endocrinol.* **73**:279.

Yalow, R.S., and Berson, S.A., 1960, Immunoassay of endogenous plasma insulin in man, *J. Clin. Invest.* **39**:1157.

Yalow, R.S., and Berson, S.A., 1966, Purification of ^{131}I parathyroid hormone with microfine granules of precipitated silica, *Nature (London)* **212**:357.

8

Freeze–Fracturing

Karl H. Pfenninger

1. Introduction

After a rather controversial childhood, the freeze–fracture technique (Moor and Mühlethaler, 1963) has grown up to fulfill a very important role in modern cell biology and membrane research. With the development of the concept of integral membrane proteins (Singer and Nicolson, 1972; Bretscher, 1973), the visualization by freeze–fracture of distinct intramembranous structures, *intramembranous particles* (IMPs), has gained special significance in the search for structure–function relationships in biological membranes.

As reviewed by Mühlethaler (1973), freeze–dry and freeze–fracture techniques were originally thought to be the only feasible approaches to the electron-optical investigation of biological samples. However, efforts to freeze and process biological material adequately were quickly overtaken by the more and more perfected technique of thin sectioning of embedded tissue. It was clear, though, that the procedures used to prepare thin sections introduced numerous artifacts that could potentially be avoided by physical fixation, i.e., freezing, of cells and tissues. Thus, the development of a freeze–fracture–replication technique retained its importance. However, adequate freezing of cells and especially of tissues is an extremely difficult task because of the formation of ice crystals that disturb native biological structure. Various approaches to solve the freezing problem have been tried, with moderate success, for many years. In the meantime, most investigators have resorted to the impregnation of specimens with cryoprotectants prior to freezing. This treatment of the biological material is usually carried out after aldehyde fixation to avoid cryoprotectant-induced artifacts. Thus, the

Karl H. Pfenninger • Department of Anatomy, Columbia University, New York, New York 10032.

original purpose of the technique, i.e., physical fixation, is defeated. Except for rare situations, the dream of studying vitrified, live tissues has not come true so far.

Yet, experience with a greatly improved freeze–fracture technique in past years and understanding of the images that it can produce have clearly demonstrated the immense value of two previously unsuspected unique properties of this method: (1) visualization of intramembranous structure and (2) exposure of large expanses of membrane amenable to *en face* examination. The impact of the data thus obtained is evident when one considers the contribution of freeze–fracture data to our knowledge of intercellular junctions (e.g., see Staehelin, 1974; Gilula, 1975). In neurobiology, freeze–fracture results have crucially influenced our understanding of synaptic mechanisms (e.g., Akert *et al.*, 1969, 1972; Nickel and Potter, 1970; Pfenninger *et al.*, 1971, 1972; Sandri *et al.*, 1972; Dreyer *et al.*, 1973; Heuser *et al.*, 1974, 1979; Landis *et al.*, 1974; Pfenninger and Rovainen, 1974; Raviola and Gilula, 1975; Hanna *et al.*, 1976; Wood *et al.*, 1977) and have permitted new insights into specialized membrane structure in retinal rods (e.g., Clark and Branton, 1968; Hong and Hubbell, 1972), myelin (Branton, 1967; Schnapp and Mugnaini, 1975), nodes of Ranvier (Livingston *et al.*, 1973; Rosenbluth, 1976; Schnapp *et al.*, 1976; Wiley and Ellisman, 1979), and growth cones (Pfenninger and Bunge, 1974). Many structures in the nervous system remain to be explored with this method.

2. Basic Background Information

As its name indicates, the main feature of the freeze–fracture technique consists of the cleavage of a frozen specimen (Fig. 1). The newly exposed surface is then replicated by shadow-casting with evaporated heavy metal at an angle of 45°. It is this metal replica that, on complete dissociation and removal of the biological material, is being looked at with a transmission electron microscope. The "depth" or three-dimensional appearance of the resulting image is the consequence solely of uneven metal deposition on the specimen surface (which produces a light–shadow effect) and thus is totally different in nature from the image obtained with a scanning (secondary) electron microscope (Fig. 2). Freeze–fracture provides valuable information because cleavage follows the specimens' natural boundaries, i.e., the membranes of cells and organelles.

Obviously, the method requires specific conditions for adequate processing of biological materials. These conditions are met in a high-vacuum chamber equipped with a low-temperature microtome and heavy-metal and carbon-shadowing units. The crucial procedural points are listed below.

1. *Ideal freezing conditions* to avoid (or at least limit) ice-crystal formation. Crystallization and recrystallization of aqueous solutions are extremely fast between 0 and $-80°C$, exceeding a growth rate of 10^7 nm/sec at $-10°C$.

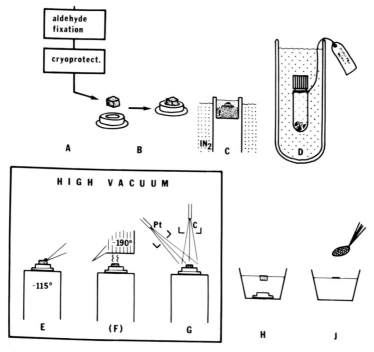

Fig. 1. Principal steps in the preparation of freeze–fracture replicas. (A) Fixation and cryoprotectant treatment of tissue; (B) mounting of tissue block on gold alloy carrier; (C) freezing in
liquid-nitrogen-cooled Freon 22 (note that a superficial layer of Freon has been thawed; the
carrier has been submersed in the liquid and pressed against the solid coolant); (D) storage of
frozen specimens in liquid nitrogen. *Inside* the freeze–fracture apparatus: (E) fracturing on cold
stage with liquid-nitrogen-cooled knife; (F) etching [optional step (see the text)]; (G) replication
by coating with platinum–carbon alloy at a 45° angle and, subsequently, with carbon at 90°. After
removal from the freeze–fracture apparatus: (H) cleaning of the replica by digestion of organic
matter and washing in distilled water; (J) mounting of replica on Formvar-coated grid.

This necessitates a critical freezing rate for crystal-free vitrification of greater
than $10^{6\circ}$C/sec for pure water, or approximately $10^{5\circ}$C/sec for biological
specimens (Moor, 1973*a*). Furthermore, heat conductance in cells and tissues,
which are water-rich, is poor. This property is actually the main limiting
factor in freezing procedures. From heat conductance and critical freezing
rate, and assuming an ideal freezing procedure, one can estimate the maximal
depth or thickness of true vitrification of a tissue block at 2–3 μm [in a
spherical droplet with its more favorable geometry, the maximal diameter
for complete vitrification reaches about 20 μm (Moor, 1973*a*)]. Only alterations of the physicochemical properties of the biological material can improve
this situation (see below). As noted above, the most commonly used solution
to the problem is the application of antifreeze agents or cryoprotectants such
as glycerol and glucose. In 20% glycerol, the critical freezing rate of tissues
can be reduced to about $10^{2\circ}$C/sec, and objects up to 0.5 mm thick can be
vitrified easily (Moor, 1964, 1973*a*).

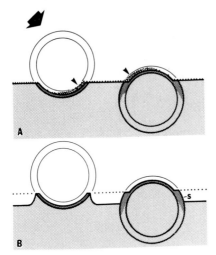

Fig. 2. Fracturing, etching, and replication. (A) The fracture follows membranes, the biological specimen's natural boundaries, and cleaves them into two leaflets so that intramembranous faces are exposed. Shadow-casting (large arrow) of the exposed surfaces at 45° results in uneven deposition of platinum to form a light–shadow pattern (arrowheads). (B) If the membranes are suspended in distilled water, this can be sublimated away (etching) so that true membrane surfaces (s) are revealed (platinum deposits omitted in this figure).

2. *Maintenance of the frozen specimens at low temperature* to avoid recrystallization. As mentioned above, recrystallization rates of water are very fast down to approximately −80°C; this process ceases to occur only at about −130°C (Moor, 1973*a*), so that storage of the specimens in liquid nitrogen (−196°C) and minimal exposure to higher temperatures are imperative. Accordingly, the fracture and shadow processes have to be carried out on a cold stage at a temperature not exceeding −90°C and with the aid of a liquid-nitrogen-cooled knife.

3. *High vacuum during the fracture process* to avoid contamination. The cold specimen, like any cold object, acts as a trap for contaminants in its surroundings, especially water and hydrocarbon vapors. Buildup of these contaminants on the freshly fractured specimen surface introduces artifactual structures and drastically reduces the resolution of the replica (Rash *et al.*, 1979; Steere *et al.*, 1979). Therefore, the fracture-replication process must be carried out under excellent high-vacuum conditions, i.e., at less than 10^{-6} torr (1.3×10^{-6} mbar). Recent studies have shown that freeze–fracturing in ultrahigh vacuum, carried out at very low temperature, produces by far the best and most spectacularly beautiful replicas (Gross *et al.*, 1978). If specimens suspended in distilled water (rather than cryoprotectant) are processed under vacuum, and if an effective cold trap (e.g., the fracture device's knife arm, which is colder than the specimen) is moved close to the fractured material, the surface can be "etched"; i.e., water can be sublimated away from it to expose actual cell surfaces (Fig. 2). This process is called freeze–fracture–etching or deep–etching.

4. *Fine-grain replication of the fractured (and etched) surface.* Heavy metal is evaporated at a 45° angle onto the fractured specimen. The resulting replica is the crucial element in the imaging process. The thickness and grain size of the replica are major determinants of the resolving power of the method.

A well-controlled evaporation unit ensures reproducible film thickness. Further important parameters that influence replica resolution are vacuum, condensation temperature, and composition of the metal vapor. The use of metal alloys such as platinum–carbon (95:5%) or tungsten–tantalum, rather than pure metals, as well as decreased specimen temperature and increased vacuum result in smaller grain size (Moor, 1973*b*).

5. *Complete dissociation of the replica from the tissue.* Since the image is produced by differential electron density of the heavy-metal deposit, it is crucial that replicas be completely freed from adhering organic and inorganic matter before examination in the electron microscope.

3. Outline of Technique

The major steps of a basic procedural scheme that applies to most nerve tissue studies are (1) fixation, (2) cryoprotectant treatment, (3) freezing, (4) fracturing and shadowing, (5) replica cleaning and mounting, and (6) electron microscopy (see Fig. 1). General references are Moor and Mühlethaler (1963), Benedetti and Favard (1973), Sandri *et al.* (1976), and Rash and Hudson (1979).

3.1. Fixation

Any aldehyde fixation (carried out at room temperature or in the cold with glutaraldehyde, paraformaldehyde–glutaraldehyde, acrolein) that produces satisfactory results in thin sections is adequate for subsequent freeze–fracture procedures. However, the tissue is not transferred into osmium tetroxide, which would interfere with the fracture process because of its reaction with membrane lipids. Instead, tissues (blocks of less than 1-mm side length), cultures, or isolated cells are rinsed in buffer (e.g., 0.15 M phosphate buffer, pH 7.3, containing 0.12 M sucrose for isoosmolarity with the fixative) for $\frac{1}{2}$ to 1 hr in the cold or at room temperature and then treated with cryoprotectant.

3.2. Cryoprotectant Treatment

A 25% glycerol solution in the buffer mentioned previously is adequate for most neurobiological situations. It is important to avoid sudden exposure of the tissue samples to this hypertonic solution. Rather, the glycerol concentration in the incubation medium is slowly increased, over a period of about $\frac{1}{2}$ hr, to the desired level of 25%. The specimens remain in this solution for an additional $\frac{1}{2}$ hr (extended glycerol treatment over hours tends to extract cytoplasmic structures). Incubation with cryoprotectant can be carried out at room temperature.

3.3. Freezing

As pointed out earlier, rapid freezing is an essential component of this technique. Accordingly, samples, i.e., tissue blocks, should be small, and an ideal cooling agent should be used. Liquid nitrogen by itself is not suitable because, under atmospheric conditions, it is at its boiling point and, therefore, immediately forms an insulating gas layer around every warm object that is immersed into it. Much superior to this is liquid-nitrogen-cooled Freon 22, which is solid at $-196°C$. Just prior to the freezing procedure, a thin, superficial layer of the Freon 22 is thawed with a metal rod; a specimen, mounted on a metal carrier, is then rapidly immersed into this liquid and pressed against the solid Freon 22 underneath. After about 30 sec, the frozen sample is quickly transferred into liquid nitrogen, where it can be stored for months without damage.

The choice of the specimen carrier depends on the apparatus used for the fracture–shadow procedure. Balzers High Vacuum Company produces two main types of gold-alloy platelets, one with a flat surface, suitable for tissue samples that are very small or flat and for cell suspensions, and a second, cupped type that is preferable for tissue blocks that can be inserted into its depression for better anchoring. In all cases, the carriers must be free of grease and completely wet with the glycerol–buffer solution. The specimens have to protrude freely above the metal surface of the carriers by about 0.5 mm and should be covered with a small amount of the liquid. They are then ready for immersion in Freon 22, specimen side up, as explained above. The freezing procedure is one of the most important steps, and the result of the procedure will depend greatly on the manual skills of the operator.

3.4. Freeze–Fracturing and Shadowing

The procedures explained herein depend greatly on the type of freeze–fracture apparatus used. I have attempted to present a neutral description, but some of the explanations may apply more readily to the Balzers unit than to others.

Frozen samples are transferred onto the precooled stage of a freeze–fracture apparatus, the cutting and shadowing systems of which have been previously readied for use. The stage has to be precooled as much as possible (cf. Pfenninger and Rinderer, 1975) and wet with cold, liquid Freon 22 for better thermal contact between stage and carrier. The transfer of the specimen must be executed as rapidly as possible, with liquid-nitrogen-precooled forceps. All these precautions are necessary to avoid warming up of the frozen sample. Furthermore, the vacuum chamber should be kept open for as short a period as possible to keep frosting of the stage to a minimum.

The chamber is then evacuated to at least 10^{-4} torr, and the specimen stage is warmed up to about $T_o = -130°C$ (which leads to temporary deterioration of the vacuum). Only when the vacuum has again reached 10^{-4} torr is the knife arm, which is being kept away from the specimens, cooled to the lowest temperature possible. Thus, it will also act as a cold trap. This is the time to turn on any other cold traps that might be installed within the vacuum chamber. The specimen is then warmed to its fracturing temperature of about $T_o = -115°C$ [which corresponds to an actual specimen temperature T_s of about $-100°C$ (Pfenninger and Rinderer, 1975)], and when a vacuum of less than 10^{-6} torr has been reached, the cutting process is started. In a substantially higher vacuum, the specimens can also be fractured at lower temperature without the danger of contamination. It is important that the specimen temperature be considerably higher than that of the knife arm; this way, the knife arm is a more efficient cold trap than the specimen, where condensation of gases from the chamber would lead to severe contamination.

During the cutting process in a Balzers freeze–fracture apparatus, flakes of frozen, brittle material are shaved off the tissue block until a macroscopically smooth surface at the desired level of the block has been achieved. This surface can then be "etched" by *backing* the very cold knife arm for 30–45 sec to a position over the specimen. During this process, which leads to somewhat crisper appearance of fine structural detail, small amounts of water are sublimated from the specimen surface and condensed on the back of the knife holder. This process of "etching" is not to be confused with the deep-etching (see below) that leads to the exposure of cell surfaces; membrane surfaces cannot be demonstrated by freeze–fracture in cryoprotectant-treated tissue blocks.

Following the "etching" step or immediately after the last cut, a 2.0- to 2.5-nm-thick layer of, for example, platinum–carbon alloy is evaporated at a 45° angle onto the cleaved specimen surface. The best results are achieved with an electron-beam-gun evaporation unit in conjunction with a quartz-crystal film-thickness monitor. These ensure a reproducible coating of fine grain and controlled thickness, parameters that greatly influence the resolution of the replica. The metal shadowing does not measurably heat up the specimen (Pfenninger and Rinderer, 1975), but warming of the most superficial layers of molecules cannot be excluded and is in fact likely to occur. Thus, it is an advantage to work at very low specimen temperature. Following the platinum–carbon, a layer of carbon about 25 nm thick is evaporated onto the specimen perpendicular to the surface. This second coat, which is invisible in the electron microscope, serves solely to reinforce the extremely thin and fragile platinum replica.

3.5. Cleaning and Mounting the Replica

At this point, the specimens, with replicas attached, can be removed from the chamber, thawed, and floated off the carriers in distilled water. The goal of the procedures described below is to completely remove biological material from the replica by digestion while keeping the replica intact and, if possible, floating on the cleaning solutions. An effective, all-round cleaning agent is the commercial bleach Clorox (mainly hypochlorite), which is added in slowly increasing amounts to the water on which the specimens are floating. Often at this point, or even earlier, in the distilled water, tissues start swelling, an effect that leads to fragmentation of the replicas. This can be suppressed at least partially by floating the samples on, for example, 1.5–2.0 M NaCl in water to which small amounts of Clorox (a few drops) are added over a period of hours. After about half a day, when most of the tissue is already dissolved, the specimens are transferred with a platinum-wire loop into concentrated Clorox. Myelin-containing nerve tissue may be very difficult to dissolve. In this case, the best procedure for tissue removal has to be determined empirically. One of the following procedures may produce satisfactory results: Clorox heated to about 70°C, lipid extraction (acetone or chloroform–methanol) preceding the Clorox treatment, 40% chromic acid preceding Clorox, or detergent treatment (e.g., 0.1% sodium dodecyl sulfate) preceding Clorox. Incubation in each solution should last for several hours, and the specimens have to be rinsed in water between the different solutions. Transfers are always done with the aid of a platinum-wire loop and, preferably, through concentration series to prevent breakage of the replica due to sudden changes in surface tension of the cleaning solutions. It is preferable to avoid the use of organic solvents because the specimens and replicas become drowned in them. Drowned replicas are usually hard to refloat on water and are therefore difficult to mount on grids.

A clean replica can be recognized under the dissecting microscope when illuminated at a low angle. Even at high magnification, there should not be any whitish, fuzzy material underlying the replica. Clean replicas are rinsed at least three or four times in double-distilled water before being mounted on Formvar-coated electron-microscope grids (preferably 150 mesh). Floating replicas can easily be picked up from above against the coated side of the grid, whereas drowned replicas have to be lifted out of the water by a grid placed underneath, a rather delicate procedure.

3.6. Electron Microscopy

Freeze–fracture replicas should be examined in the electron microscope at 80 kV. Stability of the replicas in the beam poses no problems. The anticontamination device of the microscope should always be used, especially if the replicas are not entirely clean and organic and inorganic matter is burning and evaporating in the microscope column. Because the replicas have an inherent resolution limit of about 3 nm, it is generally not useful to take pictures at magnifications greater than ×50,000. Replicas of good quality produce a tremendous amount of image contrast, so that the use of a relatively large objective aperture may be considered. Negatives should be not dark, but gray, and must be printed on low-contrast paper. Otherwise, a loss of structural detail is inevitable.

Figure 3 is an example of freeze–fractured lamprey spinal cord showing

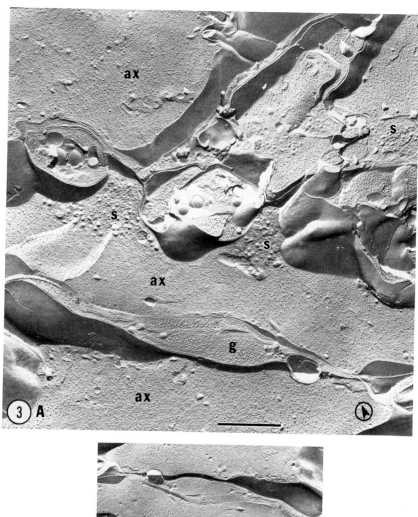

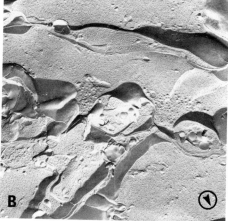

Fig. 3. Survey electron micrograph of large axons (ax) in lamprey spinal cord. Note clusters of synaptic vesicles (s) in evaginations of large axons and in nerve terminals. While (A) is shown in proper orientation (note the shadowing direction as indicated by the circled arrowhead), (B) is rotated by 180° so that the opposite three-dimensional impression is generated. (g) Glial processes. Magnification: × 17,500 (A); calibration bar: 1 μm.

large axons with *en passage* synaptic regions and neuropil. Note the orientation of this electron micrograph. Improper orientation (Fig. 3B) reverses the three-dimensional impression that one gains from the picture. The shadowing direction of platinum should always point toward the top or the upper left-hand corner of the picture so that electron-lucent (platinum-free) areas appear above the protruding elements that have caused them. With this orientation, the viewer is faced with the optical illusion that light is shining from above onto the relief of the specimen surface, the way he or she is accustomed to observing objects in nature.

4. Critical Evaluation of Technique and Interpretation of Results

Inasmuch as most authors of freeze–fracture papers present very little detailed information on the procedures they have employed, and freezing–fracturing–etching–replicating involves several important parameters that cannot be measured easily (e.g., freezing speed), an evaluation of freeze–cleave material has to be carried out on the basis of the published pictures. This can be very difficult, because it is often impossible to know what a particular type of structure should ideally look like. However, the broad experience gained with this technique in the past decade makes it possible to establish at least a few standards that are generally applicable. As explained above, the following three parameters are most critical (cf. Benedetti and Favard, 1973; Böhler, 1976; Rash and Hudson, 1979): (1) state of vitrification of the material (i.e., size of ice crystals); (2) amount of contamination on fracture face; and (3) resolution of replica.

4.1. Artifacts

While membrane structure is relatively well maintained even under unfavorable conditions, extracellular space and cytoplasm are good indicators of the freezing conditions of the specimen. Well-frozen material exhibits very finely granular texture. The coarser this texture and the larger the units, i.e., the ice crystals, in it, the farther away one is from the desired state of vitrification (Fig. 4). Small cytoplasmic structures, nonmembranous or membranous, are then no longer visible and, frequently, membranes overlying ice crystals show deformation artifacts.

Contamination artifacts (e.g., Rash *et al.*, 1979; Steere *et al.*, 1979) may be harder to spot, but usually are most obvious around intramembranous particles. These particles, as judged from unfixed, cryoprotectant-free cells that were frozen at very high speed (Plattner *et al.*, 1973; Pfenninger, 1976 unpublished) or under very high pressure (Riehle and Höchli, 1973; Moor *et al.*, 1980), as well as conventionally frozen specimens (etched or unetched), even if fractured in ultrahigh vacuum (Gross *et al.*, 1978), generally appear as sharply defined, globular structures that are partially embedded in the surface of a smooth-looking membrane matrix (Fig. 5A). These particles may

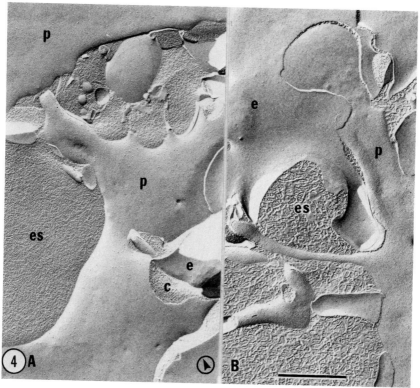

Fig. 4. Well-frozen (A) and poorly frozen (B) processes of supporting cells in nerve-tissue cultures. While membranes show little differences, the extracellular fluid (es) has formed ice crystals of considerable size in (B). (c) Cytoplasm; (p) protoplasmic and (e) external leaflets of the plasmalemma. Magnification: × 18,600; calibration bar: 1 μm; circled arrow: shadowing direction.

appear much enlarged and flattened out in contaminated specimens (Fig. 5B). In good preparations, small pits can be detected that represent sites from which particles have been removed along with the complementary membrane leaflet. These pits fill easily with contaminants so that they represent very sensitive indicators of specimen contamination.

The resolution limit of a particular replica is reached when graininess interferes with fine structure. The value of describing morphological detail at high magnification becomes even more questionable because of the so-called "decoration effect" (Moor, 1971, 1973b). This phenomenon is the result of ordered condensation of the metal alloy. If decoration occurs along existing elements in the fracture face, it may be beneficial to the image because it may enhance the visibility of certain structures. However, if the structures are near the resolution limit of the replica, one cannot be sure whether they have been generated by true biological elements or by condensation artifact.

A last possible artifact that deserves consideration when interpreting freeze–fracture results is deformation of structure as a result of the fracture

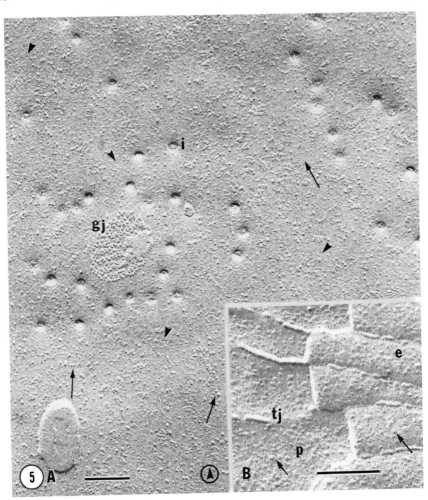

Fig. 5. Intramembranous structure. (A) Protoplasmic face of supporting cell in a nerve-tissue culture. Note various-sized intramembranous particles (arrows), a gap junction (gj), and indentations (i) caused by formation or fusion of plasmalemmal vesicles. The membrane matrix is smooth except for occasional dimples (arrowheads). Magnification: × 61,600. (B) Similar material, but contaminated preparation. Note the altered appearance of intramembranous particles (arrows). (tj) Tight junctions interconnecting adjacent protoplasmic (p) and external (e) plasmalemmal leaflets of two neighboring cells. Magnification: × 85,100. Calibration bar for both pictures: 0.2 μm; circled arrow: shadowing direction.

process (cf. Clark and Branton, 1968; Kirchanski *et al.*, 1979). Deformation has been observed in a variety of nonbiological and biological materials and can be seen frequently in collagen bundles (e.g., collagen fibrils extending freely over membrane faces). Mechanical deformation may be an important factor influencing the shape of intramembranous particles, although this cannot be determined with our present technology.

4.2. Significance

Assuming adequate freeze–fracture methodology, major problems still exist with the identification of elements and the significance of structures revealed in a replica. The reason for this is that freeze–fracture yields information (from intramembranous views) that is otherwise inaccessible and thus cannot be correlated directly with data obtained by alternate techniques. The first and usually most difficult question to answer is which type of cell a particular membrane area belongs to. Complicating factors are that: (1) the cytoplasm associated with a membrane region is often not revealed by the fracture; (2) nonmembranous cytoplasmic organelles are seen only occasionally; and (3) nonstructural parameters such as staining properties do not exist in freeze–fracture. It follows that a correlation of freeze–fracture and thin-section results is confined almost exclusively to shape and size of membrane-bounded elements (e.g., a cellular process) and their spatial relationship to other membranous structures. Obviously, extensive thin-section data are a prerequisite for successful analysis of freeze–fractured material, and this correlation almost always requires tangential or secantlike fractures that reveal plasma membrane as well as intracellular organelles of the same cell or cellular process. On the basis of the combined results from thin sections and secantlike fractures, it may be possible to define criteria for the identification of specific membrane areas and thus to assign a particular intramembranous structure to a specific cell type, cell region, or cellular organelle. A method for the thin sectioning of replicated tissues and thus for direct replica–thin-section correlation has been described (Rash, 1979), but is not likely to be widely applicable.

The problem regarding the nature of membrane faces exposed by freeze–fracture had been the subject of heated discussions until 1970, when it could be shown by freeze–fracture/deep-etch studies combined with cell-surface labeling (Pinto da Silva and Branton, 1970; Tillack and Marchesi, 1970) and by double-replica analyses (Wehrli *et al.*, 1970) that the cleavage plane splits membranes down the middle and thus reveals intramembranous structure (Fig. 6). On the basis of independent evidence, the intramembranous course of the cleavage plane was also demonstrated for nerve tissue (Pfenninger *et al.*, 1972). It follows that the same membrane can be seen in two different ways in fractured material: the *e*xternal half or leaflet (previously called the B face, now termed the *E* face) can be viewed from a vantage point inside the cellular element; conversely, the inner, cytoplasmic, or *proto*plasmic leaflet (previously the A face, now the *P* face) may be observed from a vantage point outside the cellular element (Fig. 5). In freeze–fractured intracellular organelles, cytoplasmic and luminal leaflets can be distinguished analogously. A new terminology for all membrane faces has been introduced by a group of freeze–fracture investigators (Branton *et al.*, 1975). The description of intramembranous structures deals primarily with two parameters: intramembranous particles (IMPs) and membrane matrix.

Intramembranous particles. A steadily increasing body of evidence indicates

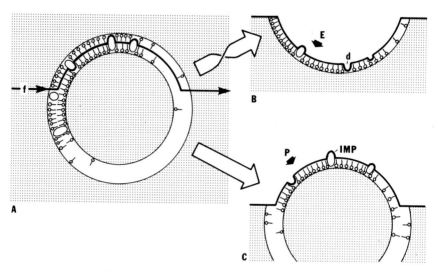

Fig. 6. Highly schematic view of exposed membrane faces in a freeze–cleaved cell. (A) Native material with a typical plane of fracture (f) indicated and resulting complementary fracture faces (B,C). (E) External and (P) protoplasmic membrane leaflets; (IMP) intramembranous particles; (d) dimples left by particles that were fractured away with the complementary membrane leaflet.

that IMPs are aggregates of integral membrane proteins. This view is likely to be accurate for membranes of nerve cells as well (e.g., Pfenninger, 1978). IMPs occur in a variety of sizes and shapes and may vary greatly in density (see Fig. 5A). Furthermore, IMP frequency and structure are usually different for complementary faces of the same membrane. The majority of IMPs are

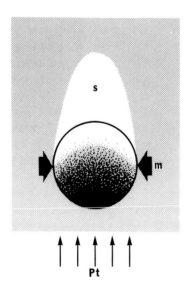

Fig. 7. Measurement (m) of the diameter of IMPs. Arrows: shadowing direction; (s) shadow (free of platinum and thus electron-lucent).

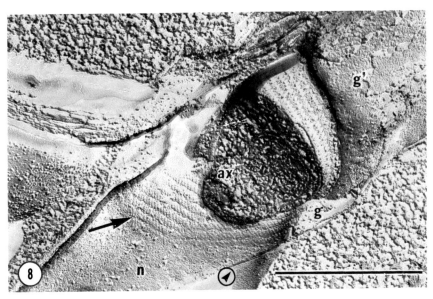

Fig. 8. Textured membrane matrix (arrow) of a glial–axonal junction adjacent to node of Ranvier in rat spinal cord. (n) Nodal axolemma, external leaflet; (ax) axoplasm; (g,g') glial loops of myelin sheath (Livingston *et al.*, 1973). Magnification: × 80,000; calibration bar: 0.5 μm; circled arrow: shadowing direction.

located on the P face of the plasmalemma (and corresponding intracellular membrane leaflets) with the exception of those in certain junctional regions. Except in these regions, particles on the E face are sparse and may, according to recent findings, result from artifactual filling (by contaminating water) of pits left by P-face particles (Moor, personal communication). Why integral membrane proteins, many of which extend carbohydrate trees into the extracellular space, are usually found on the cytoplasmic leaflet (P face) escapes explanation at present. The measurement of IMP density requires simple counting of particles in perpendicularly exposed membrane regions. IMP size, however, can be hard to assess because of its apparent dependence on the shadowing angle and the contribution of the replica thickness to the particle diameter. While the latter problem cannot be resolved, the former difficulty is avoided by measuring IMP diameter along an axis that is perpendicular to the shadowing direction (Fig. 7).

The *membrane matrix* usually exhibits smooth-surfaced structure (Fig. 5A) with some pits that probably represent IMP insertion sites, but the matrix may also be of textured appearance. Patterns "molded" into what is assumed to be a lipid surface have been observed, for example, in the paranodal junction between axon and myelin sheath (Fig. 8) (Livingston *et al.*, 1973; Rosenbluth, 1976; Schnapp *et al.*, 1976), but their nature is controversial at present.

5. Modification of the Technique and Integration with Other Methods

5.1. Vitrification

As pointed out above, an ultrarapid freezing method without the necessity for cryoprotectant treatment would greatly enhance the value of the freeze–fracture method. Thus, a great deal of effort has been made to improve freezing technology. While the vitrification of suspended particulate samples (e.g., isolated cells, subcellular fractions) can be achieved relatively easily by "jet" or "spray" freezing (e.g., Bachmann and Schmitt-Fumian, 1973*a,b*; Plattner *et al.*, 1973), considerable difficulty still exists with tissue blocks. Two major approaches exist today:

1. *Rapid freezing on a metal block.* This method, used originally by Van Harreveld and Crowell (1964) and Van Harreveld *et al.* (1974), later by Dempsey and Bullivant (1976), and more recently improved by Heuser *et al.* (1979), involves the rapid movement of fresh tissue onto the surface of a very cold (liquid-nitrogen- or liquid-helium-cooled) metal block the large heat-sink action of which results in rapid removal of heat from a superficial layer of the tissue. Excellent results have been obtained with this method, but a variety of drawbacks exist: Theoretical considerations (see above) indicate that a layer only a few micrometers thick can be vitrified by this method and thus can be considered as devoid of freezing artifacts; however, the structures of interest may not be found within this thin tissue layer, and the investigator is likely not to know the precise depth of an observed structure within the tissue block. Furthermore, dissection and mounting of the specimen prior to freezing may cause alterations (e.g., anoxia, mechanical damage) of its cells, and the best-frozen, most superficial tissue layer is the most likely to be harmed during this process. Nevertheless, important new results could be obtained with this procedure.

2. *High-pressure freezing.* The principle of this technique consists of using the cryoprotectant action of high pressure (2100 bars) to reduce the minimal freezing rate needed for vitrification. This is done by raising the pressure rapidly in a chamber containing the specimen just prior to and during freezing by a jet of liquid nitrogen (Riehle and Höchli, 1973; Moor *et al.*, 1980). However, the problems involved in tissue handling prior to freezing (e.g., mechanical damage, anoxia) apply here also and have to be considered. Furthermore, sophisticated machinery is necessary for the execution of high-pressure freezing.

As important as vitrification methodology is, it should nevertheless be pointed out that the analysis of ultrarapidly frozen material has so far confirmed results obtained from aldehyde-prefixed, cryoprotectant-treated and conventionally frozen samples.

5.2. Double-Replica Technique

There are situations in which it is of importance to achieve a precise evaluation of intramembranous particles on *complementary membrane faces*. In such cases, a double-replica technique has to be employed (Steere and Moseley, 1969; Chalcroft and Bullivant, 1970; Wehrli *et al.*, 1970). With this method, the specimen is sandwiched and frozen between two carriers; the carriers are then broken apart to achieve a cleavage plane through the biological sample, and the two halves of the sample are shadowed simultaneously. This permits the recovery of complementary faces. However, a great deal of difficulty lies in the search for corresponding areas in the two replicas, especially when tissue blocks are used. The effort needed is considerable, so that the application of the double-replica method is justified only in those cases in which it is crucial to know, for instance, whether one IMP in one membrane leaflet corresponds to a particle, to a pit, or to smooth matrix surface on the complementary face.

5.3. Deep-Etching

As mentioned before, *true cell surfaces* can be revealed by a freeze–fracture–etch process in which water surrounding cells and covering membranes is sublimated away following the fracture step. This method can be extremely useful if applied conjointly with surface labeling techniques employing, for example, antibodies against membrane components or lectins conjugated with a morphological marker such as ferritin. In the resulting preparations, in border zones between fractured (intramembranous) and etched (true surface) regions, it is possible to correlate the distribution of the surface label with that of specific intramembranous structures such as IMPs (e.g., Tillack *et al.*, 1972). However, the deep-etching procedure can be carried out successfully only on isolated cells or on cell-membrane preparations that were suspended and frozen in distilled water.

5.4. Thin-Layer Fracturing

The more and more widespread use of *tissue-culture* systems in various fields of biology, including neurobiology, has called for a special procedure that permits the preparation of freeze–fracture replicas from cells and tissues grown *in vitro*, without removal from their substrate. Different techniques have been proposed in recent years (e.g., Pfenninger and Rinderer, 1975; Collins *et al.*, 1975). In principle, they are similar in that they all employ a double-carrier system between which the cultures are sandwiched. For cleavage of the cultured cells, the two carriers are fractured apart in a manner similar to the double-replica procedure. With the aid of such procedures, it has become possible to obtain novel data on, for example, the membrane properties of nerve growth cones (Pfenninger and Bunge, 1974; Pfenninger and Rees, 1976).

5.5. Radioautography

It is of particular interest to combine the freeze–fracture method with a technique such as *radioautography*, permitting the localization of chemical

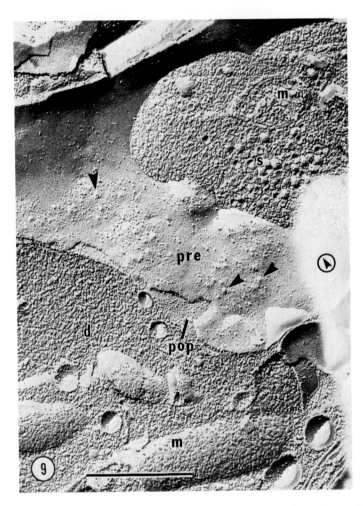

Fig. 9. Fracture through synaptic nerve terminal exposing simultaneously axoplasm with synaptic vesicles (s) clustered against presynaptic membrane [(pre) external leaflet], postsynaptic membrane [(pop) protoplasmic leaflet], and dendritic cytoplasm (d). (m) Mitochondria; arrowheads: vesicle attachment sites. Magnification: × 57,000; calibration bar: 0.5 μm; circled arrow: shadowing direction. From Pfenninger *et al.* (1972).

compounds in replicas. Attempts have been made to integrate the two methods (Branton, 1974) but remain, so far, in the experimental stage. This is due to the great technical difficulties involved in coating deep-frozen specimens with photographic emulsion under high-vacuum conditions. Coating has to be performed under these conditions because only in the frozen state, and up to the platinum- and carbon-coating steps, do the specimens maintain morphological and biochemical integrity. Later, of course, biological material must be thawed and dissolved away.

5.6. Correlative Studies

The paramount importance of the correlation of freeze–fracture analyses with thin-section studies (which might involve radioautography or cytochemistry) has already been stressed.

6. Application to a Neurobiological Problem: Freeze–Fracturing Synapses

In the late 1960's, synaptic ultrastructure was studied extensively in thin sections. An important question that remained unclear, however, was the exact relationship between synaptic vesicles and the presynaptic membrane during the process of transmitter release. This problem was, and still is, difficult to deal with in sectioned material, because thin sections, on the one hand, do not permit the viewing of entire membrane areas and, on the other hand, are too thick to allow detailed examination of the initial interaction between transmitter vesicle and presynaptic membrane. It was logical, therefore, to attempt investigation of the synapse by freeze–fracture (Pfenninger *et al.*, 1972).

A first difficulty hampering this work was the identification in freeze–fractured neuropil of the nerve terminal and, especially, of the presynaptic membrane. This membrane can now be recognized on the basis of (1) its apposition to, for example, a putative dendritic element, (2) the close association of clusters of synaptic vesicles to the same membrane region, and (3) its slightly raised appearance. This indentation into the nerve terminal corresponds to the commonly observed widening of the intercellular space at synaptic sites (Fig. 9). Obviously, these are criteria that have been identified earlier in thin-sectioned material and that are suitable for application to secant-like fractures through nerve terminals.

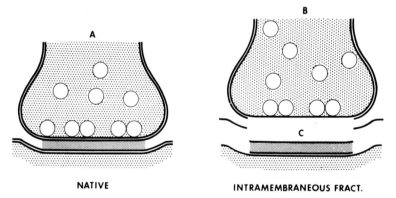

NATIVE   INTRAMEMBRANEOUS FRACT.

Fig. 10. Typical cleavage planes through synaptic membranes. For explanation, see the text. From Pfenninger *et al.* (1972).

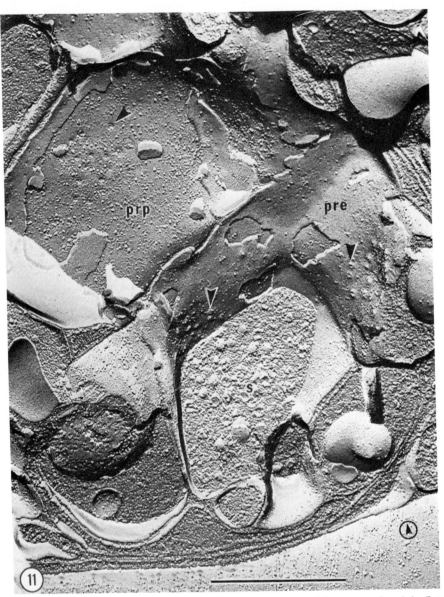

Fig. 11. Freeze–fractured presynaptic membranes [(pre) external and (prp) protoplasmic leaflets] in the pigeon optic tectum. (s) Synaptic vesicles; arrowheads: vesicle-attachment sites. Magnification: × 73,000; calibration bar: 0.5 μm. Circled arrow, shadowing direction. From Akert *et al.* (1972).

At the time of the first description of freeze–fractured synaptic membranes (Pfenninger *et al.*, 1971), the nature of fracture faces was not clear because the mechanism of cleavage was not fully understood. Analysis of different fracture planes through synaptic regions indicates that nothing other than *intramembranous fracture* can explain the membrane faces that are observed (Pfenninger *et al.*, 1972). As shown in Figs. 10 and 11, these faces are: presynaptic (external leaflet) membrane islands attached to postsynaptic (protoplasmic leaflet) membrane sheets, or presynaptic (protoplasmic leaflet) regions viewed through lacunae in the postsynaptic membrane (external leaflet). This material has thus demonstrated, in agreement with modern views of the freeze–fracture mechanism, that nerve membranes, like other membranes, are split down the middle to expose their inner faces.

Freeze–fractured synaptic membrane faces exhibit some previously unknown, characteristic structures, the most important being the vesicle attachment sites (VASs). These elements consist of small dimples in the membrane

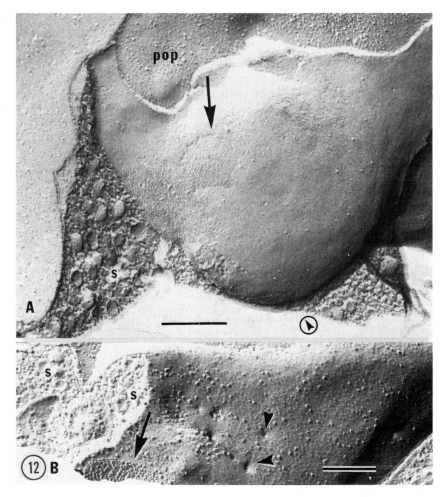

Fig. 12. Mixed chemical and electrical synapses in the spinal cord of the lamprey as seen in the external (A) and the protoplasmic (B) leaflet of the presynaptic membrane. (s) Synaptic vesicles; arrows: gap junctions; arrowheads: VASs; (pop) protoplasmic leaflet of postsynaptic plasmalemma. Magnifications: × 89,000 (A) and × 72,400 (B); calibration bars: 0.2 μm. Circled arrow, shadowing direction. From Pfenninger and Rovainen (1974).

(as seen in the protoplasmic leaflet) or protuberances (as seen in the external leaflet) and resemble, to some degree, the fractured necks of plasmalemmal vesicles in endothelial cells. Thus, VASs are good candidates for sites of interaction between transmitter vesicles and presynaptic membrane. Knowing the hexagonal arrangement of synaptic vesicles adjacent to the presynaptic membrane, one might therefore ask whether the VAS distribution, which appears unordered at first sight, is truly random or whether it can be correlated with (1) the hexagonal array of the synaptic vesicles or (2) the trigonal distribution of dense projections (Akert *et al.*, 1969; Pfenninger *et al.*, 1969). Extensive morphological analysis—by measuring distances between VASs, optical diffraction of VAS patterns, and direct observation of complete VAS arrays—has revealed that VASs indeed occupy a variable proportion of nodal points of a hexagonal pattern identical to that formed by synaptic vesicles (Pfenninger *et al.*, 1972). This evidence, albeit indirect, shows that transmitter vesicles interact with the presynaptic membrane in a specific fashion. That this interaction is the morphological basis for exocytosis of the transmitter has subsequently been established by freeze–fracture analysis of stimulated and resting synapses (Streit *et al.*, 1972; Heuser *et al.*, 1974; Pfenninger and Rovainen, 1974; Pfenninger, 1977). In summary, it has been possible to identify in freeze–fractured interneuronal synapses and neuro-muscular junctions, by careful comparison with thin-sectioned material, specific sites of vesicle–plasmalemma interaction that are stimulation- and Ca-dependent and, therefore, represent the most convincing morphological evidence for exocytosis as the mechanism of neurotransmitter release. In their recent elegant studies, Heuser, Reese, and collaborators (Heuser *et al.*, 1979) have confirmed these data in unfixed, rapidly frozen material.

Freeze–fracture not only has furthered our understanding of the biology of chemical synapses, but also represents the most conclusive approach to the identification and analysis of *electrical synapses*, which are structurally identical to gap junctions (e.g., Gilula, 1975; Staehelin, 1974). Examples of mixed electrical and chemical synapses in the spinal cord of the lamprey are illustrated in Fig. 12.

References

Akert, K., Moor, H., Pfenninger, K., and Sandri, C., 1969, Contributions of new impregnation methods and freeze–etching to the problem of synaptic fine structure, in: *Mechanisms of Synaptic Transmission* (K. Akert and P. Waser, eds.), *Prog. Brain Res.* **31**:223–240.

Akert, K., Pfenninger, K., Sandri, C., and Moor, H., 1972, Freeze–etching and cytochemistry of vesicles and membrane complexes in synapses of the central nervous system, in: *Structure and Function of Synapses* (G.D. Pappas and D.P. Purpura, eds.), pp. 67–86, Raven Press, New York.

Bachmann, L., and Schmitt-Fumian, W.W., 1973a, Spray–freeze–etching of dissolved macro-molecules, emulsions and subcellular components, in: *Freeze–Etching: Techniques and Applications* (E.L. Benedetti and P. Favard, eds.), pp. 63–72, Société Française de Microscopie Electronique, Paris.

Bachmann, L., and Schmitt-Fumian, W.W., 1973*b*, Spray–freezing and freeze–etching, in: *Freeze–Etching: Techniques and Applications* (E.L. Benedetti and P. Favard, eds.), pp. 73–79, Société Française de Microscopie Electronique, Paris.

Benedetti, E.L., and Favard, P. (eds.), 1973, *Freeze–Etching: Techniques and Applications*, Société Française de Microscopie Electronique, Paris.

Böhler, S., 1976, *Artefacts and Specimen Preparation Faults in Freeze–Etch Technology*, Balzers Aktiengesellschaft, Liechtenstein.

Branton, D., 1967, Fracture faces of frozen myelin, *Exp. Cell Res.* **45**:703–707.

Branton, D., 1974, Interpretation of freeze–etch results, in: *Proceedings of the 8th International Congress on Electron Microscopy*, Canberra, 1974, Vol. II, p. 28, Australian Academy of Science, Canberra.

Branton, D., Bullivant, S., Gilula, N.B., Karnovsky, M.J., Moor, H., Mühlethaler, K., Northcote, D.H., Packer, L., Satir, B., Satir, P., Speth, V., Staehelin, L.A., Steere, R.L., and Weinstein, R.S., 1975, Freeze–etching nomenclature, *Science* **190**:54–56.

Bretscher, M.S., 1973, Membrane structure: Some general principles, *Science* **181**:622–629.

Chalcroft, J.P., and Bullivant, S., 1970, An interpretation of liver cell membrane and junction structure based on observation of freeze–fracture replicas of both sides of the fracture, *J. Cell Biol.* **47**:49–60.

Clark, A.W., and Branton, D., 1968, Fracture faces in frozen outer segments from the guinea pig retina, *Z. Zellforsch.* **91**:586–603.

Collins, T., Bartholomew, J., and Calvin, M., 1975, A simple method for freeze–fracture of monolayer cultures, *J. Cell Biol.* **67**:904–911.

Dempsey, G.P., and Bullivant, S., 1976, A copper block method for freezing non-cryoprotected tissues to produce ice-crystal-free regions for electron microscopy, *J. Microsc. (Oxford)* **106**:251–270.

Dreyer, F., Peper, K., Akert, K., Sandri, C., and Moor, H., 1973, Ultrastructure of the "active zone" in the frog neuromuscular junction, *Brain Res.* **62**:373–380.

Gilula, N.B., 1975, Junctional membrane structure, in: *The Nervous System*, Vol. I, *The Basic Neurosciences* (D.B. Tower, ed.), pp. 1–11, Raven Press, New York.

Gross, A., Bas, E., and Moor, H., 1978, Freeze–fracturing in ultrahigh vacuum at −196°C, *J. Cell Biol.* **76**:712–728.

Hanna, R.B., Hirano, A., and Pappas, G.D., 1976, Membrane specializations of dendritic spines and glia in the weaver mouse cerebellum: A freeze–fracture study, *J. Cell Biol.* **68**:403–410.

Heuser, J.E., Reese T.S., and Landis, D.M.D., 1974, Functional changes in frog neuromuscular junctions studied with freeze–fracture, *J. Neurocytol.* **3**:109–131.

Heuser, J.E., Reese, T.S., Dennis, M.J., Jan, Y., Jan, L., and Evans, L., 1979, Synaptic vesicle exocytosis captured by quick freezing and correlated with quantal transmitter release, *J. Cell Biol.* **81**:275–300.

Hong, K., and Hubbell, W.L., 1972, Preparation and properties of phospholipid bilayers containing rhodopsin, *Proc. Natl. Acad. Sci. U.S.A.* **69**:2617–2621.

Kirchanski, S., Elgsaeter, A., and Branton, D., 1979, Low temperature freeze fracturing to avoid plastic distortion, in: *Freeze–Fracture: Methods, Artifacts, and Interpretations* (J.E. Rash and C.S. Hudson, eds.), pp. 149–152, Raven Press, New York.

Landis, D.M., Reese, T.S., and Raviola, E., 1974, Differences in membrane structure between excitatory and inhibitory components of the reciprocal synapse in the olfactory bulb, *J. Comp. Neurol.* **155**:67–92.

Livingston, R.B., Pfenninger, K., Moor, H., and Akert, K., 1973, Specialized paranodal and internodal glial–axonal junctions in the peripheral and central nervous system: A freeze–etching study, *Brain Res.* **58**:1–24.

Moor, H., 1964, Die Gefrierfixation lebender Zellen und ihre Anwendung in der Elektronenmikroskopie, *Z. Zellforsch.* **62**:546–580.

Moor, H., 1971, Recent progress in the freeze–etching technique, *Philos. Trans. R. Soc. London B* **261**:121–131.

Moor, H., 1973*a*, Cryotechnology for the structural analysis of biological material, in:

Freeze–Etching: Techniques and Applications (E.L. Benedetti and P. Favard, eds.), pp. 11–20, Société Française de Microscopie Electronique, Paris.

Moor, H., 1973*b*, Evaporation and electron guns, in: *Freeze–Etching: Techniques and Applications* (E.L. Benedetti and P. Favard, eds.), pp. 27–30, Société Française de Microscopie Electronique, Paris.

Moor, H., and Mühlethaler, K., 1963, Fine structure in frozen–etched yeast cells, *J. Cell Biol.* **17:**609–628.

Moor, H., Bellin, G., Sandri, C., and Akert, K., 1980, The influence of high pressure freezing on mammalian nerve tissue, *Cell Tissue Res.* (in press).

Mühlethaler, K., 1973, History of freeze–etching, in: *Freeze–Etching: Techniques and Applications* (E.L. Benedetti and P. Favard, eds.), pp. 1–10, Société Française de Microscopie Electronique, Paris.

Nickel, E., and Potter, L.T., 1970, Synaptic vesicles in freeze–etched electric tissue of *Torpedo*, *Brain Res.* **23:**95–100.

Pfenninger, K.H., 1977, Cytology of the chemical synapse: Morphological correlates of synaptic function, in: *Neurotransmitter Function* (W.S. Fields, ed.), pp. 27–57, Symposia Specialists, Miami.

Pfenninger, K.H., 1978, Organization of neuronal membranes, *Annu. Rev. Neurosci.* **1:**445–471.

Pfenninger, K.H., and Bunge, R.P., 1974, Freeze–fracturing of nerve growth cones and young fibers: A study of developing plasma membrane, *J. Cell Biol.* **63:**180–196.

Pfenninger, K.H., and Rees, R.P., 1976, From the growth cone to the synapse: Properties of membranes involved in synapse formation, in: *Neuronal Recognition* (S.H. Barondes, ed.), pp. 131–178, Plenum Press, New York.

Pfenninger, K.H., and Rinderer, E.R., 1975, Methods for the freeze–fracturing of nerve tissue cultures and cell monolayers, *J. Cell Biol.* **65:**15–28.

Pfenninger, K.H., and Rovainen, C.M., 1974, Stimulation- and calcium-dependence of vesicle attachment sites in the presynaptic membrane: A freeze–cleave study on the lamprey spinal cord, *Brain Res.* **72:**1–23.

Pfenninger, K., Sandri, C., Akert, K., and Eugster, C.H., 1969, Contribution to the problem of structural organization of the presynaptic area, *Brain Res.* **12:**10–18.

Pfenninger, K., Akert, K., Moor, H., and Sandri, C., 1971, Freeze–fracturing of presynaptic membranes in the central nervous system, *Philos. Trans. R. Soc. London B* **261:**387.

Pfenninger, K., Akert, K., Moor, H., and Sandri, C., 1972, The fine structure of freeze–fractured presynaptic membranes, *J. Neurocytol.* **1:**129–149.

Pinto da Silva, P., and Branton, D., 1970, Membrane splitting in freeze–etching, *J. Cell Biol.* **45:**598–605.

Plattner, H., Schmitt-Fumian, W.W., and Bachmann, L., 1973, Cryofixation of single cells by spray-freezing, in: *Freeze–Etching: Techniques and Applications* (E.L. Benedetti and P. Favard, eds.), pp. 81–100, Société Française de Microscopie Electronique, Paris.

Rash, J.E., 1979, The sectioned-replica technique: Direct correlation of freeze–fracture replicas and conventional thin-section images, in: *Freeze–Fracture: Methods, Artifacts, and Interpretations* (J.E. Rash and C.S. Hudson, eds.), pp. 153–160, Raven Press, New York.

Rash, J.E., and Hudson, C.S. (eds.), 1979, *Freeze–Fracture: Methods, Artifacts, and Interpretations*, Raven Press, New York.

Rash, J.E., Graham, W.E., and Hudson, C.S., 1979, Sources and rates of contamination in a conventional Balzers freeze–etch device, in: *Freeze–Fracture: Methods, Artifacts, and Interpretations* (J.E. Rash and C.S. Hudson, eds.), pp. 111–122, Raven Press, New York.

Raviola, E., and Gilula, N.B., 1975, Intramembrane organization of specialized contacts in the outer plexiform layer of the retina: A freeze–fracture study in monkeys and rabbits, *J. Cell Biol.* **65:**192–222.

Riehle, U., and Höchli, M., 1973, The theory and technique of high pressure freezing, in: *Freeze–Etching: Techniques and Applications* (E.L. Benedetti and P. Favard, eds.), pp. 31–62, Société Française de Microscopie Electronique, Paris.

Rosenbluth, J., 1976, Intramembranous particle distribution at the node of Ranvier and adjacent axolemma in myelinated axons of the frog brain, *J. Neurocytol.* **5:**731–745.

Sandri, C., Akert, K., Livingston, R.B., and Moor, H., 1972, Particle aggregations at specialized sites in freeze–etched postsynaptic membranes, *Brain Res.* **41:**1–16.

Sandri, C., Van Buren, J.M., and Akert, K., 1976, Membrane morphology of the vertebrate nervous system, *Prog. Brain Res.* **46.**

Schnapp, B., and Mugnaini, E., 1975, The myelin sheath: Electron microscopic studies with thin sections and freeze–fracture, in: *Golgi Centennial Symposium Proceedings* (M. Santini, ed.), pp. 209–233, Raven Press, New York.

Schnapp, B., Petacchia, C., and Mugnaini, E., 1976, The paranodal axo–glial junction in the central nervous system studied with thin sections and freeze–fracture, *Neuroscience* **1:**181–190.

Singer, S.J., and Nicolson, G.L., 1972, The fluid mosaic model of the structure of cell membranes, *Science* **175:**720–731.

Staehelin, L.A., 1974, Structure and function of intercellular junctions, *Int. Rev. Cytol.* **39:**191–283.

Steere, R.L., and Moseley, M., 1969, New dimensions in freeze–etching, in: *27th Annual Meeting, Electron Microscopy Society of America* (C.J. Arceneaux, ed.), pp. 202–203, Claitor's, Baton Rouge.

Steere, R.L., Erbe, E.F., and Moseley, J.M., 1979, Controlled contamination of freeze–fractured specimens, in: *Freeze–Fracture: Methods, Artifacts, and Interpretations* (J.E. Rash and C.S. Hudson, eds.), pp. 99–109, Raven Press, New York.

Streit, P., Akert, K., Sandri, C., Livingston, R.B., and Moor, H., 1972, Dynamic ultrastructure of presynaptic membranes at nerve terminals in the spinal cord of rats: Anesthetized and unanesthetized preparations compared, *Brain Res.* **48:**11–26.

Tillack, T.W., and Marchesi, V.T., 1970, Demonstration of the outer surface of freeze–etched red blood cell membranes, *J. Cell Biol.* **45:**649–653.

Tillack, T.W., Scott, R.E., and Marchesi, V.T., 1972, The structure of erythrocyte membranes studied by freeze–etching. II. Localization of receptors for phytohemagglutinin and influenza virus to the intramembranous particles, *J. Exp. Med.* **135:**1209–1227.

Van Harreveld, A., and Crowell, J., 1964, Electron microscopy after rapid freezing on a metal surface and substitution fixation, *Anat. Rec.* **149:**381–386.

Van Harreveld, A., Trubatch, J., and Steiner, J., 1974, Rapid freezing and electron microscopy for the arrest of physiological processes, *J. Microsc. (Oxford)* **100:**189–198.

Wehrli, E., Mühlethaler, K., and Moor, H., 1970, Membrane structure as seen with a double replica method for freeze fracturing, *Exp. Cell Res.* **59:**336–339.

Wiley, C.A., and Ellisman, M.H., 1979, Development of axonal membrane specializations defines nodes of Ranvier and precedes Schwann cell myelin formation, *J. Cell Biol.* **83:**83a.

Wood, M.R., Pfenninger, K.H., and Cohen, M.J., 1977, Two types of presynaptic configurations in insect central synapses: A freeze–fracture and cytochemical analysis, *Brain Res.* **130:**25–45.

Enzyme Histochemistry

E. Marani

1. Introduction

In the following pages, various important aspects of enzyme histochemistry will be discussed. Enzymes are proteins, and a basic understanding of enzyme histochemistry will require first a brief review of proteins including their biochemical characteristics. Special attention will be directed toward the biochemistry of enzymatic catalysis. After this introductory material, some aspects of enzyme histochemistry will be more specifically reviewed in the context of the introductory presentation of enzyme biochemistry. It should become clear that an accurate interpretation of histochemical results is possible only with the help of detailed knowledge of the biochemical properties of enzymes.

1.1. Amino Acids: The Building Blocks

Three types of amino acids can be found in nature: the α, β, and γ types. Most of the amino acids in proteins, however, are of the α type. Further, α-amino acids have an asymmetrically substituted carbon atom that implies the existence of two stereoisomers for every amino acid (except, of course, glycine), designated D and L. This D and L formalism is based on the same principles used in carbohydrate chemistry. While the naturally occurring carbohydrates belong mainly to the D class, the naturally occurring amino acids belong to the L class. By far the greatest part of naturally occurring proteins are constructed from about 20 different L-α-amino acids, which may be divided into six groups according to the nature of the substituents on the α-carbon atom.

E. Marani • Laboratory of Anatomy and Embryology, University of Leiden, Leiden, The Netherlands.

Apart from being the building blocks of proteins, some amino acids appear to function in the nervous system as neurotransmitter substances. For example, glycine is postulated as a transmitter in the spinal cord (Aprison, 1975) and proline as a transmitter in the visual system, the cerebellum, and the spinal cord (Johnston, 1975). Other amino acids participate in neurotransmission in a more indirect manner by being precursors for transmitter substances. For example, glutamate can be decarboxylated to γ-aminobutyric acid (GABA) (Albers *et al.*, 1972). Apart from this precursor function, glutamic acid is itself a neurotransmitter. In summary, then, amino acids are precursors of proteins; they can be precursors of neurotransmitter substances; they can themselves be neurotransmitters; and they may have an important role in certain metabolic pathways, known as transamination reactions, and thereby enter the tricarboxylic acid cycle. All these functions are displayed by glutamic acid.

Once taken up by a neuron, amino acids may be transported along the axon to the axon endings, the boutons, by a mechanism yet unknown. The transport may be of the amino acid, as such, or of the amino acid incorporated in a protein. This phenomenon is known as anterograde axonal transport (anterograde = away from the cell body). By injection of tritiated amino acids into selected areas of the brain, the course of the axons originating from the cell bodies in that area can be followed by means of autoradiographic techniques. This method has become an important tool in the study of neuronal interconnections (see Volume 2, Chapter 9).

1.2. Proteins and Enzymes

1.2.1. General Remarks on Proteins

Proteins are macromolecular compounds with molecular weights anywhere from ten thousand to several millions. On the basis of their solubility characteristics, they can be subdivided into several groups, of which the globular proteins are by far the most important to enzyme histochemists. The diameters of the globular proteins also show considerable variation, ranging from 50 to 5000 Å. However, there is not always a close correlation between diameter and molecular weight. Because of this, proteins cannot pass a semipermeable membrane with pores smaller than 50 Å. This property is used in biochemistry for the purification of proteins from smaller molecules and is known as dialysis. Its histochemical counterpart may be found in semipermeable-membrane techniques (Meijer, 1972, 1973; Meijer and Vloedman, 1973; Meijer and de Vries, 1974), in which a semipermeable membrane is interposed between the tissue section and the incubation medium. This prevents the diffusion of soluble enzymes into the medium.

Proteins contain charged side groups, the sign of which charge is pH-dependent. This is due to extra acidic or basic groups in certain amino acids, such as glutamic acid and arginine, respectively. Differences in the net charges of proteins may aid in their separation by electrophoresis techniques, in which proteins are caused to move down an electrical field (see Chapter 7 and Volume 2, Chapter 4).

1.2.2. Structure of Proteins

In genetically controlled protein synthesis, many amino acids are coupled together by means of peptide bonds to form a macromolecular compound: the polypeptide chain. (In proteolysis, which is brought about by the proteases, the reverse reaction takes place.) The sequence of amino acids in this chain is called the *primary structure*. With a few exceptions, this polypeptide chain does not exist in a simple linear compound, but is curved in on itself in a helical fashion, called the α-helix. This helix is not always precisely helical in all places of the molecule because this is prevented by bulky side groups (as, for example, in phenylalanine) or the presence of amino acids such as proline. The helical conformation is known as the *secondary structure*. The secondary structure is stabilized by the interactions of the $>C=O$ and $>NH$ groupings in the peptide bonds. The α-helix, again, may be turned and twisted in a particular way, yielding a spatial structure superimposed on the secondary structure. This is known as the *tertiary structure*. The tertiary structure is stabilized by weak interactions between the side groups (which all project away from the long axis of the α-helix). These interactions include ionic interactions between oppositely charged groups, hydrogen bonding, hydrophobic bonding, Van der Waals forces, and disulfide bridges. It is supposed that secondary and tertiary structures necessarily emerge from the primary structure and are thus genetically determined. Furthermore, several α-helices can interact with one another, resulting in a structure consisting of many polypeptide chains twisted and folded in a particular way and held together by the same kind of interactions that stabilize the tertiary structure. One might call this the *quaternary structure*.

1.2.3. Active Site and Denaturation

It can be easily imagined that the sometimes very extensive folding of the α-helix brings together side groups that are far away from one another in the primary structure. A particular three-dimensional orientation of side groups might, under suitable conditions, exhibit enzymatic activity; that is, a specific arrangement of side groups may favor the interaction of one or more components of a particular reaction. This special site of the protein molecule is called the *active site*.

As suggested above, the enzymatic activity of a protein is dependent on the active site, the structure of which is dependent on the tertiary (or quaternary) structure of the protein. Any factor that disrupts this structure will also modify, to a greater or lesser extent, the arrangement of the active site and hence the enzymatic activity. Disruption of the tertiary structure is known as *denaturation* and can be brought about by a change in the pH, temperature, or ionic strength of the solution. All these are changes in the physical environment of the protein molecule, and they induce physical denaturation, which in most instances is reversible. The tertiary structure of proteins can also be disrupted by chemical means, as is done by most fixatives. The denaturation brought about by chemical processes is in general *not* reversible. Unfortunately, during histochemical experimentation, it may be necessary to act on the enzyme under investigation in such a way (e.g., fixation, especially in ultrahistochemistry, or high salt concentrations, sometimes encountered in incubation media) that denaturation of the protein results. One must take these denaturing conditions into account when interpreting the results of histochemical procedures.

Sometimes, it is found that several seemingly different enzymes all catalyze the same reaction. On the one hand, this may be brought about by a rather low substrate specificity, as has been illustrated for the hydrolysis of ATP. This reaction, under suitable conditions, may be catalyzed by acid phosphatase, alkaline phosphatase, or one of the many ATPases, but these are all totally different enzymes. On the other hand, many enzymes are built up of more than one polypeptide chain. These chains, then, can be combined in various ways, as may be illustrated by lactate dehydrogenase (LDH). This enzyme consists of a combination of so-called M and H chains totaling four chains per molecule. This permits the following combinations: M_4, M_3H, M_2H_2, MH_3, and H_4. All five catalyze the same reaction but differ in their kinetic and physical properties, e.g., the net charge over the complex at a certain pH. This latter characteristic provides the basis on which the various enzyme forms may be separated from one another by gel electrophoresis (see Volume 2, Chapter 4). These five forms are known as isoenzymes or isozymes, and on occasion, they can be differentiated by the addition of certain compounds in the medium (Gollnick and Armstrong, 1976) (for a description of LDH isoenzymes in the brain, see Friede, 1966).

1.3. Coenzymes and Cofactors

In some instances, the tertiary structure alone of a protein induces its enzymatic activity. In many other instances, however, the protein needs some extra compound to perform this activity. If this compound is a complex organic structure, we speak of a *coenzyme*. The terms coenzyme and prosthetic group are usually synonymous, although coenzymes belong to parts of the enzyme that dissociate easily, while a prosthetic group is bound tightly to the enzyme. The enzyme (holoenzyme) can thus dissociate into two parts: the coenzyme and the apoenzyme. This is a reversible reaction.

The typical coenzymes could better be named cosubstrates (Karlson, 1972), since they are related to the enzymatic reaction as an energetic active component or they function as a phosphate donor (ATP delivers the phosphate for the hexokinase reaction; the coenzyme takes the hydrogen from the substrate, codehydrase I). In this sense, coenzymes are often mediators between two enzymes, and in such a way that enzyme systems can arise (e.g., in glycolysis). Prosthetic groups can be discriminated from coenzymes because they are strongly coupled to the enzyme and remain in the enzymatic reaction (Karlson, 1972).

Many enzymes need metal ions to perform an enzymatic reaction. Because in the course of tissue preparation for enzyme histochemistry these (highly soluble) ions are lost, the enzyme will not be effective unless these ions are added to the incubation medium. We might see this as activation of the enzyme; hence the name "activator," which is so often encountered in descriptions of histochemical procedures when the addition of certain metal salts is discussed. It would seem more appropriate to replace this term with "cofactor" to distinguish cofactors from coenzymes. For most enzymes, the mechanism by which these ions work is unknown. Some ions can be substituted with others yet have the same or nearly the same influence on the enzyme activity [e.g., the interchangeability of Zn^{2+}, Co^{2+}, and Mn^{2+} in yeast alcohol dehydrogenase (Sadler, 1976)].

1.4. Subdivision of Enzymes

The International Union of Biochemistry has classified enzymes in sections and in a numerical order. This subdivision is adopted here for those enzymes that are of interest in enzyme histochemistry but are still commonly called by their trivial names. Some enzymes are frequently studied histochemically [e.g., acetylcholinesterase (EC 3.1.1.7)], while others have attracted less attention in neurobiology [e.g., betainaldehyde dehydrogenase (EC 1.2.2.8) (see Fig. 1)].

1.5. Substrate Specificity

Any substrate molecule can enter, dependent on its chemical structure, into several chemical reactions. For example, an amino acid can be decarboxylated or coupled to another amino acid. Other reactions are also possible.

Because of its structure as a protein, a given enzyme can undergo only one of these reactions. This is called *Wirkungsspezifität* (Karlson, 1972). For example, a decarboxylase will catalyze only the decarboxylation reaction. Not all amino acids will be broken down by a particular decarboxylase. The choice among certain amino acids made by an enzyme (e.g., a particular decarboxylase) is called substrate specificity. The degree of this substrate specificity differs from one enzyme to another, and no general rules can be given (Karlson, 1972). The specificity of enzymes can be confusing for enzyme histochemistry; e.g., aldehyde oxidase (EC 1.2.3.1) also oxidizes derivatives of pyridine and quinoline (Chance, 1952). AMP, a specific substrate for 5'-nucleotidase (EC 3.1.3.5), can be broken down by acid (EC 3.1.3.2) or alkaline phosphatase (EC 3.1.3.1). Peptidases, on the other hand, are relatively specific (M. Dixon and Webb, 1958), while cholinesterases are less specific for substrates. The dehydrogenases are highly specific with some exceptions, e.g., alcohol dehydrogenase (EC 1.1.1.1), LDH, and aldehyde dehydrogenase (M. Dixon and Webb, 1958).

A direct consequence of the differences in substrate specificity for the

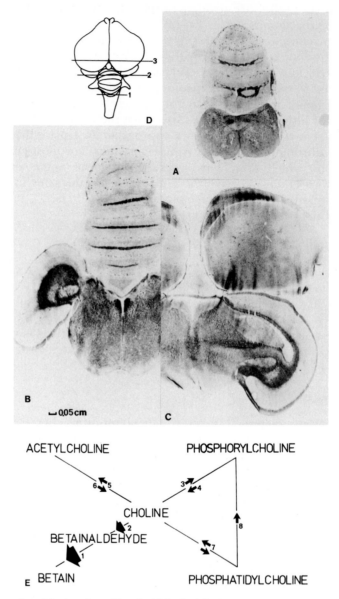

Fig. 1. Examples of the location of betainaldehyde dehydrogenase in different sections through the brainstem and cerebellum of the domestic fowl. The levels of the sections are indicated by the numbers 1–3 on a dorsal view of the brain (D). Section (A) is located at 1, section (B) at 2, and section (C) at 3. Betainaldehyde dehydrogenase is an enzyme that plays an important role in choline breakdown. (E) A scheme for the enzymes involved in the metabolism of choline, modified according to Van der Waart (1974). Enzymes: (1) betainaldehyde dehydrogenase; (2) choline dehydrogenase; (3) choline kinase; (4) alkaline phosphatase; (5) choline acetyltransferase; (6) acetylcholinesterase; (7) phospholipase via lysophospholipase and glycerophosphofinicocholinediesterase; (8) phospholipase.

enzyme histochemist is shown in the next example. Manocha and Shanta (1970) postulated a connection between the cells in the mouse cerebellum with an inhibitory function and the localization of the enzyme 5'-nucleotidase. We (Marani and Voogd, 1977) showed that the 5'-nucleotidase activity in Golgi cells and Purkinje cells (both inhibitory cells) is due to residual inherent acid phosphatase activity and not to 5'-nucleotidase. Acid phosphatase, even at pH 7.2 (the pH at which the 5'-nucleotidase reaction takes place), acts on AMP and apparently disturbs the specific localization of 5'-nucleotidase in the mouse cerebellum.

For the enzyme histochemist, it is important to know whether, and where, enzymes can disturb the desired localization of enzymes. It is obvious from the example given that erroneous conclusions can be drawn if this principle is not kept in mind.

Some types of enzymes can discriminate between optical isomers. In enzyme histochemistry, examples of this stereospecificity of certain enzymes can be noted, e.g., L-amino acid dehydrogenase (L-amino acid : NAD oxidoreductase, EC 1.4.3.2.) and D-amino acid oxidase (D-amino acid : O_2 oxidoreductase, EC 1.4.3.3.) Furthermore, some types of enzymes can discriminate between geometric isomers. Substrates can have the *cis* or *trans* configuration, and a selective attack on the *cis* isomers is possible, while the corresponding *trans* isomers are not broken down. Enzymes that interconvert the L type of the substrate into the D type or the *cis* into the *trans* type are called isomerases. For hydroxyproline-2-epimerase (EC 5.1.1.a.), an enzyme-histochemical method has been described (Adams and Norton, 1964). This enzyme catalyzes the reaction of L-hydroxyproline into D-allohydroxyproline. The *cis–trans* isomerases (EC 5.2.) cannot be demonstrated in enzyme histochemistry, but an illustrative example of this type of reaction is the conversion of 11-*trans*-retinal into 11-*cis*-retinal. This interconverting enzyme reaction is important in the biochemical pathway of vision (see Van der Meer *et al.*, 1977).

2. The "Enzyme-Histochemical" Reaction

The aim of an enzyme histochemist is an exact localization of enzymes in a biological specimen. This accurate localization is fundamental, and sometimes a conclusion can be drawn far beyond this level by comparison with other techniques, but only if precise localization is attained. In general, there are three conditions necessary for such localization. In the first place, the desired enzyme reaction must be possible under optimal conditions in the artificial environment of the incubation medium. Second, the reaction products must be precipitated on the exact spot where the enzyme action occurs. Finally, these precipitated reaction products must be visualized where they are localized.

The idea is that the enzyme can be localized by its reaction products. Although the idea and even the conditions seem obvious, it became evident in the history of enzyme histochemistry that sometimes these conditions and even the idea have self-contradictory elements. To determine an enzyme by its reaction products, many accessory products are needed, and these may affect the enzyme capacity. Or the enzyme may be soluble and may thus displace its reaction products over the studied tissues or cells. Some additional conditions are accordingly deduced from the principal conditions (e.g., see O'Sullivan, 1955). First, the enzyme *in vitro* must not be soluble and its place must not be altered by preparatory processing. Second, the capture reaction, to precipitate the reaction product, must be a fast one to overcome diffusion of the reaction product. Third, no diffusion of precipitated reaction products to other cell compartments or tissue compartments may occur during any of the additional operations needed in the capture-reaction sequence. Fourth, the color developer must not change the location of the principal product. Further conditions are known (O'Sullivan, 1955), but in the main, they fit in among these four rules.

The example given of the accessory products that affect enzyme capacity (or the presence of a certain isozyme) does not fall within the compass of the added rules. With the reasoning that some localization is better than no localization, these effects are accepted on the condition that the inhibitory effects or background effects are determined, described, and evaluated with respect to the given localization (for a dicussion of background effects, see Burstone, 1962).

Comparisons are often made with biochemical results, and Glenner (1965) gave the conditions with an agnosticism that enables as "critical an approach as possible to be made to the evaluation of histochemical enzyme systems." Glenner (1965) stated that the following assumptions should not be made unless their validity has been adequately demonstrated:

A substrate is specific for a specific enzyme or group of enzymes.

A histochemical substrate identifies the same activities as its "naturally occurring analogues."

A naturally-insoluble (undenatured) enzyme has a soluble component and vise versa.

The reagents added to the histochemical incubation medium have no effect on the characteristics of enzymic activity demonstrated.

The steps of the histochemical tissue-preparation have no effect on the localization and characteristics of enzyme activity.

Absence of enzymic activity toward a substrate in a tissue site indicated the absence either *in vivo* or *in vitro* of an enzyme capable of hydrolyzing the substrate in that site.

The advantage of the approach of Glenner (1965) is that the limitations of enzyme histochemistry are also indicated by the given assumptions.

In this chapter, discussion of the types of enzyme reactions will be based on the principal conditions already mentioned and will keep in mind the conditions pointed out by Glenner (1965). Each of the main conditions can be represented by a type of reaction: the enzyme reaction, the capture reaction, and a visualization reaction. The equations used are:

For the enzyme reaction: $\quad$ I. $E + S_1 + (S_2) \rightleftarrows P_1 + (P_2) + E$

For the capture reaction: $\quad$ I. $P_1 + (P_2) + X \rightarrow PX \downarrow$

For the visualization reaction: $\quad$ II. $PX + Z \rightarrow PZ \updownarrow$

In these equations, I signifies that the reaction is carried out in the incubation medium, while II signifies the use of a different medium. Precipitation is symbolized by $\downarrow$ and the formation of a visible product by $\updownarrow$. The other symbols used are E (enzyme), S (substrate), P (product), X (capture agent) and Z (color developer).

2.1. Hydrolytic Enzymes (EC 3)

Different types of enzyme-histochemical reactions can be illustrated in the class of hydrolytic enzymes when one varies the presence or absence of a capture or visualization reaction and the procedures that are carried out within and without the incubation medium.

2.1.1. Type 1 Reactions

Type 1 reactions are known in hydrolytic enzyme histochemistry as noncoupling reactions, altered-solubility methods, and intramolecular-rearrangement methods. They can be grouped together in the following equation:

$$\text{I. } E + S \rightarrow E + P \updownarrow (P = PX = PZ)$$

An example of this type is the alkaline phosphatase method of Friedenwald and Becker (1948). This type of reaction is seldom found in the literature, although Pearse (1968) stated with reference to intramolecular rearrangement that "future studies in this direction may be expected to yield important results."

By avoiding separate capture and visualization reactions while providing all the conditions in one and the same chemical reaction, most of the difficulties concerning diffusion of the reaction product before capture in the coupling reaction or diffusion by the visualization procedure may be overcome.

2.1.2. Type 2 Reactions

This type of reaction is called the postincubation-coupling method and is described by the following equations:

$$\text{I. } E + S \rightarrow E + P \downarrow (P = PX)$$

$$\text{II. } P + Z \rightarrow PZ \updownarrow$$

The reaction has been employed for acid phosphatase by Seligman *et al.* (1949) using the 6-bromo-2-naphthyl phosphate or the 6-benzoyl-2-naphthyl phosphate method (Rutenberg and Seligman, 1955). For the visualization reaction, a diazonium salt was used. A limitation is that the reactions can be performed only at acid pH because at alkaline pH, the released reaction products are soluble, the capture agent for these reaction products being the hydronium ion of water. With naphthyl AS phosphate, a postcoupling method was developed by Burstone (1962) for alkaline phosphatase.

2.1.3. Type 3 Reactions

These are called the simultaneous capture or coupling reactions:

$$I.\ E + S \rightarrow E + P$$

$$I.\ P + Z \rightarrow PZ\ \updownarrow\ (PZ = PX)$$

In this type of reaction, the capture agent and the color developer are identical.

Since the capture–color agents are added to the incubation medium, effects of these agents on the enzymatic activity may be expected, and these effects must be evaluated. This also holds for Type 4 and 5 reactions. [For a discussion of the kinetics of these reactions, see Holt and O'Sullivan, (1958).] The best-known examples of these reactions used the Naphthol AS Series of substrates with freshly hexazotized parafuchsin (Barka and Anderson, 1962).

2.1.4. Type 4 Reactions

Known as the multiple-sequence method, these reactions can be described as:

$$I.\ E + S \rightarrow P + E$$

$$I.\ P + X \rightarrow PX\ \downarrow$$

$$II.\ (PX + Y \rightarrow PY\ \downarrow)$$

$$II.\ PX(PY) + Z \rightarrow PZ\ \updownarrow$$

Sometimes, an extra conversion to another invisible, insoluble reaction product (the third equation) is needed.

The best-known examples of the multiple-sequence method are the lead sulfide method of Gomori (1941) and the modified technique of Wachstein and Meisel (1957). The calcium–cobalt technique of Gomori (1952) is an example of the conversion of the insoluble captured reaction product into another, invisible one. The liberated phosphate is captured by Ca^{2+}, and the calcium phosphate is first converted to cobalt phosphate before the enzyme location is visualized as cobalt sulfide.

2.1.5. Type 5 Reactions

The intermediate or postcoupling enzyme reaction uses a hydrolytic enzyme to localize another enzyme. The equation can be given as:

$$\text{I. } E_1 + S \rightarrow E_1 + P_1$$

$$\text{I. } P_1 + E_2 \rightarrow P_2 + E_2$$

$$\text{I. } P_2 + X \rightarrow P_2X$$

$$\text{II. } P_2X + Z \rightarrow P_2Z \updownarrow$$

This principle is applied in the reaction of 3′,5′-cyclic phosphodiesterase of Shanta *et al.* (1966). The reaction of phosphodiesterase cannot be shown, but the reaction product (5′-AMP) can be attacked by 5′-nucleotidase, which is added to the incubation medium. The liberated phosphate can be captured and marked with lead sulfide. In the following sections, we will discuss this kind of method in detail.

2.1.5a. An Example of a Contested Method (5′-Exonuclease). Coupling techniques using an intermediate enzyme have been developed for hydrolytic enzymes [5′-exonuclease = 3′,5′-cyclic phosphodiesterase (Shanta *et al.*, 1966)] as well as for oxidoreductases [e.g., "coupled peroxidase method for MAO" (Graham and Karnovsky, 1965*b*, 1966)]. The liberated product in such a reaction (5′-nucleotides or H_2O_2) is shown or localized by a second enzyme added to the incubation medium using the liberated product of the first enzyme reaction as substrate. The final product is shown by the lead sulfide technique for phosphodiesterase or the diaminobenzidine technique for monoamine oxidase (MAO).

Snake venom is added to the 5′-exonuclease incubation as a source of the second enzyme (5′-nucleotidase), although snake venoms may also contain 5′-exonuclease, which can lead to erroneous localizations of that enzyme in the tissue section being examined (Tatsuki *et al.*, 1975). In testing this possibility that the snake venom of *Crotalus atrox* can contain phosphodiesterase activity, we used latex particles as a substitute for tissue sections in incubations according to the technique described by Poelmann and Daems (1973). In Fig. 2A, five latex particles on which the lead reaction product is indeed present are shown.

In addition to this objection, some disadvantages have been reported by Breckenridge and Johnston (1969). For example, the incubation medium inhibits phosphodiesterase activity as much as 90%, since it contains 2 mM

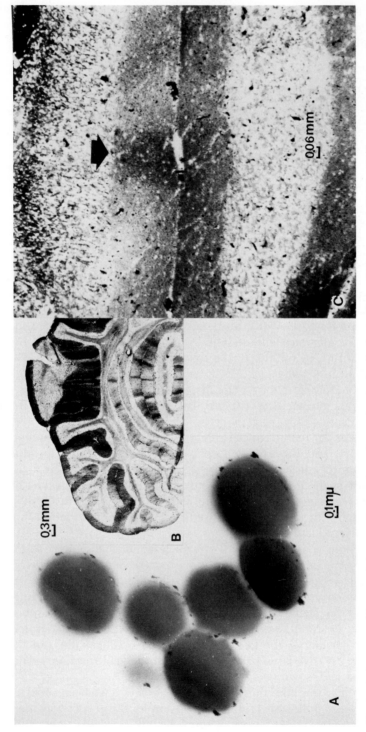

Fig. 2. (A) Latex particles containing reaction product after an incubation in the medium of Shanta *et al.* (1966). (B) Survey of a horizontal section incubated for 5′-nucleotidase after the method of Scott (1965) for 2.5 hr. (C) Detail of a section of the mouse cerebellum incubated in the medium of Shanta *et al.* (1966) for 3′,5′-phosphodiesterase activity. The arrow indicates a bandlike arrangement in the molecular layer, as can be found with the 5′-nucleotidase incubation as well (see the text).

lead acetate. It is unlikely that the snake venom 5'-nucleotidase penetrates homogeneously into the sections, which have been previously dried. Theophylline is not very suitable as an inhibitor of phosphodiesterase. Finally, phosphodiesterase activity decreases during the long incubations that are necessary for the reaction of Shanta *et al.* (1966).

In an attempt to localize phosphodiesterase at adrenergic nerve endings, Gerebtzoff and Ziegels (1974) concluded that the localization was imprecise at the light-microscopic level. They indicated that the lead precipitate was found near and not at the site of phosphate liberation and that the liberation of phosphate by 5'-nucleotidase gave a diffuse appearance in the cerebellar molecular and granular layer. The technique of Shanta *et al.* (1966) is therefore considered unreliable.

Looking more critically at this type of reaction, and assuming that, for example, a whole cerebellum is incubated, interference can occur due to endogenous 5'-nucleotides. The endogenous amounts of 5'-nucleotides, determined by different methods (e.g., enzymatic analysis, salt precipitation, and chromatography), are averaged and presented in Table 1.

The average wet weight of the mouse cerebellum is 75 mg (Marani, 1979), and an estimate of the total amount of 5'-nucleotides present would be nearly 0.1 nmol. Assuming that the added 5'-nucleotidase from snake venom can convert all endogenous 5'-nucleotides (and the ones newly liberated by 5'-exonuclease as well), and knowing from a conversion of the result of Breckenridge and Johnston (1969) that the newly liberated 5'-nucleotides are released at a rate of 4.3 nmol/min, then it follows that the 5'-nucleotides are responsible in the first minute for 2% of the amount of lead precipitate. When an inhibition of 90% of the phosphodiesterase activity by lead ions is also taken into account, then in the first minute, 20% of the captured phosphate is formed by endogenous 5'-nucleotides, or after an incubation of 20 min, 1% of the total captured phosphate is still due to endogenous 5'-nucleotides. Since a residual activity of acid phosphatase of nearly 30–40% of the optimum activity at pH 7.2 in the mouse cerebellum is present, this activity can convert 5'-nucleotides liberated by phosphodiesterase activity and introduce another error in localization.

Since, for the whole cerebellum, this optimum acid phosphatase activity is nearly 200 IU/liter at pH 5.0, its residual activity will therefore be (at pH 7.2) 60 IU/liter. At higher pHs, an additional activity from alkaline phos-

Table 1. Mean Amounts of 5'-Nucleotides in Brain Tissue[a]

5'-Nucleotide	Amount (nmol/g wet wt.)
AMP	0.956
GMP	0.08
UMP	0.08
IMP	0.08

[a] From Heald (1960).

phatase can be expected (Marani, 1979), as well as a decrease of the acid phosphatase activity. In the mouse cerebellum, a rostrocaudal pattern is present consisting of 5'-nucleotidase-positive and -negative bands (Fig. 2B). This means that an uneven distribution of endogenous 5'-nucleotidase is present, and that its measured activity for the whole cerebellum can vary around 300 IU/liter, dependent on the extraction method used in the 5'-nucleotidase procedure (Marani, 1979).

The activity of 5'-nucleotidase from *Crotalis atrox* is 2.963×10^{-3} IU/min per mg, when 1 mg/10 ml of this snake venom is added to the incubation medium. This results in an activity of 300 IU/liter in the incubation medium due to the inhibition by lead ions.

The following scheme summarizes the enzymes involved in this 5'-exonuclease reaction and their activities:

$$3',5'\text{-cAMP} \longrightarrow 5'\text{-AMP} \longrightarrow A + (PO_4)^{3-}$$

↑	↑
5'-Exonuclease	Endogeneous 5'-nucleotidase
	(300 IU/liter)
(300–430 IU/liter)	Snake venom 5'-nucleotidase
	(300 IU/liter)
	Residual acid phosphatase activity
	(60 IU/liter)

The high activity of endogenous 5'-nucleotidase competes with the added snake venom 5'-nucleotidase for the 5'-nucleotides liberated by 5'-exonuclease. This means that under certain conditions (an equal distribution of phosphodiesterase, phosphodiesterase in the same location as endogenous 5'-nucleotidase, or a diffusion of 5'-nucleotides because the activity of snake venom is low compared to the activity of phosphodiesterase), one observes the distribution of endogenous 5'-nucleotidase rather than the exonuclease phosphodiesterase. Sections incubated for endogenous 5'-nucleotidase (according to Scott, 1965) (see Figure 2B) exhibit a band pattern in the molecular layer of the mouse cerebellum quite similar to the pattern exhibited by the 5'-exonuclease (= 3',5'-phosphodiesterase) incubation (according to Shanta *et al.*, 1966) (see Figure 2C) (see also Marani and Boekee, 1973). The 5'-exonuclease technique of Shanta *et al.* (1966) can therefore be considered as a contested method. If it is used, extreme caution is needed and care must be taken to provide for controls, including those for endogenous enzymes and substrates.

In the γ-aminobutyric acid (GABA) pathways, there are several enzymes that can be shown by these multistep reactions. GABA-transaminase (Hyde and Robinson, 1976) is demonstrated by the following procedure:

I. α-Ketoglutarate + GABA ⟶⌐
  └→ glutamate + succinic semialdehyde
      ↑
  GABA transaminase

I. Succinic semialdehyde + NAD$^+$ + H$_2$O → succinate + NADH + H$^+$

Succinic semialdehyde dehydrogenase

I. NADH + PMS$^+$ → PMSH + NAD$^+$

I. PMSH + nitro-BT → nitro-BTH$_2$ (formazan) $\updownarrow$ + PMS$^+$

Using a specific inhibitor for GABA-transaminase (aminooxyacetic acid), the same location of formazan deposits was found as without an inhibitor, indicating that the technique in our hands was less than reliable (Vosseberg, unpublished results).

An analogous mechanism is used to demonstrate glutamate decarboxylase (Higashi *et al.*, 1960).

2.2. Oxidoreductases

This class of enzyme includes the dehydrogenases, diaphorases, oxidases, and peroxidases. The accuracy of dehydrogenase demonstration and the number of dehydrogenases that could be demonstrated increased due to progress in the development of tetrazolium salts in the 1950s. As a result, (nitro-BT), tetranitro-BT, MTT, and half-TNBT became available (Tsou *et al.*, 1956; Pearse, 1954; Seidler and Kunde, 1969).

While progress in the dehydrogenase field depended on a suitable electron or hydrogen acceptor, the oxidase and peroxidase techniques progressed following the discovery of a suitable hydrogen donor. Seligman *et al.* (1968, 1973) provided breakthroughs in the field of oxidase location by the discovery of the aminobenzidines. The combination of the diaminobenzidine peroxidase technique, the revelation of the uptake of horseradish peroxidase (HRP) in boutons (Holzman *et al.*, 1973), and the consequent retrograde transport (La Vail and La Vail, 1972) in nervous tissues yielded an explosive impulse to the study of connections in the central nervous system (e.g., Kuypers, 1974).

2.2.1. Oxidases and Peroxidases

The enzymes that react with oxygen can be classified into three subdivisions: oxidases, oxygenases, and hydroxylases. Of these, only the oxidases have received any attention in the field of enzyme histochemistry. The oxidases themselves can be further classified into subgroupings. Only some of these subgroupings contain more than one enzyme that has been demonstrated histochemically. It is doubtful that some enzymes such as xanthine oxidase, diaminooxidase (= histaminase), and D'-amino-acid oxidase can be demonstrated by existing oxidase techniques. Other oxidases such as cytochrome oxidase, peroxidase, catalase, dopa oxidase (tyrosinase), MAO, and uricate oxidase can be reliably demonstrated with existing techniques.

Various types of reactions can be used to demonstrate the localization of the oxidases, but all of them are based on electron transfer. Because of this, these types of reactions are not as easily grouped as are the reactions used to localize the hydrolytic enzymes. Although the grouping of these enzymes will be discussed below, the types used to classify the hydrolytic enzyme-localization reactions will not be used.

2.2.1a. Polymerization Reactions. Polymerization of reaction products was introduced by Seligman and his co-workers and is used, for example, to demonstrate peroxidase or catalase activity. The best-known substrate is 3,3'-diaminobenzidine used in HRP studies. The advantage of the polymerizing diamines lies also in their formation of electron-dense products with OsO_4. Therefore, reactions using diaminobenzidine are highly suitable for electron microscopy.

The same reaction type is involved in the older benzidine reaction studied by Lison (1936). In this reaction, the peroxidatic activity was demonstrated by the conversion of benzidine to a blue or brown product. The blue product is unstable, as is discussed by Straus (1964). As a consequence of this reaction-product instability, many procedural modifications have been investigated, as summarized by Pearse (1972). Of these modifications, the most-accepted methods using benzidine are those according to Van Duyn (1955) or the Wachstein and Meisel (1964) incubation.

2.2.1b. Indolphenol Reactions. The indolphenoloxidase or amine free-radical coupling method is used by cytochrome oxidase localization. Its equation can be described as:

$$I.\ E + S \rightarrow P + E$$

$$I.\ P + X \rightarrow PX \downarrow\ (PZ = PX)$$

The work of Person and Fine (1961) revealed that the only substance that is oxidized in this reaction is *N,N*-dimethyl-*p*-phenylenediamine. This substance is converted to semiquinones, which can form a complex with α-naphthol derivatives to form a colored reaction product. This so-called "Nadi reaction" (Lison, 1936) has evolved further with the use of other substrates. While in the original reaction dimethyl-*p*-phenylenediamine and α-naphthol were used, the first substance was replaced by 4-amino-*N,N*-dimethylnaphthylamine.

The insoluble reaction product that derived from this reaction was called indolnaphthol purple (Nachlas *et al.*, 1958), while the original reaction product was called indolphenol blue. Replacements of the second substrate, α-naphthol, by other naphthols failed (Pearse, 1972). This was attempted because the naphthol inhibits terminal respiratory enzymes and it can produce autooxidation of the diamine (Person and Fine, 1961).

Polymerizing diamines (see Section 2.1.1) are also used in combination with the α-naphthols to produce an insoluble reaction product. The best reagent for the "Nadi reactions" was shown to be *N*-benzyl-*p*-phenylenediamine (Seligman *et al.*, 1967).

2.2.1c. Tyrosinase Reactions. The catecholoxidases or tyrosinases (= dopa-oxidases) can be shown with a typical kind of reaction:

$$\text{I. } E + S + \tfrac{1}{2}O_2 \rightarrow P_1 + E$$

$$\text{I. } P_1 + E + \tfrac{1}{2}O_2 \rightarrow P_2 + E$$

$$\text{I. } P_2 \text{ spontaneous?} \rightarrow P_3 \updownarrow \ (P_3 = PZ = PX)$$

An intensive study and a critical view are given by Van Duyn (1953, 1957*a*,*b*) and Van der Ploeg and Van Duyn (1964). The degree of accuracy of localization in this reaction can be criticized (Pearse, 1972), and the studies by Van Duyn exhibit doubt as to the presence of a single enzyme complex in the formation of melanin. In the enzyme system tyrosine $\xrightarrow{1}$ dopa $\xrightarrow{2}$ intermediate → melanin, tyrosinase probably acts on steps 1 and 2. It is still uncertain whether peroxidase or a peroxidaselike activity is involved in this enzyme complex. Van der Ploeg and Van Duyn (1964) showed that one of the intermediates formed is 5,6-dihydroxyindole. This substance can easily be converted to melanin by peroxidases and H_2O_2, and it can be used as a substrate to localize peroxidase systems (Van der Ploeg *et al.*, 1973).

2.2.1d. Coupled Peroxidase Reactions. The coupled peroxidatic method is known for uricase (= urate oxidase) (Graham and Karnovsky, 1965*a*), but is also used for MAO determination. (For a discussion of the peroxidase reaction, see Section 2.2.1a.)

2.2.1e. Formazan Reactions. Methods based on the reduction of tetrazolium salts for the demonstration of oxidases and all dehydrogenases are in this group. MAO and xanthine oxidase are demonstrated with this method:

$$\text{I. } S + E \rightarrow P + EH$$

$$\text{I. } EH + X \rightarrow E + HX \updownarrow \ (HX = PX = PZ)$$

Only an indolamine oxidized by MAO has the capacity to reduce tetrazolium salts. This reduction is probably not achieved by a diaphorase system. The primary reaction product can reduce the tetrazolium salt directly (Glenner *et al.*, 1960). These results are important because reduction of tetrazolium salts by transferred electrons in the MAO reaction is not mediated by other systems. In other systems, the formazan reaction product indicates the localization of the transferring systems rather than the localization of the studied enzymes.

2.2.1f. Endogenous Peroxidase in Brain Tissue. Since the introduction of the retrograde HRP technique in neuroanatomy (Kuypers *et al.*, 1974; La Vail and La Vail, 1972; La Vail *et al.*, 1973; Kristensson *et al.*, 1971; Kristensson and Olsson, 1971), it has been taken to with such enthusiasm that one may expect renewed inquiries into the localization of the endogenous peroxidase and catalase activities.

A possible disturbing factor for an optimal localization of the HRP end product, on both the electron-microscopic and the light-microscopic level, may be the endogenous enzyme activity precipitating 3′,3′-diaminobenzidine (De Duve and Baudhuin, 1966). Although most authors agree with the conclusion that endogenous peroxidase or catalase or both do not disturb their HRP research (Kuypers *et al.*, 1974; Kievit and Kuypers, 1975, and personal communication), no paper has yet appeared that clearly documents the effects of the use of 3′,3′-diaminobenzidine cytochemistry (see Novikoff, 1973).

In the literature, some remarks on residual activity interfering with the HRP technique are available (Henkel *et al.*, 1975; Marani *et al.*, 1979). Recently, Vosseberg (personal communication) has found that prolonged incubations of more than 1 hr do give artifacts in the HRP technique. These facts indicate that a more fundamental view is needed concerning the localization of endogenous oxidase activity. In the histochemical literature, a distinction is made between catalase and peroxidase activities on the basis of the needed H_2O_2, of the use of the inhibitor aminotriazol (Heim *et al.*, 1955) or DCPIP (P. M. Novikoff and A. B. Novikoff, 1972), of their resistance to fixation, and by shifts in the pH of the buffer (Herzog and Fahimi, 1976). The inhibitors aminotriazol and DCPIP are attacked in the literature as a poor means of distinguishing between catalase and peroxidase (Herzog and Miller, 1972; Roels *et al.*, 1973; Fahimi, 1975). The H_2O_2 concentrations used in HRP studies are not always indicated or cited from the literature, even though the concentration is crucial for the peroxidase reactivity (Herzog and Fahimi, 1976).

Since most HRP sections can have a brownish background, unevenly distributed over the sections, and perhaps caused by diffusion of the HRP or endogenous oxidases, the solubilities of the enzymes involved have to be checked.

In the cerebellum, a peroxidase activity is seldom reported. For example, Arvy (1966) does not mention peroxidase in his chapter on oxidases. This section therefore contains a critical note on peroxidase techniques directed toward results obtained in cerebellar studies.

Solubility of endogenous peroxidase and catalase activities was studied in two series, a transverse one and a sagittal one, using the membrane method of Meier (1972, 1973), and making in each series parallel series without membranes. The difference between the normally incubated sections and the ones on membranes is the overall higher activity in the membrane-treated sections, especially in and around great blood vessels. An identical localization in the comparable series was found, but a higher activity was present in the membrane-treated series. Activity is found in the neuropil of the molecular layer, in Purkinje cells, and in neurons of the cerebellar nuclei. Sometimes, the row of Purkinje cells is disturbed, but this is due to lack of Purkinje cells, as is also found in normal series, rather than to the negativity of Purkinje cells. No regular lattice of these empty places was found. The granular cells are positive, but their nucleus cannot be discerned as a negative spot, as can the nuclei of all other cell types studied in the cerebellum. No positive Golgi cells could be discerned light-microscopically, and the basket cell axons and the infraganglionic plexus are negative. Bergmann's glia seemed to be negative, since they cannot be discerned in the infraganglionic plexus.

By using an excessively high concentration of H_2O_2 (0.5%), the residual activity in $3',3'$-diaminobenzidine cytochemistry can be attributed to catalase (Herzog and Fahimi, 1976; Deisseroth and Dounce, 1970) and artifacts (Novikoff, 1973). The remaining reactivity in these sections is localized at the erythrocytes and greater blood vessels (see Table 2). The fixation results are considered to be in close correspondence with those in Table 2. Even a short fixation in formaldehyde and glutaraldehyde solutions yielded total inhibition of the activity in all structures of the cerebellum except the blood vessels and the erythrocytes. Prolonged incubation in citrate buffer (for storage of sections) diminishes the activity in all structures except for the blood vessels and erythrocytes. From these results, it is clear that the residual activity present after fixation and high concentrations of H_2O_2 belongs to catalase activity, or nonspecific activity. These activities are the disturbing factor in the HRP technique, but are clearly discernible from HRP activity.

Fixations over 1 hr will destroy the main endogenous peroxidase activity in the cerebellum, and H_2O_2 percentages somewhat higher than 0.03% form an additional help for suppressing this activity (see Herzog and Fahimi, 1976; Roels *et al.*, 1973; Wisse, 1970, 1972, 1974).

Considering the results presented in Table 2, it seems unnecessary to use a mixture of glutaraldehyde and formaldehyde. Prolonged fixation with formaldehyde alone inhibits light-microscopically all endogenous peroxidase activity. Prolonged incubations, over 1 hr, after short fixation must be rejected, since $3',3'$-diaminobenzidine reaction products can appear in the cells, disturbing the desired HRP localization.

2.2.2. Monoamine Oxidase

The previous discussion has indicated that MAO determinations can be made not only with the coupled peroxidatic method of Graham and Karnovsky (1956*b*) but also with methods based on the reduction of tetrazolium salts to a formazan deposit (Glenner *et al.*, 1960). The importance of the MAO reaction using tetrazolium salts is due to the fact that the reduction of tetrazolium salts is caused, not by transferred electrons, but rather by a direct action of the products formed (aldehydes or NH_3). The disadvantages of this tetrazolium-localizing MAO technique are the inhibitory action of nitro-BT on MAO, the spontaneous reduction of tetrazolium salts by most of the catecholamines, and the nonspecific binding of nitro-BT to phospholipoproteins in membranes.

Recently, Williams *et al.* (1975*b*) developed a modification of the tetrazolium technique for the demonstration of brain MAO, using preincubation with a high salt concentration to avoid tetrazolium inhibition of MAO. The

Table 2. Effects of Different Fixations and Fixation Times on Endogenous Peroxidase and Catalase Activity[a]

Brain area	Citrate buffer (0.1 M, pH 7.2)			Formaldehyde 2% in citrate buffer			Glutaraldehyde 1.25% in citrate buffer			Formaldehyde 2% and glutaraldehyde 1.25% in citrate buffer				
	\ Time (min)													
	30	60	120	30	60	120	30	60	120	15	30	60	120	240
Blood vessels	+	+	+	+	+	+	+	+	+	+	+	+	+	+
Purkinje cells	+	+	+	−	−	−	−	−	−	−	−	−	−	−
Granular cells	+	+	±	−	−	−	−	−	−	−	−	−	−	−
Molecular layer	+	+	±	+	+	−	+	−	−	+	+	+	−	−
Erythrocytes	+	+	+	+	+	+	+	+	+	+	+	+	+	+
Neurons in the cerebellar nuclei	+	+	+	−	−	−	−	−	−	−	−	−	−	−

[a] The incubation medium contained: 0.05 M Tris, pH 7.6; 10^{-2} M 3',3'-diaminobenzidine; 0.06% H_2O_2. The temperature and time were 25°C and 1.5 hr. Results are from series 9565 and 9566.

method makes it possible to use different substrates, not only indolealkylamines but also phenylalkylamines. Since two types of MAOs are discerned biochemically on the basis of their affinity for 5-hydroxytryptamine (5-HT) and noradrenalin (MAO-A, the inhibitor of which is clorgyline) and for benzylamine and β-phenylethylamine (MAO-B, which is insensitive to clorgyline), this method opens up the possibility of discerning both enzymes histochemically. It seems to be a promising method, because after addition of 5-HT to the incubation medium and high concentrations of clorgyline, a third type of MAO (called MAO-C) can also be discerned histochemically (see Fig. 3) (Williams *et al.*, 1975*a*).

The areas containing this MAO-C seem to lie in brain areas that are localized outside the blood–brain barrier, and the proposed function of this MAO-C is a role in the modulation of the movement and the effects of specific cerebrospinal-fluid or blood monoamines in these areas.

3. Methodology

3.1. Histochemical Procedures

An extensive review of cold-knife and cryostat techniques is given by Pearse (1968). The following description of the methods used concerns experiences in our laboratory, derived mainly from experiments with brain tissue. Brain tissue must be handled with care in both light and electron microscopy, since osmotic changes will occur when brain tissue is not carefully treated before or after the removal of the brain from the cranium. The animals are anesthetized, and for this purpose all kinds of narcotics are used. In the main, chloroform and ether are given to small animals, while larger ones, such as cats and rabbits, are anesthetized intraperitoneally with, for example, barbiturates.

Since autolytic processes start immediately after the blood supply is disrupted, any part of the preparation that can precede decapitation must be performed first. After decapitation, for experiments using light microscopy, the brain is removed as rapidly as possible from the cranium, embedded, and frozen. It is preferable that the whole procedure from decapitation till the start of the freezing be carried out within 5 min for small animals and within 7–8 min for larger ones. Otherwise, there exists the possibility of autolytic changes. When preparations take longer, in our methodology, we usually reject the tissue for use in enzyme histochemistry.

The embedding takes place in a medium that stiffens at low temperature [e.g., Tissue Tek, Ames and Co.; aqua destillata (AD) is also sometimes used (El-Badawi, personal communication)]. It may be necessary to surround the tissue by reference tissue, e.g., liver or kidney. Various methods can be used to freeze samples for the cryostat procedure: gaseous or solid CO_2, or immersion in liquid nitrogen, propane or Arcton (I.C.I.). The use of liquid nitrogen for the cooling of isopropane to $-80°C$ until $-130°C$ is recom-

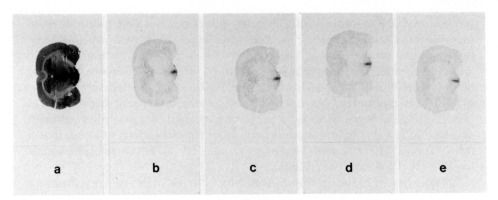

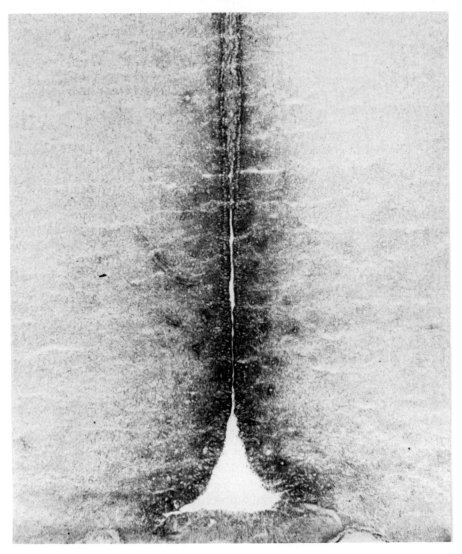

mended because, in this way, freeze artifacts (see Fig. 4) are overcome. If the holder with embedding medium and tissue is placed directly in the liquid nitrogen and not in isopropane, a vapor mantle of nitrogen is created by the relatively hot holder, which interrupts the heat exchange among holder, tissue, embedding medium, and liquid nitrogen. On the other hand, overly rapid freezing may still occur with the recommended method, but can be overcome by using talcum powder in nitrogen–propane techniques (Winckler, 1970).

Carbon dioxide freezes tissue slowly, and ice crystals will form if the tissue specimen is thicker than 1 cm (Silver, 1974). The tissue is placed in a cold microtome or cryostat, and after the temperature of the embedded tissue equals the temperature in the cryostat, sections of the desired thickness are made. Sections are intercepted on a clean glass slide (use alcohol and acetone and dry). Since the freezing technique can change the positioning of the tissue, and the cutting direction of the knife can overthrow a cutting direction perpendicular to the assumed tissue direction, and since reference lines are difficult to establish in the embedding medium or in soft, unfixed tissue, reconstruction can meet with severe difficulties afterward (See Marani, 1978). The sections are placed into the incubation medium for the desired time, with or without prefixation, depending on the enzyme studied. Sections on glass slides are preferably placed vertically into glass jars, thus preventing free reaction products from adhering to the sections. The incubation medium is stirred slowly during incubations of a large number of sections to obtain good exchange between the incubation medium and the sections. If a color reaction is to be carried out after the incubation, this is usually done after the sections are washed in a buffer or in AD, and on occasion after fixation. After fixation, sections are covered with an embedding medium. The type of embedding medium (e.g., glycerine gelatin) is dependent on the reaction carried out.

Research at the ultrastructural level requires better preservation of the submicroscopic structures of the brain tissue. Therefore, brains are normally prefixated by perfusion or, sometimes, by immersion fixation. The fixation solutions normally contain formalin or glutaraldehyde or both. Freed from the cranium, the brain is prepared and sectioned on a vibratome or a tissue-chopper to sections of 50–250 µm. The tissue-chopper method necessitates embedding in gelatin or alginate (Lewis and Shute, 1963). If necessary, immersion fixation with the same solutions may precede or follow incubation.

After the incubation for electron-microscopic procedures, the tissue sections are washed in buffer. No color reaction is carried out; as in most of the reactions developed for electron microscopy, the final reaction product

---

Fig. 3. (*Top*) MAO activity in coronal sections of rat brain with 5-HT as substrate. (a) Distribution of MAO-A activity; (b–e) MAO activity after inhibition with clorgyline: (b) 10^{-7} M; (c) 10^{-6} M; (d) 10^{-5} M; (e) 10^{-4} M. MAO-C activity is found in the region of the hypothalamus and third ventricle. ×2. (*Bottom*) MAO-C activity in the vicinity of the infundibular recess. The section was exposed to 10^{-5} M clorgyline for 5 min prior to incubation in 5-HT–NBT medium. ×480.

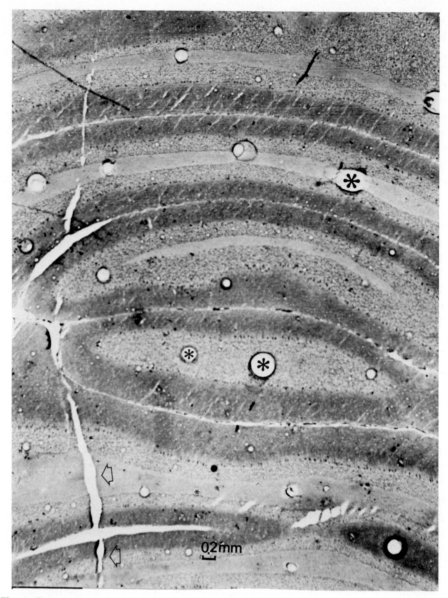

Fig. 4. Transverse section through the cerebellum of the cat. This section was incubated for glucose-6-phosphatase (Pearse, 1968). Freeze artifacts are indicated. Cracks are indicated by outline arrows, and holes by asterisks.

is electron-dense rather than colored. Hereafter, the routine procedure for electron microscopy follows that outlined in Volume 2, Chapter 7. Ultrathin sections are cut on an ultratome and are stained with Pb or U ions. Sometimes, the reaction product consists of Pb ions; in such cases, the use of sections that have not been treated with Pb ions is recommended.

The comparisons given in Table 3 of the methods for electron and light microscopy clarify a substantial number of the differences between the methods. While light microscopy must be concerned with autolytic processes, electron microscopy tries to prevent these processes by the use of perfusion fixation. A better preservation of the structure is attained with electron microscopy, but it does have the great disadvantage that fixation is necessary before any enzyme activity can be detected. Since a number of enzymes fail to react following fixation, this means that the total number of enzymes that can be localized on the ultrastructural level is less than with light microscopy. Add the consequence that not all enzymatic procedures give an electron-dense reaction product, and it is clear that the scope of electron-microscopic enzyme histochemistry is far narrower than that of light-microscopic enzyme histochemistry.

After incubation, dehydration and embedding must still be performed in electron-microscopic procedures, and even though the reaction product is captured and localized near the enzyme location, some diffusion may still occur.

The enumerated pros and cons of the electron-microscopic procedure as contrasted with the light-microscopic method deserve extra attention because a distinction at a lower morphological level, with the chance of a bad reaction product, can lead one to false conclusions.

Table 3. Summary of Procedures in Light Microscopy and Electron Microscopy for Enzyme Histochemistry

Light microscopy	Electron microscopy
Anesthetizing	Anesthetizing
	↓
	Fixation
	↓
↓	Preparation of brain tissue
Preparation of brain tissue	↓
↓	Sectioning on a vibratome or tissue-chopper
Sectioning on a cryostat	↓
↓	Prefixation (if necessary)
Prefixation (if necessary)	↓
↓	Incubation
Incubation	↓
↓	Fixation (OsO$_4$)
Fixation (formalin)	↓
↓	Dehydration
Covering glass slides	↓
	Embedding in Epon
	↓
	Ultratome sectioning
	↓
	Coloring/Pb and/or U ions
↓	↓
Examination	Examination

3.2. Lesion Techniques

3.2.1. Mechanical Lesions

One of the experimental tools used in brain research is the lesion technique. By mechanically interrupting the connections between neurons, some conclusions can be drawn concerning connections in the brain. Degeneration may be followed not only by argentophilia of boutons and axons, or by changes in the neurons, e.g., retrograde chromatolysis (Brodal, 1940), but also by changes in enzyme activity.

Table 4 shows a series of research experiments in which effects are enumerated that can occur after lesions of the cranial or peripheral nerves. Although compiled mainly from the work of the group of Shute and Lewis on acetylcholinesterase (AChE), it also shows, in particular, that recovery of the affected areas for this enzyme is possible. The same holds for butyryl-cholinesterase (BuChE) activity, as demonstrated for the cells of the hypoglossal nerve of the rat after a recovery time of 30 days. The opposite may also occur. After vagotomy at the cervical level (see Table 4), an increase in acid phosphatase activity in the dorsal motor nucleus X is found. The activity returns to normal after some time. Return of the affected cells to a normal enzymatic pattern is frequently found after the cutting of spinal roots, vagotomy, or hypoglossotomy.

A rather peculiar situation occurs in the hypoglossal area. After hypoglossotomy, Davidoff (1973) found an increase in 5′-nucleotidase activity in the hypoglossal nucleus. This activity is bound to cells that are clearly smaller than the hypoglossal nerve cells. The activity is attributed to glial cells by Davidoff (1973). While the hypoglossal neurons recover from the lesion by increasing their cholinesterase activity, simultaneously the 5′-nucleotidase in the glial cells decreases.

After lesions of the trigeminal nerve branches, Rustioni *et al.* (1971) found an acid phosphatase topical localization in the substantial gelatinosa Rolandi for the ophthalmic, maxillary, and mandibular branches of the trigeminal nerve. Studies by Coimbra *et al.* (1974) have revealed that the acid phosphatase activity is localized in C terminals between vesicles, as well as in the hyaloplasm of the peripheral elements of the glomeruli. Dorsal root sections caused complete disappearance of acid phosphatase from the peripheral profiles and the C terminals. This type of degeneration (Knyihar and Csillik, 1976) is also followed by a regenerative proliferation in the substantial gelatinosa (Csillik and Knyihar, 1975).

3.2.2. Chemical Lesions

Administration of chemical substances can affect specific brain areas. The most recently developed methods use chemical lesions for the selective destruction of the inferior olive by 3-acetylpyridine (Desclin, 1974, 1976; Desclin and Escubi, 1974). 3-Acetylpyridine is a nicotinamide antagonist that

Table 4. Effects of Mechanical Lesions

Authors	Lesion	Location	Effects[a]	Time
Navaratnam and Lewis (1975), Lewis *et al.* (1970).	Vagotomy	Cervical	Atrophy of BuChE-containing neurons in the dorsal motor nucleus of the vagus nerve	Within 3 weeks
			Reversible AChE loss in the dorsal motor nucleus of the vagus nerve	Within 3 weeks
			Reversible AChE loss in the area subpostrema	Within 3 weeks
			Reversible AChE loss in the ambiguus nerve	Within 3 weeks
			Increase of AP in the dorsal motor nucleus of the vagus nerve	After some time again normal
			Increase in AP activity in the ambiguus nerve	Within 3 weeks
Hughes and Lewis (1961)	Leg damage	Sciatic nerve	Increase in AChE in nerve, approaching the spinal cord with a velocity of 0.5 mm/day	A few hours
Flumerfelt and Lewis (1975)	Hypoglossal nerve	Cut at branching near the digastricus muscle	Disappearance of AChE activity from cells of the hypoglossal nerve	Within 2 days; normal after 20 days
			Disappearance of BuChE activity from cells of the hypoglossal nerve	Within 2 days; normal after 20 days
Filogamo and Candiollo (1962)	Right plexus brachialis	Near primary branching	AChE activity lost at C_5–C_7 in the cells and neuropil of the dorsal horn	Within 2–6 days; normal after 60 days
Shute and Lewis (1967)	Lesion in the dorsal tegmental pathway	Lesion via midbrain tegmentum	Loss of AChE activity in: ipsilateral posterior inferior colliculus, deep layers of anterior superior colliculus, superficial and deep pretectal nuclei, intralaminar nuclei of thalamus	After 18 days
	Lesion in the ventral tegmental pathway		Loss of AChE in: Caudate nucleus	After 6 days
			Ipsilateral No. oculomotor nerve	After 20 days

[a] (BuChE) Butyrylcholinesterase; (AChE) acetylcholinesterase; (AP) acid phosphatase.

may have effects on mice, rats, rabbits, and cats. In rats, especially, a total destruction of the inferior olivary complex and an almost complete destruction of the ambiguus nerve were observed, while facial and hypoglossal nuclei were only partially affected. All the enumerated nuclei, except the inferior olivary complex, are thought not to play a part in providing the cerebellum with climbing fibers (Desclin, 1974). In this sense, a specific destruction of climbing fibers is effected. Another example of selective lesions by chemical compounds is the destruction of certain brain areas by monosodium glutamate. In mice, rats, chicks, guinea pigs, hamsters, and monkeys, a degenerative effect of glutamate has been reported. The pattern of damage tends to parallel the distribution of circumventricular organs (Olney, 1969). These brain regions are all areas in which fenestrated endothelial walls are present and in which a local entry from the blood into the central nervous system is possible. Recently, the neurotoxic effects of glutamate on mouse area postrema, also belonging to the circumventricular organs, has been described (Olney *et al.*, 1977).

The effects of 6-hydroxydopamine (6-OH-DA) on catecholamine-containing systems is due mainly to a long-lasting depletion of noradrenalin-containing boutons (for a survey of the 6-OH-DA effects, see Malmfors and Thoenen, 1971). Administration of 6-OH-DA does affect central brain areas such as the locus coeruleus. The cells of this nucleus are destroyed after treatment with 6-OH-DA (Lewis and Schon, 1975). The cells degenerate, and this process may be followed by their subsequent loss of AChE activity. Cells of other nuclei in the vicinity of these coeruleus cells, which are not cholinergic yet contain AChE, are not affected, which shows the selectivity of 6-OH-DA treatment for this area.

Lesion techniques thus used in neurobiology can serve as a direct help for enzyme histochemistry to understand localizations of enzymes found in certain brain areas. Since one of the consequences of the degeneration of terminals found in, for example, trigeminal sections is a cessation of the resupply of axoplasmic enzyme proteins, including enzymes such as acid phosphatase in the substantia gelatinosa Rolandi, and the subsequent regenerative process, a way of studying axoplasmic flow by means of enzyme histochemistry is given.

3.3. Premortem and Postmortem Damage to Brain Tissue

From the moment neurobiologists started to study the brain, the problem of "dark cells" had to be resolved. An excellent review has been written by Cammermeyer (1962), who explained the dark cells as an artifact caused by the cell body's shriveling. Although Cammermeyer does not specifically mention hyperosmolality of the surrounding fixatives or dehydration medium, this is his main explanation. Another group of authors (Friede, 1964; Stensaas *et al.*, 1972) pointed to postmortem effects and premortem traumatizations. It seems that careless treatment of the brain causes the appearance of dark cells even up to 2 or 3 hr after a perfusion fixation. Dark cells (see Fig. 5) are characterized by a dark outer strip and shriveling, causing a light halo that is clearly present in light microscopy. In sections prepared for electron microscopy, a dark matrix is present throughout the whole perikaryon and its offshoots. The intercellular space is increased, and it appears as though the cell itself has withdrawn from the surrounding neuropil and cells (Fig. 5). According to the theory of Friede (1964), these cells possess a different microcellular osmolality.

In brain development, certain areas can be discerned that contain a high concentration of degenerating neuronal cells. These cells are also characterized by a dense matrix, but they differ slightly from dark cells in respect to the concentration of ribosomes, a loosening of cell contacts, and a dense karyoplasm (Maruyama and Agostino, 1967; Schlüter, 1973; O'Connor and Wyttenbach, 1974). In mature brains, it is difficult to discern the experimental degeneration that is caused in offshoots, since the same darkening of the matrix as in dark cells in early stages is produced (Alksne *et al.*, 1966).

Damaged cells may appear hyperchromic under both electron and light microscopy regardless of whether the damage is the result of genetic factors, experimental degeneration, premortem stress, or postmortem deterioration. However, nerve cells damaged around the time of death (as in the last two examples in the preceding sentence) should not differ enzymatically from surrounding, undamaged cells, since the brain is fixated at approximately the same time, i.e., just before or just after the traumatization. Cells damaged because of genetic factors or experimental degeneration would be expected to differ enzymatically from undamaged cells and should therefore be discriminable from cells damaged around the time of death. As demonstrated in Table 4 and discussed in Section 3.2.1, experimentally degenerated cells do show a pattern of enzyme activity different from that of the surrounding tissue. Contrary to the expectation discussed above, cells that are hyperchromic following traumatization around the time of death do exhibit some differences in enzyme pattern when compared to undamaged cells. Pyknomorphic, chromophilic neurons demonstrate a strong phosphorylase activity in the cell body and dendrites, as described, for example, by Friede (1959) for Purkinje cells. A loss of succinic dehydrogenase after traumatization or damage in pyramidal cells and cells of the nucleus cochlearis has also been reported. Furthermore, an increase in alkaline phosphatase in cerebral cortical cells has been reported (Miller, 1949). It seems obvious that careless handling of laboratory animals before, during, or after experiments can induce hyperchromic nerve cells that can react in unexpected ways in enzyme-histochemical procedures.

3.4. Ischemia

According to Petresco (1958), chronic ischemia produces hyperchromic neurons (see also Section 3.3). Ischemia may arise from general failure of the circulation, from local occlusion of blood vessels, or from a combination of both causes (Dixon, 1965). Ischemia includes lack of oxygen, anoxia,

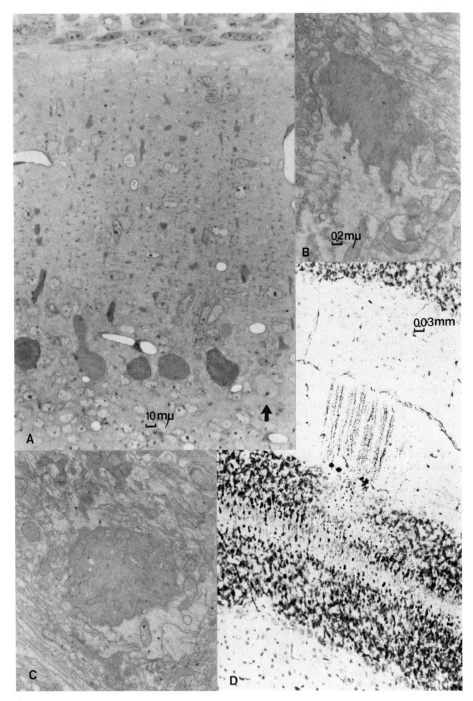

Fig. 5. (A) Transverse cerebellar section (1 μm) colored with toluidine blue. Dark Purkinje cells are seen between the granular layer and the molecular layer. The arrow points to a normal Purkinje cell. (B, C) Electron micrographs of Purkinje cell dendrites that belong to dark Purkinje cells. Note the dark cytoplasm and the shrinkage of the dendrites as compared to their

deprivation of glucose, and accumulation of potassium ions. Anoxia causes loss of mitochondrial enzymes (Becker, 1961; MacDonald and Spector, 1963) and a decrease of succinic dehydrogenase and cytochrome oxidase (Spector, 1963). These decreases in enzyme activity persist up to 24 hr after the end of a transient anoxic period. A loss of reductase activity exists 24–72 hr after the period of anoxia [this holds for both NADH and NADPH tetrazolium reductases (Becker, 1961, 1962; Becker and Barron, 1961)], as does a loss of malic, lactic, and glutamic dehydrogenase and ATPase activity (MacDonald and Spector, 1963).

In electron microscopy, but also in light microscopy, anoxia may cause ultrastructural changes (Hills, 1964a,b; Dixon, 1965). The extensive changes in anoxia that may occur both in enzyme activity and in the subcellular structure justify a careful setup of experiments for enzyme histochemistry and electron microscopy (see also Palay and Chan-Palay, 1974). Other forms of ischemia, in the main, must be induced experimentally and do not interfere with normal experimental procedures.

3.5. Fixation

Humason (1972) states:

> The use of fixed tissue for enzyme demonstration is debatable, but cutting frozen unfixed tissues may result in a greater loss of enzyme into the incubating medium and more cell damage than if fixed tissue is used.

The dilemma of fixation in enzyme histochemistry has been condensed by Humason (1972) in a nutshell. Although different techniques have been proposed for the fixation of enzymes without disturbing their activity, all the proposals may be criticized. Many enzymes will tolerate some exposure to fixatives, but others are not able to withstand their action. Aspartate aminotransferase (formerly glutamicoxalacetate transaminase) has been shown by Lee and Torack (1968) to be destroyed by 1% glutaraldehyde and by 4% formalin. Still, aspartate aminotransferase may be localized by using light formalin fixations (Martines-Rodriguez *et al.*, 1976), while acetone fixation has only a minimal effect on this enzyme (Lee and Torack, 1968).

Calcium formalin (Baker, 1944) is used at low temperatures (4°C) and is recommended for the preservation of many enzymes. Nevertheless, formaldehyde in rat brains has been shown to be a bad fixative for lipids (Deierkauf and Heslinga, 1962). This must mean that lipid-associated enzymes could be easily extracted if formalin fixation is used.

Glutaraldehyde was introduced as a fixative by Sabatini *et al.* (1964), and when used in a phosphate buffer can be tolerated by some enzymes; yet others are totally destroyed by this fixative (see Section 3.5.1). A combination of formalin and glutaraldehyde fixation is frequently used in electron-microscopic neuroenzyme histochemistry. As already pointed out, the inhi-

---

surrounding. (D) True degeneration of Purkinje cells in a transverse cerebellar section. The section was colored for normal degeneration studies (Fink and Heimer technique). Both Purkinje cell somata as well as the dendritic trees are impregnated with silver.

bitory action of these fixatives on enzymes may be severe and reduces the total amount of demonstrable enzymes for electron microscopy.

In light microscopy, prefixation frequently occurs with absolute acetone at low temperatures, but Burstone (1958) warned against its use in parenchymatous tissue, since it may cause considerable distortion. The acetone method was introduced by Gomori (1952) and has been discussed by Novikoff *et al.* (1960). In their comparison of the cold-acetone and cold-formol calcium fixation, they preferred cold-acetone fixation for light microscopy because enzyme activities for succinate, lactate, malate, β-hydroxybutyric, glutamic, and α-glycerophosphate dehydrogenases were only slightly changed. The compromise between good morphological and good enzyme-activity preservation is difficult, because each enzyme reacts differently in its environment dependent on the fixatives used. The appropriate fixation procedure can be found mainly by trial and error. But, generally, one must keep in mind that some enzymes cannot withstand fixation at all.

3.5.1. Effects of Glutaraldehyde

A review by Hopwood (1972) showed that the effect of prolonged glutaraldehyde fixation is time-dependent. Glutaraldehyde fixation, for example, has an inhibitory action on acid phosphatase that increases with the fixation time. Not all enzymes behave in the same way in their sensitivity to glutaraldehyde. Nearly total inhibition is found for Na^+- and K^+-activated ATPase and 5′-nucleotidase. Heavily affected (more than 50% reduction of the enzymatic activity after 1 hr of fixation or less) are the lysosomal enzymes, acid phosphatase and arylsulfatase; isocitric dehydrogenase, β-hydroxybutyrate dehydrogenase, and lactate dehydrogenase; and of the oxidoreductases, catalase and peroxidase. Slightly inhibited (less than 50% reduction of the activity after 1 hr of fixation or less) are alanine aminotransferase and cholinesterase (Hopwood, 1972).

A general rule for glutaraldehyde treatment of enzymes can be proposed: use short perfusion fixations for glutaraldehyde in combination with perfusion washing; eventually, use both procedures with substrate protection, since washing does increase the enzymatic activity and since perfusion allows for short fixations in comparison to immersion fixation. Sometimes, substrate protection does help the enzymatic site to survive damage by fixatives (see Hopwood, 1972). For more information on the significance and effects of glutaraldehyde fixation, consult Hayat (1970) and the special fixation issue of the *Histochemical Journal* (Volume 4, 1972).

3.5.2. Effects of Other Fixatives

Formaldehyde diffuses rapidly and fixes the tissues slowly as compared to glutaraldehyde, which penetrates slowly but fixes rapidly (Hayat, 1970). In comparison with glutaraldehyde, formaldehyde gives a much better preservation of enzyme activities. This is probably due to the slow fixation rate of formaldehyde.

Paraformaldehyde is preferred to formaldehyde because the latter can be impure. This impurity can affect or interfere with enzyme localization. Acrolein gives good morphological preservation, and enzymatic activity is well protected. On the other hand, acrolein destroys the enzymatic activity of both alkaline phosphatase and ATPase (Hayat, 1970). Hydroxyadipaldehyde is used for the demonstration of unstable enzymes such as cytochrome oxidase, glucose-6-phosphatase, and succinic dehydrogenase, although ultrastructural morphology is poorly preserved with this agent (Hayat, 1970).

3.5.3. Perfusion Fixation

The technique of perfusion fixation is used to aid histological fixation in electron-microscopic histochemical studies. Perfusion is defined as the passage of fluid or blood through the vascular bed of an organ (Ross, 1972). Vascular perfusion has often been used to obtain satisfactory fixation for histological purposes. Although a most ingenious apparatus has been developed for biochemical purposes (Ross, 1972), a rather simple method can be chosen for fixation perfusion. An extensive description of the perfusion procedure can be found in Chapter 12 or the volume *Cerebellar Cortex* (Palay and Chan-Palay, 1974).

In addition to the method of suspending flasks of fixative with the flow rate determined by atmospheric pressure, it is also possible to use a peristaltic pump that provides adjustable rates of flow. Differences in pump speed and in the diameter of the silicone rubber tubing can be used to obtain the desired flow rate. Although the pressure is difficult to measure, the needed outflow is easily adjusted for the kind of animal one is working on (see Ross, 1972). Excessive pressure or outflow affects the tissue treated. The most easily recognizable features of this phenomenon are distended blood vessels and abnormal packing of other tissue elements near the vessels. Excessive or prolonged high pressure affects the brain in a way similar to that found in hypertensive encephalopathy. Its most common pathological feature is edema. Sometimes, fixation perfusion is preceded by saline perfusion, in which case edema may also occur.

4. Topographic Enzyme Histochemistry

4.1. Acetylcholinesterase in the Cat Cerebellum

Scott (1964, 1965) was the first to report that the enzyme 5′-nucleotidase was confined to parallel longitudinal bands within the molecular layer of mouse and rat cerebellum. This arrangement has been confirmed by Marani and Voogd (1973), and in addition, Marani (in prep.) reports an identical distribution for this enzyme in the shrew (*Sorex araneus*) and the vole (*Cletrionomys glarolus*). Similarly, Ramon-Moliner (1972) provides photographs that are suggestive of a bandlike deposition of acetylcholinesterase (AChE) in the molecular layer of the cerebellum of 4-month-old cats. The summary in this section gives the distribution pattern of AChE in the cat cerebellum. The functional significance of the biochemical gradients of this enzyme within the cerebellar cortex is also discussed. In our study, immature and adult cats were used in studying AChE activity (Table 5) (see Marani and Voogd, 1977). The band pattern could be demonstrated with both of two histochemical methods used. In 19 cats, 2–4 months old, the pattern was present in the vermis of both the anterior and the posterior lobes. In older animals, the pattern was not observed.

The molecular layer throughout all lobules of the anterior lobe contained symmetrically disposed acetylthiocholinesterase (AthChE), positive areas alternating with narrow negative zones (Fig. 6). Transverse sections indicate that the AthChE activity is present in narrow striations reaching from the somata of the Purkinje cells to the pial surface (Fig. 6). The somata of Purkinje cells are negative for AthChE even at the base of AthChE-positive areas in the moecular layer (Fig. 6). The neuropil of the granular layer is strongly AthChE-positive, but the grandular cells themselves are negative. No activity is found in the white matter of the cerebellum.

The margins of a midline positive area are ill defined. The area consists of some closely packed vertically arranged striations. Negative spaces between the striations accumulate laterally to form a narrow negative band. Lateral

Table 5. Distribution Pattern of Acetylcholinesterase in the Cat Cerebellum

Age of cat	Kind of reaction[a]	Parallel series with inhibitor[b]	Concentration (M)	Incuba-tion time (hr)	IO pattern[c]	Cerebellar pattern
1–2 days	AChE—K	iso-OMPA	$10^{-3}, 10^{-4}, 10^{-5}$	$1\frac{1}{2}$	−	−
25 days	AChE—K	iso-OMPA	$10^{-3}, 10^{-4}, 10^{-5}$	$1\frac{1}{2}$	±	−
7 weeks	AChE—K	iso-OMPA	10^{-6}	$1\frac{1}{2}$	+	±
7 weeks	AChE—K	iso-OMPA	10^{-6}	$1\frac{1}{2}$	+	+
7 weeks	AChE—K	iso-OMPA	10^{-6}	$1\frac{1}{2}$	+	+
9 weeks	AChE—K	iso-OMPA	10^{-6}	$1\frac{1}{2}$	+	+
12 weeks	AChE—K	iso-OMPA	10^{-3}	$1\frac{1}{2}$	+	+
4 months	AChE—K	—	—	$1\frac{1}{2}$	+	+
4 months	AChE—G	—	—	3	+	+
4 months	AChE—G	—	—	3	+	+
4 months	AChE—K	—	—	$1\frac{1}{2}$	+	+
4 months	AChE—K	iso-OMPA	10^{-5}	$1\frac{1}{2}$	+	+
4 months	AChE—K	FDP	10^{-6}	$1\frac{1}{2}$	+	+
4 months	AChE—K	iso-OMPA	10^{-6}	$1\frac{1}{2}$	+	+
4 months	ButhChE—K	—	—	$1\frac{1}{2}$	−	−
4 months	AChE—K	Eserine	$10^{-3}, 10^{-4}, 10^{-5}$	$1\frac{1}{2}$	−	−
Mature	AChE—K	iso-OMPA	$10^{-3}, 10^{-4}, 10^{-5}$	$1\frac{1}{2}$	+	−
Lesions						
4 months	AChE—K	Cerebellum	—	$1\frac{1}{2}$	+	
4 months	AChE—K	Cerebellum	—	$1\frac{1}{2}$	+	
4 months	AChE—K	Gracilus	—	$1\frac{1}{2}$	+	
4 months	AChE—K	Cun. int. + grac.	—	$1\frac{1}{2}$	+	
4 months	AChE—K	Cun. int. + grac.	—	$1\frac{1}{2}$	+	
4 months	AChE—K	Cun. int. + grac.	—	$1\frac{1}{2}$	+	

[a] (K) Method of Karnovsky and Roots (1964); (G) method of Gomori (1952).

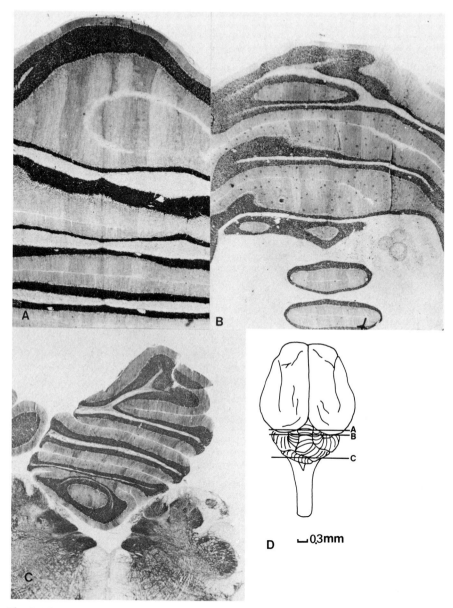

Fig. 6. Three transverse sections through the cerebellum of the cat after incubation for AChE. The sections are indicated on a dorsal view of the cat brain (D). (A, B) Sections through the anterior lobe; (C) section through the posterior lobe vermis. Note the striations in (A) and the negative Purkinje cells at the transition from the granular layer (nearly black) and the striated molecular layer. Section (B) shows the diverging path the bands take when going from lobules II and III (*bottom*) to lobule V (*top*). In (C), the positive bands are small compared to the less positive ones in lobules IX through VIII, but tend to widen in the sublobules of VII (*top*).

to these negative bands, symmetrically disposed positive bands are present. The transition between these positive areas and the medial negative band is gradual. On the lateral side, the border adjoining a second negative band is particularly sharp. Often, this lateral border is located opposite a constriction of the white matter that corresponds to the lateral border of the medial compartment A of the white matter (Voogd, 1964, 1967, 1969). The bands broaden progressively as they move laterally from lobule I to lobule V in the anterior lobe (Fig. 7). Consequently, both negative and positive bands become wider in the dorsal part of the anterior lobe, but the two negative bands never attain the width of the adjoining positive bands. Together, the three positive and four negative bands in the medial part of the anterior lobe occupy less than one third of its total width, i.e., are located in the region commonly designated as the vermis of the anterior lobe.

In sagittal sections through the anterior lobe, the distribution of AthChE activity is more uniform. The striations are either absent or much wider and less distinct as compared with transverse sections. Differences in overall enzyme activity in different parts of a folium or among different folia can be observed, but the transition between the positive and negative areas is always gradual.

In the posterior lobe, the bands have the same general appearance as in the anterior lobe. The contrast between positive and negative bands is less obvious. The Purkinje cell bodies are negative for AthChE activity. The peculiar stripes are present in both negative and positive areas. These bands are limited to the posterior lobe vermis (Fig. 8). Distinct bands could not be distinguished in the hemisphere. The AthChE activity in the posterior lobe vermis may be limited to bands corresponding to the A raphe (Voogd, 1964, 1967, 1969) except for lobules IX and X (Fig. 8).

Biochemical tests for AChE activity in the molecular layer are difficult, because the intense AChE activity in the granular layer obscures the response that might be obtained from the molecular layer. The band pattern itself interferes with manometric determinations (Austin and Phillis, 1965) and makes it impossible to assess the activity restricted to the molecular layer, since no stereotactic description of the band pattern is available now. The addition of cholinesterase (ChE) and pseudo-cholinesterase (p-ChE) inhibitors provides another method for checking whether AChE is involved in the band pattern.

Eserine (Long, 1963) inhibits ChE activity in the molecular layer. The specific p-ChE inhibitor iso-OMPA (Aldridge, 1953; Long, 1963; Robinson, 1971) was added in concentrations ranging from 10^{-3} to 10^{-6} M. The band pattern was retained regardless of the concentration of iso-OMPA. Since it has been reported (Silver, 1967) that the molecular layer of the cat cerebellum is nearly devoid of p-ChE, we anticipated that iso-OMPA would not produce any change in the AChE response, and it did not. Previous manometric measurements in the molecular layer and in the granular layer (Austin and Phillis, 1965) indicated that the total of ChE activity in the molecular and granular layers is equal. These results are not in agreement with the

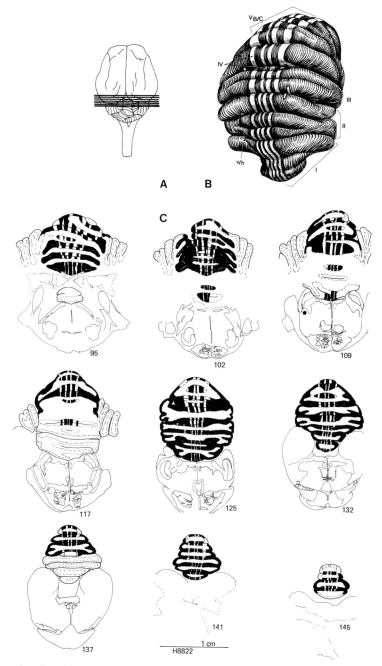

Fig. 7. A series of sections through the anterior lobe. (C) AChE-positive and -negative bands are indicated. The sections are indicated in a dorsal view of the cat brain (A). (B) AChE pattern in a reconstruction in a frontal view. (C) Sections of series H 8822 demonstrating AChE activity in black. Although the hemispheres contain activity, this activity is not shown. Open areas in the molecular area suggest the difficulty in interpretation [e.g., section 145 (*bottom right*)].

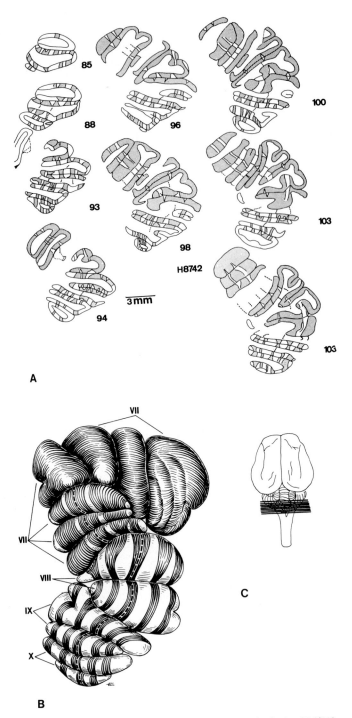

Fig. 8. Distribution of AChE activity in the posterior lobe vermis. Series H 8742 was incubated according to the method of Karnovsky and Roots (1964). (A) Sections with stippled areas

histochemical evidence that AChE activity is highest in the granular layer and, apparently, greatly exceeds that in the molecular layer (Silver, 1967). Light-microscopic histochemical analysis of AChE activity in the molecular layer therefore cannot be definitely attributed to any known structure in the molecular layer. The literature is confused concerning the presence of AChE in the molecular layer of the cerebellar cortex of the cat or sheep (Palmer and Elleker, 1961; Phillis, 1965a,b; Silver, 1967). In newborn cats, AChE activity was found in the Purkinje cells of neonates, but it had disappeared in 3-week-old cats (Sakharova, 1966). In adult cats, it is variously reported absent (Austin and Phillis, 1965; Kaśa and Csillik, 1965), to be present only after long incubations (Silver, 1967), or to be bound to Purkinje cells (Joo *et al.*, 1965; Phillis, 1965a,b). The specific distribution of the enzyme described by us and previously illustrated by Ramon-Moliner could be demonstrated only in cats 4 months old or younger. This, and the disappearance of AChE activity from the molecular layer as cats approach maturity, may explain the conflicting opinions expressed in the literature concerning the presence of AChE in the cerebellum. In the guinea pig, ChE is bound to the parallel fibers and Golgi cell dendrites; in sagittal sections, the Purkinje cell dendrites are negative (Kaśa *et al.*, 1965).

In the cat, the striated appearance of the positive areas in the molecular layer in transverse sections contraindicates that the enzyme is bound to parallel fibers. The absence of striations in sagittal sections cannot be explained when the enzyme is bound to Golgi cell dendrites or Bergmann's glia, because glial cells are supposed to contain p-ChE (Kaśa *et al.*, 1965; Silver, 1967). The unaltered appearance of the pattern of positive and negative bands in the molecular layer after iso-OMPA treatment also seems to rule out the Bergmann's glia. The present experiments indicate or may indicate that the enzyme is located in the dendritic tree of the Purkinje cell or in structures associated with it. This is in accordance with the observations in mature sheep (Phillis, 1965a,b; Suran, 1974a,b), in which the Purkinje cells were found to be definitely positive for AthChE. One objective of this study is to prove whether we are dealing with a "true" AChE. This is important because some authors (Crawford *et al.*, 1966) found "that ACh is unlikely to be an excitatory transmitter within the feline cerebellum, particularly at mossy fiber granule cell synapses, despite the presence of relatively high levels of AChE within the mossy fiber terminals." In the molecular layer, both muscarine and carbamylcholine excited Purkinje cells more strongly than did acetylcholine. Others (McCance and Phillis, 1964) determined that cells within the granular layer are excited by acetylcholine, whereas in the molecular layer, Purkinje cell excitation was a consequence of prior granule cell activation. These findings are at odds with this study that indicates the presence of AChE in the molecular layer.

---

indicating a high concentration of reaction products. The positions of these transverse sections are indicated on a dorsal view of the cat brain (C). (B) Reconstruction of the posterior lobe vermis viewed caudally. The midline is indicated by the dashed line.

4.2. Acetylcholinesterase in the Cat Inferior Olive

The inferior olive (IO) in mammals has been subdivided into the dorsal accessory olive (DAO), the medial accessory olive (MAO), and the principal olive (PO) (Kooy, 1916; Maréschal, 1934; Brodal, 1940). The IO can be further subdivided on the basis of its afferent and efferent (olivocerebellar) fiber connections. This can be illustrated particularly well for the DAO. According to Brodal (1940), the caudal and lateral parts of the DAO, which receive spinoolivary fibers (Brodal *et al.*, 1950), project to the contralateral vermis, whereas the rostromedial part of the DAO projects contralaterally to more lateral parts of the anterior lobe. Voogd (1969) and Armstrong *et al.* (1974) described a projection from the caudal one third of the DAO to the vermis and (Armstrong *et al.*, 1974) from the rostral part of the DAO to discrete areas in the pars intermedia of the anterior lobe and the paramedian lobule. Afferents from the cerebral cortex (Sousa-Pinto, 1969; Sousa-Pinto and Brodal, 1969) and the dorsal column nuclei (Groenewegen *et al.*, 1975), moreover, terminate in the rostral part of the DAO, but are absent from its caudal portion. These fibers from the gracile and internal cuneate nuclei terminate in discrete zones in the ventrocaudal and dorsomedial parts of the rostral DAO, respectively (Boesten and Voogd, 1975; Groenewegen *et al.*, 1975). These two regions in the rostral part of the DAO seem to coincide with the localization of the olivary neurons projecting to the hindlimb and forelimb areas in the projection zones of the DAO in the pars intermedia and the paramedian lobule (Armstrong *et al.*, 1974).

When it was perceived that the AChE distribution in the rostral DAO (as illustrated by Ramon-Moliner, 1972) coincided with the dual projections of the dorsal column nuclei to this same area of the DAO, the AChE distribution in the IO was investigated. Histochemical (see Table 5) and biochemical analyses led to the conclusion that the IO is compartmentalized into regions with different AChE activity. This subdivision of the IO will be compared with the connections of the IO as reported by previous investigators.

4.2.1. Distribution of AChE in the Inferior Olive

In Figs. 9–12, discussed below, enzyme activity is expressed as high (black in the figures), medium (hatched), and low (white). The pattern in the IO nucleus was determined from transverse sections. In a number of experiments, the distribution of AChE activity was determined from negative photographic enlargements of serial transverse sections through the IO. Some series were plotted independently by two observers. Three arbitrary levels of enzymatic activity were distinguished—low, medium, and high— and are indicated in the drawings of the IO. Reconstructions of the main parts of the IO were made according to the method of Brodal (1940). Enzyme activity is present in the cytoplasm and nucleus of the IO perikarya. It is also contained within the neuropil of the IO. In areas with a high or medium activity, some negative cell bodies are found against the darker background of the neuropil (Fig. 9). Where the neuropil is negative, some reactive

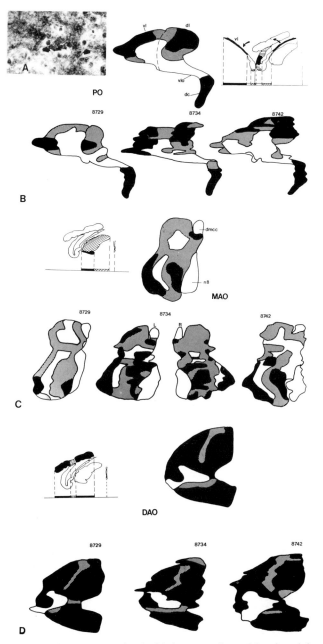

Fig. 9. (A) Cells in a section counterstained with hematoxylin and incubated for AChE. These cells are surrounded by a negative area (arrows). (B) Reconstruction of the left PO from three different series showing the distribution of AChE. The scheme at the top summarizes the results from three series for this subnucleus. The projection of the PO is indicated at the top of the figure. (C) Same schemes used in (B) applied to the MAO. (D) Comparable schemes used for the DAO. (dc) Dorsal cap; (dl) dorsal leaf of PO; (dmcc) dorsomedial cell column; (nβ) nucleus beta; (vl) ventral leaf of PO; (vlo) ventrolateral outgrowth. Regions with high, medium, and low AChE activity are indicated by black, hatching, and white, respectively.

neurons were encountered. The areas of the neuropil of the IO nucleus with high or medium activity contrast strongly with the fibers of the medial lemniscus and the reticular formation surrounding the IO, which are negative for AChE (Fig. 10).

The neuropil configuration generally conforms to the accepted subdivision of the IO in the cat. The borders are slightly wider, however, than delineated in Nissl-stained series. This phenomenon, and the use of young cats, account for the increase in the mediolateral dimension in parts of the IO made in this study, when compared with those of Brodal (1940) and others made from Nissl material. The distribution of this enzyme is described from transverse series 8729, illustrated in Fig. 11. The reconstructions of this and two other series (8734 and 8742) are depicted in Fig. 9.

4.2.2. Comparison of the Pattern with and without Iso-OMPA

No differences can be observed in AChE distribution within the PO of iso-OMPA-treated material. The pattern for the MAO is less clear. Rostral areas of high and medium activity cannot be differentiated. In one series (8897), the usual pattern of enzymatic activity was reversed, so that the more intense reaction occurred in a middle strip rather than at the periphery. Examination of reconstructions from a series of 4-month-old cats with and without iso-OMPA (Fig. 12) (see also Table 5) indicates that the lateral and medial margins of the rostral DAO can be recognized as separate regions containing much lower activity than the intense response obtained from central zones. Generally, however, the pattern remains the same prior to and following treatment with iso-OMPA, as evidenced by series 8895, in which alternate sections were incubated with 10^{-4} and 10^{-3} M iso-OMPA, and series 8945 (cat 3 months old, treated with 10^{-3} M iso-OMPA). The same areas remain positive in series treated with 10^{-6} M iso-IMPA (8809, 8897, 8903, 8919, and 8920). The ChE pattern observed in the IO, therefore, is due mainly to the AChE activity, confirming the biochemical determinations (see Section 5.2). In series of 7-week-old kittens (8903, 8919, and 8921) not treated with iso-OMPA, the lateral and medial margins of the DAO similarly show less activity than the central parts. This indicates that the p-ChE activity in 4-month-old cats appears relatively late in development.

In the synapse, AChE is usually found either in the postsynaptic membrane or inside synaptic vesicles (Robinson and Bell, 1967), but it can also be bound to the presynaptic membrane (Robinson and Bell, 1967; Koelle *et al.*, 1974). Therefore, the distribution of a positive AChE response in the IO (see Figs. 10 and 11) does not imply that AChE is primarily associated with particular afferent or intrinsic systems of the IO, nor is it necessarily produced by the perikarya of the IO.

The distribution of AChE in the rostral DAO, moreover, mirrors the projections of the contralateral internal cuneate and gracile nuclei to the medial and lateral parts of this subnucleus (Boesten and Voogd, 1975;

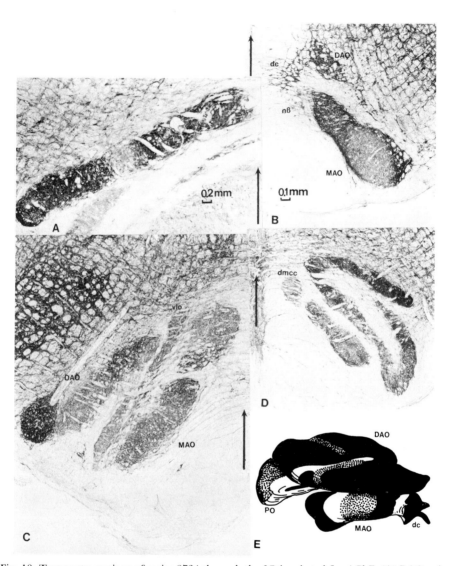

Fig. 10. Transverse sections of series 8734 through the IO incubated for AChE. (A) DAO at its rostral end. The absence of reaction product is demonstrated in the middle area. (B) Distribution of AChE in the caudal MAO, nucleus beta (nβ), and dorsal cap (dc). The caudal part of the DAO starts to appear in this section. In (C), the ventrolateral outgrowth (vlo) can be discerned. (D) The most rostral part of the IO with its dorsomedial cell column (dmcc) is shown. (E) The IO is shown in reconstruction. AChE areas are shown as viewed from oblique caudal. In all photographs (A–D), the negative areas are caused by passing hypoglossal fibers, while the midline is indicated by an arrow placed in the midline (B, D) or paralleling its direction (A, C).

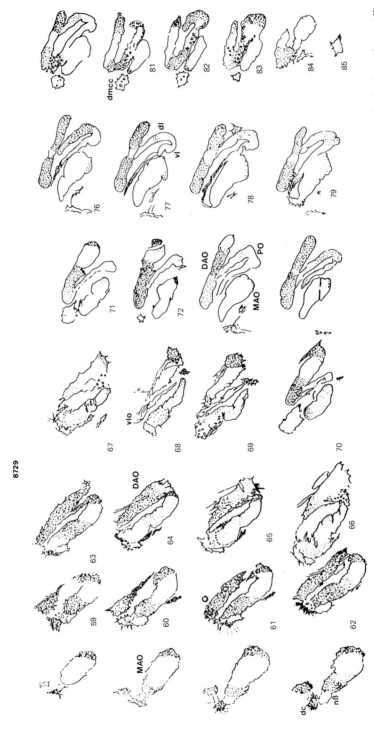

8729

Fig. 11. IO stained for AChE activity shown as a stippled drawing. The numbers of the serial brainstem sections are indicated. (dc) Dorsal cap; (dl) dorsal leaf of the PO; (vlo) ventrolateral outgrowth; (dmcc) dorsomedial cell column; (nβ) nucleus beta; (vl) ventral leaf of the PO. Regions with high, medium, and low AChE activity are indicated by intense stippling, moderate stippling, and no stippling, respectively.

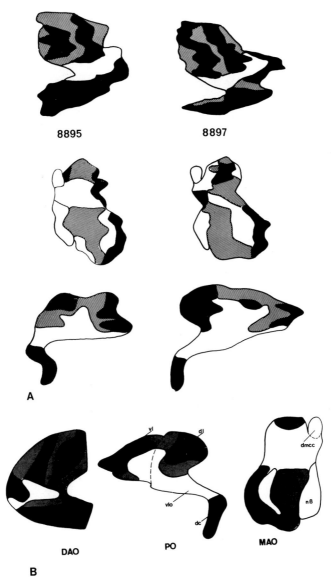

Fig. 12. (A) Reconstructions of AChE activity in the DAO, MAO, and PO after treatment with 10^{-6} M iso-OMPA (series 8897) or with 10^{-5} M iso-OMPA (series 8895). Rostral is to the top of the figure. The lateral margins of the rostral part of the DAO are less active in comparison to series that have not been treated with inhibitors. The rostral aspect of the MAO is difficult to interpret. (B) Final diagram for the distribution of AChE in the IO complex. See the Fig. 11 caption for definitions of the abbreviations.

Groenewegen *et al.*, 1975). Two AChE-positive areas with a negative area between them can be distinguished in the caudal part of the MAO. Cell group β is also negative.

When other afferent fiber systems are considered, the lumbar spinoolivary fibers project to the strongly positive areas in both the caudolateral MAO and the caudal DAO. In the DAO, however, spinoolivary projections extend onto its rostrolateral part, which shows much less activity in our iso-OMPA-treated material. Cervical spinoolivary fibers, moreover, terminate in the more rostromedial parts of the MAO (Boesten and Voogd, 1975) within regions that show low or no AChE activity, as well as in the strongly positive contralateral medial part of the opposite DAO. From lesions of the dorsal mesencephalic reticular formation, Busch (1961) identified this area as the origin of the tegmentoolivary tract. No termination pattern for this tract has been identified within the IO. Lesions of the medial longitudinal fasciculus produce intense degeneration in the medial tegmental tract that supplies the rostral half of the MAO (medium or low activity for AChE) and the ventral lamella of the PO (containing high, medium, and low AChE activity). The results of Busch (1961) do not reveal an overlap between the mesencephalic descending termination in the IO and the AChE pattern. The recent studies of Walberg (Walberg, 1974; Walberg *et al.*, 1976) provide partial agreement with the findings of Busch (1961). These projections, however, are not coincident with the AChE pattern. The projections of the nucleus ruber on the IO described by Walberg (1956) terminate mainly in the ipsilateral dorsal lamella of the PO. This projection is affirmed by the autoradiographic studies of Edwards (1972). The rubroolivary tract ends in areas with high, medium, and low AChE activity. No overlap of the projections from the ruber and the AChE pattern can be detected. The caudate and pallidal projections, as described by Walberg (1956), are difficult to interpret and will not be considered.

Preliminary experiments with chronic animals containing lesions of the dorsal column nuclei did not show a change in the distribution of AChE in the IO and (see Table 5) therefore do not favor the conclusion that AChE is bound to afferent olivary systems of the IO. As in the DAO, the distribution of AChE in the MAO and the PO leads to a compartmentalization of the IO that corresponds closely to subdivisions of the IO based on the organization of the olivocerebellar projections (Brodal, 1940; Brodal *et al.*, 1950; Armstrong *et al.*, 1974). Localization on the perikarya is supported, because in our counterstained series we found activity in most perikarya. Furthermore, following fractionation of the IO complex into P_1, P_2, an S_2 fractions (Whittaker *et al.*, 1964), the S_2 fraction (microsome fraction) contained 62% of cholinesterase activity, the P_2 (synaptosome fraction) nearly 23%, and the P_1 (nuclei and debris) 15%. In this respect, the distribution of areas projecting to discrete zones in the cerebellar cortex of the cat is of interest (see Groenewegen and Voogd, 1977).

Subdivisions present in the IO based on the AChE subdivisions mirror the IO distribution based on the climbing-fiber projections (Fig. 13). On the other hand, the cerebellar distribution of AChE in the molecular layer as

AFFERENT PROJECTION   INHERENT   EFFERENT PROJECTION
BORDERS   AChE   BORDERS
BORDERS

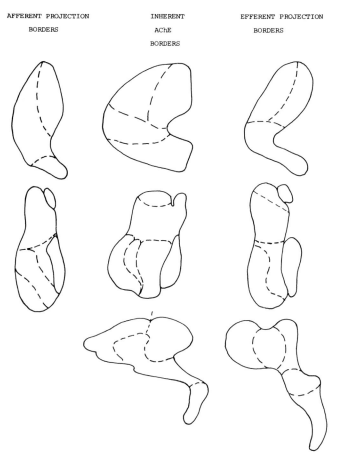

Fig. 13. Comparisons of the afferent projection borders (Boesten and Voogd, 1975; Groenewegen *et al.*, 1975), the inherent AChE borders (Marani *et al.*, 1977) and the efferent projection borders (Groenewegen and Voogd, 1977). The similarity of the borders in the subdivisions of the IO complex is evident. The DAO, MAO, and PO are shown from top to bottom.

described in Section 4.1 does not correspond with the enzymatic concentrations determined for the cells within the IO that are the source of topographically related olivocerebellar projections. The unique distribution of "true" AChE activity in the IO therefore cannot totally be explained on the basis of the known afferent and efferent connections with the IO, although inherent AChE borders are comparable with efferent and afferent projection borders (Fig. 13).

4.3. 5'-Nucleotidase in the Mouse Cerebellum

Scott (1964, 1965, 1969) described a pattern in the cerebellum of the mouse with the help of the histochemical reaction for the enzyme 5'-nucleotidase. He described in more detail the longitudinal course of five bands in the vermis of the cerebellum. No indications were given about the splitting of the bands, nor were all the bands reported in the anterior lobe. These bands can be traced all over the rostrocaudal extent of the cerebellum from the nodulus to the lingula. Voogd (1967) emphasized the importance of the rostrocaudally directed bands of the enzyme 5'-nucleotidase. He suggested that the same principle underlies this distribution of the enzyme 5'-nucleotidase in the molecular layer of the cerebellar cortex of the mouse (Scott, 1965), the bandlike distribution of spinocerebellar fibers to the granular layer [rabbit (Van Rossum, 1969), chicken and pigeon (Vielvoye, 1970, 1977)], and the division of the white matter of the folia cerebellae into parasagittal compartments in the ferret (Voogd, 1964).

Hardonk and De Boer (1968) concluded that of the several isoenzymes of the 5'-nucleotidase that occur in the mouse, only one is present in the brain of this animal. These results were confirmed in my experiments (see Section 5.4) for the cerebellum. To compare these patterns, a more complete description of the 5'-nucleotidase distribution was necessary. A description is given only of the 5'-nucleotidase band pattern in the molecular layer of the *anterior lobe* of the mouse.

4.3.1. Some Remarks on the Morphology of the Anterior Lobe

The anterior lobe in the mouse cerebellum is divided into culmen IV–V, centralis III, and lingula I–II (Sidman *et al.*, 1971). Two deep fissures divide the anterior lobe into three lobules. The precentral fissure separates lobule I–II (lingula) from lobule III (centralis), and lobule IV–V (culmen) is situated between the preculminate fissure and the primary fissure. In some cerebella, lobule I–II and lobule IV–V can be further subdivided into their separate lobules by additional sulci (Fig. 14). No discontinuities have been noticed in the molecular layer.

4.3.2. Description of the 5'-Nucleotidase Pattern

4.3.2a. Lingula I–II. In the lingula, five bands were found. They are indicated by arabic numbers in Fig. 15. Number 1 is the midsagittal band. The positions of the remaining bands are almost bilaterally symmetrical with respect to band 1. On both sides of band 1, bands 2 are situated, and laterally to bands 2, bands 3 are found. In Fig. 15, some sections of the lingula may be seen. The 5'-nucleotidase areas are indicated with dots. To make a map of the anterior lobe that would be of any importance for a topological localization of the 5'-nucleotidase-active areas, all the slices were projected on a horizontal plane, as is indicated in Fig. 15. A topological map was made by drawing connections between the projected broken lines. Some variability is noticed in the bands of the lingula. In series 8151 and 8553, bands 2 tend to split in the rostral part of the lingula. In series 8552, in the caudal part of the lingula, bands 2 and 3 split near the continuation of the lingula and the centralis. Band 3 fuses with band 2 at some places and at the top of the lobule at the right side either fuses or disappears.

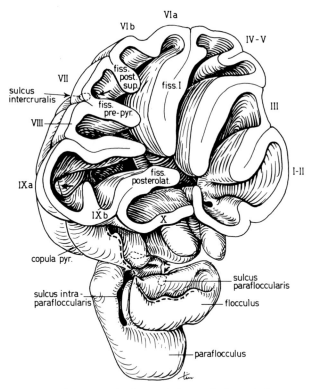

Fig. 14. Reconstruction of the molecular layer of the cerebellum of the mouse indicating the main fissures and sulci in the midsagittal plane. The main subdivisions of the anterior lobe are important for the discussion of 5′-nucleotidase in the text. The anterior lobe is divided from the posterior lobe by its primary fissure, while lobule I–II is bordered by the fourth ventricle. Subfissures divide the anterior lobe into lobules I–II, III, and IV–V.

The areas between the bands in all lobuli are "negative" for the reaction products of the enzyme 5′-nucleotidase. They show considerably less activity. Therefore, it looks as though there are negative areas between the bands in reconstruction. But in series 8552 and 8553 (with 2 hours' incubation), it is obvious that there is a weak reaction for 5′-nucleotidase, but considerably less than in the reported bands. It was not possible to discern, in these two series, more bands in the areas between the reported bands. The weak reaction might well be caused by the diffusion of reaction products from the positive bands.

4.3.2b. Centralis III. In the centralis, the band pattern is bilaterally symmetrical with respect to band 1. Eleven bands are found in the whole centralis. Bands 1, 2, and 3 are continuations of bands 1, 2, and 3 of the lingula. Band 1 is the medial one. Laterally on each side of bands 3, bands 4, 5, and 6 are found in the centralis (see Fig. 15). Band 4 has a subdivision in the rostral and caudal parts. These subbands are numbered 4a and 4b; band 4a is medial to band 4b. Band 4a sometimes fuses with band 3, passing

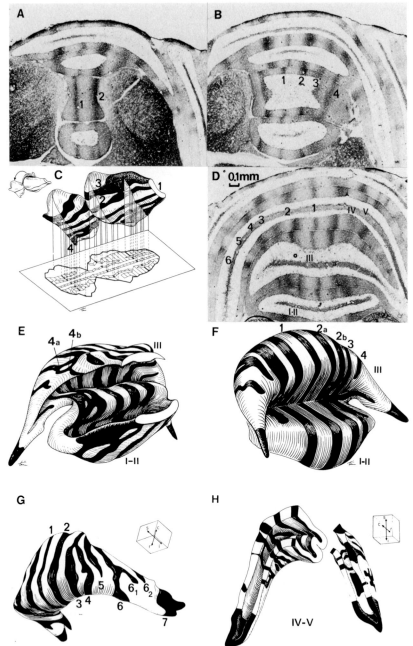

Fig. 15. (A, B, D) Sections (series 8553) through the anterior lobe from rostral (A) to caudal (D). The sections were incubated for 2 hr according to Scott (1965). The positive bands that can be discerned are indicated by numbers (see the discussion in the text). (C, E, F, G, H) Reconstructions of lobule I–II (C), lobules I–II and III in caudal view (E) and in rostral view (F), and lobule IV–V in oblique rostral view (G) and in caudal view (H). (C) Indicates how a semistereotactic map can be produced from, for example, lobule I–II. For a discussion of the continuities of the different bands, refer to the text. (r) Rostral; (c) caudal; (i) inferior; (s) superior.

from III to IV–V. The bands are localized mainly in the central part of the lobule; the most lateral part of the lobule is filled up with bands 5 and 6. In the lateral part of the centralis, as well as in the culmen, it is very difficult to distinguish bands because the differences in the areas (e.g., between bands 5 and 6) are hard to recognize after the 2-hr incubation series as well as after the 1-hr incubations.

4.3.2c. Culmen IV–V. The culmen has a subdivision at the top, but the band pattern is not disturbed by this subdivision. Thirteen bands are found in the whole culmen, as are two strongly positive areas. Band 1 is situated in the midsaggital plane. It is a continuation of band 1 of the centralis, and bands 2 and 3 are continuations of the same bands of the centralis. All bands are almost bilaterally symmetrical with band 1. Bands 3 and 4 sometimes fuse (see Fig. 15). Band 5 is located on the medial lateral bow of the lobule. The most lateral tip of the culmen is positive for the enzyme 5'-nucleotidase. Therefore, this area was called band 7. In the area between bands 6 and 7, two small positive areas were found that were numbered 6_1 and 6_2. They do not form any bands extending into the 6, while top of the lateral part of the culmen. On the caudal and rostral surfaces of the culmen, these areas fuse with band 6. Area 6_2 lies laterally to area 6_1. The components of the lateral area of this lobule are difficult to distinguish for the same reason as already discussed for the centralis.

4.3.2d. Considerations. Each band has been numbered. Numbering of bands in distinct lobules demonstrates a clear visible relationship only when bands in one lobule have been reported to have a continuation into the next lobule. Bands 1–4 continue into all the lobules, but the reconstruction methods used make it difficult to follow the continuation of laterally situated bands such as band 4. It may be that splitting of band 3 in the lingula gives rise to bands 3 and 4 of the centralis. Further complications follow because band 2 of the lingula also tends to split. Complicated schemes of band sequencing could be devised, but are not considered here. As described, laterally situated bands tend to spread in areas of the molecular layer as they pass from one lobule to another. In such areas, distinctions between bands seem to disappear. Band 3 of the lingula, as mentioned, may give rise to bands 3 and 4 and, perhaps, band 5 of the centralis. Band 6 of the culmen may arise from band 5 of the centralis. If a band splits and rejoins, it is considered as a single unit and the original numbering is retained, although letters are assigned to the split portions of the band (e.g., band 4 becomes bands 4a and 4b). Branches to positive areas arising from a band retain the number of the original band with a further specification (e.g., band 6 gives rise to bands 6_1 and 6_2). As noted in the band numbering of each lobule, the number of bands in each lobule increases as they progress from the lingula through the culmen. This increase parallels the increasing surface area of the molecular layer. Remarkably, most of the bands are located in the central portions of the lobules. Although the lobules tend to vary somewhat, morphologically, among brains, this variation did not noticeably affect the band patterning. The difference between positive bands and negative bands for the enzyme 5'-nucleotidase is not an absolute. The negative bands are considerably less positive for the 5'-nucleotidase enzyme, and each "negative band" in the lobules is limited by positive bands, but not all positive bands are limited by negative ones. Therefore, it is unlikely that a positive and a negative band together form a unit. It is possible that the importance of the pattern is conditioned, not by the positive bands, but, on the contrary, by the negative ones.

Homologies in 5'-nucleotidase patterns have been examined in an attempt to illuminate the functional role of this enzyme in the mouse cerebellum. Thus, the 5'-nucleotidase patterns in insects and the mouse were compared. Sprey (1970*a,b*) published two articles on histochemical changes in the imaginal disks of *Calliphora erythrocephala* (the blowfly). In the first article, he describes the localization of glycogen in the imaginal disks. There are areas less positive for glycogen and areas that are strongly positive for glycogen. No direct connection was found between the glycogen content and chitin formation, although some indications are present. Sprey (1970*a*) could not find a correlation between the results of glycogen and 5'-nucleotidase patterns. The combination of the presence of 5'-nucleotidase and glycogen was also found by Milaire (1969). In studying syndactylis in the rat, he found in normal embryos of the 17th day 5'-nucleotidase activity in the metapode (the diaphysairy center of phalanxes II, III, and IV). In the autopode, he found 5'-nucleotidase activity in the mesoblast and gentroapicale of the synovial sheaths. A partial coincidence of the presence of 5'-nucleotidase and glycogen can be found in the area in which bone growth appears.

A corresponding glycogen pattern in the cerebellum of the mouse has been sought, but no coincidence of glycogen or the α-glucan phosphorylase with the 5'-nucleotidase pattern could be found. After administration of 3',5'-cAMP, only an inhibition of the α-glucan phosphorylase could be found (Marani and Boekee, 1973). Thus, it is probable that the glycogen pattern is typical for chitin formation in insects and for bone growth, but has nothing to do with the role of 5'-nucleotidase in the cerebellum. In a second article, Sprey (1970b) published some data on the localization of the enzyme 5'-nucleotidase. 5'-Nucleotidase from *Calliphora* can break down AMP and ATP. The 5'-nucleotidase from mouse cerebellum can break down only 5'-nucleotides (Scott, 1965).

In general, there is one negative band in insects ranging from proximal to distal, sandwiched between two positive areas, except the antennal disk. In an article on AMPase activity in the imaginal disks of some dipteran larvae, Majoor (1973) found in some species of *Drosophila* (*araicas, guarmuni, robusta*) and in *Musca, Phormia,* and *Zaprionus* different areas of 5'-nucleotidase activity. Absence of 5'-nucleotidase activity was reported in *Drosophila hydei, D. funebris,* and *D. melanogaster.* Sprey (1970*b*) proposed that the pattern had something to do with the rigidity of the cells of the imaginal disks. On the basis of results with different species, Majoor (1973) questioned this role for the enzyme 5'-nucleotidase as proposed by Sprey. Hardonk and De Boer (1968) and Hardonk and Koudstaal (1968) suggested from their studies in rat and mouse that there is a correlation in several tissues between a low proliferation rate and a high activity of 5′-nucleotidase. Vijverberg (1973) denies the proposed connection for *Calliphora erythrocephala*. Looking for the incorporation of tritiated thymidine in the wing and leg disks, he noticed no distinct relationship between his pattern of DNA synthesis and the 5′-nucleotidase pattern of Sprey. However, in *Calliphora*, a 5′-nucleotidase pattern was found, but there was no correlation between the glycogen or DNA pattern and the 5′-nucleotidase pattern. Therefore, the functions of the enzymes as proposed by Sprey (1970*b*) and by Hardonk and De Boer (1968) and Hardonk and Koudstaal (1968) remain unsupported.

In peripheral nerves and root ganglia of the rat, Novikoff *et al.* (1966) found no activity of nucleoside monophosphatase. Only after incubation of 1 hr could a very light staining of unmyelinated fibers be demonstrated. Schlaeffer *et al.* (1969) noticed also for nucleoside phosphatase a low activity for AMP in spinal roots and ischiadic nerves of the rat. Although Schlaeffer *et al.* (1969) found that nucleoside phosphatase can break down ATP and AMP, the brain 5′-nucleotidase found in sheep is inhibited by ATP (Ipata, 1966, 1967, 1968). Gomori and Chessick (1953) found 5′-nucleotidase activity in the dog cerebellum apparently located in axons. Since the alkaline phosphatase in mouse cerebellum is located mainly in the fiber layer (Marani and Kurk, unpublished results) and Gezelius and Wright (1965) found in *Dictyostelium discoideum* an alkaline phosphatase that can break down AMP, one cannot state that the activity in the fiber layer is only 5′-nucleotidase activity. Naidoo and Pratt (1954) saw that the perikarya of Purkinje cells in the rat cerebellum were devoid of 5′-nucleotidase activity, although their molecular and granular layers have equivalent 5′-nucleotidase activity. Scott (1965, 1969) previously showed that there is a 5′-nucleotidase band pattern in the rat molecular layer. Shanta and Manocha (1970) studied 5′-nucleotidase activity in the cerebellum of the squirrel monkey and in *Macaca mulatta*. (For discussion, see Section 1.5.)

Although the number of bands and their presence in the hemisphere in some species distinguish the 5′-nucleotidase band pattern from that observed for AChE in the cat, both seem to be associated with the Purkinje cell dendritic tree (Marani and Voogd, 1973). Although the enzymatic reaction is different, the effect expected is equal, in the sense that AChE activity is thought to play an important role in transmission, while such a role is postulated for the 5′-nucleotidase (Suran, 1974*a,b*) in the trigeminal system. The effect of AMP, a substrate for 5′-nucleotidase, on the spontaneous firing capacity of Purkinje cells has been measured, after microiontophoretic administration (Kostopoulos *et al.*, 1975). This effect was different from the effect after administration of adenosine. This kind of experiment stresses the idea that 5′-nucleotidase could play a role in transmission by creating Purkinje cell dendrites with high concentrations of 5′-nucleotides or dendrites with a high concentration of reaction products of 5′-nucleotidase.

The effects of AMP have been studied in connection with noradrenergic transmission (Bloom *et al.*, 1971; Hoffer *et al.*, 1971; Siggins *et al.*, 1971), and it has recently been shown that some 5'-nucleotides can enhance glutamate-induced potentials (Ozeki and Sato, 1970). Glutamate is postulated as a climbing-fiber transmitter in the rat (Guidotti *et al.*, 1975). A direct sensitivity for glutamate of mainly the dendritic portion of the Purkinje cells has also been described after microiontophoretic administration of glutamate (Chujo *et al.*, 1975).

The existence of mediolateral gradients of 5'-nucleotidase in the cerebellar molecular layer would lead, anyhow, to alternating bands of excited and depressed Purkinje cells after stimulation of the diffusely organized noradrenergic afferent system or after stimulation of the equally well distributed climbing fibers.

In a great variety of animals, a distribution pattern similar to that of AChE is present in the molecular layer for the enzyme 5'-nucleotidase (Scott, 1964, 1965). In contrast to the AChE pattern, the 5'-nucleotidase organization in mouse, rat, and shrew consists of more bands. In the mouse, 5'-nucleotidase compartmentalization can be observed throughout the entire cerebellum. Despite the differences between the AChE and 5'-nucleotidase band positions, some similarity can be noticed. In a preliminary report (Marani and Voogd, 1973), the 5'-nucleotidase activity was assigned to the Purkinje cell dendritic tree, as is the case for the AChE pattern. The conclusion has to be that two types of structures containing 5'-nucleotidase or AChE exist in the cerebellum of the mouse, rat, shrew, vole, and cat. These types can be differentiated only by enzyme-histochemical methods. Marani and Boekee (1973) state:

> Although the enzymatic reaction is different, the effect expected is equal, in the sense that AChE activity is thought to play an important role in transmission while such a role is postulated for the 5'-nucleotidase too (Suran, 1974*a,b*), as was already reported by us.

This hypothesis stresses the idea that the band patterning in the molecular layer of different animals is a general feature, probably connected with the transmission in the molecular layer of the cerebellar cortex. In Section 4.2.1, it was reported that an agreement was found between the projection of the dorsal column nuclei, the efferent projections, and AChE deposition in the cat's IO. From a physiological point of view (Armstrong *et al.*, 1974), one can correlate these efferent areas with olivocerebellar bandlike projections within the molecular layer of the cat. The subdivisions of the olive project to the white matter in a bandlike arrangement (Voogd, 1969). Injections of tritiated leucine in the IO also reveal a bandlike distribution in the molecular layer (Groenewegen and Voogd, 1977; Voogd *et al.*, 1975). The bandlike gradient of the projections from the IO to the cerebellum is based on physiological (Oscarsson, 1973), autoradiographic, and silver techniques. It is peculiar that in the IO, there is found an AChE subdivision (Marani *et al.*, 1977) that agrees partially with a compartmentalization arrangement based on afferent and efferent projections and, similarly, that AChE activity is localized in compartments within the molecular layer, as is true for olivocerebellar connections.

4.4. Monoamine Oxidase in the Chicken Cerebellum

Recently, using the monoamine oxidase (MAO) method of Williams *et al.* (1975*b*), we demonstrated in the fowl cerebellum a topographical distribution for this enzyme. Using 5'-hydroxytryptamine as a substrate, while controls were performed with 10^{-5} M clorgyline, evidence was gathered for an uneven distribution of MAO-A in the fowl cerebellar white matter. A simple comparison could be made with the uneven distribution of fiber diameters in the white matter as described in detail by Feiraband and Voogd (1975) of our group. Although preliminary results were described and shown, it seems obvious that areas containing high MAO-A activity mirror the areas in which mainly Purkinje cell axons are localized (see Fig. 16) (Feirabend *et al.*, 1977). The unexpected localization of the activity of this enzyme at places where Purkinje cell axons are distributed is strengthened by the agreement with the already-discovered lamination in the fowl cerebellum, although no sound explanation can be given as to why these areas containing Purkinje cell axons would be provided with MAO activity. Another possibility is that thin fibers, which are always more numerous as compared to Purkinje cell axons, even in areas with thick fibers, are responsible for this MAO-A activity. These thin fibers may be mossy fibers and climbing fibers as well as locus ceruleus fibers.

5. Comparisons with Biochemical Results

5.1. Activators

No general rule can be given for the use of and the need for activators for different enzymes. The effects of monovalent and bivalent ions have been published in several studies in the literature (for a review, see Dixon and Webb, 1958). The known cations that can affect enzymes are: Na^+, K^+, Rb^+, Cs^+, Mg^{2+}, Ca^{2+}, Zn^{2+}, Cd^{2+}, Cr^{3+}, Cu^{2+}, Mn^{2+}, Fe^{2+}, Co^{2+}, Ni^{2+}, Al^{3+}. Of this group of cations, Mg^{2+} seems to be the natural activator of enzymes that convert phosphorylated substrates, although Mg^{2+} can in most cases be replaced by Mn^{2+}.

Most enzyme activators are know from the biochemical literature and are used accordingly. The next example may show how one can start by studying an enzyme localization and end up studying activators. It was preceived in our studies that the cerebellar 5'-nucleotidase pattern was not always present in different incubation media. It was observed, in checking various substances and buffers that could be used for 5'-nucleotidase determinations, that all substances that enhanced the 5'-nucleotidase band pattern and 5'-nucleotidase activity contained sodium ions. 5'-Nucleotidase from different sources can use various cations as activators, although most 5'-nucleotidases use Mg^{2+}, except the bacterial 5'-nucleotidase. The interchangeability of Mg^{2+} and Mn^{2+} as shown for 5'-nucleotidases in Table 6 could be expected for K^+ and Na^+, but Na^+ often acts as a competitive

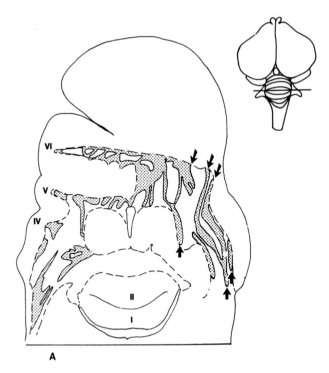

Fig. 16. Two sections through the cerebellum of the fowl. The position of the sections is indicated in a dorsal view of the brain (*top*). (A) Schematic drawing indicating the areas containing thick fibers in the cerebellar white matter (courtesy of Dr. H.K.P. Feirabend). (B) Results of an MAO incubation according to Williams *et al.* (1975*a,b*) at the same level in the cerebellum. The arrows in (A) and (B) are comparable (refer to the text for further discussion).

inhibitor for K^+ activation (e.g., in phosphotransacetylase and aldehyde dehydrogenase (Dixon and Webb, 1958). While Song and Bodansky (1967) found no sodium or potassium effect on rat liver 5′-nucleotidase, research on cerebellar 5′-nucleotidase in the mouse showed such effects. Separate administration of sodium and potassium raised the 5′-nucleotidase

Table 6. 5′-Nucleotidase Activators

Taxon	Species or tissue	Authors	Cation activators
Bacteria	*Escherichia* *Shigella* *Citrobacter*	Neu (1967*a,b*)	Co^{2+}, Mn^{2+}, Zn^{2+}
Snakes	Venoms	Sulkowski *et al.* (1963)	Mg^{2+}, Co^{2+} in low concentrations
Birds	Heart	Gibson and Drummond (1972)	Mg^{2+}
Mammals	Rat liver	El-Aaser and Reid (1969)	Mg^{2+}, Mn^{2+}, Fe^{2+}, Co^{2+}
	Rat brain	Bosmann and Pike (1970)	Mg^{2+}, Mn^{2+}, Co^{2+}
	Mouse cerebellum	Scott (1965)	Mg^{2+}, Mn^{2+}, Ni^{2+}
	Human serum	Van der Slik (1975)	Mg^{2+}

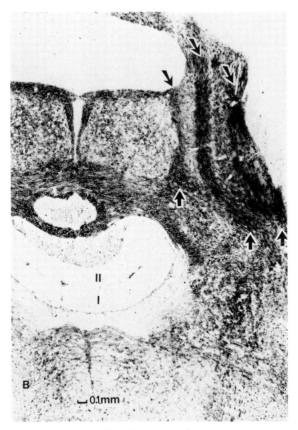

Fig. 16. (*Continued*)

activity, while combined administration in physiological concentrations of both ions gave a rather stable activity of 5′-nucleotidase, without adding Mg^{2+}, as shown in Table 7. Exclusion of different side effects for AMP deaminase, acid phosphatase, alkaline phosphatase, and the *p*-nitrophenyl activity of K^+- and Na^+-activated ATPase could be provided for. The effect of combined administration of K^+ and Na^+ was tried in order to produce an experiment imitating the different concentrations of K^+ and Na^+ that can occur when an action potential induces K^+ and Na^+ exchange. Combinations of a high Na^+ and a low K^+ concentration or a low Na^+ and a high K^+ concentration can then be expected. One can calculate the summation of the K^+ and Na^+ effects from those they have separately, and this calculation can be compared to the results of the combined administration of these ions (see Fig. 17). From the calculated graphs, a lowering of the total 5′-nucleotidase activity at combinations of a high sodium and a low potassium concentration were expected (see Fig. 17).

Under physiological conditions, however, cerebellar 5′-nucleotidase converts—under the described conditions (Marani, 1980*a*), independently of the

Table 7. Effects of K^+ and Na^+ Administration to a
K^+- and Na^+-Free Medium[a]

Concentration (M)	5'-Nucleotidase activity (IU/liter) (mean ± S.D.)
K^+ 1.8×10^{-2}	34.8 ± 3.4
K^+ 5.4×10^{-2}	63.2 ± 2.6
K^+ 9.9×10^{-2}	93.7 ± 4.3
K^+ 45.0×10^{-2}	98.8 ± 5.2
Na^+ 1.8×10^{-2}	24.4 ± 2.3
Na^+ 5.4×10^{-2}	29.9 ± 0.9
Na^+ 9.9×10^{-2}	43.3 ± 3.1
Na^+ 45.0×10^{-2}	51.4 ± 2.2

[a] The medium consisted of 0.025 M Veronal buffer, pH 7.2; 0.0025 M AMP (free acid); 0.005 M glucose-1-phosphate; and 500 IU/liter adenosine deaminase. The determinations were performed according to Persijn and co-workers (Persijn *et al.*, 1969; Persijn and van der Slik, 1970) using this incubation medium. Addition of K^+ and Na^+ was effected by using KCl and NaCl. No Mg^{2+} was added in any step of the procedure. Mg^{2+} present in the homogenates was not chelated.

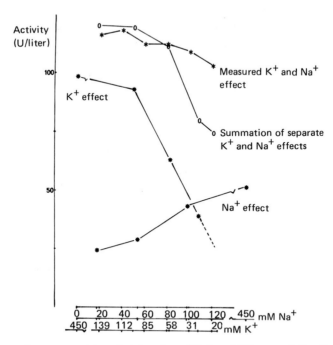

Fig. 17. Effects of separate or joint administration of Na^+ and K^+ on cerebellar 5'-nucleotidase activity in homogenates treated with 0.1% Triton × 100. The 5'-nucleotidase activity is expressed in U/liter.

Na$^+$ and K$^+$ concentration, but in the presence of both ions in physiological concentrations—the same amount of 5'-AMP. This unexpected behavior shows that cerebellar 5'-nucleotidase is peculiar not only in regard to its pattern present in rat and mouse cerebella, but also by its reaction to Na$^+$ and K$^+$ ions. In this sense, cerebellar 5'-nucleotidase behaves differently from most other 5'-nucleotidases. However, this can easily be understood in the environmental conditions of the central nervous system with its carefully protected maintenance of K$^+$ and Na$^+$ concentrations.

5.2. Inhibitors

In enzyme histochemistry (as in biochemistry), it is necessary to assure that a given incubation medium yields a reaction that is due solely to the particular enzyme under study. To accomplish this, one tries to inhibit specifically the activity of that enzyme. This can be done in either of two ways. First, one could pretreat the section in a way that is known (from other sources) to inhibit or destroy the activity of the enzyme. An example is the pretreatment with buffers at low pH for the inhibition of glucose-6-phosphatase. Another possibility arises if for the given enzyme a compound is known that specifically inhibits this (and only this) enzyme. In this case, the compound can be added to the incubation medium, and in this way a more specific test becomes possible. A well-known example is the addition of sodium fluoride in the incubation medium for acid phosphatase. It is not the best example, however, since sodium fluoride inhibits many more enzymes than just acid phosphatase. Better examples will be discussed below in connection with acetylcholinesterase (AChE) and alkaline phosphatase. It will also be shown that uncritical application of such inhibitors may yield quite unexpected results.

5.2.1. Inhibitors in AChE Histochemistry

AChE belongs to the group of carboxyesterases from which can be demonstrated by histochemical procedures the so-called A-esterases (EC 3.1.1.1), B-esterases (EC 3.1.1.2), C-esterases (EC 3.1.1.6), and the cholinesterases, which in turn are subdivided into AChE (EC 3.1.1.7) and pseudocholinesterase (p-ChE) (EC 3.1.1.8). AChE is important in neurohistochemistry in localizing sites of destruction of the neurotransmitter acetylcholine (ACh). Since ACh breakdown products cannot be precipitated, the histochemical procedure uses acetylthiocholine, which is even more readily hydrolyzed than ACh itself, as a substrate. The procedure is often used to demonstrate cholinergic pathways. Although the use of AChE for this purpose can be criticized, it will be clear that it is necessary to distinguish between the (specific) AChE and the nonspecific carboxyesterases, and in particular between AChE and p-ChE, particularly because there is a functional difference between AChE and p-ChE (Silver, 1967). This being the case, the use of inhibitors becomes more or less imperative. Unfortunately, the inhibition of carboxyesterases is difficult, but the reverse, the inhibition of cholinesterases (AChE + p-ChE) can be effected by eserine (physostigmine). Other tools in the investigations are the use of (iso-OMPA), which is considered to be a specific inhibitor of p-ChE, and—though this is not really within the scope of inhibitor studies—the use of butyrylthiocholine as a substrate. This compound is preferentially hydrolyzed by p-ChE.

The following example will illustrate the tactics employed. In our laboratory, we were studying the AChE patterns in the inferior olive (IO). Biochemical determinations of the hydrolysis of acetylthiocholine in homogenates of the IO area showed (see Fig. 18) that the addition of eserine was inhibitory over the whole range of concentrations studied. This result was confirmed in parallel histochemical experiments on that area. This meant that the acetylthiocholine-hydrolyzing activity was due to either p-ChE or AChE, or to both. Further investigations were then carried out with the p-ChE inhibitor iso-OMPA (Fig. 18). Again, acetylthiocholine hydrolysis was found to be inhibited at all concentrations used. This points to the conclusion that a substantial part of the acetylthiocholine-hydrolyzing activity in the sections was due to p-ChE, a conclusion that was confirmed by substituting butyrylthiocholine for acetylthiocholine: addition of iso-OMPA then gave a definite inhibition. This means that in studying the AChE patterns in the IO, the use of p-ChE inhibitor is most imperative. However, at iso-OMPA concentrations higher than needed for total p-ChE inhibition (with butyrylthiocholine as substrate), we *still* found inhibition with acetylthiocholine as substrate. This means that high concentrations of iso-OMPA will also inhibit AChE activity, which is supported by the results of Desmedt and La Grutta (1957).

In most neurobiological cholinesterase studies, at least these two enzymes that can convert acetylthiocholine are present. Sometimes, biochemical evidence gathered in such studies can hint at the presence of both AChE and p-ChE. Klück (1980) studied the peripheral innervation of the rat bladder and urethra with the help of the AChE reaction. In a biochemical study comparable to that used for the IO, a plateau was found in the curve for cholinesterase inhibition using acetylthiocholine as substrate (see Fig. 18F). The presence of nonspecific esterases was excluded by the eserine assays. The interpretation given was that inhibition of p-ChE was affected at low concentrations, while an effect of this substance on AChE appeared only at high iso-OMPA concentrations. The results were compared to sections treated with different iso-OMPA concentrations and the same substrate. Disappearance of cholinesterase activity in the rat bladder muscles was achieved, while at plateau concentrations, the innervation pattern in these muscles remained [see Fig. 18 and compare to the results of El-Badawi and Schenk (1966, 1967, 1968a,b, 1969, 1970, 1971a,b, 1974)].

As shown, biochemical results can sometimes mean that the enzyme histochemist will meet difficulty in localizing enzymes. On the other hand, assuming that a generally accepted inhibitor is a good inhibitor under all circumstances can be a source of error, as has been shown for the alkaline phosphatase inhibitor levamisole (Borgers, 1973).

As previously noted, acid phosphatase activity can disturb the 5'-nucleotidase localization in the mouse cerebellum at pH 7.2. Since there is no acid phosphatase inhibitor known that does not also affect 5'-nucleotidase, except perhaps for Ni^{2+} (for the effect of tartrate on brain acid phosphatase, see Felicetti and Rath, 1975), a solution for this interference could be a pH shift toward the alkaline range. Cerebellar 5'-nucleotidase pattern remains present at pH 8–9 (Marani and Kurk, unpublished results). Activity at alkaline pH range is also found for many other nucleotidases.

Pure 5'-nucleotidase activity should be revealed when levamisole is used as an inhibitor for cerebellar alkaline phosphatase. At pH 8, levamisole exerts an inhibitory effect on the alkaline phosphatase of mouse cerebellum homogenate (see Fig. 19). When β-glycerophosphate was used as a substrate for the demonstration of alkaline phosphatase activity in mouse cerebellar sections, strong deposits of the reaction product, lead sulfide, appeared in the fiber layer. However, addition of levamisole (0.5 mM) reduces but does not totally inhibit alkaline phosphatase in the cerebellar fiber layer. Although this problem has not been satisfactorily resolved, it does emphasize the need for accurate studies of inhibition, even for generally accepted inhibitors.

An additional problem arose when biochemical studies demonstrated that levamisole (0.5 mM) activated AMPase activity at pH 8–10 in mouse cerebellar homogenates (see Fig. 19). Obviously, both 5'-nucleotidase activity and AMPase activity remain in sections of the fiber layer. Such AMPase activation clearly precludes the possibility of distinguishing between 5'-nucleotidase activity and alkaline phosphatase activity in the fiber layer at pH 8–10 when levamisole is used to inhibit the latter activity. Arguments in favor of 5'-nucleotidase localization purely in axons have been based on inconclusive results in the literature (see Section 4.3).

Another example of the confusion that exists regarding inhibitors concerns the Ni^{2+} inhibition of 5'-nucleotidase. Determination of 5'-nucleotidase activity has diagnostic value in hepatic diseases and osteoblastic metastases (see Van der Slik, 1975). Campbell (1962) introduced a method for determination of 5'-nucleotidase activity. The difference in activities with and without Ni ions was assumed by Campbell (1962) to represent true 5'-nucleotidase activity. In the study of Ahmed and Reis (1958), it was generally accepted that placental 5'-nucleotidase activity was inhibited by Ni^{2+}. On the assumption that Ni^{2+} acts in the same manner in liver and bone, this method (Campbell, 1962) was generally introduced in clinics. There was already some confusion present in the literature concerning the use of Ni^{2+} for 5'-nucleotidase (see Van der Slik, 1975). It is also known that Ni^{2+} seems to be an activator for yeast, snake venom, and bull seminal 5'-nucleotidase (Drummond and Yamamoto, 1971), while Ni^{2+} is also reported as a bone acid phosphatase inhibitor (Schwartz and Bodansky, 1964, 1965). Scott (1965) published his results for the effects of various Ni^{2+} concentrations on the

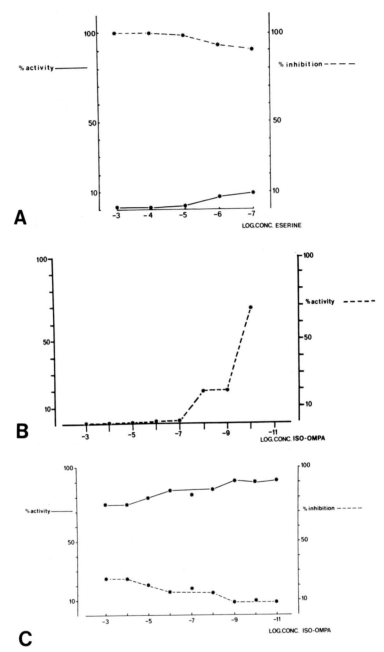

Fig. 18A–C. (A) Effect of eserine (physostigmine) on AChE activity in homogenates of the IO area of the cat. Nearly total inhibition is achieved, which indicates the absence of nonspecific esterases. Biochemical determinations were performed according to Marani (1977). (B) Effect of iso-OMPA on butyrylthiocholinesterase activity in homogenates of the IO of the cat. Nearly total inhibition is achieved at concentrations higher than 10^{-7} M iso-OMPA. (C) Effect of iso-OMPA on acetylthiocholinesterase activity in homogenates of the IO area of the cat. Partial inhibition is achieved over the whole range of iso-OMPA concentrations used.

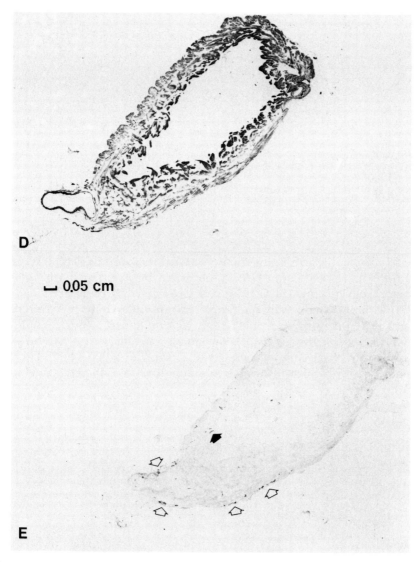

Fig. 18D,E. (D) Midsagittal sections of the rat bladder and urethra incubated for AChE activity according to Karnovsky and Roots (1964). (E) Effect of iso-OMPA administration (30 min preincubation with 10^{-4} M iso-OMPA) on AChE activity in sections of rat bladder and urethra. (⟁) Juxtamural ganglia; (⬥) an intramural activity of a nerve fiber.

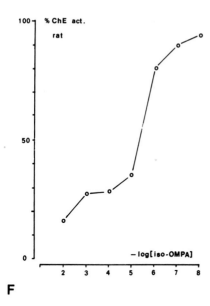

F

Fig. 18F. Effect of iso-OMPA on homogenates of the rat bladder and urethra treated with 0.1% Triton ×100. At low concentrations, iso-OMPA inhibits the acetylthiocholinesterase activity only slightly. A plateau is reached at concentrations of 10^{-3} and 10^{-4} M iso-OMPA. At higher concentrations, iso-OMPA still inhibits acetylthiocholinesterase activity. Courtesy of Dr. P. Klück.

activity of the cerebellar 5′-nucleotidase pattern. It is obvious that nickel ions do not disturb the band pattern (see Table 8) but do cause the disappearance of intrasomatal Purkinje cell activity and Golgi cell activity (determined as nonspecific phosphatase activity). Since this example shows the different effects in different tissues of one cation (Ni^{2+} inhibited 5′-nucleotidase in placenta and not in the cerebella), one must conclude that a careful and thorough study of the use of these types of inhibitors is needed. The 5′-nucleotidase determination method is now, according to Campbell, no longer commonly used in clinics.

5.3. Centrifugation Techniques

Glenner (1965) writes:

> As distinct from the biochemist, the histochemist has been primarily concerned with the description of those enzymes that are present in tissues in an isoluble state—or can be so made by a variety of denaturing techniques—and whose activity can be demonstrated by the formation of a dense precipitate at the site of the enzymic activity.

Biochemists, on the other hand, have to aid them centrifugation techniques to separate different populations of cell fractions (e.g., Gray and Whittaker,

Table 8. Activities after Ni^{2+} Administration Determined According to Scott (1965)[a]

Ni^{2+} concentration (M)	10^{-2}	10^{-3}	10^{-4}	10^{-5}	10^{-6}
5′-Nucleotidase band pattern	88	104	127	104	96
Intracellular 5′-nucleotidase	0	100	161	100	100

[a] The activity measured without Ni^{2+} is 100.

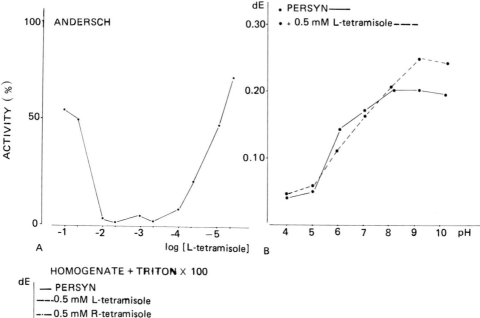

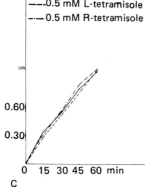

Fig. 19. (A) Influence of *l*-tetramisole on alkaline phosphatase (pH 8) activity determined according to the method of Andersch and Szcypinski (1947) in cerebellar homogenates treated with 0.1% Triton ×100 (see Marani, 1977). (B) Activation of 5′-nucleotidase by *l*-tetramisole in the alkaline range. (C) Demonstration of the lack of effect of either *l*- or *r*-tetramisole on 5′-nucleotidase.

1962). They can determine, following centrifugal separations, whether an enzyme is localized in a microsomal, a lysosomal, a mitochondrial, a synaptosomal, or another fraction (see Volume 2, Chapter 3). A rough but useful localization of enzymes can be determined. Enzymes can be examined with centrifugation techniques that are not well suited to undergo enzyme-histochemical methods.

Hardonk *et al.* (1975) used centrifugation techniques and then performed enzyme-histochemical analyses on their embedded pellets. A comparison could be made in this way between centrifugation (biochemical) techniques and enzyme histochemistry. They obtained good agreement between their biochemical and histochemical data on marker enzymes in subcellular particles such as the mitochondrial marker enzyme succinate dehydrogenase, α-glycerophosphate, the lysosomal marker enzyme acid phosphatase, and the plasma membrane markers 5′-nucleotidase and ATPase derived from rat livers, mouse spleens and ovaries, and bovine adrenals. Since it was impossible for Scott (1964), 1965) to correlate 5'-nucleotidase with any structure in the cerebellum observable with light microscopy, it is interesting to know the subcellular distribution of the 5'-nucleotidase activity. It is the more interesting because Pilcher and Jones (1970) stated in a short communication that the 5'-nucleotidase activity over the different mouse cerebellar fractions (Gray and Whittaker, 1962) has a distribution that "reflects that of specific synapses containing 5'-nucleotidase" and that "it may be possible to distinguish two types of synapses on the basis of their 5'-nucleotidase content." However, when Persijn's method (Persijn et al.,1969; Persijn and Van der Slik, 1970) of determining 5'-nucleotidase activity is used, rather than the Fiske and Subbarrow (1925) phosphate-determination method, it was difficult to replicate the Pilcher and Jones (1970) results. In fact, the distribution of 5'-nucleotidase among the various fractions differed even though the method for preparing the P_2 fraction was varied only slightly. Pilcher and Jones (1970) used the original method of Gray and Whittaker (1962), while we followed that of Israël and Frachon-Mastour (1970). The latter is essentially that of Gray and Whittaker (1962) but with a modification of preparing the P_2 fraction (Whittaker et al., 1964). The P_2 fraction in our experiments was centrifuged for 30 min at 10,000g, instead of for 1 hr at 17,000g. Our results did not confirm a distribution reflecting specific synapses containing 5'-nucleotidase, as did the results of Pilcher and Jones (1970). Accordingly, we believe only that 5'-nucleotidase must be loosely bound to one or more cell organelles or structures that appear in both the P_2 and P_3 fractions (see Marani, 1977).

As can be noted in Table 9, the specific activities are nearly equal in all the experiments. The recovery of the total 5'-nucleotidase activity in these experiments is high. The P_1 fraction, as stated previously (see Fig. 3), contains mainly nuclei and debris. Our results for the percentage of activity in the P_1 fraction of 5'-nucleotidase are in the range of Pilcher and Jones (1970), which makes it clear that a portion of the 5'-nucleotidase activity is associated with the nuclei and the debris.

As Table 9 shows, only 44% of the total activity is recovered in P_2, while Pilcher and Jones (1970) recovered 78.8 and 67.9% in their two experiments. It is unlikely that this difference can be explained by the difference in centrifugation, although Whittaker et al. (1964) stated that the difference in the method resulted in a lower yield of synaptosomes. The difference can be

Table 9. 5'-Nucleotidase Activity of Different Fractions at pH 7.2[a]

	Homogenate	P_1	P_2	S_2
Mean of Expts. 1–6	107.6 ± 2.6	23.6 ± 0.6	47.6 ± 2.0	38.5 ± 2.6
Relative %	—	21%	44%	36%
Recovery: 103%				

[a] The means and standard deviations of 6 experiments are given. The specific activity is expressed in mIU/ml per min.

*Table 10. Relative Distribution of 5'-Nucleotidase
Activity at pH 7.2[a]*

| | Experiment No. | | | |
Fraction	10	11	12	Mean
P_1	15	14	15	15
P_2	39	36	39	38
P_3	15	14	15	15
S_3	30	35	30	32

[a] The activity is expressed as percentages of the total activity.

explained by the solubility of 5'-nucleotidase. Pilcher and Jones (1970) did not find much activity in the S_2 fraction. In our experiments, nearly 36% of the 5'-nucleotidase activity of the total homogenate was recovered in S_2.

In the microsomal fraction (P_3), 15% of the 5'-nucleotidase activity of the total homogenate was recovered (Table 10). This is nearly the same activity as was recovered by Pilcher and Jones (1970).

As already described above with respect to the P_2 fraction, it is obvious— also from earlier experiments (Marani and Voogd, 1977)—that part of the 5'-nucleotidase enzyme is soluble. In these experiments, nearly 32% of the activity was recovered in the soluble fraction, S_3. Pilcher and Jones (1970) did not find any activity in the supernatant of P_3, although the same speed and time were applied. To check whether the 5'-nucleotidase activity left in the P_2 fraction is bound to a special type of synaptosome, we did three experiments using the linear-sucrose-gradient technique (Van der Krogt, 1974). The results are shown in Fig. 20. The interpretation given suggests that 5'-nucleotidase is not bound to mitochondria, because there is no overlap of the monoamine oxidase (MAO) peak and the one for 5'-nucleotidase. The cytoplasmic marker lactate dehydrogenase (LDH) overlaps the 5'-nucleotidase peak. The sharpness of the peak for 5'-nucleotidase and the overlap with the top of the LDH peak can be seen as an argument for the binding of this part of the 5'-nucleotidase activity to a certain type of particles within the synaptosome pool. At the top of the gradients, an amount of 5'-nucleotidase activity is found that is interpreted as 5'-nucleotidase in solution or bound to ruptured structures.

Considering the relative distribution of the 5'-nucleotidase activity over the sum of the acitivity of the different fractions, it is obvious that the P_2 and S_3 fractions have the same 5'-nucleotidase activity, while the P_3 fraction contains only half that of either the P_2 or the S_3 fraction. The comparison of the mean specific activities of the fractions ($P_1 = 3.3$, $P_2 = 6.5$ $P_3 = 6.2$, and $S_3 = 7.1$) does not indicate that there is a difference between the P_2, P_3, and S_3 ratio per milligram of protein. The results found in these experiments for the P_2 fraction are more in agreement with the results of Israël and Frachon-Mastour (1970) in the rat cortex than with the results of Pilcher and Jones (1970).

Using centrifugal techniques for cerebellar 5'-nucleotidase, a distribution

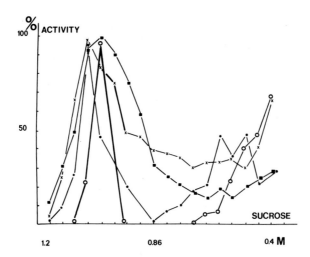

Fig. 20. Linear sucrose gradients from crude P_2 fractions. Each point in the figure is the mean of three experiments. (○) 5′-Nucleotidase activity; (★) MAO activity; (×) protein determinations; (■) LDH activity. All points are represented as percentages of the fractions with the highest content of enzyme activity or protein. The recoveries are: protein, 104%; 5′-nucleotidase, 95%; MAO, 92%; LDH, 75% (Marani, 1977).

of cerebellar 5′-nucleotidase over several fractions was found that agreed partially with the results of Pilcher and Jones (1970), except for the soluble fraction (see also Marani, 1979). Pilcher and Jones (1970) concluded from their cerebellar 5′-nucleotidase experiments an axoplasmic and a synaptic distribution for this enzyme.

Cerebellar 5′-nucleotidase is very labile and easily denatured by fixatives (Scott, 1965), but rapid preparation of ultrathin sections by a short perfusion fixation followed by a short perfusion incubation and, afterward, an immersion incubation (see Section 6.2) can show 5′-nucleotidase at the ultrastructural level with enzyme-histochemical methods. A location within parallel fibers and parallel fiber synapses was encountered, but an intradendritic location in Purkinje cell subsurface cisternae and spine apparatus was also noted. It was also clear from light microscopy that basket cell axons contain 5′-nucleotidase. Our sucrose-gradient centrifugations (see Fig. 20) showed a population within the synaptosomal fraction that contained 5′-nucleotidase activity, but the conclusions that a synaptic localization with the origin of this activity (Pilcher and Jones, 1970) could be called premature.

Although centrifugation studies are very important, sometimes conclusions are drawn that are merely postulates, as is shown for 5′-nucleotidase distribution. These postulates need an ultrastructural enzyme-histochemical check, especially in nervous tissue that is built up of elements that can form far mor particles by fractionation than are expected from, for example, liver fractionation.

5.4. Isoenzymes

Using AMP as a substrate, 5'-nucleotidase localization in the cerebellum (see Fig. 15) can be influenced by the incubation time. In incubations using short time periods, 5'-nucleotidase is present in a band pattern shown by positive and negative areas. The negative areas separating the positive ones, in contrast to the situation in the fiber layer, are not strictly negative. After longer incubation periods (>2–3 hr), the negative areas, as compared to the still-negative fiber layer, are more positive, but are less positive than the positive bands. For any functional or topographical explanation, it seems important to know whether two different isozymes are present or one type in various quantitities.

5'-Nucleotidase in brain tissue has been studied over the last few years by Scott (1964, 1965, 1969), Ipata (1967, 1968), Hardonk and De Boer (1968), Bosmann and Pike (1970), and Suran (1974a,b). In the molecular layer of the cerebellar cortex of the mouse, Scott (1965) found reaction products of 5'-nucleotidase with different 5'-nucleotides to be located in a pattern of longitudinal bands (see Section 4.3) Moreover, he concluded that different forms of 5'-nucleotidase are present in the cerebellum of the mouse. At high magnifications, products of a specific mouse brain 5'-nucleotidase, reacting with AMP, CMP, and TMP (pH optimum 7.2), were found to be located inside the perikarya of Purkinje cells, whereas the reaction products wer located outside the perikarya when the substrates UMP, GMP, IMP, and NMN were used. The results of the experiments of Scott (1965, 1969) in which various cations were added to the incubation medium furnish additional proof for the existence of different isoenzymes. Hardonk and De Boer (1968) have investigated rat and mouse 5'-nucleotidase by means of agar electrophoresis. In several organs of the mouse, various isoenzymes were found. In homogenates of whole mouse brain, however, only one form of the 5'-nucleotidase enzyme (pH optimum 7.5) was found to be present. It has not yet been determined whether several forms of the enzyme 5'-nucleotidase contribute to the formation of the longitudinal pattern in the cerebellar molecular layer or whether this is due to quantitative differences in enzyme products of only one form of this enzyme. This information may be a necessary prerequisite for quantification of the 5'-nucleotidase band pattern with the method of Scott (1969) and for ultrastructural localization of the 5'-nucleotidase.

The advantages of disk electrophoresis (Brewer, 1970) over agar electrophoresis are offset by the greater variability in the results. To overcome this deficiency, densitometric scans of samples from different mice were compared. Where large discrepancies were encountered in densitometric scans, an additional sample of the aqueous phase of the homogenized cerebella (these samples were stored at −50°C) was taken, and electrophoresis was performed again. Another cause for the variablity of our results, and possible explanation for the discrepancy between some of our findings and those reported in the literature, may be found in the extraction procedures. The specific activities of 5'-nucleotidase, acid, nonspecific, and alkaline phosphatase were therefore determined in cerebellar homogenates extracted with butanol in our initial experiments following the procedure of Hardonk and De Boer (1968). In later experiments, butanol was excluded. The homogenates obtained from one cerebellum were divided into two equal portions. One volume was subjected to one of the three extraction procedures (butanol, electrophoretic buffer, or Triton X100). The remaining volume was used to determine enzyme activity in the homogenate. The 5'-nucleotidase activity was determined with the method of Persijn and Van der Slik (1970) and nonspecific phosphatase with *p*-nitrophenylphosphatase (Andersch and Szcypinski, 1947). In addition, proteins in each sample were determined with the method of Lowry *et al.* (1951).

In Table 11, the specific activities of the butanol, electrophoretic buffer, and Triton X100 extracts are given for acid phosphatase, 5'-nucleotidase, alkaline phosphatase, and nonspecific phosphatase. These results indicate that enzyme activities of homogenates treated with butanol are consistently lower than in comparable homogenate extracts. Moreover, a strong variability in enzyme activity after butanol treatment is present. These low values cannot be explained solely by the incomplete extraction of lipoproteins by butanol (Pasquini and Sato, 1972). When the latter values are taken into account, still only 25–30% of the total homogenate activity for 5'-nucleotidase is found in homogenates treated with butanol. Since butanol may also exert an influence on the isoenzyme pattern (Kabara and Konvich, 1972), disk electrophoresis of homogenates treated with butanol was repeated with aqueous extracts and with extracts of Triton X100. No important differences were noticed.

5.4.1. Densitometric Scans of 5'-Nucleotidase

In most 5'-nucleotidase densitograms of incubations with AMP at pH 7.2, there are large overlapping peaks, and in addition a small peak may follow the large values (Fig. 21). The presence of several peaks in the densitogram means either that there are several isoenzymes of 5'-nucleotidase or that acid phosphatase also acts on AMP at pH 7.2. Should the latter be the case, the densitogram of 5'-nucleotidase actually represents the sum of the reaction products of 5'-nucleotidase and nonspecific phosphatase.

5.4.2. Densitometric Scans of Incubations for Acid Phosphatase with β-Glycerophosphate at pH 5.0

The densitometric scans generally show three peaks (Fig. 21). In most cases, the large peaks overlap; in addition, a small peak is present. The findings reported by Scott (1965) made it clear that β-glycerophosphate is not broken down by cerebellar 5'-nucleotidase. Persijn *et al.* (1969) and Belfield and Goldberg (1976) even found an inhibition of 5'-nucleotidase activity by β-glycerophosphate. Comparison of the acid phosphatase scans,

Table 11. Specific Enzyme Activities in Various Extracts

Enzyme	Butanol		Electrophoretic buffer		Triton X100 (0.1%)	
	Homogenate	Supernatant	Homogenate	Supernatant	Homogenate	Supernatant
Acid phosphatase	36.13 ± 0.86	0.86 ± 0.46	40.35 ± 0.82	14.30 ± 1.08	42.13 ± 2.91	30.43 ± 1.20
5'-Nucleotidase	231.00 ± 14.1	24.90 ± 3.2	153.16 ± 4.32	103.50 ± 2.73	322.00 ± 7.25	280 ± 16.00
Alkaline phosphatase	39.26 ± 4.2	32.13 ± 3.1	22.11 ± 0.64	12.24 ± 0.46	43.80 ± 3.4	28.33 ± 3.88
Protein (mg/ml)	8.36 ± 0.03	2.16 ± 0.06	6.98 ± 0.09	3.92 ± 0.04	8.30 ± 0.05	7.12 ± 0.11
Nonspecific phosphatase	14.56 ± 0.46	2.56 ± 0.92	16.20 ± 0.69	5.22 ± 0.20	15.13 ± 0.23	11.70 ± 0.24

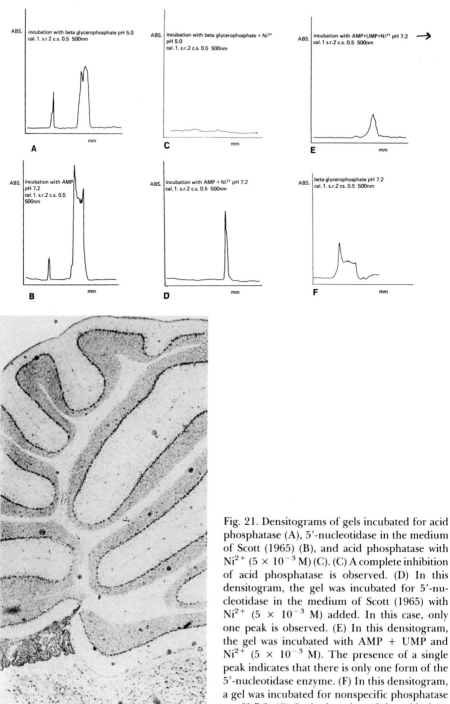

G ⌙ 0.125 mm

Fig. 21. Densitograms of gels incubated for acid phosphatase (A), 5′-nucleotidase in the medium of Scott (1965) (B), and acid phosphatase with Ni²⁺ (5 × 10⁻³ M) (C). (C) A complete inhibition of acid phosphatase is observed. (D) In this densitogram, the gel was incubated for 5′-nucleotidase in the medium of Scott (1965) with Ni²⁺ (5 × 10⁻³ M) added. In this case, only one peak is observed. (E) In this densitogram, the gel was incubated with AMP + UMP and Ni²⁺ (5 × 10⁻³ M). The presence of a single peak indicates that there is only one form of the 5′-nucleotidase enzyme. (F) In this densitogram, a gel was incubated for nonspecific phosphatase at pH 7.2. (G) Sagittal section of the acid phosphatase localization prepared according to the method of Barka and Anderson (1962).

however, strongly suggests that part of the former may in fact contain a residual activity of acid phosphatase at pH 7.2. Previously, Ipata (1966) indicated that 5'-nucleotidase coincided with nonspecific phosphatases. Hardonk and De Boer (1968), however, using β-glycerophosphate, found hardly any acid phosphatase activity in butanol extracts of the mouse brain at pH 7.2. To substantiate our findings of acid phosphatase activity with β-glycerophosphate in butanol and aqueous extracts at pH 5.0, we repeated the incubations of butanol extracts with *p*-nitrophenylphosphate as a substrate. The same densitograms resulted after incubations of the gels. Subsequently a pH–activity curve for nonspecific acid phosphatase was prepared using *p*-nitrophenylphosphate as a substrate (Fig. 22). Although the pH optimum lies at pH 5.0, we found an activity of acid and alkaline phosphatases of 30–40% of the maximum between pH 6.8 and 7.4.

We can conclude, therefore, that a residual activity of the acid phosphatase is present at the pH at which the reaction for 5'-nucleotidase is measured. There is also some alkaline phosphatase activity at this pH, but judging from the slope of the curve in Fig. 22, we assume that it will be a minor one compared with that of the acid phosphatase. To demonstrate which part of the densitogram of 5'-nucleotidase (Fig. 21) is really due to this enzyme, acid phosphatase must be selectively inhibited. Inhibition was obtained with Ni^{2+}.

According to Scott (1965), Ni^{2+} in a concentration between 10^{-3} and 10^{-4} M does not affect 5'-nucleotidase activity (see Section 5.2). We found that Ni^{2+} added in a concentration of 5×10^{-3} M to the incubation medium for acid phosphatase strongly inhibits the activity of butanol-treated homogenates. This can be shown for both the pH–activity curve (Fig. 22) and the densitometric scans of acid phosphatase (Fig. 21). In histological preparations, the Ni^{2+} concentration causes a slight reduction of the 5'-nucleotidase activity.

5.4.3. Densitometric Scans of Butanol Extracts with Ni^{2+} Added to the Incubation Medium for Acid Phosphatase

A 50–100% inhibition occurs in 80% of the mouse cerebellum when 5×10^{-3} M Ni^{2+} is added to the incubation medium for acid phosphatase (Fig. 21). It is not possible to obtain selective inhibition of acid phosphatase in 5'-nucleotidase incubations by increasing the concentrations of Ni^{2+}, because at higher concentrations the 5'-nucleotidase itself would be inhibited (Scott, 1965). The great variability in inhibition obtained in different mice may be due to the variation inherent in the butanol extraction referred to previously. In some mice, the acid phosphatase activity is completely suppressed in densitometric scans of butanol extracts incubated for 5'-nucleotidase with 5×10^{-3} M Ni^{2+}.

Variation in 5'-nucleotidase peaks occurs when (5×10^{-3} M) Ni^{2+} is added to the incubation medium. In most analyses, only one small peak is present; in others, three peaks can still be discerned. Total inhibition of the enzyme activity at pH 7.2, as can be seen in densitometer readings of acid

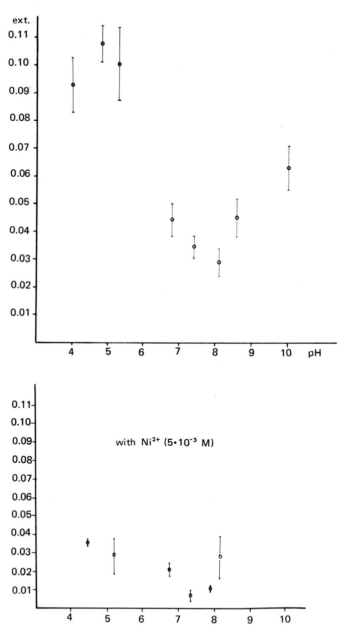

Fig. 22. Effect of Ni^{2+} on the phosphatase activity as determined according to Andersch and Szcypinski (1947).

phosphatase incubated at pH 5.0, was never observed. When densitometer readings with one small peak (Fig. 21) represent a situation in which a 100% inhibition of acid phosphatase was obtained, this may signify that only one isoenzyme of 5'-nucleotidase is present in the cerebellum of the mouse. Since Scott (1965) found different localizations for the reaction products of isoenzymes of 5'-nucleotidase reacting with two different groups of nucleotides, densitograms using a nucleotide from each group (UMP and AMP) were compared. We found identical densitometer patterns with AMP and UMP. Also, after incubation with AMP + UMP and Ni^{2+} (5×10^{-3} M), only one peak was found in the densitograms (Fig. 21). This may signify that only one isoenzyme of 5'-nucleotidase is present in the mouse cerebellum and that it reacts with both AMP and UMP.

5.5. Biological Rhythms

Rhythms in enzyme activity that are frequently encountered (see Schevin *et al.*, 1974) may be divided into exogenous and endogenous types. The light–dark cycle can induce an exogenous rhythm, for the determining factor is the "Zeitgeber" outside the organism. When the Zeitgeber is constant, the rhythm stops. Endogenous rhythms, on the contrary, are independent of factors from outside the organism, although the so-called "free-running" phenomenon may occur. Under constant environmental conditions, the free-running period normally takes on a value of 20–28 hr, after which the cycle is repeated. Such nearly 24-hr rhythms are called circadian rhythms. Day rhythms caused by exogenous or endogenous factors can influence biochemical and histochemical determinations if different experiments are compared (Mayersbach, 1964).

In our laboratory during other experiments performed on rats, we established the presence of daily fluctuations. They were verified for the dopaminergic system in striata, while in the same rats the cerebella were used to follow circadian rhythms for 5'-nucleotidase and nonspecific phosphatases. The latter enzyme activities were determined by their *p*-nitrophenylphosphatase activity. The methods used are described by Van Dijke (1975) for the dopaminergic part, while the cerebellar enzymes will be described by Marani (1980*b*). Every 2 hr, eight rats were used in the first experiment (1031/1974), and during the repetition (1218/1974), we used six rats every 4 hr. The experiments started at 9:00 AM and lasted till 9:00 the following morning. The male Wistar rats were kept at a constant temperature of 24°C, and a relative humidity of 55% was provided. The light period lasted from 6:00 till 18:00 hr, while entrance for feeding and other purposes was always at 12:00. The animals' average weight was nearly 400 g. The heads were removed with a "small-animal decapitator" and immediately frozen for 14 sec to overcome leakage of dopamine (DA) from synaptosomes and, it was hoped, to stop autolytic processes.

Figure 23 shows the effects measured for DA, tyrosine hydroxylase in the presence of the synthetic cofactor 2-amino-4-hydroxy-6,7-dimethylte-

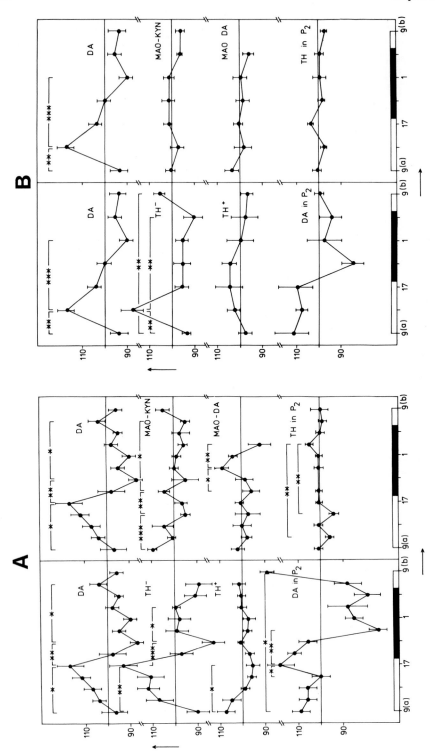

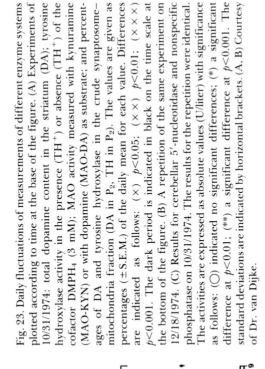

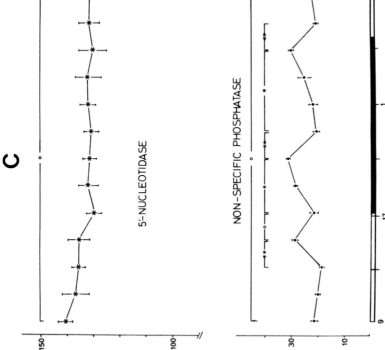

Fig. 23. Daily fluctuations of measurements of different enzyme systems plotted according to time at the base of the figure. (A) Experiments of 10/31/1974: total dopamine content in the striatum (DA); tyrosine hydroxylase activity in the presence (TH⁺) or absence (TH⁺) of the cofactor DMPH₄ (3 mM); MAO activity measured with kynuramine (MAO-KYN) or with dopamine (MAO-DA) as substrate; and percentages of DA and tyrosine hydroxylase in the crude synaptosome–mitochondria fraction (DA in P_2, TH in P_2). The values are given as percentages (±S.E.M.) of the daily mean for each value. Differences are indicated as follows: (×) $p < 0.05$; (××) $p < 0.01$; (×××) $p < 0.001$. The dark period is indicated in black on the time scale at the bottom of the figure. (B) A repetition of the same experiment on 12/18/1974. (C) Results for cerebellar 5′-nucleotidase and nonspecific phosphatase on 10/31/1974. The results for the repetition were identical. The activities are expressed as absolute values (U/liter) with significance as follows: (○) indicated no significant differences; (*) a significant difference at $p < 0.01$; (**) a significant difference at $p < 0.001$. The standard deviations are indicated by horizontal brackets. (A, B) Courtesy of Dr. van Dijke.

trahydropteridine (DMPH$_4$) (TH$^+$), and tyrosine hydroxylase in the absence of the cofactor (TH$^-$). The MAO determinations were performed with kynuramine as substrate (MAO-KYN) and with DA (MAO-DA). The total amount of DA and tyrosine hydroxylase was also determined in P$_2$ pellets of striatal tissues, but these determinations will not be discussed here. The results show a rhythm for striatal DA, and they are comparable as to minima and maxima with the results of Bobillier and Mouret [(1971) rat forebrain], Collu *et al.* [(1973) whole rat brains], and Di Raddo and Kellogg [(1975) rat telencephalon]. The results differ, however, from those of Schevin *et al.* [(1968) whole rat brains], who found a cycle of 5–8 hr.

The rate-limiting step in DA synthesis is the activity of tyrosine hydroxylase (Levitt *et al.*, 1965), while the regulation of this enzyme *in vivo* is probably determined by the cofactor tetrahydrobiopterine (Lovenberg and Victor, 1974). In the experiments of Dr. Van Dijke, the tyrosine hydroxylase activity fluctuated little in the presence of the exogenous cofactor, while high correlations were found between tyrosine hydroxylase activity in the presence of endogenous cofactor and the determined DA quantities (significant correlation of TH$^-$ and DA). The first experiments concerning MAO activity showed small fluctuations, but these could not be replicated. The results for the cerebellar enzymes are shown in Fig. 23. 5'-Nucleotidase activity in the cerebellum shows no daily fluctuations, as determined with the method of Persijn and Van der Slik (1970), while clear changes in activity at different times are found for nonspecific-phosphatase activities. The comparison of the daily fluctuations in striata and cerebella, for the enzymes studied, under a constant light-dark cycle show no correlation. It is clear that different enzymes in different parts of the nervous system can have various rhythms, although 5'-nucleotidase exhibited a constant activity, indicating that some enzyme systems do not show rhythms.

6. Neuroenzyme Histochemistry in Electron Microscopy

6.1. Introduction

The performance of neuroenzyme histochemistry on the subcellular level nearly always contains pitfalls, which are mainly inherent in the tissue studied and which may be avoided by trial and error. The difficulties that emerge from the tissue composition are due to the compactness of brain tissue, the fact that its tissue fluid is of a different composition than that of other tissues, and the fact that brain tissue is protected by a blood–brain barrier. The copper ferricyanide technique, for example, can easily be used in light-microscopic cryostat sections. However, the poor penetration capacity of copper ferricyanide becomes readily apparent in electron-microscopic studies due to the compactness of brain tissue (see also Lewis and Knight, 1977).

Substrate protection and perfusion incubation are sometimes impossible, since the substrate in these methods can be converted by enzymes of the blood–brain barrier, making it difficult to estimate the osmolality actually available for fixation in the brain tissue itself.

A good survey of problems one may meet with subcellular enzymehistochemistry is afforded by Lewis and Knight (1977), since, for most problems, examples derived from neuronal tissue are given that can be a great help in neuroenzyme histochemistry. In the following sections, two examples already extensively discussed in this chapter will be used to demonstrate the localization of a brain enzyme that cannot survive long-lasting fixations (5'-nucleotidase) and of an enzyme that is less sensitive to fixatives, such as cholinesterase.

6.2. 5'-Nucleotidase

Studies concerning 5'-nucleotidase always show the difficulties of the simultaneous demonstration of other hydrolases, as already indicated in this chapter. In addition to the presence of nonspecific phosphatases, cerebellar 5'-nucleotidase exhibits great sensitivity to fixation (Scott, 1965). On the other hand, the peculiar localization of 5'-nucleotidase in a band pattern makes one curious to know which cerebellar element or elements are responsible for its appearance. Its localization was eventually ascertained, in addition to which good preservation of the cerebellar molecular layer was attained.

6.2.1. Fixation

Sometimes, tissue can undergo a short incubation and afterward some of the tissue elements are still recognizable. The effect of a 5'-nucleotidase incubation without a preceding fixation is shown in Fig. 24. It is obvious that none of the structures can be identified on the basis of criteria derived from purely morphological studies. Long fixation times failed to show, in prolonged incubations, the distribution of this enzyme. However, 5'-nucleotidases from other sources have been demonstrated. Klaushoffer and Pavelka (1975) showed its presence (on the subcellular level) in the smooth muscle cells of the rat's gastrointestinal tube. An additional handicap in determining the 5'-nucleotidase localization was its distribution in a plexiform layer. Such layers are very sensitive to shrinkage of the tissue, resulting in wide intercellular spaces between the cerebellar elements. Partial destruction of the layers can then occur. Cerebellar 5'-nucleotidase activity can be demonstrated using a short perfusion fixation followed by an incubation period. A prolonged fixation after the incubation did promote tissue preservation (Marani, 1977). Animals were anesthetized and vascularly perfused with buffered physiological saline solution, to which sodium nitrite and heparin were added. Administration of sodium nitrite and heparin before the saline perfusion is also possible.

The following description can be used, is of a method for adapting the percentages of the fixatives for most enzymes that cannot withstand fixative

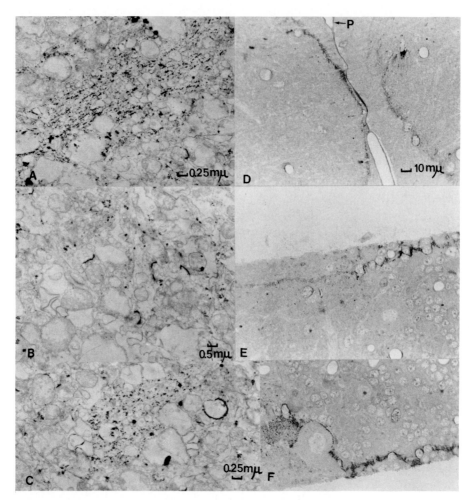

Fig. 24. (A–C) Effect of an incubation according to Scott (1965) for 5′-nucleotidase without any prefixation. Postfixation was achieved with 4% paraformaldehyde on 0.16 M cacodylate buffer. (A) The reaction product is localized in a restricted area lying obliquely from edge to edge in this photograph of the molecular layer. (B) The reaction product is shown in half-circle arrangements. (C) This photograph shows a restricted area within the molecular layer filled with reaction products. Nevertheless, it is unclear to which structures the reaction products are bound. (D–F) Results gathered with the method of Hanker *et al.* (1974) for the demonstration of AChE in 1-mμm sections. The penetration of 3′,3′-diaminobenzidine in a 100-mμm thick vibratome section is poor, and its conversion to an electron-dense product is limited to a narrow zone in this section of the cat cerebellum. (D) Shows this phenomenon at the transition of one molecular layer to another one separated by the pia mater (P). (E) Demonstrates the transition of molecular layer to granular layer, while (F) shows the same aspects around a Purkinje cell (P).

action. The use of two different fixative solutions, in our experience, has great advantages. With no pause between, two fixation solutions were used, the first of which contained 4% paraformaldehyde in 0.16 M cacodylate buffer (pH 7.3) and the second of which contained 2.5% glutaraldehyde in 0.16 M cacodylate buffer (pH 7.3) to which was added 5.4 g/liter sucrose (Rinvik and Grevova, 1970). The osmolalites were: saline solution, 300 mosmol; paraformaldehyde solution, 1400 mosmol; glutaraldehyde solution, 580 mosmol. The saline solution was perfused until all blood was replaced by it. The paraformaldehyde solution was perfused for 2 min, which was followed by nearly 5 min with glutaraldehyde. The perfusion speed was, in mice, 15 ml/min, and for cats, 50–60 ml/min. In our opinion, the high osmolality of the first fixative is needed for a good preservation of neuronal tissue, as also recorded by Carlstedt (1977) in his study of the transitional regions in the spinal cord. Our laboratory also found that separate administration of both fixatives is more advantageous than a mixture of formaldehyde and glutaraldehyde (see Carlstedt, 1977), but high percentages of fixatives in the first fixation solution are not that frequently reported in studies as producing rather good ultrastructural preservation. In the series studied, the molecular layer was always better preserved with the inclusion of a postfixation. Care must be taken that solutions are prepared fresh, as also seems to be the case for paraformaldehyde solutions (Lewis and Knight, 1977). In studies concerning normal structures in the central nervous system, Ca^{2+} ions can be prescribed in the buffers or fixatives. It must be noted that addition of cations can have a deleterious effect on the enzymes studied, resulting in an inhibition of the enzyme by Ca^{2+}.

6.2.2. Controls

As in most studies concerning enzyme registration on the subcellular level, it is necessary to use inhibition control studies. For 5'-nucleotidase studies, a general inhibitor for nonspecific phosphatases is unknown, and therefore it is always necessary to have checks on the presence of disturbing enzymes. In such cases, the best check is the localization of the disturbing enzyme itself (using β-glycerophosphate in the 5'-nucleotidase case) on the ultrastructural level, and the possibility of coincidence of localizations has to be regarded with care. Also recommended for 5'-nucleotidase is a mixture of 2' and 3' isomers of cytidine monophosphoric acid (= cytidylic acid). Sometimes, inherent controls can be present. From the light-microscopic studies, it was known that in the 5'-nucleotidase longitudinal pattern, after short incubation periods or after fixation, the negative areas contained no reaction product. These results must be confirmed by electron microscopy, and in the ultrathin sections studied, a measure of the demonstration of 5'-nucleotidase is the demonstration of the presence of negative areas.

The biochemical data on gradient-centrifugation procedures indicated a certain constituent of the synaptosomal pool with 5'-nucleotidase activity.

This result must be confirmable to the ultrastructural localization of 5'-nucleotidase. The presence of circumstantial data can be used to ascertain enzyme locations, and may be added as arguments in judgments on the reliability of such determinations.

6.2.3. Localization

The striped appearance, in transversal sections, of the 5'-nucleotidase band pattern after butanol treatment or short fixation periods did indicate a localization perpendicular to parallel fibers, but also paralleled by the distribution of Purkinje cells and climbing fibers. A rather surprising result was the discovery of the presence of 5'-nucleotidase reaction products in parallel fibers, but in accordance with the gradient-centrifugation studies, it was found here in boutons (Marani, 1977). 5'-Nucleotidase activity was also present in the Purkinje cell dendritic tree, as could be expected from the light-microscopic studies (Fig. 25). Its localization in the Purkinje cell was distributed over the dendritic tree and its appendages, the spines, but was absent in the somatas, as was most clearly demonstrated in rat Purkinje cells. The activity in Purkinje cell somatas in the mouse coincided light microscopically with nonspecific phosphatase activity and was therefore rejected as 5'-nucleotidase activity.

Within the Purkinje cell dendrite, a specialized structure is present, making identification easy. The subsurface cisternae or hypolemmal cisternae (Palay and Chan-Palay, 1974) characterize the Purkinje cell dendritic area. They are localized mainly beneath the plasmalemma and are connected to each other by branches of the agranular endoplasmic reticulum (Palay and Chan-Palay, 1974). In between the cisternae, or in the agranular endoplasmic reticulum, the 5'-nucleotidase reaction products are found within the main dendritic arborization (Marani, 1977) (see also Fig. 25). The Purkinje cell spines contain specialized structures, described by Gray (1959) for the cerebral cortex and denominated as the spine apparatus. The structure consists of two, three, or more tubular arrangements, but the dense bands between these tubular arrangements are absent in the cerebellar cortex. This apparatus is considered by Palay and Chan-Palay (1974) as a continuation of the agranular reticulum. The 5'-nucleotidase activity is found between the tubular arrangements of the spine apparatus (Fig. 25). The functional loop granular axon (= parallel fiber), in contact with Purkinje cell dendrites on its spine protrusions, seems to be determined by the presence of 5'-nucleotidase.

6.3. Acetylcholinesterase

The ultrastructural localization of acetylcholinesterase (AChE) activity has been studied in many parts of the nervous tissue. A good survey of the methods available for esterases is given by Lewis and Knight (1977). Further, the copper–glycine methods can be strongly recommended for use in the central nervous system. In the same chapter, a simplified adaptable incubation procedure is given. In sections, preferably not thicker than 50 nm, a good

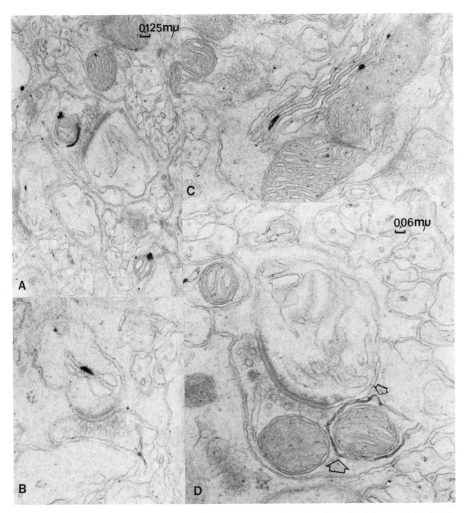

Fig. 25. Electron micrographs showing the reaction product for 5'-nucleotidase activity in the molecular layer of the mouse cerebellum. The technique used is described in the text and in Marani (1977). (A) Shows the reaction product in a parallel fiber bouton synapsing on a Purkinje cell spine. The reaction product seems to be localized in a semicircle around a mitochondrion. However, a comparison with (D) shows that the reactivity is localized in cisternae that are present around the mitochondria of the same system [i.e., a parallel fiber bouton synapsing on a Purkinje cell spine (see arrow)]. (B) Shows the presence of reaction product in the spine apparatus of the Purkinje cell spine. (C) Shows the reaction product in the subsurface cisternae of a Purkinje cell dendrite. In general, the 5'-nucleotidase activity in the parallel fiber to Purkinje cell dendrite system is localized endogenously. Compare this localization to the results for AChE in the molecular layer of the cat cerebellum in Fig. 27.

penetration of the incubation medium is obtained with this method. To prevent diffusion artifacts at the electron-microscopic level, a rather low pH (between 5 and 6) must be maintained. This is a pH well away from the pH optimum of AChE. The technique, therefore, is relatively insensitive and shows mainly spots of high activity or high quantities of enzyme.

6.3.1. Fixation

It is well established that brain AChEs are relatively insensitive toward fixation (Lewis and Knight, 1977). The AChE pattern as shown in the molecular layer of the cat cerebellum can be fixated for over 40 min in both the formaldehyde and the glutaraldehyde fixative, as described above, without an obviously severe loss of enzyme activities. Nevertheless, the incubation for esterases itself, independent of the fixation time, does easily disturb the ultrastructure of the studied tissue, since incubation times over 1 hr are needed in a medium that in most cases is hypotonic. A strong disturbance of the ultrastructure was noticed with the method of Karnovsky and Roots (1964). In our material, the maintenance of a rather good ultrastructure was effected with the Lewis and Shute method (see Lewis and Knight, 1977).

6.3.2. Controls

In cholinesterase studies, inhibitors or alternative substrates or both are used. The use of inhibitors is recommended even though the inherent specificity of the copper–glycine technique is high. In studies of nervous tissue, to be sure of the demonstration of true AChE, acetylcholine iodide is used in combination with the inhibitor (iso-OMPA) or ethopropazine. The same holds for the detection of pseudocholinesterases with butyrylthiocholine iodide and BW 284C51 or BW 62C47 (both numbered substances are inhibitors for true AChE). Since the copper–glycine technique shows areas with high enzymatic activity, a rather spotted appearance of the cholinesterase activity is demonstrated. The copper ferricyanide procedure has the advantage, in our experience, that it shows the overall location of the enzyme studied. Although the copper ferricyanide procedure has serious drawbacks, it can be used to check the overall localization in the subcellular structures, while the addition of the copper–glycine technique can be used to study specific localizations within or around subcellular structures (see Fig. 25).

6.3.3. Techniques

Some preliminary results can be given for AChE localization within the cat's cerebellar molecular layer as this relates to the techniques used (see Table 12). For the ultrastructural localization of this enzyme, we used three reactions, the Lewis ans Shute (1964, 1966), the Karnovsky and Roots (1964), and the Hanker et al. (1973) methods. The Lewis and Shute (1964, 1966) method is a modification of the copper–glycine reaction mechanism introduced by Koelle and Friedenwald (1949). It involves two steps:

1. $Cu^{2+} + 2 \text{ glycine} \longrightarrow Cu(\text{glycine})_2 + 2H^+$

   In a mild alkaline environment, the copper–glycine is stable.
   In an acid pH range (5–6), some free copper ions are present, but an excess of glycine prevents this problem.

2. AThCh + Cu(glycine)$_2$ + 2H$_2$O $\xrightarrow{\text{AChE}}$ Cu(ThCh) $\wp$ + 2 acetic acid + 2 glycine

At pH 6.0–6.5 the optimal precipitate of copper thiocholine is produced, while the optimum for the enzyme is approximately pH 8.

By the constitution of the incubation medium, an excess of $_4^{2-}$ ions is produced that forms crystals with the Cu(ThCh)$_2$ precipitate (Malmgren and Sylven, 1955). The proposed structure of these crystals is

$$[Cu^+\text{-S-}C_2H_4\text{-}N^+\text{-}(CH_3)_3]\ SO_4{}^{2-}$$

Copper itself is electron-dense and can be visualized in the electron microscope.

The Karnovsky and Roots (1964) and the Hanker *et al.* (1973) methods are closely related. Acetylthiocholine is converted by AChE to thiocholine and acetic acid by addition of Na$_3$Fe(CH)$_6 \cdot$ H$_2$O and Cu^{2+} to the incubation. Cupriferrocyanide is formed, which is known as Hatchett's brown. In the Hanker *et al.* (1973) method, the cupriferrocyanide can enhance the oxidative polymerization of 3′,3′-diaminobenzidine, and by the subsequent osmication, an electron-dense reaction product is formed. The results of the Karnovsky and Roots (1964) method for electron microscopy are shown in Fig. 26. In a small penetration area, reaction products are formed. The bad penetration is caused by the poor diffusibility of the ferricyanide ion into the tissues. Since the underlying principle of the Hanker *et al.* (1973) method is the Karnovsky and Roots (1964) method, it suffers from the same disadvantages (see Fig. 24). In the penetration area, reaction products can be studied with both techniques, although several artifacts can be recognized (Fig. 27). The localization, however, in both techniques equals the results of the Lewis and Shute (1964, 1966) technique in the cat's cerebellar molecular layer. The Lewis and Shute technique is designed to overcome some disadvantages of

Table 12. Electron Microscopy of the Acetylcholinesterase Pattern in the Cat Cerebellum

Series	Age	Kind of re-action[a]	Inhibitor (M)	Ultra-struc-ture	Enzyme activity	Remarks
E275	18 wk	K + H	10^{-4} iso-OMPA	−	+	—
E277	18 wk	K + H	10^{-4} iso-OMPA	−	+	—
E285	12 wk	L	10^{-4} iso-OMPA	+	+ +	Short rinse with succinate solution
E286	12 wk	L	10^{-4} iso-OMPA	+	+	3-hr rinse with succinate solution
E288	16 wk	L	10^{-4} iso-OMPA	±	+	Short rinse with succinate solution
E292	2 days	L	10^{-4} iso-OMPA	+	−	1-hr rinse with succinate solution
E293	3 days	L	10^{-4} iso-OMPA	+	−	1-hr rinse with succinate solution
E324	12 wk	L	10^{-4} iso-OMPA	±	+	1-hr rinse with succinate solution
E325	12 wk	L	10^{-4} iso-OMPA	±	+	1-hr rinse with succinate solution
E326	13 wk	L	10^{-4} iso-OMPA	+	+	24-hr rinse with succinate solution
E327	8 wk	L	10^{-4} iso-OMPA	+	+	24-hr rinse with succinate solution

[a] (K) Karnovsky and Roots (1964); (H) Hanker *et al.* (1973); (L) Lewis and Shute (1964, 1966).

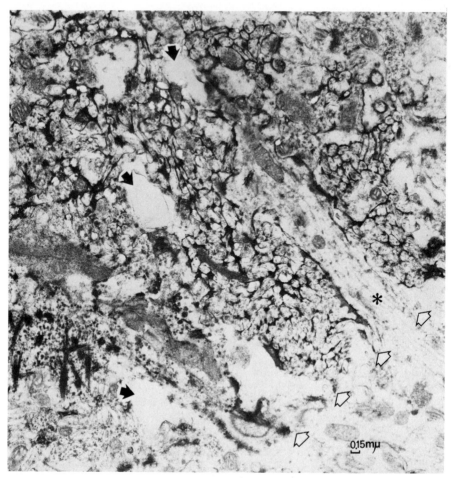

Fig. 26. Electron micrograph from a part of the molecular layer of the cat cerebellum that has been incubated after Karnovsky and Roots (1964). (△) Penetration front; (◆) holes that are frequently encountered with this technique. Nevertheless, reaction products can be found in the Purkinje cell dendrites (*) in an elongated agranular cisterna (running away from the top open arrow) and at the Purkinje cell membrane. The transversely cut parallel fibers are surrounded by reaction products of the AChE reaction.

the original copper–glycine method, (for a detailed description and comparison to other modifications, see Lewis and Knight, 1977). As shown in table 12, a rather good ultrastructural preservation is achieved, although the reaction products show a speckled effect (Fig. 28) because this method shows mainly areas with a high enzymatic activity (Lewis and Knight, 1977).

The combination of the three techniques described above allows one to study the overall distribution of the enzyme AChE, while in particular the Lewis and Shute method provides the better ultrastructural localizations with the fewest artifacts. Artifacts in the Lewis and Shute technique include reaction products in mitochondria and holes in areas with an overall reaction-product localization.

6.3.4. Localization

The description given here is a summary of the localizations found with the three techniques used as described in the previous section. AChE activity is found in the subsurface cisternae of Purkinje cell dendrites (Figs. 26–28) and in the agranular endoplasmic reticulum (not shown). As in the 5′-

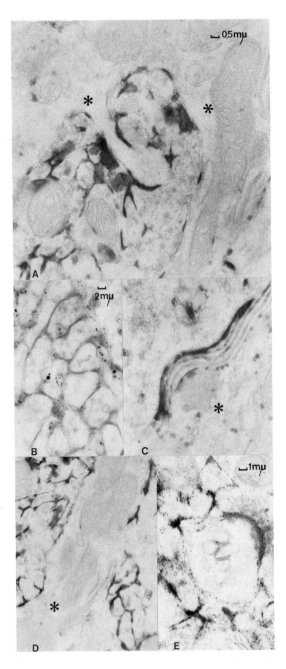

Fig. 27. Localization of the reaction products for AChE studied by the method of Hanker *et al.* (1974) (A,C,D) Asterisks (*) denote Purkinje cell dendrites. (A) The reaction products are localized around a Purkinje cell spine that extends into transversely cut parallel fibers. (C) In the Purkinje cell dendrite, the reaction products are found between the subsurface cisternae; (D) the reaction product is also confined to the cell membrane of the Purkinje cell dendrites and parallel fibers. (E) The synapse between the Purkinje cell spine and the parallel fiber is free of AChE reaction products. The reaction product can be distributed in droplets, showing an artifact of the Hanker *et al.* (1974) technique.

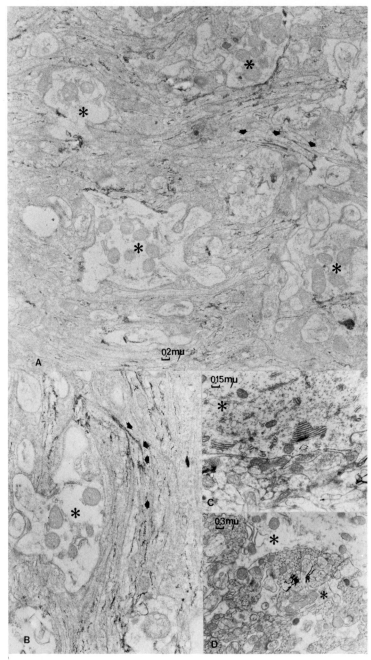

Fig. 28. (∗) Purkinje cell dendrites. (A) A horizontal section through the molecular layer of the cerebellum of the cat. Reaction product is found with the Lewis and Shute technique (Lewis and Knight, 1977) at the Purkinje cell membrane around parallel fibers (▲) that are horizontally cut in this section. (B) Identical situation but at a higher magnification. (C) Activity in the Purkinje cell subsurface cisternae (△). (D) A Purkinje cell spine intruding between parallel fibers. Both the Purkinje cell spine and the parallel fibers contain AChE activity in the intercellular space and thus probably at their cell membrane surfaces.

nucleotidase pattern, the Purkinje cells can be subdivided into two subpopulations, one characterized by the presence of the enzyme AChE, the other by its absence. In the Purkinje cells positive for AChE, a second subdivision can be made. The cell soma is negative for AChE, while the dendritic tree contains this enzyme in some organelles, or the Purkinje cell spines contain AChE activity between the cisternae of the spine apparatus, thus involving a part of the dendritic tree that is of postsynaptic origin (Figs. 27 and 28). Among other structures present in the cerebellar cortex, the parallel fibers contain activity in the intercellular spaces and in the intercellular space between the Purkinje cell spine and the parallel fiber (Figs. 26–28).

The stellate cell contains positivity in the nuclear envelope and the rough endoplasmic reticulum. Other structures are now under investigation.

7. Summary

This chapter details histochemical studies on the localization of enzymes within the mammalian cerebellum. Research in our laboratory has led to the conclusion that a longitudinal band pattern was present for climbing fibers in the cat cerebellum (Groenewegen and Voogd, 1977) as well as mossy fibers in chicken (Vielvoye, 1977) and in rabbit (Van Rossum, 1969), while the arrangement of the Purkinje cell axons confirms such a distribution for the fiber layer (Voogd, 1964; Feirabend *et al.*, 1977). An expression of this organizational principle was also found on the biochemical level for two types of enzymes, 5'-nucleotidase and acetylcholinesterase, in the molecular layer, and for monoamine oxidase in the fiber layer. In the granular layer of the rat, such a distribution for acetylcholinesterase was also present (unpublished results).

Since the presence or absence of an enzyme in one and the same cell type is directly related to differences in gene expression, it must be concluded that although cell types may have the same morphological characteristics, they differ in their biochemical content as expressed by the presence or the absence of certain enzymes. Therefore, what was considered as one uniform structure present in the whole cerebellum now has to be subdivided into two structures, alike in morphology, but different in content. Since both types of enzymes discussed in this chapter, 5'-nucleotidase and acetylcholinesterase, can break down substrates that may be involved in neurotransmission and since their localization, as studied so far, is related to pre-and postsynaptic structures within boutons or spines, a direct link with neurotransmission events is acceptable.

These facts indicate far-reaching significance for those studying the cerebellum with biochemical or physiological methods.

ACKNOWLEDGMENTS. The author would like to thank Dr. D. Williams, Department of Pathology, Welsh National School of Medicine, Cardiff, Wales, for providing Fig. 3, Dr. K. Van Dijke for part of Fig. 22, and Dr. M. Van der Waard for the use of the scheme in Fig. 1. Dr. H.K.P. Feirabend kindly provided part of Fig. 16, and I gratefully acknowledge his critical evaluation of the comparison between MAO distribution and thick-fiber distribution. The author is very grateful to Dr. R. Vosseberg for the simultaneous checks of incubations, critical reading of part of the manuscript, and the elucidating discussions, and to Dr. J.A. Van der Kroft for the use of clorgyline and the discussions on centrifugation techniques. Thanks are equally due to Dr. P.R. Lewis for introducing me to the copper–glycine techniques and for providing me in advance with the description of the Lewis and Shute method (see Lewis and Knight, 1977). The author is indebted to Dr. J. Voogd, who permitted me to spend time for finishing this chapter, and to all members of our laboratory for their cooperation. Finally, the author wishes to thank M.M. Nihom for correcting parts of the manuscript and Mrs. E. Bruyn-Van der Staay for typing this chapter and her persisting criticism on my careless wording. The cooperation of the *Journal of Comparative Neurology*, the *Journal of Anatomy*, and the *Journal of Experimental Neurology* for the use of figures is highly appreciated.

References

Adams, E., and Norton, I.L., 1964, Purification and properties of inducible hydroxyproline-2-epimirase from *Pseudomonas, J. Biol. Chem.* **239**:1525.

Ahmed, Z., and Reis, J.L., 1958, The activation and inhibition of 5′-nucleotidase, *Biochem. J.* **69**:386.

Albers, R.W., Siegel, G.W., Katzman, R., and Agranoff, B.W., 1972, *Basic Neurochemistry*, Little, Brown, Boston.

Aldridge, W.N., 1953, The differentiation of true and pseudo-cholinesterase by organophosphorus compounds, *Biochem. J.* **53**:62.

Alksne, J.F., Blackstad, T.W., Walberg, F., and White, L.E., Jr., 1966, Electron microscopy of axon degeneration: A valuable tool in experimental neuroanatomy, *Ergeb. Anat. Entwicklungsgesch.* **39**(1):1.

Andersch, M., and Szcypinski, A.J., 1947, Use of *p*-nitrophenyl-phosphate as the substrate in determination of serum acid phosphatase, *Am. J. Clin. Pathol.* **17**:571.

Aprison, M.H., 1975, Comments on potential new transmitter candidates, p. 9, ISN Barcelona Meeting.

Arvy, L., 1966, Cerebellar enzymology, *Int. Rev. Cytol.* **20**:277.

Armstrong, D.M., Harvey, R.J., and Schild, R.F., 1974, Topographical localization in the olivocerebellar projection: An electrophysiological study in the cat, *J. Comp. Neurol.* **154**:287.

Austin, L., and Phillis, J.W., 1965, The distribution of cerebellar cholinesterases in several species, *J. Neurochem.* **12**:709.

Baker, J.R., 1944, The structure and chemical composition of the Golgi Element, *Q. J. Microsc. Sci.* **85**:72.

Barka, T., and Anderson, P.J., 1962, Histochemical methods for acid phosphatase using hexazonium pararosanilin as coupler, *J. Histochem. Cytochem.* **10**:741.

Becker, N.H., 1961, The cytochemistry of anoseic and anoseic–ischemic encephalopathy in rats. II. Alterations in neuronal mitochondria identified by diphosphopyridine and triphosphopyridine nucleotide diaphorase, *Am. J. Pathol.* **38**:587.

Becker, N.H., 1962, The cytochemistry of anoseic and anoseic–ischemic encephalopathy in rats. III. Alterations in the neuronal Golgi apparatus identified by nucleoside diphosphatase activity, *Am. J. Pathol.* **40**:243.

Becker, N.H., and Barron, R.D., 1961, The cytochemistry of anoseic and anoseic–ischemic encephalopathy in rats. I. Alterations in neuronal lysosomes identified by acid phosphatase activity, *Am. J. Pathol.* **38**:161.

Belfield, A., and Goldberg, D.M., 1976, Comparison of sodium beta-glycerophosphate and disodium phenylphosphate as inhibitors of alkaline phosphatase in determination of 5′-nucleotidase activity of human serum, *Clin. Biochem.* **3**:105.

Bloom, F.E., Hoffer, B.J., and Siggins, G.R., 1971, Studies on norepinephrine-containing afferents to Purkinje cells of rat cerebellum. I. Localization of the fibers and their synapses, *Brain Res.* **25**:501.

Bobillier, P., and Mouret, J.R., 1971, The alterations of the diurnal variations of brain tryptophan, biogenic amines and 5-hydroxyindole actic acid in the rat under limited feeding, *Int. J. Neurosci.* **2**:271.

Boesten, A.J.P., and Voogd, J., 1975, Projections of the dorsal column nuclei and the spinal cord on the inferior olive in the cat, *J. Comp. Neurol.* **161**:215.

Borgers, M., 1973, The cytochemical application of new potent inhibitors of alkaline phosphatases, *J. Histochem. Cytochem.* **21**:812.

Bosmann, H.B., and Pike, G.Z., 1970, Membrane marker enzymes: Isolation, purification and properties of 5′-nucleotidase from rat cerebellum, *Biochem. Biophys. Acta* **227**:402.

Breckenridge, B.McL., and Johnston, R.E., 1969, Cyclic 3′,5′-nucleotide phosphodiesterase in brain, *J. Histochem. Cytochem.* **17**:505.

Brewer, G.J., 1970, *An Introduction to Isoenzyme Techniques*, Academic Press, New York.

Brodal, A., 1940, Experimentelle Untersuchungen über olivo-cerebellare Lokalisation, *Z. Gesamte Neurol. Psychiatr.* **169**:1.

Brodal, A., Walberg, F., and Blackstad, T., 1950, Termination of spinal afferents to the inferior olive in cat, *J. Neurophysiol.* **13**:431.

Burstone, M.S., 1958, The relationship between fixation and techniques for the histochemical localization of hydrolytic enzymes, *J. Histochem. Cytochem.* **6**:322.

Burstone, M.S., 1962, *Enzymehistochemistry and Its Application on the Study of Neoplasms*, Academic Press, New York.

Busch, H.F.M., 1961, *An Anatomical Analysis of the White Matter in the Brain Stem of the Cat*, Van Gorcum, Assen, Holland.

Cammermeyer, J., 1962, An evaluation of the significance of the "dark" neuron, *Ergeb. Anat. Entwicklungsgesch.* **36**:1.

Campbell, D.M., 1962, Determination of 5′-nucleotidase in blood serum, *Biochem. J.* **82**:43.

Carlstedt, T., 1977, I. A preparative procedure useful for electron microscopy of the lumbosacral dorsal rootlets, in: *Observations on the Morphology at the Transition between the Peripheral and the Central Nervous System in the Cat*, Acta Physiol. Scand., Suppl. 466.

Chance, B., 1952, The kinetics and stoichiometry of the transition from the primary to the secondary peroxidase peroxide, *Arch. Biochem. Biophys.* **41**:416.

Chujo, T., Yamada, Y., and Yamamoto, C., 1975, Sensitivity of Purkinje cell dendrites to glutamic acid, *Exp. Brain Res.* **23**:293.

Coimbra, A., Sodré-Borges, B.P., and Magalhoes, M.M., 1974, The substantia gelatinosa Rolandi of the rat: Fine structure, cytochemistry (acid phosphatase) and changes after dorsal root section, *J. Neurocytol.* **3**:199.

Collu, R., Jéquier, J.C., Letarte, J., Leboeuf, G., and Ducharme, J.R., 1973, Diurnal variations of plasma growth hormone and brain monoamines in adult male rats, *Can. J. Physiol. Pharmacol.* **51**:890.

Crawford, J.M., Curtis, D.R., Voorhoeve, P.E., and Wilson, V.J., 1966, Acetylcholine sensitivity of cerebellar neurones in the cat., *J. Physiol.* **186**:139.

Csillik, B., and Knyihar, E., 1975, Degenerative atrophy and regenerative proliferation in rat spinal cord, *Z. Mikrosk. Anat. Forsch.* **89**(6):1099.

Davidoff, M., 1973, Uber die Glia im Hypoglossuskern der Ratte nach Axotomie, *Z. Zellforsch. Mikrosk. Anat.* **141**:427.

De Duve, C., and Baudhuin, P., 1966, Peroxisomes (microbodies and related particles), *Physiol. Rev.* **46**:323.

Deierkauf, F.A., and Heslinga, F.J.M., 1962, The action of formaldehyde on rat brain lipids, *J. Histochem. Cytochem.* **10**:79.

Deisseroth, A., and Dounce, A.L., 1970, Catalase: Physical and chemical properties, mechanism of catalysis and physiological role, *Physiol. Rev.* **50**(3):319.

Desclin, J.C., 1974, Histological evidence supporting the inferior olive as the major source of cerebellar climbing fibres in the rat, *Brain Res.* **77**:365.

Desclin, J.C., 1976, Early terminal degeneration of cerebellar climbing fibres after destruction of the inferior olive in the rat: Synaptic relations in the molecular layer, *Anat. Embryol.* **149**:112.

Desclin, J.C., and Escubi, J., 1974, Effects of 3-acetylpyridine on the central nervous system of the rat, as demonstrated by silver methods, *Brain Res.* **77**:349.

Desmedt, J.E., and La Grutta, G., 1957, The effect of selective inhibition of pseudo cholinesterase on the spontaneous and evoked activity of the cat's cerebral cortex, *J. Physiol.* **136**:20.

Di Raddo, J., and Kellogg, C., 1975, *In vivo* rates of tyrosine and tryptophan hydroxylation on regions of rat brains at four times during light–dark cycle, *Nannyn-Schmiedeberg's Arch. Pharmacol.* **286**:389.

Dixon, K.C., 1965, Ischaemia and the neurons, in *Neurohistochemistry* (C.W.M. Adams, ed.), pp. 558–598, Elsevier, Amsterdam, London.

Dixon, M., and Webb, E.C., 1958, *Enzymes,* Longmans, Green, London.

Drummond, G.J., and Yamamoto, M., 1971, Nucleotide phosphomonoesterases, in: *The Enzymes,* Vol. IV (D. Boyer, ed.), pp. 337–352, Academic Press, New York.

Edwards, S.B., 1972, The ascending and descending projection of the red nucleus in the cat: An experimental study using an autoradiographic tracing method, *Brain Res.* **48**:45.

El-Aaser, A.A., and Reid, E., 1969, Rat liver 5'-nucleotidase, *Histochem. J.* **1**:417.

El-Badawi, A., and Schenk, E.A., 1966, Dual innervation of the mammalian urinary bladder: A histochemical study of the distribution of cholinergic and adrenergic nerves, *Am. J. Anat.* **119**:405.

El-Badawi, A., and Schenk, E.A., 1967, Histochemical methods for separate, consecutive and simultaneous demonstration of acetylcholinesterase and norepinephrine in cryostat sections, *J. Histochem. Cytochem.* **15**:580.

El-Badawi, A., and Schenk, E.A., 1968*a*, The peripheral adrenergic innervation apparatus. I. Intraganglionic and extraganglionic adrenergic ganglion cells, *Z. Zellforsch. Mikrosk. Anat.* **87**:218.

El-Badawi, A., and Schenk, E.A., 1968*b*, A new theory of the innervation of bladder musculature. Part I. Morphology of the intrinsic vesical innervation apparatus, *J. Urol.* **99**:585.

El-Badawi, A., and Schenk, E.A., 1969, Innervation of the abdominopelvic ureter in the cat, *Am. J. Anat.* **126**:103.

El-Badawi, A., and Schenk, E.A., 1970, Intra- and extraganglionic peripheral cholinergic neurons in the urogenital organs of the cat, *Z. Zellforsch. Mikrosk. Anat.* **103**:26.

El-Badawi, A., and Schenk, E.A., 1971*a*, A new theory of the innervation of bladder musculature. Part II. The innervation apparatus of the ureterovesical function, *J. Urol.* **105**:368.

El-Badawi, A., and Schenk, E.A., 1971*b*, A new theory of the innervation of bladder musculature. Part III. Postganglionic synapses in ureterovesicourethral autonomic pathways, *J. Urol.* **105**:372.

El-Badawi, A., and Schenk, E.A., 1974, A new theory of the innervation of bladder musculature. Part IV. Innervation of the vesico-urethral function and external urethral sphincter, *J. Urol.* **111**:613.

Fahimi, H.D., 1975, Fine structural cytochemical localization of peroxidatic activity of catalase, in: *Techniques of Biochemical and Biophysical Morphology* (D. Glick and H. Rosenbaum, eds.), Vol. 2, pp. 197–245, John Wiley, New York.

Feirabend, H.K.P., and Voogd, J., 1975, The efferent projection of the cerebellar cortex in the white Leghorn (*Gallus domesticus*) and its relation to the longitudinal organization of the cerebellar white matter, *Exp. Brain Res.* **23**(Suppl.):70.

Feirabend, H.K.P., Vielvoye, G.J., Freedman, S.L., and Voogd, J., 1977, Longitudinal organization of afferent and efferent connections of the cerebellar cortex of the white Leghorn (*Gallus domesticus*), *Exp. Brain Res.*, Suppl. I, p. 72.

Felicetti, D., and Rath, F.W., 1975, Zum Vorkommen und zur Isolierung einer durch Zink stark aktivierbaren sauren Phosphatase im Grosshirn der Ratte, *Acta Histochem.* **53**(2):281.

Filogamo, G., and Candiollo, L., 1962, Observations on the behaviour of acetyl cholinesterase (AChE) in the new cells of the spinal reflex arc, after section of the peripheral nerves (experimental investigations in *Lepus Cuniculus* L.), *Acta Anat.* **51**:273.

Fiske, C.H., and Subbarow, Y., 1925, The colorimetric determination of phosphorus, *J. Biol. Chem.* **66**(2):375.

Flumerfelt, B.A., and Lewis, P.R., 1975, Cholinesterase activity in the hypoglossal nucleus of the rat and the changes produced by axotomy: A light and electron microscopic study, *J. Anat.* **119**(2):309.

Friede, R.L., 1959, Histochemical demonstration of phosphorylase in brain tissue: Association of postmortal neuron changes with phosphorylase activity, *J. Histochem. Cytochem.* **7**:34.

Friede, R.L., 1963, Interpretation of hyperchronic nerve cells: Relative significance of the type of fixative used, of the osmolarity of the cytoplasm and the surrounding fluid in the production of cell shrinkage, *J. Comp. Neurol.* **121**:137.

Friede, R.L., 1964, Axon swellings produced *in vivo* in isolated segments of nerves, *Acta Neuropathol.* **3**:229.

Friede, R.L., 1966, *Topographic Brain Chemistry*, Academic Press, New York.

Friedenwald, J.S., and Becker, B., 1948, Histochemical localization of glucuronidase, *J. Cell. Comp. Physiol.* **31**:303.

Gerebtzoff, M.A., and Ziegels, J., 1974, An attempt to localize cyclic phosphodiesterase at adrenergic nerve endings, *J. Neural Transm.*, Suppl. XI, p. 181.

Gezelius, K., and Wright, B.E., 1965, Alkaline phosphatase in *Dictyostelium discoideum, J. Gen. Microbiol.* **38**:309.

Gibson, W.B., and Drummond, G.I., 1972, Properties of 5′-nucleotidase from avian heart, *Biochemistry* **11**(2):223.

Glenner, G.G., 1965, Enzyme histochemistry, in: *Neurohistochemistry* (C.W.M. Adams, ed.), pp. 109–160, Elsevier, Amsterdam.

Glenner, G.G., Weissbach, H., and Redfield, B.G., 1960, The histochemical demonstration of enzymatic activity by a non-enzymatic redox reaction: Reduction of tetrazolium salts by indo-3-acetaldehyde, *J. Histochem. Cytochem.* **8**:258.

Gollnick, P.D., and Armstrong, R.B., 1976, Histochemical localization of lactate dehydrogenase isozymes in human skeletal muscle fibers, *Life Sci.* **18**:27.

Gomori, G., 1941, The distribution of phosphatases in normal organs and tissues, *J. Cell. Comp. Physiol.* **17**:71.

Gomori, G., 1952, *Microscopic Histochemistry: Principles and Practics*, Chicago University Press.

Gomori, G., and Chessick, B.O., 1953, Esterases and phosphatases of the brain: A histochemical study, *J. Neuropathol. Exp. Pathol.* **12**:387.

Graham, R.C., and Karnovsky, M.J., 1965a, The histochemical demonstration of uricase activity, *J. Histochem. Cytochem.* **13**:448.

Graham, R.C., and Karnovsky, M.J., 1965b, The histochemical demonstration of monoamine oxidase activity by coupled peroxidatic oxidation, *J. Histochem. Cytochem.* **13**:604.

Graham, R.C., and Karnovsky, M.J., 1966, The early stages of absorption of injected horseradish peroxidase in the proximal tubules of mouse kidneys: Ultrastructural cytochemistry by a new technique, *J. Histochem. Cytochem.* **14**:291.

Gray, E.G., 1959, Axo-somatic and axo-dendritic synapses of the cerebral cortex: An electron microscope study, *J. Anat.* **93**:420.

Gray, E.G., and Whittaker, V.P., 1962, The isolation of nerve endings from brain: An electron-microscopic study of cell fragments derived by homogenization and centrifugation, *J. Anat.* **96**:79.

Groenewegen, H.J., and Voogd, J., 1977, The parasagittal zonation within the olivocerebellar projection. I. Climbing fiber distribution in the vermis of cat cerebellum, *J. Comp. Neurol.* **174**:417.

Groenewegen, H.J., Boesten, A.J.P., and Voogd, J., 1975, The dorsal column nuclear projections to the nucleus ventralis posterior lateralis thalami and the inferior olive in the cat: An autoradiographic study, *J. Comp. Neurol.* **162**:505.

Guidotti, A., Biggio, G., and Costa E., 1975, 3-Acetylpyridine: A tool to inhibit the tremor and the increase of c-GMP content in cerebellar cortex elicited by harmaline, *Brain Res.* **96:**201.

Hanker, J.S., Thornburg, L.P., Yates, P.E., and Moore, H.G., 1973, III: The demonstration of cholinesterases by the formation of osmium blacks at the sites of Hatchett's brown, *Histochemie* **37:**223.

Hanker, J.S., Thornburg, L.P., Yates, P.E., and Romanovicz, D.K., 1974, The demonstration of arylsulfatases with 4-nitro-1,2-benzenediol mono (hydrogen sulfate) by the formation of osmium blacks at the sites of copper capture, *Histochemistry* **41:**207.

Hardonk, M.J., and De Boer, H.G.A., 1968, 5'-Nucleotidase. III. Determinations of 5'-nucleotidase isoenzymes in tissue of rat and mouse, *Histochemie* **12:**29.

Hardonk, M.J., and Koudstaal, J., 1968, 5'-Nucleotidase. II. The significance of 5'-nucleotidase in the metabolism of nucleotides studied by histochemical and biochemical methods, *Histochemie* **12:**18.

Hardonk, M.J., Bouma, J.M.W., Mulder, G.J., and Konings, 1975, Enzyme histochemical evaluation of centrifugation procedures, *Acta Histochem.*, Suppl. XIV, p. 91.

Hyat, M.A., 1970, *Principles and Techniques of Electron Microscopy: Biological Applications*, Vol. I, Van Nostrand Reinhold, New York.

Heald, P.J., 1960, *Phosphorus Metabolism of Brain*, Pergamon Press, Oxford.

Heim, W.G., Appleman, D., and Pyform, H.T., 1955, Production of catalase changes in animals with 3-amino-1,2,4 triazole, *Science* **122:**693.

Henkel, C.K., Linauts, M. and Martin, G.F., 1975, The origin of annulo olivary tract with notes on other mesencephalo-olivary pathways: A study by the HRP method, *Brain Res.* **100:**145.

Herzog, V., and Fahimi, H.D., 1976, Intracellular distinction between peroxidase and catalase in excocrine cells of rat lacrimal gland: A biochemical and cytochemical study, *Histochemistry* **46:**273.

Herzog, V., and Miller, F., 1972, Endogeneous peroxidase in the lacrimal gland of the rat and its differentiation against injected catalase and horseradish-peroxidase, *Histochemie* **30:**235.

Higashi, H., Okawa, K., and Takamatsu, H., 1960, *Proc. Jpn. Histochem. Soc.*, p. 80; cited in Pearse (1972).

Hills, C.P., 1964a, Ultrastructural changes in the capillary bed of the rat cerebral cortex in anoseic–ischemic brain lesions, *Am. J. Pathol.* **44:**531.

Hills, C.P., 1964b, The ultrastructure of anoseic–ischemic lesions in the cerebral cortex of the adult rat brain, *Guy's Hosp. Rep.* **113:**333–348.

Hoffer, B.J., Siggins, G.R., and Bloom, F.E., 1971, Studies on norepinephrine-containing afferents to Purkinje cells of rat cerebellum. II. Sensitivity of Purkinje cells to norepinephrine and related substances administered by microiontophoresis, *Brain Res.* **25:**523.

Holt, S.J., and O'Sullivan, D.G., 1958, The studies in enzyme cytochemistry. Part I. Principles of cytochemical staining methods, *Proc. R. Soc. London Ser. B* **148:**465–480.

Holzman, E., Teichberg, S., Abrahams, S.J., Citkowitz, E., Crain, S.M., Kawai, H., and Peterson, E.R., 1973, Notes on synaptic vesicles and related structures, endoplasmic reticulum, lysosomes and peroxisomes in nervous tissue and the adrenal medulla, *J. Histochem. Cytochem.* **21:**349.

Hopwood, D., 1972, Theoretical and practical aspects of glutaraldehyde fixation, *Histochem. J.* **4:**267.

Hughes, A.F.W., and Lewis, P.R., 1961, Effect of limb ablation on neurones in *Xenopus* larvae, *Nature (London)* **189:**333.

Humason, G.L., 1972, Animal tissue techniques, in: *A Series of Books in Biology* (D. Kennedy and R.B. Park, eds.), pp. 415–483, W.H. Freeman, San Francisco.

Hyde, J.C., and Robinson, N., 1976, Improved histological localization of GABA-transaminase activity in rat cerebellar cortex after aldehyde fixation, *Histochemistry* **46:**261.

Ipata, P.L., 1966, Resolution of 5'-nucleotidase from non-specific phosphatase from sheep brain and its inhibition by nucleosidetriphosphates, *Nature (London)* **214:**618.

Ipata, P.L., 1967, Studies on the inhibition by nucleosidetriphosphates of sheep brain 5'-nucleotidase, *Biochem. Biophys. Res. Commun.* **27:**337.

Ipata, P.L., 1968, Sheep brain 5'-nucleotidase: Some enzymic properties and allosteric inhibition by nucleoside-triphosphates, *Biochemistry* **7**:507.

Israël, M., and Frachon-Mastour, P., 1970, Fractionnement du cortex cérébral du rat, distribution subcellulaire de la 5'-nucleotidase et des cholinestérases, *Arch. Anat. Microsc.* **59**(4):383.

Johnston, G.A.R., 1975, Proline, an inhibitory transmitter?, ISN Barcelona Meeting (1975), p. 11.

Joo, F., Savay, G., and Csillik, B., 1965, A new modification of the Koelle–Friedenwald method for the histochemical demonstration of cholinesterase activity, *Acta Histochem.* **22**:40.

Kabara, J.J., and Konvich, D., 1972, The extraction of brain isoenzymes with solvents of varying polarity, *Proc. Soc. Exp. Biol. Med.* **139**:1326.

Karlson, P., 1972, *Biochemie für Mediziner und Naturwissenschaftler*, 8th ed., George Thieme, Stuttgart.

Karnovsky, M.J., and Roots, L., 1964, A "direct-coloring" thiocholine method for cholinesterases, *J. Histochem. Cytochem.* **12**:219.

Kaśa, P., and Csillik, B., 1965, Cholinergic excitation and inhibition in the cerebellar cortex, *Nature (London)* **208**:695.

Kaśa, P., Joo, F., and Dsillik, B., 1965, Histochemical localization of acetylcholinesterase in the cat cerebellar cortex, *J. Neurochem.* **12**:31.

Kievit, J., and Kuypers, H.G.J.M., 1975, Subcortical afferents to the frontal lobe in the rhesus monkey studied by means of retrograde horseradish peroxidase transport, *Brain Res.* **85**:261.

Klaushoffer, R., and Pavelka, M., 1975, Studies on 5'-nucleotidase histochemistry. III. 5'-Nucleotidase activity in smooth muscle cells of the rat's gastrointestinal tube, *Histochemistry* **43**:373.

Klück, P., 1980, The autonomic innervation of the human urinary bladder, bladder neck and urethra: A histochemical study, *Anat. Rec.* **198**:3.

Knyihar, E., and Csillik, B., 1976, Effects of peripheral axotomy on the fine structure and histochemistry of the Rolando substance: Degenerative atrophy of central processes of pseudounipolar cells, *Exp. Brain Res.* **26**:73.

Koelle, G.B., and Friedenwald, J.S., 1949, A histochemical method for localizing cholinesterase activity, *Proc. Soc. Exp. Biol. Med.* **70**:617.

Koelle, G.B., Davis, R., Smyrl, E.G., and Fine, A.V., 1974, Refinement of the bis-(thioacetoxy)aurate I method for the electron microscopic localization of acetylcholinesterase and non-specific cholinesterase, *J. Histochem. Cytochem.* **22**:252.

Kooy, F.H., 1916, *The Inferior Olive in Vertebrates*, De Erven Bohn, Haarlem, The Netherlands.

Kostopoulos, G.K., Limacher, J.J., and Phillis, J.W., 1975, Action of various adenine derivates on cerebellar Purkinje cells, *Brain Res.* **88**:162.

Kristensson, K., and Olsson, Y., 1971, Retrograde axonal transport of protein, *Brain. Res.* **29**:363.

Kristensson, K., Olsson, Y., and Sjostrand, J., 1971, Axonal uptake and retrograde transport of exogenous proteins in the hypoglossal nerve, *Brain Res.* **32**:399.

Kuypers, H.G.J.M., Kievit, J., and Groen-Klevent, A.C., 1974, Retrograde axonal transport of horseradish peroxidase in rat forebrain, *Brain Res.* **67**:210.

La Vail, J.H., and La Vail, M.M., 1972, Retrograde axonal transport in the central nervous system, *Science* **176**:1416.

La Vail, J.H., Winston, K.R., and Tish, A., 1973, A method based on retrograde intraaxonal transport of protein for identification of cell bodies of origin of axons terminating within the C.N.S., *Brain Res.* **58**:470.

Lee, S.H., and Torack, R.M., 1968, Electron microscope studies of glutamic oxalacetic transaminase in rat liver cells, *J. Cell Biol.* **39**:716.

Levitt, M., Spector, S., Sjoerdsma, A., and Udenfriend, S., 1965, Elucidation of the rat limiting step in norepinephrine biosynthesis in the perfused guinea pig heart, *J. Pharmacol. Exp. Ther.* **148**:1.

Lewis, P.R., and Knight, D.P., 1977, Staining methods for sectioned material, in: *Practical Methods in Electron Microscopy* (A.M. Glanert, ed.), pp. 137–223, North-Holland, Amsterdam.

Lewis, P.R., and Schon, F.E.G., 1975, The localization of acetylcholinesterase in the locus coeruleus of the normal rat and after 6-hydroxydopamine treatment, *J. Anat.* **120**(2):373.

Lewis, P.R., and Shute, C.C.D., 1963, Alginate gel; An embedding medium for facilitating the cutting and handling of frozen sections, *Stain Technol.* **38**:307.

Lewis, P.R., and Shute, C.C.D., 1964, Demonstration of cholinesterase activity with the electron microscope, *J. Physiol.* **175**:5.

Lewis, P.R., and Shute, C.C.D., 1966, The distribution of cholinesterase in cholinergic neurons, demonstrated with the electron microscope, *J. Cell. Sci.* **1**:381.

Lewis, P.R., Scott, J.A., and Navaratnam, V., 1970, Localization in the dorsal motor nucleus of the vagus in the rat, *J. Anat.* **107**:197.

Lison, L., 1936, *Histochemie Animale*, Paris.

Long, J.P., 1963, Structure activity relationships of the reversible anticholinesterase agents, in: *Handbuch der experimentellen Pharmakologie*, Vol. 15 (G.B. Koelle, ed.), p. 374, Springer-Verlag, Berlin.

Lovenberg, W., and Victor, S.J., 1974, Regulation of tryptophan and tyrosine hydroxylase, *Life Sci.* **14**:2337.

Lowry, B.H., Rosebrough, N.J., Farr, H., and Randall, R.J., 1951, Protein measurement with the Folin phenol reagent, *J. Biol. Chem.* **193**:265.

MacDonald, M., and Spector, R.G., 1963, The influence of anoseia on respiratory enzymes in rat brain, *Br. J. Exp. Pathol.* **44**:11.

Majoor, G.D., 1973, AMPase activity in the wing disk and mesothoracic leg disk of some dipteran larvae, *Netherl. J. Zool.* **23**:111.

Malmfors, T., and Thoenen, H., 1971, *6-Hydroxydopamine and Catacholamine Neurons*, North-Holland, Amsterdam.

Malmgren, H., and Sylven, B., 1955, On the chemistry of the thiocholine method of Koelle, *J. Histochem. Cytochem.* **3**:441.

Manocha, S.L., and Shanta, T.R., 1970, *Macaca mulatta: Enzyme Histochemistry of the Nervous System,"* Chapter X, Cerebellum, Academic Press, New York.

Marani, E., 1977, The subcellular distribution of 5'-nucleotidase activity in the mouse cerebellum, *J. Exp. Neurol.* **57**:1042–1048.

Marani, E., 1978, A method for orienting cryostat sections for three-dimensional reconstructions, *Stain Technol.* **53**:265–268.

Marani, E., 1979, The morphology of the mouse cerebellum, *Acta. Morphol. Neerl. Scand.* **17**:33–52.

Marani, E., 1980a, K$^+$ and Na$^+$ activation of cerebellar 5'-nucleotidase, *J. Exp. Neurol.* **67**:412–422.

Marani, E., 1980b, 5'-Nucleotidase in der Molekularschicht des Rattenkleinhirns, *Acta Histochem.* Suppl. XXI, pp. 237–242.

Marani, E., The 5'-nucleotidase isoenzyme in the mouse cerebellum (in prep.).

Marani, E., and Boekee, A., 1973, Aspects histoenzymologiques de la localisation de l'adenylcyclase, de la C.3',5'-nucleotide phosphodiesterase, de la 5'-nucleotidase et de l'alpha-glucanphosphorylase dans le cervelet de la souris, *Bull. Assoc. Anat. (Nancy)* **57**(158):555.

Marani, E., and Voogd, J., 1973, Some aspects of the localization of the enzyme 5'-nucleotidase in the molecular layer of the cerebellum of the mouse, *Acta Morphol. Neerl.-Scand.* **11**(4):365.

Marani, E., and Voogd, J., 1977, An acetylcholinesterase band pattern in the molecular layer of the cat cerebellum, *J. Anat.* **124**(2):335–345.

Marani, E., Voogd, J., and Boekee, A., 1977, Acetyl cholinesterase staining in subdivisions of the cat's inferior olive, *J. Comp. Neurol.* **174**:209–226.

Marani, E., Rietveld, W.J., and Osselton, J.C., 1979, Ultrastructural localization of the endogenous peroxidase activity in the ventromedial arcuate nucleus, *I.R.C.S. Med. Sci.* **7**:501–502.

Maréschal, P., 1934, *L'olive Bulbaire: Anatomie, Ontogénèse, Phylogénèse, Physiologie et Physiopathologie*, Doin, Paris.

Martines-Rodriguez, R., Garcia-Legura, L.M., Toledano, A., and Martines-Murille, R., 1976, Aspertate amino transferase activity and glutamic dehydrogenase in the cerebellar cortex in several species of animals: A histochemical study, *J. Hirnforsch.* **17**:387–398.

Maruyama, S., and Agostino, A.N.D., 1967, Cell necrosis in the central nervous system of normal rat fetuses: An electron microscopic study, *Neurology* **17**:550.

Mayersbach, H. von, 1964, *The Cellular Aspects of Biorhythms*, Springer-Verlag, Heidelberg.

McCance, J., and Phillis, J.W., 1964, The action of acetylcholine on cells in cat cerebellar cortex, *Experientia* **20**:217.

Meijer, A.E.F.H., 1972, Semipermeable membranes for improving the histochemical demonstration of enzyme activities in tissue sections. I. Acid phosphatase, *Histochemie* **30**:31.

Meijer, A.E.F.H., 1973, Semipermeable membranes for improving the histochemical demonstration of enzyme activities in tissue sections. III. Lactate dehydrogenase, *Histochemie* **35**:165.

Meijer, A.E.F.H., and de Vries, G.P., 1974, Semipermeable membranes for improving the histochemical demonstration of enzyme activities in tissue sections. IV. Glucose-6-phosphate dehydrogenase and 6-phosphogluconate dehydrogenase (decarboxylating), *Histochemistry* **40**:349.

Meijer, A.E.F.H., and Vloedman, A.H.T., 1973, Semipermeable membranes for improving the histochemical demonstration of enzyme activities in tissue sections. II. Non specific esterase and beta-glucuronidase, *Histochemie* **34**:127.

Milaire, J., 1969, Etude morphogénétique de la syndactylie postaseiale provoquée chez le rat par l'hadacidine. I, *Arch. Biol. (Liège)* **80**:167.

Miller, R.A., 1949, A morphological and experimental study of chromophilic neurons in the cerebral cortex, *Am. J. Anat.* **84**:201.

Nachlas, M.M., Crawford, D.T., Goldstein, T.P., and Seligman, A.M., 1958, The histochemical demonstration of cytochrome oxidase with a new reagent for the Nadi reaction, *J. Histochem. Cytochem.* **6**:445.

Naidoo, D., and Pratt, O.E., 1954, The development of adenosine 5'-phosphatase activity with the maturation of the rat cerebral cortex, *Enzymologia* **5**:298.

Navaratnam, V., and Lewis, P.R., 1975, Effects of vagotomy on the distribution of cholinesterases in the cat medulla oblongata, *Brain Res.* **100**:599.

Neu, H.C., 1967a, The 5'-nucleotidase of *Escherichia coli*. I. Purification and properties, *J. Biol. Chem.* **242**(17):3905.

Neu, H.C., 1967b, The 5'-nucleotidase of *Escherichia coli*. II. Surface localization and purification of the *Escherichia coli* 5'-nucleotidase inhibitor, *J. Biol. Chem.* **242**(17):3986.

Novikoff, A.B., 1973, Studies on the structure and function of cell organelles: 3,3'-Diamino benzidine cytochemistry, in: *Electron Microscopy and Cytochemistry* (E. Wisse, W.T. Daems, I. Molenaar, and P. Van Duyn, eds.), pp. 89–109, North-Holland, Amsterdam.

Novikoff, A.B., Shin, W.Y., and Drucker, J., 1960, Cold acetone fixation for enzyme localization in frozen sections, *J. Histochem. Cytochem.* **8**:37.

Novikoff, A.B., Quintana, N., Villaverde, H., and Forschirm, R., 1966, Nucleoside phosphatase and cholinesterase activities in dorsal root ganglia and peripheral nerve, *J. Cell. Biol.* **29**:525.

Novikoff, P.M., and Novikoff, A.B., 1972, Peroxisomes in absorptive cells of mammalian intestine, *J. Cell. Biol.* **53**:532.

O'Connor, T.M., and Wyttenbach, C.R., 1974, Cell death in the embryonic chick spinal cord, *J. Cell Biol.* **60**:448.

Olney, J.W., 1969, Brain lesions, obesity and other disturbances in mice treated with monosodium glutamate, *Science* **164**:719.

Olney, J.W., Rhee, V., and De Gubareff, T., 1977, Neurotoxic effects of glutamate on mouse area postrema, *Brain Res.* **120**:151.

Oscarsson, O., 1973, Functional organization of spino-cerebellar paths, in: *Handbook of Sensory Physiology*, Vol. II (A. Iggo, ed.), Chapt. II, p. 339, Springer-Verlag, Berlin.

O'Sullivan, D.G., 1955, Diffusion and simultaneous chemical reactions. II. The equations of those systems in which transport occurs from one region to an adjoining region, *Bull. Math. Biophys.* **17**:243.

Ozeki, M., and Sato, M., 1970, Potentiation of excitatory junctional potentials and glutamate-induced responses in crayfish muscle by 5'-ribonucleotides, *Comp. Biochem. Physiol.* **32**:203.

Palay, S.L., and Chan-Palay, V., 1974, *Cerebellar Cortex: Cytology and Organization*, Springer-Verlag, Berlin.

Palmer, A.C., and Elleker, A.R., 1961, Histochemical localization of cholinesterases in the brainstem of sheep. *Q. J. Exp. Physiol.* **46**:344.

Pasquini, J.M., and Soto, E.F., 1972, Extraction of proteolipids from nervous tissue with *n*-butanol–water, *Life Sci.* **11**(Part II):433.

Pearse, A.G.E., 1954, Intracellular localisation of dehydrogenase systems using monotetrazolium salts and metal chelation of their formazans, *J. Histochem. Cytochem.* **5**:515.

Pearse, A.G.E., 1968, *Histochemistry Theoretical and Applied*, Vol. I, Churchill Livingstone, Edinburgh.

Pearse, A.G.E., 1972, *Histochemistry Theoretical and Applied*, Vol. II, Churchill Livingstone, Edinburgh.

Persijn, J.P., and Van der Slik, W., 1970, A new method for the determination of serum 5′-nucleotidase, *Z. Klin. Chem. Klin. Biochem.* **8**:398.

Persijn, J.P., Van der Slik, W., and Timmer, C.J., 1969, On the determination of serum 5′-nucleotidase activity in the presence of betaglycerophosphate, *Clin. Biochem.* **2**:335.

Person, P., and Fine, A., 1961, Studies of indophenol blue synthesis. I. The role of free radical formation by heart muscle particulates during the "G" Nadi reaction, *J. Histochem. Cytochem.* **9**:190.

Petresco, A., 1958, Les modifications de l'activité de la phosphatase alcaline dans le neurone cortical atrophié du lapin, *Ann. Histochim.* **3**:159.

Phillis, J.W., 1965*a*, Cholinesterase in the cat cerebellar cortex deep nuclei and peduncles, *Experientia* **21**:266.

Phillis, J.W., 1965*b*, Cholinergic mechanisms in the cerebellum, *Br. Med. Bull.* **21**:26.

Pilcher, C.W.T., and Jones, D.G., 1970, The distribution of 5′-nucleotidase in subcellular fractions of mouse cerebellum, *Brain Res.* **24**:143.

Poelmann, R.E., and Daems, W.T., 1973, Problems associated with the demonstration by lead methods of adenosine triphosphatase activity in resident peritoneal macrophages and exudate monocytes of the guinea pig, *J. Histochem. Cytochem.* **21**:488.

Ramon-Moliner, E., 1972, Acetylthiocholinesterase distribution in the brain stem of the cat, *Ergebn. Anat.* **46**:1.

Rinvik, E., and Grevova, J., 1970, Observations on the fine structure of the substantia nigra in the cat, *Exp. Brain Res.* **11**:229.

Robinson, P.M., 1971, The demonstration of acetylcholinesterase in autonomic axons with the electron microscope, in: *Progress in Brain Research*, Vol. 34, *Histochemistry of Nervous Transmission* (O. Eränkö, ed.), pp. 357–370, Elsevier, Amsterdam.

Robinson, P.M., and Bell, C., 1967, The localization of acetylcholinesterase at the autonomic neuromuscular junction, *J. Cell Biol.* **33**:93.

Roels, F., Wisse, E., De Prest, B., and Van der Meulen, J., 1973, Cytochemical discrimination between peroxidases and catalases using diamino benzidine, in: *Electron Microscopy and Cytochemistry* (E. Wisse, W.T. Daems, J. Molenaar, and P. van Duijn, eds.), pp. 115–118, North-Holland, Amsterdam.

Ross, B.D., 1972, *Perfusion Techniques in Biochemistry: A Laboratory Manual* Clarendon Press, Oxford.

Rustioni, A., Sanyal, A., and Kuypers, H.G.J.M., 1971, A histochemical study of the distribution of the trigeminal divisions in the substantia gelatinosa of the rat, *Brain Res.* **32**:45.

Rutenberg, A.M., and Seligman, A.M., 1955, The histochemical demonstration of acid phosphatase by a post-incubation coupling technique, *J. Histochem. Cytochem.* **3**:455.

Sabatini, D.P., Miller, F., and Barrnett, R.J., 1964, Aldehyde fixation for morphological and enzyme histochemical studies with the electron microscope, *J. Histochem. Cytochem.* **12**:57.

Sadler, P.J., 1976, Zinc in enzymes, *Nature (London)* **262**(5566):258.

Sakharova, A.V., 1966, *Tsitologiya* **8**:54; cited in Silver (1967).

Schevin, L.E., Harrison, W.H., Gordon, P., and Panly, J.E., 1968, Daily fluctuation (circadian and ultradian) in biogenic amines of the rat brain, *Am. J. Physiol.* **214**:166.

Schevin, L.E., Halberg, F., and Pauly, J.E., 1974, *Chronobiology*, Georg Thieme, Stuttgart.

Schlaeffer, W.W., La Valle, M.C., and Torack, R.M., 1969, Cytochemical demonstration of nucleoside phosphatase activity in myelinated nerve fibers of the rat, *Histochemie* **18**:281.

Schlüter, G., 1973, Ultrastructural observation on cell necrosis during formation of the neural tube in mouse embryos, *Z. Anat. Entwicklungsgesch.* **141**:251.

Schwartz, M.K., and Bodansky, O., 1964, Properties of activity of 5′-nucleotidase in human serum and application in diagnosis, *Am. J. Clin. Pathol.* **42**:572.

Schwartz, M.K., and Bodansky, O., 1965, Serum 5'-nucleotidase in patients with cancer, *Cancer* **18**:886.

Scott, T.G., 1964, A unique pattern of localization within the cerebellum of the mouse, *J. Comp. Neurol.* **122**:1.

Scott, T.G., 1965, The specificity of 5'-nucleotidase in the brain of the mouse, *J. Histochem. Cytochem.* **13**:657.

Scott, T.G., 1969, The quantitative assay of 5'-nucleotidase in brain by histophotometry, *Histochem. J.* **1**:215.

Seidler, E., and Kunde, D., 1969, Verbesserter histochemischer Dehydrogenasenachweis mit Monotetrazolium Salze, *Acta Histochem.* **32**:142.

Seligman, A.M., Nachlas, M.M., Manheimer, L.H., Friedman, Q.M., and Wolf, G., 1949, Development of new methods for the histochemical demonstration of hydrolytic intracellular enzymes in a program of cancer research, *Ann. Surg.* **130**:333.

Seligman, A.M., Plapinger, R.E., Wasserkrug, H.L., Deb, C., and Hanker, J.S., 1967, Ultrastructural demonstration of cytochrome oxidase activity by the Nadi reaction with osmiophilic reagents, *J. Cell. Biol.* **34**:787.

Seligman, A.M., Karnovsky, M.J., Wasserkrug, H.L., and Hanker, J.S., 1968, Nondroplet ultrastructural demonstration of cytochrome oxidase activity with a polymerizing osmiophilic reagent, diamino benzidine (DAB), *J. Cell. Biol.* **38**:1.

Seligman, A.M., Shannon, W., Hoshino, Y., and Plapinger, R., 1973, Some important principles in 3,3'-D.A.B. ultrastructural cytochemistry, *J. Histochem. Cytochem.* **21**:756.

Shantha, T.R., and Manocha, S.L., 1970, *Macaca mulatta: Enzyme Histochemistry of the Nervous System*, Chapter IX, The Cerebellum, Academic Press, New York.

Shantha, T.R., Woods, W.D., Waitzman, M.B., and Bourne, G.H., 1966, Histochemical method for localization of cyclic 3',5'-nucleotide phosphodiesterase, *Histochemie* **7**:177.

Shute, C.C.D., and Lewis, P.R. 1967, The ascending cholinergic reticular system: Neocortical, olfactory and subcortical projections, *Brain* **90**:497.

Sidman, R.L., Angevine, J.B., Jr., and Pierce, E.T., 1971, *Atlas of the Mouse Brain and Spinal Cord*, Harvard University Press, Cambridge.

Siggins, G.R., Hoffer, B.J., and Bloom, F.E., 1971, Studies on norepinephrine-containing afferents to Purkinje cells of rat cerebellum. III. Evidence for mediation of norepinephrine effects by cyclic 3',5'-adenosine monophosphate, *Brain Res.* **25**:535.

Silver, A., 1967, Cholinesterases of the central nervous system with special references to the cerebellum, *Int. Rev. Neurobiol.* **10**:57.

Silver, A., 1974, The biology of cholinesterases, in: *North-Holland Research Monographs: Frontiers of Biology*, Vol. 36 (A. Neuberger and E.L. Tatum, eds.), pp. 1–596, North-Holland/American Elsevier, Amsterdam.

Song, C.S., and Bodansky, O., 1967, Purification of 5'-nucleotidase from human liver, *Biochem. J.* **101**:5C.

Sousa-Pinto, A., 1969, Experimental anatomical demonstration of a cortico-olivary projection from area 6 (suppl. motor area) in the cat, *Brain Res.* **16**:73.

Sousa-Pinto, A., and Brodal, A., 1969, Demonstration of a somato-topical pattern in the cortico-olivary projection in the cat: An experimental anatomical study, *Exp. Brain. Res.* **8**:364.

Spector, R.G., 1963, Cerebral succinic dehydrogenase, cytochrome oxidase and monoamine oxidase activity in experimental anoseic ischaemic brain damage, *Br. J. Exp. Pathol.* **44**:251.

Sprey, T.E., 1970a, Morphological and histochemical changes during the development of some of the imaginal disks of *Call. erythrocephala*, *Netherl. J. Zool.* **20**:253.

Sprey, T.E., 1970b, Localization of 5'-nucleotidase and its possible significance in some of the imaginal disks of *Call. erythrocephala*, *Netherl. J. Zool.* **20**:419–432.

Stensaas, S.S., Edwards, C.G., and Stensaas, L., 1972, An experimental study of hyperchronic nerve cells in the cerebellar cortex, *Exp. Neurol.* **36**:427.

Straus, W., 1964, Factors affecting the cytochemical reaction of peroxidase with benzidine and the stability of the blue reaction product, *J. Histochem. Cytochem.* **12**:462.

Sulkowski, E., Björk, W., and Laskowski, M., Sr., 1963, A specific and nonspecific alkaline monophosphatase in the venom of *Bothrops atrox* and their occurrence in the purified venom phosphodiesterase, *J. Biol. Chem.* **238**:2477.

Suran, A.A., 1974*a*, 5'-Nucleotidase and acid phosphatase of spinal cord: Quantitative histochemistry in cat and mouse substantia gelatinosa, *J. Histochem. Cytochem.* **22:**802.

Suran, A.A., 1974*b*, 5'-Nucleotidase and acid phosphatase of spinal cord: Quantitative histochemistry in cat and mouse spinal cords and in mouse brain, *J. Histochem. Cytochem.* **22:**812.

Tatsuki, T., Iwanaga, S., and Suzuki, T., 1975, A simple method for preparation of snake
venom phosphodiesterase almost free from 5'-nucleotidase, *J. Biochem.* **77:**831.

Tsou, K.C., Cheng, C.S., Nachlas, M.M., and Seligman, A.M., 1956, Syntheses of some *p*-
nitrophenyl substituted tetrazolium salts as electron acceptors for the demonstration of
dehydrogenases, *J. Am. Chem. Soc.* **78:**6139.

Van der Krogt, J.A., 1974, Localisatie van enzymen van het catecholamine-metabolisme in
rattehersenen, Thesis, University of Leiden, The Netherlands.

Van der Meer, K., Mulder, J.J.C., and Lugtenburg, J., 1977, A new facet in rhodopsin
photochemistry, *Photochem. Photobiol.* **24:**363.

Van der Ploeg, M., and Van Duyn, P., 1964, 5,6-Dehydroxy indole as a substrate in a
histochemical peroxidase reaction, *J. R. Microsc. Soc.* **83:**415.

Van der Ploeg, M., Streefkerk, J.G., Daems, W.T., and Brederoo, P., 1973, Quantitative aspects
of cytochemical peroxidase reactions with 3,3'-diaminobenzidine and 5,6-dehydroxy indole
as substrates, in: *Electron Microscopy and Cytochemistry* (E. Wisse, W.T. Daems, I. Molenaar,
and P. Van Duyn, eds.), p. 123, North-Holland, Amsterdam.

Van der Slik, W., 1975, Determination and diagnostic value of 5'-nucleotidase, Thesis (Krips
Repro, ed.), University of London, Leiden, Meppel, The Netherlands.

Van der Waart, 1974, Choline en cholinederivaten in rattehersenen, *in vitro* onderzoek over het
metabolisme, Thesis (J.H. Pasmans, ed.), University of Leiden, Leiden Den Haag, The
Netherlands.

Van Dijke, C.P.H., 1975, Over de subcellulaire localisatie van dopamine in het striatum van de
rat, Thesis (Krips Repro, ed.), University of Leiden, Meppel, The Netherlands.

Van Duyn, P., 1953, Inactivation experiments on the DOPA factor, *J. Histochem. Cytochem.* **1:**143.

Van Duyn, P., 1955, An improved histochemical benzidone-blue peroxidase method and a note
on the composition of the blue reactive product, *Rec. Trav. Chim. Pays-Bas Belg.* **74:**771.

Van Duyn, P., 1957*a*, Histochemistry of DOPA factors. II. The nature of the localizing-
mechanisms in the DOPA-reaction, *Acta Physiol. Pharmacol. Neerl.* **5:**413.

Van Duyn, P., 1957*b*, Histochemistry of DOPA factors. III. Inactivation experiments on the
DOPA factors in neutrophilic and eosinophilic leucocytes and erythrocytes, *Acta Physiol.
Pharmacol. Neerl.* **5:**428.

Van Rossum, J., 1969, Corticonuclear and corticovestibular projections of the cerebellum, Thesis,
Leiden University, Van Gorcum, Assen, The Netherlands.

Vielvoye, G. J., 1970, Distribution and termination of spinocerebellar fibers in the cerebellum
of the pigeon (*Columbia domestica*), *Acta Morphol. Neerl.-Scand.* **7:**367.

Vielvoye, G.J., 1977, Spinocerebellar tracts in the white Leghorn (*Gallus domesticus*), Thesis (Krips
Repro, ed.), University of Leiden, Meppel, The Netherlands.

Vijverberg, A.J., 1973, Incorporation of tritiated thymidine in the wing and leg disks of Call.
erythr. M., *Neth. J. Zool.* **23:**189–214.

Voogd, J., 1964, The cerebellum of the cat: Structure and fiber connections, Thesis University
of Leiden, Van Gorcum, N.V., Assen, The Netherlands.

Voogd, J., 1967, Comparative aspects of the structure and fibre connexions of the mammalian
cerebellum, in: *Progress in Brain Research*, Vol. 25, *The Cerebellum* (C.A. Fox and R.S. Snider,
eds.), pp. 99–135, Elsevier, Amsterdam.

Voogd, J., 1969, The importance of fiber connections in the comparative anatomy of the
mammalian cerebellum, in: *Neurobiology of Cerebellar Evolution and Development* (R. Llinas,
ed.), Institute for Biomedical Research Symposium, Chicago.

Voogd, J., Groenewegen, H.J., and Boesten, A.J.P., 1975, Olivo-cerebellar connections in the
cat, The Kyoto Symposium (1975), Tokyo.

Wachstein, M., and Meisel, E., 1957, Histochemistry of hepatic phosphates at a physiological
pH: With special reference to the demonstration of bile canaliculi, *AM. J. Clin. Pathol.* **27:**
13.

Wachstein, M., and Meisel, E., 1964, Demonstration of peroxidase activity in tissue sections, *J. Histochem. Cytochem.* **12:**538.

Walberg, F., 1956, Descending connections to the inferior olive: An experimental study in the cat, *J. Comp. Neurol.* **107:**77.

Walberg, F., 1974, Descending connections from the mesencephalon to the inferior olive: An experimental study in the cat, *Exp. Brain. Res.* **20:**145.

Walberg, F., Brodal, A., and Hoddevik, G., 1976, A note on the method of retrograde transport of horseradish peroxidase as a tool in studies of afferent cerebellar connections, particularly those from the I.O.; with comments on the orthograde transport in Purkinje cell axons, *Exp. Brain Res.* **24:**383.

Whittaker, V.P., Michaelson, J.A., and Kirkland, R.J.H., 1964, The separation of synaptic vesicles from nerve-ending particles ("synaptosomes"), *Biochem. J.* **90:**293.

Williams, D., Gascoigne, J.E., and Williams, E.D., 1975a, A specific form of rat brain monoamine oxidase in circumventricular structures, *Brain Res.* **100:**231.

Williams, D., Gascoigne, J.E., and Williams, E.D., 1975b, A tetrazolium technique for the histochemical demonstration of multiple forms of rat brain monoamine oxidase, *Histochem. J.* **7:**585.

Winckler, J., 1970, Zum Einfrieren von Gewebe in Stickstoff-gekühlten Propan, *Histochemie* **23:**44.

Wisse, E., 1970, An electron microscopic study of the fenestrated endothelial lining of rat liver sinusoids, *J. Ultrastruct. Res.* **31:**125.

Wisse, E., 1972, An ultrastructural characterization of the endothelial cell in the rat liver sinusoid under normal and various experimental conditions, as a contribution to the distinction between endothelial and Kupffer cells, *J. Ultrastruct. Res.* **38:**528.

Wisse, E., 1974, Observations on the fine structure and peroxidase cytochemistry of normal rat liver Kupffer cells, *J. Ultrastruct. Res.* **46:**393.

Bibliography for Standard Incubation Methods

Recent Literature

Lodja, Z., Gossrau, R., and Schiebler, T.H., 1979, *Enzymehistochemistry*, Springer Verlag, Berlin.

Pearse, A.G.E., 1980, *Histochemistry: Theoretical and Applied*, Vol. 1, 4th ed., Churchill Livingstone, England.

Pearse, A.G.E., 1972, *Histochemistry: Theoretical and Applied*, Vol. 2, 3rd ed., Churchill Livingstone, England.

Older Literature

Burstone, M.S., 1972, *Enzyme Histochemistry and Its Application in the Study of Neoplasms*, Academic Press, New York.

Spannhof, L., 1967, *Einführung in die Praxis der Histochemie*, Gustav Fisher Verlag, Jena.

Critical evaluations on neuroenzyme histochemistry can be found in: Adams, C.W.M., 1965, *Neurohistochemistry*, Elsevier, Amsterdam.

Index